CAMBRIDGE LIBRARY COLLECTION

Books of enduring scholarly value

Botany and Horticulture

Until the nineteenth century, the investigation of natural phenomena, plants and animals was considered either the preserve of elite scholars or a pastime for the leisured upper classes. As increasing academic rigour and systematisation was brought to the study of 'natural history', its subdisciplines were adopted into university curricula, and learned societies (such as the Royal Horticultural Society, founded in 1804) were established to support research in these areas. A related development was strong enthusiasm for exotic garden plants, which resulted in plant collecting expeditions to every corner of the globe, sometimes with tragic consequences. This series includes accounts of some of those expeditions, detailed reference works on the flora of different regions, and practical advice for amateur and professional gardeners.

A Dictionary of the Economic Products of India

A Scottish doctor and botanist, George Watt (1851–1930) had studied the flora of India for more than a decade before he took on the task of compiling this monumental work. Assisted by numerous contributors, he set about organising vast amounts of information on India's commercial plants and produce, including scientific and vernacular names, properties, domestic and medical uses, trade statistics, and published sources. Watt hoped that the dictionary, 'though not a strictly scientific publication', would be found 'sufficiently accurate in its scientific details for all practical and commercial purposes'. First published in six volumes between 1889 and 1893, with an index volume completed in 1896, the whole work is now reissued in nine separate parts. Volume 3 (1890) contains entries from *Dacrydium* (a genus of coniferous trees) to *Gordonia obtusa* (a species of evergreen tree).

Cambridge University Press has long been a pioneer in the reissuing of out-of-print titles from its own backlist, producing digital reprints of books that are still sought after by scholars and students but could not be reprinted economically using traditional technology. The Cambridge Library Collection extends this activity to a wider range of books which are still of importance to researchers and professionals, either for the source material they contain, or as landmarks in the history of their academic discipline.

Drawing from the world-renowned collections in the Cambridge University Library and other partner libraries, and guided by the advice of experts in each subject area, Cambridge University Press is using state-of-the-art scanning machines in its own Printing House to capture the content of each book selected for inclusion. The files are processed to give a consistently clear, crisp image, and the books finished to the high quality standard for which the Press is recognised around the world. The latest print-on-demand technology ensures that the books will remain available indefinitely, and that orders for single or multiple copies can quickly be supplied.

The Cambridge Library Collection brings back to life books of enduring scholarly value (including out-of-copyright works originally issued by other publishers) across a wide range of disciplines in the humanities and social sciences and in science and technology.

A Dictionary of the Economic Products of India

VOLUME 3: DACRYDIUM TO GORDONIA

GEORGE WATT

CAMBRIDGE
UNIVERSITY PRESS

University Printing House, Cambridge, CB2 8BS, United Kingdom

Published in the United States of America by Cambridge University Press, New York

Cambridge University Press is part of the University of Cambridge.
It furthers the University's mission by disseminating knowledge in the pursuit of
education, learning and research at the highest international levels of excellence.

www.cambridge.org
Information on this title: www.cambridge.org/9781108068758

© in this compilation Cambridge University Press 2014

This edition first published 1890
This digitally printed version 2014

ISBN 978-1-108-06875-8 Paperback

A

DICTIONARY

OF

THE ECONOMIC PRODUCTS OF INDIA.

BY

GEORGE WATT, M.B., C.M., C.I.E.

REPORTER ON ECONOMIC PRODUCTS WITH THE GOVERNMENT OF INDIA.

OFFICIER D'ACADEMIE; FELLOW OF THE LINNEAN SOCIETY; CORRESPONDING MEMBER OF THE
ROYAL HORTICULTURAL SOCIETY, &C., &C.

(ASSISTED BY NUMEROUS CONTRIBUTORS.)

IN SIX VOLUMES.

VOLUME III.,

Dacrydium to Gordonia.

Published under the Authority of the Government of India,
Department of Revenue and Agriculture.

LONDON:

W. H. ALLEN & Co., 13, WATERLOO PLACE, S.W., PUBLISHERS TO THE
INDIA OFFICE.

CALCUTTA:

OFFICE OF THE SUPERINTENDENT OF GOVERNMENT PRINTING, INDIA,
8, HASTINGS STREET.

1890.

CALCUTTA :
GOVERNMENT OF INDIA CENTRAL PRINTING OFFICE,
8, HASTINGS STREET.

PREFACE to Vol. III.

SUBSEQUENT to the appearance of the first volume of this work, the Editor was engaged, for nearly two years, in connection with the Colonial and Indian Exhibition. On his return to India in April 1887, he resumed the Dictionary work, and the second volume was published in little more than a year from that date. During the course of preparation of that volume, however, the Government of India considered it desirable to modify materially the scope and character of the work, enlarging it in some directions and abbreviating it in others. It was, for example, deemed unnecessary to give botanical descriptions of the plants dealt with, and thought advisable to practically omit all imported articles of Indian trade, to discontinue reference to Ceylon products, when not directly connected with India, and also to reduce the number of tables given in statistical accounts of trade. A minor departure was, at the same time, enjoined in the adoption of the third person, in preference to the first, but that would, in any case, have been necessitated, for, shortly after the second volume had been completed, the Government of India was enabled to render invaluable aid by the deputation as collaborateurs of Mr. J. F. Duthie, Director of the Botanical Department, Northern India, and shortly afterwards of Dr. J. Murray, of the Indian Medical Service. The Editor has now to express his warmest thanks to these gentlemen for the able assistance they have rendered. He need only add that the respective share taken by each contributor is indicated by the appearance of his name on the right-hand top corner of the pages.

During the preparation of the third volume the Editor's task was indeed a pleasant one, for, the entire material of the Dictionary having been brought together and arranged by him some years ago, his editorial work consisted in seeing that the elaboration of the portions entrusted to his collaborateurs was on the plan laid down by the Government of India.

It may perhaps be admitted that the third (and perhaps also the second) volume manifests a considerable improvement on the first. This was to be expected, since the co-operation of Mr. Duthie and Dr. Murray ensured greater accuracy, through doubtful points having invariably been decided in consultation. A numerous circle of correspondents have also been consulted, amongst whom may be specially mentioned Dr. George King, Superintendent of the Royal Botanic Gardens, and Dr. D. Prain, Curator of the Herbarium, Calcutta; Mr. H. Medlicott (and his successor Dr. W. King), Superintendent of the Geological Survey; and the authorities of the Imperial Museum. The Directors of Land Records and Agriculture in the various provinces, by official requisitions through the Revenue and Agricultural Department of the Government of India, have given the Editor much useful information on various subjects. On trade questions invaluable assistance has been rendered by Mr. J. E. O'Conor, Assistant Secretary to the Government of India, Finance and Commerce Department, by the Chambers of Commerce, and by many mercantile experts and planters throughout the country, to all of whom the Dictionary is indebted for many of its most useful features. The official correspondence of the Government of India has also continued to be placed under free contribution, and the various branches of the Secretariat have uniformly and graciously responded to applications for assistance by placing their files on Economic Products at the disposal of the Editor.

GEORGE WATT,

Editor, Dictionary of the Economic

Products of India.

– SIMLA,

July 1890.

DICTIONARY

OF

THE ECONOMIC PRODUCTS OF INDIA.

DACRYDIUM, *Soland.; Gen. Pl., III., 433.*

A genus of coniferous trees, mostly natives of the Eastern Archipelago and the Malay Peninsula, of Fiji, New Zealand, Australia and Tasmania. They yield very beautiful woods and are highly ornamental, on account of which their cultivation is being largely prosecuted in most countries. Perhaps the species in greatest demand is **D. Franklinii,** *Hooker,* which yields the celebrated Huon Pine.

Dacrydium elatum, *Wall; Fl. Br. Ind., V., 648 ;* CONIFERÆ.   **I**

> **References.**—*Kurz, Forest Flora, Burma, II., 499 ; Gamble, Man. Timb., 394 ; Indian Forester, III., 178-9 ; VII., 362 ; XI., 106 ; XII., 282 ; Smith, Econ. Dict., 217, 353 ; Trans. Agri.-Hort. Soc. Ind., V., 110.*

Habitat.—Burma, probably Tenasserim. A tree, 30 to 60 feet in height, with dimorphous leaves. Very little is known regarding it, and it is, therefore, alluded to here more on account of the high value placed on its congeners than of any special properties reputed to be possessed by the Indian representative.

DACTYLIS, *Linn.; Gen. Pl., III., 1193.*   2

Dactylis glomerata, *Linn. ;* GRAMINEÆ.

COCK'S FOOT GRASS.

> **Syn.**—D. HISPANICA, *Roth. ;* D. GLAUCESCENS, *Willd.*
> **References.**—*Roxb., Fl. Ind., Ed. C.B.C., 114; Voigt, Hort. Sub. Cal., 717 ; Thwaites, En. Ceylon Pl., 374; Mueller, Select Ex.-Trop. Pl., 101 ; Murray, Pl. and Drugs, Sind, 14; Royle, Ill. Him. Bot., 28, 417, 423 ; Treasury of Bot., 379 ; Morton, Cyclop. Agri., 600.*

Habitat.—A tall, perennial grass, said to be common on the Himálaya of the N.-W. Provinces and the Panjáb. It receives its English name from the fancied resemblance of its flowering spikes to a fowl's foot.

Fodder.—Highly valued in Europe as a fodder grass for cattle. It forms a portion of most good pastures, especially on chalky or loamy soils. In *Morton's Cyclopædia of Agriculture* a full account of the grass is given. It is there said to be "one of the most widely distributed and valuable of hay and pasture grasses, being common in all countries of Europe south of the Arctic circle, as well as in the north of Africa, and in the corresponding latitudes of Asia and America. In Britain it forms a principal constituent of all the best natural pastures and meadows." The soil required is said to be "of a deep, rich, and moist but not saturated

FODDER.

3

B

DÆDALACANTHUS purpurascens.	Indigo-producing plants.

FODDER.

description:" "the finest developed native specimens are generally found in waste places, by the sides of hedges and dykes, on way-side banks, and in shady copses. It surpasses most of the native grasses in the enduring rapidity of its growth after being eaten or cut down, as well as in the quantity and quality of its produce; and as it is readily devoured by cattle, sheep, and horses, it became, at an early period in the history of grass culture, an object of agricultural care, having been grown in England in 1764, and that, at first, from seed received under its American name of *Orchard grass*, from Virginia, where considerable progress had been made in its cultivation." Royle alludes to it as common on the North-West Provinces and the Panjáb Himálayas, and in Atkinson's *Himálayan Districts* it is said to occur at Naini Tál, Kathi, Jalat, and Jhuni on open situations, at an altitude of 6,000 to 8,000 feet. By several writers it is spoken of as "frequent on the Himálayas," but no effort appears to have been made to cultivate the plant for fodder purposes. In the *Gazetteer* of Mysore and Coorg it is said to be cultivated in the Bangalore Gardens, but practical experiments have still to be performed to ascertain the Indian regions where its cultivation is possible. Roxburgh alludes to two plants—**D. lagopoides,** *Linn.*, and **D. brevifolia,** *Linn.*—as found on salt, sandy soil near the sea. The former is referred to by Dalzell and Gibson (*Bombay Flora, p. 298*) as common near the sea, and is said to be the **Poa brevifolia,** *Kunth.* Roxburgh placed these plants in **Dactylis,** because of Burman having done so, but was of opinion that they were more probably forms of **Poa.** At all events they are not species of **Dactylis.**

Dactyloctenium ægyptiacum, *Willd.;* Gramineæ, see **Eleusine ægyptiaca,** *Pers.*

D. scindicum, *Boiss.* see **Eleusine scindica.**

DÆDALACANTHUS, *T. Anders.; Gen. Pl., II., 1082.*

A genus of shrubs containing several highly ornamental plants, some of which are extensively cultivated in Indian gardens. They are known to afford indigo, a property possessed by many members of the family to which they belong. It is probable also that they are all, like **D. roseus,** used medicinally. The following may be specially enumerated.

[Acanthaceæ.

4 **Dædalacanthus nervosus,** *T. Anders.; Fl. Br. Ind., IV., 418;*

    Syn.—Justicia nervosa, *Vahl.,* Bot. Mag, t. 1358; Eranthemum nervosum, Br. Prod., 477.
    Vern.—*Shechin,* Nepal; *Topatnyok,* Lepcha; *Nalla nilámbari, vádámbram,* Tel.
    References.—*Gamble, Man. Timb., 280; Cat. Darj., 59; Walter Elliot, Flora Andhr., 126, 187; Bomb. Gaz., XV., 440; N.-W. P. Gaz., IV., p. lxxvi.*

    Habitat.—A frequent plant at the base of the Himálayas (1,000 to 3,000 feet) from the Panjáb to Bhután. Cultivated in most tropical countries; flowers bright blue.

    Properties of this and the other species have been described for the sake of economy under the genus.

5 **D. purpurascens,** *T. Anders.; Fl. Br. Ind., IV., 420.*

    Syn.—Eranthemum nervosum, in *Dals. & Gibs.,* Bomb. Fl., p. 195; E. pulchellum, *Roxb.,* Fl. Ind., Ed. C.B.C., 37.
    Vern.—*Kalla-jati,* Beng.; *Gul-sham,* Hind.

    Habitat.—A fairly abundant plant in the forests of the central table-land of India, at altitudes of 1,000 to 4,000 feet in Central India, Bombay

Ghâts, Belgaum, Parisnath, Assam, &c. **Roxburgh** describes it as "a most stout flowering shrub" "generally in its full beauty in February."

Dædalacanthus roseus, *T. Anders.; Fl. Br. Ind., IV., 419.*

6

Syn.—Justicia rosea, *Vahl.;* Eranthemum roseum, *Br.*
Vern.—*Dasmúli,* Mar.
References.—*Dals. & Gibs., Bomb. Fl., 195; Dymock, Mat. Med. W. Ind., 2nd Ed., 587.*
Habitat.—A shrub 2 to 6 feet in height, flowers deep blue, turning bright red as they fade. Frequent in the Western and Southern Deccan, from Bombay Ghâts to Mangalore.
Medicine.—Dymock mentions "the root boiled in milk is a popular remedy for leucorrhœa; dose, one drachm. In the Southern Concan it is given to pregnant cattle to promote the growth of the fœtus."

MEDICINE.
Root.

7

D. splendens, *T. Anders.; Fl. Br. Ind., IV., 418.*

8

Reference.—*Gamble, Man. Timb., p. 280.*
Habitat.—A handsome shrub, with long spikes of pink flowers, common in the undergrowth of *Sál* forests (*Gamble*).

DÆMIA, *R. Br.; Gen. Pl., II., 764.*

[Asclepiadeæ.

Dæmia extensa, *R. Br.; Fl. Br. Ind., IV., 20; Wight, Ic., t. 596;*

9

Syn.—Asclepias echinata, *Roxb.;* Raphistemma ciliatum, *Hook. f. in Bot. Mag., t., 5704;* Cynanchum extensum, *Ait.*
Vern.—*Sagowani, utran, jutuk,* Hind.; *Chágulbánti,* Beng.; *Uttururi,* Uriya; *Karial, siáli, trotu,* Pb.; *Utarana, kharial,* Sind; *Utarni,* Bomb.; *Utarana,* Mar.; *Nagala dudhi,* Guj.; *Utran, utarni, jutuk, jutup,* Dec.; *Utarni, vélip-parutti, uttámani,* Tam.; *Jittupáku, guruti-chettu, dushtupu,* Tel.; *Hála-kóratige,* Kan.; *Vélip-paritti,* Malay; *Yugaphala* (according to **Ainslie**), Sans.
References.—*Roxb., Fl. Ind., Ed. C.B.C., 256; Thwaites, En. Ceylon Pl., 196; Dals. & Gibs., Bomb. Fl., 150; Stewart, Pb. Pl., 145; Aitchison, Cat. Pb. and Sind Pl., 735; Grah., Cat. Bomb. Pl., 122; Sir Walter Elliot, Flora Andh., 48, 67, 75; Wight, Contrib., 59; Linn., Soc. Jour., XIX., 177; Pharm. Ind., 142; Ainslie, Mat. Ind., II., 452; Moodeen Sheriff, Supp. Pharm. Ind., 129; Dymock, Mat. Med. W. Ind., 2nd Ed., p. 523; S. Arjun, Bomb Drugs, 85; Murray, Pl. and Drugs, Sind, 161; Home Dept. Cor. Regarding Pharm. Ind., 239; Bidie, Cat. Paris Exhib.; Drury, U. Pl., 175; Lisboa, U. Pl. Bomb., 233, 274; Birdwood, Bomb. Pr., 317; Royle, Ill. Him. Bot., 272; Liotard, Paper-making Mat., 5, 15, 20; Hunter, Gaz., Orissa, II., 181; Gaz., Mysore and Coorg, I., 56; Gaz., N.-W. P., I., 82; X., 313; Jury Rep., Madras Exhib.; Spons, Encyclop., 947; Balfour, Cyclop., I., 875.*
Habitat.—A common fœtidly-scented climber; met with throughout the hotter parts of India, ascending to 3,000 feet, but does not apparently occur in Burma or the Malay Peninsula. Distributed to Afghanistan.
Fibre.—The stems yield a fibre which has been recommended as a substitute for flax. It is said to be very fine and strong; a sample shown at the Madras Exhibition, 1855, gained a medal. **Birdwood**, in his *Bombay Products,* remarks that it is the commonest weed in the Deccan, where it is called *Ootrun,* and that the late **Colonel Meadows Taylor** was the first to draw attention to its valuable fibre. Although this fibre is frequently mentioned by Indian writers, it does not appear to have been thoroughly examined. In *Spons' Encyclopædia* the statements, first published in the *Madras Jury Reports,* are reproduced. But neither **Roxburgh** nor **Royle** appear to have examined the fibre. **Balfour** says it is

FIBRE.
Stems.
10

FIBRE.

a promising substitute for flax. In a recent report furnished by the Conservator of Forests, Northern Circle, Madras, it is stated that "the plant is common in the drier districts of the Presidency. It affords a very pretty fibre which is said to be sometimes used for fishing lines."

The fibre was not shown at the Colonial and Indian Exhibition, but as the plant is extremely plentiful, there should be no difficulty in procuring a large annual supply.

MEDICINE
Plant.
II
Leaves.
12
Juice.
13

Medicine.—The PLANT has emetic and expectorant virtues, and is extensively employed by natives in the diseases of children. Ainslie says that "a decoction of the LEAVES is given to children as an anthelmintic, in doses not exceeding three table-spoonfuls; the JUICE of the leaves is ordered in asthma." The *Pharmacopœia* assigns the above properties to it, as current native opinion, but adds that, although reputed to be a cure for snake-bite, this rests on insufficient grounds. Dr. Oswald held that it was a fairly good expectorant in the treatment of catarrhal affections, in ten-grain doses, for which purpose it was used at the Pettah Hospital, Mysore. Dr. Dymock says that in "Western India the plant has a general reputation as an expectorant and emetic. In Goa the juice of the leaves is applied to rheumatic swellings." Drury adds the further fact that "the juice of the leaves mixed with *chunam* is applied externally in rheumatic swellings of the limbs."

SPECIAL OPINIONS.—§ "Used in infantile diarrhœa." (*Surgeon-Major D. R. Thomson, M.D., C.I.E., 1st District, Madras*). "The fresh leaves made into a pulp are used as a stimulating poultice in carbuncle, with good effect" (*Asst. Surgeon Sakharam Arjun Ravat, L.M., Bombay*). "Certainly valuable as an emetic with infants: the *leaves* are washed, and the juice expressed by rubbing between the palms of the hands; the leaves of the dark *Toolsi* are similarly treated, and then a mixture of the juices is given; this preparation is a stimulating emetic" (*Civil Surgeon B Evers, M.D., Wardha*).

FODDER.
Plant.
14

Fodder.—The PLANT is said to be browsed by goats.

Dœmonorhops (Calamus) Draco.—The Dragon's Blood Palm; see Vol. II., C. No. 68.

15

Dahi (DADHI, *Sans.*)

A term given to a kind of curd or rather coagulated milk. To prepare this the milk is first boiled, then soured by being thrown into an unwashed vessel in which *dahí* had been previously kept. At times, however, an acid is employed to precipitate the solid ingredients of the milk, and rennet is used by a certain limited community only. *Dahí* thus differs from curd as prepared in Europe in being practically sour boiled milk. The milk is boiled almost immediately after being obtained from the cow, and thus contains all its fat or butter. *Dahí* in the liquid state is largely consumed, so that the whey not being separated *Dahí* contains in solution all the sugar of milk. The curd or casein, even if separated from the whey, contains, however, too much fat to be made into cheese. It is, in fact, cream cheese, and on drying crumbles to a powder. The whey is separated by pressing the curd inside a cloth, and in this condition it is largely used in cookery, and is the basis of all the sweetmeats made in India. The natives of India have thus come to learn that to eat the liquid *dahí* they are consuming a wholesome mixture of the muscle-forming materials casein and fat with the heat-giving ingredient—sugar—the equivalent of starch. But to eat the curd alone to any large extent would be injurious by causing severe constipation. After being made into sweetmeats it is, however, rendered highly nutritious through having restored to it sugar,

DAHI.

and by being mixed with flour of wheat or of rice is made into an article of diet. Hence it follows that the sweetmeats so largely consumed as a midday meal in India partake of all the ingredients of food and are not mere luxuries like the sweets of Europe.

The trade in expressed *dahí* is very extensive, and within a radius around the larger cities immense quantities are carried by train—the plastic substance being contained within a cloth and resting in open baskets. The manufacture of cheese is practically unknown in India except as cream cheese, and it seems probable that by the working classes a cheap cheese to be eaten along with bread would be appreciated. But there exists the practical difficulty which, in all probability, suggested the present course, namely, that the climate of India would sour the milk before the cream could have time to rise to the surface ; hence in all probability the practice of rapidly boiling milk which is all but universal in this country.

See **Milk** and **Rennet.**

Dakh, a term applied in Hindustani to grapes, but also to raisins, currants, or the fruit of **Sageretia oppositifolia**—the *gidardák* or Jackal's vine. **16**

Dakra, a substance said to be used in Nepál to poison elephants. It is made up in balls along with rice. *Dakra, dhakka,* &c., are names given to **Elæodendron Roxburghii,** the bark of which is a virulent poison; and **Cissampelos Pareira** is said to be the *Dakh nirbisi,* or antidote to *Dakh.* The exact nature of the Nepál poison does not appear to have been made known, but it more than likely contains Aconite. **17**

Dal, a generic name for split peas, but more especially applied to the split peas of **Cajanus indicus,** the *Arhar-ka-dal* ; **Phaseolus Mungo** and **P. radiatus** are the *Mung-ka-dal,* while **Cicer arietinum** (gram) is the *Channa-ka-dal,* and **Lens esculenta** the *Masur-ka-dal.* **18**

DALBERGIA, *Linn.; Gen. Pl., I., 544.*

A genus of valuable trees comprising some 60 or 70 species ; cosmopolitan in the tropics. The generic name was given in honour of **Dalberg,** a Swedish botanist. The genus DREPANOCARPUS differs only in having versatile anthers, and in the fruit being lunate to reniform. Into that genus **Kurz** placed the following Indian species : **D. Cumingii, D. reniformis, D. spinosa,** and **D. monosperma.** These are the species which, in the *Flora of British India,* constitute the sub-genus SELENOLOBIUM, except that **D. Cumingii** does not appear to be described in the *Flora.* **Bentham** and **Hooker,** in the *Genera Plantarum,* regard DREPANOCARPUS as an American genus, with one species African but none indigenous to India. India, however, possesses, including the DREPANOCARPUS of **Kurz,** some 29 species of DALBERGIA, of which the following are the more important, and although in some cases not specially dealt with in this work, all are of considerable economic interest.

Dalbergia cana, *Grah.; Fl. Br. Ind., II., 237 ;* LEGUMINOSÆ. **19**

Habitat.—A tree 40 to 60 feet high (according to **Kurz**), a scandent plant (according to the *Flora of British India*), frequent in the tropical forests of the eastern slopes of the Pegu Yomah, and still more frequent from Martaban down to Tenasserim (*Kurz*).

Structure of the Wood.—"White, turning brownish, rather heavy, of a very coarse fibre, soon attacked by xylophages" (*Kurz, For. Fl. Burm., I., 344*).

TIMBER.
20

D. cultrata, *Grah.; Fl. Br. Ind., II., 233.* **21**

Vern.—*Yendike, yindaik, veng-daik,* BURM.

| DALBERGIA lanceolaria. | The Blackwood. |

References.—*Kurz, For. Fl. Burm., I., 342; Gamble, Man. Timb., 128; Indian Forester, I., 120; VI., 125; VIII., 416; Balfour, Cyclop., I., 878.*

Habitat.—A moderate-sized tree of Burma (Prome), in general habit resembling **D. lanceolaria**, especially in the character of the pod. **Kurz** says it is common in all leaf-shedding forests, especially in the upper mixed savannahs and *Eng* forests, all over Burma from Ava to Martaban, and down to upper Tenasserim.

RESIN.
22
OIL.
23
TIMBER.
24 **Resin.**—Exudes a red resin according to **Kurz. Mr. M. H. Ferrars** says that the Karenis use the plant for the propagation of lac.

Oil.—Balfour states that this tree furnishes a useful oil.

Structure of the Wood.—Purplish-black, with darker streaks, harder than, but in structure similar to, that of **D. latifolia.** Weight 83℔ a cubic foot. The sapwood is pale-coloured, turning pale brown, very perishable; the heartwood blackish and ebony-like, often streaked red on a paler ground, extremely durable.

DOMESTIC.
25 It is employed for wheels, agricultural implements, handles of *dahs* and spears, but especially for carving.

26 ## Dalbergia (Drepanocarpus) Cumingii, *Bth.;* as in *Kurz, For. Fl.* [*Burm., I., 336.*

Habitat.—A tree-like scandent shrub, met with in Tenasserim.

DYE.
27 **Dye.**—Kurz says this "is a dye-wood, and furnishes the *Kayu lakka* of commerce." The writer can discover no other reference to this plant than that given by **Kurz.** It is not described apparently in the *Flora of British India.* **Gamble** (*Man. Timb., 124*) simply repeats **Kurz's** words.

28 ## D. foliacea, *Wall.; Fl. Br. Ind., II., 232.*

Vern.—*Tatebiri,* NEPAL.

References.—*Kurz, For. Fl. Burm., I., 347; Gamble, Man. Timb., 129.*

Habitat.—A large, straggling shrub, met with in the Eastern Himálaya and Burma (according to **Gamble**); the *Flora of British India* mentions only Ava, Pegu, and Martaban.

TIMBER.
29 **Structure of the Wood.** —White, porous, with a small, dark heartwood, in structure resembling that of **D. stipulacea,** except that the medullary rays are broader (*Gamble*).

30 ## D. glomeriflora, *Kurz; Fl. Br. Ind., II., 236.*

Habitat.—A tree 30 to 40 feet in height, found occasionally in the upper mixed forests of the Prome Yomah, at 1,000 to 2,000 feet elevation. It flowers in March and April (*Kurz, For. Fl. Burm., I., 345*).

31 ## D. hircina, *Benth.; Fl. Br. Ind., II., 236.*

Vern.—*Saras, bandir, tantia, gogera,* N.-W. P.; *Bakalpattia, tantia,* KUMAON.

References.—*Brandis, For. Fl., 151; Gamble, Man. Timb., 124; Indian Forester, XI., 3; Atkinson, Him. Dist., 309.*

Habitat.—A small tree of the Central and Eastern Himálaya, from Garhwál and Kumáon to Bhután, ascending from the foot of the hills to altitudes of 5,000 feet. Flowering season April to May, the seeds ripening in July.

32 ## D. lanceolaria, *Linn.; Fl. Br. Ind., II., 235.*

Syn.—*D. FRONDOSA, Roxb.; D. ZEYLANICA, Roxb.; D. ARBOREA, Heyne; D. ROBUSTA, Wall; D. HIRCINA, Wall.*

Vern.—*Takoli, bithúa,* HIND.; *Chakemdia* (in Puri), BENG.; *Píri,* KOL.; *Chapot siris,* SANTAL; *Bander siris,* NEPAL; *Takoli, bithúa,* N.-W. P.,

D. 32

| The Blackwood or Rosewood. | (*G. Watt.*) | **DALBERGIA** **latifolia.** |

Pássi, RAJ., MERWARA; *Dandous,* SIND; *Takoli, harrani, gengri,* BOMB.; *Harréni,* DHARWAR; *Dándashi,* THANA; *Dandous, kaurchi, dandúsa,* MAR.; *Barbat, parbáti,* BANSWARA; *Gengri,* PANCH MEHALS; *Nal valanga,* TAM.; *Erra pachchári, pedda sópara, yerra patsaru, pasarganni,* TEL.; *Vel-urruvai* (TAM. in Ceylon), *be-lúlabba* (*Roxb.*), SING.

References.—*Roxb., Fl. Ind., Ed. C.BC., 534; Brandis, For. Fl., 151; Beddome, Fl. Sylv., 88; Gamble, Man. Timb., 128; Thwaites, En. Ceylon Pl., 93; Dals. & Gibs., Bomb. Fl., 78; W. Elliot, Fl. Andh., 53, 150; Wight & Arnott, Prod., 266; Trimen, Cat. Ceylon Pl., 27; Campbell, Econ. Prod. Chutia Nagpur, No. 8442; Duthie, Report of a Tour in Merwara; Atkinson, Him. Dist., 309; Drury, U. Pl., 175; Lisboa, U. Pl. Bomb., 61; Bomb. Gaz. (Thana Dist.), XIII., 24; (Kanara Dist.) XV., 433; Gaz., N.-W. P. (Bundelkhand), I., 80; Balfour, Cyclop., I., 878; Ajmere-Merwara, Special Report by Assist. Conservator, Forests.*

Habitat.—A deciduous tree of the sub-Himálayan tract, from the Jumna eastwards, ascending to 2,500 feet; also met with in Central and South India, and Bombay. Kurz does not mention it as met with in Burma, but the Conservator of Forests in Bengal reports that though scarce this small tree occurs in the Puri District.

Oil.—The OIL expressed from the SEED is said to be used in rheumatic affections. "The MILK which exudes from the ROOT is occasionally applied to ulcers" (*Drury*).

OIL. Seed. 33 Root. 34

Medicine.—Drury says that "the BARK in infusion is given internally in dyspepsia, and the LEAVES are rubbed over the body in cases of leprosy and other cutaneous diseases." That information, he remarks, is derived from Roxburgh, but the writer cannot find the passage referred to in Roxburgh's works, and suspects that Drury was in reality compiling from Ainslie's Dalbergia arborea, *Willd.,* which is Pongamia glabra, the seeds of which yield a well-known oil, useful in skin affections. Beddome, however, states that "the BARK and an OIL obtained from the SEEDS are in use medicinally with the natives." The Revd. A. Campbell writes that the Santals use the bark along with that of Flacourtia Ramontchi, as an external application, during intermittent fever. The leaves and the ROOT, he adds, are also employed medicinally.

MEDICINE. Bark. 35 Leaves. 36

Seeds. 37

Root. 38

Structure of the Wood.—White, moderately hard; not durable; no heartwood. Weight 62℔ per cubic foot. Beddome says the timber is useful for building purposes. In the *Bombay Gazetteer* (Konkan), it is stated that the wood is used for the handles of tools and small agricultural implements. Roxburgh observes that it is a quick-growing, large, beautiful tree, the timber of which is useful for many purposes. Similarly, Balfour reports that it affords a strong and useful timber.

TIMBER. 39

Dalbergia latifolia, *Roxb.; Fl. Br. Ind., II., 231.*

THE BLACKWOOD OR ROSEWOOD OF SOUTHERN INDIA.

Syn.—D. EMARGINATA, *Roxb.*

Var. sissoides is said, by the *Flora of British India,* to occur on the Nilgiri hills. It differs from the normal condition in having the leaflets rather narrower in proportion to their length and somewhat obtusely pointed. It is the D. sissoides, *Grah.,* and the D. javanica, *Miq.* Beddome writes, however, that this form is common in the forest about Coimbatore and at Palghát, on the Anamallays, at Madura and Tinnevelly. He adds: "the wood is generally of a redder colour, and the tree flowers in the rainy season (July) instead of in the hot weather; it is always distinguished by the Palghát axemen as the *ecruputu,* while D. latifolia is called *ectee* (Dr. Wight apparently transposes these native

40

41

names). I cannot, however, distinguish the two trees botanically; the flowers of **sissoides** are said to be rather larger and the leaves narrower, but these differences are not constant, and the same drawing might answer for either tree; I cannot, therefore, look upon **sissoides** as more than a variety of **latifolia**." Balfour remarks of this form that the wood contains much oil which unfits it for receiving paint, and he adds logs are almost always faulty in the centre.

Vern.—*Shisham*, HIND.; *Sitsál (sweta sál,* i. e., *white sál)*, BENG.; *Ruti, kiri, siso,* KOL.; *Satsaiyar,* SANTAL; *Sissua, sissa,* URIYA; *Ruzerap,* MICHI; *Serís,* GOND; *Sirás, sissú, sirsa,* MANDLA; *Sitsál,* OUDH; *Serisso,* KURKU; *Shisham,* PB.; *Bhotuk,* BHIL; *Shisham,* MERWARA, RAJ.; *Tali,* SIND; *Shisham, siras,* C. P.; *Shissam, sissu, kalaruk, tivas, shisar,* BOMB.; *Siswa, shisham, sisu, sisva, kalarukh, bhotheula, sissúi,* MAR.; *Sissu,* GUZ.; *Shisao,* KON.; *Iti, eriwadi, tótakatti, jittagé, yerugudu,* TAM.; *Irugudu, iruvudu, virugadu, jitegi, yerugudu, jitangi, jitregi,* TEL.; *Biti, thodágatti,* KAN.

Sir Walter Elliot points out that the *Símsupa* of Wilson is incorrectly applied to this tree and should be assigned to the *Sál* proper (**Shorea robusta**). *Patsa* and *yerugudu* exactly correspond with the English Black-wood. The Black-wood which **Dr. Hove** describes in his Tour in Bombay (in 1787) was most probably **Diospyros montana** and not **Dalbergia latifolia**.

References.—*Roxb., Fl. Ind., Ed. C.B.C., 532, 533; Brandis, For. Fl., 148; Kurz, For. Fl. Burm., 342; Beddome, Fl. Sylv., XXIV; Gamble, Man. Timb., 127; Dalz. & Gibs., Bom. Fl., 77; Wight, Icon., t. 1156; Econ. Prod. Chutia Nagpur by Rev. A. Campbell, No. 9454; Duthie, Report of a Tour in Merwara; W. Elliot, Fl. Andh., 71, 75, 128, 176, & 192; Mason, Burma and Its People, pp. 530, 769; Cleghorn, 164; Lisboa, U. Pl. Bomb., 60, 278; Birdwood, Bomb. Pr., 328; Royle, Ill. Him. Bot., 195; Liotard, Dyes, 33; Indian For., I., 84, 99; II., 18, 19, 412; III., 45, 201; IV., 292, 366, 411; V., 497; VI., 304; VIII., 102, 105, 125, 387, 414; IX., 356; X., 222, 309, 549, 552; XII., 188 (XXII.), 313, app. 12; XIII., 120; XIV., 159, 199, 421; Balfour, Cyclop., I., 879; Smith, Dic., 53, 357; Treasury of Bot., 380; Kew Off. Guide to Bot. Gardens and Arboretum, 45; Sind Gaz., 193; Bomb. Gaz., (Ahmedabad), IV., 23; Bomb. Gaz. (Nasik), XVI., 18; Bomb. Gaz. (Ahmednagar), XVII., pp. 18, 26; Bomb. Gaz. (Poona), XVIII., p. 52; Bomb. Gaz., XV., 33, 67; Report by Shuttleworth, Conservator, Forests, Bombay; Settlement Report, Seone, 10; of Chandwara, 110; of Nimar, 306; of Chanda, app. VI; of Upper Godavery, 37; of Bhandara, 18; Baitool, 125; Raipore, 75; Manual of Trichinopoly by Moore, 77; Mysore and Coorg Gas., I., 48, 52, 60; III., 20; Man. Coimbatore by Nicholson, pp. 401, 484; Man. Cuddapah by Gribble, 56, 71, 262; Forest Admin. Rep., Chutia Nagpur, 1885, pp. 6, 30; Settl. Rep., Lahore, 15, &c., &c.*

Habitat.—A deciduous tree, attaining a large size in South India, also found in Oudh, Eastern Bengal, and Central India. The *Flora of British India* states that it is "common through the Western Peninsula, Sikkim, and Behar, *Hooker fil.*, Bundelcund, *Edgeworth*." **Mr. A. T. Shuttleworth** writes of Bombay: "The tree grows extensively and vigorously in the Deccan, Konkan, and Guzerat forests, but does not attain to any large size." **Mr. McGregor**, Conservator of Forests, Southern Circle, gives it as "common in Kanara, Belgaum, and Dhárwar in the moist regions, not, as stated by **Brandis**, in the dry forests."

"Best reported to the Bombay Government that the tree was difficult to rear owing to the ravages of insects on the sprouting seeds. It may, however, be successfully grown during heavy rains. The seed may be sown in drills well supplied with the refuse of lamp-oil mills. It may also be grown from suckers, but the wood does not turn out so well as when reared from seed." (*Conf.* with **Drury**.) **Beddome** remarks: "It is found throughout the Madras Presidency, Mysore, Coorg, Bombay, Central

| The Blackwood or Rosewood. | (*G. Watt.*) | DALBERGIA latifolia. |

India, and parts of Bengal, Sikkim, and the Andaman Islands." "It is not found in Ceylon, nor I believe in Burma. It ascends the mountains to nearly 4,000 feet, and grows equally well in the dry deciduous forests with teak, and in the moist evergreen sholas, and it is often associated with bamboo." "It flowers in March and April." It may be raised from seed, but is a very slow grower. Colonel J. G. Macrae reports that in Sind this plant has been experimentally cultivated, but with indifferent results. The Conservator of Forests, Bengal, reports that it has been introduced into the Sitapahar reserve, Chittagong Division, and promises to succeed ; it is nowhere indigenous in the Hill Tracts and Collectorate of Chittagong. The Conservator of Forests, North Circle, Madras, reports that it is "found throughout the Presidency, and varies greatly in size according to the moisture of the locality. In Malabar, the West Nilgiri slopes, South Canara, and Travancore, it grows to a large size and furnishes splendid pieces of timber fit for export. In Ganjam, Godavery, and the Eastern Ghât forests generally, it grows fairly big, and gives a rather harder, darker wood of finer quality, while in the hills of Cuddapah, North Arcot, Bellary, and the Western Carnatic, it is smaller and gives only pieces for small furniture and carved house-posts." It is also said to be common in the deciduous forests of Coorg, the wood selling in the forests for 5 to 6 annas a cubic foot.

Gum.—The tree is said to yield a GUM (*E. A. Fraser, Assistant Agent to Governor General, Rajputana*).

GUM.
42

Oil.—The SEEDS yield OIL, of which almost nothing further than this fact is at present known ; indeed, the same doubt as has been expressed regarding the oil of **Dalbergia lanceolaria** may be viewd as applicable to the statements made by some writers regarding the oil of D. latifolia. Beddome makes no mention of the oil or of any medicinal properties as assigned to this species.

OIL
Seeds.
43

Fodder.—Mr. Shuttleworth (Conservator of Forests, Bombay) reports that the LEAVES are used as FODDER. Mr. Lisboa (quoting from Brandis) remarks that this is the case in Oudh, but he makes no mention of the practice being followed in Bombay.

FODDER.
Leaves.
44

Structure of the Wood.—Sapwood yellow, small ; heartwood extremely hard, dark purple, with black longitudinal streaks ; no distinct annual rings, but alternating concentric belts of dark and light colour, which, however, run irregularly into each other. Weight from 50 to 66℔ a cubic foot ; growth moderate, 5 to 9 rings an inch. It coppices well, is easily raised from seed, and reproduces naturally and easily.

TIMBER.
45

It is a valuable furniture wood, and is exported to Europe from the forests of Kanara and Malabar. Wood sent to London for sale in 1878 fetched £13-10s. per ton. It is also employed for cart-wheels, agricultural implements, gun-carriages, &c. It is good for carving and fancy work, and is used for the handles of knives, kukris, and other arms. It has been employed for sleepers. Nine sleepers, which had been down seven to eight years on the Mysore State Railway, were found to be, when taken up—five good, three still serviceable, and one bad. It has been grown in plantations in Malabar and Kanara (*Gamble & Brandis*). In the *Bombay Gazetteer* it is stated : "the timber is one of the most valuable in India, is strong, very hard, close-grained, and of a purple black. It takes a beautiful polish and is reckoned the best furniture wood. A seasoned cubic foot weighs 50℔." In the *Lahore Gazetteer* it is stated that a fair sized tree will fetch from R40 to R70. Kurz says the heart-wood is greenish or greyish black and often mottled or lighter veined. Used extensively in India for cabinet-work, knees of vessels, agricultural implements, combs, &c. In Trichinopoly vases and other ornamental articles are made of the wood. It is

Furniture.

Combs.
46

Vases.
47

TIMBER.	sometimes called Indian rose-wood from the resemblance when polished to the timber of that name. The planks of black-wood have one great defect—a tendency to split longitudinally when not well seasoned. **Beddome** remarks of the wood : " it differs much in colour, but is generally purple-black ; it admits of a very fine polish, and is our best furniture wood, and is extensively used for gun-carriage purposes." "It generally fetches a higher price than teak." **Roxburgh** says Bengal-grown timber is " not so heavy as that obtained on the coast of Coromandel and Malabar, though fully as beautiful." **Wight** states that the Madras plant more closely corresponds with Roxburgh's D. **emarginata** than D. **latifolia**, " but the wood of the former is not black, which I think fatal to their identity. It is possible, however, that the Malabar tree may be specifically distinct from the Bengal one." **Wight** also states that planks, often 4 feet in diameter, are obtained from Malabar, and that too after all the white external wood has been removed. **Roxburgh** alludes to a tree 20 feet in circumference.

Dalbergia Mooniana, *Thwaites* ; see **Pericopsis Mooniana,** *Thwaites* ; *Fl. Br. Ind., II., 252.*

48   **D. (Drepanocarpus,** *Kurz*) **monosperma,** *Dalz. ; Fl. Br. Ind.,* [*II., 237.*

    Syn.—D. PANICULATA, *Wall. ;* D. TORTA, *Grah.*

    Habitat.—Shores of the Western Peninsula, Ceylon, and the Malayan Peninsula (*Fl. Br. Ind.*). Tidal jungles of Upper Tenasserim (*Kurz*). A scandent bush with hooked branches (Conf. with *Gamble, Man. Timb., 124 ; Dalz. & Gibs., Bomb. Fl., 78*).

49   **D. nigrescens,** *Kurz, For. Fl., I., 346.*

    Vern.—*Thitsanweng* or *thitsawnwin,* BURM.
    References.—*Kurz, For. Fl. Burm., 346 ; Gamble, Man. Timb., 129.*

    Habitat.—A moderate-sized deciduous tree, of the dry mixed forests of Upper Burma. Leaves small, blunt or retuse, panicles dense or compact, pedicles short. The name is given on account of the leaves turning black on being dried.

TIMBER.
50
    Structure of the Wood.—Light-grey, soft ; weight 38℔ a cubic foot.
    There is some doubt about the identification of these species owing to the absence of concentric bands (*Gamble*). It is not referred to in the *Flora of British India.*

D. ougeinensis, *Roxb. ;* see **Ougeinia dalbergioides,** *Benth. ; Fl. Br.* [*Ind., II., 161.*

51   **D. ovata,** *Grah.; Fl. Br. Ind., II., 231.*
    Syn.—D. GLAUCA, *Wall.*
    Vern.—*Madama* (**Kurz**), *douk-ta-louk* (**Mason**), BURM.
    References.—*Kurz, For. Fl. Burm., 343 ; Mason's Burma, 530, 769.*

52
    Var. obtusifolia.—A form with leaflets 3-5 inches long, oblong, obtuse, emarginate ; found in Burma.
    Kurz regarded D. **ovata**, *Grah.*, as distinct from **D. glauca**, *Wall*, thus restoring two species which, in the *Flora of British India*, were reduced to one. Of **ovata**, he says, the leaflets are acuminate, and to **glauca** he assigns the characters given above to the variety **obtusifolia**. The writer prefers following the *Flora of British India* in all matters of synonymy, since he has no means of examining the plants and of thus forming a personal opinion.

Dalbergia paniculata, *Roxb.; Fl. Br. Ind., II., 236.* 53

Vern.—*Dhobein, dhohein, pássi, satpuria,* HIND.; *Pondri,* KOL.; *Surteli, passi,* BAIGAS; *Padri,* GOND, BHIL; *Dubein,* BANDA; *Katsirsa,* N.-W.-P., OUDH; *Phassi,* KURKU; *Dobein, dhobin, pássi,* C. P.; *Padri,* DHAR-WAR; *Pondarra, sheodur, topia, pási,* or *phási,* MAR.; *Passi,*MELGHAT, BERAR; *Patchalai, valange,* TAM.; *Potrum, pachchári, porilla sápara, patsuru, porilla sopara, tella patsaru, toper,* TEL.; *Hasur guniri, pachári, padri,* KAN.; *Piangani,* MALAY; *Tapoukben,* BURM.

References.—*Roxb., Fl. Ind., Ed. C.B.C., 534; Brandis, For. Fl., 150; Kurz, For. Fl. Burm., I., 345; Beddome, Fl. Sylv., 88; Gamble, Man. Timb., 129; Dals. & Gibs., Bomb. Fl., 78; Sir W. Elliot, Fl. Andh., 140, 155, 178; Dymock, Mat. Med. W. Ind., 2nd Ed., 889; Lisboa, U. Pl. Bomb., 61; Birdwood, Bomb. Pr., 328; Balfour, Cyclop., I., 879; For. Adm. Rep., Ch. Nagpur, 1883, 30; Bomb. Gaz., III., 200; XI., 26; XV., 67; Gaz., Mysore and Coorg, I., 48; Gaz., N.-W. P. (Bundel-khand), I., 80; Indian Forester, II., 18; IV., 321; IX., 357; XIII., 120; XIV., 421; Settlt. Rep. of Chanda, App. IV.*

Habitat.—A large, deciduous tree, according to **Gamble** met with in the North-West Himálaya from the Jumna to Oudh, Central and South India (quoted by **Kurz** as met with in Burma, but identification doubtful). **Balfour** states that it grows at Moulmein. By the *Flora of British India* its habitat is given as the "plains of the Western Peninsula." **Brandis** states that it occurs in "South and Central India, Gonda forests of Oudh, Siwalik tract west of the Jumna, ascending to 2,500 feet." He adds "the leaves are shed in February-March; the new foliage comes out in April and May, with the flowers." **Beddome** remarks: "This tree is common in the plains and subalpine dry forests throughout the Madras Presidency." **Mr. McGregor,** Conservator of Forests, South-ern Circle, Bombay, reports that it is common in Dharwar, Belgaum, and Kanara. **Mr. G. Greig,** Conservator of Forests, N.-W. Provinces, alludes to the tree as met with in the Banda forests. **Colonel G. J. van Someren** refers to this tree as met with in the Melghát forests of Berar. The Editor of the *Indian Forester* (XIV., 421) says: "**D. paniculata** is a moderate-sized tree, attaining a girth of 5 to 6 feet and a height of 60 to 80 feet, is widely distributed throughout South and Central India, and is also found in the Sub-Himálayan tracts to the east of the Sárda river. Unlike its allies, **D. Sissoo** and **D. latifolia,** which form dense highly-coloured useful heartwoods, the whole wood is whitish-grey and soft and abnormal in pos-sessing narrow soft layers of parenchyma, alternating with broad concen-tric masses of wood, so that planks cut out of old trees often fall to pieces."

Gum.—The tree is reported to yield a GUM.

GUM.
54

Structure of the Wood.—Yellowish or greyish white, soft, perishable; no heartwood. Structure most remarkable, entirely different from that of the other species of the genus: broad concentric masses of wood alternate with narrow, dark-coloured belts of a fibrous substance, resembling the inner bark. Wood not durable and very subject to the attacks of insects. Weight according to **Skinner** 48℔; **Gamble** 37℔; **Beddome** 60℔ un-seasoned and 48℔ seasoned, per cubic foot. Specific gravity 768. **Rox-burgh** says the wood is white and firm to appearance but less useful than some of the other species. **Beddome** remarks that it is used for building and other purposes. It affords useful fire-wood. **Kurz** affirms that it is "good for common household purposes."

TIMBER.
55

In the *Indian Forester* (XIV., 421) an interesting note is given by the Editor on a sample of coppice shoot furnished by **Mr. S. C. Moss,** Sub-Assistant Conservator, Tinnevelly, "which shows a coppice shoot springing from the zones of soft tissue between two of the concentric layers

TIMBER.

of the wood : in one specimen the shoots are from close to the centre of the stem. The stumps were 12 inches in radius, and the concentric rings vary from half an inch to a whole inch in thickness. In the case of shoots springing from near the centre of the stem, the latter appears to have been decomposed at the centre, and the shoot, which may have originated in a layer of soft tissue, has passed radially across three zones of harder and two of softer tissue. This discovery of **Mr. Moss'** appears to be a new one in vegetable physiology, as adventitious shoots generally spring from the cambium zone, or directly between the wood and bark."

DOMESTIC.
56

Domestic Uses.—Leaves and twigs are used to manure fields in Madras (*Ind. For., IX., 357*).

57

Dalbergia purpurea, *Wall. ; Fl. Br. Ind., II., 235.*

A scandent species allied to D. LANCEOLARIA.

Vern.—*Thitpôt,* BURM.

Habitat.—Martaban and Pegu : common in the mixed forests down to Upper Tenasserim.

TIMBER.
58

Structure of the Wood.—Sap-wood light, not much used : heart-wood black and ebony-like (*Kurz, For. Fl. Burm., I., 344*).

59

D. reniformis, *Roxb.; Fl. Br. Ind., II., 238 ; Wight, Ic., t. 261.*

Syn.—D. FLEXUOSA, *Grah.*; D. STIPULATA, *Wall.*; DREPANOCARPUS RENIFORMIS, *Kurz, For. Fl. Burm., I., 336* (see the note above under the genus DALBERGIA).

Vern.—*Tankma* (Kurz) and *Douk-loung* (Mason), BURM.; *Kures,* SYLHET (Roxburgh).

References.—*Roxb., Fl. Ind., Ed. C.B.C., 534 ; Mason's Burma and Its People, pp. 530 and 769.*

Habitat.—A large, crooked, bushy tree " common in the swampy forests of Pegu and Martaban down to Upper Tenasserim ; flowering in February and March, and the fruit ripening in April and June " (*Kurz*). The *Flora of British India* adds that it is found in " Sillet." Roxburgh says that in Sylhet it flowers in March and the seeds ripen in December.

TIMBER.
60
DOMESTIC.
61

Structure of the Wood.—White, turning yellow, coarsely fibrous, light, very perishable.

Domestic Uses.—Roxburgh states that "the wood yields a greenish flame, and is reckoned the best for burning limestone."

62

D. rimosa, *Roxb. ; Fl. Br. Ind., II., 232 ; Wight, Ic., t. 262.*

Vern.—*Kaogrum,* SYLHET.

Habitat.—A shrubby species, met with in the tropical zone of the Eastern Himálayas, ascending to 4,000 feet—Khásia hills, Sylhet, Assam. Brandis (on the authority of **Stewart**) says that it is also met with in the Siwalik tract and outer Himálayas west of the Jumna. Reported to be cultivated in Bangalore (*Mysore Gaz.*). (Conf. with *Gamble, Man. Timb., 124; Brandis, For. Fl. N. Ind., 148; Roxb., Fl. Ind., Ed. C.B.C., 536.*)

D. robusta, *Roxb.;* see Derris robusta, *Benth. ; Fl. Br. Ind., II., 241.*

63

D. rubiginosa, *Roxb.; Fl. Br. Ind., II., 232.*

Vern.—*Karra sirli, tella tige,* TEL. **Sir Walter Elliot** remarks that Roxburgh's name *tella tige* simply means "white climber."

Habitat.—A scandent species, to be distinguished from **D. monosperma** by the character of the stamens and ovary; according to the *Flora of British India* it is met with in the Western Peninsula. **Roxburgh's** locality

for it was the Circar mountains. It is described by **Mr.** Talbot as occurring in Kanara.

Dalbergia scandens, *Roxb.;* see **Derris scandens.** Sir Walter Elliot
remarks that this is the *Chiratala bódi* and the *surlí* in Telegu.
Rheede, VI., 22.

D. Sissoo, *Roxb.; Fl. Br Ind., II., 231.* 64

THE SISSOO.

Vern.—*Shísham, síssu, síssai, sísam, sísu,* HIND.; *Shísu (Sisu* by **U. C.**
Dutt), BENG.; *Sissu,* ASSAM; *Sísu,* URIYA; *Sissái,* OUDH; *Sisú,*
N.-W. P.; *Tálí* or *táhlí safeda, shin, nelkar, shísham, shísháí, shía,*
shewa, PB.; *Shewa* **(Gamble, Stewart),** *Zagar* **(Lace),** PUSHTU;
Shawa or *shewa* (PUSHTU) in Bannu and Peshawar Districts; *Shísham*
(MERWARA), RAJ.; *Síssu, táli,* SIND; *Síssu,* BOMB.; *Tanach,*
sísam, GUZ.; *Yette, núkku-kattái,* TAM.; *Síssú, karra,* or *síssu-karra*
(síssu by **Elliot),** TEL.; *Biridi, cishmabage,* KAN.; *Sínsapá* **(U. C.**
Dutt), *shingshupa* (Roxburgh), SANS.; *Sásam, sásim,* ARAB.

Dr. Moodeen Sheriff explains that in Dukhni the word *Shísham* is used
for any wood which is black or reddish black and heavy, whatever tree
may produce it. *Sishu-kát* is the Bengali name for the above wood,
not *Shíshu* by itself, which means a young boy. It may be added that,
according to some writers the word is *síssú,* by others it is *síssú.*

References.—*Roxb., Fl. Ind., Ed. C.B.C., 533; Brandis, For. Fl., 149;*
Beddome, Fl. Sylv., t. 25; Gamble, Man. Timb., 124; Dals. & Gibs.,
Bomb. Fl., Suppl., 25; Stewart, Pb. Pl., 65; Aitchison, Cat. Pb. and
Sind Pl., 50; Sir W. Elliot, Fl. Andh., 168; Dr. Stock, Report on
Sind; Moodeen Sheriff, Supp. Pharm. Ind., 129; U. C. Dutt, Mat.
Med. Hind., 318; Murray, Pl. and Drugs, Sind, 129; Firminger, Man.
Gar., 448; Baden Powell, Pb. Pr., 342, 577; Atkinson, Him. Dist., 734;
Drury, U. Pl., 177; Lisboa, U. Pl. Bomb., 60, 217; Royle, Ill. Him. Bot.,
8, 191, 195; Spons, Encyclop., 879; Balfour, Cyclop., 879; Smith, Dic.,
379; Treasury of Bot., 381; Kew Off. Guide to the Mus. of Ec. Bot., 45;
Kew Off. Guide to Bot. Gardens and Arboretum, 76; Jour. Agri. Hort.
Soc., 1885, Vol. VII., pt. III., New Series; Procgs. Soc., ci., 1875-78,
Vol. V., 72; Report, Colonial and Ind. Exhibn., Ind. Timbers, p. 3;
Indian Forester, III., 45; IV., 321, 366, 386, 411; V., 180; IX., 75,
92, 490; X., 60, 402; XII., app. 1, 27 XIII., 55, 339; XIV., 159, 199,
421; Bombay Gazetteers, V., 285; VI., 12; VII., 32, 35; Punjab Settl.
Rep. (Jhang), 20; (Simla), XLIII; (Dera Ghazi Khan), 4; (Hazara),
10; (Kangra), 21; (Peshawar), 13; (Guserat), 133; Punjab Gazetteers
(Ludhiana), 10; (Amritsar), 4; (Karnal), 16; (Rawlpindi), 15; (Jhang),
15; (Sialkot), 11; (Jalandhar), 4; (Shahpur), 69; (Musaffargarh), 21;
(Hazara), 13; (Bannu), 23; (Dehra Ismail Khan), 19; (Rhotak), 14;
N.-W. Provinces Settl. Reports (Shajehanpur), IX.; N.-W. P. Gaset-
teers (Meerut), 33, 248; (Bundlekhand), 80; (Agra), LXXI.; Madras
Manuals—Trichinopoly, 77; Central Prov. Settl. Report (Chanda),
108; Mysore Gas., I., 48, 60; Gazetteer of Orissa, II., 5, 179; Sind
Gazetteer, 695, Special Report by Col. J. G. Macrae, Conserv., Forests;
Special Report by Conserv., Forests, Assam; Assist. Conserv., Forests,
Merwara, and Assist. Conserv., Forests, Quetta; Trans. Agri. Hort.
Soc. Ind., VII., 129, an account of the tree in Cuttack.

Habitat.—A large, deciduous tree of the sub-Himálayan tract from the
Indus to Assam, ascending to 2,000 feet. The *Flora of British India*
states that it is found in "the plains throughout India proper" and distributed to Baluchistán and Afghanistán. The extensive list of references
given above may be accepted as indicating its distribution, and it has been
found necessary to abridge very greatly the enumeration that might have
been given. It may briefly be said to occur in every district in India, many
of its localities, however, being the result of the effort to extend its cultivation. It is probable that its indigenous habitat is very much narrower
than we are accustomed to think. Neither **Kurz** nor **Mason** make any

mention of its occurrence in Burma. **Roxburgh** regards it as a native of
Bengal "and of the adjoining provinces to the northwards." **Brandis** views
it as a native of the sub-Himálayan tract, and adds "generally gregarious,
mostly on sand or gravel along the banks of rivers or on islands extending
50 to 100 miles into the plains. Believed to be indigenous also in
Guzerat, Baluchistan, and Central India. I have never seen it really
wild outside the sub-Himálayan belt. Cultivated and often self-sown
throughout India : thrives best on light soil, and requires a considerable
amount of moisture. The old leaves turn reddish brown and begin to fall in
December, but continue to be shed up to February, when the young foliage
comes out, continuing till April." "Flowers from March to June, at times
with a second flush between July and October; the seed ripens from
November to February, and generally remains long on the tree."
Beddome says it occurs as an avenue tree in the Madras Presidency.
Mr. J. H. Lace, Assistant Conservator of Forests, Quetta, who has
given much careful study to the plants of Baluchistan, says: **D. Sissoo** is
"indigenous about the Harnai, the Mehrab-Tangi, and up to Sharigh
(4,000 feet). The Wam Tangi Forests near the Harnai is chiefly composed
of it, where it grows up to 35 feet in height." **Mr. Mann,** Conservator of
Forests, Assam, says: " It occurs naturally only in the Eastern Duars of
the Goalpara District in Assam. With the exception of a few scattered
trees in the Lakhimpur District up stream from Dibrugarh, no Sissu is
found in the Cachar or Sylhet Districts." The Assistant Conservator of
Forests, Ajmere-Merwara, writes that while **D. latifolia** and **D. lanceolaria**
are wild, **D. Sissoo** is cultivated. The Conservator of Forests says, it is
cultivated in Sind, and that plantations, 20 years old, exist. It requires a
good soil and care during its first year or two. **Stewart** regards it as
indigenous in the Kachhi Forest, Panjáb, on the islands of the Indus
opposite Bannu. The Conservator of Forests, Northern Circle, of Madras (in
a report forwarded through the Board of Revenue) says that "Sissu is only
found in cultivation in the Madras Presidency. It does well on river
banks as in the plantations on the Cauvery in Trichinopoly and fairly-
well on coast stands as at Musulipatam."

OIL.
Wood.
65
Seeds.

Oil.—The WOOD is said to yield an empyreumatic, medicinal OIL. In
a recent report from the Forest Department, North-Western Provinces, it is
stated that oil is expressed from the SEEDS.

MEDICINE.
Raspings.
66
Root.
67
Leaves.
68
Oil.
69
Mucilage.
70
Decoction,
71
Bark.
72
FODDER.
73
TIMBER.
74

Medicine.— The RASPINGS of the wood are officinal, being regarded as
alterative (*Beddome*). "It is considered by natives to be hot (*Stewart*).
Useful in leprosy, boils, eruptions, and to allay vomiting : also in special
diseases" (*Baden Powell*). The ROOTS are said to be so astringent
that they are neither eaten by rats nor ants. "The LEAVES and saw-
dust (raspings) in decoction are esteemed in eruptive and special diseases,
and to allay vomiting. The OIL is also applied externally in cutaneous
affections" (*Atkinson, Himálayan Districts*).
SPECIAL OPINIONS.—§ " The MUCILAGE of the leaves mixed with sweet
oil is a good application in cases of excoriation. A DECOCTION of the leaves
is given in the acute stage of gonorrhœa" (*Civil Surgeon J. Anderson, M.B., Bijnor*). "The BARK made into pills with aromatics such as
ginger, &c., checks cholera" (*V. Ummegadien, Mettapollium, Madras*).
Fodder.—The young trees are liable to be browsed by cattle, goats,
and camels (*Stewart*); but the arrangements for forest conservation
prevent this as much as practicable.
Structure of the Wood.—Sapwood small, white; heartwood brown
with darker longitudinal veins, close and even-grained, seasons well,
very hard. Annual rings not distinctly marked, alternating dark and
light-coloured bands, which run into each other.

| Sissoo Wood. | (*G. Watt.*) | DALBERGIA spinosa. |

The wood is very durable, seasons well, and does not warp or slip. It is highly esteemed for all purposes where strength and elasticity are required. Clifford says that "in strength it is only inferior to *sál*, while in many other useful qualities it surpasses it, and has the advantage of being lighter. For FELLOES and NAVES of wheels and carved work of every description, for framings of carriages and similar work, it is unsurpassed by any other wood, owing to its fine seasoning and standing qualities." It is extensively used for boat-building, carts, and carriages, agricultural implements, in construction, and especially for furniture.

Roxburgh's account of this timber may be here given: "this tree yields the Bengal SHIP-BUILDERS their crooked timbers and knees. It is tolerably light, remarkably strong, but unfortunately not so durable as could be wished." Formerly, it was more extensively used for GUN-CAR-RIAGES than it can be at present, owing to the comparatively small supply. With regard to its durability and strength as a wood for wheels, Clifford writes: "The WHEELS of our ordnance carriages have never failed, however arduous or lengthened the service has been on which they have been employed, of which no more striking example can be furnished than the campaign in Afghánistán, about the most trying country in the world for wheels. Some of our batteries served throughout the campaign, went to Bamían, and even to the Hindoo Koosh, and came back again to India without a break-down, while Royal Artillery wheels, built of the very best materials Woolwich could produce, specially for Indian service, almost fell to pieces after a few months' exposure and service on the plains of India."

It has been tried and found to be good for SLEEPERS, and Mr. Mc-Master, in the *Proceedings of the Institution of Civil Engineers*, Vol. XXIII., 1863, says it will be really good for that purpose. The wood makes excellent CHARCOAL. Stewart recommended the cultivation of the tree for the purpose of railway FUEL, and Mr. Baden Powell, while Conservator of Forests in the Panjáb, planted out large tracts of country for this purpose. It is much planted as an AVENUE tree all over India, and in forest plantations in the Panjáb and Bengal. At the Colonial and Indian Exhibition Conference on timbers, Sir D. Brandis is reported to have said: "The tree is chiefly found along the streams which emerge from the Himálaya. Large trees became scarce about 60 years ago, but the tree is now regularly and extensively planted. An exhaustive report was prepared in 1826 by an eminent botanist, Dr. Wallich, respecting the localities producing the *Sissu*, which showed that the supply of large timber was at that time nearly exhausted. *Sissu* can, however, be easily cultivated in India, and on a large scale, in fact almost as easily as Spruce in Europe. Very extensive plantations have already been formed, and they could be extended over a great area if a sufficient demand arose for the timber. The tree has, for example, been cultivated in the south of India, but the plantations are still too young to judge whether it will there attain any large size."

Sissu wood might be exported from Calcutta.

Sacred Uses.—The tree is planted by the Hindus, being viewed by them as sacred.

TIMBER.

Felloes.
75
Naves.
76

Ship-building.
77
Gun-carriages.
78

Wheels.
79

Sleepers.
80

Charcoal and Fuel.
81

Avenue Tree.
82

SACRED.
83

84

Dalbergia spinosa, *Roxb.; Fl. Br. Ind., II., 238.*

> Syn.—D. HORRIDA, *Grah.*; DREPANOCARPUS SPINOSUS, *Kurz, For. Fl. Burma, I., 337.*
> Vern.—*Yechinya*, BURM.
> Habitat.—A stiff, erect shrub with the branches spine-tipped, frequent on the shores of the Eastern and Western Peninsulas and at Chittagong.

DALBERGIA volubilis.	The Dalbergias.

MEDICINE.
Roots.
85

Medicine.—" The ROOTS powdered absorb alcohol, and a spoonful of the powder in a tumblerful of water is said to be sufficient to destroy in less than half an hour the effects of alcohol even in cases bordering on delirium tremens " (*Kurz*).

TIMBER.
86

Structure of the Wood.—"Soft, beautifully silvery white, close and straight-grained " (*Kurz*).

[*453.*

87

Dalbergia stipulacea, *Roxb. ; Fl. Br. Ind., II., 237 ; Wight, Ic., t.,*

Syn.—D. FERRUGINEA, *Roxb.* ; D. TINGENS, *Wall.* ; D. CASSIOIDES, *Wall.* ; D. LIVIDA, *Wall.* ; D. ROSTRATA, *Grah.*

Vern.—*Tatebiri,* NEPAL ; *Tón-nyok,* LEPCHA ; *Garodosal,* MICHI ; *Dank talaungnwi,* BURM.

References. —*Kurz, For. Fl. Burm., 346 ; Gamble, Cat. Darj. Pl., 29, 129.*

Habitat.—A large, climbing shrub of the Eastern Himálaya, ascending to 4,000 feet ; also of Assam, the Khásia Hills, Chittagong, and Burma.

TIMBER.
88

Structure of the Wood.—Soft, greenish-grey, hard, close-grained, very prettily marked with lines of different colours. Weight 48℔ a cubic foot.

89

D. sympathetica, *Nimmo ; Fl. Br. Ind., II., 234.*

Syn.—D. FRONDOSA, *Wall.* ; D. FERRUGINEA, *Hohen.*

Vern.—*Petaguli* or *pentgul, titávali, yakayela,* MAR. ; *Titábli,* GOA.

References.—*Dals. & Gibs., Bomb. Fl., 78 ; Dymock, Mat. Med. W. Ind., 2nd Ed., 236 ; Bomb. Gaz. (Kanara), XV., I., 433.*

MEDICINE.
Bark.
90

Leaves.
91

92

Habitat.—A scandent plant armed with large curved thorns, frequent on the hills of Western India. Dymock says it is common near Bombay, and Talbot that it is found at Kanara.

Medicine.—The BARK is used as a *lép* to remove pimples. The LEAVES are in Goa employed as an alterative (*Dymock*).

D. tamarindifolia, *Roxb. ; Fl. Br. Ind., II., 234 ; Wight, Ic., t. 242 ; Roxb., Fl. Ind., Ed. C.B.C., 53 ; Gamble, Man, Timb., 124.*

Syn.—D. LIVIDA, *Wall.* ; D. MULTIJUGA, *Grah.* ; D. BLUMEI, *Hassk.* ; DERRIS PINNATA, *Lour.*

Vern.—*Keti,* SYLHET ; *Damar,* NEPAL.

Habitat.—A scandent species met with in the Eastern Himálayas— Nepál, Sikkim, Sylhet, the Khásia Hills, &c., ascending to 4,000 feet. Kurz says it is not unfrequent in the Andaman Islands and in Tenasserim (*For. Fl. Burm , I., 348*). Talbot reports its occurrence in the forests of Kanara.

FODDER.
Leaves.
93
94

Fodder.—The LEAVES resemble those of the tamarind and are eaten by cattle.

D. volubilis, *Roxb. ; Fl. Br. Ind., II., 235.*

Vern.—*Bhatia, bankhara,* HIND. ; *Bir munga, nari siris,* SANTAL ; *Nubari,* URIYA ; *Rongdi,* MAL. (S.P.) ; *Bhatia,* KUMAON ; *Alei, alai,* MAR. ; *Bandigarjana, bandi guriginja* (Elliot doubts the correctness of these names), TEL.

References.—*Roxb., Fl. Ind., Ed. C.B., 536 ; Brandis, For. Fl., 152 ; Kurz, For. Fl. Burm., 346 ; Elliot, Fl. Andh., 22 ; Dymock, Mat. Med. W. Ind., 2nd Ed., 237 ; Indian Forester, X., 326 ; Gaz., N.-W. P. (Bundelkhand), 80 ; Himálayan Districts, 309 ; Bomb. Gaz., XV., I. (Kanara), 433.*

Habitat.—A large climber, met with in the Central and Eastern Himálaya, Oudh, Pegu, and Ceylon. The Conservator of Forests, Bengal, in a recent report, states, however, that it also occurs in Orissa.

MEDICINE.
Root-juice.
95

Medicine.—Dymock states that it is applied to aphthæ and is used as a gargle in sore-throat. The ROOT-JUICE with cummin and sugar is given in gonorrhœa.

D. 95

Dammar.	(*G. Watt.*)	DAMMAR.

Fodder.—According to the Rev. A. Campbell cattle and goats eat the leaves of this plant.

Structure of the Wood.—Light-brown, hard, very tough.

Dalchini, see **Cinnamomum Tamala, C. 1183.**

Dalima, a name given in Orissa to a hard stone employed for making utensils, &c.

Damasonium indicum, see **Stratiotes alismoides,** *Linn.*

Dammar.—A trade generic name for a series of resins separately recognised by specific appellations. Of these the following may be specially mentioned, the reference being given for each, to further passages in the present work, where fuller details will be found :—

1st, East Indian Dammar.—Also known as "Singapore" or "White" Dammar. This is the true Dammar and is obtained from the species of **Dammara** described below, the best known of which is the Amboyna pine (**D. orientalis**) a native of Malacca, Borneo, Java, Sumatra, &c.

2nd, Kauri or Cowdee Dammar.—A fossil resin derived from **Dammara australis,** the chief supply of which is obtained from New Zealand. An extremely fine yellow amber-like resin.

3rd, Sal Dammar.—Known in Indian commerce as *rál.* This is the stalactitic resin obtained from **Shorea robusta,** which see.

4th, Black Dammar.—The resin obtained from **Canarium strictum,** *Roxb.,* which see, Vol. II., C. No. 285. Some interesting commercial facts regarding this and other Indian gums were published by the Public Works Department of the Government of India in a special report derived from correspondence with the Local Governments. This report appeared in 1871, and the following pages deal with Black Dammar :—2, 3, 4, 6, 7, 8, 9, 10, 13, 23, 30, and 69.

5th, Rock Dammar.—This is obtained from two species of **Hopea,** *viz.,* **H. odorata,** a native of Burma, and **H. micrantha,** a native of Malacca, Borneo, and Sumatra, &c.—(See **Hopea.**)

6th, White Dammar or Dhoop resin.—This name is often applied to the first Dammar enumerated above, but also to the resin derived from **Vateria indica,** which see.

7th, Green Dammar.—A term given to the resin of **Shorea Tumbuggaia,** which see.

8th, Pwenyet (or Poon-yet) Dammar.—A resinous or waxy substance obtained from certain trees in Burma. It is the hive of a peculiar bee, but much doubt exists as to the true nature and source of the substance. See **Pwenyet** in this work, and also **Dr. Forbes Watson's** account of it in the report on Gums and Gum-resins published by the India Office (1874), page 95.

In the countries where they are obtained, the dark-coloured and impure dammars are used for caulking boats and other such purposes. The purer qualities are exported to Europe and America, where, according to their specific properties, they are used for various purposes. Nearly all are, however, employed as varnishes, the purer qualities being employed to give a gloss to cotton and other fabrics. The less pure forms are used as varnishes by coach-builders and painters. The finest quality of all is that known in the trade as *Kauri* or *Cowdee* resin. This is a fossil dammar derived from **Dammara australis,** the supply of which mainly comes from New Zealand. The exports of this substance from New Zealand average between 2,500 and 6,000 tons annually, the larger quantity either going direct or *viâ* London to the American market.

c

FODDER.
96
TIMBER.
97

98

99

East Indian.
100
Amboyna Pine.
101
Kauri.
102

Sal.
103
Black.
104

Rock.
105

White.
106

Green.
107
Pwenyet.
108

D. 108

DAMMARA, *Lamb.; Gen. Pl., III., 436; Fl. Br. Ind., V., 650.*

Lambert (*Genus Pinus, smaller edition*) accepted Salisbury's position in separating the species of Dammar from the genus Pinus; he, however, preferred the name **Dammara,** *Rumph.,* to **Agathis,** *Salisb.,* the result being that **Dammara** has become better known. In a work like the present, which is more or less of a commercial character, it has been thought desirable to preserve the older name **Dammara,** and as **Agathis** has not been dealt with in the first volume of this work, it becomes all the more necessary to give the economic information in the present place. The *Flora of British India* adopts **Agathis** in preference to **Dammara.**

109

Dammara australis, *Lamb.; Genus Pinus, t. 54.*

The Kauri Pine.

Syn.—Agathis australis, *Salisb.*

References.—*Gordon, Pinetum, 108; Gamble, Man. Timb., 394; Indian Forester, III., 177, 184; V., 104; VII., 363; XII., 476, 553; Mysore and Coorg Gaz., I., 66; Smith, Dict. Econ. Pl., 149; Royle, Productive Resources, 68; Mueller, Extra-Tropical Pl., 102; Beddome, Fl. Sylv., 227; Trans. Agri-Hort. Soc., V., 110; VI., 103—105.*

Habitat.—A native of New Zealand, now confined to the North Island, but formerly more extensively distributed. Cultivated in most tropical and sub-tropical countries. The tree is being experimentally cultivated in India, but apparently not with the vigour which the importance of the subject deserves. **Royle** alludes to a consignment of 353℔ of seed of Dammar having been consigned to India in 1796, and in Mysore the descendants (presumably) of this stock may still be seen. **Beddome** alludes to **Dammara** as represented on the mountains of Madras.

**RESIN.
110**

Resin.—In the above remarks regarding Dammar resin some of the main features of the trade have been indicated. In **Lambert's** work, quoted above, is reproduced **Rumphius'** interesting article on the subject, one of the most important which has as yet appeared. Some idea of the value of the resin may be obtained from the fact that the imports into Great Britain are stated to have been worth £200,000. The tree is rapidly being exterminated in New Zealand, as its timber is of great value, and it is problematic how long the supply of fossil resin will continue to meet the growing demand. The tree attains a height of 120 to 150 feet, with a circumference of 24 feet.

111

D. orientalis, *Lamb.; Genus Pinus, t. 55.*

The Amboyna Pine.

Syn.—D. alba, *Rumph.*; Agathis loranthifolia, *Salisb.*; *Fl. Br. Ind., V., 650.*

Vern.—*Theet-men* (according to **Mason**), Burm.

Habitat.—A large tree, native of Amboyna, and Ternate, of the islands of Molucca, Java, Borneo, &c. **Mason,** in his list of the plants of Burma, enumerates this species, but **Kurz** makes no mention of it. **Wallich** states that it is found in Tavoy, and the *Flora of British India* that it is a native of Penang and Perak.

**RESIN.
112**

Resin.—The timber is of little value, but the tree affords large quantities of a transparent resin known as Dammar. This is conveyed to most parts of the world, being used in India as incense and for medicinal purposes. In Europe it is largely used like the resin of the above species for purposes of varnishing and for waxing or polishing fabrics. (*O'Shaughnessy, Beng. Disp., p. 617.*)

D. 112

It seems probable that this species might with great ease be grown in Burma, and possibly also in the Andaman Islands.

Damson; see **Plum,** and also **Prunus communis,** *Huds.*

Dana—a grain, and especially gram, but the name is also given to many plants, such as the *Anardana, Ramdana,* species of **Amarantus;** *Shakardana,* **Colebrookia oppositifolia;** *Behdana,* **Cydonia vulgaris;** *Hazardana,* **Euphorbia thymifolia;** *Kaladana,* **Ipomœa hederacea,** &c. | 113

Dandelion; see **Taraxacum officinale,** *Wiggers;* COMPOSITÆ.

Dandy, Banghy and Palanquin Poles, Woods used for— | 114

These woods are elastic and capable of bearing a considerable weight. They might, accordingly, be employed for cart shafts. Dandy poles are used by many races of India to carry loads across the shoulder, a package being balanced at each extremity.

Acer cultratum.	Fraxinus xanthoxylloides.
A. pictum.	Grewia oppositifolia.
Bambusa arundinacea.	G. tiliæfolia.
Betula Bhojpattra.	G. vestita.
Cotoneaster obtusa.	Lagerstræmia parviflora.
Cupressus torulosa.	Quercus dilatata.
Diospyros melanoxylon.	Q. semecarpifolia.
Ficus bengalensis.	Taxus baccata.
F. indica.	Ulmus campestris.
Fraxinus floribunda.	

Danewort; see **Sambucus Ebulus.**

DAPHNE, *Linn.; Gen. Pl., III., 190.*

Daphne cannabina, *Wall.; Fl. Br. Ind., V., 193;* THYMELÆACEÆ. | 115
POPULARLY known as the NEPAL PAPER PLANT.

Syn.—DAPHNE PAPYRACEA, *Wall.;* D. ODORA and BHOLUA, *Don;* D. PAPYRIFERA, *Ham. MS.*

Vern.—*Set barúwa, satpúra,* HIND.; *Dunkotah, gande, kaghuti, bhullu soang,* NEPAL; *Dayshing,* BHUTIA; *Balwa or bhalua, chamboi, barua,* KUMAON; *Niggi, mahadeo-ka-phúl* (God's Flower), *jeku* (SIMLA), PB.; *Hsele,* BURM.

References.—*Brandis, For. Fl., 386, 577; Gamble, Man. Timb., 315; Cat. of Trees, Shrubs, and Climbers of Darjeeling, 67; Stewart, Pb. Pl., 188-9; O'Shaughnessy, Beng. Dispens., 7, 531; Baden Powell, Pb. Pr., 515; Atkinson, Him. Dist., 378, 574, 795—97; Drury, U. Pl., 178; Royle, Ill. Him. Bot., 321; Christy, Com. Pl. and Drugs, VI., 13; Royle, Fib. Pl., 311; Spons, Encyclop., 947; Balfour, Cyclop., I., 888; Treasury of Bot., 383; Kew Off. Guide to the Mus. of Ec. Bot., 47; Special Report on Nepal by Dr. Gimlette; Stewart's Report of a tour in Hazara (in Jour. Agri.-Hort. Soc. India, XIV., p. 13; Hodgson, Jour. As. Soc. Beng., I., 8; Madden, Jour. As. Soc. Beng., XVIII., 610; Asiatic Res., XIII., 385; Trans. Agri.-Hort. Soc. India, V., 220-231; Conservator of Forests, Assam, in a recent report states that the plant is wild in the Khásia hills; Conservator of Forests, N.-W. P., reports that though the plant is common in the Jaunsar Division, it is not used for paper-making; Simla Gazetteer, 12.*

Habitat.—A large shrub or small tree found on the Himálaya from the Indus to Bhután, between altitudes of 3,000 and 10,000 feet; also on

DAPHNE cannabina.	The Nepal Paper Plant.

the Khásia and Naga Hills; one of the most abundant bushes on the hills between Manipur and Burma.

Gamble remarks that this species blossoms from November to February, and that the fruits ripen and become red in May. He adds that the flowers are "exceedingly sweetly scented" (*List of Darjeeling Trees, &c., p. 67*). Brandis says it flowers in "March and April, also in autumn," but he makes no mention of its being sweetly scented. The synonym **D. odora,** *Don,* would most probably imply that the flowers were scented. In the Simla district this species flowers from the middle of December to the end of February or middle of March, but the flowers are then devoid of any smell. It is probable that under certain circumstances it may have two seasons of flowering, in one of which it may be scented. Most authors describe the plant as "a large shrub," and Brandis says it attains a height of seven to eight feet. In Simla it is one of the most abundant plants, with **Skimmia Laureola** and **Sarcococca pruniformis** forming the forest under-brushwood, but none of these plants much exceed three feet in height.

FIBRE.
Bark.
116

Fibre.—The well-known Nepál paper is said to be made from the BARK of this and the other species of **Daphne,** and of the allied plant, **Edgeworthia Gardneri.** European interest in this paper may be stated to have originated in **Lord Auckland's** enquiry regarding it in the year 1837. It was of course known to the natives of India for several centuries prior to that date, and official records, on daphne paper, dated 1817, were submitted to His Lordship for inspection. Very little has since been added to our knowledge of the subject, and the reports quoted below were first published about the beginning of the present century.

" The process of making paper from this plant is thus described in the *Asiatic Researches.* After scraping the outer surface of the bark, what remains is boiled in water with a small quantity of oak-ashes. After the boiling it is washed and beaten to a pulp on a stone. It is then spread on

Setburosa.
117

moulds or frames made of bamboo mats. The *Setburosa,* or paper shrub, says the same writer in the above Journal, is found on the most exposed parts of the mountains, and those the most elevated and covered with snow throughout the province of Kumáon. In traversing the oak-forests between Bhumtah and Ranigur, and again from Almorah to Chimpanat and down towards the river, the paper-plant would appear to thrive luxuriantly only where the oak grows. The paper prepared from its bark is

Cartridges.
118

particularly suited for cartridges, being strong, tough, not liable to crack or break, however much bent or folded, proof against being moth-eaten, and not subject to damp from any change in the weather; besides, if drenched or left in water any considerable time, it will not rot. It is invariably used all over Kumáon, and is in great request in many parts of the plains, for the purpose of writing *misub-namahs* or genealogical records, deeds, &c., from its extraordinary durability. It is generally made about one yard square, and of three different qualities. The best sort is retailed at the rate of 40 sheets for a rupee, and at whole-sale 80 sheets. The second is retailed at the rate of 50 sheets for a rupee, and 100 at wholesale. The third, of a much smaller size, is retailed at 140 sheets, and wholesale 160 sheets to 170 for a rupee " (*Drury, U. Pl., 178.*)

Another early account of **Daphne** paper and the process of its manufacture is that given by the late **Mr. B. H. Hodgson** (*Jour. As. Soc. Beng., Vol. I., 8*), then Resident at the Court of Nepál. In describing this industry (which differs but little from that pursued with ordinary paper-making in India), it may suffice to indicate briefly the main features of **Mr.** **Hodgson's** account of the process, materials, and manipulation. The reader, however, will find **Mr. Hodgson's** complete article

D. 118

FIBRE.

reproduced in Atkinson's *Himálayan Districts, page 795,* also in the *Trans. Agri.-Hort. Soc. Ind., V., 228-231.*

Mr. Hodgson says a stone mortar is required and a mallet or pestle of hard wood, proportioned to the mortar and the quantity of bark it is desired to pulp. The alkali employed is the ash of oak-wood. This is placed in a basket of close wicker-work and water allowed to percolate through; the fluid thus obtained is the alkali used. The freshly-peeled bark is then placed in an open metallic vessel (the heat necessary being too great to allow of the employment of earthen boilers), and over these is poured the alkali. Four seers of oak-ash, through which five seers of water have been slowly poured, afford the alkaline solution sufficient to do a large handful of the bark. After the solution has reached the boiling point the bark is placed in it to the extent of as much as will float in the alkaline solution.

The boiling is then continued for half an hour, when the alkaline juice will be found to be nearly absorbed and the bark quite soft. This is now carried to the stone mortar and beaten with the mallet until reduced to a pulp. It is next freely stirred in another vessel containing pure water, until it loses all stringiness and will spread itself out in the water when shaken. The pulp is now ready for the frame. This has stout wooden sides, so that it may readily float, and a bottom of cloth the meshes of which are so regulated as to retain all the pulp but allow water to pass through easily. In throwing the pulp on the frame it is passed through a sieve so as to remove the lumpy portions and impurities. The sieve is of the same size as the frame. It is placed on the top of it and both are allowed to float on the water of the cistern. When sufficient pulp has passed through to cover the frame with a layer of the desired thickness, the sieve is removed, and while holding the frame in the left hand, a dexterous movement of the water and pulp with the right causes it to diffuse uniformly over the surface of the frame. The frame is then raised carefully from the water so as to allow of drainage without disturbing the film of pulp. The paper thus made is partially dried on the frame by being exposed edgewise to a fire. It is then removed, and if desired is polished by means of a conch shell, while placed on a flat board. A peculiarity of Daphne paper consists in the fact that it may be polished until it can be used for writing on without the aid of any sizing material.

Mr. Atkinson adds in his more recent account of this paper that it is "manufactured exclusively by the tribes inhabiting Cis-Himálayan Bhot, known as Murmis, Lepchas, &c., or generically as Rongbo, in contradistinction to the Sokpo, the name given to the inhabitants of Trans-Himálayan Bhot. The manufactories are mere sheds, established in the midst of the great forests of the upper ranges, which afford an inexhaustible supply of the material as well as of wood-ashes and good water, both of which are essential to the manufacture of the raw material into the blocks from which the paper is made." Dr. Royle (*Fibrous Plants*) mentions that at the Great International Exhibition of 1851 a sample of Nepál paper was exhibited of such size as to occasion universal surprise. He continues: "This paper is remarkable for its toughness, as well as its smoothness; some of it, in the form of bricks of half-stuff, was sent to England previous to the year 1829. As the quantity was not sufficient for a complete experiment, a small portion of it was made into paper by hand. An engraver, to whom it was given for trial, stated that "it afforded finer impressions than any English-made paper, and nearly as good as the fine Chinese paper which is employed for what are called Indian paper proofs." Dr. Campbell (*see Agri.-Hort. Soc. Trans., V., 222*) repeats Mr. Hodgson's statements and describes the paper made by the Bhoteahs as strong and almost as

---

**DAPHNE
cannabina.**   **The Nepal Paper Plant.**

FIBRE.   durable as leather, and quite smooth enough to write on. For office records, he says it is incomparably better than any Indian paper. It is occasionally poisoned by being washed with preparations of arsenic in order to prevent the destruction caused by insects. Many of the books of Nepál, written on this paper, are said to be of considerable age, and the art of making the paper seems to have been introduced about 500 years ago from China, and not from India." "The paper," he continues, "is so pliable, elastic, and durable, that it does not wear at the folds during twenty years, whereas English paper, especially when eight or ten sheets are folded up into one packet, does not stand keeping in this state uninjured for more than four or five years." He then refers to a copy of a Sanskrit work which he inspected, the date of transcription of which was A.D. 1687, or 150 years prior to his writing of it, that it was in a "perfect state of preservation, having all that time withstood the ravages of insects and the wear and tear of use."

The writer had the pleasure recently to receive from **Dr. Gimlette,** Residency Surgeon, Nepál, some interesting facts regarding Nepál economic products and industries. The following passage, as supplementing the facts derived from the earlier writers (briefly reviewed above) may be here taken from **Dr. Gimlette's** account of paper-making :— "This paper, justly celebrated for its toughness and durability, is manufactured from two or three forms of **Daphne** and also from **Edgeworthia Gardneri,** the last mentioned producing the finest and whitest paper. It is manufactured by the cis-alpine Bhotias, who inhabit the mountains between Nepál proper and Thibet. The barks of the different species are generally mixed together, that of **Daphne papyracea** being seldom used alone except for cordage. *Shosho, arbadi, shedbarwa,* or *letbarwa,* are names given by the Bhotias to the **Daphne** shrubs; *Kaghuti, bara kaghuti,* and *chota kaghuti* are names also used, but all seem to be somewhat loosely applied." "The paper sells in the Katmandu bazar at the rate of six annas per twenty-four large sheets." **Dr. Campbell** reported in 1837 that the price was then 160 sheets, per Nepalese rupee, to 400; or from 9 to 13 Company's rupees per maund. The transport to Patna (a distance of 200 miles) he estimated at R 1-12, and the price in Patna only a little more than in Catmandoo. This latter fact he explains by the circumstance of there being a monopoly of the sale of paper kept up by the Nepalese Government.

Throughout the greater part of India Daphne paper may be purchased, so that the manufacture by the hill tribes must be very extensive. Around Simla it is not made; indeed, the people seem utterly ignorant of the value of the plant—one of the commonest of wild plants. They prefer to make their ropes from **Grewia oppositifolia,** and alike neglect the **Daphne** and the wild hemp. This seems to be the state of affairs on most of the outer ranges. At Nagkanda (some 40 miles to the north of Simla) the writer came across some men carrying loads of **Daphne** bark, and was told it was being carried to the east where it was made into paper. This fact is in support of **Stewart's** statement that the Panjáb Himálayan tribes do not make the paper, though it is well known to be extensively made in Kumáon. The Forest Officer of Jaunsar reports that, though the plant is very common in all the forests above 5,000 feet, paper is not made of its bark, but that the local supply used for Patwari maps, &c., is imported from Kumáon. **Mr. G. G. Minniken,** Forest Officer of Bashahr, recently informed the writer that **Daphne cannabina** was not used in his district for paper-making, though it was probably exported to be used as an adulterant.

CHEMISTRY.   Chemistry of Daphne.—In the chemical analysis of the fibres of
119   India, published by **Messrs. Cross, Bevan,** and **King, Daphne** is placed

CHEMISTRY.

at the bottom of the list, since it possesses, of all the Indian fibres examined by these gentlemen, the lowest amount of cellulose, namely, 22·3 per cent. Chemistry, in the verdict of percentage of cellulose as an indicator of merit, is thus in opposition to practical experience, for, although it would not perhaps pay to export the bark or paper half-stuff of this (or, indeed, of any other plant) to Europe for the paper-making industry, pure and simple, there can be hardly any room for doubt but that the **Daphnes** in many respects are the best of all Indian paper materials. The chemical test given above may, however, be accepted as demonstrating their unsuitability for textile purposes. In *Spons' Encyclopædia* (*Vol. I.,* 947) an old report regarding the fibre is reproduced, namely:—"The inner bark, prepared like hemp, affords a very superior paper material. The paper made from it is particularly suitable for cartridges, being strong, tough, and not liable to crack or break, however much bent or folded; it is proof against being moth-eaten, and is not affected by change in the weather; if drenched or left in water for a considerable time it will not rot. It is in universal request locally for writing deeds and records on, being quite smooth, and almost indestructible." It may, however, be pointed out that the process described above (by means of which the hill tribes manufacture their Daphne paper) is one mainly characterised by the very slight amount of alkali necessary to produce the pulp. A crude alkaline ash, with the boiling conducted for only half an hour, and that too in an open vessel, is all that is necessary. Such a treatment may not completely reduce the fibre, though it proves sufficient to produce a workable pulp. **Messrs. Cross, Bevan, and King** urge that the only safe criterion of the merits of a fibre is obtained from its percentage of cellulose, and that being so, **Daphne** would be the most worthless of Indian fibres. The writer has, on several occasions, ventured to express an opinion opposed to this somewhat sweeping conclusion, but has had to admit that he bases his comparative want of faith in the cellulose theory on practical and not chemical considerations. The present seems a strong case in point. **Daphne** fibre, as a paper material, holds the foremost place among Indian paper stuffs, in opposition to its low percentage of cellulose, and thus seems to call for extended research, since chemistry must undoubtedly be able to account for this fact. It would almost seem as if the expeditious and wholesale modern methods of paper-making, indeed of fibre extraction generally, removed the materials of vascular concretion or disturbed conditions of the ultimate cellulose fibrils that were essential to their strength as textile or paper materials. The loss by weight and the injury to strength effected by a strong boiling alkali, and under a high pressure, does not seem a conclusive proof that, with some other process, the fibre thus condemned would not be found to possess properties of great merit. At all events, Daphne paper, as made in India, will endure for many years under a treatment that, in a few weeks, days, or even hours, would render the modern papers produced in Europe perfectly worthless.

The figures of analysis published by **Messrs. Cross, Bevan, and King** regarding **Edgeworthia** curiously enough confirm, in a remarkable manner, **Dr. Gimlette's** statement that the paper made from that plant is superior to that from **Daphne cannabina.** Their analysis is as follows:—Moisture 13·6 per cent., ash 3·9; loss by hydrolysis, for five minutes, in soda alkali 21·6, for one hour 34·7; amount of cellulose 58·5 per cent.; mercerising 16·5; increase of weight on nitration 126; loss by acid purification 8·3; amount of carbon 41·8 per cent. It is to be regretted that these chemists did not furnish a similar complete report of **Daphne** so as to allow of comparison. They seem to have been so disappointed with the low per-

DAPHNE	
Mezereum.	The Nepal Paper Plant.

CHEMISTRY. centage of cellulose n **Daphne** as to have considered it not deserving of further investigation. Their published results are, however, sufficient, when taken in the light thrown on the subject by **Dr**. Gimlette, to suggest the possibility that past writers may have been in error in ascribing the high merit of the Nepal paper to **Daphne cannabina**—the D. **papyracea** of the older authors. It is just possible that to **Edgeworthia Gardneri** the merit of the Nepál paper is due. If this be so future effort should be directed towards extending our knowledge of this comparatively scarce plant and of rapidly undeceiving the public mind of a misleading error. In this consideration the curious fact may be called to mind that Nepál paper-making is confined to the Central and Eastern Himálaya (the habitat of **Edgeworthia**) and is not practised in the Panjáb, where **Daphne cannabina** is so abundant, but **Edgeworthia** absent. (**Edgeworthia** is the *Aryili* and **Daphne involucrata**, the *chhota aryili*.)

In the absence of a satisfactory investigation of the merits of the fibres obtained from the individual species of **Daphne**, the above account of **Daphne** fibre and paper, may, in the present state of our knowledge, be viewed as applicable to **D. cannabina** conjointly with that given under the species below.

SACRED.
120
Sacred Uses.—The flowers of this, and perhaps also of all the Indian **Daphnes**, are used by the Hindus as offerings to their idols.

TIMBER.
121
Structure of the Wood.—White, moderately hard. Flowers very sweet-scented.

122 ## Daphne involucrata, *Wall; Fl. Br. Ind., V., 193.*

Syn.—DAPHNE LONGIFOLIA, *Meissn.*; D. WALLICHII, *Meissn.*; ERIOSO-LENA WALLICHII, *Meissn.*; SCOPOLIA INVOLUCRATA, *C. A. Mey.*
Vern.—*Shedbarwa, chhota aryili*, NEPAL.

Habitat.—A shrub of the Eastern Himálaya, the Khásia Hills, Upper Assam, East Bengal, and Burma. Gamble says that this species flowers in January and February, and that the fruits, which are black, ripen in May. Being an East Himálayan species this is not described by Stewart nor by Brandis. Gamble, however, distinguishes between **Daphne Wallichii,** *Meissn.* (the *chhota aryili*.) and D. **longifolia**, *Meissn* (the *Shedbarwa*), and he states that while they both flower at the same time the latter does not mature its black fruits till November and December. (*List of Trees, &c., in Darjeeling District, p. 67.*)

FIBRE.
123
Fibre.—The BARK is used in the manufacture of Nepál paper.

124 ## D. Mezereum, *Linn.*

MEZEREON, *Eng.*; ECORCE DE MEZEREON, DE GARON, DE LAUREOLE, DE THEYMELEE, BOIS GENTIL, *Fr.*; KELLERHALS, SEIDELBASTRINDE, KELLERHALSRINDE, *Germ.*; MEZEREO, *It.*; MEZEREON, *Sp.*

Vern.—*Mezereon*, or *mázariyún*, ARAB.; *Masirium*, or *mazariyun*, PERS.

References.—*Brandis, For. Fl., 384; Gamble, Man. Timb., 315; Pharm. Ind., 188; O'Shaughnessy, Beng. Dispens., 530; Moodeen Sheriff, Supp. Pharm. Ind., 174; Dymock, Mat. Med. W. Ind., 2nd Ed., 673; Flück. & Hanb., Pharmacog., 540; U. S. Dispens., 15th Ed., 941; S. Arjun, Bomb. Drugs, 118; Murray, Pl. and Drugs, Sind, 109; Irvine, Mat. Med. Pat., 56, 73, 122; Birdwood, Bomb. Pr., 75; Royle, Ill. Him. Bot., 321; Spons, Encyclop., 818, 1414; Balfour, Cyclop., I., 889; Treasury of Bot., 383; Kew Off. Guide to the Mus. of Ec. Bot., 113; Year Book, Pharm., 1873, 91, 92; 1874, 628; Irvine, 56, 73, 122.*

Habitat.—A deciduous shrub with pink flowers in lateral clusters; native of North-East Europe from Italy to the Arctic regions and eastwards to Siberia, &c. The flowers appear in spring before the leaves, and

D. 124

are succeeded by red berries. Although it is occasionally met with in Britain, by most writers it is there viewed as an introduced plant.

It is said by **Mr. Murray** to be common in the Panjáb Himálaya, and to be cultivated in gardens as an ornamental shrub. It may be cultivated, but is certainly not a wild plant, on the Himálaya nor anywhere in India.

History.—The Mezereon, according to Muhammadan physicians, is a leaf, of which there are three kinds, white, yellow, and black. The white is described as the best. The word *Mázariyun* is by most authorities said to be of Greek and not Arabic origin, and the plant referred to is thought to be the **Daphne Mezereum** of botanists. At all events, that plant has held a place in European medicine for the past 300 years, but the parts used are the bark or the berries, and not the leaves as described by **Mir Muhammad Husain** and other Muhammadan writers. Thus, very considerable confusion exists, and it seems probable that the *Kamela* which **Irvine** and other modern Indian authors refer to **Daphne Mezereum** is not the Mezereon of European writers. **Irvine** remarks that "the seeds are imported from Cabul and used as an irritant." In another place he again reverts to the subject of **Daphne Mezereum**, but calls it the *Mameera*, and states that "this root is like Mezereon and used in the same way." On a still further page, and again under **Daphne Mezereum**, he gives another account, calling the plant (in the vernacular) by the names of *Uzul-ool* and *Mazrioon*. "The root," he says, "brought from Persia is used as a stimulant sudorific." (*Conf.* with remark under **D. oleoides**, para. MEDICINE). **Dr. Dymock** (under *Mazariyun*) gives an account of the drug as described by **Mir Muhammad Husain**, but makes no mention of any drugs sold at the present day in Bombay drug shops under that name. **Assistant Surgeon Sakharam Arjun**, however, says that *Mázirúm* is "the Mezerion root of the Pharmacopœia. It is chiefly used by the Unani Hakeems in venereal complaints." **Dutt** (in his *Materia Medica of the Hindus*) and **Ainslie** (*in his Materia Indica*) are silent as to Mezereon, and, while **Sir William O'Shaughnessy** gives, what appears to be an outline of the leading facts attributed to the European drug, he says nothing as to its uses in India. In the Indian Pharmacopœia both **Daphne Merzereum** (the Mezereon) and **D. Laureola** (the Spurge Laurel) are made officinal. In France and the United States **D. Gnidium** is also officinal. It thus seems probable that as all the **Daphnes** possess more or less the same chemical properties, if the *Mazariyun* of the Indian bazars is a **Daphne** at all, it will be found to be one or other of the species indigenous to India or Persia but not the **Daphne Mezereum** of Europe.

Medicine—.Since the probability exists that the *Mázariyún* of India is an indigenous species of **Daphne**, or at all events that any **Daphne** might be used as such, it may not be out of place to give here a brief review of the medicinal and chemical properties assigned to the drug in Europe. Mezereon, when taken internally, is supposed to be alterative and sudorific, and to be useful in venereal, rheumatic, and scrofulous complaints. Externally applied it is a rubefacient and vesicant, but to obtain the last effect it has to be first steeped in hot vinegar and kept in contact with the skin by means of a bandage. In English medicine it is prescribed as an ingredient of the Compound Decoction of Sarsaparilla. An ethereal extract of the BARK has been recommended, however, as an ingredient in a powerful stimulating liniment.

Chemistry.—"Mezereun contains a crystalline bitter glucoside, daphnin, which, by the action of acids, is converted into daphnetin. An acrid resin is contained in the inner bark. Daphnin is also contained in the bark of other species of **Daphne**.

HISTORY.
125

MEDICINE.
Mazariyun.
126

Sarsaparilla.
127
Bark.
128
CHEMISTRY.
129

DAPHNE pendula.	Daphne Paper Plants.

CHEMISTRY.

"Umbelliferone has been obtained by dry distillation of the resinous acid of the bark.

"A greenish-yellow oil has been extracted from the **Daphne Mezereum** seeds, which is stated to act as an irritant and vesicant" (*Prof. Warden*).

130

Daphne oleoides, *Schreb.; Fl. Br. Ind., V., 193; Royle, Ill., t. 81.*

Syn.—DAPHNE MUCRONATA, *Royle;* D. CORIACEA, *Royle;* D. BUXIFOLIA, *Vahl.;* D. ACUMINATA, *Boiss. ;?* D. CASHEMIREANA, *Meissn.*

Vern.—*Kútilál, kanthan, gandalún, shalangri, zosho, shing, mashúr, swána, jikri, dona, channi niggi, kágsari, sind, kánsian, sonái, zhi, kak,* PB.; *Laghúne,* AFG.; *Pech,* SIND.
The above vernacular names are given by most authors under the old synonym of **D. mucronata.**

References.—*Brandis, For. Fl., 385; Gamble, Man. Timb., 315; Stewart, Pb. Pl., 189; Aitchison, Cat. Pb. and Sind Pl., 130; Aitchison, Kuram Valley Flora (Jour. Linn. Soc., XVIII.), 25; Baden Powell, Pb. Pr., 577; Atkinson, Him. Dist., 574; Royle, Ill. Him. Bot., 321.*

Habitat.—A small much branched shrub, met with on the Western Himálaya, from Garhwál westward to Murree, the Suliman Range, and Afghánistán, occuring at altitudes of from 3,000 to 9,000 feet.

Regarding the season of flowering of this species there seems to be some confusion. Brandis says that it occurs in September and October; and the fruits, which are orange or scarlet, mature in May and June. As if to contrast this statement with an error made by **Stewart** he gives the following paranthetic quotation ("blossoms May-July, at times October, the fruit usually ripening June-October"—*Stewart*). Gamble refers to the plant as met with in the Simla District : if it does so, it must be extremely rare. The writer has not as yet come across it in Simla, but with reference to the season of flowering he has samples of the plant from Quetta in full flower and dated May, and from Pangi, dated June.

MEDICINE.
Roots.
131

Medicine.—Aitchison, in his *Flora of Kuram Valley,* says that the ROOTS of this plant are used internally, when boiled, as a medicine, being purgative. In another place he says : "Camels will not eat this shrub except when very hungry. It is poisonous, producing violent diarrhœa. I feel certain that much of the mortality of camels in the Kuram division was due to the prevalence of this shrub."

Bark.
132
Leaves.
133
Berries.
134

The BARK and LEAVES are used in native medicine. The BERRIES are eaten to induce nausea. **Stewart** refers to this plant as hurtful to camels, thus making the same observation as recorded by **Aitchison. Stewart** further says the bark is used by women in Kanáwar for washing their hair, and adds that it has been tried for paper-making.

It seems highly probable that the Mezereon which **Irvine** and other writers mention as imported into India from Afghánistán and Persia is this plant and not the true **D. Mezereum.**

SPIRIT,
BERRIES.
135
136

Spirit.—Brandis says that on the Sutlej a spirit is distilled from the BERRIES..

D. pendula, *Sm.; Fl. Br. Ind., V., 194.*

Syn.—DAPHNE MONTANA, *Meissn.;* ERISOLENA MONTANA, *Blume.*

Reference.—*Kurz, For. Fl. Burm., II., 333.*

Habitat.—A smaller plant in all its parts, otherwise doubtfully distinct from D. involucrata; met with on the hills between Nattoung and Moulmein, Burma. Kurz says it occurs on the damp hill forests of the Martaban east of Tounghoo at 5,000 to 6,000 feet elevation, and flowers in April.

FIBRE.
137

Fibre.—It seems probable that this plant affords the Nepál paper said to be made in Burma and the Straits.

D. 137

Daphnidium, *Nees ; Gen. Pl., III., 163.* Reduced to **Lindera,** *Thunb., Fl. Br. Ind., V., 182.*

Daphniphyllopsis capitata, *Kurz ;* see **Nyssa sessiliflora,** *Hook. f., Fl. Br. Ind., II., 747 ;* CORNACEÆ.

DAPHNIPHYLLUM, *Bl. ; Gen. Pl., III., 282.*

[EUPHORBIACEÆ.

Daphniphyllum glaucescens, *Blume. ; Fl. Br. Ind., V., 353 ;*   **138**

[1878-9.

Syn.—D. ROXBURGHII, *Baill. ;* GOUGHIA NEILGHERRENSIS, *Wight, Ic., t.*
Vern.—*Nir-chappay* (by the Badagas), NILGIRI HILLS.
References.—*Beddome, Fl. Sylv., t. 288 ; Gamble, Man. Timb., 384 ; Man. Madras Adm. Report, II., 110 ; Balfour, Cycl. Ind., I., 889 ; Thwaites, Enum. Ceylon Pl., 290.*

Habitat.—A small tree, met with in the Nilgiri and Pulney hills, South India, and Ceylon. It is a highly ornamental foliage tree, on account of which it is being cultivated in shrubberies.

Structure of the Wood.—Very inferior, but makes excellent fuel.   Timber.
139

D. himalayense, *Muell. ; Fl. Br. Ind., V., 354.*   **140**

Habitat.—A small tree, very much like the preceding, but found on the Himálayas from Kumáon to Upper Assam, and Burma, at altitudes from 4,000 to 9,000 feet.

In Atkinson's *Himálayan Districts, p. 379,* it is said to be known as   DOMESTIC.
Rakt chandan and *Rakt angliya,* and is "frequently used in marking the   **141**
tika mark on the forehead." The *Rakt chandan* of most writers is **Ptero-carpus santalinus** or **Adenanthera pavonina,** the wood of either of which is in the plains used for marking the forehead.

Dárchíní or **Dálchíní,** see **Cinnamomum Tamala,** *Fr. ;* Vol. II., C. 1183.

Darengri.—Balfour mentions this as a name given in Kashmír to a   **142**
leaf used in dyeing. The writer is unable to discover what plant is meant.

Dari, see **Carpets,** Vol. II., C. 627.

Darmá, see **Mats and Matting.**   **143**

It seems probable that the true *darmá* mat is that made of **Phrag-mites Roxburghii,** *var.* **angustifolia.** The reeds are split open and plaited into mats. Mr. T. N. Mukharji, however, in his work (*Art Manufactures of India, p. 310*) says : "Bamboo mats called *Darmá* are largely employed in Eastern Bengal for the construction of the walls of houses." The writer's experience of Bengal goes towards the conviction that, though similarly constructed and like the true *darmá* mats used in house construction, bamboo mats are not generally designated *darmá* mats.

At a conference held at the Colonial and Indian Exhibition mats of the **Phragmites** reed were shown, as also those of split bamboo ; the gentlemen who examined these were of opinion a trade might be done in the former but not in the latter. The contention as to what is or is not *darmá* is therefore of little importance as compared with the distinction urged above of sending **Phragmites** mats to Europe in preference to bamboo in any efforts towards opening up trade in these articles. This explanation has been thought necessary, since, in his chapter on Mats, Mr. Mukharji makes no mention of the reed mats here specially indicated. The plant from which they are made is abundant on all the islands and sandy river banks in Bengal, and the trade in making and selling these mats is very extensive. See **Phragmites Roxburghii**.

Dates and **Date Palm**, see **Phœnix dactylifera,** *Linn.*

DATISCA, *Linn.; Gen. Pl., I., 844.*

144   **Datisca cannabina,** *Linn.; Fl. Br. Ind., II., 656 ;* DATISCACEÆ.

   Syn.—D. NEPALENSIS, *Don.*
   Vern.—*Akalbir* or *kalbir, bhang-jala,* HIND.; *Akalber, bajr* or *bhang-jala,* N.-W. P.; *Wuftangel,* KASHMIR; *Akilbir, eqilbir, bhang jala, drinkhari, sida atsú,* PB.; *Akalbar,* HIND. IN BOMB. (*Dymock*).
   References.—*Gamble, Man. Timb., 207; Stewart, Pb. Pl., 191; Doñ, Prodr. Nep., 203; Dymock, Mat. Med. W. Ind., 2nd Ed., 355; Murray, Pl. and Drugs, Sind, 43; Baden Powell, Pb. Pr., 372; Atkinson, Him. Dist., 724, 774; Liotard, Dyes, 90, 96; Wardle, Report on the Dyes of India, p. 24; Linnæan, Soc. Jour. XIX., 4; Balfour, Cyclop., I., 897, 1005; Robinson, Gleanings from French Gardens, p. 42.*

Habitat.—A tall, erect herb, resembling hemp, hence the specific name. It is met with in the temperate and sub-tropical Western Himálaya from Kashmír to Nepál, at altitudes from 1,000 to 6,000 feet, but is by no means a plentiful plant. **Dr. Dymock** says : " The plant is a native of Sind." This seems highly doubtful.

DYE.   **Dye.**—Many writers allude to this as a special dye used in Kashmír to
145   dye silk a delicate yellow colour. Throughout the Himálaya it is more or less employed, being combined, it is said, with red colours to soften the tint, and with indigo to produce a favourite green (*pista*). **Stewart** writes : " In some of the places where it grows, the yellow root is used to aid in dyeing red, and **Cleghorn** states that it is exported from Pangi, Lahoul, and Kúllú, to Nadoun and Amritsar to be used in dyeing woollen thread. **Edgeworth** mentions that for this purpose it is combined with *asbarg* " (**Delphinium saniculæfolium** [or rather **D. Zalil,**—*Ed.*]). In a recent report furnished by the Conservator of Forests, North-West Provinces, it is stated that the dye-stuff is exported from the Himálayas to the plains to be used both as a dye and medicine.

The parts employed are the yellowish wood, bark, and root.

SPECIAL OPINIONS.—§ " Used extensively as a dye, for which purpose it is exported from Kashmír " (*Surgeon-Major J. E. T. Aitchison*). " **Datisca cannabina** (*akilbir*) is found sparsely scattered throughout the forest in upper Kunawur, and more plentifully to the west of Wangtú, particularly in the Saldung Valley. It is known as producing a yellow dye, and the roots sell at Amritsar for R14 per maund of 80℔. In August the roots are dug up (the bark peeled off), dried in the sun, and then packed for export to Rampur or Amritsar. About 200 to 300 maunds are obtained annually in Bashahr on the Sutlej. It is not known if any be sent from the Rupin or Paber Valleys. One root yields from ½ to 1 seer. The seed or flowers are of no use as far as is known " (*G. G. Minniken, Esq., Assistant Conservator of Forests, Bashahr*).

MEDICINE.   **Medicine.**—Medicinally, it acts as a sedative in rheumatism. As a
146   bitter and purgative, it is used sometimes in fevers and in gastric and scrofulous complaints. In intermittent fevers, it is administered in doses of from 5 to 15 grains (*Dymock, Mat. Med. West. Ind., 1st Ed*). In the second edition of his work **Dr. Dymock** seems to have modified slightly his statements regarding the drug, but adds that in " Khagan the bruised

| The Akalbir. | (*G. Watt.*) | DATURA. |

MEDICINE.
Root.
147
Bark.
148
CHEMISTRY.
149

ROOT is applied to the head as a sedative." **Balfour** says it is used as an expectorant in cattarrh. " The BARK also contains a bitter principle like quassia."

Chemistry.—The peculiar property of the dye principle of this plant does not seem to have been worked out, and results of some interest may be expected from its thorough examination. **Dr. Warden** (Professor of Chemistry, Calcutta), in reply to an enquiry on this subject, furnished the following brief note :—

" It contains a glucoside, datiscin, which crystallises in colourless needles or laminæ. It forms with alkalis a deep yellow solution ; and, according to **Bracannot**, it dyes fabrics, both mordanted and unmordanted. It forms yellow lakes with lead salts." **Dr. Dymock** furnishes a slightly more detailed account. He writes : " The leaves and roots contain a glucoside $C_{21} H_{22} O_{12}$, which may be obtained by exhausting them with alcohol, evaporating to a syrup, and precipitating the resin with water ; from the decanted liquid crystals may be obtained which should be redissolved in alcohol, and the remaining traces of resin removed by re-precipitation with water. Datiscin may then be obtained in colourless silky needles or scales, little soluble in cold water, and only sparingly so in warm water and ether. The crystals are neutral and have a bitter taste ; they melt at 180° C." (*Wurtz, Dict. de Chem., t. I., 1134.*)

DATURA, *Linn. ; Gen. Pl., II., 901.*

150

A genus of herbaceous plants, containing in all some 10 or 12 species. These are widely distributed throughout the tropical and temperate regions, both of the Old and the New World. They are all regarded as being highly poisonous, and from the remotest antiquity, have been used both medicinally and criminally. It seems probable, however, that they have been known in Europe, comparatively speaking, during modern times only. By some writers **Datura Stramonium** is supposed to be the στρύχνος μανικòς of **Theophrastus** and **Dioscorides** This, however, seems open to doubt, as the descriptions of the plant alluded to by these classical authors do not justify such an opinion. It is, indeed, doubtful if even **Stramonium** was known during the Roman Empire. In modern Greek it bears the name τατουλα, a word clearly derived from the Persian *Tátúlah*. The earliest Muhammadan writers on medicine, however, describe several forms under the name *Jouz-el-mathil*; and the modern Indian name *Dhatura* and the Persian *Tátúlah* come from the Sanskrit *Dhustura*, while the name given to it in Southern India *Ummettak-káy* comes from the Sanskrit synonym *Unmatta*.

The Arabic and Sanskrit literatures fully establish an ancient knowledge of the properties of the drug. But so much difference of opinion prevails amongst modern writers on the medicinal as well as non-medicinal forms, that it may not be out of place here to analyse these opinions, and to furnish, at the same time, brief contrastive descriptions that may assist in the separate recognition of the forms met with in India. We may thus be able to procure in the future more trustworthy information than we presently possess. It is customary, for example, to read of the " white-flowered datura " and of the " purple-flowered datura, " but the colour of the flower is in all probability a matter of accident or of cultivation—it is certainly not specific. Writers who speak of the purple-flowered being a more powerful poison than the white, may have originally got their ideas from an ascertained fact, namely, that a form, *one* of the characters of which was to have flowers of that colour, had the poisonous property highly developed. But any of the Indian species or varieties may have purplish

DATURA.	Colour of the Flowers of Datura.

HISTORY.

flowers. Indeed, **D. alba,** a name formerly given to what is now treated as a variety of **D. fastuosa,** has often purplish flowers. So, again, **D. Metel** has generally white flowers, but sometimes they are purplish. Either of these species may, however, have been the white-flowered datura of the early writers ; and very probably it was **D. Metel** that was their less poisonous white-flowered datura and not **D. alba** as supposed by most modern authors. It thus follows that nothing could be more misleading than to base an opinion as to the merits of a datura simply because of its flowers being purple or white. Few plants could generically be more easily recognised than the *dhaturas.* The long-plaited corolla and inflated calyx, the latter separating transversely on fertilisation so as to leave a collar around the base of the thorny fruit, are unmistakeable characters. And these characters are peculiar to all the forms, but cultivation may modify the colour of the flower or even double the flower—one corolla appearing to grow from the interior of the other. How far the chemical properties of the plants are affected by selection or care in cultivation it is impossible to discover. But one thing is certain, that the daturas have been and are to some extent cultivated, and many of the peculiar forms met with in certain localities are most probably escapes from a former cultivation. Indeed, it is scarcely possible to avoid the conviction that cultivation has had far more to say to the peculiarities of the daturas than is generally believed. In a great many Indian localities the plant appears at most only semi-wild, and has all the appearances of being the degenerated offspring of a cultivated stock, once upon a time much more generally cared for than at the present day. There are, for example, numerous forms known to the native expert that would be utterly unrecognisable in the herbarium. Like the forms of **Aconitum Napellus** some of these are poisonous and others comparatively innocuous. The shepherd will dig up and eat one form of Aconite but eschew another, recognising it as a virulent poison. But to the botanist they are indistinguishable. This same knowledge is prevalent regarding the forms of datura. That we should longer remain entirely ignorant of these facts is doubly to be regretted, since we are alike unable to check criminal abuse and to take full advantage medically of the meritorious forms.

As sold in the Indian bazars, datura should be used with the greatest caution. It would richly repay any person having the opportunity and leisure to prosecute such researches, to cultivate in India side by side all the forms known to the natives, and having critically examined and described these, to have them subjected to chemical analysis. It might then be possible to establish some more trustworthy standard by which to differentiate the daturas than we possess at present. Such a study might not reveal a more extensive series of varieties and cultivated forms than is supposed to exist ; but that the specific distinctions recognised by botanists would thereby be broken down seems highly likely. Possibly all the Indian daturas constitute but one or at most two species. The differences currently admitted are scarcely more than what in most other genera would be attributed to climatic causes. **Datura Stramonium** might be called the type of the temperate or alpine series, and **D. fastuosa** that of the tropical or plains assemblage. Some of the conditions of the former, like some of those of the latter, have blueish flowers ; certain are recognised as virulent poisons, others sufficiently less so to be employed neither criminally nor medicinally. M. Naudin devoted much careful study to the species of Datura, cultivating all those of which he could procure the seeds. It is recorded that Dr. George Bidie, C.I.E., of Madras, sent seeds of **D. alba** to **Professor Flückiger,** and that these were handed over to **Naudin.** As the result, plants, *first,* of the true **D. alba,** were obtained ; *second,* plants with flowers white inside and violet outside ; *third,* plants with double corollas of a large

HISTORY.

size and yellow colour. It is remarkable that these should all be said to
have been obtained from the seed of **D. alba**; and it would be instructive to
know (if by any chance the observation was made at the time) whether the
seeds were collected from one individual plant or from two or more plants.
It has to be admitted that the utmost we can say of the Indian daturas is
a confession of defective knowledge, and an appeal for more critical study.
The reader is referred to the remarks below (under each species) for a
brief description of the forms commonly recognised. But before passing
from this introductory account it may be as well to allude to one or two
authors whose writings deserve consideration, although it is impossible to
decide to what particular species or form they more especially allude.
Garcia de Orta visited India in 1534 and became physician to the hospital at Goa. In 1563 he published his *Coloquios*, in which much valuable
information is given regarding datura and most other Indian drugs. He
describes, at pages 83 and 84, the criminal uses of the drug in the hands of
servants and highway robbers. Shortly after, **Huyghen van Linschoten**
visited India; and the Journal of his Voyage (published 1596) gives a most
complete account of *dhatura*—the plant found around Goa—and hence
presumably a form of **D. fastuosa.** He writes: "They have likewise an
hearbe called Deutroa which beareth a seed, whereof brusing out the sap
they put it into a cup or other vessel and give it to their husbands,
eyther in meate or drinke and presently therewith, the man is as though
hee were halfe out of his wits, and without feeling, or else drunke (doing
nothing but) laugh, and sometimes it taketh him sleeping (whereby he
lieth) like a dead man, so that in his presence they may doe what they
will and take their pleasure with their friends, and the husband never
know of it. In which sort he continueth foure and twentie hours long,
but if they wash his feete with colde water hee presently reviveth and
knoweth nothing thereof but thinketh he had slept." Commenting on
Linschoten's account of the drug his contemporary **Paludanus** states that
"*Deutroa* of some called *Tacula*" (a misprint for *tatula*), "of others
Datura, in Spanish *Burla Dora*, in Dutch *Igell Kolban*, in Malaha
Vumata Caya, in Canara *Datura*, in Arabic, *Marana*" (the Arabian name
is *Jauz masal*) "in Persian and Turkie *Datula*. Of the description of
his hearbe and fruit you may read in the *Herballes*, if any man receaveth
or eatheth but half a dramme of this seed, hee is for a time bereaved of
his wits and taken with an unmesurable laughter." **Linschoten** frequently recurs to datura. "This hearbe," he says, "groweth in all places
in abundance, and although it is forbidden to be gathered or once used,
never-the-less those that are the principal forbidders of it, are such as
dayly eat thereof, &c." It is somewhat remarkable, however, that while he
enlarges at great length on the various criminal uses of the drug, he makes
no mention of the medicinal.

The *Makhzan* recommends preference to be given to the purple-flowered datura, and the author adds as his reason that all the parts of the
plant are powerfully intoxicating and narcotic. He gives the following
account of datura intoxication:—"Everything the patient looks at appears dark; he fancies that he really sees all the absurd impressions of
his brain; his senses are deranged; he talks in a wild disconnected manner,
tries to walk but is unable; cannot sit straight; insects and reptiles float
before his eyes, he tries to seize them and laughs inordinately at his failure. His eyes are bloodshot, he sees with difficulty and catches at his
clothes and the furniture and walls of the room. In short, he has the appearance of a madman." According to **Dutt**, "Sanskrit writers do not
make any distinction in the properties of the two varieties of datura, and in
practice both are indiscriminately used. Sometimes the white-flowered

DATURA fastuosa.	The Black Datura.

variety is specified, as, for example, in a prescription for insanity, quoted below. *Dhaturá* leaves are used in smoking by debauched devotees and others accustomed to the use of *gánjá.* The seeds are added to the preparations of *bháng* (leaves of **Cannabis sativa**) used by natives for increasing their intoxicating powers. The use of the powdered seeds in sweetmeats, curry powder, &c., for the purpose of stupifying travellers and then robbing them is well known." Further on, Dutt says of the habit of smoking the leaves as a cure for spasmodic asthma : " I have not met with any written prescription for it in Sanskrit or vernacular medical works, nor does the *Taleef Shereef* allude to the practice as known to the Mussulman Hakims. It would seem, therefore, that this use of the drug is of recent origin." Smokers of *gánjá,* however, as is well known, suffer from violent fits of a kind of false asthma, so that the habit of smoking the leaves by devotees, &c., to which Dutt alludes is practically a recognition of the property the knowledge of which he excludes the early Sanskrit authors from possessing. In the passage quoted above it may be doubted whether Dutt is narrating his own knowledge of the modern employment of the leaves or is quoting the opinions of Sanskrit writers. The point is of considerable historic interest. Ainslie found that the natives of South India, during his time (1820), were unacquainted with the value of the leaves in the cure of asthma, and it is commonly stated by writers on the subject that the discovery of this property is due to European medical science.

151 **Datura fastuosa**, *Linn. ; Fl. Br. Ind., IV., 242 ;* SOLANACEÆ.

THE BLACK DATURA.

Syn.—DATURA HUMMATU, *Bernh. ; Dalz. and Gibs., Bomb. Fl., 174.*

Vern.—*Kala dhatúrá*, HIND. ; *Kala dhuturá*, BENG. ; *Dhatura*, SANTAL ; *Khunuk* (according to Irvine), BEHAR ; *Toradana* (Peshawar District), PUSHTU ; *Dhaturo* (there are two kinds—*acho*, white, and *káro*, black, Stocks), SIND ; *Kala dhatúrá*, BOMB. ; *Kálá-dhatúrá, údah-dhatúrá,* DEC. ; *Kala dhatúrá,* or *kálo-dhatúro,* GUJ. ; *Karu-vúmattai* (**Moodeen Sheriff**), *Karu-umate, karú umatay* (Ainslie), TAM. ; *Nolla-ummetta* (**Elliot**), TEL. ; *Karu-ummatta* (**Moodeen Sheriff**), *rotecubung, kechubung* (Ainslie), MALAY ; *Pa-daing-ame, padáyinkhatte,* BURM. ; *Attana* (Trimen), *Kalu attana* and *antenna* (Ainslie), SING. ; *Dhattúra, dhustura, unmatta, kála-hémiká* (**Moodeen Sheriff**), *krishna dhattúra* (Ainslie), SANS. *Jouz-massel* (Avicenna states is more correctly **D.** Metel, but that name is now given to this species) ; *Jouz-másle asvad, jouz-másame-asvad,* ARAB. ; *Kechu-búh* (according to Ainslie), ARAB. in Egypt ; *Tátúrahe-siyáh* (Nabrak according to Stocks), *guzgiah* (Ainslie), *kaiz-másale-siyáh, kouz kunáe-siyah,* and *Tátúrahe-siyáh* (**Moodeen Sheriff**), PERS.

References.—*Roxb., Fl. Ind., Ed. C.B.C., 188 ; Dalz. & Gibs., Bomb. Fl., 174 ; Flora Andhrica, 128 ; Mason's Burma, 488, 798 ; Report on the Botany of Merwara by J. F. Duthie ; Voyage of John Huygen van Linschoten to the East Indies, I., 210-211 and II., 68 ; Garcia de Orta, Coloquios, pp. 83-84 ; Ainslie, Mat. Ind., I., 442, 636 ; O'Shaughnessy, Beng. Dispens., 59 ; Moodeen Sheriff, Supp. Pharm. Ind., 130 ; U. C. Dutt, Mat. Med. Hind., 207 ; Dymock, Mat. Med. W. Ind., 2nd Ed., 518 ; Flück. & Hanb., Pharmacog., 462 ; U. S. Dispens., 15th Ed., 1364 ; Bent. & Trim., Med. Pl., 192 ; S. Arjun, Bomb. Drugs, 97 ; Murray, Pl. and Drugs, Sind, 155 ; Waring, Bazar Med., 52 ; Irvine, Mat. Med., Patna, 27 ; Hummatu in Rheede's Hort. Mal. ; Baden Powell, Pb. Pr., 297, 363 ; Atkinson, Him. Dist., 735 ; Drury, U. Pl., 188 ; Birdwood, Bomb. Pr., 209 ; Balfour, Cyclop., I., 897 ; Smith, Dic., 152 ; Med. Top. Ajmir, 133 ; Mysore and Coorg Gaz., I., 56, 63 ; Gazetteers (Kanara), XV., 439 ; (Gujrat) 11 ; Peshawar 26 ; Orissa II., 180 ; Special Report from the Government of Burma where it is* *said to occur in Chindwin Valley, Kyaukpyu, Mandalay, Upper Burma, Toungoo, Ruby Mines, and Bhamo districts ; Special Report from Bengal, where the plant is said to be often grown in gardens.*

Habitat.—A small shrub found all over the tropical parts of India, in waste places. There are said to be two if not three recognisable forms of this plant, the type being ascertained by the following characters : Flowers white or purplish, large ; corolla often 7 inches long with a spreading mouth which is sometimes 5 inches in diameter ; teeth 5 or 6, but the flower frequently seen to be double, one corolla within another. Fruit roundish (not ovate), spinose all over ; stalk recurving with maturity until the fruit becomes pendent. When the seeds are ripe the fruit opens irregularly, forming a few short valves.

This is very generally reputed to be the most virulent form of the Indian daturas ; but in few cases, indeed, do authors distinguish it from the variety **alba** described below, so that the statements of medical uses given both in this place and under **D. alba** may fairly well be regarded as dealing jointly with one and the same species. As already quoted, the *Makhzan* gives preference to the purple-flowered datura (presumably **D. fastuosa**) ; but according to Dutt the Sanskrit writers do not make any distinction in the properties of the two plants, though the white form (**D. alba**) is recommended to be used for insanity. **Dalzell** and **Gibson** say there are several well-known varieties of **D. alba.** They, however, make no mention of these being used medicinally. Of **D. Hummatu** (= **D. fastuosa**) they remark that in Bombay it is almost as common as the preceding. They then add : "These plants are intoxicating and narcotic ; the root is used in violent headaches and epilepsy ; poultices are made of the leaves for repelling cutaneous humours ; the bruised seeds are applied to boils." The *Pharmacopœia of India* makes **D. alba** officinal, but says of **D. fastuosa**: "It is generally thought to be the more powerful of the two, but there is no evidence of this being the case. The probability is that they possess equal powers as a narcotic and anodyne ; but clinical observation is wanting to confirm this." It is only necessary to say in order to confirm this confession of ignorance, that, while the active principle of **D. Stramonium** has been isolated, and its properties determined, neither **D. alba** nor **D. fastuosa** have as yet been critically examined, and it is therefore practically by a comparison only with the therapeutic actions of these and of **D. Stramonium**, that we are enabled to infer that they contain the alkaloid *Daturine.*

But it may be added that **D. fastuosa** is so universally believed to be stronger than **D. alba** or **D. Metel**, that it is preferentially used by the criminal classes.

Criminal Purposes.—Considered by some of the native doctors a better variety than the white ; the *Pharmacopœia of India* affirms that there is no foundation for this opinion. The SEEDS constitute a favourite poison for criminal purposes. These seeds or a preparation from them are generally employed by the Indian road-side poisoners, not for the purpose of destroying life, but simply to stupify their victims with the view of easy committal of theft. Death may follow as a consequence of overdose. (*See* Chevers' *Jurisprudence.*) The seeds are also in Bengal employed to render liquor more intoxicating, and for this purpose they are burned upon charcoal, the vessels being inverted to catch the smoke. The seeds may also be used in the form of a powder for the same purpose when a stronger intoxicant is desired. When the vessels are full of smoke the liquor is thrown into them and the mouth covered over for a night. It seems remarkable that when thus burned, the smoke should retain its poisonous and intoxicating properties. **Dr. Dymock** states that in

CRIMINAL PURPOSES.

Seeds.
152

DATURA fastuosa.	The Black Datura.

CRIMINAL PURPOSES.

Bombay the smoke from the seeds burned over charcoal is also used to make liquor intoxicating. **Mr. H. Sewell,** Collector of Cuddapah, reports: "This is known as *Umatai* in Tamil. There are two species—white and black. Both grow wild and are not cultivated. The former is not used for any purpose. For mixing with intoxicating beverages, for instance toddy, the latter is useful. Its seeds are soaked in that liquor along with a quantity of poppy seeds ground to a paste. The mixture is then strained and mixed with fresh-drawn toddy, which gives the latter intoxicating power. It is not possible to estimate the quantity of datura seed consumed in this way." **Mr. Baden Powell** (*Pb. Prod., I., 297*) alludes to a series of samples shown in the Lahore Museum as illustrative of the criminal methods of using the drug in Upper India. He says (quoting from a report on these written in 1863): "The series consists of the seeds of the plant in their raw state, seeds roasted, essence of the seeds, atta (flour) drugged with the poison, sugar ditto, and tobacco ditto." He then remarks this is the agent used by the "Thugs" to stupify their victims. "Both kinds of the dhatura, the white and the purple, are used, but the white (*sic*) is considered the most efficient." "For poisoning purposes the seeds are parched and reduced to a fine powder: thus it is easily mixed with sugar, atta, tobacco, &c. Also the professionals distil the seeds with water, forming a powerful essence; ten drops of this is sufficient if put into a *chillam* of the *huka* to render a man insensible for two days. The taste is acrid and bitter, and soon followed by a burning suffocating sensation. It is very difficult to detect in a *post mortem* examination. The victims are usually discovered in a state of insensibility, and breathing hard and heavily; if removed, care should be taken not to expose them to the heat of the sun, which is fatal. The action of the poison is quicker in the hot weather than in the cold; much, of course, depends on the individual constitution of the victim, but usually in hot weather it begins to work in five minutes, coma supervening within the hour. In cold weather it begins to act in a quarter of an hour or twenty minutes."

MEDICINE.
153

Medicine.—For Medical uses proper, see under *var.* **alba.**

SPECIAL MEDICAL OPINIONS RECORDED UNDER **D.** fastuosa.§—"The form of datura with blue flowers is considered stronger than the white kind. No doubt this drug prevents hydrophobia. There are persons here and there in this district who are considered professors in curing hydrophobia. But none of them will reveal the secret of the medicine used. With great pains and labour I discovered this remedy. I have myself treated many cases successfully, and some of my pupils have been equally successful. My treatment consists in giving the medicine previous to the time of the development of hydrophobia."

"It is usually found that hydrophobia comes forty days after the patient has been bitten by the mad dog (except some rare cases which I have known to happen within two or three weeks). My treatment is to give the following medicine two weeks after the patient has been bitten, *i e.,* between the fifteenth and twenty-fifth days. In the morning after the fifteenth day of the bite, about six o'clock, give a dessert spoonful of tea wood-charcoal powder. (This seems to be given lest the poison of the juice overcomes the patient.) Half an hour after give an ounce of the JUICE of the black datura leaves. Soon after follow with Palmyra jaggery or something else in order to check vomiting. Then, bind the person lest he does mischief to others, and keep him in the sun for four or five hours, until midday. Then, the person gradually becomes mad, and does many things like the mad dog (when these symptoms appear, it is evident that the patient had been really bitten by a mad dog, and that he will totally recover). In the afternoon pour many pots of cold water over the head. This causes

The Black Datura.	(*G. Watt.*)

MEDICINE.

great annoyance to the patient and he resents it to the utmost, protesting loudly. Food should now be given such as pork, salt-fish, brinjal, horse-gram, Bengal-gram, &c., &c. The patient may be considered out of danger and should receive simple light diet.

"If you were to treat a person already suffering from hydrophobia, then you must scratch the front part of his head with a lancet so as to make it bleed a little and rub in the ground LEAVES of the black datura as well as give the juice internally" (*V. Ummegudien Mattapollian, Madras*).

Leaves.
154

The above has been given as a sample of many similar violent remedies, the writer has received, from native practitioners, in all of which datura is recommended in the cure of supposed mad dog bites. The English of the original has been slightly altered and superfluous matter removed, but the principle and method of treatment has been faithfully preserved. [*Ed., Dictionary, Economic Products.*]

"I have used this drug pretty extensively. In painful swellings I apply the JUICE of fresh leaves, or make a poultice of them. The fresh juice in ophthalmic pain I find very useful; it checks the inflammation if there be any. Inhalation of the smoke of the burning DRY LEAVES and TWIGS is always useful in asthmatic fits. Smoking the powdered dry leaves and twigs relieves the spasm, but when smoked in excess brings on giddiness and fainting. The seeds are said to be useful in cases of hydrophobia, and the anther in cholera" (*Civil Surgeon D. Basu, Faridpur, Bengal*). "The dried ROOT of the plant I have frequently used, as smoking, to relieve fits of asthma" (*Nundo Lall Ghose, Bankipur*). "In ear-ache the fresh juice of the leaves is useful, a drop or two poured inside the ear" (*Assistant Surgeon T. N. Ghose, Meerut*). "The dried leaves are smoked in cases of asthma. The expressed juice of the leaves is used as an external application to relieve the pains of gout and rheumatism, and in cases of glandular inflammation and enlargement. The leaves are also employed as poultices to check inflammation of the breast and excessive secretion of milk in cases where an abscess is threatened" (*Civil Surgeon J. H. Thornton, B.A., M.B., Monghyr*). "When in Jessore, about five years ago, in two separate instances, a batch of men were sent to me by the police, all with well-marked symptoms of dhatura-poisoning, and some proved fatal" (*Civil Surgeon G. Price, Shahabad*). "The leaves constitute an anodyne poultice. The SEEDS are mostly used in medicine. They are believed to be aphrodisiac and are also employed for cough, diarrhœa, asthma, intermittent fevers" (*Surgeon-Major Robb, Civil Surgeon, Ahmedabad*). "Smoking of leaves is a useful antispasmodic in asthma and chronic bronchitis. I found the juice of fresh leaves efficacious when applied over painful glandular swellings" (*Assistant Surgeon Shib Chandra Bhattacharji, Chanda, Central Provinces*). "Dry root of the above, in about half-grain doses, is given by the Hakims of the N.-W. Provinces to take with betel leaves in syphilitic diseases. The seeds are also employed by them for impotence in the following way: seeds of 15 fruits dried and pounded are well boiled in ten seers of cow's milk; out of this milk as much ghee as possible is made; this ghee is believed to contain strong aphrodisiac properties, and is rubbed on the genitals twice a day, to stimulate them, and about four grains of the ghee is also given internally once a day (*Assistant Surgeon Nobin Chunder Dutt, Dhurbhanga*). "In Mysore the juice of the leaves is given once daily with curdled milk for gonorrhœa" (*Surgeon-Major John North, I.M.S., Bangalore*). "Have used the leaves warmed over a fire nightly as an external anæsthetic in rheumatism" (*Dr. Picachy, Civil Medical Officer, Purneah*). "The leaves are useful as a local application in rheumatism. The concentrated juice of the leaves is prescribed in mumps as a

Juice.
155
Dry Leaves.
156
Twigs.
157

Root
158

Seeds.
159

D 2

DATURA fastuosa.	The White Datura.

MEDICINE.

local application, and has a marked effect in reducing the swelling and tenderness" (*Narain Misser, Kathe Bazar Dispensary, Hoshangabad, Central Provinces*). "An extract made from the seeds is a good mydriatic, and the leaves are used as emollient and suppurative" (*Honorary Surgeon Easton Alfred Morris, in Medical charge, Tranquebar*). "The leaves of this plant are boiled, made into a poultice, and applied locally to boils and abscesses to relieve pain and hasten suppuration" (*Surgeon W. F. Thomas, Mangalore*). "A few seeds with *úqarqarha* (**Anacylus Pyrethrum**) root and cloves are chewed as an aphrodisiac" (*Dr. Emerson*). "A paste composed of datura and turmeric is useful in checking inflammation of the breasts" (*Civil Surgeon J. Anderson, M.B., Bijnor*). "The juice of the leaves is a good substitute for Belladonna" (*Surgeon Major P. N. Mookerjee, Cuttack, Orissa*).

160

Var. alba, *Nees ; Fl. Br. Ind., IV., 243.*

WHITE DATURA.

Syn.—D. ALBA, *Nees.*

Vern.—*Saféd-dhatúrá, sádah-dhatúrá,* HIND. ; *Dhútúrá,* BENG. ; *Dather,* KASHMIR ; *Dhotará,* MAR. ; *Ujlá-dhatúrah,* DEC. ; *Dholo dhatúro,* GUJ. ; *Umattai,* TAM. ; *Ummetta, duttúramu,* TEL. ; *Ummatte-gidá,* KAN. ; *Ummatta, ummam,* MALAY. ; *Padáyin-phiu,* BURM. ; *Sudu-attana,* SING. ; *Ummatta-vrikshaha,* SANS. ; *Jous-másal* or *Jous-másle-abyaz,* ARAB. ; *Kouz-másale-saféd, tátúrahe-saféd,* PERS.

NOTE.—It is doubtful how far the vernacular names given by authors for D. fastuosa and D. alba can be regarded as specific, since either forms may have white or blue flowers. Indeed, these plants have more generic than specific names, the simple equivalents of *Datura* of the plains as the names given by the hill tribes are but further synonyms, though given to the form met with in the higher regions, *viz.,* D. Stramonium.

References.—*Flora Andhrica, 48, 186 ; Mason's Burma, 488, 798 ; Pharm. Ind., 175, 460 ; Ainslie, Mat. Ind., I., 442 ; O'Shaughnessy, Beng. Dispens., 469 ; Moodeen Sheriff, Supp. Pharm. Ind., 130 ; U. C. Dutt, Mat. Med. Hind., 207 ; Dymock, Mat. Med. W. Ind., 2nd Ed., 518 ; S. Arjun, Bomb. Drugs, 96 ; Murray, Pl. and Drugs, Sind, 155 ; Year Book Pharm., 1880, 250 ; Baden Powell, Pb. Pr., 363 ; Atkinson, Him. Dist., 735 ; Drury, U. Pl., 188 ; Lisboa, U. Pl. Bomb., 268 ; Bomb. Gaz., V., 27 ; Balfour, Cyclop., I., 897 ; Home Dept. Cor. regarding Pharm. Ind., 222, 230, 321 ; Madras Man. Admin., II., 65 ; Man. Cuddapah, 200 ; Orissa Gaz., II., 180 ; Gaz., Mysore and Coorg, I., 56 ; Gaz., N.-W. P. (Meerut), II., 506, III., 81.*

Habitat.—A large, spreading annual, two to four feet high, found like the type form of the species throughout the warmer parts of India, though it only rarely ascends above 3,000 feet. This form doubtfully deserves the rank of a variety. The characters of the flower and fruit are almost identical with that already given, except that they are smaller, the teeth of the calyx being less than half the size of those in D. fastuosa, and almost lanceolate-acuminate instead of ovate-acuminate. Flowers white or slightly bluish outside.

If anything this is even more abundant, and fortunately so, for it is very generally reputed to be less virulent than the black dhatura.

MEDICINE.
Seeds.
161
Leaves.
162

Medicine.—The properties of the Indian plains *Datura* are supposed to be practically identical with those of D. Stramonium and analogous to those of Belladonna. The officinal parts are the SEEDS and the LEAVES : of the former a tincture, an extract, and a plaster are prepared, and of the latter a poultice, but the dried leaves are also smoked to relieve urgent symptoms in spasmodic asthma, the dyspnœa of phthisis, emphysema of the lungs, or even in chronic catarrh. The tincture and extract are sedative and narcotic ; the former preparation by many writers is recommended as a useful and cheap substitute for opium, 20 drops of the tincture being equal to a grain of opium. The latter has been frequently employed as a convenient substitute for extract of Belladonna, the dose being a quarter of a grain, increased gradually to a grain and a half, thrice daily. Dr. Bidie suggests an extract from the leaves, and in the *Pharmacopœia* this preparation is well spoken of : "In a case of phthisis, in which it was employed in two-grain doses, it acted favourably on the dyspnœa, and produced much the same effect as extract of Belladonna, in doses of a third of a grain." Dr. Bidie considers that the larger dose in which it can be administered is an advantage. The plaster and the poultice are effectual local anodynes in case of nodes, rheumatic enlargements of the joints, painful tumours or external piles. The plaster is frequently used on the chest in asthma and chronic pulmonary affections, but neither should be applied to ulcerated surfaces owing to the risk of absorption of the poison. Amongst native women a poultice of *datura* leaves is a favourite method of arresting the secretion of milk in cases of painful breasts. The active principle *daturine* in place of atropia has been proposed for ophthalmic purposes, but with comparatively little success. The effect of the administration of *datura* is to produce dilatation of the pupil : "should it become very large and dilated this may be taken as a sign that the medicine has been carried as far as it can with safety, whether it has produced its other intended effects or not" (*Waring, Baz. Med.*, 53).

Waring recommends it to be tried in tetanus (lock-jaw consequent on a wound) when other better remedies are not procurable. "A poultice of the leaves, renewed three or four times a day, should be kept constantly to the wound, which should be further cleansed if covered with thick discharge or slough by the process of irrigation of tepid water. The tincture of datura, doses of 20 to 30 drops in water, may also be given internally three or four times a day. The dose must be regulated by the effect produced, but it may be continued (unless the spasms previously yield), till it produces full dilatation of the pupil with some degree of giddiness, drowsiness, or confusion of ideas, beyond which it is not safe to carry the medicine. If the spasms abate, *i.e.*, if they recur at more distant intervals, and are less severe and prolonged when they do occur, the medicine in smaller doses, at longer intervals, may be continued till the spasms cease altogether ; but if, under the use of the remedy, after it has produced its specific effects on the system, the spasms show no sign of abatement, no good, but perhaps harm, will result from continuing it. In addition to the above means datura liniment should be well rubbed in along the spine several times daily. The patient should be confined to a dark room, and protected from cold draughts of air ; the bowels should be opened, if necessary, by turpentine enemas. The strength should be supported by strong beef-tea or mutton broth, by eggs beaten up with milk, and by brandy or other stimulants" (*Bazar Medicines*, 56).

The above may be viewed as a brief abstract of the current European medical uses of datura, but by the natives of India the drug is highly spoken of in the treatment of insanity, and of the painful headaches which often precede epilepsy and mania ; and Ainslie mentions that Muhammadan doctors especially prescribe for these purposes a powder of the ROOT in very small doses not exceeding a quarter of a grain increased to three grains. Ainslie adds that "Berguis and Stoerck ordered the inspissated juice of the leaves of D. Stramonium in epilepsy." Indeed, the modern use of datura may be said to date from Baron Stoerck's success with it in the treatment of mania and epilepsy. Though still occasionally employed for these diseases, its use might be said to be almost confined to neuralgic and rheumatic affections, dysmenorrhœa, syphilitic pains, cancerous sores, and

Root.
163

DATURA fastuosa.	The White Datura.
MEDICINE.	

spasmodic asthma, but most of all for the last complaint. **Waring** affirms that in some few cases it is not serviceable in asthma, but so frequently is it of great benefit that patients subject to chronic asthma should always keep a pipe filled and ready to light. **Dymock** remarks that " Sanskrit writers describe the plant as beneficial in mental derangements, fever with catarrhal and cerebral complications, diarrhœa, skin diseases depending upon the presence of animal parasites, painful tumours, inflammation of the breasts, &c. A pill made of the pounded seeds is placed in decayed teeth to relieve toothache, and the leaves are smoked along with tobacco in asthma. According to Dutt no mention of the latter use of the plant is to be found in old Hindu books. Muhammadan writers also are silent upon this point. Ainslie found upon enquiry that the physicians of Southern India were unacquainted with the value of datura in spasmodic asthma, but he tells us that his friend, **Dr. Sherwood** of Chittore, noticed the smoking of **D. fastuosa** as a remedy in that disease. In the Konkan the juice of **D. alba** is given with fresh curds in intermittent fever to the extent of one tolá during the intermission, and at least two hours before the fever is expected." Ainslie (p. 637) mentions a case in which great relief was obtained in sciatica from an extract administered in one-eighth of a grain to grain doses. **Drury** states that a preparation of the leaves in oil is used in the cure of itch and rheumatic pains by being rubbed on to the part affected. In Sind, it is said that a poultice of the bruised leaves and rice flour is believed to relieve the pain and hasten the expulsion of guinea-worm. "The leaves of the white variety are sometimes chewed with the same object " (*Murray*). " In Rájputana mothers smear their breasts with the juice of the leaves to poison their new-born female children" (*Drury*). **Mr. H. Z. Darrah**, Director of Land Records and Agriculture, Assam, has furnished the following information regarding the daturas of the Assam Valley : "The Assamese *dhutara* is probably the **D. Stramonium.**"—[This is most unlikely, since we have no knowledge of that species existing so far to the east ; it is more than probable that one or two forms of **D. fastuosa**, or possibly also **D. Metel**, constitute the *dhutá-rá* of Assam.—*Ed.*] " The white flower, the purple flowering, and also the yellowish-tinged variety, are met with growing wild in villages and waste places. A few plants are specially protected for medicinal use. It is known as a strong poison and to cause delirium. The dried leaves are rarely smoked, and then only as a remedy in illness, but the leaves are used as a paste and applied : the seeds are taken internally with other articles as a medicine, and sometimes the root is used. It is not used as an intoxicant. It is said, according to a popular idea, to be put as an ingredient into a medicine used to prevent hydrophobia after the bite of a mad dog, but is given carefully and in sufficient quantity only to produce delirium or madness, which is thought to take the place of the madness of the hydrophobia. It is said to be ineffectual when hydrophobia has begun." **Mr. Sewell**, Collector of Cuddapah, Madras, writes that two forms grow wild in his district—the white and the black ; the former is not used for any purpose, but the latter is employed for making toddy intoxicating. "The leaf is smoked along with tobacco by asthmatic patients." **Mr. H. Willock**, Collector of Trichinopoly, while stating that one or two forms grow " in back-yards and gardens," adds " it is never smoked."

SPECIAL MEDICAL OPINIONS COMMUNICATED REGARDING **D. alba**§—"The juice of the leaves I have frequently used to dilate the pupils with success " (*Nundo Lall Ghose, Bankipur*). " The leaves are employed as an external application in rheumatism, the joint being enveloped in the leaves of the castor oil plant afterwards" (*Lal Mahomed, Hospital Assistant, Main Dispensary, Hoshangabad, Central Provinces*). "I have

| The White Datura. (*G. Watt.*) | DATURA Metel. |

used a pulp of the leaves, made with water, as an application in sweating of the feet with success" (*Civil Surgeon L. Cameron, M.D., Nuddea*). "The juice of the leaves is used as an antiperiodic in intermittent fevers" (*Surgeon-Major D. R. Thomson, M.D., C.I.E., Surgeon, 1st District, Madras*).

Datura Metel, *Linn. ; Fl. Br. Ind., IV., 243.*

164

The origin of the name Stramonium is obscure, but it appears to have been first given to this species—a plant which, as a matter of history, is known to have been cultivated at Venice under that name about the middle of the sixteenth century. **D. Stramonium** reached Europe some short time after, and taking more kindly to its new home spread rapidly, and in time came to bear the name Stramonium, which botanists have given to it as its classical specific designation. Another curious fact is vouched for by Avicenna, namely, that originally **D. Metel** was the *Massel* or *Mathil* of Arabic writers, although by modern usage that classical name has been assigned to **D. fastuosa.** Of **D. Metel** the *Flora of British India* affirms that the flowers are whitish purple, but **Ainslie** states that **D. fastuosa** is the **D. rubra** of **Rumphius** "and is distinguished from **D. Metel** by having dark-coloured flowers, while those of **D. Metel** are white." He then proceeds to further distinguish these species by their foliage—**D. fastuosa** having the leaves ovate, angular ; while **D. Metel** has cordate, almost entire leaves, and is pubescent. He adds that **D. Metel**, according to **Forskahl** (*Flora Arabiæ Felicis*), has three Arabic names, and that it is the **D. alba** of **Rumphius** and the *Humatu* of **Rheede**. Dealing apparently with **Rheede's** *Humatu*, the *Flora of British India* refers it to **D. fastuosa**, and a doubtful variety of that species, based on **Rheede's** drawing, in which the fruit is shown as smooth instead of spinescent. On the other hand, **Roxburgh's D. Metel**, which he states to be **Rheede's** *Hummatu* (*Hort. Mal., II., 47, t. 28*), is reduced both by **Dunal** (*DC., Prod. XIII., Pt. I., 542*), and the *Flora of British India*, to the variety **D. alba**, described above. It would thus seem that a considerable amount of difference of opinion prevails amongst botanists, and it is therefore not to be wondered at that writers on the medical properties of these plants should have got confused as to the "white datura." The name *Metel* would indicate that the plant first so named came to Europe through the Arabs, and *Humatu* is doubtless a mistake for *Unmatta* or *Ummatta*, the Sanskrit and South Indian name for any datura. It is to be regretted, therefore, that such names should have been adopted, in botanical literature, as the classic names of species to which they only very doubtfully belong.

Vern.—There are no specific vernacular names intended in India to denote this species. All the names given above might be applied to it, but more especially those recorded under **D. alba.** Indeed, the writer strongly suspects that the "white dhutúrá" of the early Sanskrit and Arabic writers was **D. Metel**, as now known to botanists, and not the variety of **D. fastuosa** known as **alba.** This suggestion seems at least worthy of being tested chemically, and if **D. Metel** should be found to contain less of the poisonous principle than **D. fastuosa**, it might be held as partly confirmed. The most trustworthy modern writers hold that there is no difference between **D. fastuosa** and **D. alba**, whereas for centuries the purple datura has been held to be much more poisonous than the white.

References.—*Mason's Burma, 488, 798 ; Ainslie, Mat. Ind., I. 443 ; U. C. Dutt, Mat. Med. Hind., 297 ; Birdwood, Bomb. Pr., 60, 209; Smith, Dic., 152 ; Mysore, Cat. Cal. Exh., 21 ; Fleming, Med. Pl. and Drugs in As. Res., Vol. XI., 165.*

DATURA **Stramonium.**	**Stramonium or Thorn Apple.**

Habitat.—A herbaceous plant, found in the Western Himálaya and mountains of the West Deccan Peninsula; probably introduced into India. Fleming (*As Res.*) in the passage quoted below affirms that this is a native of the Himálayas, and is the species met with in Kashmír. It is widely naturalised in the Old World and produces flowers and seeds the whole year.

This is a much more temperate species than the preceding, but in shape of flower and character of fruit can with difficulty be distinguished. The corolla possesses, however, 10 instead of 5 teeth or petals; the leaves are pubescent, and show a pronounced tendency to be cordate at the base. The stems are almost sub-villose, a character by which the plant may be recognised in the bazar product from all the other Indian daturas. It is a much smaller species than any of the others, its pubescence and 10-petalled corolla being its characteristic features.

MEDICINE.
165

Medicine.—Sir George Birdwood mentions this plant in his list of drugs of Bombay as if it were *the datura.* It possesses properties similar to those of the other species. Fleming (*As. Res., XI., 1840*), gives it the names of *D'hatura,* HIND., and *D'hustura,* SANS., and refers to *Murray, I., 670,* and *Woodville, II., 338,* works which the writer has not the opportunity of consulting. In a further passage (quoted in full under **D. Stramonium**) Fleming holds that this is a native of India, and seems to concur with Linnæus, that it might be used in preference to Stramonium.

166

Datura Stramonium, *Linn.; Fl. Br. Ind., IV., 242.*

STRAMONIUM or THORN APPLE.

Syn.—DATURA FEROX, *Nees* (the plant described by **Madden** as the *Kála dhatúra* of Kumaon, *Atkinson, Him. Dist., p. 370*); D. WAL-LICHII, *Dunal;* STRAMONIUM VULGATUM, *Gærtn.*

Vern.—By many writers this bears the same popular names as have already been given under **D. fastuosa,** *var.* **alba,** and if the suggestion that **D. Metel** is the white *datura* proves incorrect, this is much more likely to be the plant meant than the **D. alba** of Botanists. *Sada dhútúrá,* BENG.; *Tattur, dattura,* PB.; *Kachola, datura,* AFG.; *Umatai,* TAM.; *Ummetta,* TEL.; *Datturi gida,* KAN.

References.—*Stewart, Pb. Pl., 156; O'Shaughnessy, Beng. Dispens., 59; Balfour, Cyclop., I., 897; Smith, Dic., 152; Kew Off. Guide to the Mus. of Ec. Bot., 100; Fleming, Med. Pl. and Drugs in As. Res., Vol. XI., 165.*

Habitat.—The temperate Himálaya from Baluchistan and Kashmír to Sikkim. It is distributed east and west along the outer Himálayas and thus covers a region of over 1,000 miles. Taking the neighbourhood of Simla as fairly representative of that area, it is very abundant around Simla, and is met with everywhere on the march north to Upper Kulu (a distance across the outer ranges of perhaps 150 miles); but everywhere it frequents road-sides and village sites, and but rarely is seen in the forests or on the wild uncultivated hills. In the deep valley of the Sutlej it is particularly plentiful, miles of country, as at Rampur, being literally covered with **Cassia Sophora, Cannabis sativa,** and **Datura.** It is often, however, difficult to say in these lower warm valleys, whether **D. fastuosa** or **D. Stramonium,** is the species present, since one plant may be found with the erect and the next with the nodding fruit. On the higher slopes no doubt need be entertained, as the plant there met with has the characteristic ovate, erect fruit bursting regularly into four valves for half of the entire length of the capsule. Although thus very abundant on the Himálayas, **Stramonium,** like the daturas of the plains of India, exists in an isolated or disconnected manner from the surrounding vegetation, or forms compact formidable clumps, to the exclusion or extermination of all other plants, attitudes which, to the writer, are suggestive of aggressive invasion and conquest. Dr. Aitchison mentions D. Stramonium as met with in Afghánistan, but doubts its being indigenous, and Mr. J. H. Lace has kindly furnished the author with a specimen from Quetta, which he remarks is plentiful in or about cultivation up to a height of 7,000 feet. D. Stramonium is peculiarly the Himálayan representative of the genus from 3,000 up to 9,000 feet. The *Flora of British India* regards it as indigenous to India, but M. Alphonse de Candolle (*Geographie Botanique, II., 1855, p. 731*), comes to the conclusion that D. Stramonium, *L.*, appears to be indigenous to the Old World, probably the borders of the Caspian Sea or adjacent regions, but is certainly not a native of India; that it is very doubtful if it existed in Europe in the time of the ancient Roman Empire, but that it appears to have spread itself between that period and the discovery of America. At the same time he holds that D. Tatula (a form most writers express the strongest hesitation in accepting as specifically distinct from D. Stramonium) is a native of Central America. If the account of the peculiar attitude here given of D. Stramonium be accepted as supporting M. de Candolle's emphatic statement that it is not a native of the Himá-layas, then must the further opinion be held that all the species of datura met with in India are introduced and acclimatised plants.

The botanical characters by which D. Stramonium may be recognised have been partly indicated above, but it may be as well to repeat these more fully. It is a more compact plant than D. fastuosa, more succulent and of a considerably paler green than the plant of the plains. The flowers are also much smaller, being only 1 to 3 inches in diameter, but the fruit is longer, being ovate and *sitting permanently erect in the bifur-cations of the stem*, instead of recurving on maturity. It also bursts open regularly, forming four valves, which split for half or the entire length of the capsule. Except in the variety described below (Tatula), the flowers are always white, but the most important characters are those given above for the fruit, which should be compared with the description of the fruit of D. fastuosa (see page 33).

Towards the close of the sixteenth century D. Stramonium was culti-vated in England by Gerarde, who received the seed from Constantinople. In his *Herbal* he calls it *"The Thorny Apple of Peru,"* and says " it is a drowsy and numbing plant with properties resembling the Mandrake (Atropa Mandragora), a plant, which gets its name from Atropos, the eldest of the all-powerful Parcæ, the arbiters of life and death."

Medicine.—It seems probable that on the Himálaya D. Stramonium is used for all the purposes indicated under D. fastuosa and D. alba. Stewart says : "The SEEDS are used in poisoning, and are given medicinally in asthmatic complaints, being sometimes smoked with tobacco thus, and for vicious indulgence. The LEAVES are applied to boils and ulcers, and are also smoked with tobacco for asthma." Mr. Baden Powell states that in the Panjáb (here he probably means the plains, and hence D. fastuosa and not D. Stramonium would be indicated) "the drug has its medicinal uses, and its value as a curative in asthma is known both to Europeans and Natives, who smoke the seed in their *hukas* when so afflicted."

Fleming (*As. Res., XI., 1840, p. 166*) says : "The D. Stramonium, *Linn.*, which is the species used in medicine in *Europe*, is not found in

MEDICINE.

Seeds.
167
Leaves.
168

DATURA **Stramonium.**	Stramonium or Thorn Apple.

MEDICINE.

Hindústǎn,* but the **D. Metel** grows wild in every part of the country. The soporiferous and intoxicating qualities of the seeds are well known to the inhabitants, and it appears, from the records of the native Courts of Justice, that these seeds are still employed for the same licentious and wicked purposes as they were formerly in the time of **Acosta** and **Rumphius** (See *Rumph., Amb. V., 242*). I do not know that either the seeds, or the

Juice.
169

extract prepared from the expressed JUICE of the plant, are used in medicine here; but those who place any faith in the accounts given by **Baron Stœrck**, and **Mr. Odhelius** (*vide* **Murray** and **Woodville**) of the efficacy of the extract of the **Stramonium**, in the cure of mania, epilepsy, and other convulsive disorders, may reasonably expect the same effects from the extract of **Metel**, the narcotic power in the two species being perfectly alike. **Linnæus**, indeed, has given a place, in his *Materia Medica* to the *Metel*, in preference to the **Stramonium**."

SPECIAL OPINIONS REGARDING **Datura Stramonium** §.—" I have

Fruit.
170

used the FRUIT as a poultice and anodyne in whitlow" (*Dr. Picachy, Civil Medical Officer, Purneah*). "A good anodyne application is made by preparing a warm infusion of the leaves, and this is effective in inflammatory pains; the crude juice of the leaves mixed with opium and rock salt makes a good local anodyne preparation when applied hot in rheumatism" (*Surgeon Edward S. Brander, M.B., F.R.C.S.E., I.M.D., Rungpore*). "The leaves, made into cigarettes, are smoked to relieve asthma. The smoke is inhaled into the lungs" (*E. G. Russell, Superintendent, Asylums, at Presidency General Hospital, Calcutta*).

For the European uses of the drug the reader is referred to works on Materia Medica.

CHEMISTRY.
171

Chemistry.—It has been stated that it is presumed that chemically the Indian forms of datura differ among themselves and from Stramonium more in degree than in quality. The active principle is the alkaloid *daturine*, a substance practically identical with *atropine*. The experiments of **Schroff**, however, would indicate that *atropine* has twice the poisonous energy of *daturine*, although the two alkaloids agree in composition, possess the same qualities in regard to solubility and fusing point, and have the same crystalline form. The identity of *daturine* with *atropine* has been maintained by several chemists, while the admission of the greater poisonous property of the latter is opposed to such an opinion. **Ladenburg** states that **D. Stramonium** contains two alkaloids, which he designates as heavy and light daturine. **Pochl** affirms that solutions of *daturine* are levogyrate, while those of atropine exhibit no rotatory power. It is probable that the light daturine if isolated would bear a much closer approximation to atropine than the mixture of the two.

The leaves contain the alkaloid in a much smaller proportion than the seeds, and even the latter possess only $\frac{1}{10}$th per cent. In the seeds it is said to be combined with malic acid. According to **Joubert** datura for opthalmic purposes is more powerful and lasting than atropia.

§ **Dr. Warden** (Professor of Chemistry, Calcutta) has kindly furnished the following note regarding the chemistry of datura :—"The alkaloids, atropia and datura, contained, respectively, in **Atropa Belladonna**, and **Datura Stramonium**, are either identical or agree very closely in chemical

---

* "In the *Asiatic Researches, VI., 351*, **Colonel Hardwicke** enumerates the **Datura Stramonium** among the plants which he found in the *Sirinagur* country; but he afterwards ascertained that the plant which he met with was the **Datura Metel**; and has candidly authorised me to notice the mistake" (*Foot-note by Dr. Fleming*).

| Gharbhuli or Tatula Apples; Carrot. (*G. Watt*) | DAUCUS Carota. |

composition. Both widely dilate the pupil when applied locally to the CHEMISTRY. eye or introduced into the system: 100 parts of the different portions of the plant in the dried state yielded to Gunther the following results:—

Datura Stramonium.		*Atropa Belladonna.*	
Seed	·318 to ·365	Leaves	·833
Stalks	·063	Stalks	·146
Leaves	·169 to ·307	Fruit, ripe	·813
Root	·065	„ unripe	·955
		Seed	·407
		Root	·810

Gharbuli.
172

Var. Tatula, *Willd.*; *Fl. Br. Ind., IV., 242; flowers purple.*

The young fruits, strung on threads and imported into India from Persia, seem to be those of this variety. It is said to be common everywhere around villages in Afghánistan. The name by which these young fruits are sold is *gharbhúli* in Bombay, and *maratia múghú* in Madras (*Ainslie, Mat. Ind., II., 185*). They are regarded as sedative and slightly intoxicating. The writer is by no means sure that he has been able to identify this form, but from the descriptions published by botanical authors it cannot be regarded as more than a darker coloured state of the **D. Stramonium** commonly met with. The name given to it—*Tatula*—is the Turkish corruption of *dhatura*, through the Persian, the Sanskrit being *dhattura* or *dhustura*; it would be equally applicable to any form of datura. Further, it cannot be affirmed that the identification of the Persian article, *gharbhúli*, with the Madras, *maratia múghú*, is anything more than a suggestion, still less can it be held that these have been satisfactorily determined to be the young fruits of **D. Stramonium** *var.* **Tatula.** (*Conf.* with *Moodeen Sheriff, Supp. Pharm. Ind., 131.*) O'Shaughnessy says:—"It is a native of North America, very nearly the same as **D. Stramonium** but is a larger plant with purple stems, and the corolla similarly stained at the edges." But in this opinion he was most probably in error, the plant he regarded as **D. Tatula** being more likely a cultivated state of **D. Metel.** M. DeCandolle appears, however, to consider **D. Tatula** to be of Central American origin, and if that be so its Turkish name would be a most misleading accident, and its identity with the Persian *gharbhúli* highly problematic.

SPECIAL OPINIONS.—§ "Enters into aphrodisiac preparations" (*T. Ruthnam Moodelliar, Native Surgeon, Chingleput, Madras Presidency*). "Sometimes produces almost magical effects in asthma, and in paroxysmal neuralgias, even when **D. Stramonium** has failed" (*E. G. Russell, Superintendent, Asylums, at Presidency General Hospital, Calcutta*).

DAUCUS, *Linn.; Gen. Pl., I., 928.*

Daucus Carota, *Linn.; Fl. Br. Ind., II., 718;* UMBELLIFERÆ.

173

THE CARROT, *Eng.*; CAROTTE, *Fr.*; GEMEINE MOHRE, GELBE RÜBE,*Germ.*; CAROTA, *It.*; LANAHORIA, *Sp.*; MORKOV, *Rus.*

Vern.—*Gáger, gájar,* HIND.; *Gágar,* BENG.; *Mor múj, bul muj, kách,* KASHMIR; *Gájar,* PB.; *Zárdák,* AFG.; *Pétaigágar* (Stocks says that *gájjar* alone is the sweet potato), SIND; *Gásara,* MAR.; *Gájar,* GUZ.; *Gájjara kelangu, manjal-mullángi, kárttu-kishangu,* TAM.; *Gajjara gedda, pita-kanda, pach-cha-mullangi, shikha-múlamu,* TEL.; *Gajjari,* KAN.; *Garjara, shikha-múlam,* SANS.; *Jasar,* ARAB.; *Zardak, gasar,* PERS.

NOTE.—The *Talif Seriff* gives *Seali* as the name for the Carrot. The *Ain-i-Akbari* describes a creeper having a long edible conical root under the name *Séáli*, and Brandis gives *Siáli* as the Panjábi for **Pueraria tuberosa.** Dr. Dymock informs the writer that *Shaqáqul* (translated wild carrot in the *Ain-i-Akbari*) is **Trachydium Lehmanni,** *Bénth. et Hook., f.* Dr. Aitchison, in his report on the Botany of the

DAUCUS Carota.	The Carrot.

Afghán Delimitation Commission, calls that plant *Shahk-ukhal,* and says it is a very common annual in the loamy soil of the Badghis, the roots of which are collected and exported to India *viá* Herat. The *Shaqáqul* of the *Ain-i-Akbari* was a vegetable apparently regularly eaten in the time of the Emperor, and Trachydium is certainly not so in India at the present day.

References.—*Roxb., Fl. Ind., Ed. C.B.C., 270 ; Dalz. & Gibs., Bomb Fl. Suppl., 41 ; Stewart, Pb. Pl., 105 ; Aitchison, Cat. Pb. and Sind Pl., 68 ; Flora Andhrica, 57 ; Stocks' account of Sind ; Darwin, Animals & Plants under Domestication, I., 326, II., 31, 33, 113, 277, 311 ; Ainslie, Mat. Ind., I., 56; O'Shaughnessy, Beng. Dispens., 368 ; Moodeen Sheriff, Supp. Pharm. Ind., 131 ; U. C. Dutt, Mat. Med. Hind., 298 ; Dymock, Mat. Med. W. Ind., 380 ; U. S. Dispens., 15th Ed., 1598 ; Bent. & Trim., Med. Pl., II., 135; S. Arjun, Bomb. Drugs, 64 ; Murray, Pl. and Drugs, Sind, 200 ; Müeller, Sel., Ex-trop. Pl., 104 ; Johnston's Chem. Com. Life, 60, 86, 158 ; Johnston, How Crops Grow, 155-156 ; Anderson, Agri. Chemistry, 286 ; Baden Powell, Pb. Pr., 351 ; Atkinson, Him. Dist., 355, 703, 735 ; Lisboa, U. Pl. Bomb., 161 ; Birdwood, Bomb. Pr., 161 ; Royle, Ill. Him. Bot., 228-9, 231 ; Atkinson, Economic Products, Pt. V., 13, 18 ; Bomb. Gaz., V, 26 ; VII, 40 ; Folkard, Plant Lore, 270 ; Firminger, Man. Ind. Gard., 93, 100-101, 168 ; Spons, Encyclop., 1432 ; Balfour, Cyclop., 590, 898 ; Smith, Dic., 94 ; Treasury of Bot., 386 ; Morton, Cyclop. Agri., 407, 632 ; Kew Off. Guide to the Mus. of Ec. Bot., 77 ; Fleming, Med. Pl. and Drugs, in As. Res., Vol. XI., 166 ; Jour. Agri. Hort. Soc., 1875-78, Vol. V., 39, 1871-74, Vol. IV, 14 ; Report, Saharunpore Bot. Gardens, 1884, 6 ; Report, Lucknow Gardens, 1885, 5 ; Famine Com. Rept., App. to Parts I. and II., p. 87 ; Report by Sir E. C. Buck (then Director of Agri., N.-W. P.) dated 16th Oct. 1878 ; Annual Report, Sett., Port Blair, 1870-71, p. 43 ; Bomb. Gas. (Kathiawar), Vol. VIII., p. 183 ; Special Report from Director, Land Records and Agri., Burma ; Quarterly Journal of Agriculture (Vol. XI.) 1840-41, p. 268 ; Vol. III., 1847-49, p. 163 ; Vol. VI., 1853-55, p. 217 ; Vol. XI., 1863-65, p. 229 ; Adams, Wanderings of a Naturalist, 299 ; Ain-i-Akbari, by Abul Fazl (Transl. by Blochmann) pp. 63, 64, & 67.*

Habitat.—According to the *Flora of British India* the Carrot is a native of Kashmír and the Western Himálaya, at altitudes of from 5,000 to 9,000 feet. Stewart says its range in Kashmír is from 3,200 to 5,000 feet, and Adams alludes to the bear as feeding on the carrot and strawberry root. Dr. Johnstone has, in his herbarium of Simla plants, a specimen collected on Murale hill which has large fleshy roots. Of this he remarks that it is a favourite food with bears.

Throughout India the carrot is cultivated, by the Europeans, mostly from annually imported seed, and by the natives, from an acclimatised if not an indigenous stock. In many parts of the country a greenish white carrot is preferred as being very hardy and productive. This rises some two or three inches above the soil, is a coarse root which possesses little of the flavour of the European carrot, but it is able to withstand the extreme heat of summer, and may be raised in some parts of the country throughout the year. It thus produces a return at seasons when other tubers or roots are scarce or not available. This is particularly the case in Behar (Patna) and some parts of the North-West Provinces. Sir Edward Buck, while Director of Agriculture in these Provinces (1878), wrote a long and interesting note on carrot cultivation as a means of human food in periods of threatened scarcity or famine. The arguments then advanced have given to the subject of carrot cultivation in India an interest which, as an ordinary garden crop, it did not previously possess. The present account deals, therefore, more fully with the subject than most persons acquainted with Indian agriculture might be prepared to expect, and it is hoped, should necessity ever arise for strenuous efforts being made to produce food, that a compilation like the present, from all existing sources of Indian information, may prove useful.

D. 173

| The Carrot. | (*G. Watt.*) | **DAUCUS Carota.** |

History of the Carrot.—Besides its Indian habitat the Carrot is a native of Europe (with the exception of the extreme north), of Abyssinia and North Africa, of Madeira and the Azores, and eastwards through Northern Asia to Siberia and Kamtschatka. By some writers it is held to be also a native of America; by others is regarded as but an introduced plant that has there become completely naturalised. In its wild state, while in foliage, flower, and fruit, it can with difficulty be distinguished from the cultivated plant, but it has never been observed to produce in Europe the succulent root for which it is famed as a cultivated plant. It is well known, however, that care and a liberal supply of nourishment, will produce conditions both in animals and plants that become hereditary, and which once acquired wi'l long continue even under the cruellest treatment. Darwin states that "the experiments of Vilmorin and Buckman on carrots and parsnips prove that abundant nutriment produces a definite and inheritable effect on the so-called roots, with scarcely any change in other parts of the plant." Conversely, neglect or the consequences on a succulently developed plant running wild would naturally be to reduce the edible property until in time it might ultimately disappear. This retrogression is, however, much less than is commonly supposed; indeed, amongst scientific writers the belief prevails that plants or animals long domesticated (such as the horse or wheat), if we knew their ancestral forms, would probably be found to never completely revert under any treatment. This may in fact be said to be the explanation of the very word "acclimatisation." Darwin mentions many instances where the seeds of English annuals failed completely in the plains of India until they had been first successfully grown in Darjeeling and acclimatised seeds produced. Speaking of carrots he refers to a case where a consignment of English seed was sent both to Madras and to Hyderabad. The former failed, but the latter was found to furnish a seed stock that succeeded admirably afterwards in Madras. It seems likely that in the wild state, the tendency to produce a succulent root might more readily occur in a warm than in a cold country, and that hence, in all probability, the natives of higher Asia may have first thought of cultivating the carrot. At all events Stewart states that in one part of Kashmír he found that the people eat the wild carrot, a circumstance confirmed in a measure by the observation that bears eat it. Indeed, it seems highly improbable from the simple examination of the wild carrot as met with in Europe, that the idea could ever have occurred of cultivating it, in the hope of producing an esculent root. Aitchison found the carrot wild in the Kuram valley, but of the Hari-rud (Afghánistan) he says the *sárdák* is not indigenous but a weed and an escape from cultivation, in cultivated land; also that the carrot is very extensively cultivated both in Afghánistán and in Persia. According to most writers the δαῦκον of Theophrastus was the carrot, and from that word the generic name as given by botanists has been derived. The early Muhammadan physicians, who, in many respects, give indications of an intimate knowledge of the contemporaneous Greek medical science, have handed down to the drug-seller of the Indian bazárs the word *Dúkús*. The *Makhzan*, under that name, describes three umbelliferous seeds, one of which may be **Daucus Carota**. At the same time it is known that the Greeks actually cultivated the Carrot in classical times, though not perhaps to any great extent. The root seems, however, to have been associated with indecency from a very early period, and similarly, the Hindus present the carrot or the radish along with fruit to their friends at the Makar Sankránti. The Greeks often talked about κέρατα ποιεῖν τινί and the individual so favoured was a χερασθορός: carrots and horns are, in fact, closely associated. The Greeks called the plant Phileon, because of its supposed connec-

HISTORY.
174

DUACUS Carota	The Carrot.

HISTORY.

tion with amatory affairs. The word Daukon was given to an umbelliferous plant, but not necessarily the carrot, though generally accepted as such. Carota, the Latin name, was perhaps derived from *caro*, flesh, and *carota* is mentioned by Apicius, the celebrated author on cookery (A. D. 230). Some writers, however, say that it is derived from *car*, the Celtic for red. This seems highly improbable, as it is doubtful if the Celtic race cultivated any vegetable so far back as the date given above for Carota, from which Carrot is doubtless derived. The Persian names for the root are *Zardak* and *Gazar*, its Sanskrit *Garjaru*, and its Arabic *Jegar*, words in all likelihood obtained from one source, and that probably the Sanskrit. Persian scholars, at all events, accept *Gazar* as a simple Sanskrit word, and not a derived one, but the modern *Zardak* is said to come from *zar*, golden, or *zard*, yellow. The resemblance of the Kashmír name *Mor múj* to some of the European names, noteably the Russian *Morkov*, is remarkable.

Carrots appear to have been regularly used in India in the time of Akbar (corresponding to the period of Queen Elizabeth in England). They are alluded to among the vegetables and pickles used by the Emperor, but there occurs also the word *Shaqáqul*, which both Gladwin and Blochmann have translated "wild carrots," though, as already shown, this translation is most probably incorrect. While much reliance cannot be put on names of plants, as historic evidences, it is significant that throughout the languages of India, indeed, from Central Asia to Cape Comorin, there should prevail in every language a name for the carrot which seems to have come from a common source. To that name is frequently added a further word meaning "root" or "tuber." Thus, in Tamil, it is the *Gajjara-kelangu*, the *Kartu-kizhangi*. Whether or not we view *Kartu* as an approximation to the European derivatives Carota, Carrot, &c., the further explanatory word simply means tuber or bulb. In this connection it may be added that Ainslie, who wrote of Madras at the beginning of the present century, gives the Tamil for Carrot as *Cárrot-kálung*. The Telegu language, among many other names for the carrot, has the following : *Gajjara-gedda*, *pita-kanda*, and *shikhá-mulamu*. Here, again, the terminal words *gedda* (or rather *gadda*) and *mulamu* (the Sanskrit *mulam*) simply mean root or rhizome, and are the equivalents of *jar* in Hindustani, *vér* in Tamil, and *véru* in Telegu. The derivation of the Latin name *Carota* mentioned above, as is customary with writers on this subject, has been given as *caro*, flesh, but the evidence of cultivation would almost lead to the inference that the carrot spread from Central Asia to Europe, and if so it might be possible to trace from the Indian, Sanskrit, and Persian names those of Europe. Ainslie has no hesitation in affirming that India obtained the carrot from Persia, but in the *Ain-i-Akbari* Abul Fazl makes no mention of carrot as having been introduced. While he goes into details regarding many of Akbar's fruits and vegetables, specially mentioning those, such as the pine-apple, which were less known, he treats carrots as a matter of course. The Muhammadan invaders of India were perhaps, for centuries before Akbar's time, equally familiar with the *Garjara, Gazar, Gajir, Zardak*, the golden root, and thus, as a regular vegetable, it was grown and eaten in India when in Europe it was scarcely known as more than a wild plant. As a somewhat curious historic fact it may in conclusion be stated that in the reign of James the First ladies adorned their head-dresses with carrot leaves, the plant having begun to be cultivated in England during Queen Elizabeth's time. It was largely grown in many parts of the Continent of Europe some time before it found its way to England. Belgium and Holland may especially be mentioned, since in these countries, even at

D. 174

The Carrot.	(*G. Watt.*)	**DAUCUS Carota.**

the present time, it is a recognised field crop, whereas in England as a whole it has not left the domain of garden production.

ABSTRACT OF THE PUBLISHED STATEMENTS REGARDING CARROT CULTIVATION IN INDIA.

Bombay Presidency.—Of Gujarat it has been said that carrots of two kinds are cultivated, "the long-rooted" and the "blunt spindle form." These are "grown at various times in different parts of the province. Generally, they are grown in garden beds from seed sown broadcast, and are sometimes transplanted from the nursery in the *rabi* season like onions, from which their cultivation does not differ, except that a light and rich soil is preferred for carrots, and great care is necessary so as not to break the roots in transplanting. The space between each plant is a full span. They take three months to mature, but nipping or removing the heads prolongs the growth, so that a supply can be ensured several months after the ordinary time of maturity. The young plants are also taken up in their half growth for the market. The produce is from 5,000 to 10,000℔. " The carrot is further stated to be sown in Gujarat from August to May, and the crop gathered four months later. Of Cutch it is reported that carrots are "much grown as a field crop." "Cutch is famous for its carrots." It is said of Poona that "with the help of water and manure, carrots are grown in large quantities in good black soil in the east of the district. The root is eaten as a vegetable, both raw and boiled. It is also slit and dried in the sun, when it will keep five or six months. When sun-dried it is called *usris*, and has to be boiled before it is eaten." In garden lands the carrot may be sown in Poona at any time, but in dry crop lands in July or August only. In Khandesh " the carrot is widely grown and with great success. The chief Khandesh carrot is long and reddish, in flavour not much inferior to the best European kinds. The seed is always sown on the third or fourth day before the *amavasya* (*e.g.*, the last day of the Hindu month), as it is believed that the woody heart of the carrot will thus be reduced to the smallest possible size." Of Ahmadnagar, a curious process is reported of obtaining carrot seed, which brings to mind the Panjáb method of cultivating a form of radish that has resulted in the production of a new vegetable, namely, the plant known as **Raphanus sativus** *var.* **caudatus.** Instead of the root being eaten the treatment followed in the Panjáb has resulted in the production of a pod of great length, which is eaten as a vegetable. The Ahmadnagar carrot seed is thus produced : " When the crop is ready the husbandman cuts off a thick slice from the crown end of the root of the carrot. This he puts two fingers deep below the soil in any place where there is a liberal supply of water. After a few weeks the roots shoot into vigorous flower stems, the seed of which is gathered four or five months after they have been planted. There are thus two crops in the year—one, the root produced from the seed, the other the seed produced from the root." In Kolhapur carrots are sown in September to November, and the crop obtained three months later. " During the first two months the crop is watered every ten days. In the third month the root begins to ripen and watering is stopped. A full-sized carrot is four or five inches long and weighs about two ounces.

Hyderabad, Sind.—In the experimental farm various kinds of imported carrots have been experimentally grown. The Altringham was found to give the best results, the yield having been in 1885-86 (*Farm Report, page 35*) 7,360℔ an acre.

Mysore and Coorg are stated to produce a very good quality of carrots ; in the *Central Provinces* occasional references occur to carrots as

DAUCUS Carota.	The Carrot.

CULTIVATION.

a garden crop. In the *Bengal Gazetteers* and other publications mention is also made of carrots. In Rungpore, for example, they are said to be sown in October and November, the crop being gathered in March and April, but in Patna they are said to be sown in July and harvested in December and January. Of the *Panjáb* brief notices are made of carrot cultivation.

Panjab.
180

In Jhang, for example, it is said that "the zamindar's food consists largely of carrots" (*Replies to the Famine Commission, page 228*). In Sialkot carrots "are grown all over the district, but the superior kinds of English carrots are little known or appreciated." In Hazara the carrot is sown in September and October and gathered in December and January.

N.-W. P. and Oudh.
181

The N.-W. Provinces and Oudh.—It has been estimated that there are 30,962 acres under carrots in these provinces, 2,557 acres being dry land and the remainder irrigated. In Oudh 35,721 acres, of which 3,599 are on dry land. Similar figures for the other provinces of India are not available; *but as the carrot is very nearly cultivated to the same extent in most provinces an approximate idea of the total area under carrots may be assumed.* In the N.-W. P. Agricultural Farm Report (*1884-85, page 17*) useful information is given regarding experiments made in the cultivation of European and the so-called indigenous carrot. The following results were obtained :—

	Outturn per acre in maunds.	Manure.	Ploughings.	Weedings.	Waterings.
Belgian Carrots— On ridges On lines . .	153·5 113·8	Poudrette 250 maunds per acre.	5	6	7
Country Carrots— On ridges On lines . .	355·3 315·5				

The country thus gave at least double the return of the imported. The seed was sown in September and October, and the crop obtained in November and December. Of Meerut it is said that carrot cultivation is becoming more general. In 1870 there were 250 acres, and in the replies to the Famine Commission it was contended that carrots were " most useful under failure of *kharif*."

Assam.
182
Burma.
183
Madras.
184

Of *Assam, Burma,* and *Madras* little can be learned regarding carrot cultivation, and it seems probable that, in these provinces, the root is only raised as a garden vegetable. Of Burma the Director of Land Records and Agriculture reports :—" It is planted at the beginning and reaped at the end of the rainy season. The soil required for its cultivation is a porous, moist, sandy one." It is only grown in Burma to a small extent.

FAMINE CROP.
185

ARGUMENTS REGARDING CARROT CULTIVATION AS AN EMERGENT CROP AT SEASONS OF THREATENED SCARCITY OR FAMINE.—The following may be given as a brief abstract of the leading facts and arguments advanced in Sir E. C. Buck's report on this subject, to which reference has been made above :—" In the half-yearly agricultural report in the N.-W. Provinces, published in the local Gazette of September last, I adverted to the extension of carrot cultivation which had taken place in consequence of the failure of the *kharif* in 1877." The replies received from enquiries instituted all over these provinces " corroborate the ideas which had been formed of the reliance placed by the agricultural population upon carrots, and to a less extent upon radishes, under failure of the ordi-

| The Carrot. | (*G. Watt.*) | **DAUCUS Carota.** |

nary autumn harvest. The reason is simple. When the *kharif* grain crops fail and food-stocks are reduced to a low ebb, the people have little to depend upon for food, unless purchased at a ruinous price, until March, when the spring harvests are gathered. But carrots and radishes can be raised in a hurry, and being sown in September or October, supplement the food-supply in the winter months at a time when scarcity is greatest." "A large weight can be produced from a small area." "Irrigation, which cannot be easily spread over a large extent in a year of drought, is concentrated on a minimum space and can be utilised to its maximum power. The facts gleaned by the enquiries are, that in the upper portions of the provinces where the failure of the *kharif* was greatest, the cultivation of the carrot rose to three or four times the ordinary extent, and would have increased much more had seed been obtained. The price of seed rose from R7 or 8 a maund to R30 or 40 a maund, and in some instances to a very much higher rate, especially in the Central Doáb, where the price ranged above R50. There is no doubt whatever that the carrot crop fed thousands of the starving poor." It has been shown that "allowing four seers a head per diem, the carrots of an acre of land would support ten persons for 200 days, could they eat carrots alone. Carrots must, however, be supplemented by some grain, but it may be presumed that a supply of grain (the outturn of about two acres) sufficient for ten persons for 200 days could be made to satisfy twenty persons for the same time if supplemented with the outturn. The above estimate is framed on the estimate of an outturn of 200 maunds an acre, which is much below the possible maximum, an outturn of over 300 maunds an acre not being uncommon, in addition to about 50 to 100 maunds of nourishing fodder for cattle. The English outturn runs up to 20 to 30 tons or from 500 to 700 maunds an acre." "I was informed that it was in the year of scarcity, 1869, that the Rohilkhand population first took the idea extensively from the cultivators of *the* Meerut side, and I am convinced from private enquiries that the practice is less common in the south than in the north of the provinces." Mr. T. N. Mukharji, who, under Sir Edward Buck's directions, instituted enquiries into the subject of carrot cultivation during the period of threatened famine referred to above, gives some instructive facts which were brought to his notice. He sums up the benefits (and these *are existing* benefits) from carrot cultivation during such an emergency, thus :—*1st*, carrots give a large amount of food in a small area; *2nd*, they afford food to both men and cattle; *3rd*, they save the ryots from the hands of the baniyas to whom they are bound to give up all grain; the baniyas will not take carrots on account of their not keeping.

In concluding this brief notice of the existing information regarding the carrot as a famine food, it may be said that some of the issues raised in connection with the enquiry have been since solved. The experimental farms have, for example, established the fact that imported seed, even were it procurable in sufficient quantity on a sudden demand in September, would only give about a third of the return of acclimatised. The suggestion therefore may be offered that an effort might be made to improve and extend the cultivation of an acclimatised stock, so that in the hands of as many ryots as possible there would always exist a certain amount of good seed. The effort might also be made to ascertain how far the carrots of one province could be cultivated in another, so that if the N.-W. Provinces were threatened with famine, it would be known what particular forms of Bombay or of Madras seed might advantageously be sent to these Provinces, or sent from the N.-W. to the Panjáb, to Central India, to Bombay, or to Bengal. Cross breeding of Indian with European and of interprovin-

E

DAUCUS Carota.	The Carrot.

cial stocks might be carried on alongside of continuous efforts to acclimatise European seed of good quality. Sir Edward Buck's farther remarks regarding the discovery of what parts of the south of Europe could afford seed suitable to India might also receive consideration, for it is clearly a desirable feature of a subject, like that of extended carrot cultivation, to know the producers to whom application should be made for seed, and this can only be learned after extensive comparative tests have been carried out.

Oil.—Carrot SEED yields a medicinal oil; this is obtained by distillation. It is a pale-yellow volatile oil and may be said to be the chief property of the seeds. It has a strong penetrating odour and a warm and somewhat unpleasant taste.

Medicine.—The carrot is not officinal in the English nor Indian Pharmacopœias, but by the natives the SEEDS are considered a nervine tonic. Boiled with honey and fermented, they produce a spirituous liquor. A decoction of the LEAVES and seeds is said to be used as a stimulant to the uterus during parturition. The ROOTS are made into a marmalade which is considered refrigerant. Dr. Dymock writes that "in the Concan a poultice of carrots and salt is used in tetter, and the seeds are eaten as an aphrodisiac." Formerly the carrot seeds (fruits) were used in European medical practice, and they are so still in America. They possess aromatic, stimulant, and carminative properties, and were used in diseases of the kidney, flatulent colic, dropsy, &c. A poultice made of the roots is even at the present day resorted, in domestic medicine, to correct the discharge from ill-conditioned sores. The raw rasped root is also deemed useful as a stimulating application, and is made into an ointment with lard.' This is used in burns and scalds to good effect. Pickled-carrots are much lauded by Persian writers as a cure for spleen. In the *American Dispensatory* it is stated that the wild root may be substituted for the seeds. It is whitish, hard, branched, and possesses a disagreeable smell.

SPECIAL OPINIONS.—§ "The crushed roots form the vehicle for many medicines used by native Hakims, and have the reputation of having tonic properties" (*Narain Misser, Kothe Bazar Dispensary, Hoshangabad, Central Provinces*). "The raw carrot when eaten acts as a mechanical anthelmintic" (*Surgeon Major D. R. Thomson, M.D., C.I.E., Surgeon, 1st District, Madras*). "Poultice of the root is useful in chronic and fœtid ulcers" (*Surgeon Major George Cumberland Ross, Delhi*). "Boiled and given to cattle with the view of making them fat" (*Assistant Surgeon Annund Chunder Mookerji, Noakhally*). "The seeds are used to bring about abortion. The roots are used as poultice" (*Surgeon Major Robb, Civil Surgeon, Ahmedabad*). "Used in dysentery and enlargement of spleen" (*John McConaghy, M.D., Civil Surgeon, Shahjahanpore*).

Chemistry of the Carrot.—"The constituents of the root are crystallisable and uncrystallisable sugar, a little starch, extractive, gluten, albumen, volatile oil, vegetable jelly or *pectin*, malic acid, saline matters, lignin, and a peculiar crystallisable, ruby-red, neutral principle, without odour or taste, called *carotin*. This latter principle has been well studied by Husemann, who gives it the formula $C_{18}H_{24}O$. Husemann has also described a colourless compound, *hydrocarotin* $C_{18}H_{30}O$, which exists with *carotin* in the juice of the carrot, and is probably changed into the latter by oxidation as the plant develops in growth." "The substance called *vegetable jelly* was by some considered a modification of gum or mucilage, combined with a vegetable acid. Braconout found it to be a peculiar principle, and named it *pectin* from the Greek πηκτις expressive of its characteristic property of gelatinising. It exists more or less in all vegetables, and is

	The Carrot.			(G. *Watt*.)

abundant in certain fruits and roots from which jellies are prepared. It may be separated from the juice of fruits by alcohol, which precipitates it in the form of a jelly. This, being washed with weak alcohol, and dried, yields a semi-transparent substance bearing some resemblance to *ichthyocolla*. Immersed in 100 parts of cold water, it swells like bassorin, and ultimately forms a homogeneous jelly." "A striking peculiarity is that, by the agency of a fixed alkali or alkaline earthy base, it is instantly converted into pectic acid, which unites with the base to form a pectate."

The following table (abstracted from an extensive series of analyses published in *Anderson's Agricultural Chemistry*) exhibits the comparative value of carrots with five other articles of human and cattle food :—

	Nitrogenous compounds.	Oil.	Respiratory compounds.	Fibre.	Ash.	Water.
Oats . . .	11·85	5·89	57·45	9·00	2·72	13·09
Wheat . . .	11·48	...	73·52	0·68	0·82	13·50
Hay . . .	9·40	2·56	38·54	29·14	5·84	14·30
Carob bean . .	3·11	0·41	62·51	18·60	2·80	12·57
Carrot . . .	1·87	...	9·91	3·07	1·11	86·04
Turnips . . .	1·27	0·20	4·07	1·08	1·71	91·47

Mr. Horsford gives the analysis of the carrot as 10·66 nitrogenous matter, 84·59 non-nitrogenous ingredients, and 5·77 inorganic constituents. These figures seem to conflict somewhat with Professor Anderson's table given above, especially in the ash, but turning to Johnson's *How Crops Grow* the ash is shown to vary from 5·1 to 10·9 per cent. The average of ten analyses gave the ash as 7·5 of the total weight of root which was composed as follows : 37·0 Potash, 20·7 Soda, 5·2 Magnesia, 10·9 Lime, 1·0 Oxide of Iron, 11·2 Phosphoric acid, 6·9 Sulphuric acid, 2·0 Silica, and 4·9 Chlorine. Professor Anderson's table is doubtless comparatively correct, and it therefore shows the value of carrots relative to the other foods there presented.

Carrots contain starch, the granules of which are very small and round, and in some cases muller-shaped, with distinct central hilums (*Bell*).

Food and Fodder.—The so-called ROOT, as produced in garden cultivation, constitutes an important item in the supply of the markets frequented by the European community. Although certain classes of Hindus in Bengal object to eat the carrot on account of some fanciful resemblance to beef, or because of its smell, still the natives of India, as a whole, are year by year taking more freely to it. At the same time it must be added that, although by the Muhammadans and certain classes of Hindus the carrot has been cultivated for ages, it is only within recent years that it has become a recognised article of diet. By certain classes the young carrots are only used as pickles. By others " the root is first boiled in water, then squeezed out and cooked in *ghi*." This latter practice accords with the scientific injunctions of the chemist, *viz.*, that the turnip, carrot, and other such roots, being deficient in fat, can only become staple articles of human diet if combined with fat or oil. Carrots are generally cooked with fat in Europe, and perhaps the grain with which they are eaten in India supplies, even in famine time, enough fat to sustain life.

In Europe carrots have become a recognised article of cattle food. In India the opinion prevails that to give horses a daily small allowance of carrots improves the gloss of the coat. Carrot tops afford a useful fodder for cattle, and the contention that carrots should be resorted to in times of famine is strengthened by this fact.

FOOD.
Root.
192

FODDER.
193

E 2

DEBREGEASIA hypoleuca.	The Debregeasia Fibre.

Coffee is often largely adulterated with carrots, and the reputed use of carrots as an adulterant in marmalade, doubtless rests on the presence of the vegetable jelly referred to above under the paragraph—" Chemistry of the Carrot."

DOMESTIC.
Tooth-sticks.
194
195

Domestic Uses.—The peduncles and flower-stalks are used by the hill-tribes as tooth-sticks.

Davallia, *Smith ; Hooker and Baker, Synopsis Filicum, 88 ; Beddome, Ferns, British India and Ceylon, p. 58.*

A genus of handsome ferns named in honour of the Swiss botanist Davall, the chief characters of which are the creeping rhizome and the involucre impressed in the substance of the margin of the frond so as to form an urceolate cyst like a miniature capsule. The economic history of Ferns is extremely imperfect. The above brief notice has been thought desirable so as to assign a place and number in the present work for one of the most extensive and most elegant of the genera, in the hope that with the advance of knowledge, we may be able to mention the uses to which some of the species are put.

Deadly Nightshade, see **Atropa Belladonna,** *Linn.* ; Vol. I., No. 1614.

Deal, *see* Fir, Pine, and Pinus.

DEBREGEASIA, *Gaud. ; Gen. Pl., III., 390.*

196

Debregeasia hypoleuca, *Wedd. ; Fl. Br. Ind., V., 591 ;* URTICACEÆ.

Syn.—DEBREGEASIA BICOLOR, *Wedd., in DC. Prod.* ; URTICA BICOLOR, *Roxb.* ; BOEHMERIA SALICIFOLIA, *Don.* ; B. HYPOLEUCA, *Hochst.* ; MISSIESSYA HYPOLEUCA, *Wedd., in Ann., Sc. Nat.* ; MOROCARPUS SALICIFOLIUS, *Blume.*

Vern.—*Puruni,* N.-W. P. ; *Tashiári* or *tashari siar,* KUMAON ; *Sihárú,* KANGRA ; *Kharwala, shakai,* TRANS-INDUS and AFG. ; *Chainchar, chairyili* or *chenjúl, amrer, sandári,* JHELUM ; *Sansárú, súss,* CHENAB ; *Siárú, talsiari, thána,* RAVI ; *Pincho, prin, siárú,* SUTLEJ.

References.—*Roxb., Fl. Ind., Ed. C.B.C., 656 ; Brandis, For. Fl., 405 ; Gamble, Man. Timb., 326 ; Stewart, Pb. Pl., 215 ; Aitchison, Cat. Pb. and Sind Pl., 136 ; Atkinson, Him. Dist., 317, 798 ; Report on Fibres shown at the Colonial and Indian Exhibition by Cross, Bevan, King, and Watt, p. 52 ; Special Report furnished by the Conservator of Forests, N.-W. P.*

Habitat.—A large shrub of the western temperate Himálayas, from Kashmír and the Salt range to Kumaon, altitude 3,000 to 5,000 feet. Distributed to Afghánistán and Abyssinia.

FIBRE.
197

Fibre.—All the species of **Debregeasia** afford strong and useful fibres, which are more or less extracted by the hill tribes and used for ropes and cordage. Our knowledge of these fibres is, however, too imperfect to allow of separate accounts being given, in which the comparative merits of the fibres from the various species would be discussed. It has, therefore, been thought desirable to draw up in one place a brief review of all the opinions which have been published regarding these fibres, but it must be added that, should hopes ever be entertained of the utilisation, commercially, of **Debregeasia** fibre, the first step would naturally be to have the individual properties of the species thoroughly investigated. In general terms it may be said that writers on Panjáb products, who refer to a **Debregeasia** fibre, are speaking of **D. hypoleuca** ; descriptions dealing with the Central Himálayas (*e.g.,* Kumaon, Garhwal, and Nepal) refer to **D. hypoleuca** and **D. velutina** ; of the Eastern Himálayas (*e.g.,* Sikkim, Assam, and Burma

D. 197

FIBRE.

to **D. velutina** and **D. Wallichiana**; while the **Debregeasia** fibre of the mountains of Western and Southern India is exclusively **D. velutina.**

Of Panjáb writers **Stewart** says: " In the eastern part of the Panjáb, its bark appears, as in the North-Western Provinces, to be used for making ropes, but it is not generally employed in this way." **Mr. Baden Powell** remarks of **D. hypoleuca** that it is " not yet recognised as a merchantable commodity. The fibre is valued for net ropes, on account of its resisting the action of water. The fibre, it would appear, is prepared by the hill people without steeping. It is merely dried, and when brittle is beaten, the fibre separates easily; the plant is cut in October." " But **Dr. Royle** quotes **Capt. Rainey**, then Political Agent at Sabáthú, who describes the process of preparation as laborious. The plant being cut is exposed one night in the open air. The stalk is then stripped of its leaves and dried in the sun; when dry, it is placed in a vessel with water and wood-ashes, and boiled for 24 hours. After boiling the fibre is well washed in a stream. The fibre is then sprinkled with flour of the grain *kodra* (**Paspalum scorbiculatum**) and left to dry: it is then ready for spinning." **Capt. Huddleston** (*Trans. Agri.-Hort. Soc., VIII., p. 275*), in his paper on the Hemp of the Himálayas, appears, under *jur kundalú, kundalú*, and *kubra* to be alluding to this fibre. He says—" It grows chiefly in the northern parts of the district, in great quantities: it also grows in the middle ones; and from its fibres the natives make rope for tying up their cattle and snow-sandals. One bundle will produce about a seer of fibre, but it is not collected for sale. The plant grows about eight or nine feet high, and the stalks are about the size of a finger in thickness. It is cut in the cold season, and the stalks are soaked a few days in water before the fibre is stripped off from the thick end like hemp." Passing further East **Mr. Atkinson** writes of Kumaon (and under the name **D. bicolor**): " The *tushi-yára* is very common all over the lower hills, ascending as high as 7,000 feet, and is particularly abundant in the Siwaliks. It yields a very strong cordage fibre." **Brandis** (in his *Forest Flora of North-Western and Central India*) says—" Twine and ropes are made of the fibre." **Gamble** also repeats this statement, but on the other hand the Conservator of Forests of the North-Western Provinces, in a recent report, writes of Jaunsar forests that **D. hypoleuca** is " not used for fibre."

Considerably more information is available regarding **D. velutina.** In the Madras Manual of Administration (*Vol. I., 313*), it is mentioned as one of the chief fibre plants of the Presidency. The Manager of the Glen Rock Fibre Company, Wynaad, 's reported to have sent a consignment, presumably, of this fibre to London. It was valued at £70 per ton. Of the Madras Presidency it is commonly stated that it is much used both by the natives generally and the managers of coffee estates. **Mr. J. Cameron** (Superintendent of the Botanic Gardens, Bangalore), in a note communicated to the writer, states that " this is one of the commonest and most conspicuous plants in the Wynaad and Nilgiri sholas. Its fibre is used for bow-strings, and it would only appear to require to be better known to be much appreciated." **Dalzell** and **Gibson** describe the plant (the *capsi*) as met with in the Concan and Ghát jungles, but make no mention of its fibre. **Mr. W. A. Talbot** also alludes to it as found in Kanara (*Bombay Gazetteer, XV., 444*), and **Mr. Lisboa** (*Useful Plants of Bombay*) says it is " common at Mahábleshwar and the Konkan jungles. The inner bark yields a fibre which, in Ceylon, &c., is used for cordage and fishing lines."

Of **D. Wallichiana**, **Mr. Gamble** makes the statement that it yields a " fibre used sometimes for cordage."

The reader is referred for further particulars to the Selections of the

DECAMALI **Gum.**	**The Debregeasia Fibres.**

Records of the Government of India, in the Department of Revenue and Agriculture (Vol. I., No. 18 of 1888-89), where, under the heading of Rhea and allied Rhea fibres, the writer endeavoured to clear up the ambiguity that prevails regarding **Boehmeria, Villebrunea,** and **Debregeasia.**

FODDER.
198

Fodder.—Stewart mentions that the leaves are eaten by sheep.

TIMBER.
199
200

Structure of the Wood.—Soft and grey : of no value.

[1959.

Debregeasia velutina, *Gaud.; Fl. Br. Ind., V., 590; Wight, Ic., t.*

> **Syn.**—DEBREGEASIA LONGIFOLIA, *Wedd., in DC. Prod.;* MISSIESSYA VE-LUTINA, *Wedd., in Ann. Sc. Nat.;* MOROCARPUS LONGIFOLIA, *Blume;* URTICA LONGIFOLIA and ANGUSTIFOLIA, *Blume; Burm.; ?* URTICA BICO-LOR, *Wall.;* U. VERRUCOSA, *Moon.;* CONOCEPHALUS NIVEUS, *Wight, Ic., t. 1952; Dals. and Gibs., Bomb. Fl., 239.*
>
> **Vern.**—*Tashiari,* NEPAL; *Kamhyem,* LEPCHA; *Kapsí,* BOMB.; *Kapsí,* KAN.; *Pwot-chaubeng, putchaw, Burm.*
>
> **References.**—*Brandis, For. Fl., 405; Kurz, For. Fl. Burm., II., 428; Beddome, Fl. Sylv. (Man. 226, t. 26, f. 5); Gamble, Man. Timb., 326; Dals. & Gibs., Bomb. Fl., 239; Lisboa, U. Pl. Bomb., 126, 234; Madras Man. Adm., I., 313.*

Habitat.—A tall shrub of the sub-tropical Himálaya, from Kumaon to Sikkim, Assam, the Khásia Hills, Tenasserim, the Deccan Peninsula from the Concan to Cape Comorin; altitude on the Himálaya from 2,000 to 5,000 feet, on the Nilghiri hills 7,000 feet.

FIBRE.
201

Fibre.—See the paragraph above under **D. hypoleuca.**

TIMBER.
202
203

Structure of the Wood.—Heartwood reddish-brown, hard; sapwood white.

D. Wallichiana, *Wedd.; Fl. Br. Ind., V., 591.*

> **Syn.**—DEBREGEASIA LEUCOPHYLLA, *Wedd., in DC. Prod.;* MOROCARPUS WALLICHIANUS, *Kurz;* MISSIESSYA WALLICHIANA, *Wedd.;* URTICA LEUCOPHYLLA, *Wall.*
>
> **Vern.**—*Púrúni,* NEPAL; *Senén,* LEPCHA.
>
> **References.**—*Kurz, For. Fl. Burm., II., 428; Gamble, Man. Timb., 326; Thwaites, En. Ceylon Pl., 262.*

Habitat.—A small tree (20 to 30 feet in height) met with in the Eastern Himálaya from Sikkim to the Khásia hills, Pegu, and Tenasserim; altitude 2,000 to 4,000 feet. (*Fl. Br. Ind.*) **Gamble** says it even ascends to 7,000 feet.

FIBRE.
204

Fibre.—See the paragraph above under **D. hypoleuca.**

TIMBER.
205
206

Structure of the Wood.—Annual rings distinctly marked by a white line. A very pretty plant with round leaves of the purest white beneath.

DECAISNEA, *Hook. f. & Thoms.; Gen. Pl., I., 42.*

Decaisnea insignis, *Hook. f. & Th.; Fl. Br. Ind. I., 107;*

[BERBERIDEÆ.

> **Syn.**—SLACKEA INSIGNIS, *Griff. Itn. Not., 187.*
>
> **Vern.**—*Lúdúma,* BHUTIA; *Nomorchi,* LEPCHA.
>
> **References.**—*Hooker's Him. Jour., II., 197; Balfour, Cyclop., I., 902; Treasury of Bot., 388.*

Habitat.—An erect shrub, which inhabits the eastern parts of the Himálaya, in Bhutan and Sikkim, in altitudes between 6,000 to 10,000 feet.

FOOD.
Fruit.
207

Food.—Produces a very palatable FRUIT, which ripens in October, and which is eaten by the Lepchas of Sikkim.

Decamali Gum—see Gardenia lucida.

D. 207

| The Deer. | (*G. Watt.*) | DEER. |

DEER: *Jerdon, Mammals of India, p. 248.*

The name Deer is applied to a group of Ruminant Mammals characterised by possessing osseous solid horns or antlers, which are shed
annually, at a period contemporaneous with the renewal of the hair; also
by the absence of a gall-bladder. They constitute the family CERVIDÆ,
the BOVIDÆ (or oxen, buffalos, sheep, goats, and antelopes), having a bony
prolongation from the skull, a core, encased by a hollow perennial horn
which grows from the base, throughout the life of the animal. The position of the Musk-deer and of the Deerlets is open to considerable difference of opinion. Most authors place them in one or in two families
intermediate between the CERVIDÆ and the BOVIDÆ. Others regard the
musk-deer as representing an aberrant genus of the BOVIDÆ (on account
of its possessing a gall-bladder), and view the Deerlets as transitionary
forms of CERVIDÆ approaching the antelopes and the musk-deer.

The classification of even the more highly developed CERVIDÆ is admittedly imperfect, but most authors recognise the following sub-families
and genera to which, in a popular work such as the present, may be sub-
joined the MOSCHIDÆ or Musk-deers and the TRAGULIDÆ or Mouse-
deers :—

FAMILY **CERVIDÆ.**

    SUB-FAMILY I.—**Cervinæ.**—THE STAGS PROPER.

    Genus CERVUS.—(1) **C. Wallichii**, the Kashmír Stag; and (2) **C.
affinis**, the Sikkim Stag.

        SUB-FAMILY II.—**Rusinæ.**—THE RUSINE STAGS.

    Genus RUCERVUS.—(3) **R. Duvaucelli**, the Swamp Deer; and (4) **R.
Eldii**, the Manipur Stag.

    Genus RUSA.—(5) **R. Aristotelis**, the Sambar Stag.

    Genus AXIS.—(6) **A. maculatus**, the Spotted Deer; (7) **A. porcinus**,
the Hog Deer.

    Genus CERVULUS.—(8) **C. aureus**, the Rib-faced or Barking Deer.

FAMILY **MOSCHIDÆ.**—THE MUSK DEER.

    Genus MOSCHUS.—(9) **M. moschiferus.**

FAMILY **TRAGULIDÆ.**—THE CHEVROTIANS OR DEERLETS.

    Genus TRAGULUS.—(10) **T. Napu**, the Javan Deerlet.

    Genus MEMINNA.—(11) **M. indica**, the Indian Mouse-deer.

Reference has been made above to the difference of opinion that prevails among zoologists as to the true position of the Musk-deer and the
Deerlets. These agree with each other in having no horns, and in possessing long canine tusks, and also in being higher over the croup than the
shoulders. But the Barking-deer has also long canine teeth and is considerably higher on the hind quarters. At the same time it has been
contended that some of the members of the genus AXIS have been found to
possess a gall-bladder. **Professor Flower** is disposed to regard the absence
of horns as an argument in favour of a Cervine position, since the males
of none of the Bovine animals are hornless. The most natural arrangement would therefore appear to be that given above, which would pass
into the BOVIDÆ with the Nilgai, the Antelopes, and the Goral, to the Goats,
Sheep, and Oxen.

The above brief indication of the classification (in a work on Economic
Products) has been deemed necessary from the difficulty that exists in
grouping the skins, horns, antlers, musk, &c., according to some stand-

ard that would be found to correspond with the arrangement adopted in the classification of the animals in purely Zoological Museums.

It is not intended, in this work, to deal with sport, but it may be worth mentioning (in connection with the subject of domestication) that in the *Ain-i-Akbari* an interesting account will be found of the fighting-deer kept by the Emperor, also of the sport of hunting for these animals by means of snares attached to trained deer. It is stated that "His Majesty had 12,000 deer" kept for these purposes.

The following paragraphs will be found to contain the vernacular names of the commoner species, their habitat and other peculiarities; but for further particulars regarding the Economic Products derived from these animals, the reader is referred to the following subject-headings :—"Hides and Skins," "Leather," "Horns, Antlers, and Ivory." For the Bovine animals to "Oxen," and to "Sheep and Goats."

Axis maculatus (*Jerdon, Mam., 260*).

212

THE SPOTTED DEER.

Vern.—*Chital, chitra* (*chritri-jhánk*, male), HIND.; *Boro-khotiya* (RUNG-PORE), *chatidah* (BHAGULPORE), *buriya* (GORAKHPORE), BENG.; *Sárang,* BELGAUM; *Polli maun,* TAM.; *Dupi,* TEL.; *Lupi,* GOND.; *Sarga, jati, mikka,* KAN.; *Zubbi,* ARAB.; *Gousun,* PERS.

Habitat.—Throughout the greater part of India, except the Panjáb, but apparently not found east of the Bay of Bengal. It is met with abundantly on the lower and outer slopes of the Himálaya, and immense herds may be seen in the Sunderbuns. It frequents forests bordering on streams, and is gregarious, very often occurring in herds of 30 to 40 or even 100. The most elegant and graceful of Indian deer, it is said to be found only in fascinating bits of country, its dappled hide being seen to sparkle in sunlight, of the mixed bamboo glades, as it bounds from the intruder on the slightest indication of danger.

See an interesting account of this deer in the Kanara Gazetteer, page 101. It is there stated to be rapidly being exterminated.

Skin.
213
Antlers.
214

SKIN AND ANTLERS.—The skin, a yellowish or rufous fawn, spotted with white, is much admired for ornamental purposes. The antlers have the tres-tines longer than the royals or posterior tines. They are shed in February and March, and are commercially in considerable demand, but actual statistics cannot be obtained. Liverpool is said to have imported from 1851—55, 20,000 of these antlers, and during the same period 700 of the skins. The following note, furnished by **Major A. E. Ward**, will be read with interest :—"There is a considerable trade in the horns as well as in the skins of the spotted deer. Formerly, in the times when this deer was plentiful, some of the Cawnpore leather firms gave contracts to men who supplied *shikáris* with powder and ball, and thus ruined the shooting in many parts of the Terai and the Duns. One firm gave a wholesale price of R50 per hundred skins; and at this rate attracted many offers of sale. The flesh is exchanged by the hunters for flour, &c. The tanned leather does not wear well."

"The spotted deer is very irregular in its breeding habits. It accordingly sheds its horns at no absolutely fixed period. The horns may thus be seen to be in velvet on some individuals and quite hard in others at almost any season of the year."

FOOD.
215

Food.—The flesh "is very good eating in the cold weather months." Ainslie (*Mat. Ind., I., 110*) says, "as venison it is not worth much, unless when caught young and fed properly." In Kanara an animal on account of its flesh is said to sell for R5-8.

D. 215

The Hog and Rib-faced Deer ; the Sikkim Stag. (*G. Watt.*)	DEER.

Axis porcinus (*Jerdon, Mam., 262*).

THE HOG-DEER.

216

Vern.—*Párá,* HIND.; *Nuthrini haran,* BENG.; *Khar-laguna,* NEPAL TARAI, but *Sugoria* is also sometimes given to it.

Habitat. —Throughout India, though less frequent in the central parts; abundant in Assam, Burma, and Ceylon. It is seldom found in forest land, preferring open grassy jungle. It lies all day in sheltered thick parts and only rises when run upon by the sportsman or his beaters. It gets its name of Hog-deer on account of its awkward gait. Major Ward writes that it leaves the thickets for "swampy ground directly the hot weather comes on, and may often be found in snipe jheels in the cold weather."

Skin and Antlers.—According to the same authority the skin of this species is not in much demand.

Skin.
217

Food.—The meat is said to be fair by some writers, but Major Ward is of opinion that it cannot be recommended. His words are: "This deer suffers greatly from internal parasites. And although the flesh is at times fairly good, what between these intestine parasites and the fact that the skin is often pierced by the grub of the *Bot,* I think the meat cannot be recommended."

FOOD.
218

Cervulus aureus (*Jerdon, Mam., 264*).

219

THE RIB-FACED DEER, the BARKING-DEER or MUNTJAC of India; the RED HOG-DEER in Ceylon.

Vern.—*Kakar, bherki, jangli-bukra,* HIND. ; *Maya,* BENG. ; *Ratwa,* NEPAL; *Karsiár,* BHOTIA ; *Siku, suku,* LEPCHA; *Kondákuri,* BELGAUM ; *Advikuri,* KAN. ; *Gutra, gutri,* GOND.; *Bekra baikur,* or *Kekar,* MAR. ; *Kuka-gori,* TEL.; *Gee,* BURM.

Habitat.—India, Burma, Ceylon, the Malay Peninsula, Sumatra, Java, Borneo, &c. Sportsmen describe this as a retiring little forest animal; generally found alone or at times in pairs, "creeping," as Hodgson remarks, through the tangled jungle or under fallen trees. It is said in Kanara to love the dense shade of the *Kárvi* (Strobilanthus) that covers Sahyádri slopes (*Kanara Gazetteer, page 102*).

Skin and Antlers.—Major Ward says that "the skin of the barking deer is very largely in demand, as it is very tough when tanned. Shoes and leather socks are made in great numbers from it. Saharanpur, Meerut, and Dehra *mochees* are the principal dealers in this hide." " The horns are too small to be of value."

Skin.
220

Antlers.
221

Food.—"It is excellent venison, but rarely carries any fat." This statement is confirmed in the *Gazetteer of Ratnagiri,* but the venison is said to be all the more appreciated in a district where mutton is scarcely attainable.

FOOD.
222

Cervus affinis (*Jerdon, Mam., 251*).

223

THE SIKKIM STAG.

Vern.—*Shon,* TIBETAN.

Habitat.—The Eastern Himálaya (Sikkim side of Tibet, Chumbi Valley). Major Ward is, however, very doubtful if this stag is to be found at all in the Chumbi Valley. "Mr. Ney Elias tells me," he writes, "that it is scarcely known in those parts even as an animal which exists."

Antlers.—According to Major Ward, the antlers are very large, a pair in Simla measuring 54 inches in length. He adds:—" A magnificent pair of antlers which I have at home quite dwarfed the pair of Kashmir stag's horns, 47 inches long, now in my possession at Simla."

Antlers.
224

| DEER. | The Kashmir Stag and Musk Deer. |

225

Cervus Wallichii (*Jerdon, Mam., 250*).

THE KASHMIR STAG.

Vern.—*Barasingha*, HIND.; *Hangul* or *Honglu*, KASHMIR.

Habitat.—Kashmír, the Sind valley, to Budrawar and Kishtar eastward, inhabiting pine forests at altitudes of 9,000 to 12,000 feet, descending to lower levels in autumn and winter. The larger stags, Major Ward writes, seldom come below 7,000 feet. " In the spring this animal migrates from the valleys of Kashmír and wanders far, often crossing the lower passes, *viz.*, the Mingan, the Togila, &c. It clings, however, to country that is fairly wooded. It is rapidly decreasing in number."

Antlers.
226

Antlers.—" The horns form a portion of the tribute paid by the shikáris to the Máhárája of Jummu. The best are sold at high prices from R15 to R30 per pair, and are bought by taxidermists and collectors of horns."

227

Moschus mochiferus (*Jerdon, Mam., 266*).

THE MUSK DEER.

Vern.—*Kastura*, HIND.; *Rous, rús, kasturé*, KASHMIR; *La-lawa*, TIBET; *Rib-jo*, LADAK; *Bena*, KANAWAR; *Mussuck-nabu*, PAHARI.

Habitat.—Found throughout the Himálaya at elevations above 8,000 feet: distributed to Central and Northern Asia and Siberia. The musk deer is a forest-loving animal, keeping much to one locality. It is wonderfully sure-footed and is able to leap and bound over the steepest and most broken ground. Colonel Markham (*Four. Sporting Adventures and Travel in Chinese Tartary and Thibet*) says: " On a gentle slope I have seen them clear a space of more than sixty feet at a single bound, for several successive leaps, and spring over bushes of considerable height at the same time." It is an exceedingly shy animal of nocturnal habit, and not much larger than a greyhound. Of all ruminants it is reported to eat the least, and although no connection can be traced between the nature of the food it eats and the production of musk, it is a common opinion among traders that those reared in forest-clad countries are better than those met with in open rocky regions. It is said to eat the tangled grey lichen (*Usnea*) that hangs from trees everywhere on the higher Himálaya, and the leaves of various shrubs, as also grasses, roots, &c. Colonel Markham alludes to a popular opinion that it eats the leaves of a laurel (*kedar pattu*), probably Litsæa umbrosa, a small tree or bush frequent up to about 7,000 feet, but certainly not common at the altitudes where the musk-deer lives. Major Ward, however, repudiates this statement, characterising it as a native absurdity.

228

MUSK.

MUSC, GRAINE D'AMBRETTÉ, *Fr.*; MOSCHUS, BIZAM, *Germ.*; MUSCHIO, *It.*; ALMIZELE, *Sp.*

Vern.—*Kastúri, mushk*, HIND.; *Kashtúri*, BENG.; *Kasturi*, MAR., TAM., TEL., MAL.; *Misk, mishk, mushk*, ARAB.; *Mushk*, PERS.; *Mushk náfá*, PB.; *Mriganábhi, kasturi*, SANS.; *Kado*, BURM.

References.—*Sterndale, Mam. of Ind., 494; Piesse, Art Perfumery, 246; U. C. Dutt, Mat. Med., 279; Moodeen Sheriff, Supp. Pharm. Ind., 177; Pharm. Ind., 282; U. S. Dispens., 15th Ed., p. 962; Baden Powell, Pb. Prod., 189; Ainslie, Mat. Ind., I., 228; Ure, Dict. Arts, &c., III., 213; Balfour, Cycl. Ind., 1021; Spons, Encycl., 1524; Davies, Trade and Resources of the N.-W. Boundary of India, CCXXXVII.*

DESCRIPTION
229

Description.—" The musk is milky for the first year or two, afterwards granular; the dung of the males smells of musk, but the body does not, and the females do not in the slightest degree." " The musk-deer is

D. 229

Musk and the Musk Deer.	(*G. Watt,*)	DEER.

much sought after for its musk, many being shot and snared annually. **DESCRIPTION.**
A good musk-pod is valued at from 10 to 15 rupees. The musk as sold
is often much adulterated with blood, liver, &c. One ounce is about the
average produce of the pod." A few anatomical details of interest (by
Dr. Campbell) may be here given:—"The musk-bag lies at the end of
the penis, and might be termed a prœputial bag. It is globular, about 1½
inch in diameter, and hairy, with a hole in the centre about the diameter
of a lead pencil, from which the secretion can be squeezed. The orifice of
the urethra lies near this, a little posteriorly. Round the margin of the
opening of the gland is a circle of small glandular-looking bodies. The
musk when fresh is soft, not unlike moist gingerbread. The anus is surrounded by a ring of soft hairs, the skin under which is perforated by
innumerable small pores, secreting an abominably offensive stuff, which
pressure brings out like honey. The scrotum is round and naked. There
is, besides, a peculiar organ or gland on the tail, which indeed is composed
almost wholly of it. The tail of the male is triangular, nude above, thick,
greasy, partially covered with short hair below, and with a tuft of hairs at
the end, glued together by a viscid liquor. It has two large elliptic pores
beneath, basal and lateral, the edges of which are somewhat mobile, and
the fluid, which appears to be continually secreted, has a peculiar and
rather offensive odour." (*Jerdon's Mammals, p. 268.*) Colonel F. Markham thus describes the preparation of the so-called "musk-pod" : " The
musk-pods which reach the market, through the hands of the native hunters, are generally enclosed in a portion of the skin of the animal, with the
hair or fur left on it. When they have killed a musk-deer they cut round
the pod, and skin the whole of the belly. The pod comes off attached to
the skin, which is then laid, with its fleshy side, on a flat stone previously
heated in the fire, and thus dried without singeing the hair. The skin
shrinks up from the heat into a small compass, and is then tied or stitched
round the pod, and hung up in a dry place until quite hard. This is the
general method of preparing them; but some put the pod into hot oil
instead of laying it on a hot stone; but either method must deteriorate the
quality of the musk, as it gets either completely baked or fried. It is
best both in appearance and smell if the pod is at once cut from the skin,
and allowed to dry of itself." Mr. F. Peak (of Peak Allen & Co.) wrote
to Mr. Piesse (*Art of Perfumery, p. 256*) that " The thin bladder-like
skin dries in the sun in a few hours; that in the hair pods, on the contrary,
gets quite roasted in the process of preserving and preparing." " I sent
both kinds home to ascertain which was best, and that in the pods without
the hairy-skin was declared to be far superior." Referring to the process
of drying skin around the pods he adds: " By the continued heat much
of its odour is driven off, and it is consequently deprived of its qualities as
a remedial agent, and for the use of the perfumer is greatly deteriorated."
(*See also Peak P. J. Tr., Feb. 1861.*)

    Adulteration.—The extent to which musk adulteration has been carried **ADULTER-**
seems natural enough, especially at the present time, when nearly every com- **ATION.**
mercial article is counterfeited to some extent. The high price paid for the **230**
perfume, the uncertainty of the supply, and the difficulty of detection must
have all naturally tended to suggest a certain amount of adulteration.
Colonel Markham writes: " I have often seen pods offered for sale which
were merely a piece of musk-deer skin filled with some substance, and tied
up to resemble a musk-pod, with a little musk rubbed over to make it smell.
These are easy to detect, from there being no navel on the skin, it being
cut from any part of the body. But the musk is sometimes taken out of
real pods, and its place supplied by some other substance, and these are
difficult to detect even if cut open, as whatever is put in is made to resem-

DEER.	Musk and

ADULTER-ATION.

ble musk in appearance, and a little genuine added makes it smell nearly as strong. Some have only a portion of the musk taken out, and its place thus supplied; and others have all the musk left in, but something added to increase the weight." The above description of the process and materials of adulteration differs but little from that written in 1596 by John Huygen van Linschoten. That early traveller was misinformed when he stated that the true musk was the testicles of the animal, and this mistake his contemporary, Dr. Paludanus, corrected. He wrote: "Some are of opinion that muske groweth at certaine times of the yeare about the navell." But Linschoten's account of the Chinese adulteration of the article, traded in by the Portuguese at so early a date, is worth quoting in this place. He says that having killed the animals "they let them lie and rot, blood and flesh together: which done they cut them in pieces both skinne, flesh and blood, all mixed togeather, and thereof make divers purses, which they sow (in a round forme) and are in that sort carried abroad and sold. These purses are commonly of an ounce waight the peece, and by the *Portingales* are called *Papos,* but the right *Papos* and perfect *Mosseliat* is the bullockes or stones of the beast, the others, although they passe among them for *Mosseliat,* are not so good as the stones : therefore the Chinaes (Chinese) who in all thinges are very subtill, make the purses cleane round, like the stones of the beaste, therewith to deceive the people, and so the sooner to procure them to buy it." So again he says: The Chinaes are very deceitful in selling of *Mosseliat* (or Muske), for they folsife it verie much, sometimes with oxen and cowe's livers, dried and beaten to powder, and so mixed with the *Mosseliat,* as it is dayly found by experience in searching of it."

COMMERCIAL FORMS.
Cabardien.
231
Assam.
232
Tonquin.
233

Commercial Forms of Musk.—Piesse says there are three kinds, *viz.*— "The *Cabardien,* or Russian Musk, which is rarely, if ever, adulterated; from its poorer fragrance, however, it does not fetch more than 8*s.* an ounce in the pod. The *Assam Musk* is next in quality : it is very strong, but has a rank smell: the pods are very large and irregular in shape : it fetches about 24*s.* per ounce in the pod. The *Tonquin* or Chinese Musk yields the kind mostly prized in England, and is more adulterated than the former; market price from 26*s.* to 32*s.* per ounce in the pod." Further on Mr. Piesse again refers to the Assam Musk : —" The musk of Assam and South Thibet reaches Europe by way of Calcutta. It is sent in bags enclosed in a chest of wood or tin-plate, which holds about two hundred pods. The form of this musk is more valuable than that of the Nankin musk." Although Mr. Piesse publishes extracts from his correspondence with Peak, Allen and Co. regarding Himálayan musk, he does not (*in his Art of Perfumery*) furnish any information as to the comparative value of Himálayan and Assam. Dr. U. O. Dutt says that, according to Sanskrit literature, there are three kinds of musk—" The *Bhávaprákása* describes three varieties of musk, namely, *Kamrupa, Nepála,* and *Káshmira* musk. Kámrupa musk is said to be of black colour and superior to the others. It is probably China or Thibet musk brought to India *viâ* Kamroop in Assam. Nepála musk is described as of a blueish colour and intermediate quality. Káshmira is of inferior quality."

Panjab Musk.
234

The following note, regarding Panjáb musk, has been obligingly placed at the writer's disposal by G. G. Miniken, Esq., Deputy Conservator of Forests :—

"In Bashàhr on the Sutlej and on the Rupin and Paber rivers, the Musk-deer was at one time plentiful, but it is generally stated that it is not now so numerous.

"The right of hunting the musk-deer belongs to the Rajah, and he employs trained shikáris to hunt them; but this right is in truth not respected.

| The Musk Deer. (*G. Watt.*) | DEER. |

Villagers all over the country shoot for themselves, and the pods obtained are sold to chemists at Simla and Masouri. The Rajah's shikáris use nets which are set up across some gap or glade in the forest, and with dogs drive the deer into the nets where they are shot, and the pod extracted from the male while it is hot. The musk is said to be of better quality if the pod be taken out at this time. The musk is sometimes adulterated by mixing with it the blood of the slain deer, and reduced by boiling to a soft mass. The test of genuine musk is made by passing a thread through asafœtida (*Hing*) and then through the pod. If after this, the smell of the *Hing* remains the musk is not genuine.

Musk is used as medicine. It is said to be useful in venereal diseases, and for wounds. In the first case a small pill is taken once a day for two or three days; in the second case, a bit about the size of a grain of rice is applied; but if too much is put on the wound, the flesh swells. Musk mixed with '*ghi*' called in the plains *Hawan samaghri* is used to scent rooms, and to keep off bad air. It is also burnt as incense in temples. Bushahiris smoke it mixed with tobacco, and it is said to have a mild intoxicating effect. But it is especially prized for its stimulative action when taken internally; particularly for incompetence. It is useful for pains in the back, which it also strengthens.

About R5,000 worth is sold annually in Bushahir, and it is bartered in the Rampur bazár for down-country produce. Its price averages R20 per ounce. A good deal of musk is brought from Kulu and native Garwahl to Rampur.

Indian Trade in Musk.—The extent of the internal trade in musk cannot be discovered, but as the animal is systematically hunted all over the region where it occurs, and the so-called musk-pods are to be had in every drug-seller's shop, the consumption must be very extensive.

Mr. Baden Powell says that "about 100 musk-bags are imported from Chang Thán *viá* Yarkand, of which about 40 go to Yarkand, the rest to Kashmír and Jammu, and are taken by Yarkandi pilgrims to Mecca or for sale in India and other Asiatic countries; they are produced in the north-west of Rodokh and Nepál; value at Leh R7 to 15, or at Yarkand from R21 to 26. In former times musk-bags from the Dasht-i-Khattan or Great Tartar desert, were in high repute, and fetched at the least R42; but all supply from that quarter has long ceased." In many of the reports of external (or trans-frontier) land trade mention is made of musk, but not in such a manner as to allow of a trustworthy statement being compiled of the total imports in any one year. Indeed, the animal is so very generally found throughout immense portions of the British Indian Himálaya (the produce of which would not appear in reports of trans-frontier imports) that even a compilation from all the reports on Indian foreign trade by land would by no means convey a definite conception of the total trade. The imports of musk into Bengal from Sikkim and Tibet were valued in 1883-84 at R2,563, in 1884-85 at R84,100, and in 1885-86 at R55,265. During the same periods Bengal received from Bhutan musk to the value of R5,913, R8,344, and R6,624. During the last of these years (1885-86) it also obtained, from Nepal, musk to the value of R5,235, so that by these foreign sources alone India obtained R67,124 worth of musk and the previous year the imports appear to have been considerably larger. The Assam imports, not consumed in the province, must be also carried into Bengal and be distributed from Calcutta all over the country, and doubtless also a very considerable amount of the imports into the North-West Provinces and the Panjáb find their way to Calcutta. But, as stated, an elaborate compilation from all the Trans-frontier Land Trade Reports, Railway and River-borne Trade, and of all other such sources of information,

COMMERCIAL FORMS.

TRADE.
235

DEER.	The Musk Deer; the Swamp Deer.

TRADE.

would fall short of the actual mark, since a small expensive article like that of musk must be extensively trafficked in outside the limits of possible commercial statistics. Calcutta is, however, the chief emporium of the trade, and some conception of its total extent may be gathered from the figures of Foreign Exports by Sea from India, which, it may be repeated, represent the surplus over and above Indian consumption. Last year (1887-88) India exported 2,144 ounces valued at R72,116, and of that amount only R20 worth left Bombay, the rest being exported from Bengal, and R61,226 worth were consigned to the United Kingdom. The exports in 1886-87 were valued at R70,913, the smallest amount since 1878-79. The average exports for the past ten years may be taken to have been valued at R1,11,750. The total amount of musk exported from India during these years was 44,195 ounces, valued at R11,17,519. Each animal contains only one musk-pod, the average weight of which is about one ounce of musk, so that the above figures would represent an annual slaughter of about 4,500 male animals to obtain the musk exported from India. These are of course not all killed within British territory, the traders bring a large proportion from the regions on the north of the Himálayas. But on the other hand the internal or Indian consumption is not estimated for, so that it is probable the Indian trade (internal and exported) represents a slaughter of little short of 10,000 musk-deer annually. And doubtless a large number of females are caught in the snares by which the natives capture the animal, so that it is probable that nearly 20,000 are actually killed by the traders and sportsmen combined. This wholesale extermination doubtless has something to say to the visible decline in the supply and to the decrease in the exports, but it is also probable that other animal and even vegetable sources of supply are yearly coming into greater importance.

The value of the musk-pod is said to average from R10 to R15. For further particulars, in continuation with this account of the Perfume musk, see **Musk** in another volume, in which will be found the medicinal and chemical properties of the substance and its applications in the art of perfumery, together with information regarding the other sources of supply.

Skin.
236

Skin.—The skin of the musk-deer does not appear to be of any value. It is covered with rigid porcupine-like hairs.

FOOD.
237

Food.—The flesh of the young animal is reported to be tender and well flavoured. The female does not produce musk, but even, in the male, while the animal smells strongly and the dung also is musk-scented, the flesh is perfectly devoid of the odour, not even the stomach, nor the contents of the stomach, removed after death, partake of the characteristic smell.

238

Rucervus Duvancellii (*Jerdon, Mam., 254*).

THE SWAMP DEER.

Vern.—*Bara-singha*, HIND.; *Baraya* or *maha*, NEPAL TARAI; *Jhinkar*, KYARDA DOON; *Potiya haran*, MONGHYR; *Goen* or *goenjak* (male), *gaoni* (female), CENTRAL INDIA.

Habitat.—The forest lands at the foot of the Himálaya from Kyarda Doon to Bhotan. It is very abundant in Assam, inhabiting the churs and islands of the Brahmaputra down to the Sunderbunds. It also occurs at Monghyr, and extends sparingly to Central India. It lives in great herds, preferring the open forest land in the vicinity of rivers. According to Major Ward, it is common in Nepal, and is still to be found on the banks of the Sardah river and the islands intersecting its course near Moondea Ghát, in which neighbourhood he has shot several. Major Ward adds that years ago it used to be found in the Dehra Duns, but that none are at

Products of India.   63

The Eld's·Deer; the Samber Stag.   (*G. Watt.*)   DELPHINIDÆ.

present met with in those parts except considerably to the westward of Philibeet.

Rucervus Eldi (*Jerdon, Mam.*, 255).   239

THE MANIPUR or BURMA STAG; THE BROW ANTLERED or ELD'S DEER.

Vern.—*Thamin*, BURM.; *Sungrai* or *sungnai*, MANIPUR.

Habitat.—The Eastern Himálayas, Manipur, Burma, Siam, and the Malay Peninsula. It is essentially a plains-loving species, and though it frequents open tree jungle, it never ventures into dense tangled brushwoods, and on being alarmed takes to the open.

Rusa Aristotalis (*Jerdon, Mam.*, 256).   240

THE SAMBER STAG.

Vern.—*Sambar*, HIND.; *Jerai, jerao*, in the 'HIMALAYA; *Maha* in the TARAI; *Meru* or *Kadavi*, MAR.; *Kadivi*, BELGAUM; *Maoo*, GOND; *Kannadi*, TELEGU; *Ghous, gaoj*, EASTERN BENGAL; *Schap*, BURM.

References.—*The account given in the Gazetteer of Kanara District will be found interesting, p. 100.*

Habitat.—Throughout India, from the Himálaya to Cape Comorin, and through Assam and Burma to the Malay Peninsula and Ceylon. In the *Kanara Gazetteer* it is said of that district that the Samber is nowhere so numerous as it was ten or fifteen years ago. The cause of this is said to be the great increase of guns. There is scarcely a village that has not its gun or guns licensed or unlicensed. The practical extermination of the animal in Kanara is feared likely to soon occur.

Skin and Antlers.—**Major Ward** communicates the following note :—   **Skin.**
"Hide greatly in demand in India. A hind's skin will now sell for R3 to   **241**
R4, and when tanned for R7 to R10. Used for gaiters, boots, bags, &c. If dressed well, with a mixture of linseed oil and mutton fat, it will stand wet fairly well; but if not so dressed, it hardens on drying."

Food.—"The flesh of the Samber is rather coarse, and rarely fat, but   **FOOD.**
sometimes well tasted." The marrow bones and tongue are saleable. In   **242**
Kanara the natives sit on the wild fruit trees and shoot the samber when it comes to feed, or they lie in holes dug near tanks of water. The fruits on which it specially feeds are said to be **Phyllanthus Emblica, Dillenia pentagyna, Terminalia bellerica,** and **Spondias mangifera.**

DELIMA, *Linn.; Gen. Pl., I., 12.*

Delima sarmentosa, *Linn.; Fl. Br. Ind., I., 31; DILLENIACEÆ.*   243

Syn.—TETRACERA SARMENTOSA, *Willd.*

Vern.—*Mon kyourik*, LEPCHA; *Korasa-wel*, SINGH.

References.—*Roxb., Fl. Ind., Ed. C.B.C., 449; Kurz, For. Fl. Burm., I., 22; Gamble, Man. Timb., 2; Thwaites, En. Ceylon Pl., 2; Gamble, List of Trees and Shrubs, &c., of Darjeeling, p. 2; Royle, Ill. Him. Bot., 58; Balfour, Cyclop., 910; Treasury of Bot., 390.*

Habitat.—A woody climber met with in Eastern Tropical India, from Darjeeling and Assam to Singapore. **Kurz** says it is frequent in the mixed forests of Burma from Chittagong and Pegu down to the Andamans: also in Ava.

Domestic Uses.—The leaves of the plant are universally employed, in   **DOMESTIC.**
the countries where the plant occurs, in place of sand-paper to polish wood   **244**
and even metal articles.

Delphinidæ, the Whale family; see **Whale.**

D. 244

DELPHINIUM Brunonianum.   **The Larkspurs.**

DELPHINIUM, *Linn.; Gen. Pl., I., 9, 953.*

A genus of annual or perennial herbs, containing some 40 species, which are distributed throughout the north temperate zone and on the temperate tracts of lofty mountains in the southern zone. The generic name, derived from the Greek *Delphinion,* arose from the somewhat fanciful resemblance of the flower-bud to the head of the Dolphin, and the English name Larkspur was doubtless occasioned through the long spur-like prolongation at the base of the flower. The common Larkspur, **Delphinium Ajacis,** takes its specific botanical name from the supposition of its petals denoting the letters, A. I. A., the initials of Ajax, the Greek Trojan hero. The Larkspur is a favourite garden annual in India. On the Himálaya it shows a distinct desire to leave the restricting influence of cultivation, and even in some parts of the plains manifests a tendency to become perennial. Withstanding the intense summer's heat of the drier areas, it may sometimes be seen to flower during winter and spring for several successive seasons. In such cases, however, it assumes a rigid bushy habit, and has small pale-coloured flowers. In fact, it alters its faces so far as to largely lose its accepted specific characteristics, and assumes some of those of **D. orientale.** Firminger remarks that he had "failed completely to germinate imported Larkspur seed in the plains of India." The plant must be first acclimatised in the temperate regions of India, and be brought gradually down to the plains. The stock found in the plains consists of **D. Ajacis** and **D. consolida.** The latter having larger flowers on longer peduncles and the segments of the leaves broader than the former. Firminger speaks of both collectively as "a poor weedy worthless thing." In a further passage he concludes :—"If the ground where Larkspurs have grown one season be left undisturbed, an abundant crop of self-sown plants will spring up the following November and December." In Bankipur (Behar), the writer carefully marked several individual plants and found that they continued to grow throughout the year, and even formed flowers during the hottest months, provided they were watered and had the partial shade from trees. In the same way a crop of lettuce was obtained at any season, and both Larkspur and lettuce produced from self-sowings the stock of seedlings for almost any month of the year. The Larkspur was thus acclimatised to one of India's dry hot tropical climates, and had practically lost its character as a temperate-loving plant. In most parts of India (preferentially the dry or non-inundated areas) it is practically a cold season garden weed, its single, faded, purplish flowers being unworthy of care and attention.

245

[LACEÆ.

Delphinium Brunonianum, *Royle; Fl. Br. Ind., I., 27;* RANUNCU-

Vern.—*Nepári,* KUMAON; *Kastúri,* GHARWAL; *Sapfalú* (RAVI), *laskar, spet, panni, supalú, ruskar, liokpa* (SUTLEJ), PB.; *Ládara,* LADAKH; *Laskara,* SIMLA; *Múndwál,* PANGI.

References.—*Stewart, Pb. Pl., 3; Aitchison, Kuram Valley Flora (Jour. Linn. Soc. XVIII., pp. 25, 30); Atkinson, Him. Dist., 412, 735; Royle, Ill. Him. Bot., 56; Balfour, Cyclop., I., 911; Gazetteer, Simla Dist., p. 12.*

Habitat.—A very abundant plant on the higher Western Himálaya and Tibet, at altitudes of 13,000 to 17,000 feet.

MEDICINE.
Juice.
246
Leaves.
247

Medicine.—This plant is prized for its strong scent of musk. It is offered to the presiding idol of the hill temples. **Aitchison,** in his *Flora of the Kuram Valley,* remarks that the juice of the leaves of this plant are used in Kuram to destroy ticks in animals, but chiefly when they affect sheep. This is a curious fact, pointing to Stavesacre (**D. Staphisagria,** *Linn.*), which is now very largely used in Europe, and was employed both by the Greeks and Romans for a similar purpose, *viz.,* the destruction of vermin.

SPECIAL OPINIONS.—§ "In Leh it is considered so poisonous that the dew from the leaves falling on grass is said to poison cattle and horses" (*Surgeon-Major J. E. T. Aitchison, Simla*).

D. 247

The Nirbisi or Jadwar. (*G. Watt.*)	DELPHINIUM denudatum.

Perfumery.—Used as a substitute for Musk (which see). Atkinson (*Him. Dist., p. 735*), says "it is exported from the Kumaon Himálaya on account of its musk-scented leaves."

PERFUMERY.
248

Delphinium cœruleum, *Jacq.; Fl. Br. Ind., I., 25.*

249

Vern.—*Dakhangú*, Pb.

References.—*Stewart, Pb. Pl., 3; Atkinson, Him. Dist., 328, 412.*

Habitat.—A slender plant with light blue flowers, met with on the alpine Himálayas; common in the Sutlej basin from 8,000 to 17,000 feet.

Medicine.—The ROOT is applied to kill the maggots in the wounds of goats (*Stewart*).

MEDICINE.
Root.
250

D. cashmirianum, *Royle; Fl. Br. Ind., I., 26.*

251

Vern.—*Amlin* (in RAVI BASIN), Pb.

Habitat.—An alpine herb met with in the Western Himálaya, Kashmír, and Thibet, at altitudes of 10,000 to 16,000 feet.

Medicine.—Stewart says this is strongly scented like **D. Brunonianum.** Atkinson (*Him. Districts, p. 745*), alluding to the necessity of a thorough investigation of the roots, &c., exported from the hills under the names of *bikh* and *nirbisi*, after mentioning **Pæonia Emodi, Aconitum ferox, Polygonatum verticillatum,** and **Smilacina pallida,** adds : "The cylindrical tuberous roots of **Delphinium kashmerianum,** *Royle,* found at Pindari in Kumaon and Bhojgara, on the south side of the Kawari pass in Garwahl (11,000 to 14,000 feet), are absolutely identical with the ordinary *nirbisi* roots. (*See Madden, An. Hag. N. H., 2nd Ser., XVIII., 445.*") *Conf.* with **D. denudatum.**

MEDICINE
252

D. denudatum, *Wall ; Fl. Br. Ind., I., 25.*

253

Syn.—DELPHINIUM PAUCIFLORUM, *Royle* (*not of* Don).

Vern.—*Nirbisi* (according to Dymock), *judwar* (according to Murray), HIND.; *Nilo bikh*, NEPAL; *Nirbisi* (of the BHOTIAS), EAST HIMALAYA; *Múnila* (SIMLA), PB.; *Jadwár, mahferfin* (according to Dymock), ARAB.

Compare the above vernacular names with the remarks under **Curcuma aromatica, Vol. II., p. 656.**

References.—*Stewart, Pb. Pl., 3; Dymock, Mat. Med. W. Ind., 2nd Ed., 11 ; Murray, Pl. and Drugs, Sind, 74; Royle, Ill. Him. Bot., 55.*

Habitat.—An annual herbaceous plant, common on the outer ranges of the Western Himálaya, from Kashmír to Kumaon ; altitude from 5,500 to 8,500 feet. A denizen of the drier warm-temperate tracts of the Himálaya, especially on grassy slopes, where occasional brushwood occurs on southern exposures. (*Conf.* with **D. vestitum.**)

Medicine.—Only one modern author records the observation that the natives of India use this **Delphinium** medicinally—Madden wrote that the ROOT is chewed on Sundays by the people of Bashahr for toothache. It would appear to be one of the roots occasionally collected in order to be used as an adulterant for Aconite. The trade in the article is, however, extremely limited, and naturally so, since it nowhere grows in the region where the Aconites are found. It bears the name of *Nirbisi*, with the Bhotias of Nepal, and on this account alone it would appear to have been lagged into the controversy as to the root which should be accepted as the *Nirbisi* (or rather *nir-visha*) of Sanskrit writers and the *Jadwár* of the Arabic. Dr. F. Hamilton was the first to make known the existence in Nepal of various species of **Aconitum.** These he incorrectly assigned to the genus **Caltha,** but gave useful information regarding their

MEDICINE.
Root.
254

F

D. 254

DELPHINIUM denudatum.						The Nirbisi or Jadwar.

MEDICINE.
Singya bikh.
255
Bikh.
256
Bikhma.
257
Nirbisi.
258

poisonous properties. He described four forms—(1) *Singya bikh*, (2) *Bish* or *bikh*, (3) *Bikhma*, and (4) *Nirbisi*. *Bikhma*, he explained, was a powerful bitter, and Wallich subsequently identified this as **Aconitum palmatum.** *Nirbisi*, Hamilton affirmed, to be devoid of poisonous property, while he announced *Singya* to be the root of a **Smilax**, and *Bish* or *Bikh* to be a virulent poison. More recent writers have extended the list of vernacular names given to the poisonous Aconites. Thus *Singyi* or *Singyá-bish* (the horny *bis*) and *mitha-zahar* (the sweet poison) are given to two forms of **Aconitum ferox**, the separate properties of which are recognised by the Indian drug-sellers. Both Hindu and Muhammadan writers on Materia Medica refer to many forms of poisonous and non-poisonous aconites. Some of the former are so poisonous as to have obtained the fabulous reputation of proving fatal to the touch. Of the latter many forms are mentioned, the names given expanding until they include an extensive series of tonic medicines, many of which are in no way related to **Aconitum.** In a like manner the word *Bish* or *Bikh* simply means poison, the *Visha* of Sanskrit; but it became specifically restricted as a proper name to Aconite, the most poisonous of all the poisons—*Bikh* or *Bis, the* poison. So also *Bikhma* or *Bishma* would mean "*bikh*-like," and might be supposed to have been first applied to the less poisonous forms of Aconite, until, in the descending scale of transitions, the innocuous forms of Aconite were embraced by it, and in time also the root, or collection of roots, that ultimately received the designation of *Nirbisi*, with its synonyms in Arabic of *Jadwár* and *Mahferfin*, and in Persian of *Zadwár*. Whether or not the word *Nirbisi* means antidote, if a synonym for *Jadwár*, the root referred to must have been used as a drug to strengthen the system against poison—the alexipharmic of ancient writers. **Royle** wrote :—" The term *Nirbisi*, as observed by **Mr. Colebrooke**, implies that the drug is used as an antidote to poison, being composed of the privative preposition *nir* and *bis* poison; and in the *Makhzan-ul-Adwiya*, it is further explained as repelling from and purifying the body from poison." Commenting on the above opinion held by **Mir Muhammad Husain, Dr Dymock** says— " The Indian name *Nirbisi*, he (**Mir Muhammad**) explains incorrectly as *Nir*, the antidote to *Bish*, the poison. *Nirvisha* is a Sanskrit adjective meaning ' not-poisonous,' and *nirvisha* or *nirvishi* is never applied to Aconite by Hindu medicine writers, but denotes a peculiar sedge used as an antidote to certain poisons, *viz.*, **Kyllingia monocephala**, *Linn.*" According to most writers the *Jadwár* possesses alexipharmic properties, and **Dr Moodeen Sheriff** says—" *Jadwár* is the only safe word to use in ordering the non-poisonous aconites." He, however, remarks, *Nirbisi* is often confounded with the Sanskrit name *nir-visha*, " and this is partly from the partial analogy that exists between their pronunciation, and partly from their literal and general meaning being nearly the same. *Free from* or *without poison* is the literal meaning of *Nir-vishani* or *Nir-visha*, and the meaning generally attached to it in books is an *antidote*. The only difference between the above meaning and the meaning of *Nir-bisi* is, that the Sanskrit word *Visham* or *Visha* is the common name for any poison, whatever it may be, while *bis* in Hindustani is the name of a particular vegetable poison, *viz.*, the root of **Aconitum ferox**."

An antidote to Aconite poison would be a diffusible stimulant, and thus, as time went on, discovery after discovery would doubtless have expanded the list of drugs that might each deserve the name of *Nirbisi* or *Jadwár*. It may thus be safely assumed that every region and age had its favourite *Nirbisi*, and that special preparations of certain diffusible stimulants came to take the place of some particular root—the *Nirbisi* of the earlier authors. The writer had a sample of the sacred

Jadwar.
259

Costus root (the root of **Saussurea Lappa**) sent him from Assam as the
antidote used by the Akas against Aconite poison. This fact is of considerable interest as manifesting a knowledge in the properties of a
Kashmír diffusible stimulant, which, perhaps, far surpasses in its efficacy
all the indigenous antidotes met with in the Aka country. It must, therefore, be either carried from the one extremity of the Himálayas to the
other, passing from village to village and hand to hand over a wild
mountainous country of perhaps several thousand miles, or be imported
into the highland home of the savage Aka from the plains of India. But
the interest in this incident, namely, the knowledge of the properties of
a drug, does not rest here. The Akas do not import their Aconite.
They possess an indigenous species quite as virulent as the Nepal root,
which finds its way all over Asia. The Akas recognise in the supposed
cure the identity of the poison, and we have thus a flood of light thrown on
the subject of the *Bikh*, *Bikhma*, and *Nirbisi* of the ancient Sanskrit
writers, which justifies the caution that a too literal interpretation or
application of these words assigning them to this individual species and that,
may miss the mark and only multiply ambiguity with the obscurity of antiquity. This caution is rendered all the more forcible when it is added
that botanists have established the fact that, under **Aconitum ferox** and
A. Napellus—the most poisonous species of Aconite—there are forms
known to the shepherds of the higher Himálaya, which, like **Aconitum
heterophyllum**, may be eaten with impunity, or used as tonic or antiperiodic medicines. The *Makhzan-el-Adwiya* states that the only plant
that can grow near the *Bikh* is the *Jadwár*. This may be a mere
tradition, but if it be accepted as carrying any meaning with it, all idea
of the *Jadwar* being Zedoary would have to be completely set aside.
Dr. Moodeen Sheriff, indeed, urges that much unnecessary ambiguity
has been caused through an early error of regarding the word Zedoary
as derived from *jadwar* and *zadwar*. The Sanskrit scholar, the late
Mr. Colebrooke, identified *nirbisi, jadwar,* and *zadwar,* as synonymous
terms, and suggested that these were most probauly given to a species
of **Curcuma,** but he added, if this be not so, they would have to be collectively assigned to the root of some other plant. Ainslie contended that
the *nirbishie* of **Dr. Hamilton** "must not be confounded with the word
nirbisi, which is the Sanskrit for **Curcuma Zedoaria**." **Dr. Dymock** and
many other modern writers, however, assign these classical names to
Delphinium denudatum, not because of the roots of that plant agreeing
with the descriptions given by early authors, or of their being used (at
the present day) or known to possess the property of an antidote to
poison, but because the hill tribes, on a restricted portion of the Himálaya,
are stated to give it the local name of *Nirbisi*. The writer suggested to
Dr. Gimlette, Residency Surgeon, Nepal, the desirability of his institut-
ing certain enquiries into the subject of the Nepal Aconites. As the
result samples of a number of plants and roots, together with their vernacular names and notes as to uses, were communicated. The *Kala bikh*   **Kala bikh.**
of the Nepalese, for example (the *Dulingi* of the Bhotias, who make a   **260**
trade in collecting and selling these roots), was reported to be a very
poisonous form of **Aconitum ferox**, so poisonous, indeed, that the Katmandu drug-sellers will not admit they possess any. *Pahlo* (yellow)   **Pahlo bikh.**
bikh, a less poisonous form of the same plant, known to the Bhotias as   **261**
Holingi, while *Setho* (white) *bikh* (the *Nirbisi sen* of the Bhotias), was   **Setho bikh.**
A. Napellus, and *Atís*, **A. heterophyllum**. The Aconite adulterants or   **262**
plants used for similar purposes were found to be **Cynanthus lobatus**, the **Adulterants.**
true *Nirbisi* of Nepal, the root of which is boiled in oil, thus forming a   **263**
liniment which is employed in chronic rheumatism. **Delphinium** denu-

DELPHINIUM
denudatum.

The Nirbisi or Jadwar.

MEDICINE.
Nilo bikh.
264
Ratho bikh.
265

datum, the *Nilo* (blue) *bikh* of the Nepalese, and the *Nirbisi* of the Bhotias, Dr. Gimlette reported to be used by the *Baids* of Nepal for the same purposes as the *Setho* and *Pahlo bikh*. **Geranium collinum** (*var*. **Donianum**) was found to be the *Ratho* (red) *bikh* of the Nepalese, and the *Nirbisi num* of the Bhotias, and like the *Setho bikh* was stated to be given as a tonic in dyspepsia, fevers, and asthma. Lastly, a plant, never before recorded as used medicinally, namely, **Caragana crassicaulis**, was sent to the writer under the name of *Artiras* of the Nepalese and the *Kúrti* of the Bhotias; it was reported to afford a root employed as a febrifuge. The Nepalese name, *Artiras*, may be admitted as recalling *Atis* (**Aconitum heterophyllum**), and the Bhotia *Kúrti*, as bringing to mind *kutki* (**Picrorhiza Kurroa**), two drugs which, like *Nilo-bikh* (or *Nirbisi*) and the *Setho* or *Pahlo-bikh*, are employed as tonics and antiperiodics. (*Conf.* with **Coptis Teeta, Vol. II., No. 1792, p. 522.**)

Munila.
266

Delphinium denudatum inhabits the southern warmer slopes of the Himálaya, descending to lower levels than any of the aconites, though in its higher areas it becomes intermixed with **Aconitum heterophyllum**. Around Simla and extending into Kumáon and Kullu, it is known as *múnila*, but it neither bears the name of *nirbisi*, nor has assigned to it any medicinal properties. It would not be difficult to suppose that if the original *nirbisi* or *nirvisha* (for the difference may after all be but the result of modern specialisation) was obtained from the Himálayas, and was also known as the *jadwár*, it may have been some of the tonic and febrifugal roots already alluded to, if it be not, as **Moodeen Sheriff** thinks, the non-poisonous forms of aconite. This supposition would give meaning to **Father Ange's** statement (in the *Persian Pharmacopœia*, published 1681) that the root, though poisonous when fresh, was perfectly innocuous when dried, and that when mixed with food and condiments it acted as a restorative. The *nirbisi* of the plains of India—the rhizomes of **Kyllingia monocephala**—may have come to be so called from their resemblance to Zedoary, the *Judwár* of some writers. In purusing such an opinion one might be almost pardoned the speculation that in the earlier ages of medical knowledge, the strength-giving bitter roots would have been likely to attract attention and to obtain a high reputation long before the less evident, and more hypothetical remedies of modern times became known. Since these tonics abound in the higher temperate regions of Asia, they would likely enough have continued, with the migrations of the people southwards, to be carried all over the fever-stricken plains that possess but few good tonic and febrifugal drugs. The property of an antidote to poison, if ever assigned to these drugs, might fairly well have depended upon their tonic action in strengthening the system against the effect of poison. The literature of *Nirbisi* is not so complete as that of *Jadwár*, but accepting the usual assumption as correct that these are mere synonyms, the present review of this subject may be concluded with a reference to the writings of Muhammadan physicians on *Jadwár*. Under that drug **Mir Muhammad Husain** mentions *Antila* as its Arabic name, and *Sáturyús* as its Greek. **Dioscorides** refers to two forms of the aphrodisiacal drug σατύριον, but both these are most probably the saleep tubers which, in consequence of the superstitious doctrine of signatures, have for ages enjoyed in Asiatic countries the reputation of being stimulants to the generative organs. Muhammadan writers allude to saleep under the name of *Khusyu-uth-thaalab* (Foxes' testicles), and the odour of the fresh root is said to resemble that of semen. Saleep has in India the reputation of being a nervine restorative and aphrodisiac. Here then we have another link between the early *nirbisi* and the more recent Zedoary, which might serve to connect the rhizomes of the medicinal sedges **Kyllingia monocephala** and **Cyperus rotundus**. But

MEDICINE.

Mir Muhammad Husain mentions five kinds of *Jadwár*: the first and most valuable of all—the *Khatai*—is said to be black externally, purplish brown internally, and knotted. It tastes sweetish at first, but is afterwards very bitter (? **Cyperus**). The second and third come from Tibet, Nepal, Rungpore, &c. The fourth is said to be blackish, to be very bitter, and of the size of an olive: it is reported to come from the Deccan hills, and thus can be neither a **Delphinium,** nor an **Aconitum.** The fifth is the Spanish drug known as *Antila.* **Dr. Moodeen Sheriff** states that there are in the bazárs of South India three kinds of *Jadwár*, all in his opinion non-poisonous aconites.

The writer does not venture to suggest what each of **Mir Muhammad's** forms of *Jadwár* may have been, but he accepts the general inference from **Mir Muhammad's** account as confirmatory of the views already expressed, namely, that it would be unsafe to regard *Nirbisi* and *Jadwár* as more than ancient names for a drug or drugs which, with the extinction of the Arabian school of medicine, lost any specific signification they ever possessed. (The reader is referred to **Aconitum Vol. I., p. 84**; to **Curcuma Vol. II., p. 656,** and also to **Bombax, Eulophia,** and **Saleep.**)

Delphinium saniculæfolium, *Boiss.; Fl. Br. Ind., I. 25.*

267

Habitat.—An erect herbaceous rigidly-branched plant, met with in the Western Himálayas, frequenting dry hills from the Indus to the Jhelum, and distributed to Afghánistan. Racemes long, composed of many pale blue flowers, each less than half an inch in size.

History.—It has been customary to read, in works on Indian Economic Products, that from this plant is obtained the dye and medicinal flowers known as *asbarg.* The writer had occasion to examine a large sample of these flowers and twigs in connection with the preparation of the collections for the Colonial and Indian Exhibition. It was then noted that the *asbarg* flowers would not answer to the description given by botanists for **D. saniculæfolium,** and that, as a ready eye-mark, the *asbarg* flowers were clearly yellow instead of blue when fresh. At that time the enquiry was carried very little further, but **Dr. Stewart's** description was consulted when it appeared subsequent authors had disregarded the doubt indicated by the qualification "perhaps." **Stewart's** words are " a considerable import takes place from Afghánistan into the Panjáb in the flowers of perhaps the species named (D. saniculæfolium)." Then again: " **Mr. Edgeworth** first brought this substance to notice many years ago, and supposed these were the flowers of **D. altissimum,** *Wall.*; but it does not appear to grow so far west." The writer has had the pleasure to examine a plant collected by **Dr. Aitchison** in Afghánistan (**D. Zalil,** *Aitch. and Hemsl.*) and to compare it with the *asbarg* flowers sold by Indian drug-sellers. As the final result he has no hesitation in affirming that the economic facts given by all Indian writers under **D. saniculæfolium** should be carried to **D. Zalil.** (*Conf.* with that species below.)

HISTORY.
268

D. vestitum, *Wall.; Fl. Br. Ind., I., 26.*

269

Vern.—*Juhí,* SIMLA.

Habitat.—West and Central Himálayas, at altitudes from 8,000 to 12,000 feet. In the lower portion of its region, it occurs sparingly in mixed forests, is a coarse plant, attaining a height of 3 to 4 feet, and has large deeply lobed, sharply serrate leaves, and a spike of dirty purplish-blue flowers. On the higher area where it is met with on exposed grassy hills, it is extremely abundant, miles of country being covered with it along with **Achillea millefolium, Tanacetum longifolia,** &c. It is here more stunted,

D. 269

approaching the type of **D. Brunonianum** and **D. cashmirianum**. Has roundish leaves 5-9 lobed, and almost dentate instead of serrate. Flowers larger than those of the lower altitude, opening up more pronouncedly and pale blue coloured. This plant commences to appear where **D. denudatum** disappears, and ascends to the altitude where **D. cashmerianum** and **D. Brunonianum** occur.

MEDICINE.

Medicine.—On questioning hill people, who were found collecting **Jurinea macropcephala** (the roots of which are used as incense, under the name of *dhup*), and also the medicinal rhizomes of **Picrorhiza Kurrooa**, as to any uses of the roots of this **Delphinium**, the writer was informed that they were not collected, nor were they known to possess any medicinal virtues. The **LEAVES** were said, however, to be poisonous to goats. Neither the leaves nor flowers have the musk odour of **D. Brunonianum**. This negative information is alluded to here in consequence of the writer's conviction that authors who attribute medicinal properties to **D. denudatum** are most probably in error. If any **Delphinium** was a regular article of trade (medicinally) the present species might be expected to be so, far rather than the scarcer plant **D. denudatum**, which at most (though widely distributed at altitudes between 5,000 and 8,000 feet) occurs only here and there, and yields a small inert root.

Leaves.
270

271　**Delphinium Zalil**, *Aitch. et Hemsl., Botany of the Afghan Delimitation Commission, published in the Trans. Linn. Soc. (2nd series), Vol. III., 20, 30.*

Vern.—*Asbarg,* HIND.; *Asbarg* (the dye), and *ghafis* (the medicine), PB.; *Zalil,* KHORASAN; *Trayamán, gul-jalil,* BOMB.; *Asfrak, asperag, trayamán,* PERS.; *Zarir,* ARAB.

Habitat.—A perennial plant, throwing up a spike of bright yellow flowers two feet in height. Dr. Aitchison says of it: "This plant forms a great portion of the herbage of the rolling downs of the Badghis: in the vicinity of Gulran it was in great abundance, and when in blossom gave a wondrous golden hue to the pastures: in many localities in Khorasan, about 3,000 feet altitude, it is equally common." At another place he alludes to it again as with its showy blossoms covering the downs "which they illuminate with their brilliant colouring, affording a sight never to be forgotten."

DYE.
272

Dye.—The dried flowers and fragments of flowering stems are brought from Afghanistán to Multan and other Panjáb towns, from which they are conveyed all over India. In Multan, as in most other places, they are used along with *Akalbér* (**Datisca cannabina**) and alum, to dye silk a yellow colour. Sir E. C. Buck, in his *Dyes and Tans of the North-Western Provinces,* says of *Asbarg:* "A yellow dye extracted from the stalks and flowers of a species of **Delphinium**. "The flowers and stalks are imported into these provinces from Kabul and Khorasan *viâ* the Panjáb. A decoction made from them is much used in silk dyeing, giving the sulphur yellow colour known as *gandhaki*. It is also used in calico-printing. Its price is R27·5 per cwt." This dye is also alluded to by Mr. Liotard, by Dr. McCann, and by Mr. Wardle, but under the name of **D. Ajacis**.

MEDICINE.
Flowers.
273

Medicine.—The **FLOWERS** are bitter, and are said to be used medicinally as a febrifuge. Dr. Dymock publishes the following early account of the drug, being a translation from the *Makhzan-el-Adwiya:* "*Zarir* grows in the Khorján hills, and is called *Asfrak* by the people of Shiráj, and *Arjikan* by the Greeks; the stem is about a span high, flowers yellow, like those of *Asfar-i-barri,* surrounded by a few soft prickles, leaves yellowish, small, root more than a span long. *Asfrak* is cold and dry, with slight

D. 273

heating properties; also detergent, anodyne, and diuretic; it is useful in spleen, jaundice, and dropsy; mixed with barley meal, it forms a poultice, which is of much service in inflammatory swellings; its ashes are useful in itch; maximum dose 5 dirhems" (240 grains, 24 hours in decoction); "it is also used as a yellow dye." The reference to its use in itch is interesting, as showing a similar property to that of the European plant known as Stavesacre (**Delphinium Staphisagria,** *Linn.*) (*Conf.* with the remarks regarding the medical uses of **D. Brunonianum,** p. 64).

MEDICINE.

DENDROCALAMUS, *Reed; Gen. Pl., III., 1212.*

A genus of bamboo or arborescent grasses, distinguished from BAMBUSA by the pericarp of the fruit being coriaceous or hard, and by the flowers having six instead of three stamens (*Conf. with* **Bambuseæ,** *Vol. I., No. 69, page 371*). Very little of a definite nature can be written regarding the individual properties of the species of DENDROCALAMUS. All are of course used by the people in the localities where they occur, and like those of Bambusa, are collectively designated Bamboo (*Conf. with the* **Economic Uses of Bamboo,** *Vol. I., p. 387*).

Dendrocalamus Brandisii, *Kurz;* GRAMINEÆ.

Syn.—For BAMBUSA BRANDISII, *Munro; See Vol. I., p. 391.*

274

D. calostachyus, *Kurz, For. Fl., II., 62.*

Habitat.—Ava, at Bhamo, and on the Kakhyen hills east of it, at 3,500 feet elevation (*Kurz*).

275

D. criticus, *Kurz.*

Habitat.—Found in Pegu, altitude 3,000 feet; stems 15 to 30 feet. Kurz says that it is apparently restricted to the shady side of the summit of the Kambalatoung, Prome Yomah.

276

D. giganteus, *Munro.*

Syn.—BAMBUSUS GIGANTEA, *Wall.*
Vern.—*Wakli, waya,* BURM.
References.—*Gamble, Man. Timb., 430; Mueller, Sel. Extra-Trop. Pl. (7th Ed.), 132; Spons, Encyclop., 921; Balfour, Cyclop., 914; Kew Off. Guide to Bot. Gardens and Arboretum, 41.*

Habitat.—Met with in Tenasserim; stems attaining a height of 100 feet and often 26 inches in girth.
This is one of the largest (indeed next to **Bambusa Brandisii** the largest) of bamboos. It is much used in Burma for POSTS and RAFTERS in rural house-building.

277

DOMESTIC.
Posts.
278
Rafters.
279

D. Griffithianus, *Kurz, For. Fl. Burm., II., 563.*

Syn.—BAMBUSA GRIFFITHIANA, *Munro.*
Habitat.—Ava.

280

D. Hamiltonii, *Nees.*

Vern.—*Kokwa,* BENG.; *Tama,* NEPAL; *Pao,* LEPCHA; *Wah,* MICHI; *Wahnok,* GARO; *Pa-shing,* BHOTIA.
References.—*Brandis, For. Fl., 570; Gamble, Man. Timb., 430; Hooker, Himálayan Journal, I., 155; Indian Forester, I., 221, 226; VII., 49; VIII., 293; XIII., 522; XIV., 112, 114; Mueller, Select Ext. Trop. Pl., 7th Ed., 132; Balfour, Cyclop.; 914.*

Habitat.—A common bamboo in the Eastern Himálaya, from Kumáon to Assam. Generally a tall grass 40 to 60 feet in height, but sometimes found as a long and tangled bush.

281

DENDROCALAMUS strictus.	The Male Bamboo.

**FOOD.
SHOOTS.
282**

**TIMBER,
283**

Food.—The young SHOOTS are boiled and eaten in Sikkim, Bhutan, and Assam.

Structure of the Wood.—The halms are large, 3 to 6 inches diameter, rather hollow and not always straight, but they are used for every variety of purpose. The bamboo grows gregariously, on hill-sides, up to 3,000 feet, and the stems are 40 to 60 feet high. They frequently grow low and tangled, instead of straight; indeed, this bamboo may at times be recognised by this character and by the very thick shoots which grow out at the nodes (*Gamble*).

Mr. **F. B. Manson**, in an article in the *Indian Forester*, alludes to the utility of this bamboo to the tea planter in shading his estate from hot and violent winds. He then refers to the discussion as to its flowering. "I have noticed, he remarks, that the forest bamboo of the Terai is flowering pretty generally this year (1882); but the phenomenon does not universally affect all bamboos. I have also noticed clumps of this bamboo in a languishing condition which had lately flowered." **Hooker**, in his *Himálayan Journal*, says: "it. flowers every year, which is not the case with all others of this genus; most of them flower profusely over large tracts of country once in a great many years and then die away."

284

Dendrocalamus Hookeri, *Munro.*

Vern.—*Ussey, assey, denga, ukotang,* Ass.

Reference.—*Brandis, For. Fl.,* 570.

Habitat.—An allied species to **D. Hamiltonii**, but with larger leaves (15 inches long and 3-4 inches broad), met with in the Eastern Himálayas, Assam, and the Khásia hills.

**TIMBER.
285**

Structure of the Wood.—Stems 50 feet in height, and like the other species put to many useful purposes.

286

D. longispathus, *Kurz, For. Fl. Burm., II., 561.*

Vern.—*Wa-ya,* BURM.

Habitat.—Frequent along the *chongs* in the moister upper mixed forests, and also in the tropical forests of Arracan, Pegu, and Martaban (*Kurz*).

**TIMBER.
287
288**

Structure of the Wood.—Stems from 40 to 60 feet in height.

D. membranaceus, *Munro.*

Vern.—*Wa-yoi,* BURM.

Habitat.—A native of Burma.

**TIMBER.
289
290**

Structure of the Wood.—Stems 40 to 50 feet.

D. Parishii, *Munro.*

Habitat.—Brandis remarks that this species is described from specimens said to have been collected in the Panjáb Himálaya. It is closely allied to **D. Hamiltonii**, differing in its ovate laneeolate acute spikelets.

291

D. serviceus, *Munro.*

Habitat.—Found on Parisnath, Chutia Nagpur.

292

D. strictus, *Nees.*

THE MALE BAMBOO.

Syn.—BAMBUSA STRICTA, *Roxb.*

Vern.—*Báns, bans kaban, bans khúrd, kopar,* HIND.; *Karail,* BENG.; *Mathan, saring, burumat,* KOL.; *Buru mat,* SANTAL; *Bukhar* (for the Clump), PALAMOW; *Halpa, veddar, vadur,* GOND; *Bhiru,* BAIGAS;

D. 292

| The Male Bamboo. | (*G. Watt.*) | DENDROCALAMUS strictus. |

Bas, udha (kaban, bassa or *vassa,* Lisboa), BOMB.; *Bhovarlit,* MAR.; *Kark,* PANDRATOLA; *Kanka, sádhanapu venduru* (Elliot), TEL.; *Myinwa,* BURM.

References.—*Roxb., Fl. Ind., Ed. C.B.C., 304; Voigt, Hort. Sub. Cal., 718; Brandis, For. Fl., 569; Beddome, Fl. Sylv., 235, t. 325; Gamble, Man. Timb., 430; Stewart, Pb. Pl., 71; Aitchison, Cat. Pb. and Sind Pl., 171; Flora Andhr. by Sir W. Elliot 165; Mueller, Select Extra-Trop. Pl., 7th Ed., 132; Atkinson, Him. Dist., 391, 632, and 735; Econ. Prod. N.W.-P., V., 90; Lisboa, U. Pl. Bomb., 137, 188, 209, 238, 277; Liotard, Paper-making Mat., 72, 73; The Fodder Grasses of Northern Ind by J. F. Duthie, p.71; Spons, Encyclop., 921; Balfour, Cyclop., 914; For. Admn. Report, Chutia Nagpur, 1885, 34; Bombay Gazetteer, XI., 30; Indian Forester, I., 233, 255, 268, 336, 346, 359; II., 19; III., 205; IV., 229, 321; VII., 163; VIII., 106, 123, 271, 301, 369, 411, 415, 416, 418; IX., 529 to 539; X., 134, 359, 548; XII., 203, 312, 413, 418; XIII., 55, 115, 121, 513, 522, 523; XIV., 419; Manual of the Madras Presidency, II., 27.*

Habitat.—Met with throughout India, but most abundantly in the plains and lower hills of Northern and Central India, ascending to 3,000 feet. **Kurz** says it is a xeroclimatic species, common on the Continent of India, but does not go further south than Upper Tenasserim. He describes it as a bushy plant, from 20 to 30 feet in height. **Dr. King** remarks that it is the only bamboo found on Mount Abu. It is scarce in Banda, but in the drier districts of Central and Southern India, it affects the cooler northerly and westerly slopes. In Bengal and along the foot of the Himálaya where the climate is damp, it occurs chiefly on the warm southerly faces of the hills. It has often deciduous leaves, and the stems, which frequently attain a height of 100 feet, are strong, elastic, and nearly solid.

CULTIVATION.

FLOWERING, &c.—This species is sometimes said to flower gregariously, but more frequently single clumps are found to do so. **Mr. Gamble** publishes an account of its flowering along the base of the hills in the North-West Provinces. **Mr Greig** (the Conservator) reported: " I have observed numbers with one or two stems of a clump in flower; in some places as many as 5 per cent. of the clumps have flowering stems, and in others I have only found ten clumps with flowering stems out of several thousands examined. Between Kolidwara and Haldu Khata whole clumps over large areas have seeded and died, and the ground is now a dense thicket of young clumps of from 10 to 30 feet high. The seeding commenced here in 1869 or 1870, and has been going on ever since." Whole areas, he continues, in Palim, Kansore, &c., seeded and died in 1877-78. (*Man. Timbers, 430.*) **Mr. Brown** writes of the flowering of this species in the North-Western Provinces: "As an example of great vitality in certain bamboos, I may mention here that on the same road along which **Bambusa arundinacea** was growing, a clump of **Dendrocalamus strictus** flowered in 1881 and sent forth new but thin shoots in 1882. These flowered again in 1885, and now new scraggy and thin shoots are pushing up in the midst of the old clump."

" With respect to **Dendrocalamus strictus**, although the flowering is not so general as with other species, yet large areas become fertile at one time. The curious point about the flowering of this bamboo in the Siwálik Forests of the Dún and Saharanpur is that the fertility seems to spread onwards gradually and year by year. For instance, in 1883, most of the clumps in the Charkhari block flowered. In 1884, the Maiapur block, Saharanpur division, became fertile. Then the Rampur block was attacked in 1885, and this year, 1886, the bamboo in Rauli block seeded.

" Thus the seeding began in the south-east corner of the Dún, turned the corner of the Siwaliks at Hardwar, and fertility is now apparently

CULTIVA-
TION.
Flowering.
293

DENDROCALAMUS strictus.	The Male Bamboo.

FLOWERING.

gradually spreading westward among the southern face of the Siwaliks. It remains to be seen whether this gradual march will continue along the rich bamboo forests of the eastern and central ranges of the Saharanpur division. Elsewhere I have seen this species flowering only sporadically." *The Seeding of Bamboos, by A. F. Brown, Esq., published in the Ind. For., Vol. XII., p. 413.*)

A long and interesting " Note " on the cultivation of this species in the Central Provinces will be found in the *Indian Forester.* But the following brief passages may be here republished :—" In every forest producing this species, a certain number of stems flower and seed annually, but a general seeding is only an occasional occurrence. Regarding the time or conditions of seeding, nothing definite is at present known, but it is evident that general seedings are associated with a short rainfall. In general seedings all clumps of the same age appear to seed within the two years over which the seeding generally seems to extend. It is the opinion of natives, and one which is believed in by many forest officers and others, that seeding is prevented or retarded by heavy working of the clumps ; the opinion is doubtless to a certain extent correct, but it is improbable that cutting will have effect if deferred till the clumps begin to flower. It is not an uncommon thing to find small one-year old shoots from clumps entirely cut over, producing seed." " It has been observed that a poor and unfavourable soil is conducive to the production of seed." " Probably the real cause of seeding is exhaustion of the soil accessible to the roots of the clumps ,which is felt the more the dryer the season : a supposition further supported by the fact that seeding is more common on poor than on rich soils. Stems that flower casually yield hardly any fertile seed, and hardly any seed at all, whereas in the general seedings the yield is very large and of excellent quality, especially in the first year " (*Ind., Forester, IX., 531*).

Speaking of the shedding of the leaves, **Mr. Kurz** remarks that it " becomes often evergreen in damper climates, or when grown in moister localities." With reference to the flowering, he remarks that this occurs when the plant is between 25-30 years old. A man who has seen two flowerings is considered old. It is generally followed by the death of the clump, but " exceptional cases are known to me where a shoot was thrown up and grew and formed a new stock." He states that the seedlings grow from 1 to 1½ feet in height during the first year and not more than 4 feet up to the third. **Brandis** says " the stems attain their height in a few weeks at the commencement of the rains ; in the Panjáb they do not harden fully during the first year." **Stewart** also remarks that according to the natives it accomplishes its growth in two or three weeks. Owing to the annual shedding of the leaves, there is always a large amount of dry foliage on the ground which makes forests of this bamboo liable to large and very destructive fires. The writer in the *Indian Forester* quoted above, regarding the cultivation of this species in the Central Provinces, remarks :—" It is probable that as a living plant this bamboo will come into use for the consolidation and support of embankments ; the complete and endless network of rootlets which develop around every clump and extend from the surface to 9 or even 12 inches below, binds the whole surface soil into a solid mass which can be cut into blocks with a spade, but is not easily broken until the rootlets die or decay " (*Indian Forester, IX., 529*).

SOILS.
294

Soils suitable for D. strictus.—" Widely as the species is distributed it is not to be found in all localities nor on all soils. The slopes of hills, ravines, and the banks of *nalas* are the favourite localities. In the plains it occurs forming dense masses and covering large areas, but on sandy

D. 294

| The Male Bamboo. | (*G. Watt.*) | DENDROCALAMUS strictus. |

soils only. A rich and free soil, good drainage and plenty of moisture, are favourable if not essential to its production; though, as already stated, it is found forming dense masses in the sandy plains; in such places it only flourishes on the banks of *nalas*, or where there is a good deposit of vegetable mould. On a considerable area of poor sandy soil it abounds without attaining any size, and in such cases its existence can only be attributed to conditions being favourable to germination, and to the protection afforded to the young plants by tree vegetation.

"In clay soils, and the combinations of clay and lime (*kankar*), not unfrequently met with, the species refuses to grow. In the black cotton soils of the plains, and even in very wet soils, it will grow luxuriantly when once thoroughly established, but young plants soon succumb to excessive moisture.

"Though not very productive, pure bamboo forests exist in several places in the Central Provinces; the species thrives best when associated with tree vegetation. It is more or less shade-bearing according to age as a young seedling; except under artificial cultivation it will not without shade live through a single hot season, while even with mature clumps light tree shade appears favourable to the plant, and under the latter condition the yield of individual clumps is greater and finer than in pure bamboo forest."

REPRODUCTION.—This is "secured by seed and by rhizomes with rootlets and portions of the stems attached. In the early stage of existence the rhizomes are larger in proportion to the stems and have greater vital powers. It is also probable that the little shoots, resembling seedlings in appearance, which are occasionally produced in dense masses at each node, would take readily if planted, and that shoots hid under ground, with portions of the leaf-bearing branches above, would take root and produce shoots at each node.

"The artificial cultivation of this species has, in the Central Provinces, only been carried on since 1875, and as might be expected, there is much yet to learn on this subject; nevertheless, a certain amount of information and experience has been gained which it would be useful to place on record.

"In propagating by sets from existing clumps, it is advisable that three or four shoots with their rhizomes should be taken, together with their roots, for each pit to be planted, and that as much of the soil as possible should be preserved above the roots. The stem should be cut back immediately above joints to a length of five or six feet; the sets should be planted as quickly as possible, six to eight inches of stem being placed below ground. The first burst of the monsoon is the most favourable time for this operation; in the absence of rain the water-supply must be kept up artificially till foliage is developed; if the soil is good, further tending will be unnecessary; clumps thus raised on good free soil produce marketable shoots in five years.

"In propagating by seed sowings may be made *in situ*, or seedlings may be raised in nurseries and transplanted. Of the former method experience is confined to the result of one experiment, in which the area dealt with was 50 acres situated on the slopes of hills. The soil was not good, though not extremely poor, but there was a little cover on the ground; the sowings were in prepared lines, but no manure of any kind was applied. The seed was put down in July but sown too thickly, and at the end of the rains the plants averaged 18 inches, or four times the height of natural seedlings of the same age, but the plants were weak. Had the soil been rich and the sowing less thick, or had the plants been properly thinned on appearance above ground, it is more than probable

CULTIVATION.

Reproduction. 295

Propagation. 296

Shoots. 297

Seeds. 298

DENDROCALAMUS strictus.	The Male Bamboo.

CULTIVA-TION.

that the growth would have been really vigorous." "It is probable that excellent results may be obtained by sowing in pits three feet in diameter and one foot deep, filled with good rich mould, provided the plants are thinned till when four feet in height; not more than four plants should stand in each pit." "If the seed be good more than 10 seers to the acre is not likely to be necessary." "As bamboos need not, as a rule, be planted nearer than 15×15 feet, an acre of nursery will suffice for planting about 80 acres."

Thinning or Cutting.
299

THINNING OF CLUMPS AND CUTTING FOR THE MARKET.—" As regards cutting or thinning, it is obviously essential to preserve, in a vigorous condition, those eyes whose turn it is next to produce shoots; it has already been indicated that, after clumps have produced full-sized shoots, reproduction is generally from rhizomes of two years old, though occasionally it proceeds from those of greater age. It is, therefore, obvious that to secure a maximum production no shoot should be cut until the end of the second monsoon succeeding that in which it was itself produced, unless increased production is rendering the forest too dense, a condition which cannot be said to exist as long as there is ample space for the full development of foliage on all standing stems, and clear space for the upward course of new shoots."

" The maintained production of shoots must prevent general seedings, which only succeed the cessation of the production of shoots. It is also probable that the complete removal of the older shoots will result in the decay of the rhizomes attached to them, and that thus the stems left will become independent of the old parent-root, and be less likely to seed than if their connection were maintained. As long, therefore, as the production of shoots does not annually increase, and there is no indication of the standing crop being too dense, all shoots should be preserved till the dry season following the second rains after that in which they were produced, when they should be cut and removed." The author of the interesting article on this bamboo from whom the above passages have been abstracted proceeds to state that, where a demand exists for green stems, a limited amount may be cut from each clump, but that, unless the reproduction be vigorous, they should not be cut off close to the ground but two feet above, "thus leaving eyes for the development of branches and foliage to preserve the vigour of the root."

Season of Cutting.
300

SEASON OF CUTTING AND PERIOD WHEN THE CLUMPS COME INTO BEARING.—" With the view to production, the best season for cutting is from the time the leaf begins to fade, up to the time the clumps become leafless." The period before a wild or cultivated forest may be expected to come into bearing has been variously stated. " The number of years necessary for the production of full-sized shoots is undetermined, but is known to vary greatly according to the conditions under which the plants have grown up. In natural forests there is reason to believe that full-sized shoots are not produced until the clumps are about twelve years old, but in really successful artificial plantations the time will probably be reduced to six years. **Sir D. Brandis**, in the passage already quoted, states that the shoots attain their full height in a few weeks, but in the Panjáb they do not harden during the first year. This of course refers to the formation of shoots on a clump in full bearing condition. **Dr. Schlich**, in his Forest Administration Report of the Central Provinces, says eight years may be taken as the time in which artificially-raised bamboos of this species will, under ordinary circumstances, come into bearing."

FIBRE.
301

Fibre.—The fibre from the stem is suitable for the manufacture of paper, but its high value prevents it from being so used. **Kurz** remarks that the natives of Behar "employ the *jungli bans* (**Dendrocalamus strictus**) for

		DENDROCALAMUS
The Male Bamboo	(*G. Watt.*)	strictus.

making neatly-worked plates, hand-fans, &c., which are generally sold in the towns through the whole of India."

Medicine.—The silicious matter found near the joints in this and most bamboos (*tabashír*) is used as a cooling, tonic, and astringent medicine. It has not been satisfactorily proved, however, that **D. strictus** does actually produce *tabashír*; but **Mr. Atkinson** affirms that it does. The leaves are given to animals during parturition, from a supposition that they cause a more rapid expulsion of the placenta (*Dr. Emerson*). For this purpose it is said to be used by native women both criminally and in ordinary midwifery practice.

SPECIAL OPINIONS.—§ "A decoction of the leaves is given to women after delivery to put the uterus in order" (*Assistant Surgeon T. N. Ghose, Meerut*). "The juice of the leaves in about two-ounce doses taken frequently is used in certain parts of the North-Western Provinces for causing criminal abortion" (*Assistant Surgeon Nobin Chunder Dutt, Durbhanga*). "I have seen the leaves used to aid parturition" (*Civil Surgeon S. M. Shircore, Moorshedabad*). "The joints when made into a decoction are used as a medicine to procure abortion" (*Surgeon Major A. S. G. Jayakar, I. M. D., Muskat, Arabia*). "The leaves are given to horses when suffering from cough, and the leaves boiled in water for convalescents to bathe in" (*Honorary Surgeon P. Kinsley, Chicacole, Ganjam, Madras*).

Food and Fodder.—The LEAVES are eaten by buffaloes and are fairly good fodder for horses. **Duthie** remarks the foliage affords abundant fodder for elephants, and **Lisboa** that the leaves are eaten by cattle. The SEEDS are eaten by men in times of famine. "The relative value of this food may be estimated by the fact that while wheat, the principal food-grain, sold at 12 seers for the rupee, bamboo seed sold at from 40 to 50 seers" (*Indian Forester, IX., 529*).

Structure of the Wood.—This is the male bamboo of most writers, a name given to it because when fully developed it becomes practically solid. It would appear, however, that in certain localities and soils it does not show so pronounced a tendency to do so as in others, the central canal often remaining fairly large. The outer shell is, however, hard and strong, yet elastic, and hence this is for its size one of the most useful of bamboos. It is employed for a variety of purposes, such as spear handles, and all the requirements of native house-building and for basket-work. The following passage from the *Indian Forester* enumerates some of the uses:—

"In the Central Provinces this bamboo is used as a substitute for timber, for rafters and battens, spear and lance shafts, walking sticks, whip handles, ploughman's driving sticks and spade handles, stakes to support sugar-cane, on light soils, stakes for *pán* plants and for construction of *jaffries* for *pán* gardens, for the construction of strong fencings to resist wild animals, the manufacture of small mats used like slates in roofing, mats for floors, covers of carts and various other purposes, sieves, hand-punkahs, umbrellas, light chairs and sofas, drenching horns, vessels for holding grease and oil, specially for lubricating cart wheels, bows, arrows, and cordage, and for the manufacture of many other minor articles. It is also used for the buoyage of heavy timbers in rafting, and when converted into charcoal, is in request for the finer smith's work. Dry stems are also used for torches and the production of fire by friction" (*Indian Forester, IX., 529*).

Trade in Male Bamboos.—Very little can be learned of the trade in this most valuable article. The reports that exist deal with limited tracts and for different seasons, so that a combined statement for all India cannot be drawn up. The value of the bamboo varies according as the culms are green (*e.g.*, young) or dry and seasoned. "In the vicinity of large towns

Marginal notes:

MEDICINE.
Tabashir.
302
Leaves.
303

FOOD AND FODDER.
Leaves.
304
Seeds.
305

TIMBER.
306

Spear Handles.
307

Rafters.
308
Battens.
309
Walking Sticks.
310
Stakes.
311
Torches, &c.
312

TRADE.
313

DENDROCALAMUS strictus.	The Male Bamboo.

TRADE.

and markets the higher value generally attaches to green bamboos, being sometimes as much as twice that of dry bamboos. As regards seasoning the preference is in some places given to bamboos that have been soaked in water for a length of time, while in others bamboos thus seasoned will not command a market. The chief use of water-seasoning appears to be the destruction of the insects which attack the bamboo when cut out of season." "Bamboos cut in the rains are always liable to speedy decay."

Hyderabad.
314

Particulars were called for (in connection with the preparation of the present article) as to the trade in this bamboo, its price and other such information. The reports received from the various provinces of India may be here summarised. Of Hyderabad (Berar) **D. strictus** is stated to occur chiefly in the hills of the Gawilgarh Range. It is said also to be common in Melghat—plentiful in the reserves, though disappearing from the forests. The total exports during the past ten years are reported to have been valued at R2,54,885, or a mean annual value of about R25,000. The local price is returned as R1-8 per hundred.

Coorg.
315

Of Coorg it has been reported :—"Chiefly used in roofing, fencing, baskets, &c.; annual sales from Government forests in Coorg 2 lakhs; price in the forest R1 to R1-8 per hundred. Probably 8 or 10 lakhs could be cut yearly from the Coorg forests without diminishing the supply. From a forest point of view it is desirable to diminish the number to a large extent and allow timber to take the place of the bamboo. If it were not for the periodic seeding and dying off of the bamboos they would gradually cover the whole forest to the exclusion of tree growth, as tree plants seldom get up where bamboos are thick. The seeding of the **D. strictus** usually takes place by clumps. Every year scattered clumps seed."

Madras.
316

Two reports from Madras may be here given : – Of the Northern Circle it has been said—"This is the 'male bamboo.' Universally found on the drier slopes of hills, and occasionally in ravines, where, as in the Nilghiris, it often attains a large size, even 3-4 inches in diameter. It is in general use for all the purposes for which bamboo is required. The annual production cannot well be given, as the supply is so much greater than the demand, but the amount exported from the Government Forests is very considerable (*See Annual Report*)." Of the Southern Circle **D. strictus** is reported to be "common in dry forests up to 3,000 feet. It is universally used for building purposes and is in demand for spear shafts and the like. It is impossible at present to say what the annual production and amount available may be. The Government seigniorage is R1-4 per cart-load of half a ton, the collection and transport of which costs the purchasers from R4 to R5 and fetches from R6 to R9.

Ajmere.
317
Bombay.
318

The Conservator of Ajmere-Merwara writes that **D. strictus** is scarce in his district, selling for R10 per hundred. **Mr. McGregor** (Conservator of Forests, Southern Division, Bombay) reports that this bamboo occurs chiefly in the drier forests, but is very local. "The rate charged is one rupee per 100 stems." **Mr. A. T. Shuttleworth** (Conservator, Northern Division) remarks that it is very abundant in the forests, but is disappearing in parts owing to its being overworked. "It is used largely in connection with betel vine cultivation in the Thana District as props or supports." From the North-West Provinces several communications have been received. Of the Dehra Dún Division it is said to be the chief wild bamboo. "It is found in large quantities only at the eastern end of the district near and on the Siwaliks. With regard to the market, this bamboo is classified into six kinds. These are as follows :—

N.-W.
Provinces.
319

    "(1) *Saráncha.*—A hollow bamboo 6" to 9" girth, 12' long. Used for
        chicks, baskets, shouldari poles, &c. Annual export from

| The Male Bamboo. (*G. Watt.*) | DENDROCALAMUS strictus. |

Dehra Dún Division into or through Hardwár about 700 scores, at 5 annas per score.

" (2) *Rakmi or Chaniju.*—Hollow or solid bamboos up to 6″ girth and 10′ long. Mainly used for thatching. Annual export as above 28,000 scores at 2½ annas per score.

" (3) *Láthi.*—The hollow or solid lower thick end of the bamboo used for sticks. Annual export as above about 35,000 scores at 2 annas per score. Chiefly sold to pilgrims in Hardwár.

" (4) *Kain.*—The 'branches' of the bamboo used for fences and in thatching small houses. About 100 headloads are exported annually.

" (5) *Poochli.*—The upper portion of the bamboo, above the *Sarancha*, used for thatching purposes about 9′ long. Annual export about 5,000 scores at 1 anna per score.

" (6) *Dry bamboos.*—Ten feet long are used for thatching. About 11,000 scores annually exported at 1½ annas a score."

Of the Saharanpur Division it has been reported that "in Hindustani **Dendrocalamus strictus** is called *Mooger*, its girth is from 8″ to 9″, and height 70′ to 80′. It germinates in March and August, being the Hindi months *Chait* and *Sawán* respectively. In the first year, it grows thirteen times its original girth; in the second and third years, three times its girth. After three years it ceases to grow any higher. It grows (but very scarce) in some places in Garhwál and Rampur, and is cut during January and November in the year. It is used for four purposes, *viz.*, the topmost portion for fishing rods, the second portion for lance staves, the third for making *charhao* or *phar* of carts, and the fourth for making baskets, &c. It is sold at R40 per score, and is available in Garhwál and Rampur."

Of Bengal (Chhutia Nagpur Division) the Deputy Conservator of Forests reports: "Found in the Singbhoom District. Wood used for building, fencing, baskets, mats, walking sticks, spear shafts, axe handles, &c., also building houses. It is plentiful, and is sold at 4 to 8 annas per 100 in the forests."

"The male bamboo is also found in the Hazaribagh forests and in the Angul forests of Orissa. Specimens have been sent from both forests to the military authorities at Calcutta; from the former for lance staves and from the latter for army signalling. Canes were, however, pronounced more serviceable for signalling as being lighter. The annual production at Koderma is two to four in each clump. Price 12 annas per hundred; 10,000 male bamboos are available in the Koderma range."

Mr. A. Smythies (*Indian Forester, VII., 163*) furnishes some interesting facts regarding the Central Provinces. He asks the question—why is **Dendrocalamus strictus**, *Nees.*, called the "male bamboo?" He presumes this is because of its reputation of having a solid stem, but he adds "I have never myself seen a stem *entirely* solid, though I have no doubt there are such. I have seen many with a very small cavity, and many more with a large cavity. The Members of the Nagpur Hunt Club in my time were wont to use as spear shafts *almost solid* stems of D. strictus, as solid as they could get them, and I remember, in 1877, supplying the local but celebrated spear-maker, Boput of Nagpur, with about one hundred shafts of the almost solid stems of this bamboo; they came from the Moharli Forests of the Chanda Division, but *there* they were only found in one particular tract, on Vindhyan sandstone, which had been preserved from fire for some years previously. Boput told me at the time that the only other place where he could procure sufficiently solid stems was a certain forest in the Chindwara District, the name of

Marginal notes:
TRADE.

Fishing Rods.
320
Lance Staves.
321
Bengal.
322

Central Provinces.
323

Spear Shafts.
324

DERRIS elliptica.	The Male Bamboo.

TRADE.

which I forget. This tends to show that the solid, or almost solid, stem of this bamboo is not common, at least near Nagpur.

"It is this kind of stem, used for spear-shafts, which I have always understood to be the male bamboo. How is it therefore that the name of male bamboo is applied to the entire species?"

The Editor of the *Forester* in a foot-note to the above passages suggests that **Boput** might try the solid bamboos procurable in Chhutia Nagpur (Palamow or Koderma).

Solid Bamboos.
325

The writer had numerous applications while on duty at London (in connection with the Colonial and Indian Exhibition) as to the best course to be pursued in the effort to establish an agency to supply English manufactures with solid bamboos. One dealer was desirous of procuring a regular supply suitable for lance shafts, another maker wished to obtain bamboo suitable for splitting up and afterwards consolidating the strips in the construction of fishing rods. The writer was unable to furnish the desired information, but is in hopes that the present general compilation from all available sources of information may suggest the most likely localities from which supplies might be drawn. From the above quotations, mainly from the *Indian Forester*, it would seem pretty certain that **D. strictus** in any or every locality will not do. It is necessary to select a particular area where the bamboo is known to produce stems of the required degree of solidity. This fact suggests an enquiry that would seem worthy of the attention of persons who may have the opportunity of following it out, *viz.*, as to the peculiar climate, soil, and exposure that is found to produce the more soild condition of stem. Possibly it may be found that, although belonging to the species **D. strictus**, there is a recognisable variety that possesses the desired property. From some such enquiry results of great value might be expected, such as the propagation under the required climatic condition or on the necessary soil, or if climate and soil be found of minor consideration, a wider distribution of the superior stock might be encouraged so as to establish plantations of solid bamboos 'n accessible regions.

Dendrocalamus Tulda, *Nees*, see **Bambusa Tulda,** *Roxb.*

Deodar. See **Cedrus Deodara,** *Loudon* (now recognised by **Sir J. D.** Hooker as **C. Libani,** *Barrel*, var. **Deodara,** *Hook.*) ; CONIFERÆ—see *Vol. II., No. 846, p. 235 of this work.*

DERRIS, *Lour.; Gen. Pl., I., 549.*

A genus of arborescent climbers or trees, embracing some 40 species, abundant in India, but according to the *Flora of British India*, found "belting the world in the tropics." **Thwaites** remarks that in Ceylon the barks of the species there met with are used by the Singalese for making ropes. Very little of an economic nature has been recorded regarding the Indian species, and only one or two need therefore be here mentioned.

326

Derris elliptica, *Bth.; Fl. Br. Ind., II., 243 ;* LEGUMINOSÆ.

Syn.—PONGAMIA ELLIPTICA, *Wall.; Wight, Ic., t. 420.*
Vern.—*Tubah,* MALAY PENINSULA.
References.—*Roxb., Fl. Ind., Ed. C.B.C., 539 ; Kurz, For. Fl. Burm., I., 340 ; Christy, Com. Pl. and Drugs, No. 10, 1887, 39 ; Kew Reports, 1887, p. 43.*

Habitat.—A large, handsome climber, met with in Martaban, Burma, Penang, Malacca, and Siam, &c.

POISON. Roots.
327

Poison.—According to the Kew Report of 1877, the ROOTS of this plant, steeped in water, afford a useful insecticide for gardening purposes.

D. 327

It is also used to kill fish. No Indian author appears to allude to this fact. The Malays use the bark as one of the ingredients in their Ipoh arrow-poison.

Derris robusta, *Bth. ; Fl. Br. Ind., II., 241.* | **328**

> **Syn.**—DALBERGIA KROWEE, *Roxb., Ed. C.B.C., 535 ;* BRACHYPTERUM ROBUSTUM, *Dals. & Gibs., Bomb. Fl., 77 ;* DALBERGIA ROBUSTA, *Roxb., Hort. Beng., 53.*
> **Vern.**—*Mowhitta,* ASSAM ; *Bolkakarú,* GARO ; *Krowee,* SYLHET ; *Gumbong,* MAGH. ; *Buro,* KUMAON.
> **References.**—*Brandis, For. Fl., 154 ; Kurz, For. Fl. Burm., I., 339 ; Gamble, Man. Timb., 133 ; Atkinson, Him. Dist., 344 ; Indian Forester, XIV., 298 ; Balfour, Cyclop., I., 879.*
> **Habitat.**—A deciduous tree (30 to 40 feet in height) of the outer Himálaya, from the Ganges eastward, to Assam, Eastern Bengal, and down to Pegu.
> **Structure of the Wood.**—Light-brown, hard. It may be used for teaboxes. Roxburgh says it "grows quickly to a large size, yielding timber of a dark brown colour, and rather too porous for furniture, but seems very fit for various other purposes." Kurz writes the wood is "red-brown, hard, and close-grained, of a short coarse fibre, but soon attacked by xylophages." | **TIMBER.** **329**

D. scandens, *Benth. ; Fl. Br. Ind., II., 240 ; Wight, Ic., t. 275.* | **330**

> **Syn.**—D. TIMORIENSIS, *DC. ;* PONGAMIA CORIACEA, *Grah. ;* BRACHYPTERUM SCANDENS, *Dals. & Gibs., Bomb. Fl., 76.*
> **Vern.**—*Noalatá,* BENG. ; *Golari, potra, nalavail,* GOND. ; *Gunj,* PB. ; *Cheratali badu* (or *chiratala bódi*), *nala tige, motta sirli,* TEL. ; *Tupail,* MALAY ; *Migyaungnwe* (*meekyoung-nway*), BURM.
> **References.**—*Brandis, For. Fl., 154 ; Kurz, For. Fl. Burm., I., 339 ; Gamble, Man. Timb., 133 ; Dals. & Gibs., Bomb. Fl., 76 ; Elliot, Fl. Andh., 41, 117, 171 ; Bombay Gas. (Kanara), XV., I., 433.*
> **Habitat.**—A handsome, climbing shrub, met with in the Eastern Himálayas and the Western Gháts, passing round the coast to Chittagong and Siam.
> **Fibre.**—The bark affords a coarse rope fibre. | **FIBRE.** **331**

Desmanthus cinereus, *Willd.* (alluded to by *Ainslie in Mat. Med., II., 458*), is now known as **Dichrostachys cinerea,** *W. & A.,* which see.

D. nutans, *Willd.* (*Roxb., Fl. Ind., Ed. C.B.C., 420*) ; see **Neptunia olerace[]- our.**

DESMODIUM, *Desv. ; Gen. Pl., Vol. I., 519, 1002.*

A genus of shrubs or herbs, embracing 120 species, which are cosmopolitan in the tropics ; 49 met with in India. The generic name is derived from *Desmos,* a bond, in allusion to the union of the stamens. Very little of an economic nature is known regarding these plants. The bushy species seem all to contain fairly good fibres, which, in some cases, are used for paper-making. The following is a brief enumeration of the more common Indian members of the genus.

Desmodium Cephalotes, *Wall. ; Wight, Ic., tt. 209 and 373 ; Fl. Br. Ind., Vol. II., 161 ;* LEGUMINOSÆ. | **332**

> **Syn.**—HEDYSARUM CEPHALOTES and UMBELLATUM, *Roxb.* (*Fl. Ind., iii., 360*) ; DESMODIUM CONGESTUM, *Wall.*
> **Vern.**—*Bir jhawar,* SANTAL ; *Bodle kúrú,* NEPAL ; *Maniphtyol,* LEPCHA ; *Chetenta,* TEL.
> **References.**—*Voigt, Hort. Sub. Cal., 221 ; Kurz, For. Fl. Burm., I., 386 ; Beddome, Fl. Sylv., 87 ; Gamble, Man. Timb., 121 ; Dals. & Gibs., Bomb. Fl., 66 ; Campbell, List of the Economic Products of Chutia Nagpur, No. 9848 ; Bombay Gas. (Kanara), XV., Pt. I., 432.*

G

DESMODIUM latifolium.	**The Desmodium Fibres.**

FOOD and FODDER.
333
TIMBER.
334
335

Habitat.—A shrub of the Eastern Himálaya, Central Bengal, Western Gháts, South India, and Burma, ascending to 3,000 feet.

Food.—According to the Rev. Mr. Campbell, the Santals eat the fruit of this plant. He also says cattle and goats eat the leaves.

Structure of the Wood.—Yellowish, in structure resembling **D. tiliæ-folium.**

Desmodium diffusum, *DC.; Fl. Br. Ind., II., 169.*

Habitat.—A herbaceous plant, one to two feet in height, found in the plains of the Western Peninsula, Bengal, Orissa, Bundelkhand, and Burma.

MEDICINE.
336

Medicine.—Sir Walter Elliot (*Fl. Andh., 16, 36*) enters into a discussion as to the plant meant by the Telegu name *Cheppu tatta*, the *Antintulu* of some writers. In his experience these names denote **Desmodium diffusum**, but Beddome found the former given to **Coldenia procumbens**, and Ainslie assigns it (*Mat. Med., I., 23*) to **Asarum europœum**. It seems desirable to prevent confusion between these two plants, especially as the latter is a drug of some importance (*Conf.* with **Asarum, Vol. I., No. 1545, page 337**).

FODDER.
337

Fodder.—Roxburgh says the foliage of this species is eaten by cattle.

338

D. floribundum, *G. Don.; Fl. Br. Ind., II., 167.*

References.—*Kurz, For. Fl. Burm., 387; Atkinson, Him. Dist., 342, 456.*

Habitat.—A shrub met with throughout the Himálaya, up to 5,000 feet, also in the Khásia Hills. In Sikkim it is common in old cultivated lands at 3,000 to 5,000 feet.

[*II., 168.*

339

D. gangeticum, *DC.; Wight, Ic. tt. 271 & 272, now 270; Fl.Br.Ind.,*

Syn.—HEDYSARUM GANGETICUM, *Willd; Roxb., Fl Ind., Ed., C.B.C., 575.*

Vern.—*Sarivan, salpan, salún,* HIND.; *Salpáni,* BENG.; *Tandi bhedi janetet',* SANTAL; *Pústbœni,* N.-W. P.; *Shál purni ?* (Bazar name for the leaves), PB.; *Salparni, sálwan, dáye,* BOMB.; *Gita naram, koláku ponna,* TEL.; *Sála parni,* SANS.

References.—*Voigt, Hort. Sub. Cal., 221; Stewart, Pb. Pl., 67; Sir W. Elliot, Fl. Andh.. 60, 92; Campbell, List of Econ. Pl., Chutia Nagpur, No. 9275; U. C. Dutt, Mat. Med. Hind., 145, 316; Dymock, Mat. Med., W. Ind., 2nd Ed., 222; Irvine, Mat. Med., Patna, 100; Atkinson, Him. Dist., 342, 456; Botanical Tour to Hazara by Stewart (Journ. Agri. Hort. Soc. Ind., XIV., 43); Indian Forester, VIII., 101, 407-8, 417; XII; App. II.; Gazetteer of Bundelkhand, 80; Gazetteer, Kanara, 432.*

Habitat.—A common species on the lower hills and plains throughout India. On the Himálayas it ascends to 5,000 feet and is distributed east to Pegu and Ceylon.

MEDICINE.
340

Medicine.—This shrub is regarded as a febrifuge and anti-catarrhal; it is one of the chief ingredients of the Hindu preparation *dasamula koatha* so frequently alluded to in Sanskrit works. The reader is referred to *U. C. Dutt's Mat. Med. of the Hindus, p. 145,* for a full account of the preparation, or to *Dymock's Mat. Med. West. India,* where that article is reproduced.

SPECIAL OPINION.—§ " Is one of the ten roots (*Dasha mula*) of the Hindu Materia Medica " (*Assistant Surgeon Sakharam Arjun Ravat, L. M., Gorgaum, Bombay*).

341

Desmodium latifolium, *DC.; Fl. Br. Ind., II., 168; Wall., Cat., 5692; Wight, Ic., t. 270.*

Vern.—*Sim matha sura,* SANTAL; *Gába,* TEL.; *Kinbun,* BURM.

D. 341

References.—*Voigt, Hort. Sub. Cal.*, 221 ; *Kurz, For. Fl. Burm.*, 385 ; *Sir W. Elliot, Fl. Andhr.*, 55 ; *Atkinson, Him. Dist.*, 342, 456.

Habitat.—An erect undershrub (3-6 feet high) found on the Eastern Himálaya to Burma, Siam, and Ceylon.

Fibre.—It affords a strong paper fibre.

FIBRE.
342
343

Desmodium parvilolium, *DC. ; Fl. Br. Ind.*, *II.*, *174.*

Vern.—*Tandi chatom arak', tandi sunsuni*, SANTAL ; *Khet sunsuni*, HIND. (in Chutia Nagpur).

Habitat.—A small densely cæspitose and much-branched plant, common everywhere on the plains of India, and from the Himálaya to Ceylon, ascending to 7,000 feet in altitude.

Food and Fodder.—The Santals appear to eat this plant as a green vegetable. **Mr.** Duthie remarks that it is eaten by cattle, camels, and goats in Jeypur State.

FOOD AND FODDER.
344

[*II.*, *171.*

D. polycarpum, *DC. ; Wight, Ic., t. 406 (non-Wall) ; Fl. Br. Ind.*,

345

Syn.—D. ANGULATUM, *Wall.* ; D. OVALIFOLIUM, *Wall.* ; D SILIQUOSUM, *DC.*; D. HETEROCARPUM, *DC.* ; D. RETUSUM, *Don.* ; D. GYROIDES, *Hassk.*; D. PATENS, *Wight* ; HEDYSARUM PURPUREUM, *Roxb.* ; H. RETUSUM, *Don* ; H. PATENS, *Roxb.*

Vern.—*Bæphol*, SANTAL.

References.—*Dalz. & Gibs., Bomb. Fl.*, 66 ; *Roxb., Fl. Ind., Ed. C.B.C.*, 578, 579 ; *Rev. A. Campbell, Econ. Prod., Chutia Nagpur, No. 7833.*

Habitat.—An erect or sub-erect undershrub found throughout the Himálaya and everywhere in Burma : distributed to Malacca, Ceylon, Zanzibar, Philippines, China, Japan, and Polynesia.

Medicine.—The Santals are said to use a preparation of the plant in fainting and convulsions.

MEDICINE.
346
347

D.pulchellum, *Benth. ; Fl. Br. Ind.*, *II.*, *162.*

Syn.—HEDYSARUM PULCHELLUM, *Roxb.* ; DICERMA PULCHELLUM, *DC.* ; *Wight, Ic., t. 418.*

Vern.—*Birkapi*, SANTAL ; *Karra antinta*, TEL. (so called from the pods catching like burs) ; *Toung-ta-min*, BURM.

Habitat.—A shrub (3-6 feet high) met with in the Eastern Himálaya and throughout India to Burma, Ceylon, &c.

D. tiliæfolium, *G. Don.; Fl. Br. Ind., Vol. II.*, *168 ; Wall., Cat.*, *5707.*

348

Syn.—DESMODIUM NUTANS, *Wall.* ; D. ARGENTEUM, *Wall.* ; HEDYSARUM TILIÆFOLIUM, *Don.*

Vern.—*Sambar, shamru, chamrá, chamyár, chamkát, chamkúl, martan, motha, gurshagal, pri, marára, múss, múrt, laber* (according to Gamble), HIND. ; *chamyár, chamrá, marára, gur kats, dúd shambar, pírhí, kathi, laber, kálí mort,* PB. ; *Bre, kuthi,* KANGRA (most of the above Hind. names are given by Stewart as Panjábi names) ; *Laber,* SIMLA ; *Kalanchi,* MURRI.

References.—*Gamble, Man. Timb.*, 120 ; *Stewart, Pb. Pl.*, 67 ; *Baden Powell, Pb. Pr.*, 516, 577 ; *Atkinson, Him. Dist.*, 342, 456, and 793 ; *Balfour, Cyclop.*, 92 ; *Ind. For., Jany. 1885, Vol. XI.*, 3.

Habitat.—A large, deciduous shrub of the Himálaya, from the Indus to Nepál, found between 3,000 and 9,000 feet. It is also said to be met with in Tavoy.

Fibre.—The BARK yields an excellent FIBRE, extensively employed for rope-making, and in many parts of the Himálaya is used also in paper manufacture. **Mr.** Atkinson remarks that a trade is done in exporting this paper material to Tibet from Kumáon. Stewart, in his account of Hazara,

FIBRE.
Bark.
349
Paper.
350

G 2

FIBRE.

reports having found it being utilised for paper and textiles. In the *Kangra Gazetteer (p. 30)* it is stated that "the bark is used for paper-making in the jail at Dharmsála." The twigs are employed for tying loads; Stewart remarks of the form known as **argenteum** that the ropes made in Kanáwar were not lasting, but when fresh are very strong, and when platted as thick as the wrist, were found to stand under a heavy temporary strain when English ropes snapped.

MEDICINE.
Roots.
351

Medicine—The ROOTS are considered carminative, tonic, and diuretic; they are used in bilious complaints (*Dr. Emerson*).

FODDER.
352

Fodder.—The leaves afford a useful fodder (*Simla Settlement Report*).

TIMBER.
353

Structure of the Wood.—Yellowish-brown, with a darker centre.

354

[*5734; Wight, Ic., t. 292.*

Desmodium triflorum, *DC.; Fl. Br. Ind., II., 173; Wall., Cat.,*

Syn.—D. HETEROPHYLLUM, *Wall.*; HEDYSARUM TRIFLORUM, *Linn.*; H. STIPULACEUM, *Burm.*

Vern.—*Kodalia*, BENG.; *Kudaliya*, N.-W. P.; *Jangli* or *ran-methi*, BOMB.; *Munta mandu*, TEL.

References.—*Roxb., Fl. Ind., Ed. C.B.,C. 577; Voigt, Hort. Sub. Cal., 223; Thwaites, En. Ceylon Pl., 86; Mueller, Select Extra-trop. Pl., 7th Ed., 132; Sir W. Elliot, Fl. And., 120; S. Arjun, Bomb. Drugs, 197; Atkinson, Him. Dist., 342, 458, and 735; Royle, Ill. Him. Bot., 194; Balfour, Cyclop., 922; Kanara Gazetteer, 432; Mysore and Coorg Gaz., I., 60.*

Habitat.—A small, much branched, slender, trailing plant; found everywhere in the plains throughout India, ascending to 4,000 feet in Kumaon, and 6,000 to 7,000 feet in Kashmír and on the Chenáb.

MEDICINE.
Leaves.
355

Medicine.—The fresh LEAVES are applied to wounds and abscesses that do not heal well (*Wight*). Thwaites remarks that in Ceylon it is valued as a medicine in the cure of dysentery.

FODDER.
356

Fodder.—Roxburgh says this is "very common on pasture ground, and helps to form the most beautiful turf we have in India;" further, that "cattle are very fond of it." Müeller, in his *Select Extra-tropical Plants*, recommends its cultivation in regions too hot for clover. Col. Drury informs us that it springs up on all soils and situations, supplying there the place of **Trifolium** and **Medicago**.

357

Detergents and Soap Substitutes.

Medically the word "Detergent" would be given to any substance which had the power of cleansing wounds, ulcers, &c. While the lists of detergents given below embrace the better known substances of that nature they have been made to include also materials employed in place of soap, either from cheapness or because of reputed special properties. A complete list of the herbs used by the natives of India as detergent poultices, or even of those employed to cleanse the hair, would indeed be voluminous. The present account of detergent materials must therefore be viewed more as suggesting the position of such articles than as an exhaustive account of them.

Perhaps the most important of the soap substitutes are the species of **Sapindus**, the fruits of which are extensively employed to purify fabrics before being dyed. It seems probable that some of these detergents exercise a chemical influence not possessed by soap. At all events, it is often contended that certain peculiar results in dyeing can be obtained only when the fabric has been first washed with certain detergent vegetable substances, and that the same result cannot be brought about if soap be used. Speaking of the fruits of **Sapindus Mukorossi**, *Gærtn.* (= S. **detergens**, *Roxb.*), and of S. **trifoliatus**, *Linn.* (=S. **emarginatus**, *Vahl.* & S. **laurifolia**,

Detergents and Soap Substitutes. (*G. Watt.*) DETERGENTS.

Vahl.), Brandis says : "The pulp makes a lather with water and is used extensively for washing, either by itself, or mixed with soap. For flannel and Kashmír shawls it is greatly preferred to soap, and some varieties are specially esteemed for washing silk." Brandis adds that the subject of these detergent nuts would "repay further study." It seems highly probable that the natives of India recognise special forms, as having definite properties, under each of the species formed by modern botanists. In the literature of the subject considerable confusion exists. Dymock gives Sapindus trifoliatus, *Linn.*, as the true *Ritha* or soap nut : Roxburgh assigns that name to the plant now known as S. Mukorossi, *Gærtn.* It is probable that the former is the *Ritha* of Bombay and South India, and the latter of Bengal and Northern India. Whether the one is superior to the other or not, does not seem to have been investigated, and both trees are met with under cultivation throughout the greater part of India. Gamble makes practically the same remark under both species, *viz.*, that the chief value of these trees lies in their saponaceous berries, which are largely used and exported as soap substitutes. Mr. Baden Powell remarks : " For finer washing and dyeing purposes, the skin or shell surrounding the seeds of the soap-nut tree is often used. When mixed up with warm water a fine lather is soon produced, and the most delicate fabric may be washed, and even silks, without destroying the colour, which would yield to a coarse alkaline soap. The nuts are produced in parts of the hills, and are called *ritha* or *harita*. These nuts contain the principle termed Saponine. Several species have in their bark and roots saponaccous properties." Dr. J. F. Royle points out that "the exact nature of the principle might be advantageously investigated by chemists favourably situated in the native countries of the plants, and the nature of the changes ascertained which takes place, from the unripe and acrid to the bland and saponaceous ripe fruit." (*Conf.* with Dr. Dymock's abstract of the chemistry of this substance, *Mat. Med. West. Ind., 2nd Ed., p. 190.*) Many of the CARYOPHYLLACEÆ have saponaceous properties, one genus more especially, *viz.*, Saponaria—S. officinalis is the soap-wort of European writers. Baron F. von Mueller says of it that it possesses considerable "technolo gical interest, as the root can be employed with advantage in some final processes of washing silk and wool, to which it imparts a peculiar gloss and dazzling whiteness, without injuring in the least any subsequent application of the most sensitive colours." In India Saponaria Vaccaria, *Linn.*, is a common weed of cultivation throughout the plains of India, ascending the hills to 7,000 feet in altitude. It does not appear generally to have assigned to it the saponaceous properties which its congener enjoys, but Murray mentions that in Sind the mucilaginous sap *is* used by the natives in place of soap for washing clothes. The writer recently questioned the cultivators in the Dhamí State, Simla, as to the properties of the Saponaria which was found as a troublesome weed in their wheat fields. They said that it often proved poisonous to young cattle, but that older animals would not eat it. They were ignorant of its saponaceous properties. By the hill tribes of the Himálaya, however, two other Caryophyllaceous plants (Lychnis indica and Silene Griffithii) are known to be useful soap substitutes.

Under Acacia concinna, *DC.* (*Vol. I., p. 45*), will be found the main facts known regarding the detergent properties of the pods of that tree. These pods are, perhaps, next to the Sapindus berries, the best known and most useful detergents. A very considerable foreign trade is now done in both these products, but in India many others, though mostly of considerably less merit, are also extensively employed. The most general hair purifyer in the hands of the natives of India is the unctuous mud found on

DETERGENTS.   Detergents and Soap Substitutes.

river banks. Some of the clays met with locally possess so high a reputation as to constitute regular articles of trade, for example, the *Multánimatí* of the bazars of India or the Fuller's-earth of European commerce.

I.—HAIR-WASHES AND DETERGENTS EMPLOYED TO REMOVE VERMIN.

The following enumeration exhibits some of the chief articles used by the people of India as hair-washes. With some of these it may be a matter of question whether they are resorted to as simple detergents, as insecticides, or as perfumes :—

Acacia concinna, *DC.* The pods.
A. Intsia, *Willd.* The bark, used in Sikkim.
Ajuga bracteosa, *Wall.* Employed to kill lice.
Albizzia amara, *Boivin.* Leaves used in South India.
Allium sativum, *Linn.* Applied along with vinegar to prevent the hair turning grey.
Andropogon Schœnanthus, *Linn.* Used to promote the growth of hair.
Anona squamosa, *Linn.* Powdered seeds along with gram used as a hair-
Bassia latifolia, *Willd.* Oil-cake used as hair-wash.   [wash.
B. longifolia, *Linn.* Regular trade done in the oil-cake as a hair-wash.
Begonia Rex, *Putzeys.* The juice : also employed to kill leeches.
Clay (see remark above ; also Vol. II., 361).
Cuscuta reflexa, *Roxb.* The seeds.
Cyperus scariosus, *R. Br.* The rhizomes.
Daphne oleoides, *Schreb.* The bark, used in Kanáwar.
Entada scandens, *Benth.* The seeds, used in Nepál.
Haloxylon multiflorum, *Bunge.* The stems and leaves.
Indigofera aspalathoides, *Vahl.* The ashes, used as a wash to remove dandriff.
I. tinctoria, *Linn.* A strong infusion of the root said to destroy vermin.
Lawsonia alba, *Linn.* Hair dye.
Malva parviflora, *Linn.* The root.
Melia Azadirachta, *Linn.* The seeds.
Nardostachys Jatamansi, *DC.* Said to promote the growth of hair.
Peganum Harmala, *Linn.* The root, applied to kill lice.
Phyllanthus Emblica, *Linn.* Fruits largely employed.
Picrasma quassioides, *Benn.* The bark, an insecticide.
Pithecolobium bigeminum, *Benth.* A decoction of the leaves is employed to promote the growth of hair.
Prunus Armeniaca, *Linn.* The kernels (? or the oil expressed from them), used in the Panjáb as a hair-wash.
Quercus incana, *Roxb.* The galls.
Sapindus Mukorossi, *Gœrtn.* and **S. trifoliatus,** *Linn.* The fruits.
Saussurea Lappa, *Clarke.* The root largely used as a hair-wash.
Sesamum indicum, *Linn.* A decoction made from the leaves and root is employed as a hair-wash and is supposed to blacken the hair.

NOTE.—Medicinal insecticides will be found in list III.

II.—SOAP SUBSTITUTES.

The list of substances used directly as detergents in cleansing fabrics or as soap substitutes in personal ablution is less extensive than those employed for washing the hair. The following may be specially mentioned :

A. arabica, *Willd.* Decoction of the bark (used in Bengal, Sind, &c).

SOAP SUBSTITUTES.

Acacia concinna, *DC.* The pods.

A. Intsia, *Willd.* The bark.

Adansonia digitata, *Linn.* The ashes of the fruit and bark mixed with oil. The fruit is used in Africa as a soap substitute.

Agave americana, *Linn.* The juice.

Avicennia tomentosa, *Roxb.* Ashes of wood employed in Madras to wash cotton cloths.

Balanites Roxburghii, *Planch.* The pulp of the fruit for silk.

Carica Papaya, *Linn.* Leaves used by the Negros to wash linen.

Casuarina equisetifolia, *Forst.* Ash.

Clay. Several clays are stated to be used by washermen, such as that obtained from Western Sind (see *Vol. II., p. 363*; see also under **Barilla,** *Vol. I., p. 396*). Dhobies' earth.

Convallaria multiflora, *Linn.* Powdered root, used in Lahoul.

Dioscorea deltoidea, *Wall.* Shawls washed in Kashmír with the tubers.

Gardenia campanulata, *Roxb.* Used to wash out stains from silk.

Haloxylon multiflorum, *Bunge.* Used to wash cloths.

Hedychium spicatum, *Ham.* Rhizomes used in Garhwál "to wash the newly married."

Limonia acidissima, *Linn.* Pulp employed in Java as a soap substitute.

Lychnis indica, *Benth.* Roots and leaves used as soap in Lahoul.

Malva parviflora, *Linn.* The root is employed in Kanáwar to cleanse woollen cloth.

Phaseolus Mungo, *Linn.* Flour used in place of soap.

Sapindus Mukorossi, *Gœrtn.* & **S. trifoliatus,** *Linn.* The fruits extensively resorted to in place of soap to wash silken and woollen goods. (**Brandis** attributes the property to the "pulp," **Stewart** to the "large seed," and **Baden Powell** to "the skin or shell that surrounds the seed." **Dymock** says that in Bombay soap-nuts sell for R2½ to R3 for 35℔.

Saponaria Vaccaria, *Linn.* Juice reported to be used in place of soap.

Silene Griffithii, *Boiss.* Root and leaves used in Lahoul.

NOTE.—The above list of Detergents does not of course include the oils employed in soap-making, and only one or two ashes have been mentioned, because these are held to possess special merit. Alkalies obtained either from the soil or from plants (*Conf.* **Alkaline Earths, Vol. I., p. 167,** and with **Barilla, Sajji, Vol. I., pp. 394—399**), is made into native crude soap along with certain vegetable oils. See also under **SOAP.**

III.—MEDICINAL DETERGENTS.

MEDICINAL DETERGENTS.
360

Or Substances Employed to Cleanse Foul Sores and to Promote Healthy Action.

The list here given has been drawn up so as to exclude, as far as possible, external applications employed for other purposes than the above.

Acacia arabica, *Willd.* A poultice of the bruised tender leaves is applied to ulcers.

Adiantum venustum, *Don.* Applied to bruises.

Ægle Marmelos, *Correa.* Leaves made into poultice and employed in ophthalmia, &c.

Agave americana, *Linn.* Fleshy leaves used as poultice.

Albizzia amara, *Boivin.* Poultice to ulcers.

A. odoratissima, *Benth.* Bark efficacious in leprosy and inveterate ulcers.

Alstonia scholaris, *R. Br.* Milky juice applied to ulcers.

Anamirta Cocculus, *W. & A.* An ointment employed as an insecticide to destroy pediculi, &c., and in obstinate skin diseases.

DETERGENTS.	**Detergents and Soap Substitutes.**

MEDICINAL DETERGENTS.

Argyreia speciosa, *Sweet.* Root used by the Santals in the cure of running sores.

Artemisia vulgaris, *Linn.* An infusion of the leaves applied as a fomentation in ulcers.

Artocarpus integrifolia, *Linn.* The young leaves used in skin diseases and the juice applied to abscesses to promote suppuration.

Asagræa officinalis, *Lyndl.* A decoction used to destroy pediculi.

Avicennia tomentosa, *Jacq.* Unripe seeds used as a poultice to hasten suppuration.

Balsamodendron Mukul, *Hook.* Resin used in preparation of an ointment for bad ulcers.

B. Myrrha, *Nees.* A detergent to cold tumours.

B. Opobalsamum, *Kunth.* Resin made into a paste with lard is applied in scrofulous and cancerous sores.

B. pubescens, *Stocks.* Resin in form of ointment may be applied to cleanse and stimulate ulcers.

Bauhinia variegata, *Linn.* Bark is useful in scrofula, ulcers, &c.

Boswellia serrata, *Roxb.* An ointment of the resin is applied to ulcers, &c.

Calophyllum inophyllum, *Linn.* Resin used for indolent ulcers.

Capparis horrida, *Linn. f.* Cataplasm of leaves useful in boils, swellings, and piles.

Cassia alata, *Linn.* Leaves used for ringworm and other skin diseases.

C. Fistula, *Linn.* Bark and leaves used in skin diseases.

C. occidentalis, *Linn.* Same as above.

C. Sophora, *Linn.* Bark, leaves, and seeds with sandal-wood regarded as a specific in ringworm.

C. Tora, *Linn.* Bark, leaves, and seeds used in ringworm.

Cedrus Deodara, *Loud.* The oil from wood used as a remedy for ulcers, &c., and for sore-feet in cattle.

Cerevisiæ Fermentum (Yeast). Used as poultice.

Ceriops Candoleana, *Arnott.* Decoction of bark applied to malignant ulcers.

Citrus Aurantium, *Linn.* Poultice of oranges is recommended in skin affections.

Colchicum autumnale, *Linn.* Used in obstinate skin diseases.

Conium maculatum, *Linn.* An extract used in tumours.

Cordia Myxa, *Linn.* Kernels employed in ringworm.

Curcuma longa, *Roxb.* A paste made of the flowers is used in ringworm and other parasitic diseases.

Cycas Rumphii, *Miq.* Resin applied to malignant ulcers; it excites suppuration in a very short time.

Cynometra ramiflora, *Linn.* Lotion of the leaves in milk applied to skin diseases.

Delphinium cœruleum, *Jacq.* Roots applied to kill maggots in the wounds of goats.

Desmodium triflorum, *DC.* Fresh leaves applied to wounds, &c., that do not heal well.

Dioscorea bulbifera, *Linn.* Powdered tuber applied to ulcers. This remark is applicable to most yams.

Diospyros montana, *Roxb.* The fruit, placed by *Bhistis* on the boils which generally appear on their hands.

Dipterocarpus turbinatus, *Gærtn.* Wood oil applied to ulcers, ring-worm.

Embelia Ribes, *Burm.* Fruits made into various remedies for ring-worm and skin diseases.

Ervum Lens, *Linn.* Poultice applied to ulcers and in small-pox, &c.

Eugenia operculata, *Roxb.* Leaves used by the Santals in dry fomentation to sores.

Detergents and Soap Substitutes.   (*G. Watt.*)   **DETERGENTS.**

Ferula Narthex, *Boiss.* or **F. alliacea**, *Boiss.* The resin employed as a paste in ringworm.

Ficus bengalensis, *Linn.* Heated leaves applied as a poultice to abscesses.

F. Carica, *Linn.* Fruit used as a poultice.

F. Cunia, *Buch.* A bath made of the fruit and bark is regarded as a cure for leprosy.

Flemingia congesta, *Roxb.* Santals use the root as an application to ulcers and swellings on the neck.

Garcinia indica, *Chois.* Kokum butter is employed in indolent sores.

Gardenia gumnifera, *Linn. f.* Gum used to keep off insects from sores on cattle.

Grewia asiatica, *Linn.* Leaves applied to pustular eruptions.

Gynandropsis pentaphylla, *DC.* Ointment made of the plant with Sesamum oil, is used in skin diseases.

Gynocardia odorata, *R. Br.* Oil used extensively in skin diseases, scrofula.

Helicteres Isora, *Linn.* Fruits made into a liniment for sores in the ear.

Heliotropium brevifolium, *Wall.* Juice used to promote suppuration.

Hibiscus esculentus, *Linn.* Fresh capsules are employed as a demulcent and emollient poultice.

Hiptage Madablota, *Gærtn.* Leaves esteemed in skin diseases.

Holarrhena antidysenterica, *Wall.* Fruits made into a paste to allay pain in wounds.

Hydrocotyle asiatica, *Linn.* Leaves applied to ulcers and skin diseases.

Hydrolea zeylanica, *Vahl.* The leaves beaten into a pulp are considered efficacious in cleaning and healing bad ulcers.

Indigofera aspalathoides, *Vahl.* Leaves and flowers are applied in leprosy and cancerous affections.

I. tinctoria, *Linn.* An ointment is made from the extract which is used in sores. The dry powder is sprinkled over foul ulcers to cleanse them.

Jasminum humile, *Sims.* The root has been found useful in ringworm.

J. officinale, *Linn.* Same as above.

Jatropha Curcas, *Linn.* The milky juice is said to be detergent.

Kalanchoe spathulata, *DC.* Leaves in Kangra are burned and applied to abscesses.

Lagenaria vulgaris, *Seringe.* The pulp used as a poultice.

Lawsonia alba, *Lam.* A decoction of the leaves applied to ulcers, sores.

Lepedieropsis orbicularis, *Müll.* The bark is used by the Santals in skin diseases.

Lepidagathis cristata, *Willd.* The ashes are used by the Santals in the cure of sores.

Linum usitatissimum, *Linn.* Seeds employed as a poultice.

Luffa acutangula, *Roxb.*, var. **Amara.** Leaves applied to sores in cattle.

Lycopodium clavatum, *Linn.* Applied to boils, carbuncles, and papular eruptions, &c.

Malva rotundifolia, *Linn.* Seeds employed in skin diseases.

Mangifera indica, *Linn.* The gum-resin mixed with lime-juice or oil is applied to cutaneous affections, scabies, &c.

Melia Azadirachta, *Linn.* Leaves made into poultice are applied to ulcers and skin diseases of long standing. An oil is also similarly used.

M. Azadarach, *Linn.* Leaves and bark made into poultice, which is employed in leprosy and scrofula. A poultice of the flowers is said to kill lice and to cure eruptions of the scalp.

Mesua ferrea, *Linn.* A paste of the flowers with butter and sugar is used in piles.

Millettia auriculata, *Baker.* Root applied to sores on cattle to kill vermin.

MEDICINAL DETERGENTS.

DETERGENTS.   Detergents and Soap Substitutes.

MEDICINAL DETERGENTS.	**Mirabilis Jalapa,** *Linn.* The leaves used as a poultice to promote suppuration.

Mirabilis Jalapa, *Linn.* The leaves used as a poultice to promote suppuration.

Momordica Charantia, *Linn.* Whole plant powdered and applied in leprosy and malignant ulcers.

Morinda citrifolia, *Roxb.* The leaves used to promote healthy action in wounds, ulcers, &c.

Nelumbium speciosum, *Willd.* The root is used as a paste in ringworm, &c.

Nerium odorum, *Soland.* The root is said to be highly efficacious in skin diseases.

Nigella sativa, *Linn.,* var. **indica.** The seeds in combination with sesamum oil are used for skin eruptions.

Nyctanthes Arbor-tristis, *Linn.* The powdered seeds are used to cure scurfy affections of the scalp. The Santals employ a preparation of the root to cure goose-skin.

Ocimum Basilicum, *Linn.* The juice of the leaves useful in ringworm.

O. canum, *Sims.* The leaves made into a paste are used by the Santals in the cure of parasitic skin diseases.

Odina Wodier, *Roxb.* A decoction of the bark is useful in old ulcers.

Olea europea, *Linn.* The oil is applied to skin diseases.

Oroxylum indicum, *Benth.* A powder made from the bark is employed in the cure of sore-backs of horses.

Oxystelma esculenta, *Br.* The milky sap is used in Sind for ulcers.

Oryza sativa, *Linn.* Rice poultice, largely used as a substitute for linseed.

Pedalium Murex, *Linn.* Leaves employed as a useful poultice.

Penæa mucronata, *Linn.* The gum applied to sloughing ulcers.

Peucedanum graveolens, *Benth.* Leaves moistened with oil are used as a poultice or suppuracive.

Phyllanthus simplex, *Linn.* Root applied to mammary abscesses.

Pieris ovalifolia, *D. Don.* The young leaves and buds are used to kill insects, and an infusion is employed in cutaneous diseases.

Pinus longifolia, *Roxb.* Resin used as a plaster to abscesses in order to cause suppuration.

Pistacia Terebinthus, *Linn.* The turpentine is considered very valuable in cancer.

Pongamia glabra, *Vent.* A poultice of the leaves is applied to ulcers infested with worms : the juice of the root is used as a wash for foul sores; the oil is one of the best native remedies for cutaneous diseases.

Rhinacanthus communis, *Nees.* Root-bark used in dhobi's itch.

Saponaria Vaccaria, *Linn.* Juice used as a detergent and in the cure of itch.

Sesamum indicum, *Linn.* A poultice of the seeds applied to ulcers.

Sesbania ægyptiaca, *Pers.* Leaves, as a poultice to promote suppuration.

Tamarindus indica, *Linn.* Poultice of the seeds is applied to boils, &c., and of the leaves and pulp of the fruit to inflammatory swellings.

Tamarix gallica, *Linn.* Strong infusion of galls applied to foul ulcers.

Terminalia Arjuna, *Beddome.* Decoction of bark used in ulcers and cancers.

Thespesia populnea, *Corr.* The yellow juice of fruit is used in cutaneous diseases.

Trichosanthes dioica, *Roxb.* The root is resorted to in treatment of leprosy.

Vallaris Heynei, *Spreng.* Milky juice applied to wounds and sores.

Vernonia anthelmintica, *Willd.* Seeds of great repute in Sanskrit Materia Medica for white leprosy and other skin diseases.

Vitex Negundo, *Linn.* The juice of the leaves has the property of removing foetid discharges from ulcers.

D. 360

Woodfordia floribunda, *Salisb.* The powdered flowers are sprinkled over ulcers to promote granulation.

MEDICINAL DETERGENTS.

Zizyphus vulgaris, *Lam.* The bark is used to clean wounds and sores.

IV.—DENTIFRICES AND TOOTH-BRUSHES.

DENTIFRICES.
361

Materials used to clean the teeth may, as a matter of convenience, be given here under Detergents.

The following list indicates those most frequently mentioned by authors:—

Abutilon indicum, *Don.* A decoction of the bark is used as a mouthwash in toothache.

Acacia Catechu, *Willd.* Cutch is recommended as a dentifrice along with charcoal.

A. ferruginea, *DC.* A decoction of the bark is employed as a toothwash.

A. modesta, *Wall.* Twigs used by the Panjábís as tooth-brushes.

Areca Catechu, *Linn.* The burnt nuts reduced to a powder have been recommended as a dentifrice.

Aristida setacea, *Retz.* According to **Roxburgh** the culms are used in South India as tooth-picks. [brushes.

Calotropis gigantea, *R. Br.* & **C. procera,** *R. Br.* Twigs used as tooth-

Cassia auriculata, *Linn.* Twigs used as tooth-brushes. A considerable trade is done in these: they are esteemed as preferable to the tooth-brushes obtained from any other plant.

Citrullus Colocynthis, *Schrad.* Fresh root used as tooth-brush.

Cuttle-fish (or Sea-foam). Employed in the manufacture of tooth powder.

Datura alba, *Nees.* Powder of the seeds used to deaden pain.

Daucus Carota, *Linn.* Leaf-stalks employed by the hill tribes.

Euphorbia antiquorum, *Linn.* Juice given in toothache.

Ficus bengalensis, *Linn.* Juice given in toothache.

Indigofera aspalathoides, *Vahl.* Root chewed in toothache.

I. paucifolia, *Delile.* Used in Sind by Hindús.

Jasminum grandiflorum, *Linn.* Leaves chewed in ulcerations of the gums.

Juglans regia, *Linn.* Bark exported to the plains, used as a dentifrice.

Mangifera indica, *Linn.* Leaves, stalks, and twigs, used as tooth-brushes.

Melia Azadirachta, *Linn.* Twigs used as tooth-brushes.

Moringa pterygosperma, *Gærtn.* The bark employed in toothache.

Pistacia Lentiscus, *Linn.* The mastich dissolved in alcohol is employed for filling up cavities in teeth.

Plumbago rosea, *Linn.* Root applied in the cure of toothache.

Pontederia vaginalis, *Linn.* The root chewed in toothache. [tifrice.

Prunus Amygdalus, *Baillon.* The powdered charred shell used as a den-

Pterocarpus Marsupium, *Roxb.* Gum employed in toothache.

Rumex vesicarius, *Linn.* Juice given in toothache.

Salvadora persica, *Linn.* Twigs used as "tooth-cleaners" in the Panjáb.

Solanum indicum, *Linn.* The root employed in toothache.

S. xanthocarpum, *Schrad & Wendl.* The fruits boiled in *ghí* are used by the Santals for toothache. Fumigation with burning seeds is in great repute for toothache.

Streblus asper, *Lour.* Twigs employed as tooth-brushes.

Ventilago calyculata, *Tulasne.* Tendril worn by the Santals as a ring on the finger intended as a charm against toothache.

Wrightia tinctoria, *R. Br.* The fresh leaves when chewed are said to relieve toothache.

Xanthoxylum alatum, *Roxb.* Twigs used as tooth-brushes and to cure toothache.

DIAMOND.	**The Diamond.**

DENTIFRICES.

Zizyphus rugosa, *Lamk.* The powdered bark is employed by the Santals as a cure for toothache.

362

DEUTZIA, *Thumb. ; Gen. Pl., I., 642.*

A genus of highly ornamental shrubs, belonging to the SAXIFRAGEÆ, which have come into much favour by European gardeners, on account of their bunches of handsome white flowers. The rough star-shaped hairs on the leaves are serviceable in place of sand-paper, and the timber is used as fuel. The two Himálayan species are D. **corymbosa**, *Brown* (the *Daloutchi*), and D. **staminea**, *Brown* (the *Muneli* of KUMAON ; *Deutsch*, SIMLA ; *Phul Kanri*, HAZARA ; *Phurilé*, KASHMIR ; *Sai*, CHUMBA ; and the *Aruchi* or *Deús* of BASHIRH).

Devil's Tree and **Dita Bark,** see **Alstonia scholaris,** *R. Br.,* **Vol. I.,**
[No. 870.

363   **Dextrine** or **British Gum.**

A chemical substance present in most grains, having the formula $C_{12}H_{10}O_{10}$. Wheat contains 4·5 ; wheat-bran, 5·52 ; barley, 6·55 ; rye-bran, 7·79 ; malt, 8·2?. In commerce the term is applied to the substance artificially produced by the transformation of starch—the granules on bursting under the influence of heat constitute British Gum. This is largely used in calico-printing, paper-glazing, gumming envelopes and postage stamps. It seems probable that a very large proportion of the Rice exported from India to Europe is employed in the manufacture of Dextrine. See **Oryza sativa.**

Dhal, see **Cajanus indicus,** *Spreng.,* Vol. II., No. 49.

Dhourra, a name often given to the millets collectively.

Dhub or **Dub,** see **Cynodon Dactylon,** *Pers.,* Vol. II., No. 2558.

364

DIAMOND, *Man. Geology, Ind., III., pp. 1-50, IV., p. 8.*

DIAMANT, *Fr., Germ., Dutch ;* DIAMANTE, *It., Sp. ;* ALMAS, *Russ.*
Vern.—*Hirá,* HIND. ; *Almás,* ARAB. and PERS. ; also in PERSIAN *Mas. ; Hiráka,* SANS. ; *Adamas,* GREEK and LATIN.
References.—*Records of Geol. Survey of Ind., II., 9 ; V., 27 ; X., 58, 186 ; XVIII., 24; XIX., 109, 208 ; Mem. of G. S. Ind., II., 65 ; VII., 113 ; VIII., 106, 267 ; XII., 144 ; XVI., 253 ; Jour. As. Soc. Bengal, II., 403 ; V., III ; VIII., 379, 1057 ; XI., 399 ; XIII., 859 ; XV., 390 ; XXXIV.. Pt. II., 13 ; XL., Pt. I., 11 ; L., 39, also Pt. II., 31 ; Jour. Royal As. Soc., VII. (Old Series), by Capt. Newbold,pp. 226, 233 ; VII. (New Series), 125 ; Trans. R. A. S., I., 277 ; As. Res., XV., 120, 125 ; XVIII., 100 ; Madras Jour. Lit. & Sc., III., 120 ; VI., 47 ; Trans. Med. & Phys. Soc., Calcutta, II., 261, 264 ; Trans. Geol. Soc., London, 2nd Series, V., pp. 541, 568 ; Jour. Geol. Soc., London, XI., 355 ; Voyage, John Huyghen van Linschoten, in 1596 (Trans. by Hakluyt Soc.), II., 136 ; Tavernier (1665-1669), Voyages, II. ; Cæsar Frederick, 1570 (Hakluyt's Voyages) ; Marco Polo (13th century), Ed. by Col. Yule, Vol. II , 295 ; New account of the East Indies by Capt. Hamilton (1688-1728), Vol. I., XXIX., 306 ; Ain-i-Akbari by Abul Fazl (1590), Trans. by Gladwin, II., 7, 11, 32, 59 ; Blochmann's Trans., p. 480 ; Tuzuk-i-Jahángiri, pp. 154-155 ; Mustapha (1758), Oriental Report, London, 1799 ; Dr. Heyne (1814), Tracts, London, p. 92 ; Capt. Burton (1876) ; Quart. Jour. Sc., New Series, Vol. VI., 351 ; Mani Mala by Raja Sourendro Mohun Tagore ; Kelsall (1872). Bellary Dist. Man., p. 24 ; Jenkins, Report of Nagpur ; Temple, Adm. Rep., C. P., 1861-62, p. 124 ; C. P Gazetteer ; Dr. Shortt (1855), Selections, Records, Beng. Govt., Vol. IV., No. XXIII., p. 182 ; Sel. Records, Madras Government, No. XIV;*

D. 364

| The Diamond. | (*G. Watt.*) | DIAMOND. |

Statistical account of Bengal by Sir W. W. Hunter, XVII., p. 190; Atkinson, N. W. P. (1874), Panná District, p. 565; Mason's Burma and Its People, pp. 573, 731; U. C. Dutt, Mat. Med. Hind., p. 92; Man. Cuddapah Dist. by Gribble, p. 24; Settle. Report, Chanda Dist., p. 4, &c., &c.

Where found.—In India Diamonds occur over three wide areas, in each of which several limited localities are more especially famed. These may be briefly stated as, *first*, the eastern side of the Deccan from the Penner to the Sone; *second*, the Madras Presidency, as near Cuddapah, Karnul, Ellore, but more especially in the Kistna and Godavari basins (the former of which probably afforded the Golconda diamonds, a name given to them from the ancient kingdom of Golconda), and, *third*, Chutia Nagpore and the Central Provinces to Bundelkhand.

It is somewhat remarkable that the Indian diamonds have not as yet been found in what can be called their original matrix. Recently, however, they have been reported to have been discovered in the Madras Presidency in a peculiar rock answering somewhat to the " blue rock " (Peridolite) of South Africa. As matter of practical experience, they are found chiefly in alluvial deposits, such as in beds of sand and clay, in ferruginous sandstones, or in conglomerates. The best diamonds are said to be those from the Kistna district and from Panna in Bundelkhand. A further locality has been reported, namely, on the Himálaya near Simla, and this might be the *Haima* of ancient writers. The discovery of diamonds on the Himálaya has not, however, been confirmed by geologists, and although, if established, it would prove of the greatest interest geologically, the reported occurrence has not as yet been productive of practical results. It may be added that none of the Indian diamond fields can, at the present day, be viewed as of commercial importance, and it is difficult, if not impossible, to identify, for certain, all the localities alluded to by classical writers. Both practical and scientific European opinion is, however, in favour of the explanation that the lessened trade in modern times is more due to the conservative character of the diggers, in keeping their art a secret, or to the exhaustion of the surface workings which their appliances and means are alone suitable for, than to the complete exhaustion or non-existence of fairly rich unexplored diamond beds. Some few centuries ago, diamonds were undoubtedly more extensively produced in India than at the present day. India was, indeed, the first, and for a long period the only, source of diamonds known to the European nations. The decline which has since taken place may be due, in addition to the above explanation, to the discovery of the stone elsewhere, and to the application of cheaper methods of working diamond mines in other countries than are known to the people of India.

For centuries the Indian mines have been held by poor workmen who, unaided by science, have had to depend on their hereditary skill while battling against the adversity and persecution engendered through national disturbances that shook the empire, particularly from about the period of the Brazilian discovery (1727) down to the completion of the industrial settlement under British rule. It seems probable that when peace and security were restored in India, the art of diamond washing had to a large extent been lost. At the same time it should not be forgotten that the diamonds which found their way all over the civilized world from the Indian mines—the *Adamas* of the Greek and Latin writers—may have largely represented the surplus accumulation of gems collected during many previous centuries.

Some of the oldest Sanskrit writers allude to the diamond; and it appears to have been worn by the nobility of India long anterior to the earliest European mention of it. At the same time it is significant, as Mr. Ball points out (*Economic Geology of India, p. 3*), that the

WHERE
FOUND.
365

DIAMOND.

The Diamond.

HISTORY.

Hindus are not now and probably never were professional diamond diggers. The greater part of the Indian mines are worked by Gonds or Kóls; for, as **Mr. Ball** adds, "the miners in South India, though some of them are said to be Hindus and others are simply described as low out-castes, all probably came from the same family. It may, of course, be said in answer to this that the mining and washing would naturally fall to the lot of Helot races; but in some of the localities it is doubtful whether the Aryans ever held paramount power." It would not, therefore, be a great stretch of imagination to picture the aboriginal races of India using diamonds as playthings prior to the Aryan invasion, as putting in fact little more value on them than the Negroes of Brazil did, who employed their diamonds as counters in games of cards. The Aryans bringing with them wealth and enlightenment might be supposed to have soon given to the Indian gems their true value, while leaving the art of digging in the hands of the aborigines in whose country they were found. Everything, therefore, points to India having always had a limited and conservative diamond-mining community with whom it might be easily supposed the art, under adversity, would not have continued to prosper. Even assuming that the first washings of the surface beds afforded a richer yield than the subsequent re-washings of the same materials (and this is admittedly what has actually taken place), there still remains the fact that few gems have in modern times been discovered that are in any way equal to those now in the possession of the great monarchs of the world—gems, the individual histories of which are lost in the obscurity of a remote antiquity. The view may thus be admissible that the Royal Diamonds have been handed down from generation to generation, and that each represents an accidental discovery, a it marks as period of human history. The prevalent opinion, advanced by the early writers and held still by the modern Indian diggers, as to "diamonds growing," accounts for the persistence with which the same materials have been searched over and over again; and it has its explanation in the fact that the natural disintegration of the matrix brings to light stones not discovered in a former washing, from their having been closely encrusted by earthy materials. But the theory of growth has been exploded both by the chemist and the European digger. The diamond is now known to be a crystalline state of pure carbon, formed under geological influences, of which analytical research may be said to have established the rationale, but which constructive or synthetical efforts have at most only approximated towards demonstrating. We may decompose the diamond but cannot make it.

One of the older European writers who visited India and wrote of the diamond (**John Huyghen van Linschoten**) describes it as growing :—

"Diamonds, " he says," by the Arabians and Moores called *Almas* and by the Indians, where they grow, *Iraa*,* and by the Malayans, where they are likewise found, *Itam*."† "They grow in the countrie of Decam behinde Ballagate, by the towne of Bisnagar, wherein are two or three hills from whence they are digged, whereof the King Bisnagar doth reape great profitte : for hee causeth them to be straightly watched, and hath farmed them out with this condition that all Diamonds that are above twenty-five Mangelyns in waight are for the King him selfe (every Mangelyn is foure graines in waight),‡ and if anie man bee found that hideth anie such, he looseth both life and goods. There is yet another

---

* *Hirá*, SANS.
† The Malay name *Iutan* comes from the Javanese *Huiten*, which, again, is derived from the Sanskrit.
‡ According to Mr. Ball in Tavernier's time (*Econ. Geol.*, *p. 21*) "a Mangelin = 1¼ carats or 7 grains at Raolconda and Coulour ; the rati being ⅘ of a carat or 3½ grains. But 1·843 grains is more probably the correct equivalent of the Hindú *rati*."

The Diamond.	(*G. Watt.*)	DIAMOND.

HISTORY.

hill in the countrie of Decam, which is called Velha, that is, the old Rocke " (= Rocha velha—*Ed.*), "from thence come the best Diamonds, and are sold for the greatest price, which the Diamond grinders, Jewellers and Indians, can very well discerne from the rest."

"These Diamonds are much brought to sell in a Faire that is holden in a Towne called Lispor,* lying in the same countrie of Decam between Goa and Cambaia, whither the Banianes and Gusurates of Cambaia doe goe and buy them up bringing them to Goa, and other places. They are very skilfull in these matters, so that no Jeweller can goe beyond them, but oftentimes they deceive the best Jewellers in all Christendome. In this Roca Velha, there are Diamonds founde that are called *Nayfes* ready cut, which are naturall and are more esteemed than the rest, especially by the Indians themselves." "In the Straight called Tania-pura, a countrie on the one side of Malacca,† there is likewise an old rocke, which also is called Roca Velha, where diamonds are found, that are excellent : they are small, but verie good, and heavie, which is goode for the seller, but not for the buyer. Diamonds are digged like gold out of Mynes, and where they digge one yeare the length of a man into the ground, within three or foure years after, there are diamonds founde againe in the same place which grow there. Sometimes they find Diamonds of one hundred and two hundred Mangelyns, and more but verie few."

It may here be suggested that it is curious **Linschoten** did not learn of the discovery in the Deccan diamond area of any exceptionally large stones such as the Great Mogul or Koh-i-nur. His remarks are of a general not a specific character. The above passage has, however, been reproduced in full from **Linschoten's** Journal of his travels in India, because **Ball** and other writers on Indian Diamonds do not appear to have consulted that work. The explanatory notes are mostly those given by **Burnell** and **Tiele** in their revised translation, published by the Hakluyt Society. The original Dutch Edition of **Linschoten's** Journal is dated 1596, and the account given by **Tavernier** in his *Voyages*—a writer to whom most modern authors assign the first place among diamond explorers—was published about 1669. It is, indeed, often stated that **Tavernier** first made the Indian diamond famous in Europe, but **Marco Polo** in the thirteenth century wrote of them, and even **Tavernier** speaks of a trade existing in these gems in his time, while a century before **Linschoten**, in the passage quoted above, published the fact that the Christians of Goa traded in diamonds. **Tavernier** was, perhaps, the first European, however, who travelled over India with the express purpose of inspecting the diamond mines. As the result, much more precise information became current in Europe after the publication of his voyages than before. He visited the Emperor Aurangzeb on the 1st November 1665, and on the next day was permitted to examine and weigh the Court jewels. The largest diamond shown him he appears to have named 'The Great Mogul.' This, he was informed, had been obtained from the Coulour mines (Kollur, in the Kistna district, Madras)—mines opened out, as he affirms, only a hundred years before the date of his visit to India This would correspond with the date of **Linschoten's** visit. **Ball** and other writers suggest that the Great Mogul was most probably known originally as the Kollur diamond, but that, in conformity with an Asiatic practice of corrupting meaningless names into something understood while preserving the original sound, it became *Koh-i-nur* or, 'mountain of light.' **Mr. Mallet**

The Great Mogul. 366

Koh-i-nur. 367

---

* Probably Elichpur, the old capital of Berar.
† "Tandjong Pura, the old capital of Matan on the west coast of Borneo. It is mentioned by **Castanheda** and others as a town from which came diamonds."

DIAMOND	The Diamond.

refers to the most recent large diamond found in India (1881). It came from the Bellary district, and was purchased by the firm of Messrs. P. Orr & Sons of Madras. When cut as a brilliant of the purest water it weighed 24⅝ carats. This, as a kind of parody on *Koh-i-nur*, received the name of *Gor-do-Norr* in honour of the senior partner of the firm, Mr. Gordon Orr. On the other hand, many writers hold that the *Koh-i-nur* was so named by Nadir Shah (the Persian invader of India in 1739) from whose successor in 1813 it passed into the hands of Runjit Sing. In 1849 on the annexation of the Panjáb it again changed hands, and was presented shortly after to Her Majesty the Queen-Empress of India.

On the other hand, a much greater antiquity is sought to be established for the *Koh-i-nur*. A legend asserts that it was found in one of the mines in the Kistna district, and was worn 5,000 years ago by Karna, one of the heroes celebrated in the *Mahábharata*. It is then said to have passed through many hands until presented to Bábar, the founder of the Mogul dynasty in 1526, and thus descended to Aurengzeb, son of Shah Jahán. Tavernier, however, expressly states that it came to the Mogul Emperors in the time of Shah Jahán, and it is significant also that Abul Fuzl, in his *Ain-i-Akbari*, while dwelling at length on the high personal character and great wealth of Akbár (the great-grandson of Bábar), makes no mention in the list of Court jewels of any diamond that could compare with either the Great Mogul or the *Koh i-nur*. In the *Tuzuk-i-Jahángiri* some interesting facts are given regarding the Court jewels in Jahangir's time (son of Akbár), but no mention is made of the Great Mogul, so that Tavernier's statement may be accepted as correct that it came into the hands of the Mogul Emperors during the reign of Shah Jahán (son of Jahangir). This would not, however, preclude the possibility of its having been in the possession of the Kings of Golconda for many previous generations, or even detract from the probable accuracy of the tradition that it was once worn by Karna. Indeed, the king of a region from which the majority of the great diamonds were obtained, might fairly well be expected to have retained in his own family some of the best gems ever found. This is the more easily admissible when it is recollected how futile had been the efforts to conquer the diamond king, and that even Shah Jahán owed some degree of his ultimate success to the treachery of Mirimgola.

A far greater difficulty exists in tracing the Mogul diamond after the date of its having been inspected and weighed by Tavernier. On the death of Aurangzeb, the Mogul Empire rapidly fell, and from 1720 it may be said to have begun the final stage of breaking up. In 1739, the Persian invader Nadir Shah overthrew what vestiges remained of the Great Muhammadan Empire—an empire that had lasted for over two hundred years, *viz.*, from Bábar to Muhammad Bahádar Shah, the last of the race of Timur. The Persians sacked the city of Delhi and carried off money and treasure to the value of 32 millions sterling, including the Great Mogul Diamond.

Tavernier does not, however, say that that gem was found a hundred years before the date of his visit to Aurangzeb, but that the Coulour mines were opened out then. The great diamond might have been picked up centuries before, although, as pointed out above, Linschoten's silence as to the existence of any one exceptionally large diamond might be accepted as leading to an opposite inference. Some capital has been made out of Tavernier's contradictory statements regarding the weight of the gem when presented to Shah Jahán—in one place 900 ratis = 787½ carats, in another 907 ratis = 793 ¹⁄₁₆ carats. But it should be borne in mind that that was only the weight he was told it then possessed, and he may be

| The Diamond. | (*G. Watt.*) | DIAMOND. |

pardoned a discrepancy which after all is not of serious consideration. When shown to him it had been cut, and he is perfectly consistent in stating, wherever he alludes to it, that on his weighing it he found it to be 319½ ratis, or 280 carats. A good deal of discussion has taken place also as to whether Tavernier's Great Mogul Diamond of 280 carats was one and the same with the Koh-i-nur. Some writers affirm that during the time it was in the possession of the Persians, it was cut or broken by cleavage, and that from the Great Mogul was derived the Orloff diamond, and also a gem still in the possession of Persia. The Orloff diamond, now in the sceptre of the Emperor of Russia, is in the form of a half pigeon's egg, and weighs 194¾ carats. The Koh-i-nur, when it came to England, weighed 186$\frac{1}{16}$ carats, and might be described to have been a defective half egg. It has since been cut in the rose, and weighs 106$\frac{1}{16}$ carats. If the removed portion from the top of the Koh-i-nur were accepted as corresponding with the Orloff gem, the latter should have weighed considerably less than the former, and if a lower portion of the Koh-i-nur gave origin to the Orloff gem, it would be difficult to account for its shape as that of a half egg. But reasoning on these lines goes on the assumption that to account for the Orloff and Koh-i-nur as parts of one original diamond, they were parts of the diamond as seen and figured by Tavernier. It would seem that this idea has so pervaded the writings of authors who have treated of this subject that the fact that the stone presented to Shah Jahán had been reduced from 787 to 280 carats has been lost sight of. It is just possible that the severe treatment bestowed on the Venetian, Hortensio Borgio, who cut the stone for the Mogul Emperor, was because of a well-founded suspicion that he had cut off large pieces which were never accounted for. If this supposition be admissible, then the Great Mogul gem, with small pieces chipped off it while in Persia, might easily be accepted for the diamond known as the Koh-i-nur, while the somewhat mythical story of the Orloff having been picked off a Hindú idol might be viewed as the manner in which the largest of all diamonds was again restored to public notice. The person who cut the Great Mogul in the form of a half egg might have followed the same method in forming the Orloff. All this is, however, pure speculation, and the main interest rests in the fact that the Koh-i-nur, the Orloff, the Pitt or Regent, and most of the other great historic diamonds, have been obtained from India.

PRESENT POSITION AND FUTURE PROSPECTS OF THE INDIAN DIAMOND FIELDS.—It has already been stated that large diamonds, in any way comparable to those discussed above, have not been found for many generations. Various reasons have been suggested for the decline of the Indian industry, and it is perhaps only necessary in this place to state that the subject seems likely to attract much greater attention in the future. An expert has recently been examining the Hyderabad diamond fields, and while a definite report has not as yet been issued by him, the Deccan Company have had a hopeful forecast placed before them. It is perhaps unnecessary to quote here a complete series of notices regarding the diamond fields that are actually being worked. A few may, however, be mentioned, premising that nearly every writer states that the trade is unimportant, the contractors often losing heavily, and the labourers earning only a precarious livelihood.

MADRAS.—In the *District Manuals* and the *Imperial Gazetteer* brief notices occur regarding the diamonds found at the present time in the Madras Presidency. These seem to be summarised in the following passage taken from the *Manual of the Madras Administration* for 1885:—
" The diamond-bearing sandstones and conglomerates are of considerable

H

DIAMOND.	The Diamond.

extent in the Kurnool basin, especially on its western side. They have been mined at Bunganapully, Moonimadoogoo, and Goorramcondah in the latter district. At Ramalcottah and several other places in Kurnool district diamonds were, and are still, obtained by washing local alluvia formed of the *débris* of the diamond conglomerate. At and near Chenur, in Cuddapah district, the gravel beds in the alluvium of the Pennair river, which consists largely of *débris* of rocks belonging to the Karnool system, were formerly washed on a large scale, though now almost abandoned. Considerable tracts of the diamond conglomerate—the 'Bunganapully conglomerates' of the Geological surveyors — have been left untried as yet by the native miners. Conglomerate beds belonging to the Cuddapah system were formerly mined for diamonds in the Kistna district, where deserted villages occur in great numbers to the north and west of Chintapully. To this set of mines belonged the old workings at Collor, on the Kistna, which has been identified, on good grounds, with the Gani Coulour of **Tavernier**, where the Koh-i-nur was obtained. The Ramalcottah and Bunganapully mines and workings appear still to yield a remunerative supply of small and rough diamonds ; the right to mine being sold at a yearly auction. The so-called Golcondah mines, either of Gollapully near Ellore or in some parts of the Golcondah range of the Eastern Ghauts north of Rajahmundry, have been long deserted."

NIZAM'S DOMINIONS.—"When the Nizam ceded the Northern Circars to the British, he was permitted to retain possession of all the village lands of this area in which diamond mines were situated, and these villages now stand isolated in the British Kistna and Godavari districts. The revenue derived from them by the Nizam at present from ordinary agricultural resources is not inconsiderable ; but the diamond mines yield little or nothing. Eighty years before **Heyne's** visit, or about the beginning of the eighteenth century, they belonged to a powerful zamindár called **Ooparow**, but on his discovering the diamonds they were taken possession of by his sovereign the Nizam." Some of these mines have already been alluded to, such as the Kollur (Color). In that mine it is now generally believed the Koh-i-nur was found, and not at Partial, though it seems fairly certain the Pitt gem was found at the latter. The expert presently examining the mines in Hyderabad has published certain facts of interest. His communication has been discussed as follows in the *Pioneer* : —"The workings are very extensive, some being five miles in length. They are all of a superficial character, not extending below 15 feet from the surface. Wherever water or rock was met the native workers could not compete with the difficulty. The soil indications are said to be extremely satisfactory, and in many places similar to those found at Kimberley and elsewhere in South Africa. Although the diamond workings have not been carried on since the beginning of the century, a few individuals still employ themselves in re-washing the old *débris*, and the expert was shown one or two small diamonds found by them of fairly good colour." The report alluded to describes the primitive method pursued in washing and sifting for the diamonds, the information given being concluded with the following : —"By the 26th January the expert had again started from Secunderabad for Purtyal with a convoy of 80 bullock carts carrying all the necessary machinery for testing and working the different places described by him. He states that he hopes to be able shortly to send a further report in the shape of a parcel of diamonds." He adds : —"It is of course not in my power to say with any certainty that I shall find diamonds in payable quantities, but I do not suppose for one moment that the diggings are worked out, particularly as the natives have not worked the ground regularly but have left ground untouched between

| The Diamond. | (*G. Watt.*) | DIAMOND. |

all the pits which is of the same soil and, therefore, just as likely to be diamond-bearing as the pits themselves." He concludes: "I have every confidence in the venture, but do not like to be over-sanguine, and as it will not be very long before the ground will be thoroughly tested, I prefer to confine myself to saying that the chances are very much in favour of everything turning out satisfactorily. It may be of interest to you to know that in all the Kistna villages excepting Purtyal, which is on the high road, there has never, in the memory of living men, been a white man, so *that* proves plainly that no prospecting or anything of that kind has taken place within the last 80 or 90 years. With regard to working any of these places, there are no difficulties of any kind; labour can be very easily obtained, also fuel and water; and should the pits, full now, be required at once, it would be an easy matter, comparatively, to drain and pump them dry."

CENTRAL PROVINCES, SAMBULPUR DISTRICT.—"Some uncertainty," Mr. Ball states, "exists as to how far the early notices of the diamond-bearing localities of Gondwana are applicable to those situated in the Mahanadi basin." "The first visit to Sambulpur, of which there is any published account, is described in the narrative of a journey which was undertaken by Mr. Motte in the year 1766. The object of this journey was to initiate a regular trade in diamonds with Sambulpur, Lord Clive being desirous of employing diamonds as a convenient means of remitting money to England. His attention had been drawn to Sambulpur by the fact that the Rajah had, a few months previously, sent a messenger, with a rough diamond weighing $16\frac{1}{4}$ carats, as a sample, together with an invitation to the Governor to depute a trustworthy person to purchase diamonds regularly. The Governor proposed to Mr. Motte to make the speculation a joint concern, in which," writes the latter, "I was to hold a third; he the other two; all the expenses to be borne by the concern. The proposal dazzled me and I caught at it without reflecting on the difficulties of the march or on the barbarity of the country, &c." "In spite of his life being several times in danger from attacks by the natives, the loss of some of his followers by fever, and a varied chapter of other disasters, Mr. Motte was enabled to collect a considerable amount of interesting information about the country. Owing to the disturbed state of Sambulpur town, however, he was only able to purchase a few diamonds."

"The next account is Dr. Voysey's, who visited the diamond washings in Sambulpur in 1823, when on his last journey from Nagpúr to Calcutta. He states that diamonds were only found below the junction of the Ebe river with the Mahanadi, but other authorities place the limit much further up, namely, at the junction of the Mand and Mahanadi rivers. The miners were at work in the channel between the island and the right bank, about 10 miles above Sambulpur." In the Medical Topography of the districts of Ramgurh, Chutia Nagpúr, Sirgooja, and Sambulpur (dated 1825) further additional information is given regarding the Sambulpur diamonds, which fixes the diamond region on the north side of the river. A large stone is said to have been found in 1809 in these mines. This is reported to have weighed 210·6 carats and to have fallen treacherously into the hands of the Mahrattas. Nothing further has been heard of this stone, but it is presumably one of the great gems the history of which is lost. The *Central Provinces Gazetteer*, upon what authority is not known, affirms that the diamonds of Sambulpur are flat and thin, and have flaws in them. Some of the older writers on the contrary state that along with the Chutia Nagpúr stones they were of the best quality and the purest water. In the *Imperial Gazetteer* it is simply stated that diamonds are occasionally found near an island called Hírákudá or diamond island. When Sambul-

H 2

DIAMOND.	The Diamond.

POSITION AND PROSPECTS.

pur was finally taken over by the British in 1850, the Government offered to lease out the right to search for diamonds, and in 1856 a notification appeared in the *Gazette* describing the prospects in somewhat glowing terms. For a short time the lease was held by a European at the low rate of R200 a year, but it was soon given up. Mr. Ball adds that, though reports are often made of diamonds found at Sambulpur, "recent local inquiries failed to elicit a single authentic case, and the gold-washers asserted that these statements were incorrect."

Of the mines in the Chanda district it may be said that, although these are of considerable extent and are most probably the Bairagarh mines mentioned in the *Ain-i-Akbari*, Mr. R. Jenkins, in his report on the territories of the Rajah of Nagpúr, states " that they were formerly celebrated, but in his time did not yield sufficient returns to make them worth working."

Bundelkhand.

374

BUNDELKHAND, PANNA.—In the *North-Western Provinces Gazetteer* (*Vol. I., Bundelkhand, p. 565*) will be found a detailed account of the past and present of the Panna diamond mines. This has been condensed and reproduced in the *Imperial Gazetteer* as follows :—" The ground on the surface and for a few feet below," says Mr. Thornton from whom this paragraph is compiled, "consists of ferruginous gravel mixed with reddish clay ; and this loose mass, when carefully washed and searched, yields diamonds, though few in number and of small size. The matrix containing in greater quantity the more valuable diamonds lies considerably lower, at a depth varying generally from 12 to 40 feet, and is a conglomerate of pebbles of quartz, jasper, hornstone, Lydian stone, &c. The fragments of this conglomerate, quarried and brought to the surface, are carefully pounded ; and after several washings to remove the softer and more clayey parts, the residue is repeatedly searched for the diamonds. As frequently happens in such speculative pursuits, the returns often scarcely equal the outlay and the adventurers are ruined. The business is now much less prosperous than formerly, but Jacquemont did not consider that there were in his time any symptoms of exhaustion in the adamantiferous deposits, and attributed the unfavourable change to the diminished value of the gem everywhere. The rejected rubbish, if examined after a lapse of some years, has been frequently found to contain valuable gems, which some suppose have in the interval been produced in the congenial matrix ; but experienced and skilful miners are generally of opinion that the diamonds escaped the former search in consequence of encrustation of some opaque coat and have now been rendered obvious to the sight from its removal by fracture, friction, or some other accidental cause. More extensive and important than the tract just referred to is another extending from 12 to 20 miles north-east of the town of Panna and worked in the localities of Kamariya, Brijpur, Bargári, Maira, and Etwa. Diamonds of the first water, or completely colourless, are very rare ; most of those found being either pearly, greenish, yellowish, rose-coloured, black or brown." Sir W. W. Hunter adds that, according to Pogson, "inexhaustible strata producing diamonds exist here." " None of the great diamonds now known appear to be traceable to the mines in Panna, and Tieffenthaler mentions it as a general opinion that those of Golconda are superior." During the prosperity of the mines a tax of 25 per cent. was levied on their produce, but the tax now imposed is stated to exceed this rate. The revenue is divided in proportions between the Rajahs of Panna and Charkhári. The value of the diamonds still found in the mines is estimated at £12,000 per annum." Mr. Ball gives a brief account of these mines written by Mr. Medlicott and a picture of the miners at work in a shaft as seen by the late Mr. Jules Schaumburg.

D. 374

Products of India. 101

The Diamond; Clove Pink and Carnation.   (*G. Watt.*)   DICHOPSIS.

BENGAL, CHUTIA NAGPUR.—Repeated reference has been made to the diamonds found in Chutia Nagpúr. Space cannot, however, be afforded to deal in full with the mines that are said to have existed, nor even to do justice to the historic references to them. But they are not generally regarded as of much importance. Mr. Blochmann's paper on the subject of Kokrah (= the ancient name of Chutia Nagpúr) is, however, of very great interest. The diamonds possessed by Akbár and his son Jahangir are said to have been largely drawn from the mines in Chutia Nagpúr. The reader is referred to Mr. Ball's detailed account of Indian diamonds in the *Manual of the Geology of India, Vol. II., pp. 1—58.*

POSITION AND PROSPECTS. Bengal. 375

Medicine.—DIAMOND dust is known to be a powerful mechanical poison. In Hindu practice it is, however, to some extent, used as a drug. Dutt says that, according to Sanskrit authors, the diamond for medicinal purposes is purified by being enclosed within a lemon and boiled in the juice of the leaves of **Sesbania grandiflora.** It is reduced to a powder in the following manner: "A piece of the root of a cotton plant is beaten to a paste with juice of some betel leaves. Both these vegetables should not be less than three years old. The diamond is enclosed within this paste and roasted in a pit of fire. The process is repeated seven times, when the stone is easily reduced to a fine powder. Another process consists in roasting the diamond enclosed in a paste made of horn-shavings, for three times in succession. The diamond thus prepared is said to be a powerful alterative tonic that improves nutrition, increases the strength and firmness of the body, and removes all sorts of disease. Dose about one grain."

MEDICINE. 376

SPECIAL OPINION.—§ "Employed as a poison, it is administered in the shape of dust, as in the late celebrated case when the Resident of Baroda, Sir Arthur Phayre, nearly lost his life" (*Surgeon-Major J. E. T. Aitchison, Simla*).

DIANTHUS, *Linn.; Gen. Pl., I., 144.*

Dianthus Caryophyllus, *Linn.; Fl. Br. Ind., I., 214.*

377

THE CLOVE PINK and CARNATION.

Habitat.—In the *Flora of British India,* the Panjáb, at Attok, is mentioned doubtfully as a locality for this plant.

The Pink and Carnation are cultivated all over India in gardens, especially on the hills, and **D. chinensis,** *Linn.,* is practically a naturalised weed of cultivation, springing up in native gardens from self-sowings, all over the plains. The young flower-buds of these plants, from their resemblance to a nail (*Clou,* FR.; *Clout,* ENG.), were early known as cloves, and the leaves being like those of a CAREX obtained the name CARYOPHYLLUS from their cutting the hand and giving origin to caries or sores. The cloves of modern commerce by a play on these names became **Caryophyllus aromaticus,** which see, Vol. II., p. 202.

Diaphoretics, see Medicine.

DICHOPSIS, *Thw.; Gen. Pl., II., 658.*

378

A genus of trees or shrubs containing some 30 species, natives of South India, the Malay peninsula and islands, with one species in Samoa. India, as accepted by Sir J. D. Hooker in the *Flora of British India,* possesses fifteen species, of which only three, or perhaps four, are natives of India proper, the others being either Malacca or Ceylon plants. By tapping these trees a gum-like juice is obtained, the better qualities of which constitute the Gutta-percha of commerce (see **D. Gutta**). It may here be added that, while the more elastic substance—

India-rubber—is obtained from several widely different plants, Gutta-percha proper is only obtained from the Sapotacea family and mostly from one or two species of DICHOPSIS; the inferior forms obtained from other plants can at most be called Gutta-percha substitutes.

379   **Dichopsis elliptica,** *Benth. ; Fl. Br. Ind., III., 542 ;* SAPOTACEÆ.

Syn.—BASSIA ELLIPTICA, *Dals.;* ISONANDRA ACUMINATA, *Drury, Useful Plants (not of Gardner).*

Vern.—*Panchoti palu,* BOMB.; *Panchoti pala,* TAM.; *Panchonta,* KAN.

References.—*Beddome, Fl. Sylv., t. 43 ; Gamble, Man. Timb., 242; Dals. & Gibs., Bomb. Fl., 139 ; Cleghorn, Memorandum on Panchotee or the Indian Gutta tree; Drury, U. Pl., 260 ; Lisboa, U. Pl. Bomb., 90 ; Cooke, Oils and Oilseeds, 8 ; Balfour, Cyclop., I., 289 ; II., 387 ; Indian Forester, III., 24 ; VIII., 208 ; Kew Report for 1881, p. 44 ; Man. Coimbatore Dist., 41 ; Madras Man. of Administ., Vol. II., 105 ; Tropical Agriculturist, 1883, p. 960.*

Habitat.—A large tree of the Western Ghâts, extending from Bombay to Kanara, and ascending to an altitude of 4,000 feet. **Beddome** says it is a common tree in all the moist sholas of the Western Ghâts, also in the Wynaad, Coorg, Travancore, &c.

GUM.   **Gum.**—This tree yields the Indian Gutta-percha or *pálá* gum, a sub-
380   stance which has attained a certain amount of popularity as an adulterant for Singapore Gutta. It is stated that as much as 20 to 30 per cent. may be used without the characteristic properties of the Gutta-percha being destroyed. To Mr. **Lascelles** and **General Cullen** should be attributed the honour of having brought this substance prominently before the public; the latter gentleman recommended, amongst many other uses, its adaptability as a cement. **Balfour** describes the juice as obtained on tapping the trees—a process quite different from that resorted to in the Malay Peninsula with Gutta-percha. The following passage from *Drury's Useful Plants of India* gives a full account of this substance :—

" The exudation from the trunk, which has some similarity to the gutta-percha of commerce, is procured by tapping, and the quantity is not inconsiderable; but it would appear that the tree requires an interval of rest of some hours, if not days, after frequent incision. ' In five or six hours,' says **General Cullen,**' ' upwards of 1½℔ (more than a catty) was collected from four or five incisions in one tree.' Again, he writes in the same month (April) : ' Incisions were made in forty places, at distances nearly 3 feet apart, along the whole trunk. The quantity produced was 2½ *dungalies* (a *dungaly* is about half a gallon), the reeds were placed again, but in the evening no more milk was found; but the bark is thin, and the juice soon ceases to flow, although there is plenty of it in the tree.' The gum when fresh is of a milky white colour, the larger lumps being of a dullish red. Specimens of the gum were forwarded to England to be reported on by competent persons, and on an analysis of its properties, Messrs. **Teschemachar & Smith** stated : ' It is evident that this substance belongs to the class of the vegetable products of which caout-chouc and gutta-percha are types, and that it greatly resembles ' bird-lime ' in its leading characteristics, but in a higher degree. It is evident that, for water-proofing purposes, it is (in its crude state) unfit ; for, although the coal-tar, oil of turpentine paste, might be applied to fabrics, as similar solutions of caoutchouc now are, and a material obtained impervious for a time to wet, yet that, owing to the capacity of this substance to combine with water and become brittle in consequence at ordinary temperatures, such a water-proofed fabric would become useless very quickly. We do not of course in any way imply that, in the hands of some inventors, this and other difficulties to its useful application may not be over-

D. 380

The Gutta-percha.	(*G. Watt.*)	DICHOPSIS Gutta.

GUM.

come. Although unfit for waterproof clothing, moveable tarpauling, and the like, yet it might be usefully employed to waterproof fixed sheds, or temporary erections of little cost covered with calico or cheap canvas; but there are already a numerous class of cheap varnishes equally adapted for such a purpose, so that as a waterproofing material, it is but advisable, for the present, to look upon it as useless.

"Its perfume when heated might possibly render it of some value to the pastille and incense-makers.

"Its bird-lime sticky quality might be made available by the game-keeper and poacher in this country for taking vermin and small birds; we almost doubt whether a rabbit, hare, or pheasant could free itself, if hair, feathers, or feet came in contact with it. We think it might be use-fully and more legitimately employed by the trapper for taking the small fur-bearing animals; turpentine would cleanse the soiled furs. The only extensive and practical use, however, in this country to which we at pre-sent think it may probably be with advantage applied, is as a sub-aqueous cement or glue. We beg to forward you some deal-wood glued together with this substance melted and applied hot, which we have now kept under water for several days, and two fragments of glasses which have been similarly treated. You will observe that the cement has hardened at the edges, but probably without injury to its cementing properties. We have no reason to think that it would not rot under water more rapidly than wood does, but experience must be the sole guide here. We have reason to think such a glue or cement would be readily tried, and if found good, employed by joiners and others."

Oil.—It yields the "Gutta-percha Seed Oil."

Structure of the Wood.—Beddome says the timber is hard and not unlike Sál in its grain; it takes a good polish, is much employed by planters for building purposes, and might be used for furniture.

OIL.
381
TIMBER.
382

383

Dichopsis Gutta, *Bth. & Hook. f.; Fl. Br. Ind., III., 543.*

GUTTA-PERCHA.

Syn.—ISONANDRA GUTTA, *Hook.*

Vern.—*Niatú, taban,* MALAY.

References.—*Brandis, For. Fl., 286; Gamble, Man. Timb., 242; Christy, Com. Pl. and Drugs, 1885, No. 8, p. 17; Cooke, Oils and Oilseeds, 14; Balfour, Cyclop., II., 388; Smith, Dic., 204; Kew Off. Guide to the Mus. of Ec. Bot., 38; Kew Off. Guide to Bot. Gardens and Arboretum, 69; Madras Manual of Administration, Vol. I., 360; Indian Forester, VIII., 205; Journal, Agri. Hort. Soc. Ind., Vol. II., 101 (Analysis of Gutta-tuban); III., 146; Vol. IV., 59; app., Vol. IV., 221; VI. app., 50; Vol. X., Correspondence and Selections, p. 13.*

Habitat.—A tree attaining a height of 40-80 feet, met with in Malacca and Singapore, and distributed to Sumatra. It is said to flourish best on the hill-sides around Perak, but it is rapidly being exterminated from all accessible situations. Since the process of extracting the sap necessitates the killing of the tree unless practised under the most scientific system of forest conservancy, in which periodic renewal accompanies felling, exter-mination becomes a matter of time, and it is feared this is what to a large extent has actually taken place.

Gum.—This is said to afford the best quality of Gutta-percha. The following brief abstract will be found to set forth the main facts known regarding this substance, and to exhibit the plants which either yield the commercial article or which might be utilised as substitutes. Most of these are either grown in India or might easily be introduced.

Oil.—The oil from this plant was reported on by the Madras Jurors at

GUM.
384

OIL.
385

D. 385

TIMBER.
386

387

the Exhibition of 1857. A vegetable butter is said in Sumatra to be prepared from the seeds.

Structure of the Wood.—Soft, fibrous, spongy, of a pale colour, and marked with black lines.

GUTTA-PERCHA.

References.—*Kew Report for 1881 gives a long account of Gutta-percha, which has been freely consulted in drawing up the present abstract; Spons' Encyclopædia; Journal of the Agri. Hort. Soc.; Government of India Proceedings; Baden Powell, Panjab Products; Indian Forester, Vol. VIII., 205-209; Encyclopædia Britannica, Vol. XI.; Tropical Agriculturist (numerous articles in the volumes for the past four or five years); British Manufacturing Industries (Stanford's series) by Collins; Society of Arts for 1844; Dr. Montgomerie's Lecture on the Discovery of Gutta-percha; Balfour, Cyclopædia of India; M. C. Naudin, in Bulletin, Minist. de L'Agri., Paris, Dec. 1888, &c., &c.*

A commercial term for the inspissated milky sap of several plants, of which nearly all (or at least all the important ones) belong to the natural order SAPOTACEÆ. The word gutta-percha is of Malayan origin; it signifies the gum or *gutta* of the tree known as *percha*. The gutta-percha of commerce is, however, chiefly the *gutta-taban* or **Dichopsis Gutta**, a tree of Perak. As it reaches the market the gum is largely adulterated, often consisting of the inspissated saps of some five or six different plants mixed together, of which a fig and a bread fruit tree, which yield inferior India-rubbers, are probably the most frequently used. Gutta-percha seems to have come into commercial notice in Europe in the year 1845 (from the Straits), its important uses soon causing an immense demand. It was probably known as *maser-wood* at a much earlier date, and in 1822 **Dr. W. Montgomerie** experimented with it, and in 1844 read a paper on the subject before the Society of Arts, London. From that date it became a regular article of commerce. It is principally employed in coating telegraphic cables, owing to its being a perfect insulator, while it is of such a nature as to withstand, in a remarkable degree, the action of water. It is in fact much more durable when entirely submerged than when exposed to a moist atmosphere. About 10 years have been stated to be the period it will withstand the variations of climate in the air; 20 years if enclosed in iron tubes; but 20 years, when it has been submerged, have no appreciable effect upon the article. This is due to the fact that under the influence of light and air it slowly becomes oxidised, being converted into a brittle resin soluble in hot alcohol. This is the great defect of Gutta-percha, for, when oxidised, it loses its plastic nature. Under water and at great depths in the sea, it is, however, very durable, hence its value as an insulator for submarine cables. Chemically, gutta-percha is almost identical with India-rubber, but it differs physically, being tough and inelastic.

Since the date Gutta-percha was made known to Europe, perhaps no substance has developed more rapidly, and, with India-rubber, its uses may be said to be so many and so important as to make it perfectly indispensable to commerce.

The immense demand has caused an extended enquiry all over the globe with the view of expanding the field of supply or discovering substitutes in sufficient abundance likely to meet the demand without endangering the extermination of the supply of plants. As far as Gutta-percha is at present concerned, there cannot be a doubt but that a few years more will suffice to eradicate the supply from the Straits Settlements. It has been estimated that to meet the shipments of gutta-percha from Sarawak alone during the years 1854-75 over 3,000,000 trees were felled. Great Britain imported in 1880 from the Straits Settlements 62,862 cwt. of gutta-percha, valued at £505,821. The expansion of the trade may be said

Gutta-percha Substitutes.　　(*G. Watt.*)　　**DICHOPSIS.**

to be demonstrated by the fact that in 1876 the imports were only 19,665 ┃ **Inspissated**
cwt., but were two years later 49,387 cwt. The present total annual trade ┃ **Sap.**
in gutta-percha has been estimated at 10,000,000℔. The future prospects
are alarming, and such that, not only should the Colonial Government
take the most decided steps within its power for the preservation of the
plants, but a response to the demand should, if possible, be made in India.
There does not, for example, seem any very great reason why our coast
forests should not, to some extent, be made to yield gutta-percha. There
is nothing to show that the plant would not thrive in many parts of India
if once successfully introduced. Gutta-percha sells at from 6*d.* to 3*s.* and
6*d.* a ℔.

Another interesting feature, which the increasing demand for Gutta-
percha must solve, is the possibility (in a simple way) of transforming the
milky saps of some of the numerous wild plants of India so as to render
these serviceable as gutta substitutes. It need only be here added that the
difference between Gutta-percha and India-rubber is of a practical more
than chemical nature, and consequently from the juices not having been
severally tested and reported on, it is impossible to draw up a list of plants
of the former that may not hereafter be found to include some of the latter.
The reader should, therefore, consult the account given under India-rubber,
as well as the detailed descriptions furnished of each plant in their respective alphabetical places in this work. The following abstract may, however,
prove useful :—

1. **Achras Sapota,** *Linn*; SAPOTACEÆ. (*See Vol. I., A., No. 376, page 80.*)
THE SAPODILLA or SAPOTA TREE.
Largely cultivated on account of its fruit in Bengal; yields the Mexican chicle-gum, a substance closely resembling gutta-percha. In the
Journal of the Agri.-Horticultural Society of India, Vol. III., 147, a long
account of this Gutta-percha will be found including its chemical analysis.
A passage from the account there given may be here reproduced :—"Its
juice differs very remarkably by the absence of adhesiveness, to which
peculiarity, indeed, it owes its value. This promises to be considerable;
for a vegetable product which softens by hot water, while at the same
time it is capable of being moulded into any shape, that afterwards hardens
(in which state it is not acted on by a hot or moist climate), so as to be preferable to horn for the handles of axes, is capable of extensive application."

2. **Alstonia scholaris,** *R. Br.*; APOCYNACEÆ. (*See Vol. I., A., No. 872,*
page 198.)
One of the many forms of this tree has recently been discovered to be
the source of the *Gutta-pulei* of Singapore. The *Satian* has long been
known in India as yielding an inferior India-rubber, but it is doubtful if
this could be regarded as anything more than an adulterant for Gutta-
percha.

3. **Bassia Mottleyana,** *De Vriese*; SAPOTACEÆ. (*See Vol. I., B., No. 281,*
page 416.)
A tree of Malacca and Borneo, known in the vernacular as *kotian.*
Mr. Mottley says that this tall and straight tree, when wounded, yields a
copious flow of milky juice which hardens to a brittle, waxy resin, readily
softened by heat. This has been described as an inferior kind of guttapercha.

4. **Calotropis gigantea,** *R. Br.*; ASCLEPIADEÆ. (*See Vol. II., No.*
171, page 35.)
The *madar* or *akanda*, a plant scarcely to be distinguished from the
following species, the properties and uses of which are identical, and these
plants may therefore be discussed jointly. **C. gigantea** is most abundant
in the Lower Provinces and Eastern India, while **C. procera** is the species
chiefly met with in Upper or Northern and Central India.

D. 387

DICHOPSIS
Gutta.

Commercial Gutta-percha.

---

5. **Calotropis procera,** *R. Br.*

Reference.—*Agri-Hort. Soc. Ind., VIII., 107, 226, 231.*

The inspissated and sun-dried milky sap from the stem resembles Gutta-percha. The *madar* is, in fact, the most interesting and most hopeful plant not belonging to the natural order SAPOTACEÆ, which can be said to yield a substance resembling Gutta-percha ever likely to obtain a commercial reputation as a Gutta-percha substitute. **Mr. Liotard** publishes, in his Memorandum on the Materials in India suitable for the Manufacture of Paper, the opinion of **Professor Redwood** upon *Madar-gutta.* The Professor considers that it possesses many properties in common with the Gutta-percha of commerce. The specimen so reported upon was collected by **Captain G. E. Hollings,** Deputy Commissioner, Shahpur (in the Panjáb) in the year 1853, little more than one year after the date of the original discovery of this Gutta. We have learned nothing further in 30 years, and uncountable riches of fibre and gum may have all the while been wasting along every roadside and over every rubbish heap.

6. **Dichopsis elliptica,** *Benth.*; SAPOTACEÆ.

The *panchoti,* a large tree of the Western Ghâts; yields the Indian gutta-percha.

7. **D. Gutta,** *Benth. & Hook.*

It is said that the finest quality of all the guttas is the *Gutta-susu,* obtained from a botanically undetermined plant. This is very scarce, but the best commercial quality is that obtained from **D. Gutta.**

There are two forms, one with red flowers known as *tuban-merut,* and the other with white flowers, *tuban-pateh.* The young trees require shade and a rich well-drained soil, hence the preference for hill-sides. No special period is observed for collecting the gutta, but it is said to be generally collected at the close of the rains. Full-grown trees, say, 20 years old, are hewn down and tapped all along at distances of 18 inches. The yield is so variously stated that it does not seem desirable to quote the contradictory reports. A mistake seems often to have been made between the yield of sap, the yield of fresh gutta-percha, and the yield of dry gutta-percha. The weight of sap would of course be far greater than that of gutta-percha, and on drying the commercial article loses as much as 30 per cent. of its weight. It seems probable that the yield of dry gutta-percha per tree may average from 2 to 14℔. The sap is of course drawn from the middle layer of the bark, the region of laticiferous vessels. The fresh milk or latex appears under the microscope as an emulsion, a clear liquid having in it minute globules of caoutchouc. It is supposed that the caoutchouc is held in suspension in the juice through the agency of ammonia. At all events, many of the fresh milky saps like that of gutta-percha have an ammoniated odour and the addition of a little ammonia prevents the natural coagulation due to evaporation. The value of a Gutta-percha or India-rubber depends on the proportion of caoutchouc granules which it contains and on the relative absence of certain oxidised, viscid, resinous substances, soluble in alcohol. The formation of such materials is greatly prevented by a rapid evaporation of the milk. The crude sap, if in small quantities, may be concreted by rubbing between the hands, but it is more expeditiously accomplished by boiling.

Singapore and Penang are the chief collecting depôts.

8. **D. obovata,** *Clarke.*

An evergreen tree of Tenasserim, extending to Malacca and Penang. According to **Kurz** it yields gutta-percha.

9. **D. polyantha,** *Benth.*

Vern.—*Tali,* BENG.; *Sill-kurta,* CACHAR.

D. 387

Milky Saps.

A tree, 30 to 40 feet in height, occurring in Sylhet, Chittagong, and Pegu. Kurz remarks that it produces a good quality of gutta-percha in large quantities.

10. **Gutta Sundek,** the second best commercial form of gutta-percha, is at present un-identified. It occurs abundantly in the Malay Peninsula. M. Beauvisage named it as **Keratephorus Leerii,** *Husk.*, but the Kew authorities regard this as incorrect, and **Dr. Trimen,** who, in the Ceylon Botanic Gardens, has succeeded in obtaining young seedlings, thinks it may prove a species of PAYENA.

11. **Dyera costulata,** *Hook. f.*; APOCYNACEÆ, and

12. **D. laxiflora,** *Hook. f.*

Trees which inhabit the forests of Malacca, Singapore, and Sumatra. They are said to yield the *gutta-jelutong* of commerce, a form of India-rubber.

13. **Euphorbia trigona,** *Haworth*; EUPHORBIACEÆ.
    Syn.—E. CATTIMANDOO, *Elliot; Fl. Br. Ind., V., 256.*
    Vern.—*Katimandu,* TAM.

This yields the *Catimandu* cement of the Madras Presidency, used to fasten knive handles. It contains sufficient caoutchouc to make it a profitable source of supply, if not of india-rubber, at least of gutta-percha. Specially recommended by **Sir Walter Elliot** at the Great Exhibition, 1851, where a medal was awarded to the exhibitor.

14. **E. neriifolia,** *Linn.*
    Syn.—E. LIGULARIA, *Roxb.; Fl. Br. Ind., V., 255.*
    Vern.—*Mansa-sij* or *Sij.*

Yields a milky sap which, on drying, much resembles gutta-percha, and for which there seems every probability of its being used as a substitute. See a long account of the properties of this gutta-percha in the *Jour. Agri.-Hort. Soc.,* VIII., pp. 223—226.

15. **E. pulcherrima,** *Willd.* (= **Poinsettia pulcherrima,** a common garden plant with large red bracts.)

Dr. Riddell recommends this, as also the next species, as suitable for the preparation of gutta-percha.

16. **E. resinifera** (described in *Smith's Dictionary of Economic Plants*).

This plant yields the gum known as **Euphorbium,** now largely employed as an anticorrosive paint for the bottoms of ships; it comes chiefly from Morocco and Barbary. Its resisting the action of water depends upon its resemblance to gutta-percha.

17. **E. Tirucalli,** *Linn.; Fl. Br. Ind., V., 254.*
    Vern.—*Lanka sij,* BENG.; *Sehud,* HIND.; *Tiru kalli,* MAL.; *Sha-soung-leknyo,* BURM.

A small tree cultivated throughout India and used as a hedge. **Dr. Riddell** states that this yields a fairly good gutta-percha.

18. **Mimusops Balata,** *Gærtn. f.*

This tree is somewhat allied to the SAPOTA, but it yields more freely a gutta-percha sap. It is a native of British, French, and Dutch Guiana, British Honduras and Brazil, flourishing best on river-banks. It is said to afford the best of all known substitutes for the true gutta-percha of commerce and to be especially useful for submarine cables. The sweet milky sap, obtained from it, was at first used as food, by the natives, but in 1860 it was employed in the preparation of its contained caoutchouc, since which date a considerable trade has developed in the article. See *Tropical Agriculturist, 1883, p. 959; Indian Agriculturist, Nov. 20th, 1886; Jour. Soc. of Arts, Feb. 26th, and March 4th, 1864; Bulletin, Ministére de L' Agri., Paris, Dec. 1888.*

DICHOPSIS poyantha.	Gutta-percha.

GUM.

19. **Payena Maingayi,** *C.B.C.*; SAPOTACEÆ.

A tree of Malacca and Penang, said by Maingay to abound in gutta-percha; also **P. Leerii,** from which it is stated the *Gutta-Sundek* is obtained.

388

Dichopsis Helferi, *Clarke ; Fl. Br. Ind., III., 542.*

Habitat.—A closely allied tree to **D. obovata,** and may be the plant referred by **Kurz** to that species. It is a native of Tenasserim and Tavoy.

GUM.
389

Gum.—Is reported to yield a good quality of Gutta-percha.

390

D. obovata, *Clarke ; Fl. Br. Ind., III., 542.*

Syn.—ISONANDRA OBOVATA, *Griff.*

References.—*Kurz, For. Fl. Burm., II., 120; Balfour, Cyclop., II., 387.*

Habitat.—A large tree which **Kurz** says occurs in the Tropical forests of Tenasserim, but to which the *Flora of British India* assigns the habitat of Malacca and Singapore, remarking that imperfect specimens of what appears to be the plant were collected by **Falconer** at Moulmein.

GUM.
391

Gum.—Kurz writes that it yields a fair sort of Gutta-percha.

392

D. polyantha, *Benth. and Hook. f.; Fl. Br. Ind., III., 542.*

Syn.—BASSIA POLYANTHA, *Wall;* ISONANDRA POLYANTHA, *Kurz (ii., 119).*

Vern.—*Táli,* BENG.; *Sill-kurta,* CACHAR; *Thainban,* MAGH.

References. –*Gamble, Man. Timb., 242; Ind. Forester, IX., 427; XI., 319.*

Habitat.—A moderate-sized evergreen tree, met with in Cachar, Chittagong, and Arakan.

GUM.
393

Gum.—Kurz says it produces a good quality of Gutta-percha in large quantities—probably little inferior to that of Singapore. The Conservator of Forests, Assam, in a letter to the Inspector General, dated 10th November 1884, reported that this tree was well known to the people of Cachar and Sylhet, but although he had "often asked the people about its yielding Gutta-percha," he had "never heard of it being extracted or made use of, except that it is mixed sometimes with India-rubber, and in doing so the people of course sell themselves, as they always get much less for mixed rubber than for pure. I have referred the matter to the Deputy Commissioner of Sylhet and the Cachar Forest Officer to make sure." "I have ordered the Cachar Forest Officer to make an experiment to ascertain how much a tree will yield and to let me have the stuff collected to allow of its being valued.

"The following is the result of the above experiment, but the writer has not been able to discover the report, if obtained, of the commercial value of the Gutta-percha collected in Cachar : —

Yield.
394

"I had 36 trees tapped, giving a yield of 15 pounds of dry Gutta-percha. To ascertain the yield per tree I have recorded the yield of six trees, the tapping of which was personally superintended by me. The milk was weighed directly it was taken from each tree separately. Then the whole was boiled down in an iron pan over a slow fire. The result is that 6 seers 11 chattacks of milk yielded 2½ seers of Gutta-percha or one-third the weight of milk." The Forest Officer seems thus to have tapped the trees after the same manner as with India-rubber trees, whereas in the Gutta percha-producing regions the trees are felled. It is probable that a much larger yield would have been obtained had the Straits method been followed. This is not however mentioned by way of recommending the destructive system of felling the trees, but only to prevent unfavourable comparisons being drawn as to the yield.

D. 394

A domestic febrifuge—Dichroa, (G. Watt.)	**DICHROSTACHYS cinerea.**

It does not appear how often the trees were tapped, in other words, whether they yielded all that it was possible for them to do. At the same time the above experiment is instructive, each tree having on an average yielded a little over 2 seers of milk, one-third of which consisted of Gutta-percha. The average yield of true Gutta-percha from the felled trees has been variously stated, but it may be said to vary from 2, 4 to 7℔ per tree, the maximum recorded yield being 25℔ according to some writers, 50 according to others, and even 100 is given by one author. This seems highly improbable. (See the remarks, p. 106, regarding mistakes of yield arising from the' milk being spoken of in some reports, in others the fresh rubber, in a third the dried rubber.)

Food.—The FLOWERS are said to be eaten.

Structure of the Wood.—Red, hard, much valued in Cachar and Chittagong. Mann says it does not float, but he is probably referring to green wood. Major Lewin remarks that it is used in Chittagong for making beds, tools, &c., and is sawn into boards for the Calcutta market. For further information regarding Gutta percha see INDIA RUBBER.

Milky Saps.

FOOD Flowers. 395 TIMBER. 396

DICHROA, *Lour.; Gen. Pl., Vol. I., 641.*

Dichroa febrifuga, *Lour.; Fl. Br. Ind., Vol. II., 406;* SAXIFRA- [GACEÆ.

397

Syn.—ADAMIA CYANEA, *Wall.*|(*t. 213*); A. VERSICOLOR, *Fortune.*

Vern.—*Basak*, HIND.; *Basak, bansúk* (Gamble); *aseru* (Gimlette), NEPAL; *Gebokanak*, LEPCHA; *Singnamúk*, BHUTIA.

In an interesting report on the Economic Products of Nepal, **Dr. Gimlette** gives the above so-called Nepalese names (as in **Gamble**) as the Hindi names for this plant, and *Aseru* as the Nepalese.

References.—*Voigt, Hort. Sub. Cal.. 267; Gamble, Man. Timb., 172; Cat. Trees, Shrubs, and Climbers of Darjeeling, 38.*

Habitat.—An evergreen shrub, common in the forests of the Eastern Himálaya (5,000 to 8,000 feet), from Nepal to Bhutan, and in the Khásia Hills, above 4,000 feet.

Medicine.—The SHOOTS and the BARK of the roots are made into a decoction and used as a febrifuge by the Nepálese (*Gamble*). Dr. Gimlette says this drug is given in doses of five *mashas.*

Structure of the Wood.—White, moderately hard, with small pores and moderately broad, to very fine medullary rays.

Domestic Uses.—Employed by the Bhutias and Lepchas to burn at religious ceremonies.

MEDICINE. Shoots. 398 Bark. 399 TIMBER. 400 DOMESTIC. 401

DICHROSTACHYS, *DC.; Gen. Pl., I., 592.*

Dichrostachys cinerea, *W. & A.; Wight, Ic., t. 357; Fl. Br.* [*Ind., II., 288;* LEGUMINOSÆ.

402

Syn.—MIMOSA CINEREA, *Linn.; Roxb.;* DESMANTHUS CINEREUS, *Willd.;* ACACIA CINEREA, *Spreng.;* A. DALEA, *Desv.*

Vern.—*Vurtuli*, HIND.; *Kunlai, kunrat, kheri*, MHAIRWARA; *Kunlai, kanlai*, MERWARA; *Kheri*, AJMERE; *Khen*, RAJ.; *Segum kati*, MAR. & GOND.; *Vadatalla, vadatara (vedittalung kolindu*, in Ainslie), TAM.; *Veturu (velliluru konalu), yeltu (venuturu, veluturu, néla jammi, vanuturu* according to Elliot) TEL.; *Andara*, SING.; *Viravriksha* (according to Ainslie), SANS.

References.—*Roxb., Fl. Ind., Ed. C.B.C., 422; Brandis, For. Fl., 171; Beddome. Fl. Sylv., t. clxxxv.; Gamble, Man. Timb., 148; Thwaites, En. Ceylon Pl , 99; Dals. & Gibs., Bomb. Fl., 84; Aitchison, Cat. Pb. and Sind Pl., 53; Sir W. Elliot, Fl. Andh., 40, 131. 190 91, W. & A. Prod. (864), p. 278; Ainslie, Mat. Ind., II., 458; Drury, U. Pl., 181;*

D. 402

Royle, Ill. Him. Bot., 182; Liotard, Dyes, 33; Watson's Report, 18; *Balfour, Cyclop., 946; Raj. Gaz., 29; Indian Forester, Vols. III., 202; IV., 232; VIII., 30; XI.,466; XII., 33; App., 2; Gazetteer, N.-W. P. (Bundelkhand), Vol. I., 80; (Agra), Vol. IV., LXXI.*

Habitat.—A thorny shrub or small tree of the dry, stony hills of the N.-W. Provinces, Western and Central India, Rajputana, Madras, Ceylon, &c. Distributed to the Malay Islands, Northern Australia. Doubtfully distinct from D. nutans, a native of Tropical Africa.

GUM.
403
DYE.
404
FIBRE.
405
MEDICINE.
Shoots.
406
FODDER.
407
TIMBER.
408

Gum.—It is said to yield a gum, but of this nothing is known.

Dye.—The lac insect is often found on the tree.

Fibre.—Mr. **J. W. Cherry** of Salem, Madras, sent to the Calcutta International Exhibition a sample of a yellowish white good bast fibre which was said to have been obtained from this plant.

Medicine.—The young SHOOTS are bruised and applied to the eyes in cases of ophthalmia.

Fodder.—The leaves are mixed with corn and given to riding horses (*Ainslie*). It is supposed to free them from both *bots* and worms.

Structure of the Wood.—Heartwood red, extremely hard; weight 70 to 80℔ a cubic foot. Used for walking-sticks. It is, however, too small to be of much use, but is much valued for tent pegs.

DICLIPTERA, *Juss.; Gen. Pl., II., 1120.*

Several species of this genus are alluded to in the Gazetteers and other descriptive works on India. Some are cultivated in gardens, while others are referred to as wild. (*See Agra Gazetteer*, p. lxxvi; Sir **W. Elliot's** *Flora Andhrica*, pp. 38 and 183, for D. **parvibracteata**, the *Chiku velaga* of Telegu; Stewart's *Account of Hazara*, where D. **Roxburghiana** is said to be one of the more remarkable of the herbaceous plants (also *Bundelkhand Gazetteer*, p. 83, &c., &c.).

[THACEÆ.

409

Dicliptera Roxburghiana, *Nees, Fl. Br. Ind., IV.,* 553 ; ACAN-

Vern.—*Kirch, somni, lakshmana* (bazar name), PB.; *Bouna*, SIMLA.

References.—*Roxb., Fl. Ind., Ed. C.B.C., 42; Voigt, Hort. Sub. Cal., 492; Dalz. & Gibs., Bomb. Fl., 196; Aitchison, Cat. Pb. and Sind Pl., 113; Atkinson, Him. Dist., 373; Balfour, Cyclop., 946.*

Habitat.—According to the *Flora of British India* there are two forms of this plant— the one met with on the plains of India, the other on the hills. Regarding the former there seems little doubt, but with the latter it is quite otherwise. It is the hill plant alone which requires to be dealt with in this work, and this fact has necessitated the writer's examining the specimens in his private herbarium with as much care as the time at his disposal would admit of. A sample of the plant collected at Simla was by him sent to the authorities of the Royal Herbarium, Kew, the result being that it was pronounced "Dicliptera Roxburghiana, *Nees*, var.?" Presumably, it may be the plant described in the *Flora* of British India as var. **bupleuroides** (*sp. Nees in Wall., Pl. As. Rar.,III., p. 111*). The writer would be more disposed, however, to place the Simla plant in another genus than to amalgamate it with **D. Roxburghiana.** The following are the chief characteristics of the two plants as recognised by the writer :—

410

a D. **Roxburghiana,** *Nees.*

Syn.—This is apparently not the **Justicia chinensis,** *Linn.*, as described by **Roxburgh**, since that plant is said to have, among other distinctive characters, cordate leaves.

A tropical species, specimens of which in the writer's herbarium are in flower, and dated February to May. Leaves with a short petiole ($\frac{1}{4}$ to $\frac{1}{2}$ inch), nearly glabrous.

| Dicoma—A strong bitter Febrifuge. | (*G. Watt.*) | DIDYMOCARPUS aromatica. |

Flower-clusters, sessile ; bracts obovate apiculate *tricostate*. Fruit long, flattened in *the plane of the septum* ; on dehiscence, severing into two valves, each with a portion of the ruptured septum down the middle which is seen to support the seeds.

β **D. bupleuroides,** *Nees* (the Simla plant).   411

A warm, temperate plant, ascending the hills to 6,000 feet in altitude, and flowering in August to October. Leaves with the petiole 1 to 1½ inches long; all parts very hairy or hirsute. Flower-clusters pedunculate ; bracts lanceolate-acuminate, the inner ones awl-shaped. Fruit not half the length of that of the above, *flattened at right angles to the plane of the septum* ; on dehiscence, the septum separates from the valves and rising up ejects the seeds as in **Rungia**.

Medicine.—The drug sold in Upper India under the name of *laksmana* MEDICINE. is the form β. It is said to be a useful tonic.   412

DICOMA, *Cass. ; Gen. Pl., II., 492.*

Dicoma tomentosa, *Cass. ; Fl. Br. Ind., III., 387 ;* COMPOSITÆ.   413

Vern.—*Navananji-cha-pála,* BELGAUM.

References.—*Dals. & Gibs., Bomb. Fl., 182; Aitchison, Cat. Pb. and Sind Pl., 81 ; Royle, Ill. Him. Bot., 248 ; Indian Forester, XII., app. 15.*

Habitat.—An herb or low shrub, with the branches clothed with white, cottony wool. It is met with in North-West India, the Western Peninsula, and Sind to Ava.

Medicine.—Dr. **Peters** of the Bombay Medical Service has kindly   MEDICINE. favoured the writer with a note on the medicinal uses of this plant. It is,   414 he writes, an agreeable strong bitter, used in Belgaum as a febrifuge, especially in the febrile attacks to which women are subject after child-birth.

DICTAMNUS, *Linn. ; Gen. Pl., I., 287.*

Dictamnus albus, *Linn. ; Fl. Br. Ind., I., 487 ;* RUTACEÆ.   415

Syn.—D. FRAXINELLA, *Pers. ;* D. HIMALAYANUS, *Royle, Ill., 156, t. 29.*

References.—*U. S. Dispens., 15th Ed., 1634; Royle, Ill. Him. Bot., 156, t. 29.*

Habitat.—A strong smelling shrubby plant, met with on the temperate Western Himálaya from Kashmír to Kunawar (6,000 to 8,000 feet); very common in Pangí.

Medicine.—Indian writers do not appear to have paid much attention   MEDICINE. to this plant. The bark of the root was once upon a time a favourite aro-   416 matic bitter. **Storck** prescribed it for most nervous diseases, also for intermittent fever, amenorrhœa, hysteria, &c. The writer has repeatedly been told by the hill people that the plant was used medicinally, but could never discover for what purpose

DIDYMOCARPUS, *Wall. ; Gen. Pl., II., 1021.*
[GESNERACEÆ.

Didymocarpus aromatica, *Wall. ; Fl. Br. Ind., IV., 347 ;*   417

Vern.—*Kumkuma,* HIND. ; *Kumkuma, ranigovindhi,* NEPAL.

References.—*Thwaites, En. Ceylon Pl., 207; O'Shaughnessy, Beng. Dispens., 478 ; Atkinson, Him. Dist., 368 ; Royle, Ill. Him. Bot., 294.*

Habitat.—A succulent herbaceous plant, met with in Nepál and Kumaon.

Perfumery.—The whole plant is said to be used as a perfume. No PERFUMERY. subsequent author has alluded to this fact since **Wallich** first made it   418 known, and it may therefore be added as a caution against possible errors that the word *Kum-kuma* is the Sanskrit for saffron (**Crocus sativa**).

DILLENIA aurea.	Dillenia.

MEDICINE.
419

Medicine.—Wallich wrote that it was used in Nepál as an aromatic medicine, but Dr. Gimlette, who furnished the writer with a most interesting collection of the Nepál medicinal plants, was apparently unacquainted with this drug, from which circumstance it may at least be assumed to be unimportant.

DIGERA, *Forsk; Gen. Pl., III., 28.*

420

Digera arvensis, *Forsk.; Fl. Br. Ind., IV., 717; Wight,*
[*Ic., t. 732;* AMARANTACEÆ.

Syn.—D. MURICATA, *Mart.*
Vern.—*Luta mahawria, gungatiya,* BENG.; *Kari gandhari,* SANTAL; *Das,* BIJNOR; *Tartara, tandala, leswa,* PB.; *Tandala,* SIND; *Getan,* BOMB.; *Chenchali kúra, chanchali kúra,* TEL.
References.—*Roxb., Fl. Ind., Ed. C.B.C., 226; Voigt, Hort. Sub. Cal. 114; Thwaites, En. Ceylon Pl., 249; Dals. & Gibs., Bomb. Fl, 218.; Stewart, Pb. Pl., 182; Aitchison, Cat. Pb. and Sind Pl., 129; Flora Andhrica by Sir Walter Elliot, 34, 36; Dymock, Mat. Med. W. Ind., 2nd Ed., 889; Murray, Pl. and Drugs, Sind, 102; Lisboa, U. Pl. Bomb., 361; Atkinson, N.-W. P. Econ Prod., Pt. Foods, 91, 97; Indian Forester, XII., App. 20.*
Habitat.—A small annual herb of the plains of Bengal and North-West India, South Deccan, Concan, Mysore, and the Carnatic, to Peshawar and the Salt Range. Distributed on the one side through Burma to Ceylon, and on the other to Beluchistan, Afghanistan, Arabia, and Africa.

FOOD.
421

Food.—It serves as a pot-herb. Leaves and tender tops are also used by the natives in their curries.

FODDER.
422
423

Fodder.—Used as fodder in South Baluchistan.

Digitaria.—A genus of grasses, the species of which have been reduced to **Panicum,** *Linn.* Several species are alluded to as met with in the Banda District, and **D. sanguinale (Panicum sanguinale,** *Linn.*) is specially alluded by **Stewart** in his account of Hazara.

Dikamali (or Decamali) Resin; see **Gardenia lucida,** *Roxb.*

Dilivaria ilicifolia, *Nees;* see **Acanthus illicifolia,** *Linn.;* ACANTHACEÆ;
[*Vol. I., A., No. 324.*

Dill, see **Peucedanum graveolens,** *Benth.;* UMBELLIFERÆ.

DILLENIA, *Linn.; Gen. Pl., I., 13.*

424

Dillenia aurea, *Smith; Fl. Br. Ind., I., 37;* DILLENIACEÆ.
[OBOVATA, *Blume.*

Syn.—D. ORNATA, *Wall.;* D. SPECIOSA, *Griff., Notul. IV., 703;* COLBERTIA
Vern.—*Dheugr,* NEPAL; *Chamaggai,* N.-W.P.; *Byúben (sen-bwon,* according to **Mason**), BURM.
References.—*Brandis, For. Fl., 2; Kurz, For. Fl. Burm., I., 20; Gamble, Man. Timb., 3; Mason, Burma and Its People, 408, 532, 741.*
Habitat.—A large tree of Nepál, Bhután, Bengal, Burma, and the Andaman Islands; distributed to Java, Borneo &c. Mason speaks of this tree at Maulmain as being highly ornamental. The visitor in February, he says, has "his attention arrested by a tree without leaf, but covered with large gaudy yellow flowers."

TIMBER.
425

Structure of the Wood.—Grey, beautifully mottled, hard, close-grained; weight from 45 to 49℔ a cubic foot.

426

D. bracteata, *Wight, Ic., t. 358; Fl. Br. Ind., I., 37.*

Syn.—D. REPANDA, *Roxb., Fl. Ind., Ed. C.B.C., 452;* WORMIA BRAC-TEATA, *Beddome, t. 115.*

D. 426

			DILLENIA
Dillenia, the Chalta.		(*G. Watt.*)	parviflora.

Habitat.—A handsome tree of the Western Peninsula, especially at Mysore and Coimbatore.

Properties and Uses.—Practically the same as those recorded under the other species.

USES.
427

Dillenia indica, *Linn.; Fl. Br. Ind., I., 36.*

428

Syn.—D. SPECIOSA and ELLIPTICA, *Thunb.; Beddome, t. 103.*

Vern.—*Chálta,* HIND.; *Cháltá, hargesa,* BENG.; *Korkot,* SANTAL; *Chilta,* MONGHYR; *Panpui,* GARO; *Chalita, otengah,* ASSAM; *Rai, oao,* URIYA; *Ramphal,* NEPAL; *Phamsikol,* LEPCHA; *Thapru, chauralesi,* MAGH; *Mothe karamala, mota karmel, karambel,* BOMB.; *Mota karmal, karmbel,* MAR.; *Uva,* TAM.; *Uva, pedda kalinga (kalinga,* Elliot), TEL.; *Bettakanagala, kadkanagula,* KAN.; *Syalita,* MALAY; *Thabyú,* BURM.; *Carllow,* TALEING; *Hondapara,* SING.; *Bhavya* (according to Dutt), *ruvya* (Birdwood), SANS.

References.—*Roxb., Fl. Ind., Ed. C.B.C., 451; Brandis, For. Fl., 1; Kurz, For. Fl. Burm., I., 19; Gamble, Man. Timb., 2; Dalz. & Gibs., Bomb. Fl., 2; Elliot, Fl. Andh., pp. 79, 187, 148; Rev. A. Campbell, Econ. Prod. of Chutia Nagpur, No. 8782; Mason, Burma and Its People, pp. 532, 740; U. C. Dutt, Mat. Med. Hind., 294; Dymock, Mat. Med. W. Ind., 2nd Ed., 890; Lisboa, U. Pl. Bomb., 1, 143; Athinson, Econ. Prod., Pt. V., 43; Smith, Dic., 154; Jour. Agri. Hort. Soc., 1885, Vol. VII., Pt. III., New Series, 276; Vol. XIII., 345; Gasetteer of Orissa, II., 179, App. VI.; Mysore and Coorg, I., 57; N.-W. P., IV., lxvii., X., 716; Indian Forester, I., 86; V., 214, 497; VI., 240; VIII., 415, 438; X., 33; XI., 230; XIV., 297; Official Note on the Condition of the People of Assam.*

Habitat.—A large evergreen tree of Bengal, Central and South India, and Burma; often planted. Distributed through the Eastern Peninsula from Sylhet to Singapore. Rare on the plains of Northern or Western India, but occurs along the base of the hills from Kumáon and Garhwál eastward, and becomes plentiful from South Kanara southwards.

Fibre.—In the Hazaribagh District the *Tasar* silk-worm is said to feed on this plant, and in an article on the trees of Cachar (*Agri. Hort. Soc. Jour., XIII., p. 345*) the *Atlas* silk-worm is also said to feed on these leaves.

Silk-worm.
429

Medicine.—The JUICE of the fruit, mixed with sugar and water, is used as a cooling beverage in fevers, and as a cough mixture. The BARK and the LEAVES are astringent and are used medicinally. The FRUIT is slightly laxative, but is apt to induce diarrhœa if too freely indulged in. (*Roxburgh, Royle, Drury, &c.*)

MEDICINE.
Juice.
430
Bark.
431
Leaves.
432
Fruit.
433

Food.—The fruit is large, about 3 inches in diameter, and is surrounded by fleshy accrescent calyces, which, when the fruit is full grown (in February), have an agreeably acid taste, and are eaten by the natives, either raw or cooked—chiefly cooked in curries. They are also made into a pleasant jelly. The acid juice sweetened with sugar forms a cooling drink.

FOOD.
434

Structure of the Wood.—Red with white specks, close-grained; moderately hard. It is used to make helves and gunstocks, and in construction; and is said to be durable under water. It makes good fire-wood and charcoal. Weight 40 to 45℔ a cubic foot.

TIMBER.
435

D. parviflora, *Griff.; Fl. Br. Ind., I., 38.*

436

Vern.—*Lingyau,* BURM.

Habitat.—A tall, deciduous tree, met with in the forests of Tenasserim, Mergui, Pegu, and the Andaman Islands.

Properties and Uses.—Same as those recorded under the other species.

USES.
437

I

DINDIG A Gum.	Dillenia.

438

Dillenia pentagyna, *Roxb.; Fl. Br. Ind., I., 38.*

Syn.—D. AUGUSTA and PILOSA, *Roxb.;* COLBERTIA COROMANDELINA, *DC.;* C. AUGUSTA, *Wall.*

Vern.—*Agar,* MONGHYR; *Karkotta,* BENG.; *Korkotta, rai,* KOL. and MAL. (S. P.); *Korkot,* ORAON and SANTAL; *Akshi, daine-oksi, okoi,* ASSAM; *Rai,* URIYA; *Tatri,* NEPAL; *Shukni,* LEPCHA; *Akshi,* MICHI; *Akachi, uchkai,* GARO; *Kallei,* GOND; *Mirchi,* BAIGAS; *Aggai,* OUDH; *Suaruk,* MELGHAT; *Pashkouli,* RAJBANSHI; *Kallai, suha-rúk,* (Bori), C. P.; *Karamala, kanagalu,* BOMB.; *Kanagalu, karamal,* or *karmal,* MAR.; *Kanagala,* KANARA; *Malé geru,* COORG; *Rai, pinnai, noi-ték,* TAM.; *Rawadan, chinnakalinga,* TEL.; *Machil, kaltega, kadkanagola, kanagole,* KAN.; *Zambrún,* MAGH.; *Zimbyun, zengbywoon,* BURM.

References.—*Roxb., Fl. Ind., Ed. C.B.C., 451; Brandis, For. Fl., 2; Kurz, For. Fl. Burm., I., 21, 22; Gamble, Man. Timb., 3; Dals. & Gibs., Bomb. Fl., 2; Elliot, Fl. Andh., pp. 40, 163; Mason, Burma and Its People, 532, 741; Lisboa, U. Pl. Bomb., 1, 143; Burm. Gas., 2, 126; Mys. Gas., I., 48; III., 16; Forest Admn. Report, Chutia Nagpur, 1885, 28; Bombay Gazetteers, XIII., Pt. I. (Thana), 25; XV., Pt. I., 67; XVII., 25; Indian Forester, Vol. I., 78, 79, 84, 87, 88; II., 18; III., 200; IV., 292; VI., 125; VIII., 412; X., 325, 326; XI., 252, 485; XII., 311; XIII., 119; XIV., 199, 297.*

Habitat.—A deciduous tree of Oudh, Bengal, Behar, Assam, Central, South, and Western India, and Burma. In young trees the leaves are sometimes as much as two feet in length. The tree flowers in March and April, and is frequently associated with *sál* in the forest-clad lower hills of the Central table-land.

FIBRE.
439
Fibre.—Cordage is said to be made of the bark. (*Lisboa, U. Pl., Bomb., 2.*)

FOOD.
440
Food.—The flowers, buds, and fruit when green are eaten by the natives. The berry is said to have an agreeable acid flavour, resembling that of **Grewia asiatica.** The fruit is also greedily eaten by animals. (*Kanara Gazetteer.*) In Thana the deer are said to be specially fond of these fruits.

TIMBER.
441
Structure of the Wood.—Rough, moderately hard, reddish-grey; apt to split, warp, and crack; strong, heavy, durable, handsomely marked on a vertical section by the darker-coloured medullary rays which appear as broad plates. Weight 41 to 50℔ a cubic foot; growth moderately fast.

It is used for construction, ship-building, rice-mills, and for charcoal, which is of good quality. In Kanara it is considered useless except for burning; in Thana the wood is also regarded as worthless. **Lisboa,** however, in his special botanical volume to accompany the *Bombay Gazetteers,* says: "the wood is very strong, hard, heavy, porous, coarse-grained, durable, &c." He adds that it is "used for house and ship-building, buggy-shafts, rice-mills, and charcoal."

DOMESTIC.
Thatching.
442
Domestic Uses.—The leaves are sold in the bazar at Poona as a substratum for thatching. (*Dalzell and Gibson.*) It is stripped of its leaves and pollarded, in Kanara, to afford leaf-manure. The old rough leaves are employed to polish ivory and horn.

443

D. pulcherrima, *Kurz.; Fl. Br. Ind., I., 37.*

Vern.—*Byú,* BURM.

Habitat.—A large tree in the tropical forests of Pegu, Prome, and Martaban, ascending to 1,000 feet in altitude.

TIMBER.
444
Structure of the Wood.—Hard and strong, used for rice-mills, but the trunk usually remains low and crooked (*Conf.* with *Indian Forester, VIII., 416*).

Dindiga Gum; see **Anogeissus latifolia,** *Wall.;* Vol. I., No. 1149, p. 256.

DINEBRA, *Jacq.; Gen. Pl., III., 1171.*

Dinebra arabica, *Beauv.; Duthie, Fodder Gr., N. Ind., 55 ;* GRAMINEÆ.   445

Syn.—CYNOSURUS RETROFLEXUS, *Vahl. ;* LEPTOCHLOA ARABICA, *Kunth. ;* DACTYLIS PASPALOIDES, *Willd. ;* DINEBRA RETROFLEXA, *Pans. ;* D. ÆGYPTIACA, *Jacq. ;* ELEUSINE CALYCINA, *Roxb., Fl. Ind., Ed. C.B.C., 116.*
Vern.—*Bara sarpot, maljhanji,* C. P. ; *Wadata-toka gadi,* TEL.

Habitat.—A tufted annual grass growing near bushes around fields. Met with in the Panjáb, Rájputana, the Central Provinces, Bundelkhand, Bengal, and South India.

Fodder.—If cultivated might prove a useful fodder : it occurs too sparsely to be regarded as a valuable wild fodder.

FODDER.
446

DINOCHLOA, *Gen. Pl., Vol. III., 1214.*

Dinochloa andamanica, *Kurz ;* GRAMINEÆ.   447

References.—*Kurz, For. Fl. Burm., II., 570 ; Indian Forester, I., 242 ; Gamble, Man. Timb., 431.*

Habitat.—A lofty scandent bamboo, met with in the tropical forests of the Andaman Islands.

D. Maclellandii, *Kurz.*   448

Syn.—BAMBUSA MACLELLANDII, *Munro.*
Vern.—*Wa-nway,* BURM.
References.—*Kurz, For. Fl, Burm., II., 571 ; Indian Forester, I., 242, 264 ; Gamble, Man. Timb., 431.*

Habitat.—A native of Burma and Chittagong ; stems scandent, 60 to 100 feet.

DIOSCOREA, *Linn.; Gen. Pl., III., 742.*   449

History of Yams.—Modern usage has assigned to the tuberous roots of the various cultivated species of **Dioscorea** the name of Yam. It is generally stated that Yam is derived from a West Indian or an American word, passing through the Spanish and Portuguese *inhame* and the French *igname.* Thus, Purchas, in his *Pilgrimage* (published in 1625), says :—"There are great stores of *Iniamas* growing in Guinea." So also Godinho de Eredia (*L'Inde Meridionale, et le Cathay,* 1613) remarks :— "Moreover, it produces great abundance of *inhames,* or large subterranean tubers of which there are many kinds, like the *camottes* of America, and these *inhames* boiled or roasted serve in place of bread." The *inhames* alluded to (according to most writers) are the yams of the present day, the word "Yam" being a consequence of the Anglo-Saxon spirit of brevity, and *camottes* is generally accepted as sweet-potatoes. On the other hand, Ramusio (*Vol. I., p. 117, Ed. 1613*) may probably be referring to the same root when he remarks—"The root which, among the Indians of Spragnuola Island, is called *Batata,* the Negroes of St. Thomé call *Igname,* and they plant it as the chief staple of their maintenance ; it is of a black colour, *i.e.,* the outer skin is so, but inside it is white and as big as a large turnip with many branchlets ; it has the taste of a chestnut, but much better." In 1583 John Huyghen van Linschoten visited India, and, during a residence of some years, collected material for a most instructive book of travel, which was ultimately published in 1596. There can be little doubt that at least 300 years ago **Dioscoreas** of various species were regularly cultivated in India, but **Linschoten** draws a comparison between

HISTORY
OF YAMS.
450

Inhames.
451

DIOSCOREA.

Dioscoreas—Yams.

what he calls "*Iniamos*" and "*Batatas*." He speaks of them as "fruits" —a mistake made by most early writers, and which, indeed, has not been completely eradicated from popular fancy even at the present day, for the words vegetable and fruit are often used as synonyms. Speaking of yams, however, **Linschoten** says: "These Iniamos are as bigge as a yellow roote" (is carrot meant?), "but somewhat thicker and fuller of knots and as thicke on the one place as in the other, they grow under the earth like earth Nuts, and of a Dun colour, and white within like earth Nuts but not so swéete. The Batatas are somewhat more red of colour and of fashion almost like the Iniamos but sweeter, of taste like an earth Nut. These two fruits are verie plentiful, specially Iniamos, which is as common and necessarie a meat as the Figges" (he alludes here and throughout his work to 'Plantains,' which by mistake was translated 'Figges'), "they eate them for the most part rosted, and use them commonly for the last service on the boorde, they sieth them likewise in another sort of porrage, and sieth them with flesh like Colwartes or Turnops, the like doe they with Batatas." Now to remove all doubt as to the Indian so-called Iniamos, which were, 300 years ago, so "verie plentifull" and "necessarie" as plantains to the people of the west coast of India, a further quotation may be given from **Linschoten**, where he returns again to the subject of Iniamos. Speaking of the Azores he writes: "They have likewise in that island a certaine fruite that groweth under the earth, like Radishes or other roots, but the plants are trees like vines but different in leaves, and groweth longwise upon the grounde; it beareth a fruit called Batatas, that is very good, and is so great that it weigheth a pound, some more, some lesse, but little esteemed, and yet it is a great sustenance and foode for the common sort of people. It is of good account in Portingall, for thether they used to bring it for a present, and those of the Ilande by reason of the great abundance doe little esteeme it." Now, whatever may have been the *Batatas* as distinct from the *Iniamos*, one point is certain—neither can for a moment be supposed to have been Aroids. They were "vines" with leaves of a peculiar form (a remark that suggests the striking foliage of most **Dioscoreas**), and they were "vines" so large that he pronounced them "trees like vines." The use of the word "trees" almost precludes the idea of an **Ipomœa** (or **Convolvulus**), and the habit of creeping on the ground gives weight to the

supposition that by Batatas he meant simply another form of Iniamos, the more so since the natives of India and of many other countries regularly allow one or two species of yam (or **Dioscorea**) to grow along the ground instead of affording them the means of twining. The sub-woody stems of most **Dioscoreas** might fairly well be viewed as admitting the expression 'trees,' whereas that word would be wholly inapplicable to the sweet-potato. One can hardly presume **Linschoten** to have failed to specialise the foliage if it had (as in the case of the sweet-potato) a resemblance to a plant with which he was probably quite familiar—the **Convolvulus**. It is much more natural that he used the word Batatas because that word had just then reached Europe. He evidently preferred to adopt words known in Europe to giving the native names; had he mentioned the Indian vernacular names also instead of using foreign words only— *Iniamos* and *Battas*—much of the confusion that now exists would have been saved. Dr. **Paludanus**, in a foot-note to **Linschoten's** account of yams (or yams and sweet-potatoes), says: "Iniamas were this year brought hether out of Guinea as big as a man's legge and all of a like thickness, the outward part is dun-coloured, within verie white, rosted or sodden they are verie pleasant of taste, and one of the principal meates of the Black [?Moores]." In 1597, the very time **Paludanus** wrote this note, **Gerarde** was experimentally cultivating the sweet-potato in his garden in Holborn, London. **Paludanus** appears to be alluding (as he affirms) to the yam and not to the sweet-potato, but there is every probability that for a time Iniamas and Batatas were viewed as different kinds of the same tuber, at least until the independence of the plants that yielded them had been discovered. In the passage quoted from **Ramusio**, *Batata* and *Igname* are said to be synonymous. Moreover, there is nothing to show that the sweet-potato is indigenous to the Azores. If, on the other hand, it be admitted to be a native of America, the sweet-potato, at the date referred to by **Linschoten**, could not have become so abundant in the Azores as to be little valued by the people. We must, therefore, either suppose the Batata of the Azores was a form of Yam, or that **Ipomœa Batata** is a native of that group of islands (in mid-Atlantic and in the latitude of Spain), while the other known home of the plant is in tropical South America. As far as can be made out the sweet-potato, however, is a native of South Brazil and Chili only; but there are various yams known at the present day so sweet as to be often described as scarcely distinguishable from the sweet-potato (*Conf.* with **D. aculeata**). May not these have been the Batatas of China and Japan, to which many authors allude, as in a measure, establishing an Asiatic as well as an American origin for the sweet-potato? At all events, **Linschoten's** description of his Batatas as being like the Iniamos "but sweeter of taste" and "more red of colour," &c., agrees far more with some of the purple-coloured yams than with any known sweet-potato. If this view be accepted the puzzle (regarding *rukt álu* being a Sanskrit name for the sweet-potato), which seems to have perplexed **M. A. deCandolle**, is at once explained. The red-coloured *álu*, which is also the sweetest of all yams, was at first compared with the modern or introduced sweet tuber until in time *rukt álu* became the name for the sweet-potato. (*Conf. with p. 126.*) The writer has, however, failed to find any author who, under that name, alludes to the Aroid (**Amorphophallus campanulatus**) as stated by **Adolphe Pictet**. **Roxburgh** refers to the two chief forms of sweet-potato—the red and the white. To the former he gives, as the Bengali name in use in his day, the compound of *Lál-shakar kandá-álú* or the red-sweet-arum-like-yam. This translation assumes the interpretation of "*álu*" as equivalent to "yam," and gives "arum-like" as the English for "*kandá*." As a matter of fact, *kandá* is a Sanskrit name chiefly assigned to **Amorphophallus campanulatus**, so that the **Roxburghian** Bengali name would seem to be a combination of all the resemblances probably suggested to the mind of the first possessor of a sample of this valuable new tuber. Moreover, although called a Bengalí name for the plant, it is more a combination of Hindustaní and Sanskrit words, and is not, strictly speaking, Bengalí in its origin. There is no Indian name for the sweet-potato that has any stronger claim to being original and specific in its nature, so that there seems no doubt as to the sweet-potato being a modern introduction into India, and most probably considerably after the date of **Linschoten's** visit to the Portuguese possessions on the western coast. The name **BATATA** gave origin to the word **POTATO**, and there seems no reason for doubting but that both these terms should have been assigned to the **SWEET-POTATO** and not to the tuber which is now known as '**THE POTATO.**' The latter tuber was taken to Europe *after* the sweet-potato, and in the words of **Yule** and **Burnell** (*Glossary of Anglo-Indian Terms*) may be said to have robbed it of its name. This is more than an interesting historic fact. The frequent allusion to the *inhames* being used "in place of bread" suggests the possibility of the sweet cassava (**Manihot Aipi**) being the plant referred to in such passages; and if so the European form of the word (yam) may have been applied to the wrong plant. Thus sweet-potatos, cassava, and yams may be accepted as confused with one another in the writings of early travellers;

HISTORY.

Batata or
sweet potato
Conf. with
p. 121.
453

Rukt alu.
454

Kanda.
455

Potato.
456

HISTORY.

and, indeed, the literature of even the potato has to some extent got mixed up with that of the above allied tubers. A similar confusion exists in India. The potato and sweet-potato are undoubtedly modern introductions, but there are many wild species of **Dioscorea** scattered over the whole of Peninsular India, some peculiar to the hot, damp, tropical plains, others to the drier tracts, while a third set ascend to temperate regions and are met with throughout the Himálaya at altitudes up to 9,000 feet above the sea. That being so, the presumption may be admissible that most of the forms of yam met with under cultivation in India are probably indigenous also, even though in some cases they are not now known in a wild state. The name *álú* by universal modern acceptance is given, however, to the potato, and that word is by no means recent in Indian classical literature. It can be traced into some of the early Sanskrit works. It is there applied to a

Conf. with p. 120.

farinaceous tuber—perhaps to several such tubers—being, in its earliest meaning, generic more than specific. **Sir Walter Elliot** (*Flora Andhrica*) gives the Telegu words *A'llu* to the millet—**Paspalum scrobiculatum ;** *kand-ulu* to the pulse—**Cajanus indicus ;** *alachand alu* to the pulse—**Vigna Catiang ;** and *Machi-kanda* and *Páti-kanda* to **Amorphophallus.** We have no alternative, therefore, but to assume that the potato on being brought to India received an ancient name, and further, that the superior properties of the introduced tuber gradually displaced the indigenous one from public favour. According to **Professor Wilson,** and more recently to **Yule** and **Burnell,** the *álú* of early Indian classics was an aroid and most probably **Amorphophallus campanulatus** (see *Dict. Vol. I.,* A. 996). But **Amorpho-**

Conf. with p. 122.

phallus is a distinctly tropical genus of **Aroideæ** and does not extend beyond the Himálaya. If therefore *álú* was originally given to an aroid it may have denoted a species of **Arisæma** (to some extent an extra-tropical genus), but in India the members of that genus are all poisonous and none of them are cultivated.[*] **Mr. John Crawford,** in a most valuable paper on the *Migration of Cultivated Plants in reference to Ethnology* (*Selections in the Agri.-Hort. Soc. of India Jour., Vol. I., New Series, 1868, p. 15*), says, that the word *álú* had a generic meaning with the Hindús of Upper India, being equivalent to *kalangku* with the Tamil-speaking people and *ubi* with the Malayan nations. He is also disposed to view *álú* as having been originally applied to an aroid and "not to the yam with which as an extra-tropical people the Sanskrit-speaking race must have been unacquainted." While not prepared to dispute an opinion held by so many high authorities, the writer may be permitted to say that he would be more disposed to accept a **Dioscorea** as having been the plant to which the word *álú* was applied when it first received (and probably in India) a restricted or specific meaning. But even the idea of "an extra-tropical people" can have little weight, since various species of **Dioscorea** are temperate Himálayan plants and are distributed even to Afghánistan. **Aitchison,** for example, found **D. deltoidea** in low scrub from 7,000 to 8,000 feet in altitude, near Shálizan in

Olla. 457

Afghanistan.[*] Unless *álú* be taken as a form of *olla* (a Sanskrit name for **Amorphophallus,** from which the modern Bengali term *ol* is derived), there are probably no other references in Sanskrit literature to *álú* that could with any degree of certainty be accepted as alluding to an aroid. But the question may fairly well be asked, as against the argument resting on the Sanskrit-speaking people having been "extra-tropical"—how did the tropical **Amorphophallus campanulatus** come to get its early Sanskrit

Surana. 458

names ? There are several well-known Sanskrit synonyms for that aroid, such as *surana, kanda,* and *arsaghna.* In none of the modern languages of India does **Amorphophallus** bear a name traceable to *álú,* while the Sans-

---

[*] Conf. with **Dr. D. Prain's** remarks on page 122.

The Alu or Yam.   (*G. Watt.*)   DIOSCOREA.

krit synonyms given above clearly afford many colloquial names. Thus the plant is the *ol* in Bengal,* *suran* in Bombay, *kanda* in the Deccan, and in Madras (among the Telegu-speaking people) that is also one of its names. *Zamin khand* is a Panjáb name, apparently for the tubers of **Dioscorea bulbifera.** Sir Walter Elliot gives the Telegu word *Konda* as a prefix to a very large series of names of plants many of which yield farinaceous substances :—*Konda-gummudu* (**Dioscorea pentaphylla**); *Kondajíluga* (**Caryota urens**); *Konda-jonna* (**Sorghum vulgare**); *Konda-kalava* (**Kœmpferia rotunda**); *Konda gurava tíge* (**Smilax ovalifolia**), &c., &c.

The writer is aware that one cultivated aroid (*viz.*, **Colocasia antiquorum**) is in some dialects of India called *álú* or rather *kachálú*, but its Sanskrit name is *kachchi* or *katchú*, and there is much in favour of the opinion that the word *álú* is a modern addition to the Sanskrit name of that plant. If the *kachú* be the *álú* of Sanskrit writers, it is somewhat significant that it should have carried with it, not the name *álú*, but *kachú*, from which, through the Arabs and the Egyptians, the Greeks probably obtained their *colcos* or *kulkas* (Conf. with **Colocasia**, *Dict. Vol. II.*, C. No. 1731), and ultimately the name **Colocasia** itself. Mr. Baden Powell mentions a curious fact, namely, that in the bazars of the Panjáb the tubers of **Dioscorea deltoidea** are viewed as the *kachálú* collected from old plants. On the other hand, in nearly every dialect of India, *álú* is, at the present day, the generic name for the **Dioscoreas**, the species being indicated by qualifying or descriptive prefixes. It is noteworthy also that the Tamil name *kalangu* and the Malay name *ubi*—the synonyms of *álú*—should now be applied to **Dioscoreas**, while so accurate a writer as Ainslie should give the potato the name *Wallarai-kilangu* (*i.e.*, English Yam). He tells us that at the time he wrote (1813) the potato had just then been introduced into Madras. De Candolle is doubtless in error (in assuming the correctness of **Roxburgh** and **Piddington**) when he says there are no Sanskrit names for the species of cultivated **Dioscoreas** (*Origin Cult. Plants, p. 77*). Dr. U. C. Dutt (in his *Materia Medica of the Hindus* compiled from Sanskrit Medical Works), gives the following Sanskrit names :—*Dandálu* (**Dioscorea alata**, *Linn.*); *Madhválu* (**D. aculeata**); *Pindálu* (**D. globosa**); and *Raktálu*, (**D. purpurea**). The *Ain-i-Akbari* (*Blochmann's Transl., p. 71*) describes *Pindálú* as reared on lattice work and to have leaves that resemble those of the betel-leaf. It was, therefore, a climber, and the description of the leaf precludes it from having been the sweet-potato. In another page both the *Kachálú* and the *Súran* are described, so that the *Pindálú* could not have been an aroid. Throughout the whole of M. deCandolle's most instructive work, **Piddington's** popular Index to Indian Names of Plants has unfortunately been accepted as a standard authority. Thus, speaking of the absence (or rather supposed absence) of Sanskrit names for the cultivated **Dioscoreas**, M. deCandolle says :—"This last point argues a recent cultivation, or one of originally small extent in India, arising either from indigenous species as yet undefined or from foreign species cultivated elsewhere. The Bengali and Hindu general name is *álú* preceded by a special name for each species or variety; *kamálu*, for instance, is **Dioscorea alata**. The absence of distinct names in each province also argues a recent cultivation." A reference to the vernacular names given under each species (in the pages which follow) will show that even this last conclusion is not well founded. In most of the languages of India there are several names for the species (wild and cultivated) of **Dioscorea**—every shade of cultivated form frequently having its distinctive name.

Conf. with
p. 121.

* Conf. with Dr. Dymock's remarks on page 121.

DIOSCOREA.	Kriss—the Yam.

HISTORY.

Piska.
Conf. with p.
136, the
Kullu.
460

Turar.
461

Conf. with
p. 118.

It may be admissible that most of the cultivated plants which the invading Sanskrit-speaking people found in Hindústan, and which were unknown to them before their arrival in India, received from them Sanskrit names (or Sanskritised forms of their aboriginal names), or had old generic Sanskrit names in time restricted or specifically applied to them. It would, indeed, be unsafe to assume that the pre-Aryan people of India did not possess some wild or cultivated plants of value which were unknown to their conquerors. The Santals and other aboriginal races have names for most of the **Dioscoreas** which have no connection with the Sanskrit names for these plants. Thus the **Rev. A. Campbell** tells us that D. bulbifera is known to the Santals as *Piska*, and that its tubers are considered a "great delicacy." In Bombay that plant is known as *mar-páshpoli*. **D. glabra,** **Mr. Campbell** also says, affords edible tubers, which are known as *Ato sang*, and **D. globosa** the *Bengo nari*. In many parts of Western India **D. pentaphylla** is known by the name *úlsi* or *úlasi* (*ufi*, *ubi*, *papa*, are American names). The name *kríss* seems to be the most general for the species of **Dioscorea** in the Panjáb, but that name is practically confined to the hill tribes; the tubers when offered for sale are known as *tarar*, a name that at once brings to mind the Malay *tallus*, *tales*, the Otahitan *tallo* or *tarro*, and the Bombay *terem*, names for **Colocasia antiquorum** (*Conf.* with *Dict. Vol. II.,* **C.** 1731). The *Ain-i-Akbari* (1590 A.D.) describes what appears to be a Yam under the name of *Tarri*. It is said to grow mostly in the mountains and to produce tuberous roots so large as to suggest the resemblance to a "mill-stone." It is spoken of as a creeper with leaves resembling those of the water-melon (*Blochmann's Transl., p. 71*). Atkinson affirms that in Kumáon **D. sagittata**, *Royle,* is known as *tair* and *turar*, words which (with the Panjáb *tarar*) may be admitted as recalling one of the generic names,—*tega*—which **Sir Walter Elliot** states is, with the Telegu-speaking people, a synonym for the Sanskrit *pindáluh* (see also *Wilson, Sanskrit Dict., p. 384,* and *Heyne, Tracts on India, p. 55*). The commonest Telegu generic name for all the **Dioscoreas, Elliot** informs us, is *pendu lam*, which he points out is the equivalent to the Sanskrit *pindáluh.* Under *Chára kanda* (**Colocasia nymphæifolia**), **Sir Walter** draws attention to the fact that **Brown**, in his *Dictionary,* is incorrect in assigning **Wilson's** account of the *pindálu* to a **Colocasia**; **Elliot** makes no mention of *álú* or any form of that word as applied to any aroid. Thus, it may safely be affirmed that throughout India there are names for the **Dioscoreas** that belong to the languages of the aboriginal races as well as those that are of a more classical character, the latter most frequently approaching to the word *álú*, while a third set of names may have been derived from foreign and mostly American sources.

It seems probable that, as articles of food, the **Dioscoreas** of the world were cultivated at a much later date than most other vegetables, probably on account of the fact that without the trouble of cultivation they afforded an unfailing supply of food. The ancient Egyptians were not aware of the value of these tubers, and, indeed, they are doubtfully cultivated (or only to a small extent) in the Egypt of to-day. The volume of existing evidence on the origin of the cultivated **Dioscoreas**, however, points to a possible independent discovery in Asia, America, and Africa. Interchanges from these regions might easily have taken place down to the present time, and the knowledge of the properties of these useful cultivated tubers may be assumed to have thus become widely and rapidly diffused. From the Indian source cultivated forms may have been carried to the West Indies, or others brought from the West Indies to India during the period of Dutch and Portuguese influence. **DeCandolle** states that in Mauritius a yam bears the name of *Cambare marron.* "Now, he adds, *Cambare* is something like the Hindu name *kam,* and *marron* indicates a plant escaped from cultivation." But this derivation ignores the fact that *Camotes* is a Mexican name for the sweet-potato, so that the syllable "Cam" in American words and *Khám* in Indian, need be viewed as nothing more than a coincidence. On the other hand, many writers contend for a more pronounced Asiatic origin for the yams, holding that they are indigenous to the Burma-Malay Peninsula, and that India mainly furnished the stock now cultivated in the West Indies. In part support of this opinion it may be added that the Indian wild species increase in number and abundance on passing into the Eastern Peninsula. They are, however, nearly as abundant on the Western coast, but become rare in the central table-land, and are scarcely represented in the northern dry areas. Thus the head-quarters of the Indian **Dioscoreas** is in the very region where the early European settlers first established themselves.

Considerable ambiguity, it must be admitted, has crept into the Anglo-Indian literature of Yams, and this has greatly increased the labour of establishing satisfactorily the names to be given to the species of **Dioscorea.** In popular works of travel, official reports, and other such publications, the word "Yam" is practically used to embrace all starch-yielding tubers other than potato and arrowroot. Thus aroids, yams proper, sweet-potatos, &c., &c., are all designated yams. This fact has compelled the writer, in drawing up the present account of the species of **Dioscorea,** to reject all descriptions of yams in which he could not, from internal evidence, discriminate the plant or plants referred to. This has naturally deprived him of more than half the economic information that exists on the subject of yams, and has reduced very considerably the list of vernacular names that doubtless might be given for the species of **Dioscorea.** It may here also be admitted that the present account is at most but a compilation of economic information. The writer has been unable to personally verify the facts given or to work out the scientific synonyms of the plants. The references to the authors consulted will, however, it is believed, afford a key to any re-arrangement that may be found necessary on the Indian species of **Dioscorea** being re-determined and described. Even in cases where no doubt existed that a yam proper is referred to, it has often been impossible to relegate to individual species the economic information available. This has necessitated to a large extent the production of the present collective article on yams instead of following the usual course pursued in this work, of giving a separate detailed account of each species in which is exhibited its vernacular names, structural and economic peculiarities, and modes of cultivation, &c., &c.

SPECIAL OPINIONS.—Since writing the above account of yams, and in reply to letters conveying the main ideas here advanced, the writer has had the pleasure to receive two instructive communications—one from **Dr. W. Dymock** of Bombay, and the other from **Dr. D. Prain,** Curator of the Calcutta Herbarium. Instead of incorporating the facts, thus obtained, in the general statement, it has been deemed preferable to give these communications in this place. **Dr. Dymock** writes :—

"It is very difficult to say what the original *alú* of Sanskrit writers was; in Hindi it means yam, sometimes as *rat-alu* (red yam); in Marathi it is applied to **Colocasia, Calladium,** and similar plants. *Rukh-alu* or *tree-alu* is **Remusatia vivipara,** which has leaves and flowers like **Colocasia,** but sends up curious spikes covered with minute bulbs. This word is purely Sanskrit and is current in Marathi. *Kachu,* the Hindi and Bengali name for these plants, is also Sanskrit. *Kachora* and *kachu* are applied to the **Curcumas** in Hindi, Marathi, and Bengali. **Amorphophallus** is

Camotes.
462
Kam
Conf. with pp. 125 and 135.
463

Species imperfectly determined. Conf. with p. 125.

Alu.
464

Kachu.
465

DIOSCOREA.   Arso-ghna, the destroyer of piles.

Surana.
Conf. with
p. 118.
466

Súrana, and *ola* or *olla* in Sanskrit and in the vernaculars; it is also *Kand* in the vernaculars as *Jini-kand* (Edible tuber). *Súrana* is derived from *súr* to hurt; all the wild kinds are very irritating. In Marathi the name *Súran* seems to be applied to all Aroids having the peculiar shaped tuber of **Amorphophallus.**

" As regards yams we find purely Sanskrit names in the vernaculars, such as *Pásh-poli,* from *pásh* a noose, and *poli* a cake, Anglic—strangle-cake, since the wild kinds irritate the throat and cause a sensation of strangling. This name is current in Marathi for wild yams, which are also called *Manda,* a Marathi name for a cake like a mincecake, full of *Rawa* and sugar instead of mince-meat.

" Cultivated yams are called *kon* and *konphal* in Marathi, *kand* in Hindi and Guzerathi, from the Sanskrit *Kanda.* This word spelt *Kanda* is applied to all tuberous roots and bulbs in Marathi just as it is in Sanskrit. Another Marathi name for cultivated and uncultivated yams is *Cháin* or *Cháyen.* The Sanskrit names *alu* and *kanda* are about as vague as the *Bolbos* of the Greeks, but we know that *surana* means **Amorphophallus** from the epithet *Arso-ghna* or 'destroyer of piles' applied to it, for it is still used as a remedy for piles."

Dr. D. Prain's note has special reference to the distribution of the species of **Dioscorea** and of **Arisæma** :—

" I have gone through the whole of the works dealing with the ' Orient' of **Boissier** and the ' Central Asia' of the Russian writers, and I find that none of them mention either **Colocasia** or **Amorphophallus.** **Boissier's** Supplement brings our knowledge down to 1888, and Traut-vetter's Incrementa to 1884, while **Maximowicz's** last paper (*Mel. Biol., XII.*) brings us also down to March 1888.

" **Arisæma** finds no place in the pages of Russian authors. In the *Flora Orientalis, V., 43,* **Boissier** describes **A. abbreviatum,** but he states that he had not seen the plant, and he gives only **Aitchison's** Kuram Valley Flora locality. In the App. to V. (p. 734) he adds **A. Jacquemontii,** also from **Aitchison's** Kurram plants, remarking that he has seen the plant himself. He does not, however, give any other locality for either species, nor does he mention any economic uses.

" No **Dioscorea** appears in *Ledebour,* the standard Russo-Siberian Flora. **Maximowicz** gives **Dioscorea** 5-loba ir *Prim. Flor. Amur. et Mands.* He mentions no other in any of his ' diagnoses' (of which he publishes the 12th paper this year), nor does **Trautvetter** *l. c.,* who brings the Russian Flora up to date at intervals.

" **Boissier** notes (under **Tamus**), on Baker's authority, that **Aitchison** had found **D. deltoidea** in Kurram, but for all that he does not provide a description of it in his Flora. I find no mention under the families you refer to of any edible tuber, either in **Boissier** or in the Russian authors."

ULTIVA
TION.
467

CULTIVATION OF YAMS IN INDIA.

The yam may be propagated by means either of the ærial tubers which form on the stem or by means of small under-ground tubers or portions of large ones. The ærial tubers are not, however, often used, as the plants in that case require two years to reach maturity. The usual process is to dig the ground to a considerable depth and to manure it fairly well ; then form pits near trees, so as to afford support for the climbing stems, and deposit in each pit a small tuber or portion of a large one. Lateral shoots should be nipped off, otherwise large tubers will not form. If it be desired to plant an area of ground entirely with yams, the ground should be trenched and the tubers deposited two feet apart along the furrows. To afford support for the plant a trellis work should be constructed between the ridges. The Chinese

Method
of planting.
468

Dioscoreas—Yams. (*G. Watt.*) **DIOSCOREA.**

have a method of producing a large crop of small tubers without having to incur the expense of providing a support for the plants. The tubers are planted 3 or 4 feet apart along the crest of ridges, and the stems made to take root at various points by being pinned down on the ridges. By this means a crop of tubers very much like potatos is produced, the tubers forming at each point where the stem is pinned down into the earth. It is often the case in India that the long stems are not provided with a support nor are they pinned down. They are simply allowed to spread over the ground. This is a slovenly practice and results in much smaller tubers than when the stems are provided with a means of climbing. Firminger states that yams should, in the plains of India, be planted in April and the crop taken up in December. A recent report from Madras says of the Arcot district that "the yam is cultivated in January and harvested in August and September. The cost of cultivation is R20 per acre, and the profit about R30. The soil required is red loam, richly manured. The tubers are planted between shallow trenches for retaining water." The varying periods peculiar to the better known species will be found indicated in the pages that follow, but the imperfect knowledge which exists on the subject of Indian yams precludes more being done than to bring together the scattered notices.

Mr. R. Mitchell, Emigration Agent for British Guiana, Calcutta, has for some years past been prosecuting most successfully the effort to introduce improved forms of yam into India. He writes: "The Jamaica yams appear to have become thoroughly acclimatised, and yielded in 1885 at the rate of about eight tons to the acre, and last year, when they were planted rather too late, gave about five. As far as I have been able to determine, the Jamaica yams with which these experiments have been made are of three distinct kinds—the smooth white, the rough white, and the rough variety slightly tinged with pink. The last mentioned is coarser than the other two, but good for food. It also bears seeds on the vine which make excellent plants.

"The other yams I imported from British Guiana comprise four varieties—The Kush-Kush or Buck-yam, the most delicate of all the species, the White Hunt, originally from Barbadoes, also a great favourite, the Stack, and the Lisbon yam.

"The only plant of Buck-yam which reached me alive was about the size of an almond. I planted it carefully, but more than once its growth was checked by accidents. When the vine withered, I dug it up and found the produce about the size of a lady's thimble—by no means an encouraging result. It was planted again and tended still more carefully, and last year produced a yam large enough to give seven plants, fully as big as I have ever seen it in the West Indies, for it gives the smallest return of any, but is by far the most delicate. Two of the seven plants are dead, but the remaining five appear to be thriving well. The other yams yielded little more than plants, and were too small almost to eat, so they have been planted again.

"I am unable to say why they have not yielded a better return; possibly they may not have been long enough acclimatised, or were planted too late last year. They appear, however, to be thriving now, judging by the growth of the vines, which are luxuriant.

"The extent of land on which these experiments have been carried out is small, about $\frac{1}{15}$th of an acre, the soil light, and of reasonably good quality.

"I have taken so much pains to acclimatise the West Indian yam, because the varieties I have tasted here are almost uneatable, very stringy and hard, with an unpleasant earthy flavour, such as may be met with in some of the coarser West Indian varieties. I may add that the yams I am experimenting with are all water yams, and most of them very delicate

CULTIVATION.

Conf. with p. 116.

Seasons of Planting and Reaping.
469

Introduction of Improved Forms.
470

Buck-Yam.
471
White Hunt.
472
Stack.
473
Lisbon Yam.
474

DIOSCOREA.   West Indian Yams.

CULTIVATION.	and delicious food. In fact, the Buck-yam is superior, in the opinion of many, to the best potatos.

and delicious food. In fact, the Buck-yam is superior, in the opinion of many, to the best potatos.

"I am also conducting experiments in the Hill near Hope Town at different elevations with these yams, with the kind co-operation of **Messrs. Johnstone** and **Calvert**, who have planted several at lower elevations. The first I planted in March, at 5,200 feet above sea level, rotted almost without exception before the roots and germinal shoots from the bark had time to appear. I, therefore, waited until May and planted out fifty more which had already sprouted, and most of them grew up and looked healthy.

Introduced Forms not suited to the hills.
475

"The result of this hill experiment will also be communicated to you. I fear, however, that at elevations above 3,000 feet, the yam can scarcely prove a valuable addition to the root crops of the hills, being too much of an exotic. In the plains, however, where the rainfall is not less than fifty inches and fairly distributed, and the soil arenaceous and moderately good, the West Indian yam should prove a valuable addition to the food of the people. It requires from seven to eight months to attain to its full growth, and should be provided with sticks to run on, although a small patch, which was not supplied with poles, produced about as much as the rest of the ground where a bamboo trellis was provided to support the vines.

Water Yam.
476

"My father and I have experimented on yams for many years. I obtained from Barbadoes some of the finest varieties of the water yam, which in that island seldom exceed eight or ten pounds in weight. In Trinidad, however, the plants throve amazingly, and the weight was increased. In one instance a yam which I had dug and weighed myself turned the scale at 82lb, and many others averaged from 30 to 50lb;

82 lb. in weight.

moreover, these enormous yams did not in any way deteriorate in flavour, and proved quite as delicate as the original stock.

"Should you arrive in Calcutta before the vines wither, about X'mas time, I should have much pleasure in showing you the patch of yams at present under experiment.

Manure.
477

"The cultivation is simple, but laborious. The ground is dug into trenches about two feet wide by eighteen inches deep, three feet apart from centre to centre. The trenches are filled with leaves, weeds, and garden rubbish of all kinds, which admit of the free expansion of the tubers. The mould from one trench forms the bed over the preceding one. The soil being uniformly good it does not matter much where the sub-soil is placed; if poor I should put it beneath in direct contact with the accumulated vegetable matter and out of the way of the young rootlets, seeking the nourishment which they would be more likely to find among the humus of the surface soil.

Tubers.
478

The yams, cut to about the size of an ordinary man's fist, care being taken to preserve as much of the outer surface as possible, are planted at intervals of three feet in the beds. The lops in the absence of seeds are the legitimate plants, and rarely fail to grow.

Seed.
479

Seeds where procurable made excellent plants. There are seasons, however, when few, if any, appear—why I am unable to say—although this has also been a subject of experiment. As far as I have observed, however, the coarser varieties bear seeds most abundantly and frequently.

Mode of sprouting.
480

"I prefer sprouting the pieces of yam cut for plants under a layer of grass and mould a few inches thick, which should be watered at intervals. The seed bed should be prepared in the end of March, or early in April, when the seasons are similar to the weather in Calcutta. In from 3 to 6 weeks the yams commence to sprout and should be planted out at once, as the young shoot, which grows with great vigour and rapidity, even in the corner of a dark room on a stone floor, soon weakens the parent."

The Prickly-stemmed Yam.	(*G. Watt.*)	**DIOSCOREA aculeata.**

The writer would wish it to be understood that since a monograph of the Indian species of **Dioscorea** has not as yet been published, the following notes must be viewed as an attempt to relegate existing information to the species probably concerned. When the species have been examined, however, considerable departures from the present arrangement may doubtless be found necessary, the names and synonyms having to be changed.

Conf. with p. 121.

Dioscorea aculeata, *Linn.; Roxb., Fl. Ind., Ed. C.B.C., 728; Wight, Ic., 2060;* DIOSCOREACEÆ.

481

PRICKLY-STEMMED YAM or GOA POTATO; KAAWI YAM; the GUINEA YAM.

Vern.—*Man-álu,* HIND.; *Mou álu* (or *mauálu*), BENG.; *Kánta,* or *Kántékángi* (Bazar name *botat*), BOMB., GOA; *Chhota-pindálu,* DEC.; *Kantúkelangú, sirru-vullie-kelangu,* TAM.; *Kata-kelenga, kummara baddu,* TEL.; *Genasu,* KAN.; *Pudie-kelengu,* MALAY; ? *Kuhu-kukulalu* (according to Balfour), SING.; *Madhválu,* SANS. This appears to be the *Bir sangi* of the SANTALS.

References.—*Voigt, Hort. Sub. Cal., 652; Dals. & Gibs., Bomb. Fl. Supp., 92; Drury's Hand-book, Indian Flora, vol. III., 276;* is probably the *Katta kelangu* of *Rheede, Hort. Mal., VII., 37;* the *Cumbilium* of *Rumph. Amb., V., t. 126; Sir W. Elliot, Flora Andh., 102; U. C. Dutt, Mat. Med. Hind., 308; Dymock, Mat. Med. W. Ind., 2nd Ed., 842; Müeller, Select Extra-Trop. Pl., 106; Mysore Gaz., II., 11; Lisboa, U. Pl. Bomb., 178; Birdwood, Bomb. Pr.,178; Balfour, Cyclop., 950; Jour. As. Soc., Pt. II., 2, 1867, 82.*

Habitat.—A native of Central and Southern Bengal and of Western and Southern India : cultivated more or less in most provinces.

Description.—The tubers are described as oval or oblong, composed of a delicate white and rich starch. They are generally about two pounds in weight, but often considerably smaller. The stems are not winged or angled, but possess numerous prickles, and are sufficiently stout to allow of the plant being cultivated without the aid of stakes. Leaves are alternate or sub-opposite, the base deeply cordate; nerves 7 to 9. Flowers minute on panicles; base of the fruit tapering.

TUBERS.
482

Dalzell and Gibson appear to have been the first authors to observe this plant under cultivation. To Roxburgh it was only known as a wild product. The Father of Indian Botany says that the tubers in his time were "dug up in the woods, for it is not cultivated, and carried for sale to the market at Calcutta." He adds that this occurs "during the cold season." Dalzell and Gibson say: "It is the smallest of the cultivated species, but it is also the most delicate." They add that it is common in Bombay, but was originally "imported from Goa." In his *Useful Plants of Bombay,* Lisboa writes of this plant that it grows "in very good soil to a very large size; white and mealy, and is much appreciated." Dymock says that it is less common in Bombay than **D. sativa** and **D. globosa,** "but deserves to be more generally known, as it is the whitest and most delicate of the species and is dry and mealy and quite free from the mucilaginous taste which makes the common yam cloying to the appetite; it is planted in June, just before the rains, and in November yields a cluster of tubers similar in shape and size to kidney potatos." Firminger (*Manual of Gardening for India*) is silent as to this species, a fact to be perhaps accounted for by his knowledge being chiefly local (namely, Bengal), and this, together with Roxburgh's statement that it is not cultivated, may be accepted as establishing this as a more peculiarly Bombay and Madras cultivated plant. DeCandolle lays stress on the fact that none of the Indian cultivated yams have been found in a wild state. But this statement should have been that none of the Bengal cultivated species had

DIOSCOREA
alata.   Wing-stalked or White Yam—the Ubi.

been observed by **Roxburgh** as existing in a wild state. **Roxburgh** was practically ignorant of the Flora of Western India, and, had a contemporary of anything like **Roxburgh** in ability, described the plants of Bombay, it is probable that on many debatable points of plant distribution, we should have held very different views than those current in such subjects. There is nothing in **Roxburgh's** statements to disprove the opinion that **Dioscorea aculeata** was not a wild as well as a cultivated plant at the very time at which he wrote. **Mueller** (*Extra-Tropical Plants*) seems to hold that it is a native of " India, Cochin-China, and the South Sea Islands."

Conf. with p. 117.

This may be so, but the writer can discover no mention of its occurring either in Eastern Bengal, Assam, or Burma, where (if a native of Cochin-China) it might have been looked for. According to many writers it is one of the best of yams, being often very sweet, almost like a sweet-potato. **Dr. Seeman** regarded it as one of the best of esculent tubers.

Roxburgh's account of the tuber having in his time been collected in the " woods," and taken to the " Calcutta market " is interesting. The writer is aware that the rural population and aboriginal hill tribes regularly gather this wild tuber, but he does not recollect of ever having seen it offered for sale in Calcutta. Indeed, Bengal is not now a large consuming province for yams, the potato having gained too strong a position for that. Such yams as are to be found in the bazars appear all to be regularly cultivated. It seems probable that a far larger number of yams are consumed in Bombay than in Bengal. The principal species grown in Bengal is **D. globosa**, and after that the next most important is **D. alata**. In Bombay **D. sativa** (the long yam) and **D. globosa** (the round yam) are about equally abundant, and next to these comes **D. aculeata ; D. alata** is apparently but rarely met with in Western India.

FOOD.
Tubers.
483
484

Food.—Tubers extensively eaten. In the *Mysore Gazetteer* it is stated that *Genasu* is " one of the chief cultivated products in these districts."

Dioscorea alata, *Linn. ; Roxb., Fl. Ind., Ed. C.B.C., 727 ; Wight, Ic., 810.*

Kam.
Conf. with pp. 121 & 135.
485

YAM, or WING-STALKED YAM. This is the WHITE YAM of Ainslie and the UVI or UBI of Australia, Java, &c. Sometimes called BARBADOES YAM.

Vern.—*Khám-álu,* or simply *khám,* HIND. and BENG.; *Gorádu,* KHANDESH; *Chin,* KOLHAPUR ; *Kam álu,* BOMB. ; *Kon,* POONA ; *Rat álu,* GUZ. ; *Kalung, Katsjút-kelangu* (*Yamskalung* according to Ainslie), *perum vullie kalangu,* TAM. ; *Gudimi donda pendalam, niluvu-pendalum,* TEL. ; *Pendalam,* NELLORE ; *Perinvullie kélunghu* (Ainslie), MALAY ; *Kirri-kondol, kahata-kondol, lingúrella,* SING. ; *Dandálu,* SANS.

References.—*Voigt, Hort. Sub. Cal., 652 ; Thwaites, En. Ceylon Pl., 326 ; DC., Origin Cult. Pl., 77 ; Drury, Hand-book of Indian Flora, III., 274 ; Katsji-kelengu of Rheede, Hort. Mal., VII., t. 38 ; Mueller, Extra-Trop. Pl., 106 ; Ainslie, Mat. Ind., I., 329 ; U. C. Dutt, Mat. Med. Hind., 296 ; Year Book, Pharm., 1878, 275 ; Bomb. Gaz. (Poona), XVIII., 56 ; Bom. Gaz. (Khandesh), XII., 171 ; Nellore Manual, 405 ; Banda (N.-W. P.) Gaz., 85 ; Indian Forester (III.), 236 ; Firminger, Man. Gard. Ind., 121 ; Lisboa, U. Pl. Bomb., 178 ; Birdwood, Bomb. Pr., 178 ; Balfour, Cyclop., 951 ; Sir W. Elliot, Fl. Andh., 56, 134.*

Habitat.—This species is much cultivated in various parts of India, and is one of those that **Roxburgh** had not seen in a wild state, and to which **DeCandolle** makes special reference. **Lisboa** (*Useful Plants of Bombay*), however, remarks that it is " wild in the Konkan," and **Drury** that it occurs in both Konkans "flowering in the rainy season." **Ainslie** writes of the white yam that " The plant is the **Dioscorea alata,** *Linn.,*

| The Dark Purple Yam. | (*G. Watt.*) | DIOSCOREA atropurpurea. |

and is indigenous in the Indian islands, where however, though the yam often grows to a large size, it is not so delicate a root as in India. In the western parts of the Archipelago it is called *ubi*; in Ternate *ima*; in Macassar *lami*; in Amboyna *heli*; and in Banda *lutu*." Roxburgh remarks that this is the only species cultivated for food on the Coromandel Coast; flowering time the close of the rains. **Wight** that this species is "universally cultivated in the Carnatic, being that which produces the yam." **Thwaites** mentions it as cultivated in Ceylon.

Description.—The tubers are oblong, white internally, brown on the surface. They often attain a great size, being sometimes 8 feet long, and weigh 80 to 100lb. The stems are four-angled, with wide-spreading wings along the margins; they are rarely prickly, only a few prickles occurring on the lower portion of the stem.

There are numerous recognisable cultivated forms of this plant, some approximating to **D. purpurea.** According to **Roxburgh** this species is extensively cultivated in India, and holds the second place in popular estimation, following **D. globosa** and preceding **D. purpurea.**

Food.—The tubers are extensively eaten. "The common yam is grown in small quantities without water or manure in the hilly waste round the edges of fields or in house yards. If left to grow till December the root attains 2 feet long and 8 inches across. The plant, which is a creeper with longish pointed leaves, bears two to five tubers or roots which, when boiled, make an excellent vegetable" (*Bomb. Gaz., XVIII., 56*). "There are two or three cultivated kinds, and several wholesome wild yams are gathered both for food and medicine" (*Bomb. Gaz., XII., 171*). Of Nellore it is reported that there are two forms of yams grown—a white and a red. "The land is well ploughed, then holes are dug and manured, and cuttings of the yam are put into each, and watered till they sprout about the eighth day. The plant spreads over the ground if there is no frame at hand for it to climb. If trained to a frame the roots are much larger than when the plant trails along the ground" (*Conf. with page 122*).

TUBERS.
486

FOOD.
Tubers.
487

Dioscorea anguina, *Roxb.; Fl. Ind., Ed. C.B.C., 728.*

Vern.—*Kúkúr-álu,* BENG.

Habitat.—Roxburgh speaks of this species as a native of the neighbourhood of Calcutta; flowering at the close of the rains.

Food.—Tuber not esteemed, though eaten by the poor when hard-pressed.

488

FOOD.
Tubers.
489
490

D. atropurpurea, *Roxb.; Fl. Ind., Ed. C.B.C., 728.*

THE DARK PURPLE YAM: MALACCA YAM; sometimes sold in Calcutta under the name of RANGOON YAM.

Vern.—*Myouk ni,* BURMESE.

References.—*Kurz, Pegu Report, XXII.; Mason's Burma, 465; Balfour's Cyclopœdia, 951; Firminger, Man. Gar., 122.*

Habitat.—Kurz enumerates this species as one of those collected by him in Pegu. Roxburgh remarks: "This is the species so extensively cultivated in Malacca, Pegu, and the Eastern Islands." It is grown both by the Karens and the Burmese, and **Mason** says of it that it is "one of our best yams."

Description.—Roxburgh referring to the tuber of this plant says it grows so large that it may be seen through the cracks in the dry ground caused by the growth of the tuber. The tubers are irregular, round, smooth, and like the stems of a deep purple colour. The stems are mostly four-winged, or often five or even seven-winged. The leaf-stalks are also winged and stem clasping.

Rangoon Yam.
491

TUBERS.
492

D. 492

DIOSCOREA bulbifera.	**Bulb-bearing Yam.**

493

Dioscorea Batatas, *Decaisne.*

Vern.—*Sain-in,* CHINESE.

References.—*DeCandolle, Origin Cult. Pl., 78 ; Agri.-Hort. Soc. Ind., IV., New Series (1874), p. 40.*

Suitable for hill cultivation.

Habitat.—Extensively cultivated in China, but not hitherto found wild in that country. Was early introduced into Europe, where it may occasionally be found under cultivation, though it has not attained a high reputation as a vegetable ; but it is the most temperate of the cultivated species hitherto brought to notice, and is thus the one which might succeed best in the higher alpine regions of India.

According to some writers this is only a cultivated state of **D. glabra**, *Roxb.*, and on this account mainly is it mentioned in this work, since apparently the cultivated plant as described by **Decaisne** is not grown in India.

494

D. bulbifera, *Linn. ; Wight, Ic., t. 878.*

BULB-BEARING YAM.

Syn.—HELMIA BULBIFERA, *Kunth.* ; DIOSCOREA TAMNIFOLIA, *Salisb.* ; D. PULCHELLA, *Hokenhacker* ; D. PULCHELLA, *Roxb.*

Vern.—*Zamín kand,* HIND. ; *Piska,* SANTALI ; *Zamín khand* (the tubers), PB. ; *Karinda* (or *hadu-karanda*), BOMB. ; *Karanda,* POONA ; *Karu-karinda,* DEC. ; *Kurú kanda,* CHANDA ; *Kathálu, patni-alu, mati álu,* ASSAM ; *Maláká-káya-pendalam* (a form introduced from the Straits), *chedu paddu dumpa* (according to **Elliot,** for **D. pulchella,** *Roxb.*), TEL. ; *Katu-katsjil,* MALAY ; *Panúkondol,* SING.

References.—*Thwaites, En. Ceylon Pl., 326 ; Dals. & Gibs., Bomb. Fl., 247 ; Stewart, Pb. Pl., 229 ; Drury, Hand-book, Indian Flora, III., 277 ; Rheede, Hort. Mal., VII., t. 36 ; Kunth, Enum., V., 435 ; Trimen, Cat. Ceylon Pl., 93 ; Year-Book, Pharm., 1878, 275 ; Bomb. Gas. (Poona Dist.), XVIII., 56 ; Baden Powell, Pb. Pr., 259 ; Drury, U. Pl., 182 ; Lisboa, U. Pl. Bomb., 179 ; Birdwood, Bomb. Pr., 178 ; Sir W. Elliot, Fl. Andh., 110.*

See p. 130.

Habitat.—Wild in Sylhet, Chittagong, and throughout the Western Ghâts to Bombay. Cultivated in the Western Presidency, especially in the Konkan. It is also mentioned by **Aitchison** as occurring at Moradabad and Simla, and is enumerated by **Strachey** and **Winterbottom** as met with at Almora, flowering in July. **Baden Powell** refers to samples of this tuber having been sent to the Lahore Exhibition from Simla, Kashmír, and Hushyarpur. (*Conf.* with the tuber described under **D. deltoidea,** which is specially mentioned by **Aitchison** as collected at Hushyarpur.) **Baden Powell** accepts the tubers sold under the name of *Zamín khand* as being those of this species. Of the *karanda* yam of Poona it is said that it differs from the *kon* (**D. alata**) in having rounder leaves and in bearing bulbs on the stems as well as on the roots. **Elliot** says that **D. bulbifera** was introduced into India from the Straits.

MEDICINE. Tubers. **495** Leaves. **496**

Medicine.—The TUBERS are applied to ulcers after being dried and powdered. In the plains of the Panjáb the LEAVES of a species of **Dioscorea** (most probably this) are used medicinally and sold under the name of *tarar puttr.* **Baillon** (*Dictionnaire de Botanique, Vol. II., p. 437*) alludes to the known febrifugal property of the leaves of certain species of **Dioscorea,** rendering them useful in the treatment of intermittent fevers.

FOOD. Tubers. **497**

Food.—The bulbules on the stems and the tubers under ground are used as vegetables (*Birdwood*). The latter are bitter, but are rendered eatable by being covered with ashes and steeped in cold water. In a report on Assam it is stated that the tubers are boiled in *khar* water, or in an acid before being eaten. The **Rev. A. Campbell** says that cattle eat the leaves.

Dr. Stewart, under the name of **D. deltoides,** *Wall,* "(and **D. bulbifera,** *L.?*)" gives the following vernacular names : *kníss, kríss, tar, kithí, tardí,*

D. 497

Products of India.   129

| Yams used as Detergents.   (G. Watt.) | DIOSCOREA deltoidea. |

FOOD.

gúngrú, kaspat, &c., and states that the plant he refers to grows abundantly in many parts of the Panjáb Himálaya ; that the root (several pounds in weight) is largely eaten, cooked, by various classes in parts of the Siwaliks and outer hills, *after steeping it in ashes and water to remove acridity.* It is difficult to know to which of the Panjáb species **Stewart** more particularly refers. He says that in Kashmír the roots are employed for washing pashm and wool cloth. **Vigne (Stewart** further informs us) affirms that a smaller kind is used to wash cotton cloth, and a third reputed to be used for silk. **Honigberger** states that a yam is also employed in dyeing *nafarmání* (blue). **Baden Powell** describes the *Zamín khand* tubers as follows :— "When the root is cut open it is yellowish inside ; at first it is very bitter and requires to be boiled several times, and sometimes also with lime water, before it is fit to be eaten. It is used also as pickle. For this purpose it is cut into little pieces and fried in oil till it becomes of a red colour, and then it is put into vinegar, &c., or in the mixture of mustard seeds ground up with salt, &c., in water or oil, which is sometimes used as a preservator."

DOMESTIC.
498

Domestic Uses.—"The tubers of this species are employed by the Singhalese for attracting fish to certain spots where they can be easily caught. The tubers are broken into pieces and thrown into the water daily for some time " (*Thwaites*).

Dioscorea crispata, *Roxb. ; Fl. Ind., Ed. C.B.C., 728.*

499

Vern.—*Rudraksha pendalam, káya pendalam,* Tel. ; *Myouk-kya, ká-bunway, ta-tway-u,* Burm.
References.—*Kurz, Report on Pegu, XXII. ; Drury, Hand-book, Flora of India, III., 276 ; Voigt, Hort. Sub. Cal., 652 ; Elliot, Fl. Andh., 90, 165.*
Habitat.—" A native of the interior of Bengal " (*Roxb.*), and according to **Kurz**, of Pegu in Burma ; flowering time, the rainy season.

D. dæmona, *Roxb. ; Fl. Ind., Ed. C.B.C., 729 ; Wight, Ic., t. 811.*

500

Syn.—Helmia dæmona, *Kunth, Enum., V., 439 ;* Ubium silvestre, *Rumph., V., t. 126.*
Vern.—*Pedumpa, puli dumpa,* Tel. ; *Kyway, kyway-nway, kyway-pin,* Burm.
References.—*Kurz, Pegu Report ; Voigt, Hort. Sub. Cal., 653 ; Dalz. & Gibs., Bomb. Fl., 247 ; Atkinson's Him. Dist., 602 ; Hooker, Niger Flora, 538 ; Elliot, Fl. Andh., 147, 158.*
Habitat.—**R**oxburgh mentions this as a native of the Gorakhpore forests as well as of the Moluccas. **Kurz** describes it as occurring in Pegu.

It is also enumerated in the list of plants collected by **Strachey** and **Winterbottom** in Kumáon, where it is stated to occur at altitudes of 2,500 feet. **Dalzell** and **Gibson** write that it is met with " at Vingorla : Hills in Concan, rare." It is also found in Africa, a specimen being described (in *Hooker's Niger Flora*) as " apparently the same as the widely-diffused East Indian plant."

Nauseous Tubers.
501

Food.—The tuber is very nauseous, even after being carefully boiled.

D. deltoidea, *Wall.*

FOOD. Tubers.
502
503

Vern.—*Gún,* Kumaon ; *Kniss, kriss, tar, kithí, khelí, dharúr, tardi, tharrí, káns, gungrú, kaspat, parwatti* ; and the bazar name for the medicinal leaves *tarar pattr,* Pb. **Baden Powell** gives this species the Panjábi names *tarar, krish.*
References.—*Stewart, Pb. Pl., 229 ; Baden Powell, Pb. Prod., 259, 378.*
Habitat.—A common Himálayan species, occurring between 3,000 and 8,000 feet : flowering time, May. It often attains a considerable size, spreading over trees and bushes for 15 to 20 feet. Frequent in the vicinity

K

DIOSCOREA
fasciculata.   The Kidney-shaped Yam or Karen Potato.

---

of Simla and throughout Kulu, being distributed into Afghánistan. The deltoidly often trilobed leaves of this plant are very characteristic, but the small flowers occur on a lax, elongated axis in a manner peculiar to this plant.

The above Panjáb vernacular names are taken from **Stewart's** *Panjáb Plants*, but it seems probable they are the names not for one species of **Dioscorea** but for all those met with in the Panjáb Himálaya. These names are, from a historic point of view, of great interest, and are reproduced here (and also under **D. bulbifera**) in the hope that future writers may be able to properly distribute them to the species to which they more especially belong. **Aitchison**, in his *Catalogue of Panjáb and Sind Plants*, mentions three species—**D. sagittata, D. bulbifera,** and **D. sp.**, undetermined. Regarding the last he says that it is found near Hushyarpur and

Conf. with
p. 128."yields a large tuber, much sought for at certain Hindu feasts. The ground where this grows is one mass of pits from the continuous digging for the tubers called *Thurí*." It has been repeatedly urged that much damage is done to young seedling trees by the deep pits dug throughout the forests while searching for yams.

DETERGENT.
504
Detergent.—**Stewart**, in his Notes on a Tour in Khághán, mentions the *kriss* tubers as used for washing silk.

MEDICINE.
Leaves.
505
Medicine.—The LEAVES of this species are sometimes spoken of as employed medicinally, having febrifugal properties assigned to them. The TUBERS are detergent and are utilized in dyeing (*Conf.* with **D. bulbifera**).

Tubers.
506

507

Dioscorea fasciculata, *Roxb.; Fl. Ind., Ed. C.B.C., 728.*

THE KIDNEY-SHAPED YAM; KAREN POTATO.

Vern.—*Súsní álu, sathui,* BENG.; *Suthni,* BIHAR; *Kangar,* POONA; *Kadwæ-oo, tá-tway-u,* BURM.

References.—*Voigt, Hort. Sub. Cal., 652; Mason's Burm., p. 464, 813; Kurz, Prelim. Report, Pegu, XXII; Grierson, Bihár Peasant Life, 250; Firminger, Man. Gard., 122; Bomb. Gaz., XVIII., 56.*

Starch made
from the
Tubers.
508
Habitat.—"Cultivated to a considerable extent in the vicinity of Calcutta, not only for food but to make starch of the roots." (*Roxb.*) It is cultivated extensively by the Karens, and, being more like the potato than the yam, has acquired the name of the Karen potato or Tavoy potato. **Mason** adds: "I am not aware that it is ever found wild on the coast." **Kurz** mentions this species, in his *List of Pegu Plants*, but does not state whether gathered from a wild or cultivated stock. According to a writer in the *Bombay Gazetteer*, this species, known as *kangar*, is grown in Poona.

TUBERS.
509
Description.—The root consists of several small, smooth, light-coloured tubers, which are used by the natives for food and for the manufacture of starch. (*Roxb.*)

Mason describes this species as follows: "This is a small yam, not much larger than a kidney potato, which it much resembles both in appearance and taste." "It is the best vegetable we have, but unfortunately it can be obtained during a few months only in the year." The Poona

Conf. with
p. 117 and
p. 126.yam referred to above is described as "closely resembling the *kon* (**D. alata**) and the *karanda* (**D. bulbifera**). It is found in the hills. Its bulbs, which form below the ground, are like a small sweet-potato in size and shape. The flesh is white and sweet" (*Poona Gaz., XVIII., 56*). **Firminger** remarks that the *súsní álú* is "a very distinct kind of yam; the tubers are about the size, form, and colour of large kidney potatos, and when well cooked bear a greater resemblance, in mealiness and flavour, to the potato than any other yam I know."

D. 509

Products of India.

131

The Chinese; the Common; and the Tivoli Yams. (*G. Watt.*) | DIOSCOREA nummularia

Dioscorea glabra, *Roxb.; Fl. Ind., Ed. C.B.C., 729.*

510

THE CHINESE YAM.

Vern.—*Eddu tóka dumpa, nára tega,* TEL.; *Ato sang,* SANTALI; *Káteh, myouk-mway,* BURM.

Sir Walter Elliot (*Fl. Andh.,* 129) remarks that the Telegu name means "Bullock's-tail root," owing to the length of the tubers. He further adds that *téga* is the tuberous root, *tíge* the twining stem.

References.—*Kurz, Pegu Report; Sir Walter Elliot, Flora Andh.,* 49, 129; *Voigt, Hort. Sub. Cal.,* 653; *Drury, Hand-book Fl. Ind., III.,* 274.

Habitat.—A native of Lower Bengal and Sylhet.

Food.—The Rev. A. Campbell says this species is largely cultivated by the Santals of Chutia Nágpur, and that the ærial bulbs as well as the under-ground tubers are eaten. Mueller calls this the Chinese yam, and adds that the tuber is known to attain a length of 4 feet, with a circumference of 14 inches, and weight of about 14℔. "The inner portion of the tuber is of snowy whiteness, of a flaky consistence, and of a delicious flavour; preferred by many to potatos, and obtained in climates too hot for potato crops. The bulbilles from the axils of the leaf-stalks, as in other **Dioscoreas,** serve as sets for planting, but the tubers from them attain to full size only in the second year, the produce being in proportion to the set planted."

FOOD.
TUBERS.
511
Bulbs.
512

D. globosa, *Roxb.; Fl. Ind., Ed. C.B.C.,* 727; *Wight, Icon., t.* 812.

513

THE COMMON YAM.

Vern.—*Chúpri álu,* HIND. and BENG.; *Bengo nári,* SANTALI; *Chaina, chopri álu, khaun phal, safed kanphal, myouk-phal,* BOMB.; *Gúna pendálam,* TEL.; *Pindálu,* SANS.

References.—*Voigt, Hort. Sub. Cal.,* 652; *Dals. & Gibs., Bomb. Fl., Supp.,* 92; *Müeller, Extra-Trop. Pl.,* 106; *Firminger, Man. Gard.,* 121; *Atkinson, Econ. Prod. N. W. P., V.,* 21; *Drury, U. Pl.,* 183; *Lisboa, U. Pl. Bomb.,* 178; *Elliot, Fl. Andh.,* 65.

Habitat.—This species is largely cultivated, especially in parts of Bengal. The tubers are roundish, sometimes very large, inside pure white. No writer appears to have noticed the plant in a wild state in India, though it is common under cultivation; the *Bengo-nári* of the Santals may not be this plant, but if it so it is wild at the foot of Parisnath. It flowers during the middle of the rains, the spikes of flowers being verticelled on a long axis.

Food.—The TUBERS of this species are the most esteemed of all the yams; they are eaten by the natives, and are also much liked by Europeans in India. Firminger does not agree, however, with Roxburgh in this high opinion regarding the tuber sold under the name of "*choopree álú*" as being the best form of Yam. "Others appear to me," he adds, "to be superior." The natives eat them in curries, and also after being boiled. They are often baked into bread. Atkinson says the fruits of this species are edible. "Formerly largely used by the natives as a vegetable, but now almost entirely displaced by potatos" (*Gaz. Beng., I.,* 139). They are said to be planted in June and dug up in January and February.

FOOD.
Tubers.
514

D. nummularia, *Lamarck; Roxb., Fl. Ind., Ed. C.B.C.,* 729.

515

THE TIVOLI YAM.

Vern.—*Bhora álu,* HIND. and BENG.; *Káru pendalam,* TEL.

Habitat.—A native of the neighbourhood of Calcutta; flowering time the close of the rains. Root tuberous, small.

K 2

D. 515

FOOD.
Tuber.
516

Food.—The tuber appears to be unfit for human food, according to Roxburgh; but Mueller calls this the Tivoli Yam and describes it as "A high climbing prickly species, with opposite leaves. Roots cylindrical, as thick as an arm; their taste exceedingly good." These writers in all probability allude to different plants. **Sir Walter Elliot** (*Fl. Andh., 86*) says the Telegu name given above is really applied to any wild species.

[*Wight, Icon., 813.*

517

Dioscorea oppositifolia, *Linn.; Roxb., Fl. Ind., Ed. C.B.C., 729;*

Vern.—*Már-páspoli,* or *márapasapoli,* BOMB.; *Piska,* SANTALI; *Are tige* or *tégálu, avatenga tige,* TEL.; *Hiri-tala,* SING.

References.—*Voigt, Hort. Sub. Cal., 653; Dalz. & Gibs., Bomb. Fl., 247; Lisboa, Useful Pl. Bomb., 179; Thwaites, Enum. Ceylon Pl., 326; Sir Walter Elliot, Flora Andh., 16, 18; Rev. A. Campbell, Econ. Prod., Chutia Nágpur; Müeller, Extra-Trop. Pl., 107; Drury, Hand-book of Indian Fl., III., 275; Trimen, Cat. Ceylon Pl., 93.*

Habitat.—A native of the east and west coasts of Southern India, extending north to Khandalla; of frequent occurrence in the sub-alpine forests. Flowering time the rainy season. Stems smooth, round, slender, twining, annual; leaves opposite, petioled, ovate-lanceolate, slightly cordate; inflorescence panicelled. Distributed to Ceylon and China.

FOOD.
Roots.
518
Tubers.
519

Food.—The Rev. Mr. Campbell reports:—"The ROOTS and the ærial TUBERS are eaten, the former being considered by the forest tribes as a great delicacy." This remark probably refers to **D. bulbifera**—one of Mr. Campbell's specimens of *piska* is certainly D. bulbifera—Dymock enumerates this plant among those used during famine time in Bombay, both the tubers and the flowers being eaten.

FODDER.
520
MEDICINE.
Root.
521
522

Fodder.—In the Santal country cattle are said to eat the leaves.

Medicine.—"The ROOT, ground and heated, is applied to reduce swellings: it is also used in snake-bite and scorpion sting."

[*Ic., 814.*

D. pentaphylla, *Willd.; Roxb., Fl. Ind., Ed. C.B.C., 730; Wight,*

THE KAWAN YAM OF FIJI.

Vern.—*Kanta-álu,* HIND.; *Kanta álu,* N.-W. P.; *Taguna, takuli, magina muniya,* KUMAON; *Kanta-alu, ulsi, shendorwel,* BOMB.; *Chataveli, úlasi,* MAR.; *Vullie, kattu-vullie-kelangu,* TAM.; *Mullu pengdalam, pandi mukku dumpa, konda-gummudu,* TEL.; *Nureni-kelangu,* MALAY.; *Pwá-sá-o, pho-sáo,* BURM.; *Katu-wala,* SING.

References.—*Voigt, Hort. Sub. Cal., 653; Thwaites, En. Ceylon Pl., 325; Dalz. & Gibs., Bomb. Fl., 247; Rheede, Hort. Mal., VII., 34, 35; Sir Walter Elliot, Flora Andh., 95, 115, 116, 118, 126, 131, 142, 144; Dymock, Mat. Med. W. Ind., 2nd Ed., p. 843; Atkinson, Him. Dist., 389 & 602; Drury, U. Pl., 183; Lisboa, U. Pl. Bomb., 179; Birdwood, Bomb. Pr., 179.*

Habitat.—Common in the jungles, on low hills throughout the greater portion of India.

Conf. with
p. 135.

Wight remarks: "A sufficiently common species in jungles on low hills, &c., but never, so far as I have seen, cultivated, which is the more remarkable as I have always found the natives dig the tubers, whenever they had an opportunity to dress and eat them." **Roxburgh** says he has only seen this plant in its wild state. It occurs on the lower Himálaya ascending to 6,000 feet in altitude; is common a little below Simla.

Description.—The stems are prickly and furrowed, rather hairy; leaves digitately 5-divided, downy, segments oblong acuminate. The tubers oblong, large, and white, considered wholesome and palatable.

FOOD.
Tubers.
523
Flowers.
524

Food.—Affords large edible TUBERS, eaten all over India. The FLOWERS are also eaten as a vegetable; they are sold in the bazars of Bom-

The Purple and Common Yams. (*G. Watt.*)	DIOSCOREA sativa.

bay during the rainy season and are considered wholesome. The LEAVES are also sometimes eaten during times of scarcity and famine.

Dioscorea pulchella, *Roxb.; Fl. Ind., Ed. C.B.C., 728; Fl. Andh., 35;* see **D. bulbifera.**

FOOD.
Leaves.
525

D. purpurea, *Roxb.; Fl. Ind., Ed. C.B.C., 727.*

526

THE PURPLE YAM.

Vern.—*Rukto-gurániya álu, Lal-gurania-álu,* HIND. & BENG.; *Désaváli-pendalam* (= Country or Common Yam), TEL.

References.—*Voigt, Hort. Sub. Cal., 652; DC., Origin Cult. Pl., 77; Elliot, Flora Andh., 46; Müeller Extra-Trop. Pl., 107; Firminger, Man. Gard., 121; Ainslie, Mat. Ind., I., 330; Atkinson, Econ. Prod., V., 22.*

Habitat.—Cultivated in most districts of Bengal. Ainslie says that many writers consider this only a purple form of the white yam (**Dioscorea alata**). Firminger writes that this appears to be the tuber brought by a Mr. McMurray from Mauritius, and which is there cultivated as extensively as the potato is in England. "The tuber is of a dull crimson red outside, and of a glistening white within."

Food.—The TUBERS are oblong, throughout of a lighter or darker purple, but always considerably deep in the tinge. (*Roxb.*)

It is reckoned by the natives as the third best among the yams, D. globosa being considered the first and **D. alata** the second.

FOOD.
Tubers.
527

D. quinata, *Wall.*

528

Vern.—*Magiya, muniya,* KUMAON.

Habitat.—Said to be met with in the North-West Himálaya at altitudes of 6,000 feet.

Food.—Yields white edible tubers.

FOOD.
Tubers.
529
530

D. rubella, *Roxb.; Fl. Ind., Ed. C.B.C., 727.*

Vern.—*Guraniya álu, goran álu,* BENG.

References.—*Voigt, Hort. Sub. Cal., 652; Firminger, Man. Gard., 122; Gaz. Bengal, V., 307; Atkinson, Econ. Prod. N.-W.P., V., 22.*

Habitat.—Much cultivated in parts of Lower Bengal, especially about Calcutta. Stems twining 6-winged; leaves opposite sagittate cordate, 5-7-nerved with subulate points. Tubers oblong, sometimes 3 feet long, deeply tinged with red under the epidermis.

Food.—Held fourth in estimation, by the Bengalis, and employed by them as food. "Fleshy and farinaceous, and may be used as an excellent substitute for potato, towards the close of the rains, when they come to perfection and the latter vegetable is dear" (*Gaz. Beng., V., 307*). Firminger describes this as a "common but very excellent yam, as good as any perhaps in cultivation. The tuber is of a great size, crimson red on the outside, and of a glistening white within."

FOOD.
Tubers.
531

D. sagittata, *Royle (non Poir.).*

532

Vern.—*Tair, tarur, tagur,* KUMAON; *Tarúr,* N.-W. PLAINS.

Habitat.—Met with in the North-West Himálaya.

Food.—The TUBERS are edible.

FOOD.
Tubers.
533
534

D. sativa, *Linn.; Kunth, Enum. Pl., V., 341.*

COMMON YAM.

Vern.—*Rátálú,* HIND.; *Ato sang,* SANTALI; *Ratálú,* PB.; *Chiná, gordi-kaun-phal,* BOMB.; *Gorkan, gorádu,* MAR., DEC.; *Ratálu (dhola* = cultivated white form, *al* a coloured), *yamskollung,* GUZ.; *Heggenasu,* KAN.

Grierson (*Bihár Peasant Life*, 250) gives the following Bihárí names for what appears several species of Yam, though by him placed under **D. sativa**:—*Latar*, north of the Ganges; *ratar* to the west; *atár*, north-west; *kathár*, south-west; in Patna and Gaya *ratálu*; in Tirhoot *phar*; and to the east *khamharna*. While regretting inability to arrange these names under their correct botanical synonyms, there seems no doubt they are names for yams, and it is interesting to have to add that **Mr. Grierson** furnishes the Bihárí names for **Colocasia antiquorum** (the *kachchu*), as distinct from those above, which he says are yams.

References.—*Dalz. & Gibs., Bomb. Fl. Supp., 92 ; Stewart, Pb. Pl., 229 ; Drury, Hand-book Fl. Ind., III., 274 ; Year Book, Pharm., 1878, 275 ; Beyt's Gujarat Agri., p. 53, appears to be describing three cultivated forms of this species ; Lisboa, U. Pl. Bomb., 178 ; Birdwood, Bomb. Pr., 179 ; Bomb. Manual, Rev. Accounts, 101 ; Gaz. of Mysore and Coorg, I., 55 ; Smith, Dic., 444.*

Habitat.—Cultivated over the greater part of India. **Dr. Dymock**, in a letter to the author, says there are several distinct varieties of this plant in Bombay. **Lisboa** remarks that it is " wild and cultivated in India and the Archipelago." **Dalzell** and **Gibson** state that it is "the most common species cultivated." Cultivated in Chutia Nagpur by the Santals.

MEDICINE.
Tubers.
535
FOOD.
Tubers.
536

Medicine.—In the form of a powder it is used as an external application for ulcers.

Food.—The TUBERS are eaten cooked as a common article of diet. In Bombay this is the most extensively cultivated species ; of Gujarat it is said : " Three varieties of yam (*Ratálu*) are grown here, a long and a small white tuber—both called *dhola*, and a purple tuber, *lal*. They are alike generally raised on the ridges of ginger fields, irrigated and manured. Duration in the ground from seven weeks to two months." It is alluded to as cultivated in Rewa Kantha and Kathiawar. The "*ratálu*" is also said to be grown in Cawnpore, the seed for an acre being two maunds, and the produce 200 maunds. In the Hardoi Settlement Report (*p. 16*) it is stated that yam lands in the vicinity of towns fetch R50 an acre.

[*Wight, Icon., t. 815.*

Dioscorea tomentosa, *Kœnig. ; Roxb., Fl. Ind., Ed. C.B.C.,*729 ;

Syn.—HELMIA TOMENTOSA, *Kunth ;* and probably also D. TOMENTOSA, *Hohenhacker.*

Vern.—*Chenyel, cháyena,* BOMB.; *Subba dumpa,* TEL.; *Kyway pin,* BURM.; *Úyala,* SING.

References.—*Kurz, Prel. Report, Pegu, XXII. ; Dymock, Mat. Med. W. Ind., 2nd Ed., 843 ; Sir Walter Elliot, Flora Andh., 169 ; Trimen, Cat. Ceylon Pl., 93.*

Habitat.—According to **Roxburgh** this is a native of the Circars, appearing in the rains. If the above synonyms be correct, it is possibly also a native of the West Coast of India at Travancore, Mangalore, &c.

Food.—The young SHOOTS are eaten as greens in Bombay.

D. triphylla, *Linn. ; Kunth, Enum., 392.*

Vern.—*Mándá,* MAR. ; *Ts-iagri-nuren,* MAL.

References.—*Rheede, Hort. Mal., VII., t. 33 ; Dals. & Gibs., Bomb. Fl., 247 ; Dymock, Mat. Med. W. Ind., 2nd Ed., 843 ; Birdwood, Bomb. Prod. ; Year Book, Pharm., 1878, p. 275.*

MEDICINE.
Tubers.
540
FOOD.
Tubers.
541

Habitat.—Common in the Concan and in Malabar.

Medicine.—Dymock says the TUBERS are sometimes used to disperse swellings.

Food.—The TUBERS are, according to **Graham**, intensely bitter and intoxicating. They are said to have been eaten in Bombay during times of scarcity and famine. They have to be well boiled to destroy the acridity, but are even then not over-wholesome.

Dioscorea versicolor, *Wall.*

542

Vern.—*Genthi, gajir, ganjira,* HIND. ; *Dola álu, dudha álu, kanri, genti,* CHUTIA NAGPUR; *Githi* (Bijnor), N.-W. P. ; *Genthi* or *genti gajir, ghanjín,* KUMAON.

References.—*Voigt, Hort. Sub. Cal.,* 653; *Beng. Gaz., XVI., p. 53,* may be alluding to this species; *Atkinson, Him. Dist.,* 735; *Econ. Prod., V., 98.*

Habitat.—A native of the Eastern Peninsula, from Monghyr to Kumáon, Nepál, Assam, Prome, and Tavoy.

Food.—Atkinson says of a wild yam found in Kumáon (which he may be correct in assigning to this species) that it "occurs in these provinces (N.-W. P.); the BULB on the stem and ROOT is eaten as a vegetable; flowers in the rains. This climber is of considerable interest, as its large tubers furnish the deliciously fragrant yam which supplies a great part of the food of the Bhuksas in Moradabad and Bijnor when grain is scarce. The plant is common throughout the forest; and its tubers, which grew to several pounds weight, are got at by digging from 2 to 6 feet. To remove their original acridity, they are always steeped for a night in ashes and water before being cooked."

FOOD.
Tubers and Bulbs.
543

CONCLUDING REMARKS REGARDING YAMS.

Since the above was written, some interesting facts, in reply to a circular issued by the Government of India, have been received from the various local Governments. In most of the reports furnished, the attempt has not, however, been made to distinguish the species. From Burma it has been reported that in the Upper Chindwin District yams are cultivated and sold for 3·65℔ (one viss) for 1 to 2 annas. Information is also furnished regarding Pegu, where several forms are grown. Of the following districts it has been reported that Yams are cultivated :—Tharawaddy, Bassein, Shwegyin, Amherst, and Upper Burma. Of Hanthawaddy it is stated that the cultivation of an acre costs R11 and the profit amounts to R50. Of Bhamo it is remarked, two kinds occur—a white and a red. "The white kind grows wild and the red is cultivated." In the Transactions of the Agri-Horticultural Society, Vol. III., 10, Col. Burney reported that in his time Yams were scarce in Burma.

BURMA.
Price.
544

Of Madras similar reports have been received : of the Kistna District it is stated that the crop is planted out in January and harvested in August. The cost of cultivation is R20 per acre and the profit R30. The Collector of Cuddapah says that yams in his district are known as *nilwa pendalam,* and that they both grow wild and are cultivated. A further report has just been received from the Conservator of Forests, Southern Circle, Madras, a passage from which may be here given : "There are a number of species of Yam, both wild and cultivated, in Southern India. Most of them bear only underground tubers, but there are two species in Malabar and Travancore bearing tubers on the stem and branches : but these are much smaller than the underground tubers, which are as much as 4 to 5 feet in length and nearly a foot in diameter. The axillary tubers borne on the stems and branches are ovate in shape and vary from the size of a pea to about 3 inches in length. These tubers are also eaten but are chiefly used for seed. The species to which they belong appears to be **D. alata,** called in Tamil *Katsjal-kelangu* or *Perum-vullie-kalangu*; in Hindustani *Kam*; in Telugu *Nilunu.* There are also several armed species in all forests in South India—notably **D. pentaphylla** and **D. triphylla.** These bear thin, long, underground tubers, which are very

MADRAS.
545

Nilwa pendalam.
546

Kam.
Conf. with p. 121 and p. 126.
547

DIOSPYROS assimilis.	**The Ebony-wood.**

stringy or fibrous but contain a lot of starch. The Yam is available only for three months in the year, and is not depended upon in the Arcot district as food to any great extent."

BENGAL.
Conf. with
Santal names,
p. 120.
548
Pichku.
549
Byong.
550
Kullu.
551

From Bengal a note has been received to the effect that the hill tribes of Chittagong cultivate several forms of Yam. The whole plant is said to be eaten : no statistics are available however, either as to the cultivated plants or the extent to which wild yams are eaten. Of Chutia Nagpúr it is reported that a " **Dioscorea**—the *Byong-kullú*, or *Pichkú* of the Kóls— is found in the Singbhoom District. The Kóls of Saranda eat the underground tubers called *byong* after cooking them in fire. They depend to a considerable extent on this food. The people of Saitba also eat yams called *Kullú* and *Pichka* after being roasted."

552

DIOSPYROS, *Linn.; Gen. Pl.*, *665*.

The name DIOSPYROS (celestial-fruit),was doubtless suggested for this genus of plants in allusion to the supposition that the fruit of **D.** Lotus may have been the oblivion-fruit of the ancients. The utmost difference of opinion, however, prevails as to the fruit which the heroes of the Odyssey ate, and perhaps DeCandolle is near the point when he says that, after all, Homer's Lotus-plant may have existed only in the fabled garden of the Hesperides.

The Sacred Lotus Flower, which plays so important a part in Oriental literature, is of course a water-lily and had nothing whatever to do with the Lotus-tree.

All the species of DIOSPYROS yield useful timbers, the best being the forms of Ebony. Gamble says of them collectively that all the species yield timber " with small pores, often in radial lines, and fine, very numerous, uniform and equidistant medullary rays, often closely packed. In most species there are numerous wavy, concentric lines across the rays. In several respects the structure of the ebonies resembles the structure of SAPOTACEÆ." The following enumeration of the better known species has been drawn up more on account of the timbers than the other economic facts attributed to them, and it has often been impossible to say more regarding even the wood than that it is used by the people and is viewed as good but inferior to certain other species. It is hoped, however, that the habitats will indicate the distribution of these useful trees, and that in future it may be possible to furnish definite information regarding each species (*Conf.* with the remarks under **D. Ebenum**).

553

Diospyros affinis, *Thwaites; Fl. Br. Ind., III., 566*; EBENACEÆ.

References.—*Beddome, Fl. Sylv., Man., 145 ; Ic. Pl. Ind. Or., t. 127 & p. 26 ; Thwaite:, En. Ceylon Pl., 179 ; Trimen, Sys. Cat. Ceyl. Pl., 52.*

Habitat.—A middle-sized tree met with in the Tinnevelly hills and in Ceylon.

TIMBER.
554

Structure of the Wood.—Thwaites states that the timber obtained from this species is suitable for building purposes. Beddome remarks that it is a good building timber.

555

D. assimilis, *Bedd.; Fl. Br. Ind., III., 558*.

Syn.—D. NIGRICANS, *Dalz. & Gibs., Bomb. Fl., 141 (not of Wall.);* D. EBENUM, *Hiern.* (in part) ; *Bedd., Fl. Sylv.,* 65, under D. EBENUM.

Habitat.—A small tree of the Malabar coast and Kánara.

TIMBER.
556

Structure of the Wood.—This is said to closely resemble **D. Ebenum,** so much so as to have been confused with that species by Hiern. The plant is very imperfectlyknown, and little can, therefore, be ascertained regarding its economic properties. The notices under **D.** nigricans in *Lisboa's Useful Plants of Bombay,* and in the *Bombay Gazetteers,* should be read as descriptive of this species. Beddome remarks : " **D. assimilis** is very nearly allied to, if distinct as a species from, **D. Ebenum** ; it differs,

D. 556

| The Ebony-Woods. | (G. Watt.) | DIOSPYROS cordifolia. |

however, in each of the stamens in the male bearing 4-6 anthers instead of generally only 2, and the stamens in the female flowers being single instead of double; its leaves turn very black in drying; it is called *Kárámárá* in the South Kánara forests, where it is very common, both in the heavy forests, in the plains, and on the Ghâts."

Diospyros Brandisiana, *Kurz ; Fl. Br. Ind., III., 570.*   557
Reference.—*Kurz, For. Fl. Burm., II., 138.*
Habitat.—An evergreen tree met with in Upper Tenasserim.

D. burmanica, *Kurz ; Fl. Br. Ind., III., 565.*   558
Vern.—*Te (Tai-pen)*, BURM.
References.—*Kurz, For. Fl. Burm., II., 133 ; Gamble, Man. Timb., 248 ; Indian Forester, VIII., 416.*
Habitat.—A large tree (attaining a height of 60 feet) characteristic of the *Eng* forests of Burma.

D. Candolleana, *Wight, Ic., tt. 1221-2 ; Fl. Br. Ind., III., 566 ; non Thwaites, En. Cey. Pl., 181.*   559
Syn.—DIOSPYROS HIRSUTA, *Hiern.* (in part); D. ARNOTTIANA, *Miq.;* D. CANARICA, *Bedd., Ic. Pl. Ind. Or., t. 134,* and *For. Man., 145;* D. OLIGANDRA, *Bedd.* This tree has, by many writers, been confused with two Ceylon trees, *viz.,* D. HIRSUTA, *Linn. f.,* and D. THWAITESII, *Bedd.* (the D. CANDOLLEANA, *Thwaites*).
References.— *Beddome, Fl. Man., 144* (D. canarica, *Bedd.*), *Ic. Pl. Ind. Or., t. 134 ; Dalz. & Gibs., Bomb. Fl., 142 ; Lisboa, U. Pl. Bomb., 94 ; Bombay Gazetteer (Kánara), XV., I., 437.*
Habitat.—A small tree or large shrub frequent in the Deccan Peninsula, from the Bombay Ghâts to the Wynaad and Courtallum. It flowers in the hot season, the fruits ripening in the rains. Talbot describes it in Kánara as "A large tree with coriaceous leaves found near Siddápur and elsewhere in North Kánara." It flowers in the hot season, the fruits ripening in the rains.

D. Chloroxylon, *Roxb. ; Fl. Br. Ind., III., 560.*   560
Syn.—D. TOMENTOSA, *Lamk. (non Roxb.) ;* D. CAPITULATA, *Wight, Ic., tt. 1224, 1588.*
Vern.—*Anduli*, GOND ; *Ninai*, BOMB. ; *Illinda, pedda illinda, togarike, aulanche, ullinda, ulimera*, TEL.
References.—*Roxb., Fl. Ind., Ed. C.B.C., 415 ; Corom. Pl., I., 38, t. 49 ; Voigt, Hort. Sub. Cal., 344 ; Brandis, For. Fl., 297 ; Gamble, Man. Timb., 248 ; Dalz. & Gibs., Bomb. Fl., 140 ; Sir Walter Elliot, Flora Andh., 70, 148, 150, 183, 186 ; Indian Forester, II., 179 ; XII., 313 ; Lisboa, U. Pl. Bomb., 93, 164 ; Royle, Ill. Him. \Bot., 262 ; Balfour, Cyclop., 952.*
Habitat.—A middle-sized tree met with in the Deccan Peninsula to Orissa. Particularly common about Surat and in the district of Násik. Wight mentions the Balaghát mountains, Madras. Is said to be a shrub in the Godavari forests.
Food.—Fruit globose, size of a large pea; eaten when ripe, and is reported to be very palatable (*Lisboa*).   FOOD. 561
Structure of the Wood.—Hard, durable, and yellow-coloured. Lisboa says it is useful for various purposes. Balfour remarks that this tree affords a very hard useful wood.   TIMBER. 562

D. cordifolia, *Willd. ; Gamble, Man. Timb., 251.*   563
Syn.—D. MONTANA, *Roxb.* (in part) which see.

564 | **Diospyros crumentata,** *Thw.; Fl. Br. Ind., III., 567.*

Vern.—*Chemel-paniché* (TAM. in Ceylon).
References.—*Thwaites, En. Ceylon Pl., 179 ; Balfour, Cyclop., 952.*
Habitat.—A large tree met with in Ceylon, and probably only a variety of **D. affinis.**

565 | **D. dasyphylla,** *Kurz ; Fl. Br. Ind., III., 554.*

Reference.—*Kurz, For. Fl. Burm., II., 139.*
Habitat.—An evergreen tree with the branches densely fulvously villous. Met with on the Martaban hills, Burma, at a height of 4,000 feet in altitude.

566 | **D. densiflora,** *Wall. ; Fl. Br. Ind., III., 570.*

Reference.—*Kurz, For. Fl. Burm., II., 134.*
Habitat.—An evergreen glabrous tree, met with occasionally in the tropical forests of Moulmein, Arakan, Martaban, and Tenasserim.

567 | **D. discolor,** *Willd. (A.DC., Prod., VIII., 235).*

Syn.—D. MABOLA, *Roxb.*
Habitat.—A small tree, native of the Philippine Islands, introduced into India and cultivated in gardens, especially in Vizagapatam.

FOOD.
Fruit.
568 | Food.—The FRUIT is like a large quince, and in some places is called Mangosteen: its proper name should be the Mabola fruit. It is agreeable, and has a pink-coloured fleshy rind.

569 | **D. Ebenum,** *Kœnig. ; Fl. Br. Ind., III., 558 ; Wight, Ic., t. 188.*

EBONY.

Syn.—D. EBENASTER, *Roxb., Fl. Ind., non Retz.*
Vern.—*Ebans, abnús, tendu,* HIND.; *Kendhu,* or *khenda,* URIYA; *Tendú, temrú,* C. P.; *Tai, tendu, abnús,* MAR.; *Acha, nullúti, tumbi, shengú- tan, kaka-tati, tai,* TAM.; *Tuki, tumbi,* TEL.; *Kare,* KAN.; *Karunkáli, chara, acha* (TAM. in Ceylon); *Kaluwara,* SING.; *Mallali,* MAN- JARABAD. This is the εβέvos of the Greeks and the *ebenus* of the Latins. (*Conf. with Dios., I., 114 ; Pliny, 16, 40.*)
References.—*Roxb., Fl. Ind., Ed. C B.C., 412 ; Voigt, Hort. Sub. Cal., 344 ; Brandis, For. Fl., 296 ; Beddome, Fl. Sylv., t. 65 ; Gamble, Man. Timb., 251 ; Thwaites, En. Ceylon Pl., 180 ; Grah., Cat. Bomb. Pl., 108 ; Müeller, Select Ex-trop. Pl., 108 ; Trimen, Sys. Cat., Ceylon Pl., 52 ; Moodeen Sheriff, Supp. Pharm. Ind., 132 ; Dymock, Mat. Med. W. Ind., 2nd Ed., 485 ; Indian Forester, III., 23, 203 ; VII., 128 ; VIII., 29 ; IX., 349 ; X., 23, 31 ; XIII., 172 ; Royle, Ill. Him. Bot., 262 ; Reprint No. 25, Records of Pub. Works Dept., 1871, p. 48 ; Balfour, Cyclop., 952 ; Treasury of Bot., 411 ; Kew Off. Guide to the Mus. of Ec. Bot., 37 ; Kew Off. Guide to Bot. Gardens and Arboretum, 43, 68 ; Settlement Reports: Chindwara, C. P., 110 ; Nimar, 305 ; Shajehanpur, IX.; Gazetteers: Kánara, XV., 67 ; Mysore and Coorg, I., 46, III., 16 ; Madras, Man. Adm., I., 313 ; Man. Cuddapah, 262 ; Orissa, II., 5.*

Habitat.—A large tree of the Southern Konkan (Kánara) to Madras (Circars and Carnatic), Ceylon, and the Malaya. The utmost confusion exists in the writings of popular authors regarding this tree, it being generally assigned to Ceylon, and the Indian Ebony spoken of as the product of **D. melanoxylon** or of **D. Ebenaster,** thus implying that the true Ebony does not occur in India. By botanists **D. Ebenaster,** however, is viewed as a synonym for **D. Ebenum.** Lisboa (*Useful Plants of Bombay*) makes no mention of **D. Ebenum** or **Ebenaster** as occurring in

D. 569

the Bombay Presidency, and **Dr. Gray** says it is a native of Ceylon. On the other hand, **Mr. Talbot** describes the plant as a forest tree in Kánara, and adds that it is "one of the trees which are not allowed to be cut." **Mr. MacGregor,** in a recent report, affirms it is found in the Kánara forest, but not abundantly. In the Manual of Madras Administration it is given as "one of the best trees of South Kánara." It is also mentioned as met with in Cuddapah, in Mysore, in Chhindwara and Nimar. **Beddome** says of Madras that "this valuable tree is not uncommon in our mountain forests on both sides of the Presidency."

Gum.—In a report on the Gums of India, issued by the Public Works Department (1871), it is stated that the gum obtained from the ebony tree is "used in medicine for removing obstruction from any cause in the vision." But as this fact is given in a table of Hazaribagh gums, it may be doubted whether the above passage should not be recorded under some other species. It is said to be the *Kendka-gand.*

Medicine.—Muhammadan writers speak of the *Abnus* as astringent, attenuant, and lithontriptic, very much after the same fashion as it was used by the Greek physicians (*Conf.* with the *Makhzan-el-Adwiya*).

Structure of the Wood.—Grey, with irregularly-shaped masses of jet-black ebony near the centre, frequently with lighter-coloured streaks. Roxburgh says of Ebony: "There are many species of this extensive genus, which yield a hard, black wood—I mean pure intensely black (not variegated), to all of which we give the general appellation Ebony: my **D. melanoxylon** I am now describing" (D. **Ebenaster,** *Willd.*) "a second ; **Ebenus,** *Rumph., Amb., Vol. 3, Pt. I., t. 1,* seems a third. From all these I know that of the Mauritius differs essentially by the entire fruit, with ripe seed, just received from that island, and now before me. The mountains of Bengal, Bhután, and Nepál produce at least another very distinct species, *viz.,* my **tomentosum.**" In modern commerce there are perhaps five if not six very distinct timbers that pass by the name of Ebony. **D. Ebenum**—the true Ebony—obtained chiefly from Ceylon but also from Madras; **D. melanoxylon,** from India; **D. reticulata,** Mauritius Ebony; **Brya Ebenus**—Jamaica or West Indian Ebony—a tree belonging to the LEGUMINOSÆ, the wood of which is more nearly allied to the Indian black wood (**Dalbergia latifolia**), and lastly, **Melhania melanoxylon,**—St. Helena or African Ebony,—a tree of the Cocoa family. To this list might doubtless be added several of the other Indian species, many of which form dark, brown, or black central wood. The trade in the Indian Ebonies appears to have been much neglected. The impression seems to have become general that Ceylon and not India has to be looked to for a commercial supply of the timber. In the Madras Manual of Administration it is stated that, while the true Ebony occurs, "that given by **D. melanoxylon** in the Circar forests and **D. tomentosa** in those of the Deccan are more largely used, though but little is really exported, but the **D. melanoxylon,** *Roxb.*—or Coromandel Ebony—is very largely used." (*Conf.* with the account under **D. melanoxylon** on a further page.)

The writer has failed to discover any report dealing comparatively with Indian, Ceylon, Mauritius, and other Ebonies. It would, for example, be instructive to learn whether **D. Ebenum,** as grown in India, affords a superior or inferior Ebony to that obtained from the same species in Ceylon. **Brandis** says the best Indian Ebony is obtained from **D. Ebenum,** a large tree of South India and Ceylon. The *Indian Forester,* while making repeated references to Ebony, does not appear to have taken up the subject and dealt with it from this point of view. In a paper, however, on the "*Development of the Trade in Indian Woods,*" it is stated that

GUM.
570

MEDICINE.
571

TIMBER.
572

573

TRADE.
574

DIOSPYROS
Ebenum.   The Ebony-Woods.

TRADE.

there are five commercial Ebonies, and "of these the Mauritius wood is esteemed the best, and is probably the produce of the same tree as the Indian Ebony, *viz.*, **Diospyros Ebenum.** [This would not seem to be correct.—*Ed.*]. " The chief Indian and Ceylon Ebonies are **D. melanoxylon,** common in most provinces; **D. Ebenum,** of South India and Ceylon; **D. Kurzii,** the Andamanese Marble-wood, or *Teakah,* and **D. quæsita,** the Calamander wood of Ceylon. The price of Ebony in England is usually about £10 to £20 per ton, equivalent, at 50 cubic feet per ton, to four to eight shillings a cubic foot. The difficulty about Ebony is the small quantity of black heart-wood given by the common **D. melanoxylon,** and it is probable that the sale of whatever Ebony may be available will be better made at the Indian ports than in Europe. But the Andaman marble-wood should sell well if sent to Europe, and more especially as it can probably be obtained of larger size than ordinary Indian kinds " (*Vol. VII. (1881-82), p. 128*). The above, which is the only lengthy reference in the *Indian Forester* on the subject of Ebony, falls, as it seems to the writer, considerably short of the chief point to be aimed at in an effort to develop a trade in this valuable timber from India. It does not appear enough to say that the difficulty about Ebony is the small quantity of black heart-wood. It is necessary to know if this is a more adverse circumstance in India than in Ceylon and other ebony-producing countries. Unless it be contended that the produce of **D. melanoxylon** is commercially of equal value to that of **D. Ebenum,** that timber should be excluded from consideration, and the possibilities or otherwise of India becoming a source of true Ebony ought to be disposed of first, before the attempt is made to place on the market Ebony substitutes. Indeed, the various opinions recorded by authors regarding the Ebony from **D. melanoxylon** do not favour the idea that it is of equal value to that from **D. Ebenum.** This subject is worthy of consideration in its widest aspect. When it has been demonstrated that **D. Ebenum** produces an inferior or less remunerative Ebony in India than it does in other countries, it might then become desirable to foster the cultivation and trade in some of the other Ebony-yielding species. But until this has been demonstrated it would seem opposed to a possible trade in Indian Ebony to palm off all available black-woods as Ebonies. These may in themselves be good timbers for some purposes, even superior to Ebony, but success is more likely to follow an accurate declaration of their characters and properties than by taking advantage of commercial generalisations. But it cannot be stated that the timbers from Indian species of **Diospyros** have as yet been critically examined, and as Ebony is daily becoming more scarce in the European markets, it might be in the interests of this country to attempt the protection and possibly the cultivation of good species either indigenous to or found suitable for India.

DOMESTIC
USES.
Turnery.
575
Inlaying.
576
Cabinet-
Work.
577
Piano-Keys.
578
Furniture.
579

It is scarcely necessary to mention here the numerous purposes to which Ebony is put. It is in great demand for ornamental turnery, for inlaying in fancy articles, cabinet-work, and for the keys in pianos, &c., &c. Gamble gives the weight of the Indian true Ebony as ranging from 61 to 70℔ a cubic foot. Beddome regards it as considerably higher. He says : " **D. Ebenum** yields the best kind of Ebony, generally jet-black, but sometimes slightly streaked with yellow or brown; it is very heavy, close, and even-grained, and stands a high polish; unseasoned it weighs 90—100℔ the cubic foot and 81℔ when seasoned, and has a sp. gr. 1·296; it is used for inlaying and ornamental turnery, and sometimes for furniture, but there is not much demand for it in this presidency. The sap-wood is white, hard, close-grained and strong but not durable; at the same time it is used by the natives for various purposes. It is called *Nalluti* in Cuddapah and the Karnúl hill forests, where the tree is very common and well-known." In *Spons' Encyclopædia* it is stated that the approximate London market values are £5—20 a ton for Ceylon and £3—12 for Zanzibar, &c.

Diospyros ehretioides, *Wall. ; Fl. Br. Ind., III., 559.*   580

Vern.—*Ouk-chingsa, aukchinsa,* BURM.

References.— *Kurz, For. Fl. Burm., II., 129 ; Kew Off. Guide to the Mus. of Ec. Bot., 93.*

Habitat.—A large tree frequent in Burma.

Structure of the Wood.—Dark-grey, with darker streaks, moderately hard, even-grained (*Gamble*). Kurz says the sapwood is yellowish white, of a very coarse granular appearance, rather hard, the heartwood heavy, brittle, close-grained, brown or beautifully white and black mottled. The weight has been given at 41℔ to 54℔ a cubic foot. It is used for house-posts.   TIMBER. 581

[*843, 844.*

D. Embryopteris, *Pers. ; Fl. Br. Ind., III., 556 ; Wight, Ic., tt.,*   582

Syn.—D. GLUTINOSA, *Kœn. ; Roxb., Fl. Ind., Ed. C.B.C., 413 ;* EMBRYOP-TERIS GLUTINIFERA, *Roxb., Cor. Pl., I., 49, t. 70.*

Vern.—*Gáb, makur-kendi, téndú,* HIND., BENG. ; *Makarkenda (garate-rél?),* SANTAL ; *Gusvakendhu,* URIYA ; *Kendu,* ASSAM ; *Kúsi,* BUNDEL-KHAND ; *Kúsi,* BANDA ; *Timbori, gab, kúsi, téndú, timbiri,* BOMB. ; *Tem-burni, timbwini,* MAR. ; *Zeeberwo,* GUZ. ; *Tembhuran,* KHANDESH ; *Tum-bika, pani-chika, tumbilik-kay,* TAM. ; *Holle-tupra,* COORG ; *Tumil, tinduki, tumika, tumki, jumika, tubiki,* TEL. ; *Kusharta,* KAN. ; *Panich-chi, vananchik,* MALAY ; *Panichekai, tumbika* (TAM. in Ceylon); *Timbiri,* SING. ; *Tinduka, sindika,* SANS. ; *Abnúse-hindi,* ARAB., PERS. *Abnús* is the name for D. Ebenum, but it is sometimes applied to this species also.

References.—*Roxb., Fl. Ind., Ed. C.B.C., 413 ; Coro. Pl., I., t. 70 ; Voigt, Hort. Sub. Cal., 344 ; Brandis, For. Fl., 298 ; Kurz, For. Fl. Burm., II., 128 ; Beddome, Fl. Sylv., t. 69 ; Gamble, Man. Timb., 252 ; Thwaites, En. Ceylon Pl., 178 ; Stewart, Pb. Pl., 136 ; Grah., Cat. Bomb. Pl., 107 ; Rheede, Hort. Mal., III., t. 41 ; Trimen, Sys. Cat. Ceylon Pl., 51 ; Rev. A. Campbell, Report on Econ. Prod., Chutia Nagpur, Nos. 9404 and 9464 ; Sir W. Elliot, Fl. Andh., 182, 184 ; Mason, Burma, 542, 782 ; Pharm. Ind., 131, 455 ; Ainslie, Mat. Ind., II., 278 ; O'Shaughnessy, Beng. Dispens., 428 ; Moodeen Sheriff, Supp. Pharm. Ind., 132 ; U. C. Dutt, Mat. Med. Hind., 321 ; Dymock, Mat. Med. W. Ind., 2nd Ed., 483, 888 ; Flück. & Hanb., Pharmacog., 403 ; Bent. & Trim., Med. Pl., 168 ; Bidie, Cat. Raw Pr., Paris Exh., 11 ; Year Book Pharm., 1875, 217 ; K. L. Dey, Indig. Drugs, India, 49 ; Atkinson, Him. Dist., 364 ; Lisboa, U. Pl. Bomb., 165 ; Birdwood, Bomb. Pr., 166, 312 ; Royle, Ill. Him. Bot., 13, 262 ; Christy, Com. Pl. and Drugs, V., 44 ; Cooke, Gums and Gum-resins, 105 ; McCann, Dyes and Tans, Beng., 134-35, 138, 150, 159 ; Cooke, Oils and Oilseeds, 42 ; Wardle, Report on Dyes of India, 14 ; Reprint No. 25, Records in Pub. Works Dept., 1871, pp. 13, 44, 45, 46 ; Spons, Encyclop., 1684 ; Balfour, Cyclop., 952 ; Treasury of Bot., 411 ; Kew Off. Guide to the Mus. of Ec. Bot., 92 ; Kew Off. Guide to Bot. Gardens and Arboretum, 43, 68 ; For. Adm. Report, Chutia Nagpur, 1885, 32 ; Jour. As. Soc. Pt. II., No. 11, 1867, 80 ; Home Dept. Cor. regarding Pharm. Ind., 223, 237 ; Jour. Agri. Hort. Soc. Ind. (New Series) I., 1867, p. xxxvii ; Indian Forester, II., 179 ; III., 203 ; V., 13 ; VIII., 29 ; XI., 231 ; XII., 312 ; Settlement Re-port of Raipur, C. P., 76-77 ; Gazetteers, Mysore and Coorg, I., 52 ; Bundelkhand, I., 82 ; Kanára, I., 437 ; Note on Cotton Cultivation in Assam by the Director of Land Records and Agriculture, p. 33 ; Special Report, furnished for this work ; by Prof. Hummel of Leeds, and by Mr. Shuttleworth, Conservator of Forests, Bombay.*

Habitat.—A dense evergreen small tree, with dark-green foliage and long shining leaves ; common throughout India and Burma, except the arid and dry zones in the Panjáb and Sind.* Distributed to Ceylon, Siam, and the Malayan Peninsula. Very abundant in Bengal. The Conservator of Forests, Northern Division, Bombay, states that in his division D. Embryopteris is not known. Wight's figures were made from specimens "found in Malabar." The Madras Board of Revenue, in a report furnished for this work, reports that, while the tree is known in some of the forests of the Presidency, especially the Northern Circars, it is not supposed to be used for tanning purposes or to pave the bottoms of boats.

**GUM.
583**

Gum.—Considerable confusion exists on the subject of the gum or resin obtained from this tree. Cooke describes it as "dark brown, rather earthy looking, with a bright resinoid fracture, not unlike some light-coloured varieties of the black dammar. A similar specimen in the reference collection is a brown resin, with a shining fracture, externally coated with a greyish white or brownish crust. It is somewhat stalactitic as if it had trickled down the tree as it was exuded." Again : "The resin in the collection referred to this source was derived from Bhaugulpore; it is soluble in turpentine or benzole, forming a limpid varnish." Although many writers speak in general terms of a gum as obtained from this plant, they do so in such a manner as to suggest a possible confusion between a true gum and a prepared extract from the fruit. The writer cannot recollect having observed a gum exuding through the bark of D. Embryopteris, and neither Stewart, Brandis, Gamble, Beddome, nor Kurz mention a gum, though of course they all allude to the "viscid pulp, which is used as gum in book-binding, and in place of tar for paving the seams of fishing-boats. Its use for 'gabing' boats is general throughout the rivers of Lower Bengal and Assam. An infusion is used to render fishing-nets durable. It is full of tannin and is used in medicine as an astringent" (*Gamble*). Dymock writes : "The extract of the fruit is of the colour and consistence of shell-lac." An extract is thus prepared in some parts of the country which might, as in the case of Cutch, closely resemble a resin. It thus may be the case that the passage quoted above from Dr. Cooke's Report on Gums and Gum-resins, &c., refers to the extract, but what seems more likely the case, it is the description of a true gum but not a product from D. Embryopteris. Some time ago information was called for by the Department of Public Works on the subject of gums and the replies obtained from every Province were in 1871 printed in the form of a combined report. From nearly every district of India some mention was made of the substance now under consideration, but in no instance was the tree stated to yield a gum. Of Assam it was stated : "The *Kendú* tree fruit yields a resinous gum : it is used for caulking the small canoes. The Dome fishermen colour their nets with it, and thereby render them more durable in wear and tear." So again the Commissioner of Dacca wrote : "There is no resinous product which is extensively used except the preparation of the *Gáb* tree." The Commissioner of the Presideney Division, Bengal, stated that "A decoction of the unripe gáb fruits is used for caulking boats." The above passages are representative of the whole series of replies. No mention is made of a gum but the use of the decoction, extract, or fresh pulp of the unripe fruit is general, at least over the Lower Provinces. Mr. Baden Powell, in a review of Dr. Cooke's report on Gums (*Indian Forester, II., 179*), after discussing the properties of the viscid pulp, adds : "If the fruit is meant in the text, we have neither gum nor resin properly so called to deal with. Forest Officers and others who have the opportunity of studying D. Embryopteris would render a valuable service by looking up the matter of gum or no gum, and thus remove a perplexing

**RESIN.
584**

---

* It flowers in March to May and the fruits ripen about December.

RESIN.

cause of ambiguity from the literature of this subject." After having ascertained whether or not the tree exudes from its stem-bark or rind of the fruit a gum which is regularly or not collected by the people, attention might next be turned to the subject of the manner in which the viscid pulp is employed for caulking boats. According to some writers it would seem as if the pulp was used direct from the fruits; according to others an extract is reported to be prepared, or only a decoction. The former would mean that the fruits were boiled and the juice thus obtained boiled down to a tarry or resinous consistence. Of Orissa (Khorda sub-division) it has been reported that the fruits do not seem to be in demand. No use is made of the fruit in Coorg. **Dr. Buchanan-Hamilton**, one of the earliest and most accurate observers on economic subjects, thus describes the process as pursued in Dinajpur, Bengal:—

"It is a beautiful tree, common near the villages of Bengal: the fruit is eatable, but excessively sour. Its principal use is for paving the bottom of boats. It is beaten in a large mortar, and the juice expressed. This is boiled, mixed with powdered charcoal, and applied once a year to the outside of the planks. A good tree will give 4,000 fruits, worth R2 and will be in full bearing in eight years from the time when it was planted. The wood is of little value" (*Buchanan's Statistics of Dinajpur, p. 152*). **Thwaites** says of Ceylon: "The juice of the unripe fruit is employed for paving the seams of fishing-boats."

For the use of the juice in tanning lines see the para. below under **Dye.**

Dye and Tan.—The FRUIT is largely used as a TAN, being a powerful astringent. By simply steeping the half-ripe fruits in water a brownish liquid is obtained, which is sometimes used in dyeing a brown colour. This is made into a good black by being combined with Myrobalans (**Terminalia Chebula**) and Proto-sulphate of iron (*hirakash*). As a dye this substance is, however, of small importance. **Wardle** (in his recent report on the Dyes of India, p. 14) says: "The dried fruit and calyx of the plant contains a small amount of colouring matter which is soluble in boiling water." This colouring agent he experimentally employed in various processes, obtaining both with silk and cotton light drab results. **Dr. McCann**, in his work on the Dyes and Tans of Bengal, says that "according to the Collector of Hughli the preparation from the fruit imparts a brown colour, dark, inclining to red, to the object to which it is applied, and also protects the timber or fibre from the action of water. It is more prized for this latter quality than as a colouring matter." From Rájsháhí it is also reported that 'an infusion of the unripe *gáb* fruits is used for steeping fishing nets, and the astringent viscous mucus of the fruit is used everywhere for painting the bottoms of boats. *Gáb* fruit is but sparingly used for the brown dye which it yields. The fruit is simply pounded and boiled and the cloth steeped in the liquid, no auxiliary being employed." The chief use as a dye is in what might be called the tanning of fishing nets and lines. A very considerable trade is done in cutch exported to Europe to be used by the home fishermen for a similar purpose. The comparative merit of *gáb* and cutch does not seem to have been worked out, but it is somewhat significant that the Indian fishermen prefer the *gáb* decoction when they might even more conveniently obtain cutch. A sample of *gáb* fruits was (in connection with the Colonial and Indian Exhibition) given to **Professor Hummel** of Leeds. The report which the Professor has furnished exhibits these fruits as possessing 15·0 per cent. of tannic acid, the acid being of a pale red colour. It is said to have a value per cwt. of 4s. and 0$\frac{3}{4}$d., as compared with Divi Divi; of 7s. and 7$\frac{1}{4}$, with Valonia; of 10s. and 2$\frac{1}{4}$d., with Ground Sumach; and 4s. and 10$\frac{1}{2}$d., as compared with Ground Myrobalans. But the Professor adds that the

DYE.
Fruit.
585
TAN.
586

dried fruits are somewhat tough and consequently difficult if not impossible to grind. An extract should be prepared in India if the attempt be thought desirable to introduce the tan to European commerce.

For further chemical analysis and percentage of tannic acid, see concluding sentences of the paragraph below on the Medicinal properties of *gáb* fruit.

OIL.
587

Oil.—An oil, extracted from the seed by boiling, is used in native medicine.

MEDICINE.
Fruit.
588
Bark.
589
Juice.
590
Oil.
591

Medicine.—The FRUIT and the BARK both possess astringent properties. The JUICE of the unripe fruit makes a good application to fresh wounds. It is rich in tannin, and is therefore a useful domestic astringent, and the tree is so plentiful as to be at the door of even the poorest cotter. An OIL extracted from the seeds is also employed in native medicine. It is used in dysentery and diarrhœa with success. The infusion of the fruit is given as a gargle in aphthæ and sorethroat (*Kani Lal De Bahadur*). Dr. Dymock gives the following brief history of the medicinal uses of *gáb* : " The circumstance that the unripe fruit abounds in an astringent viscid juice, which is used by the natives of India for daubing the bottoms of boats, was communicated by Sir William Jones to Roxburgh in 1791. The introduction of the fruit into European practice in India is due to O'Shaughnessy. In 1868 it was made officinal in the *Pharmacopœia of India*. In Bombay the fruit is eaten by the poorer classes. The SEEDS are preserved by the country people and given as an astringent in diarrhœa. The testa is the astringent part, the albumen being almost tasteless." Ainslie, speaking of the fruit, says : " On being punctured, it gives out a juice of peculiar astringency, and which the Hindú doctors sometimes employ as an application to fresh wounds." The *Pharmacopœia* recommends the drug to be exhibited in the form of an extract prepared from the expressed juice over a water-bath. It is "of a reddish brown colour, in flexible plates, and readily soluble in water. It is an excellent astringent and very useful in diarrhœa and chronic dysentery. A solution of two drachms in a pint of water is a valuable vaginal injection in leucorrhœa." In a further page the *Pharmacopœia* explains that the reputed value of this drug rests mainly on the testimony of Sir W. O'Shaughnessy, but it adds : " There is little doubt that it possesses powerful astringent properties, and is deserving of a high place amongst indigenous remedies of this class." Dr. Dymock points out that the Diospyros fruit is intensely astringent until quite ripe, when it suddenly becomes mawkish and sweet without a trace of its former astringency. This is noticed in the ' Pharmacographia,' but not in the Indian Pharmacopœia, where unripe fruit should have been ordered." This fact is of very considerable importance and must account for the varying percentage of tannin found in the fruits by different chemists. O'Shaughnessy is said to have found 60 per cent.; Professor Hummel (see paragraph on Dye and Tan above) found only 15 per cent. It would be most instructive to have the fruits in all stages of their growth chemically analysed, since it would seem that both as a drug and a tan opinions founded on imperfect materials have become current in the literature of this subject.

Seeds.
592

SPECIAL OPINIONS.—§ " According to O'Shaughnessy the juice of the unripe fruits contains 60 per cent. of tannic acid. The bark has been used with doubtful results in the treatment of intermittent fevers " (*Dr. Warden, Prof. Chemistry, Calcutta*). " The expressed juice abounds in tannin and is used in cases of diarrhœa, dysentery, and hœmorrhage from internal organs " (*Civil Surgeon J. H. Thornton, B.A., M.B., Monghyr*). " The pulp of the ripe fruit is sweet and astringent in flavour, and may be moderately used in ordinary cases of diarrhœa and dysentery " (*Civil Surgeon Bankabehary Gupta, Pooree*).

D. 592

	DIOSPYROS
The Chinese Fig. (*G. Watt.*)	**Kaki.**

Food.—Produces a round fruit as big as a middle-sized apple, green when unripe, rusty yellow when ripe; and in the latter stage contains a somewhat astringent pulp in which the seeds are embedded.

The FRUIT when green is commonly used in caulking the bottom of boats; when ripe it is eaten by the natives, but is not very palatable. The LEAVES are also eaten as a vegetable. Dr. Dymock enumerates this tree in a list of plants which were eaten during the Kandesh famine of 1877-78. "Although the ripe fruit is very sweet, insects will not touch it."

Structure of the Wood.—Grey, moderately hard, close-grained. Beddome says the timber is only of average quality and is used for building purposes. Also that masts and yards of country vessels are made from this tree in Ceylon

Domestic Uses.—Ainslie mentions that the carpenters of the Malabar coast use the juice of the fruit "as an excellent glue."

FOOD.

Fruit.
593
Leaves.
594

TIMBER.
595

DOMESTIC.
596

Diospyros exsculpta, *Ham.,* see **Diospyros tomentosa,** *Roxb.;* D. exsculpta, *Bedd.,* see **D. melanoxylon,** *Roxb.,* var. **Beddomei;** and D. exsculpta, *Dalz.,* see **D. Tupru,** *Buch.-Ham.*

D. flavicans, *Hiern.; Fl. Br. Ind., III., 562.*

Habitat.—A small tree, native of Mergui, Tavoy, Tenasserim, and Malacca.

597

D. foliosa, *Wall.; Fl. Br. Ind., III., 556.*

598

Syn.—DIOSPYROS CALYCINA, *Bedd.*
Vern.—*Vellay toveray,* TAM.
References—*Bedd., Fl. Sylv., t. lxviii.*

Habitat.—A middle-sized tree, native of the Western Ghâts, Madras. Beddome says it occurs in the Tinnevelly District and southern portions of Madura. Very abundant in the Ghât forests from the foot up to 3,000 feet elevation.

Structure of the Wood.—Yields a valuable light-coloured wood, which is much in use in the Tinnevelly District (*Beddome*).

TIMBER.
599

D. glutinosa, *Roxb.,* see **Diospyros Embryopteris,** *Pers.*

D. Kaki, *Linn. l.; Fl. Br. Ind., III., 555; Wight, Ic., t. 415.*

600

Sometimes called the CHINESE FIG and PLUM or the KEG FIG of Japan.

Syn.—DIOSPYROS CHINENSIS, *Bl.;* EMBRYOPTERIS KAKI, *Don.;* D. COSTATA, *Rev. Hort.*
Vern.—*Tay, tee, teh,* BURM. Roxburgh says that the Chinese gardeners employed in the Calcutta Botanic Gardens call this plant *Chin.*
References.—*Roxb., Fl. Ind. Ed. C.B.C., 412; Mueller, Sel. Ex-trop. Pl., 108; Voigt, Hort. Sub. Cal., 344; Royle, Ill. Him. Bot., 262; Balfour, Cyclop., 953; Treasury of Botany, 411; Kew Off. Guide to Bot. Gardens and Arboretum, 68, 143; Grah., Cat. Bomb. Pl., 107; Smith, Dict. Econ. Pl., 151; Mason, Burma and Its People, 463, 782; Indian Forester, I., 113; VI., 25; XIII., 76.*

Habitat.—A small tree, native of the Khásia hills, Upper Assam, and Burma. Mr. M. H. Ferrars (in his *Journey of a Tour into the Kareni Country*) says: "Large areas are covered almost exclusively with *Teh* (D. Kaki), and an undergrowth of some CYPERACEÆ." Roxburgh first thought it was a native of China and Japan, but subsequently he added that it had been found in "the mountains of Nepál to the northward of Bengal." He also states that plants grown in the Botanic Gardens, Calcutta, had in twelve years only attained a height of 12 to 15 feet.

L

DIOSPYROS **Lotus.**	Andamanese Marble Wood; Green Ebony-tree.

DYE.
601

Dye.—In Japan black dyes are produced from the fruit of this tree with sulphate of iron.

FOOD.
Fruit.
602

Food.—Cultivated on account of its FRUIT which is about the size of a small apple, the better qualities being described as delicious and often made into preserves. Roxburgh says: "The fruit is tolerably pleasant, though by no means equal to a good apple, but what is worse, the trees about Calcutta are uncommonly unproductive." Balfour writes that the fruit in Japan and China "attains the size of an orange, and is frequently sent to Europe in a dried state. Preserved in sugar, it is a large thin-skinned, juicy fruit of an orange yellow colour, with a sweet taste."

603

Diospyros Kurzii, *Hiern.; Fl. Br. Ind., III., 559.*

ANDAMANESE MARBLE WOOD.

Vern.—*Teakah, thitkya,* BURM.; *Pecha-da,* AND.
References.—*Kurz, For. Fl. Burm., II., 231; Gamble, Man. Timb., 249; Indian Forester, IV., 292; V., 186; VII., 128; X., 532; Smith's Dict. Econ. Pl.; 267; Balfour, Cyclop., 953; Kew Off. Guide to the Mus. of Ec. Bot., 93; Kew Bulletin, Sept. 1887, p. 19.*

Habitat.—A large evergreen tree of the Andaman and Nicobar Islands.

TIMBER.
604

Structure of the Wood.—Handsome, streaked with black and grey; the grey wood hard; the black wood very hard, with alternate streaks of black ebony and grey. The mass of ebony occupying the centre of the tree is large and very irregular in outline, and frequently encloses interrupted concentric belts of light-coloured wood. At the Conference on Indian Timbers held in the Commercial Rooms of the Colonial and Indian Exhibition, Sir D. Brandis explained the properties of this timber. To the question as to the sizes of logs available, he replied that planks of 12 to 15 inches would, in his opinion, be the maximum. The gentlemen present

Furniture.
605

at the conference thought that if the supply were regular a good demand for the wood might be expected. Pieces of furniture made of it were exhibited, which seemed to justify the idea that it might come into use for fancy cabinet-making.

Cabinet-Work.
606

In India it is sparingly employed for furniture, but a considerable demand exists for walking sticks made of this wood. It seems to deserve to be better known as a substitute for the Ceylon Calamander wood, which it resembles in appearance very much. It is said by Major Ford to be

Sheaths.
607

used in the Andamans for handles and sheaths of blades, and for furniture. Home's survey gave 224 trees, or one tree per acre; so that it is fairly abundant.

608

D. lanceæfolia, *Roxb.; Fl. Br. Ind., III., 562.*

Syn.—DIOSPYROS MULTIFLORA, *Wall.*
Vern.—*Gúlul,* SYLHET (see under **D. Toposia**).
References.—*Roxb., Fl. Ind., Ed. C.B.C., 414; Brandis, For. Fl., 297; Atkinson, Him. Dist., 364, 524; Kurz, For. Fl. Burm., II., 136.*

Habitat.—A fairly large tree met with in Sylhet, the Khásia hills and Moulmein; flowers in March.

FOOD.
Fruit.
609

Food.—Roxburgh says the FRUIT is eaten by the natives.

TIMBER.
610

Structure of the Wood.—The tree furnishes the natives with hard durable timber used in the construction of their habitations (*Roxburgh*).

611

D. Lotus, *Linn.; Fl. Br. Ind., III., 555.*

[EBONY TREE.

THE EUROPEAN DATE PLUM: Sometimes also called the GREEN

Vern.—*Amlok, malok,* HIND.; *Amlúk (or amlok), malúk* (Jhelum Valley), *bissarhi pála* (Bias Valley), PB.

D. 611

The Coromandel Ebony.	(*G. Watt,*)	**DIOSPYROS** **melanoxylon.**

References.—*Brandis, For. Fl., 297, t. 36; Stewart, Pb. Pl., 136; Aitchison, Cat. Pb. and Sind Pl., 86; Müeller, Sel. Extra-trop. Pl., 109; Aitchison, Kuram Valley Flora (Linn. Soc. Jour., XVIII.), 78; Baden Powell, Pb. Pr., 270, 578; Royle, Ill. Him. Bot., 261-262; Indian Forester, II., 179; V., 181; Journ. Agri-Hort. Soc., XIX., 16; Balfour, Cyclop., 953; Smith, Dict., 152; Kew Off. Guide to the Mus. of Ec. Bot., 93; Kew Off. Guide to Bot. Gardens and Arboretum, 143; Settlement Report of Kohat, 30; of Hazara, 10, 95.*

Habitat.—A middle-sized tree of the northern parts of the Panjáb, ascending the Himálaya, and extending into Kashmír, Afghánistan, and Beluchistan. Probably only cultivated in India. **Stewart** says it is a handsome little tree not uncommon in the western part of the Jhelum basin, from 2,500 to 6,000 feet, and appears to be common in some parts of the northern Trans-Indus Hills. There are three trees at Juggatsakh (6,000 feet) in Kúllú, the largest of which has a girth of 12 feet. In parts of Hazára the male plant is called *Gwalidar* and the female *Amlok*.

Resin.—**Baden Powell** mentions this species along with some others as possibly resin-bearing (*Review of Dr. Cooke's Report on Gums and Resins, &c., Indian Forester, II., 179*).

Food.—The FRUIT, when ripe, is sweetish, and is eaten, either fresh or dried, by Afgháns and other tribes. The former bring quantities of it to the Peshawar bazars. It is sometimes also used in *sherbat*.

This small fruit is supposed by some to be one of the fruits which were eaten by the Lotophagi. In Southern France it is eaten when half-rotten like the Medlar (*Gamble*). Dr. **Stewart**, in his report on a visit to Hazára, says that he was familiar with the fruit in the bazars at Peshawar, but had not seen the plant up to the date of his visit to Hazára. He continues: "Subsequently, the writer got it growing abundantly in many places throughout Hazára, from 3,500 to 6,000 feet, and in 1860 on the return journey from Kashmír, in the Upper Jhelum valley below Baramúla. The fruit is sweetish and pleasant enough in taste when eaten fresh, but would hardly be valued in Europe." In the Hazára *Gazetteer* it is stated that about 2,000 maunds of the fruit are produced annually, and in the Rawalpindi *Gazetteer* it is reported that the tree is plentiful in gardens. In his *Panjáb Plants* Stewart states that, according to **Bellew**, the fruit is eaten plain or with rice or is used in *sherbats*. "I presume Irvine was mistaken in stating that spirits are in the Panjáb distilled from the fruit." **Aitchison** remarks that this is "a large tree, extensively cultivated in the Kuram district for its fruit; not met with in a wild state. It does not occur in the Hariáb District." In a paper in the *Indian Forester*, Dr. **Aitchison** adds that the Lotus fruit is considered next in value to the walnut. It is said to be purple in colour and to be about the size of a cherry or pigeon's egg. (*Conf.* with generic introduction and the remarks under D. **melanoxylon**).

Structure of the Wood.—Grey, moderately hard, close-grained. Stewart remarks that the tree is nowhere in India so abundant as to afford timber.

[III., 564.

Diospyros melanoxylon, *Roxb.; Fl. Ind., ii., 530; Fl. Br. Ind.,*

COROMANDEL EBONY.

Syn.—DIOSPYROS WIGHTIANA, *Wall.*; D. DUBIA, *Wall.*; *Wight, Ic., t. 1223.* By **Brandis, Gamble, Atkinson,** and many other writers this is made to include D. TOMENTOSA, *Roxb.,* the result being that in all probability the economic facts and vernacular names given by these authors for the species met with in Bengal and the N. W. Provinces should be removed from D. MELANOXYLON, and carried to D. TOMENTOSA. Not being able to verify each fact the writer has felt it preferable, however,

RESIN.
612

FOOD.
Fruit.
613

TIMBER.
614

615

L 2

---

to leave for future enquiry the dissection and redistribution of the information here given and that which will be found under D. TOMENTOSA.

Vern.—*Tendu, kendu, temru, abnús,* HIND.; *Kend, kyou (kiu),* BENG.; *Gora tiril,* KOL.; *Terel, kiril,* SANTAL; *Kendhu,* URIYA; *Kend,* MAL (S. P.); *Tumri, tummer, tumki,* GOND.; *Tendu,* BAIGAS; *Tendu,* C. P.; *Timberni, temru, tumri, tumbúrni* (Beddome), BOMB.; *Tumri, temru, timburni,* MAR.; *Támrug* (Baroda), *timburni* (Panch Mahals), *támrug,* GUZ.; *Balai,* KANARA; *Tembhurni,* THANA; *Tumbi, tumbali, karunthumb,* TAM.; *Tumi, tumki,* or *tunki* (Beddome), *tumida, timmurri, damádi,* TEL.; *Balai,* KAN.; *Ouk-chin-ya* (according to Balfour), BURM.; *Kenduka,* SANS.; *Abnús,* ARAB.

Abnús is given in the Jangira *Gazetteer* as the Konkan name for this species; it is called *temburni* in Kolába. Mr. J. F. Duthie points out that the *Tendu* of Northern India is D. tomentosa, *Roxb.*

References.—*Roxb., Fl. Ind., Ed. C.B.C., 412; Voigt., Hort. Sub. Cal., 344; Brandis, For. Fl., 294, in part not Blume; Beddome, Fl. Sylv., t. 67; Aitchison, Cat. Pb. and Sind Pl., 86; Pharm. Ind., 132; O'Shaughnessy, Beng. Dispens., 428; Dymock, Mat. Med. W. Ind., 485; Bidie, Cat. Raw Pr., Paris Exh., 33; Atkinson, Him. Dist., 364; Lisboa, U. Pl. Bomb., 92, 164; Birdwood, Bomb. Pr., 332; Royle, Ill. Him. Bot., 262; McCann, Dyes and Tans, Beng., 128, 135-36, 152, 159; Buck, Dyes and Tans, N.-W. P., 44, 79; Balfour, Cyclop., 953; Treasury of Bot., 411; Kew Off. Guide to the Mus. of Ec. Bot., 93; Journ., 1867, Pt. II., 2, 80; Bomb. Gaz., XV., 67; VII., 36; For. Adm. Report, Ch. Nagpur, 1885, 6, 32; Gazetteers, Bombay (Panch Mahals), III., 201; VII., 31, 32, 36; XI., 25 and 404; XIII., 27; XV., Pt. I., 67¦; XXV., 92, 164, 348, and 389.*

Habitat.—A large tree, attaining a height of 60 to 80 feet, sometimes seen as a shrub. Roxburgh mentions it as a native of "most woody mountainous countries in India, *viz.*, Ceylon, Malabar, Coromandel, Orissa, &c." It is said not to be met with in Burma (*Kurz, Brandis*). It seems probable that this species should be described as the Western representative, just as D. tomentosa might be said to be the Eastern and Northern Indian form. The *Flora of British India* retains these Roxburghian species as distinct, but many Indian authors view them only as forms of one and the same species. Lisboa says of D. melanoxylon that it "is not uncommon in Bombay, North Kánara, and the Madras forests, extending northwards as far as the Raví. It attains a height of 30 to 50 feet, sheds its foliage in the cold season, renews and flowers in the hot season, and ripens its fruits during the rains." The plant met with in the Raví basin is most probably D. tomentosa not D. melanoxylon. According to Wight the latter plant is met with on the Nilghiris and Serramallí hills near Dindigul. The Madras Board of Revenue assign it, however, to the Eastern Gháts and hills of Kurnool, so far as the Madras Presidency is concerned. It is also said to be very common in Thana in Bombay. A recent report furnished by the Forest Department of Bengal speaks of it as found in Singbhoom. It attains a girth of about 6 feet. The timber might be obtained from Madras.

616

Var. **Beddomei**=D. exsculpta, *Bedd.* (not *Ham.*). Some doubt exists regarding this plant. The description of it given in the *Fl. Br. Ind.* is based on Beddome's plate (*Fl. Syl., 66*). Beddome remarks that it is found in the Cuddapah, Salem, and Karnul forests, and probably elsewhere in the Madras Presidency. He adds that it is found also in Bengal and Bombay, but this last statement most probably rests on the identification of the Madras variety with D. exsculpta, *Ham.*, a synonym for D. tomentosa, *Roxb.*

**GUM.
617**

Gum.—Like D. Embryopteris, this species is reported to yield a gum, but nothing of a definite nature can be discovered regarding that gum and its existence may be doubted. Indeed, in the writer's opinion, it is highly

| The Coromandel Ebony. | (*G. Watt.*) | DIOSPYROS melanoxylon. |

doubtful if any **Diospyros** yields a gum or resin, though the fruits of all would most probably afford a resinoid juice.

Medicine.—The BARK possesses astringent properties, and is used in decoction in diarrhœa and dyspepsia as a tonic. In a dilute form it is used as an astringent lotion for the eyes. Roxburgh says that the bark, powdered and mixed with pepper, is given for dysentery.

MEDICINE.
Bark.
618

Food.—The tree flowers in April and May, and produces a FRUIT which, when ripe, is eaten by the natives. It seems probable that the fruit of this tree is often confused with that of **D. tomentosa.** Roxburgh's description of these two fruits may be here given. **D. melanoxylon:** berry round, of the size of a small apple, yellow, pulpy, seeds as many as 8, immersed in the pulp, kidney-shaped, sharp on the inner edge. **D. tomentosa:** berry ovate, as large as a pigeon's egg, covered with a smooth hard bark, which becomes yellow when ripe, and is filled with a soft, yellow edible pulp. The fruit of **D. melanoxylon** by some writers is said to have an astringent taste, not palatable. But by others: "The fruit when perfectly ripe has a pleasant taste and is much liked" (*Kánara*). "The fruit is of the size of a plum." It is "gathered chiefly by the Náikdás and brought into the bazar at Godhra and Halol for sale" (*Bomb. Gaz., XXV., 389*). "The pulp is yellowish, sweet, soft, and highly astringent; it is much appreciated during the hot months. Douglass mentions a variety without a stone which is cultivated in the Central Asian highlands" (*Lisboa, U. Pl. Bomb., 93*). The above remark about a stoneless form of **Diospyros** is curious in connection with the controversy regarding the Lotus fruit of the ancients, a stoneless form of that classic fruit having been alluded to by some of the early writers. In several of the District Settlement Reports of the Central Provinces reference is made to this fruit. It is, for example, said to be in ordinary use in Bilaspore. Of Hoshangabad it is remarked that the fruit is baked and eaten, but is not generally sold. It is a common expression among the hill-people that they live on *Mohwa* and *Tendú*.

FOOD.
Fruit.
619

Structure of the Wood.—Hard, of a light-pink colour, with irregular-shaped masses of black ebony in the centre. The ebony is jet black with purple streaks, extremely hard, pores and medullary rays difficult to distinguish. It is used for buildings, shoulder poles, and carriage shafts, and the ebony for all purposes of fancy-work and carving.

TIMBER.
620

The following passages from the Gazetteers, Settlement Reports, and other official publications may be here given regarding the timber obtained from this plant: The Assistant Conservator of Forests, Ajmere-Merwara, says, this is the only **Diospyros** met with in his district, and that it is known as *temru*, the wood being used for legs of beds, &c. It is reported to be one of the best timbers in Seone (Central Provinces) and to be plentiful in all parts of the Upper Godavery District. In Raipore, there are stated to be "noble forests of *tendú*." This is viewed as one of the best timber trees of Puri. **Dr. Gray** says that it is plentiful in the Gujarat and Satpuda forests and occasional in Kánara; and that it yields a kind of ebony (*Bombay Gaz., XXV., 348*).

"The wood is used for building and is fairly durable. Blocks of ebony are found in the centre of old trees from 12 to 18 inches in diameter, and on an average weighing from 75 to 80lb the cubic foot" (*Baroda Gaz.*). "Only the heartwood of old trees contains ebony, and even that is streaked with dull yellow lines. The wood, though strong, tough, and fairly durable, is not held in much esteem. A seasoned cubic foot weighs 50 to 70lb" (*Kánara Gazetteer*, the Forest Chapter, written by **W. A. Talbot, Esq.**). "This is a valuable timber tree, the wood being whitish or with a yellowish or brown tinge outside and the core jet black.

DIOSPYROS montana.	The Ebony Trees: Cigarette Covers.

TIMBER.

It is heavy, close, and even-grained, and takes a fine polish " (*Lisboa*). "The black heartwood of old trees is used for cart-wheels and for bracelets."

DOMESTIC.
621

Conf. with B 327.
Bidis.
622

Domestic Uses.—The black wood, in place of "Sandal-wood, is ground into a paste and smeared over the face and body after worshiping the gods. The LEAVES, like those of the *ápta*, are so much used in rolling cigarettes that ship-loads are every year sent to Bombay " (*Thana Gaz.*, *XIII.*, *Pt. I.*, *27*). In a recent report on the subject of BIDIS (these cigarette covers) Mr. A. T. Shuttleworth says that " a considerable revenue is realised by their sale."

623

Diospyros microphylla, *Bedd.; Fl. Br. Ind., 559.*

Syn.—DIOSPYROS BUXIFOLIA, *Hiern.*; D. VACCINIOIDES, *Wall.*; LEUC-OXYLUM BUXIFOLIUM, *Bl.*

References.—*Balfour, Cyclop., 954; Bombay Gazetteer (Kánara), XV., I., 437; Beddome, Ic. Pl., pl. 133.*

Habitat.—" A large tree " (*Flora Br. Ind.*); " an immense evergreen tree, very common in North Kánara " (*Talbot*). Beddome and several other writers allude to this box-wood-like-leaved species as met with in Mysore, the Anamallay hills, and as being distributed to Malacca, Java, and Borneo. Beddome says it is " very common in South Kánara forests of the Gháts and plains." It flowers in the cold season.

624

D. mollis, *Griff.; in Jour. Agri.-Hort. Soc., III., 145; Kurz, For. Fl. Burm., II., 130.*

This plant is not alluded to in the *Flora of British India*, unless it be D. martabanica, *Clarke*, under which is given as a synonym **Gunisanthus mollis**, *Kurz*. Turning to Kurz's paper in the *Journal of the Asiatic Society of Bengal*, where he first described the plant, he is found to state that D. mollis, *Kurz MS.*, is a synonym for **Gunisanthus mollis**, *Kurz*; but in his later publication (*For. Fl. Burma, II., 126*), he gives G. mollis, *Kurz*, as a distinct plant from D. mollis, *Griff.* Under the last mentioned name he says—" The berries produce the so-called black dye of the Shans." Mason mentions the same fact, and states that the tree grows on the mountains which separate the province of Tavoy from the Siamese territory. He calls it the *ma-kleu*. This same economic fact has been subsequently reproduced by various writers, none of whom have apparently identified the plant. Thus Colquhoun (*Among the Shans*) mentions this plant as yielding a black dye. Balfour, Liotard, and others describe it as the " celebrated vegetable black dye."

BLACK DYE.
625

It would be interesting to have this matter cleared up by original investigation. Are there, for example, two trees, *viz.*, **Gunisanthus mollis**, *Kurz*, and **Diospyros mollis**, *Griff.* (as in *Kurz, For. Fl.*); if not, are both these synonyms for D. martabanica, *Clarke* (as in the *Flora of British India*)? The dye obtained is so highly spoken of, that this subject would seem to call for early investigation. Black dyes are reported to be obtained from the shoe-flower (**Hibiscus rosa-sinensis**), from the juice of the cashew tree (**Anacardium occidentale**), from the fruit of various species of Melastoma (hence the name, because of the fruits staining the mouth black), and from this imperfectly known **Diospyros**—the Shan black-dye. In his report on Pegu Kurz says the wood of **Gunisanthus mollis** is of a red brown colour, rather heavy, has a short fibre, is close-grained and soft. It is soon attacked by xylophages.

Other Black Dyes.
626

TIMBER.
627

628

D. montana, *Roxb.; Fl. Br. Ind., III., 555; Wight, Ic., t. 1225.*

Syn.—D. CORDIFOLIA, *Roxb.*; D. RUGOSULA, *Br.*; D. BRACTEATA, *Roxb.*; D. HETEROPHYLLA, *Wall.*; D. SYLVATICA, *Wall.*, not of *Roxb.*; D. PUNCTATA, *Dcne.*; D. GOINDU, *Dalz.*; D. WALDEMARII, *Klotzsch.*

D. 628

Vern.—*Tendu, dasáundu, lohari, bisténd,* HIND.; *Ban-gáb* (*e.g.,* wild **D. Embryopteris**), BENG.; *Gada terel,* SANTAL; *Makar-tendi,* BANDA; *Hirek, keindu, temru, pasendu, kendú,* PB.; *Ambia,* BANSWARA; *Pasend,* BHURTPUR; *Temru,* MEYWAR; *Kanchan, kadal, pattewar, patwan,* C. P.; *Hádru,* PANCH MAHALS; *Goindú, kundu, temru,* BOMB.; *Timru, timbúrni, tembhurni, goundhan,* MAR.; *Timra* (Panch Mahals), *timru* or *timbaroa* (Palanpur), GUZ.; *Tendu,* KANARA; *Goinda,* KONKAN; *Muchi tanki, kaka ulimera* or *nalla ulimera, yerragoda* (or *erragoda*), *micha tummurra,* TEL.; *Goindú, kala goindú, balkuniki, jagalagante,* KAN.; *Gyútbeng* (*chop-pin*), *taubot* (**Kurz** for **D. cordifolia**), BURM.; *Kethi-kanni, vukkana* (TAM. in Ceylon), SING.; *Tumala* (according to **Roxburgh** for **D. cordifolia**), SANS.

The *Ahmadnagar Gazetteer* gives the Deccan name *Gondhan* to **D. cordifolia,** and *Timbhurni* to **D. montana**; the above Burmese name is also assigned to **D. cordifolia,** but not to **D. montana.** Similar vernacular names distinguishing the two forms exist in other parts of India, particularly with the Telegu people.

References.—*Roxb., Fl. Ind., Ed. C.B.C., 415; Corom. Pl., I., 37, t. 48; Voigt, Hort. Sub. Cal., 344; Brandis, For. Fl., 296; Kurz, For. Fl. Burm., II., 130; Beddome, Fl. Sylv. Man., 143; Thwaites, En. Ceylon Pl., 178; Dals. & Gibs., Bomb. Fl., 141 & 142; Stewart, Pb. Pl., 137; Aitchison, Cat. Pb. and Sind Pl., 86; Grah., Cat. Bomb. Pl., 107; Kurz in Jour. As. Soc., 1877, Pt. II., 235; Wight, Ic., t. 148; Trimen, Sys. Cat. Ceylon Pl., 51; Fl. Andh. by Elliot, 52, 77, 128; DC. Prod., VIII., 230; Baden Powell, Pb. Pr., 578; Atkinson, Him. Dist., 364; Lisboa, U. Pl. Bomb., 93, 164; Royle, Ill. Him. Bot., 262, Balfour, Cyclop., 954; Journ. Agri-Hort. Soc., 1875-78,\Vol. V., 74; Indian Forester, III., 203; V., 187; VI., 125 (*D. cordifolia*), II., 179; III., 203; IV., 228; XII., App. 16; Gazetteers, Bombay (Panch Máhals), III., 201; IV., 23; V., 285; XI., 26; XII., 23; XV., Pt. I., 67; XVII., pp. 24, 26; XXV., 93, 164 and 389.*

Habitat.—A tree often spinose, met with "from the Himálaya (on the Raví eastward, *Brandis*) to Ceylon and Tenasserim: common." "**D. cordifolia** has the female peduncles ½ to ¾ in., the leaves and calyx sub-persistently pubescent: **D. montana** has them glabrescent. Beddome maintains these as distinct species" (*Fl. Br. Ind.*). Roxburgh distinguishes his two plants thus:—**D. montana**: leaves ovate oblong smooth, male flowers numerous; female with only 4 sterile stamina. **D. cordifolia**: leaves linear cordate, downy; male peduncles 3-flowered: female, with 12 sterile stamina. As already remarked, the vernacular names compiled above under one species are by the natives distributed under two, and it seems probable that little advantage is gained by the combination. De **Candolle** in the *Prodromus Systematis Nat. Regni Vegetabilis* follows **Roxburgh** in regarding **D. montana** as distinct from **D. cordifolia.** But for the fact that the *Flora of British India* is the accepted botanical standard in this work, the writer would personally have been more disposed to accept **Beddome's** position. Of the Panch Mahals it is said to be common only in the Pávágad forests. **Wight** mentions the Courtallum sub-alpine forests of Madras as a locality. In India, as a whole, one or the other form occurs in most provinces if not both.

Medicine.—The FRUIT is supposed to be poisonous. The *bhistis* (water-carriers) apply it to the boils which generally appear on their hands and which give them much pain and trouble.

Food.—As with the products of nearly every other species of **Diospyros** the most conflicting statements have been published regarding the fruit of this plant. In the Gazetteer of the Pálanpur State (Bombay), it is enumerated as one of the chief fruits. In a list of Gujarat fruit trees (compiled from material furnished by **Mr. G. H. D. Wilson, C.S.,** and **Lieut.-Col. J. G. McRae**) published in the *Bombay Gazetteer (XXV., p. 389)*, it is

MEDICINE.
Fruit.
629

FOOD.
630

stated that "The fruit is eaten and much relished by the forest tribes, but is seldom offered for sale in the market;" "except in the size of its fruit, which is as large as an apple, it is scarcely distinguishable from D. melanoxylon, *tamrug*, whose fruit is of the size of a plum." On the other hand, **Mr. J. C. Lisboa** (*Useful Plants of Bombay, page 93*) describes **D. montana**, *Roxb.* (giving "**D. Goindu**, *Dalz.*" as a synonym), and states that the "fruit is bitter, not eaten." At page 164 he gives an account of "**D. Goindu**, *Dalz.*" (placing **D. montana**, *Roxb.*, under *it* as a synonym), and remarks that "The fruit globose, size of large cherry, yellow when ripe, is said by **Dr. Birdwood** to be eaten as a fruit." In the *Ahmednagar Gazetteer* it is stated that the *Gondhan* (**D. cordifolia**) bears an edible fruit. Of the Panjáb, **Stewart** writes: "The fruit is not eaten, and I have heard it called 'poisonous.'" This confusion may perhaps be accounted for by the amalgamation of **D. montana** and **D. cordifolia** into one species.

FODDER.
631

Fodder.—"Leaves used as fodder in Oudh" (*Brandis*).

TIMBER.
632

Structure of the Wood.—Yellowish-gray, soft, no heartwood, no annual rings. It is durable and would be good for furniture.

"The wood is used for making carts (*Panch Mahál Gaz.*, 201). **Dr. Hove** (a Polish botanist who visited Gujarat in 1787) describes a forest of "black-wood" and the process adopted for darkening the timber. Modern writers are disposed to regard the "black-wood" described by **Hove** as having been **D. montana** and not **Dalbergia latifolia**. **Hove** states that the wood was buried in a swamp, and after soaking till it was black was sent for sale to Surat and other places on the coast (*Conf.* with **Dalbergia latifolia**, p. 9). "A small tree of the ebony kind with black and variegated streaks towards the heart. The wood is pretty strong but is not much used" (*Kánara Gaz.*, 67). "Wood is dark-brown, mottled with white, hard, close-grained, takes a fine polish, and is used for furniture" (*Lisboa*). In Kolába it is said to be used in hut and cowshed building. "It has a very hard strong wood, but, except for cart poles, is not much used. The fact that it is not durable, that it suffers from the attacks of insects and cannot be crested, takes away from the value of its timber. The centre or heartwood, which is very small, is ebony of an inferior kind; but except that it is turned into wooden bracelets, it is apparently not known in trade" (*Bombay Gaz., XXV., 389*). In Ahmednagar it is said that the wood of this species is good fuel, and that it is used chiefly for making field and other tools. **Balfour** affirms that the tree is not uncommon along the Siwalik tract up to near the Ravi, and occasionally out in the plains westward from near Delhi to Sirsa. The wood, dark and strong, is fitted for agricultural implements, in door work, &c.; does not bear exposure, and could not be creosoted.

633

Diospyros nigricans, *Wall.; Fl. Br. Ind., III., 557.*

References.—[*Beddome, For. Man., 144*, so quoted in *Fl. Br. Ind.*, seems to the writer to be the description of the plate *124, Ic. Pl. Ind. Or.* (D. NIGRICANS, *Dalz.*), that description being simply reprinted in the *Icones*. **Beddome** probably never saw D. NIGRICANS, *Wall.*, a Khásia Hill plant, so that his plate and description appear to refer to the Konkan species, D. ASSIMILIS, *Bedd.*]

Habitat.—A tree attaining a height of 50 feet, met with in the Khásia Hills and Sylhet.

TIMBER.
634

Structure of the Wood.—Nothing is known as to the special properties of the timber of this tree. It doubtless possesses the characteristic features of the other species, but it is alluded to in this place chiefly with the view of correcting a mistake current in Indian botanical works, and

The Ebony Woods.	(*G. Watt.*)	DIOSPYROS pyrrhocarpa.

which took its origin with the late **Mr. Dalzell**, *viz.*, of mistaking **Beddome's D. assimilis** for **D. nigricans**, *Wall.* (*Conf.* with **D. assimilis**, *Bedd.*).

Diospyros oleifolia, *Wall. ; Fl. Br. Ind., III., 567.*   635

> Habitat.—A large tree, attaining a height of 60 feet, met with in Amherst, Pegu, Martaban, and Tenasserim ; frequent.
> Structure of the Wood.—Sapwood white or yellowish white, heavy, fibrous, close-grained, soft (*Kurz*).

TIMBER.
636

D. oocarpa, *Thw. ; Enum., 180 ; Fl. Br. Ind., III., 560.*   637

> Vern. - *Vellai-karun káli* (TAM. in Ceylon) ; *Kalu-kadumberiya*, SING.
> References.—*Gamble, Man. Timb., 250; Thwaites, En. Ceylon Pl., 180; Trimen, Sys. Cat. Ceylon Pl., 52; Indian Forester, X., 31 ; Balfour, Cyclop., 954.*
> Habitat.—A middle-sized tree of the Konkan, Mysore, and Ceylon.
> Structure of the Wood.—Purplish-brown, with black streaks, moderately hard ; a handsome wood. Often spoken of as one of the Calamander woods.

TIMBER.
638

D. ovalifolia, *Wight ; Ic., t. 1227 ; Fl. Br. Ind., III., 557.*   639

> Vern.—*Vedú-kunari* (TAM. in Ceylon), SING.
> References.—*Beddome, For. Man., 143; Thwaites, En. Ceylon Pl., 181; Trimen, Sys. Cat. Ceylon Pl., 52; Indian Forester, III., 203; VIII., 29; X., 31 ; Balfour, Cyclop., 954.*
> Habitat.—A middle-sized tree met with in the South Deccan Peninsula, Coimbatore and Nilgiri Hills, Anamallay and Tinnevelly Hills.

[*III., 570.*

D. paniculata, *Dalz., in Hook. Kew Journ., IV., 109 ; Fl. Br. Ind.,*   640

> References.—*Bedd., Ic. Pl. Ind. Or., t. 125, and For. Man., 144; Dalz. and Gibs., Bomb. Fl., 141 ; Lisboa, U. Pl. Bomb., 94; Kánara Gazetteer, 437 ; Balfour, Cyclop., 954.*
> Habitat.—A large tree of the Deccan and Konkan. **Mr. Lisboa** mentions the Chorla Gháts and Raighát and **Mr. Talbot** the Sahyádris near Mavimone. **Beddome** quotes the Carcoor Ghát (Wynaad) at elevations of 2,000 to 3,000 feet.
> It flowers during the cold season.

[*III., 553.*

D. pruriens, *Dalz., in Hook. Kew Journ., IV., 110 ; Fl. Br. Ind.,*   641

> References.—*Bedd., Ic. Pl. Ind. Or., t. 129, and For. Man., 144; Dalz. and Gibs., Bomb. Fl., 141 ; Lisboa, U. Pl. Bomb., 95 ; Balfour, Cyclop., 954; Bombay Gazetteers (Kánara), XV., Pt. I., 437 ; XXV., 348.*
> Habitat.—A small or middle-sized tree, found on the Western Gháts from Bombay and Kánara southwards to the Wynaad and Tinnevelly ; altitude 1,000 to 3,000 feet. It is said to be specially abundant on the Nilkund Gháts. It flowers in the cold season.

D. pyrrhocarpa, *Miq. ; Fl. Br. Ind., III., 571.*   642

> Vern.—*Tay, té*, BURM.
> References.—*Kurz, For. Fl. Burm., II., 136; Gamble, Man. Timb., 252.*
> Habitat.—An evergreen tree of the Andaman Islands. The *Flora of British India* alludes to this as a doubtfully Indian species, but both **Gamble** and **Kurz** mention it as met with occasionally in the Andaman Islands.

D. 642

DIOSPYROS stricta.	The Ebony Woods.

DYE.
643
FOOD.
Fruit.
644
TIMBER.
645
646

Dye.—Major Ford says the Burmese extract a red dye from the fruit, and that the Chinese umbrellas are dyed with that substance, which has the property of rendering them waterproof.

Food.—The FRUIT is said to be eaten by the Burmese.

Structure of the Wood.—Reddish-brown, moderately hard to hard; weight 52℔ per cubic foot.

Diospyros quæsita, *Thwaites; Fl. Br. Ind., III., 560.*

THE CALAMANDER WOOD.

Vern.—*Pú-karunkáli* (TAM. in Ceylon); *Kalumediriya*, SING.
References.—*Bedd., Ic. Pl., pl. 128; Brandis, For. Fl., 296; Thwaites, En. Ceylon Pl., 179; Tennet, Nat. Hist. Ceylon, 118; Trimen, Sys. Cat. Ceylon Pl., 52; Yule-Burnell, Gloss. Anglo-Ind. Terms, 110; Balfour, Cyclop. Ind., 954; Treasury of Botany, 411; Indian Forester, VII., 128; VIII., 29; Gamble, Man. Timb., 250.*

Habitat.—A large tree of Ceylon, alluded to in this place because of the high esteem in which the wood is held. It is at most only experimentally cultivated in India.

TIMBER.
647

Structure of the Wood.—Hard, consisting of irregular alternate layers of black ebony and greyish-brown wood. The most valuable ornamental wood in Ceylon; it is now scarce, and is much in demand. But **D. Kurzii** is often spoken of as a good substitute of this wood (which see).

648

D. ramiflora, *Roxb.; Fl. Br. Ind., III., 569; Wight., Ic., t. 189.*

Vern.—*Uri-gáb* or *gúlul,* BENG.
References.—*Roxb., Fl. Ind., Ed. C.B.C., 414; Voigt., Hort. Sub. Cal., 344; Kurz, For. Fl. Burm., II., 132.*

Habitat.—A large tree, native of East Bengal, Tipperah, the Khásia hills, &c. **Kurz** says it occurs in the tropical forests of Arakan and Chittagong.

FOOD.
Fruit.
649
TIMBER.
650

Food.—Roxburgh remarks that the FRUIT is as large as an orange, takes twelve months to ripen, and is "replete with yellowish edible pulp."

Structure of the Wood.—Roxburgh says that in the eastern frontier of Bengal this tree grows to a great size and "supplies the natives with very strong hard wood."

651

D. Sapota, *Roxb.; D.C. Prod., VIII., 228.*

Roxburgh, Voigt, and others allude to this species as met with in gardens in India. Roxburgh specially mentions that it grows most luxuriantly in the Botanic Gardens and blossoms in the hot season, but has not yet perfected its fruit in Bengal. It is a native of Mauritius and the Philippine Islands. Hyder Ali is said to have introduced the tree into his gardens in 1804. The *Mysore and Coorg Gazetteer* calls it the Date Plum, and describes it as a handsome evergreen tree.

652

D. sapotioides, *Kurz; Fl. Br. Ind., III., 562.*

Habitat.—A tree attaining a height of 50 feet; frequent in the Pegu Yomah, Burma.

653

D. stricta, *Roxb.; Fl. Br. Ind., III., 563.*   [345.

References.—*Roxb., Fl. Ind., Ed. C.B.C., 415; Voigt., Hort. Sub. Cal.,*

Habitat.—A tall, slender, conical tree, with straight trunk; met with in Eastern Bengal, Sylhet, and Tipperah, &c. **Kurz** adds that it also occurs in Chittagong.

TIMBER.
654

Structure of the Wood.—Of superior quality: its straight stems are taken advantage of in house-building.

D. 654

The Ebony Woods.	**DIOSPYROS tomentosa.**

Diospyros sylvatica, *Roxb. ; Fl. Br. Ind., III., 559.*

655

Vern.—*Tella goda,* TEL. ; *Kalúchia,* URIYA.

References.—*Roxb., Fl. Ind., Ed. C.B.C., 415 ; Thwaites, En. Ceylon Pl., 178 ; Voigt, Hort. Sub. Cal., 344 ; Balfour, Cyclop., 954 ; Trimen, Sys. Cat. Ceylon Pl., 52 ; Indian Forester, III., 203 ; X., 34 ; Elliot, Fl. Andh., 58 ; Bedd., Ic. Pl. Ind. Or., t. 121, and For. Man., 143.*

Habitat.—A medium-sized tree of the Deccan Peninsula from Bombay and the Circars to Ceylon ; ascends to 3,000 feet in altitude.

D. tomentosa, *Roxb. ; Fl. Br. Ind., III., 564 ; Wight, Ic., tt. 182, 183, not of Poir ; D. exsculpta, Ham., and of A. DC., Prodr. VIII., 223, not of Dalz. nor of Bedd.*

656

Vern.—*Tumal, mitha tendu,* HIND. ; *Kyou, kend,* BENG. ; *Tiril,* KOL. ; *Kendhu,* URIYA ; *Kinnú, kendu, tindú,* PB. ; *Chilta tumiki,* TEL. ; *Kaulay* (according to **Balfour**), KAN. ; *Kakinduka,* SANS. ; Panjáb bazar name for the medicinal raspings, *búra de abnús.*

References.—*Roxb., Fl. Ind., Ed. C.B.C., 413 ; Voigt., Hort. Sub. Cal., 343 ; Brandis, For Fl., 295 ; Stewart, Pb. Pl., 137 ; Elliot, Fl. Andh., 44 ; Indian Forester, X., 543 ; XII., App. 16 ; Baden Powell, Pb. Pr., 578 ; Royle, Ill. Him. Bot., 262 ; Jour. Agri.-Hort. Soc. (1867), p. 80 ; Balfour, Cyclop., 953, 954.*

Habitat.—A small crooked tree, found in the northern parts of Bengal, Behar, Chutia Nagpur, Bundelkhand, and Oudh—extending as far north as to the Siwalik tracts of the Panjáb. It produces whitish yellow wooly flowers, which appear in April, and small berries, which ripen in June. This would appear to be the eastern and northern representative of D. **melanoxylon,** a western Gháts species. In a recent report from the Madras Board of Revenue this species is said to occur (as far as Madras is concerned) chiefly in the ceded districts and hills of the western Carnatic : D. **melanoxylon** occurring in the eastern Gháts. **Balfour** writes that it attains its full size in 60 years. Length of trunk to first branch 8 to 10 feet and girth 4 feet (*Conf.* with the remarks under D. **melanoxylon**).

Medicine.—RASPINGS of the wood are officinal in the Panjáb, being given as an alterative (*Stewart*).

MEDICINE.
Raspings.
657

Food.—When ripe the BERRIES are yellowish, and are filled with a soft yellow, sweetish, astringent pulp, eaten by the natives. (*Conf.* with D. **melanoxylon.**) Stewart, who in all probability is correctly alluding to this species, since D. **melanoxylon** does not occur in the Panjáb, says the FRUIT, which is reported to ripen in June with the mango, is eaten, being sweetish and astringent and not unpleasant. But one authority reports its pulp as bitter, fœtid, and emetic.

FOOD.
Berries.
658
Fruit.
659

Structure of the Wood.—Roxburgh says it is black, hard, and heavy and is in short the ebony of Bengal. Of the Panjáb Stewart remarks that this plant is scarce, but that the wood "is fine, black, hard, and somewhat brittle. It carves well, and insects are said not to touch it. Mr. Watson, Madhopur workshops, informs me it is good for cogs if it could be got, though hardly so strong as **Olea**. Inland near the Rohilkhand Siwálik tract, where the tree is more common or better looked to, handsome work-boxes, &c., are constructed of the wood ; combs are made from it in the Ambala District. In Kangra it "is used for ploughs, in house-building, and for small boxes." **Balfour** remarks that the wood of young trees is white, but that of old trees is black : the heart-wood is fine, extremely hard, but somewhat brittle, and is used by agriculturists for ploughs, and for the wood-work of their houses.

TIMBER.
660
Work-boxes.
661
Combs.
662
Ploughs.
663

D. 663

664 **Diospyros Toposia,** *Ham. ; Fl. Br. Ind., III., 556.*

Syn.—D. RACEMOSA, *Roxb., Fl. Ind., Ed. C.B.C., 414 ; Wight, Ic., t. 416;* D. LANCEOLATA, *Wall , Cat., 4122;* EMBRYOPTERIS LANCEOLATA, *Don.* According to **Brandis, Gamble,** and other writers, this is reduced to D. MELANOXYLON, along with D. TOMENTOSA, but by the *Flora of British India* all three are retained as separate species.

Vern.—*Gŭlul,* SYLHET ; *Kaha-kála,* SING.

References.—*Roxb., Fl. Ind., Ed. C.B.C., 414 ; Voigt, Hort. Sub. Cal., 345 ; Kurz, For. Fl. Burm., II., 128 ; Beddome, Ic. Pl. Ind. Or., t. 122 & For. Man., 144 ; Thwaites, En. Ceylon Pl., 179 ; Trimen, Sys. Cat. Ceylon Pl., 52 ; Indian Forester, X., 34 ; Royle, Ill. Him. Bot., 262 ; Balfour, Cyclop., 954.*

Habitat.—A large tree met with in Sylhet, Cachar, and Chittagong. Roxburgh gives this the same vernacular names as recorded under his **D. ramiflora** and **D. lanceæfolia,** and as these trees are all found in the same region, it is probable the natives do not distinguish the one from the other.

FOOD.
Fruit.
665
Food.—FRUIT ripens in November and is eaten by the natives (*Roxburgh*).

666 **D. Tupru,** *Buch.-Ham. ; Fl. Br. Ind., III., 563.*

Syn.—DIOSPYROS RUBIGINOSA, *Roth. ;* D. MELANOXYLON, *Hiern.,* in part.

References.—*Brandis For. Fl., 295 ; Bedd., Fl. Sylv., t. 66 ; Dals. & Gibs., Bomb. Fl., 142 ; Bombay Gazetteer (Kanara), XV., Pt. I., 437.*

Habitat.—A small tree of the Western Deccan Peninsula from the Concan to Mysore.

667 **D. undulata,** *Wall. ; Fl. Br. Ind., III., 568.*

Habitat.—A large tree of Amherst, Mergui, and Malacca, mistaken by some writers for **D. lucida,** *Wall.,* a Singapore and Malacca species. According to **Kurz, D. undulata** occurs in the tropical forests of Martaban, Tenasserim, and the Andaman Islands. It flowers in April and May, and the fruit ripens in October to February.

668 **D. variegata,** *Kurz ; Fl. Br. Ind., III., 557.*

Habitat.—A large tree (attaining a height of 70 feet) found fairly abundantly in Assam, Pegu, and Martaban, ascending to altitudes of 1,000 feet.

TIMBER.
669
Structure of the Wood.—Sapwood white, turning greyish, heavy fibrous but close-grained, soft (*Kurz*).

DIPLOSPORA, *DC. ; Gen. Pl., II., 97.*

670 **Diplospora apiocarpa,** *Dalz. ; Fl. Br. Ind., III., 123 ;* RUBIACEÆ.

Vern.—*Panigara,* MAR. ; *Bachange,* KAN.

References. —*Beddome, Fl. Sylv., t. 223 ; Ic. Pl. Ind. Or., t. 40 ; Dals. & Gibs., Bomb. Fl., 120 ; Bomb. Gaz., XV., Pt. I., p. 68.*

Habitat.—A small tree of the Western Peninsula from the Concan southwards ascending to 5,000 feet.

TIMBER.
671
Structure of the Wood.—Used to make combs and toys (*Bomb. Gaz., XV., I., 68*).

672 **D. singularis,** *Korth. ; Fl. Br. Ind., III., 123.*

Vern.—*Thittú,* BURM.

Habitat.—A small tree distributed from the Khásia hills to Pegu, Tenasserim, Amherst, Sumatra, Borneo, &c.

D. 672

| The Garjan Oil Trees.   (*G. Watt.*) | DIPTEROCARPU alatus. |

Structure of the Wood.—Rough with numerous prominent medullary rays: weight 36℔ a cubic foot (*Kurz, Fl. Brit. Burm., II., 50; Gamble, Man. Timb., 119*).

TIMBER.
673

DIPLOTAXIS, *DC.; Gen. Pl., I., 84, 967.*

Diplotaxis Griffithii, *H.f. & T.; Fl. Br. Ind., I., 157; * CRUCIFERÆ.   674

Vern.—*Sisgai, mole,* TRANS-INDUS; *Baráni múli, bibúcha, chinaka,* (Sind Sagar Doab), PB.; *Parjan?* MERWARA.

Habitat.—A robust herb 1—3 feet high, found on the Salt Range in the Panjáb, and distributed thence through Baluchistan to Afghánistan. **Mr.** Duthie alludes to a species of **Diplotaxis** as collected by him in Merwara, and the vernacular name there given to it has provisionally been included with the above.   If this prove correct, the area of the species should be given from Merwara.

Food.—Eaten as a pot-herb.

FOOD.
675

DIPTEROCARPUS, *Gærtn. f.; Gen. Pl., I., 191, 981.*

A genus of lofty trees embracing some 50 species, natives of Tropical East Asia. Of these India (as accepted by the *Flora of British India*) possesses 17, of which 6 occur in India proper.   The others are Ceylon species, or appear in Burma and are distributed to Malacca. The generic name has been given in allusion to the winged condition of the fruit, due to the accrescent calyx.

Dipterocarpus alatus, *Roxb.; Fl. Br. Ind., I., 298;* DIPTEROCARPEÆ.   676

Syn.—DIPTEROCARPUS COSTATUS, *Gærtn. f.*

Vern.—*Garjan* (*batti-sal* according to **Balfour**, *shweta-garjan,* according to **Birdwood**), BENG.; *Kanyinbyu* (=white *Kanyin*), BURM.; *Horagaha* (according to **Birdwood**), SING.

References.—*Roxb., Fl. Ind., Ed. C.B.C., 439; Kurz, For. Fl. Burm., I., 116, 117; Gamble, Man. Timb., 33; O'Shaughnessy, Beng. Dispens., 224; Dymock, Mat. Med. W. Ind., 2nd Ed., 88; Year Book Pharm., 1877, 155; Birdwood, Bomb. Pr., 257; Cooke, Gums and Gum-resins, 114; Report on the Gums and Resins of India published by the P. W. D., pp. 19, 20, 31, 35, 37, & 62; Indian Forester, I., 365; VI., 125; VIII., 416; Balfour, Cyclop., 956; Kew Off. Guide to the Mus. of Ec. Bot., 17.*

Habitat.—A large tree met with in Chittagong, Burma, and the Andaman Islands; distributed to Siam.

Oleo-resin.—Kurz says this tree yields a WOOD-OIL in great quantity and exudes a dirty-brown resin.   The oil and resinous thicker substance are at first mixed together; this mixture is strained through a cloth whereby the clear oil separates itself from the resinous portion.   According to Roxburgh this species affords the wood-oil of Pegu.

OLEO-RESIN.
Wood-oil.
677

In a recent correspondence with **Mr. J. W. Oliver**, Forest Department, Burma, this species is given (along with **D. lævis** and **D. turbinatus,** &c.) as one of the trees that yields the thin oil which in Burmese trade reports is designated *Kanyin*-oil or Burmese wood-oil.   The thick oleo-resinous substance known in Burma as *in*-oil is obtained from **D. tuberculatus.** It is probable that the latter substance is that which sometimes bears, in India, the name of Garjan-oil, but this point has not been satisfactorily determined, and it seems likely that the Garjan-oil of European and Indian commerce may in reality be any one or a mixture of all the *Kanyin* and *in* oils, but chiefly of the former.   For particulars as to the extraction of *Kanyin*-oil see a further page under **D. turbinatus.**

DIPTEROCARPUS lævis.	**The Garjan Oil Trees.**

TIMBER.
678

Structure of the Wood.—Sapwood white; heartwood reddish-grey, moderately hard, smooth, mottled, takes a fine polish. Weight from 38 to 50℔ a cubic foot. Used for house-building and canoes, but is not durable; if exposed to wet it decays rapidly, the canoes made of it lasting only three to four years.

679

Dipterocarpus angustifolius, *W. & A.; Fl. Br. Ind., I., 299.*

Syn.—Dipterocarpus costatus, *Roxb. (not of Gærtn. f.).*
According to **Roxburgh** this species is a native of Chittagong. By the *Flora of British India* it is viewed as doubtfully distinct.

680

D. Griffithii, *Miq.; Fl. Br. Ind., I., 299.*

Syn.—Dipterocarpus grandiflorus, *Griff. (not of Wall),*
References.—*Kurz, For. Fl. Burm., I., 116; Report on Gums and Resins issued by P. W. D., pp. 34, 62, 64.*

Habitat.—A tree of the Mergui and South Andaman Islands. **Kurz** says it is common in the tropical and moister upper mixed forests of the Andamans and also in Tenasserim.

TIMBER.
681

Structure of the Wood.—Yellowish-grey, rather coarsely fibrous, close-grained, and heavy (*Kurz*).

682

D. incanus, *Roxb.; Fl. Br. Ind., I., 298.*

References.—*Roxb., Fl. Ind., Ed. C.B.C., 439; O'Shaughnessy, Beng. Dispens., 224; Dymock, Mat. Med. W. Ind., 2nd Ed., 88; Report on Gums and Resins issued by the P. W. D., pp. 19, 20, 31, 35, 37; Cooke. Gums and Gum-resins, 114; Agri-Hort. Soc. of India Journ., Vol. IV., 15; Spons' Encyclop., 1651; Balfour, Cyclop., 956, 1087.*

Habitat.—A tree of Chittagong (*Roxburgh*), but according to **Kurz** it occurs also in Pegu.

OLEO-RESIN,
Wood-oil.
683

Oleo-resin.—It yields a wood-oil or balsam. **Roxburgh** says this is the *garjun* tree of Chittagong " where the tree grows to a great size and is said to furnish the largest proportion of the best sort of wood-oil or balsam mentioned in my description of **D. turbinatus.** Flowering time November and December, and the seed ripens in April." **Balfour** seems to be mistaken when, after enumerating **Dipterocarpus alatus, D. costatus, D. incanus, D. lævis,** and **D. turbinatus** as yielding wood-oil, he adds "but **D. incanus** is supposed to yield the best sort and in the greatest quantity."

MEDICINE.
Oil.
684

Medicine.—Dymock also includes this plant along with **D. turbinatus** and **D. alatus** in his account of the medicinal Gurjun-oil, but it is certainly far less important commercially than Kanyin-oil-yielding trees of Burma.

D. indicus, *Beddome*; see under **D. lævis,** *Ham.*

685

D. lævis, *Ham.; Indian Forester, X., iii., 131 ; IX., 216.*

The lofty tree so named—a native of the tropical forests of Burma—is, by the *Flora of British India,* reduced to be a synonym for **D. turbinatus,** *Gærtn. f.* It has been the custom followed by the writer to accept the *Flora* as the standard on all botanical points, the endeavour being made in the present work to compile the economic information regarding plants under the names as established by **Sir J. D. Hooker.** Gamble, Kurz, and other Indian botanists do not, however, accept the above reduction as correct, but prefer to regard these names as belonging to distinct trees. Should this latter opinion be confirmed, the information given under **D. turbinatus** would probably, to some extent, have to be rearranged. Gamble, however, affirms that the Garjan-oil tree is **D. turbinatus,** although under

D. 685

Products of India. 159

The Male In or Inbo Tree.   (*G. Watt.*)   DIPTEROCARPUS
pilosus.

D. lævis he makes the remark that "it yields copiously a resin and a wood-oil used for painting." According to some of the more recent writers garjan and wood-oil are distinct, though both are obtained from several trees. If this be so, a rearrangement would probably not seriously affect what has been given below. It may serve a useful purpose, therefore, to mention in this place the Burmese name given to **D. lævis,** *Ham., viz., Kanyin-ni* (*e.g.,* red *Kanyin*), while **D. alatus** is known as *Kanyin-byu* (*e.g.,* white *Kanyin*). **Gamble** points out that, according to the *Flora of British India,* **D. indicus,** *Beddome, t. 94,* may be reduced either to **D. turbinatus** or **D. lævis.** He appears, however, to view it as a distinct species, a native of the western Ghâts, which is there known as *Guga* and *Walivara* in Kánarese. The Garjan-oil reported to be made in South India would, accordingly, be the produce of **D. indicus.**

Resin.—The authors who recognise this as a distinct species say that it yields a RESIN similar to that of all the other species.

Oil.—For information as to the wood-oil obtained from this plant, see under **D. turbinatus.**

Structure of the Wood.—Sapwood white; heartwood rough, reddish, soft; is rarely used, but is occasionally employed for planking and rafters; weight 43—49℔ a cubic foot.

REISN.
686

OIL.
687

TIMBER.
688

Dipterocarpus obtusifolius, *Teysm.; Fl. Br. Ind., I., 295.*

689

This is in Burma called the male *In* tree or *Inbo.*

Vern.—*Inbo, kanyin-kok* (according to **Gamble**), BURM.
References.—*Kurs, For. Fl. Burm., I., 115; Gamble, Man. Timb., 32; Indian Forester, VIII., 416.*
Habitat.—A large, deciduous tree of the Eng (*In*) forests of Prome and Martaban, ascending to 3,000 feet. It is commonly found forming small patches in the *In* forests.

Resin.—This tree is said to afford a clear white or yellow resin, not an oil. This is reported to burn readily, but is not used for any purpose.

Structure of the Wood.—Heartwood reddish-brown, rough, moderately hard. Pores large and moderate-sized. Weight 59℔ per cubic foot (*Gamble*). **Kurz** says it is "of the quality of that of *Eng.*"

RESIN.
690
TIMBER.
691

D. pilosus, *Roxb.; Fl. Br. Ind., I., 296.*

692

Vern.—*Hollong,* ASSAM.
References.—*Roxb., Fl. Ind. Ed., C.B.C., 440; Kurs, Fl. Burm., 115; Jour. As. Soc. Bengal, 1870, II., 65; also 1874, p. 98; Forest Fl. Burm., I., 115; Gamble, Man. Timb., 31.*
Habitat.—A large evergreen tree met with in Assam, Chittagong, Pegu, Arracan, hills of Martaban and Tenasserim, and also the Andaman Islands. Distributed to Sumatra.

Oleo-resin.—Mr. Oliver, in the report below and accompanying correspondence, suggests that perhaps some of the Kanyin oil of Tenasserim may be obtained from this tree.

Structure of the Wood.—"Of a reddish-brown colour, close and pretty straight grain; it does not warp or split much but quickly deteriorates unless kept in a dry and ventilated place; is attacked by nearly all the timber insects. Notwithstanding its large size, it is of little or no use except for temporary purposes and for packing boxes; it must, however, be borne in mind that in Assam this latter use forms a very important business, as not less than 400,000 boxes for packing tea are used yearly, the making of each one requiring about 1·50 cubic feet of rough timber" (*Paganini, in Indian Forester*).

OLEO-RESIN.
693

TIMBER.
694

Packing
boxes.
695

D. 695

**DIPTEROCARPUS
tuberculatus.**   The Eng or In Tree.

---

696   Dipterocarpus tuberculatus, *Roxb.; Fl. Br. Ind., I., 297.*

THE ENG (or, as it is now spelt, IN) TREE.

In a passage quoted below this is said to be known to the Burmans as the
female *In* (or *Inma*). It is reported to yield a thick oleo-resinous substance.

Syn.—D. GRANDIFLORUS, *Wall.*

Vern.—*Eng*, or *in*, BURM.; *Sooahn*, TALEING.

References.—*Roxb., Fl. Ind., Ed. C.B.C., 440; Brandis, For. Fl., 27;
Kurz, For. Fl. Burm., I., 113; Gamble, Man. Timb., 32; Special Report by Mr. Alpin, Deputy Conserv. Forests, Burma (Tour with Southern
Shan Force, 1887-88); Cooke, Gums and Gum-resins, 115; Indian Forester, I., 107, 362, 363; II., 178, 181; VIII., 113, 416; IX., 14; X., 131,
134; XIII., 56; Balfour, Cyclop., 957; Ind. For., X., iii., 131.*

Habitat.—A large, deciduous, gregarious tree, forming the "*In* forests"
of Burma and Chittagong. Distributed to Siam.

OLEO-RESIN.   Oleo-resin.—According to Roxburgh, Gamble, Kurz, and other
697   authors, this tree does not yield a wood-oil, but exudes a clear yellow resin.
Mr. J. W. Oliver, Deputy Conservator of Forests, informs the writer,
however, that it does yield an oil, but an oil of a considerably thicker substance (an oleo-resin) than the kanyin-oils described under **D. turbinatus.**
In a further page, under **D. turbinatus,** will be found a general account
of Gurjun and Wood-oil. The former appears to be the crude product,
the latter the liquid oil, obtained after the subsidence of the heavy resinous matter. This takes place on *Kanyin* and *In* oils being set aside for
a few days. Mr. B. Ribbentrop, Inspector General of Forests, on being
asked as to the difference between *Kanyin* and *In* oils replied that there
is no doubt the *In* tree affords an oleaginous substance, but whether chemically different from *Kanyin* he was not prepared to say. One point in
favour of its being different consists in the fact that it flows freely from a
wound, and practically without requiring the aid of fire (the tree being rarely
charred. The *Kanyin* oils, on the other hand, are obtainable only after the
cut surface of wood has been charred. In both cases Mr. Oliver believes,
however, that the thick dry deposit that forms on the wood clogs the
pores and prevents the escape of the oil, and that this is fired in preference
to being chipped off as a matter of convenience. It burns readily and
quickly, thus exposing the pores, whereas it would take some time to effect
the same result by chipping or paring the surface. Mr. Oliver reports
as follows: "*In* oil.—This is the produce of **D. tuberculatus** (Burmese
In or *Inma* female *In*, which is the most common species in Burma), and
is always found on laterite, gravel or clay, very often forming pure forests.
The process of extraction practised in the Prome and Tharrawaddy
districts is as follows:—a deep semi-circular niche, with a convex roof,
is made through the sap-wood near the foot of the tree, extending round
one-third of its circumference with a hollow in the lower portion of the
cut to receive the oil. After a few days the oil is collected and the wood
on the upper surface of the incision chipped away so as to expose a fresh
surface of sap-wood. This chipping has frequently to be repeated, as the
pores of the wood become clogged with congealed oil. In many cases
fire is also applied to the cut, but this appears to be not absolutely
necessary. The object of firing is probably the same as that of chipping,
viz., to remove the congealed oil. The latter is very inflammable, and
the cut surface invariably gets burned during the jungle fires, whether fire
is used in collection or not, so that between chipping and burning, a
wound, some 6 feet long by 2 feet wide, is formed in the side of the tree.
The tree thus gets gradually cut or burned through, and falls over by its
own weight. The oil is collected from four to ten times a month. A man

and boy can look after 300 trees, which yield about 20 viss a month. The time of collecting lasts from August to February. At the end of the season the congealed oil or resin which remains in the hollow is scraped off and used for TORCHES which are made of rotten wood mixed with oil and resin and rolled up in the leaves of the *satthwa*—a species of screw-pine common along the banks of streams in *In* forests. The oil is also largely used for water-proofing bamboo-baskets, for well-buckets, &c. The selling price of oil in the Prome and Tharrawaddy districts in 1882 was 5 to 7 viss for the rupee." In the *Indian Forester* (1875) Sir D. Brandis contributed a paper on the Black Burmese Varnish (obtained from **Melanorrhœa usitata**) in which he gives some particulars regarding *In* oil. He remarks that the oil exudes from the outer layers of wood. He describes the process thus :—"Deep semi-circular niches are cut into the wood, the first cut is about 4 to 6 inches deep and 12 to 18 inches wide, the bottom of the niche being slightly hollowed out to receive the oil. It oozes out and collects at the bottom of the niche about three days after the cut has been made. The surface is then charred with fire, after which the oil runs for three days. This process is repeated four times, and at the end of fifteen days the surface of the niche is cut afresh, the old charred wood being cut away and the niche enlarged. After the oil has run for three days, the surface is again charred and the original process repeated. The *Eng* tree yields oil throughout the year, and one tree often yields oil from several niches at the same time. I saw a tree with six niches,* two of which were yielding oil at the same time. One man can make 2,000 to 3,000 torches in a year, and 100 torches require about 10 viss (36℔) of oil, which is mixed with touch-wood and neatly wrapped up in the leaves of palms or of the *tsathoaben*, a species of **Pandanus**, so as to form cylinders about 20 inches long and 2 inches in diameter. They are tied with thin strips of bamboo, generally *tinwa* **(Shizostachyum pergracile)**; elsewhere in the Hlaine district the leaves of the *Zalooben* (**Licuala peltata**) are used for this purpose. This is the information which was given me in the *Eng* forest of Tyemyouk, and if it is correct, a man can collect about 700 to 1,000℔ of wood oil in a year. These torches are sold at R3-8 or R4 a 100 near the forests. The wood oil of the *Kanyin* tree is collected precisely in the same manner."

Medicine.—Mason says that the oleo-resin of this tree "is used with asafœtida and cocoa-nut oil as an application for large ulcers."

Structure of the Wood.—Brown with darker coloured heart-wood, rather heavy and loose-grained, sometimes used for canoes, but more generally for planking.

TORCHES.
698

MEDICINE.
699
TIMBER.
700

Dipterocarpus turbinatus, *Gærtn. f.; Fl. Br. Ind.,* I., 295.

701

KANYIN OIL.

Syn.— D. LÆVIS, *Ham.*, as established by the *Flora of British India.*

The term WOOD-OIL, given sometimes to the oleo-resin obtained from this plant, should be distinguished from the fatty oil (also called WOOD-OIL) which is obtained from **Aleurites**, see **Vol. I., No. 740.**

Vern.—*Garjan, tihya gurjun,* BENG.; *Kanyoung,* MAGH.; *Gurjun,* GUZ.; *Challani,* KAN.; *Kanyin-ni* (if D. **lævis** be distinct from this species they would appear to both bear the same vernacular names), BURM. Mason says the Burmese distinguish two forms of this plant—*Kanyin-(ni)* red and *Kanyin-(phu)* white; but the latter, according to modern writers, is D. **alatus.**

---

* May this not rather have been a *Kanyin* than an *In* tree ? The process here described appears to be that given by Mr. Oliver for *Kanyin.*

M

DIPTEROCARPUS turbinatus.	The Garjan or Kanyin Oil.

References.—*Roxb., Fl. Ind., Ed. C.B.C., 439 ; Kurz, For. Fl. Burm., I., 114, 115 ; Gamble, Man. Timb., 31, 32 ; Mason's Burm. & Its People, pp. 493, 516, 527, 757 ; Hooker, Him. Jour., II., 348 ; Report & Gazetteer of Burma by Major Douglas Macneill (prepared for Q. M. G.'s Dept.), Vol. II., 228 ; O'Shaughnessy, Beng. Dispens., 12, 222 ; Dymock, Mat. Med. W. Ind., 2nd Ed., 88 ; Flück. & Hanb., Pharmacog., 88 ; U. S. Dispens., 15th Ed., 1779 ; Extra Pharm. by Martindale & Westcott, p. 92 ; Year Book Pharm., 1875, 503 ; Royle, Prod. Res. of India, 77 ; Birdwood, Bomb. Pr., 11, 257 ; Cooke, Gums and Gum-resins, 113 ; Report on Gums issued by the P. W. D., pp. 19, 41 ; Spons' Encyclop., 1651 : Balfour, Cyclop., 957, 1087 ; Home Dept. Cor., 225, 230, 232, 290 ; Trans. Agri.-Hort. Soc., VIII., 345 ; Jour. Agri.-Hort. Soc., Vol. IV., 14.*

Habitat.—An evergreen tree of Eastern Bengal, Chittagong, Burma, and (according to Gamble) of the Andaman Islands. Distributed to Singapore.

It is said to be one of the loftiest of Indian trees, individual specimens being sometimes seen 250 feet in height, but **D. lævis** is generally reported to be the higher form, **D. turbinatus** rarely exceeding 200 feet. Hooker, referring to **D. turbinatus** in his account of Chittagong, says : " This is the most superb tree we met with in the Indian forests; we saw several species, but this is the only common one here ; it is conspicuous for its gigantic size, and for the straightness and graceful form of its tall unbranched pale grey trunk, and small, symmetrical crown : many individuals were upwards of 200 feet high, and 15 in girth."

OLEO-RESIN.

PROCESS OF
EXTRACTION
702

Oleo-resin.—Considerable confusion exists in the literature of Garjan and Wood-oil. Apparently several species of **Dipterocarpus** yield balsamic products to which it would seem the name *garjan* oil is assigned. In Burma one set of oils is, however, collectively spoken of as *Kanyin*-oils, another as *In*-oils. The term garjan appears to be unknown to the Burmans.

IN BURMA.
703

A reference having been made by the Revenue and Agricultural Department to the Government of Burma for particulars, to be inserted in the present publication, as to the " various species of **Dipterocarpus** that yield wood-oil (*garjan*)," the following instructive reply was obtained : The passage here quoted is in continuation of that given above under **D. tuberculatus :**—" Kanyin oil is the produce of **D. lævis** (*Kanyin-in* = Red *Kanyin*) and **D. alatus** (*Kanyin-byu*=White *Kanyin*) which are common in evergreen forests, and probably of other species of similar habitat. The oil is generally collected only in the dry weather (November to May). It is obtained by cutting two or three deep pyramidal hollows (the apex pointing towards the interior of the stem) near the foot of the tree and by applying fire to the upper cut surfaces. The oil then collects at the bottom of the hollow, which is emptied every three or four days. Fire is applied every time the oil is removed, and the upper surfaces of the hollow are rechipped three or four times during the season. In Tharrawaddy district, where trees are not very plentiful, twenty are about as many as one man can attend to. The yield of twenty trees would be about 100 viss for the season, worth R25. In Prome district oil only comes into the market in the form of torches, which are made of rotten wood steeped in oil and rolled up in *Salu* leaves (**Licuala peltata**). The exports of Kanyin oil from Burma ports during 1887-88 were as follows :—

							R
Rangoon .	.	18,826 gallons valued	.	.	.	.	16,302
Moulmein .	.	782 ,, ,,	.	.	.	.	575
Mergui .	.	55,470 viss ,,	.	.	.	.	9,394

D. 703

The Garjan or Kanyin Oil. (*G. Watt.*)	DIPTEROCARPUS turbinatus.

The exports of torches were :—   **R**

Tavoy . . 2,000 valued at . . . 30
Mergui . .850,225 ,, ,, . . . 22,372

PROCESS OF EXTRACTION.

Collectors do not keep the oil from the different kinds of Kanyin trees separate, consequently the oil that comes into the market is the produce of different species mixed in varying proportions. The Mergui Kanyin tree seldom exceeds 6 feet in girth, and is probably distinct from the species found in Pegu and North Tenasserim which commonly attain a girth of from 15 to 25 feet.

Two other communications, procured through the circular letter alluded to above, may be here published. The Conservator of Forests in Bengal reported : " **Dipterocarpus turbinatus**, the *Teli-gurjun* of Bengali, is found in the Chittagong District. It is prohibited to tap in the Collectorate owing to the large number of trees already killed by tapping. This prohibition does not extend to the hill tracts. As much of the oil exported from the hill tracts is shoulder-borne, the total amount exported cannot be definitely stated. But the total amount carried past the revenue stations and which paid a royalty of 10 per cent. *ad valorem* in 1883-84 to 1887-88 may be said to have been as follows : 1883-84, 355 maunds, 1884-85, 125 maunds, 1885-86, 96 maunds, 1886-87, 60 maunds, and 1887-88, 51 maunds.

IN CHITTAGONG.
704

"The mode of tapping is to cut a deep hollow in the tree, and keep live charcoal in it at night. The oil is removed in the morning, and fresh live charcoal put in again at night. It is repeated till the oil ceases to flow. Three, four, or more such deep hollows are often cut in the same tree, with the not surprising result that the tree is killed. The falling-off in exports is most probably due to most of the trees having been already killed by the tapping." A sketch was furnished along with the above report, in which the notch made in the trees was shewn to be the same as that described in Burma by the above passage. In this way a cavity is formed with a flat bottom on which it would be possible to deposit live charcoal, but it may here be added that in Burma charcoal does not appear to be used.

The other communication was from the Conservator of Forests, Coorg, which gives an account therefore of the wood-oil prepared in that portion of the west coast of Southern India (*Conf.* with remarks under **D. lævis** regarding **D. indicus**). The Conservator writes : " We have two oil trees in the Western Ghât Forests of Coorg. Both I believe are species of **Dipterocarpus**, but have not been able to get the flowers to identify them. The oil is contained in the pores of the wood, and is collected by cutting a hole into the centre of the tree. One species yields a yellow oil and the other a dark red. The former is sold in the bazaars mixed with dammar (the produce of **Vateria indica**) as varnish at 5 annas a bottle. The latter also makes a fair varnish. It has a strong cupaiba-like smell, and would probably be useful in medicine."

IN COORG.
705

Varnish.
706

During a conversation on this subject with the writer, **Mr. Ribbentrop** remarked that it was impossible to mistake the trees that yield *Kanyin* and *In* oils. **D. tuberculatus**, the *In*, was a low growing tree, found only on the *indaing* soils, and forming dense forests somewhat like its associate the *sál* (**Shorea obtusa**). This was in his opinion the chief if not sole source of the oil which was collected in the autumn and cold season, flowing from a wound *without the aid of fire*. The other trees alluded to were **D. turbinatus**, **D. lævis**, and **D. alatus**. These are very lofty, occur in mixed forests, and tower above the surrounding trees. They frequent deep rich soils and yield in spring their oleaginous products on being charred. **Mr. Ribbentrop** regards **D. lævis** as distinct from **D. turbina-**

M 2

tus; the former being a much loftier tree than the latter. **Kurz** mainly distinguishes these plants by the former being glabrous, while the latter is hairy.

VARIETIES.
707

Varieties of Garjan Oil.—The writer can discover no author who has separately distinguished the oleo-resins described above; indeed, in all the published accounts, which he has been able to consult, the substance described appears to be that obtained after charring the trees—the Kanyin oils. Thus **Roxburgh** wrote of **D. turbinatus** that " To procure the balsam a large notch is cut into the trunk of the tree, near the earth (say, about thirty inches from the ground), where a fire is kept up until the wound is charred, soon after which the liquid begins to ooze out. A small gutter is cut in the wood to conduct the liquid into a vessel placed to receive it. The average produce of the best trees during the season is said to be some- times forty gallons. It is found necessary, every three or four weeks, to cut off the old charred surface, and burn it afresh; in large healthy trees abounding in balsam, they even cut a second notch in some other part of the tree, and char it as the first. These operations are performed during the months of November, December, January, and February. Should any of the trees appear sickly the following season, one or two more years' respite is given them." **Lieut. Hawkes** published, in his report on the Oils shown at the Madras Exhibition of 1855, an account of the extraction of this oil by charring, the operation being performed in " March or April." But **Lieut. Hawkes** was apparently, like **Roxburgh,** ignorant of the oil extracted from **D. tuberculatus** with or without the aid of fire. **Sir J. D. Hooker** (*Him. Journals, Vol. II., 348*) gives a brief note regarding the oleo-resin obtained in Chittagong from **D. turbinatus.** He says: " A fragrant oil exudes from the trunk, which is extremely valuable as pitch and varnish, &c., besides being a good medicine. The natives procure it by cutting transverse holes in the trunk, pointing down- wards, and lighting fires in them, which causes the oil to flow." **Mason,** than whom few more trustworthy authors on Burmese subjects could be found, attributes wood-oil to **D. lævis** and **D. turbinatus,** but says of **D. grandiflora** (a synonym for **D. tuberculatus**) that " the gum of this species, as well as that of the preceding, is used by the natives to make torches." It is, however, significant that **Mason** should not have described the pro- cess of extraction of his " wood-oil " or of the " gum," nor even mentioned the seasons at which these products are obtained. **Dr. Cooke,** in his report on the Gums, Resins, and Oleo-resins of India, quotes **Roxburgh's** description of the process of extraction, and reviews the opinions advanced by **Lieut. Hawkes** under **D. turbinatus,** *Gærtn. f.*; but under **D. tubercu- latus,** *Roxb.,* he simply remarks: " A wood-oil, under the name of *Eng,* is said to be the produce of **D. tuberculatus**; this was sent to London from Burma (May 1874) for valuation and report." **Flückiger** and **Hanbury** (in their *Pharmacographia*) follow the same course, but seem not to have heard of an oil extracted without the aid of fire, such as the thick oleo-resin known in Burma as *In*-oil.

In a further paragraph will be found the opinions of medical writers regarding Garjan oil, in which it is held that there are different qualities, some of very considerably higher medicinal merit than others. This fact would point to the desirability of a thorough investigation into the oleo- resins obtained from all the species of **Dipterocarpus** in which the chemical properties and industrial merits of each should be separately estab- lished. With this in view experiments might be conducted in order to ascertain if **D. tuberculatus** is the only species that affords the oil on being simply tapped, or whether **D. turbinatus** and **D. lævis** might not also do so, and lastly what action or influence the charring process exercises. It

The Garjan or Kanyin Oil. (*G. Watt.*)	DIPTEROCARPU turbinatus.

seems probable that, assuming that the oleo-resins from all the species of **Dipterocarpus** are chemically identical, that obtained during a different season of the year, and by a different process may be distinct or have its properties changed from what might be called the normal secretion of the **Dipterocarpi.** In concluding this brief review of the literature of *garjan* oil, it may be as well to point out that, according to the report above, the *Kanyin* oil (or that produced by charring the trees) comes into the market mainly as torches. From this fact the inference might be deduced that the *garjan* oil of commerce was obtained from **D. tuberculatus,** and not from **D. turbinatus** and **D. lævis,** the species to which the oil has hitherto been attributed. The writer must, however, suggest caution in accepting this inference, but it may safely be assumed that at least the thick honey-like form of *garjan* oil is the *In* oil of Burma.

CHEMICAL PROPERTIES OF GARJAN OR WOOD-OIL.—Lieut. Hawkes (in his report on the Oils shown at the Madras Exhibition) says that this class of substances, called wood-oils, forms the connecting link between the oils and resins of the vegetable kingdom. They consist of a volatile oil holding in solution a resin, and are generally classed under the head of balsams. It is commonly stated that the oil if set aside for a time sub-sides into two substances, *viz.*, a clear thin liquid, floating above a thick mass known as *guad.* One of the most remarkable properties attributed to this oil is the fact that it is reported to act as a solvent to caoutchouc. This was apparently discovered at the beginning of the present century by **Mr. Laidlay,** and his experiments will be found in the *Transactions of the Agri-Horticultural Society of India* (*Vol. VIII. 345*); also repro-duced in *Mason's Burma.* **Mr. Laidlay** directs that the caoutchouc should be dropped into the garjan oil in small pieces. In a few hours it swells and must then be frequently stirred to facilitate the process. If heat be applied, complete solution is speedily effected. The solution obtained may be spread on cloth which is said to be thus rendered water-proof. This fact appears to have been practically lost sight of, while it might prove the key to an industrial utilisation of the substance, since such water-proofings would, from the property of the *garjan* oil, be at least proof against the attacks of insects, if they were not found in addition to possess other useful properties.

A sample of *garjan* oil obtained from Moulmein was examined by **Flückiger** and **Hanbury.** Space cannot be here afforded to reproduce their report on the substance. The reader is referred to their *Pharmaco-graphia,* p. 88, or to *Dr. Dymock's Materia Medica of Western India,* where, however, the account of the chemistry of this substance, as given by the authors of the *Pharmacographia,* is reproduced. By simple distillation with water they obtained 37 per cent. of an essential oil, leaving in the still a dark viscid liquid resin. The sp. gr. of this essential oil was found by **Flückiger** and **Hanbury** to be 0·915, but by **O'Shaughnessy** it is given as 0·931, and by **De Vry** as 0·928. One of the most remarkable physical properties of this oil is the fact that at a temperature of 130°C. it becomes gelatinous, and on cooling does not recover its fluidity. The learned authors of the *Pharmacographia* found the resin to contain, like that of copaiba, a small proportion of a crystallisable acid which may be removed by warming it with ammonia in weak alcohol. The portion of the resin which they found to be insoluble even in absolute alcohol was uncrystallisable. **Werner,** however, found a sample of *garjan* oil examined by him (as well as its resin) to be entirely soluble in boiling potash lye. The crystallisable acid extracted from the resin **Werner** called Gurjanic ($C_{44}H_{68}O_8$): it is soluble in alcohol 0·838 but not in weaker. It is dissolved also by ether, benzol, or sulphide of carbon.

VARIETIES.

CHEMISTRY.
708

DIPTEROCARPUS turbinatus.	The Garjan or Kanyin Oil.

CHEMISTRY.

The amorphous resin, which forms the chief bulk of the substance obtained after the removal of the essential oil, has not as yet been definitely analysed. Flückiger and Hanbury found, however, that after complete desiccation it was not soluble in absolute alcohol. These authors add that a sample of *garjan* balsam of unknown origin yielded a crystallisable substance answering to $C_{28} H_{46} O_2$: and this was devoid of acid character. They would thus appear to have inferred that the *garjan* oil of commerce is not a substance of uniform chemical character, hence they conclude by recommending that "a comparative examination of the product of each of the above named species of **Dipterocarpus** would be highly desirable." **Dr. Dymock**, while not materially enriching the chemical knowledge of this substance, gives much interesting information as to the medical opinion held regarding the properties of the drug. The admission of different chemical and medicinal results confirms to a large extent the contention advanced in this work, *viz.*, that there are at least two widely different substances sold in the markets of India under the name of Garjan oil, the Kanyin, and In Oils of Burma.

TRADE.
709

TRADE IN GARJAN OR WOOD-OIL.—The above special reports regarding the *garjan* oil of Burma and Chittagong make the usual admission that, owing to the cheap price of kerosine, the trade in wood-oil has very considerably declined. It is now mainly used for torches and in waterproofing, &c. The trade in the medicinal *garjan* oil must be very limited indeed. It appears to be mainly obtained from the Andaman Islands and to be the produce of **D. alatus**, and possibly **D. turbinatus.**

Garjan
Balsam.
710

Flückiger and Hanbury (*l. c.*) state that the world's supply is obtained from "Singapore, Moulmein, Akyab, and the Malayan Peninsula, and is a common article of trade in Siam." (*Conf.* with **Mr. Oliver's** opinion above as to the plant which yields the Tenasserim oil) "It is likewise produced in Canara in South India." (*Conf.* with remarks regarding **D. indicus.**) "It is occasionally shipped to Europe." The Burma oil is most probably obtained from **D. turbinatus** and **D. alatus** (*Kanyin*) and from **D. tuberculatus** (*In*). **Dr. Dymock** remarks : "Garjan Balsam is not an article of commerce in Bombay; small quantities may be sometimes obtained in the native drug shops. The Government supplies have been obtained from the Andaman Islands." **Dr. Moodeen Sheriff** (in his new work on the Materia Medica of South India, of which proofs have been kindly furnished to the author) writes that in Madras "wood-oil is pretty common in most large bazars." He describes several forms and gives their prices :—"Of the black or dark brown variety—wholesale, R12 per maund ; retail or bazar, annas 10 per pound. Of the red or reddish-brown variety—wholesale, R24 per maund; retail or bazar, R1-4 per pound. Of the pale white or grey variety—wholesale, R18 per maund ; retail or bazar, R1 per pound." He adds : "There are several varieties of *garjan* or wood-oil, but out of these, three are generally met with in the bazars, which are known as *Suféd Garjan-ká-tél* or *Suféd Lakrí-ká-tél* (the pale white or grey variety), *Lál Garjan-ká-tél* or *Lál Lakrí-ká-tél* (the red or reddish-brown variety), and *Kálá Garjan-ká-tél* or *Kálá Lakrí-ká-tél* (the black or dark-brown variety)."

Black
Variety.
711
Red variety.
712
Grey variety
713

Fully fifty years ago hopes were entertained that *garjan* oil would become an article of European trade, meeting a demand in the arts. **Dr. Royle** wrote on this subject, and a member of the Agri-Horticultural Society of India consigned five hundred gallons to London. The effort proved futile, as **Dr. Royle** reports, because the Custom-house officers refused to pass it except at the highest rate of duty, namely, that for a manufactured article. It seems probable that this obstruction prevented the industrial enterprise of the British manufacturer from being able to discover a use

The Garjan or Kanyin Oil.   (*G. Watt*)   **DIPTEROCARPUS turbinatus.**

for an article which has in consequence remained at a nominal value. (*Conf.* with p. 164.)

MEDICINE.

Garjan balsam does not appear to have been used medicinally by the early Hindus. It does not bear any Sanskrit, Arabic, or Persian names. In Muhammadan works on Materia Medica it is first mentioned in the *Makhzan* under the name of *Duhn-el-Garjan.* Ainslie was the earliest European medical writer to mention it, and that in his *Materia Medica of Hindústan*—a work published in 1813. A prior notice occurs, however, in a work by **Francklin** (*Tracts on the Dominions of Ava, p. 26*) published in 1811. But Ainslie does not seem to have continued to value the drug, since in his larger and final work—the *Materia Indica*—published in 1826, he makes no mention of it. **Sir William O'Shaughnessy** in 1841 (*Bengal Dispensatory, 222*) recommended the balsam to the consideration of European physicians. He wrote: "The *garjan* balsam varies in consistence from that of a thick honey to a light oily liquid. The colour of a fine specimen of thick *garjan* obtained from **Captain Jenkins** of Assam was pale grey; specimens sent from Rangoon by **Mr. Speir** were a light brown. As found in the bazar, this substance generally occurs as a brown oily-looking, semi-transparent liquid, in odour strongly resembling a mixture of balsam of copaiba with a small portion of naphtha." After giving the results of his chemical examination or division of the substance into its essential oil and resin, he continues: "The close resemblance in the chemical properties of this *garjan* and copaiba' balsam led to the institution of an extensive set of experiments on the medicinal effects of the former in the treatment of gonorrhœa. The results, which have been laid before the profession, and which have been confirmed by trials made by other practitioners, seem perfectly conclusive that in the treatment of gonorrhœa, gleet, and similar affections of the urinary organs, the essential oil of *garjan* is nearly equal in efficacy to the South American drug. The essential OIL may be given in 10 to 30 drop doses in mucilage, milk, rice-water, or thin gruel, and repeated thrice or still more frequently daily. It generally causes a sensation of warmth at the epigastrium, eructations, and sometimes slight purging. It communicates a strong smell of turpentine to the urine, which it increases remarkably in quantity. Some obstinate cases of chronic gonorrhœa and gleet, which had long resisted copaiba and cubebs, have been cured by this remedy in the course of the experiments alluded to." "For additional suggestions relative to the mode of administering this remedy, see **Copaiba.** In the *Pharmacopœia,* we have given a formula for a solution of the essential oils of *garjan* and cubebs in sulphuric ether, which affords a cheap but perfectly efficacious substitute for the celebrated 'Frank's Specific.' "

Pursuing, in order of publication, the Indian works which treat of this substance, the *Pharmacopœia of India* in 1868 made it officinal. It is in that work described as a "stimulant of mucous surfaces, particularly that of the genito-urinary system, diuretic," and in a further page the results of various experiments with this substitute for copaiba are given. **Dr. T. B. Henderson** of Glasgow is said to have used it only when copaiba failed, and with remarkably good results. **Dr. H. B. Montgomery** found that it is apt to produce "an eruption of a character similar to that occasionally following the use of copaiba." **Dr. Kanny Lall Dey, C.I.E.** (*Indigenous Drugs of India, p. 51*) republishes the facts given above regarding the use of the drug in the treatment of gonorrhœa, but adds that "it is also used externally as a stimulating application to indolent ulcers." **Waring** (*Bazar Medicines, p. 56*) says it has the odour

MEDICINE.
Balsam.
714

OIL.
715

| DIPTEROCARPUS turbinatus. | The Garjan or Kanyin Oil. |

MEDICINE.

and taste of copaiba, but is less powerful. "It has been used as a substitute for this latter drug in the treatment of *gonorrhœa*, and trials with it in the hands of Europeans have shown that it is a remedy of no mean value in that affection. It is only advisable in the advanced stages, or when the disease has degenerated into gleet. In the latter affection it is stated to prove most useful. It is also well worthy of a trial in *leucorrhœa* and other *vaginal discharges.*" **Dr. Waring** then proceeds to say that "great success has been found to attend its employment, both internally and externally, in the treatment of *leprosy*." He then quotes **Dr. J. Dougall's** proposed treatment for leprosy (*Indian Medical Gazette, February 2nd and March 2nd, 1874*) as follows: "Rise at day-light and wash the body thoroughly, using dry earth as a detergent, in which character it is more efficient than soap or bran. After this is completed, at 7 A.M. a dose of the emulsion is given, and for the next two hours the patient himself should perseveringly rub in the ointment over his whole body. This is a point of importance, not merely smearing it in here and there but using thorough and continuous friction over the whole surface for a couple of hours. This prolonged rubbing is not only insisted upon for the sake of the action of the ointment upon the skin, but because it is considered that any gentle employment, combined with exercise, proves beneficial both physically and mentally. After this inunction, breakfast may be taken, and some light employment followed during the day. At 3 P M., a second dose of the emulsion is given, followed by another two hours' friction. Should the emulsion act too freely on the bowels the dose should be diminished. In none of the cases treated by Dr. Dougall was there any change from the ordinary native diet, but we may reasonably expect even better results where a liberal supply of good and nourishing food is allowed. The success which has attended this treatment is very marked and encouraging and is fully confirmed by **Dr. A. S. Lethbridge**" (*Indian Medical Gazette, 1st July 1874*).

Use in Leprosy.
716

On the other hand, **Dr. Dymock** says of **Dr. Dougall's** reported success in the treatment of leprosy :--" In order to test the correctness of this statement, large quantities of the Balsam have been distributed by the Indian Government, but as far as I have heard the new treatment is not likely to prove successful. Dr. Dougall's directions for carrying out the treatment of leprosy by *Garjan* Balsam include frequent ablutions with dry earth and water, and strict attention to the hygienic condition of the patient ; it seems probable that he has attributed effects to the balsam which are in reality due to cleanliness and an improved hygienic condition. Within the last two years several tons of the drug have been distributed in the Bombay Presidency."

Dr. Moodeen Sheriff, the most recent writer on the subject of the properties of *Garjan* Balsam, says : "All the varieties of *Garjan* oil are equally useful as a local stimulant, but the red or reddish-brown and the pale-white or grey varieties are the best for internal use. The best medical properties of this oil are its usefulness in gonorrhœa and gleet, and in all forms of psoriasis, including lepra-vulgaris. In gonorrhœa and gleet it is at least equal to copaiba, and the only difference between these two drugs is that the former requires to be used in a much larger dose (2 drachms to 3 drachms) to produce the same effect as the latter. As *Garjan* balsam is always used in the shape of emulsion with mucilage the largeness of its dose is no disadvantage. With regard to its usefulness in psoriasis and lepra-vulgaris, I am not aware of any other local stimulant which is more efficacious in those diseases than this drug. I have either cured or relieved many cases of the above affections by the use of this drug, with little or no assistance of internal remedies. The internal use of wood-oil is also

| The Garjan or Kanyin Oil. *(G. Watt.)* | DIPTEROCARPUS turbinatus. |

attended with benefit in some cases of true leprosy in its early stage; but its efficacy in this respect is greatly enhanced with the addition of from five to ten drops of *Chaulmugra oil* to each drachm of it. If well mixed in the above proportions, the combination of *Chaulmugra-oil* cannot be detected. Some years ago I had received a bottle of *Gurjan-oil* of this kind from a medical friend, which proved itself more useful in a case of true leprosy than all its varieties in the bazaar, but I did not know the existence of *Chaulmugra-oil* in it until I was informed of it." **Martindale** and **Westcott** say : "It is very florescent, has an opaque, dingy, greenish grey colour seen by reflected light, yet is transparent and reddish-brown in strong day-light; it has the weak aromatic odour and bitterish aromatic taste of copaiba without the acridity—has been used as an adulterant of copaiba. It is not completely soluble in either ether or alcohol; emulsified with mucilage of **Acacia**, it is used with success like copaiba for gonorrhœa; and, in the East, as a remedy for leprosy, an emulsion is made of equal parts of the balsam and lime water, which is used freely as a liniment and given to the extent of 4 drachms three times daily."

SPECIAL OPINIONS COMMUNICATED FOR THIS WORK.—§ "Used in leprosy" *(Surgeon-Major J. B. Thomas, Waltan, Vizagapatam).* "Very effectual in relieving true leprosy. Dose internally as in the *Pharm. Ind.*; for an ointment take of the oil 1, lime water 3 parts; useful for chronic skin diseases and true leprosy" *(Thomas Ward, Apothecary, Madanapalle, Cuddapah).* "Gurjun oil is of undoubted efficacy in tuberculous leprosy" *(Civil Surgeon R. D. Murray, M.B., Burdwan).* "Used also in leprosy" *(G. A. Watson, Allahabad).* "Very useful in cases of leprosy. Externally the oil should be well rubbed into the affected parts. Internally it is taken in doses of 3 drachms or 1 drachm mixed with lime water or Liqr. Potassæ" *(Civil Surgeon J. Anderson, M.B., Bijnor).* "In leprosy it was found beneficial. It was given internally and rubbed externally in the form of an emulsion with lime water" *(Surgeon T. N. Ghose, Meerut).* "Is very useful in leprosy, used both externally and internally. A case of elephantiasis now under observation is being treated with gurjun oil. It appears to be useful, though the case is too recent for any certainty" *(Surgeon-Major E. Sanders, Chittagong).* "I have tried it frequently in cases of leprosy; it is a good dressing, and heals the ulcers as well if not better than any other application, and the inunction of the oil does the sufferer good constitutionally, but it is certainly not a specific for leprosy, nor does it stop the nerve disease" *(Surgeon-Major C. W. Calthrop, M.D., 4th Bengal Cavalry, Morar).* "The oil with a little corrosive sublimate and sulphur is a capital remedy for ringworm" *(Surgeon-Major P. N. Mookerji, Cuttack, Orissa).* "I used this oil for two years in the treatment of leprosy, but found it perfectly useless" *(Brigade-Surgeon C. Joyut, M.D., F.K.Q.C.P.; Poona).* "Gurjun tél,— The Andaman oil is the best, and useful in leprosy. Taken internally and applied externally too" *(Civil Surgeon C. M. Russell, M.D., Sarun).* "Is a good dispensary substitute for copaiba in gonorrhœa and mucous discharges. Its internal and outward use in leprosy is highly recommended" *(Dr. Picachy, Civil Medical Officer, Purneah).* "I experimented for two years with *gurjun* oil as a cure for leprosy in the lepra ward at Burdwan in 1875-76-77. It is useless as a specific, which it was claimed to be, but the ointment is a fairly good application for leprous and other ulcers" *(Civil Surgeon C. H. Joubert, M.B., Darjeeling, Bengal).* "The balsamic exudation of **D. turbinatus** or Gurjun-balsam is a very valuable external and internal stimulant. It exercises more or less beneficial influence over all skin diseases, but its curative effect in those of a scaly nature as lepra-vulgaris and psoriasis is highly satisfactory. Many a case of the

DIPTEROCARPUS **turbinatus.**	The Garjan or Kanyin Oil.

MEDICINE.

last-named disease has been relieved by its external use with little or no assistance of internal remedies. I have also employed it pretty extensively in the cure of gonorrhoea, and quite agree with what is; already mentioned on this point. There are several varieties of gurjun-balsam, but the thin and reddish brown variety is about the best" (*Honorary Surgeon Moodeen Sheriff Khan Bahadur, Triplicane, Madras*). "Useful application in scabies. It did not prove so useful in my hands in gonorrhœa when administered alone. Combined with liqr. pot. and other medicines in dram doses it has been found to be efficacious in certain cases" (*Assistant Surgeon Shib Chundra Bhuttacharji, Chanda, Central Provinces*). "Recently much praised as a cure for leprosy. I have not been able to obtain any remarkable effects from its use" (*Civil Surgeon G. Price, Shahabad*). "The oil prepared into an ointment for external application, and given internally in leprosy in early stage of the disease, undoubtedly arrests further progress, and affords great relief in advanced cases. The ointment is prepared by mixing the oil with lime water in equal parts, and churning it into a creamy substance. It should be well rubbed into the affected parts, for at least 15 minutes, every morning and evening. The oil given internally from one to ten drops, morning and night, in cold water" (*Civil Surgeon S. M. Shircore, Moorshedabad*). "It is a stimulant diuretic used in gonorrhœa and discharges from the genito-urinary organs, also in leprosy, both internally and externally, with lime water" (*Bolly Chand Sen, Teacher of Medicine*). "Gurjun oil was used extensively at the penal settlement of the Andamans in the treatment of leprosy. After long trial it was found to act beneficially in many cases as a palliative remedy, but as a specific for the cure of leprosy it completely failed" (*James Reid, Principal Medical Store-keeper to Government, Fort William*). "Has been used both internally and externally in leprosy with apparent benefit" (*Civil Surgeon J. H. Thornton, B.A., M.B., Monghyr*). "It is a very good application for various kinds of skin diseases" (*Doyal Chunder Shome*).

TIMBER.
717

Structure of the Wood.—Rough, moderately hard; heartwood reddish grey. It is used for house-building and for canoes in Burma. The best Burmese charcoal is made from this and **D. lævis**. (*Gamble.*)

Heavy, rather close grained, the sapwood pale brown, narrow, the heart wood darker brown; takes a fine polish (*Kurz*).

DOMESTIC AND INDUSTRIAL USES OF GARJAN-OIL.

PROPERTIES AND USES.

It is extensively employed by the Burmans as torches, but now-a-days to a limited extent only is it used as a lamp oil. It is largely employed in preserving bamboo wicker-work from the attacks of insects and in paving the bottoms of boats. It is also used as a varnish. It is reported to be useful as an ingredient in lithographic ink. In European medicine it is mainly utilised as an adulterant for Copaiba. But it is commonly held that if a process could be discovered of causing it to dry more rapidly, it would come largely into European use as a varnish. It has been suggested that this might be effected by mixing it with some good drying oil, or by evaporating away the essential oil. It seems to the writer, however, that a far more important way of utilising the article might be found in taking advantage of **Mr. Laidlay's** discovery that it acts as a solvent to caoutchouc. A thick coating of India-rubber is of course perfectly water-proof, but the way in which India-rubber sheetings, over-coats, &c., harden, dry, and crack at one season of the year or stick together at another under the tropical climate of India, would recommend the experiment being made to ascertain if this would be also the case with a water-proofing

Varnish.
718

Water-proofing.
719

D. 719

material made of a solution of India-rubber in¡ *Garjan* oil. The merits of *Garjan* oil have at all events not received sufficient attention by the manufacturer, and **Sir William O'Shaughnessy's** opinion may be here quoted in favour of the desirability of the matter being looked into in the future. **Sir William** wrote fifty years ago that *Garjan* was "likely to be found a perfect substitute in the arts for the expensive balsam of copaiba, now much used in the preparation of colourless varnishes and drying paints. In the coarser kinds of house and ship painting, *garjan* balsam is used as an excellent substitute for linseed oil." **Dr. Wight** also speaks highly of the property of *garjan* in preserving wood, &c., from the attacks of insects ; its ¡defects are slowness of drying, thin body when dry, and liability to being brittle.

PROPERTIES AND USES.

TESTS FOR GARJAN, COPAIBA, AND HARDWICKIA BALSAMS.—Dr. Watson says : "Its entire solubility in coal naphtha proves the absence of any of the soft resin which exists in most of the copaiba of commerce." It may be distinguished from Copaiba or the balsam of **Hardwickia** thus : shake up 1 drop of the balsam with 19 of carbon bisulphide, add one drop of nitro-sulphuric acid and agitate : Copaiba will show faint reddish-brown with a deposit of resin on the sides of the tube : *garjan*, intense purplish-red, soon becoming violet : while **Hardwickia** will not alter from its pale greenish yellow.

GARJAN TESTS. 720

Divi-divi, or **Libi-dibi,** see **Cæsalpinia Coriaria,** *Willd. ; Vol. II., p. 6 ;* LEGUMINOSÆ.

DOCYNIA, *Dcne.* (not described in *Genera Plantarum*).

Docynia indica, *Dcne. ; Fl. Br. Ind., II., 369 ;* ROSACEÆ.

721

    Syn.—PYRUS INDICA, *Roxb. ; Wall., Pl. As. Rar., II., 56, t. 173¡;* CYDONIA INDICA, *Spach.*

    Vern.—*Sopho,* KHASIA ; *Mehul, passy,* NEPAL ; *Likúng,* LEPCHA.

    References.—*Roxb., Fl. Ind., Ed. C.B.C., 406 ; Kurz, For. Fl. Burm., I., 441 ; Gamble, Man. Timb., 161 ; Cat. Trees, Shrubs, and Climbers of Darjeeling, p. 37.*

    Habitat.—A small tree of the Eastern Himálaya, from Sikkim (4,000 to 6,000 feet), Bhutan (7,000 feet), and Assam, the Khásia Hills, Manipur (5,000 feet) to Burma.

    Food.—Produces a FRUIT which is yellow green with orange patches ; is 1 to 1½ inches in diameter, and rounded at the base. When ripe the fruit has a slight quince flavour, and is eaten when half ripe by the hill tribes. The ground is often literally covered with the fruits of this tree, and in that state they are largely eaten by wild animals. They ripen in September, whereas those of the allied plant, the Quince (**Cydonia vulgaris**), begin to fall from the trees in April.

FOOD. Fruit. 722

Dock, see **Rumex.**

Dodder, see **Cuscuta reflexa,** *Roxb. ;* Vol. II., No. 2508, p. 671.

DODECADENIA, *Nees ; Gen. Pl., III., 160.*

Dodecadenia grandiflora, *Nees ; Fl. Br. Ind., V., 181 ;* LAURINEÆ.

723

    Syn. —TETRANTHERA GRANDIFLORA, *Wall. ; ?* LAURUS MACROPHYLLA, *Don ; Prod. Nepal, 64.*

    References.—*Brandis, For. Fl., 381 ; Kurz, For. Fl. Burma, II., 304 ; Gamble, Man. Timb., 304.*

DODONÆA viscosa.	Dodonæa—the Switch Sorrel.

TIMBER.
724

Habitat.—A moderate-sized tree of the Temperate Himálaya from Kumáon eastward to Burma.

Structure of the Wood.—Not known to be of any important use.

DODONÆA, *Linn.; Gen. Pl., I., 410 & 1000.*

A genus of some 40 shrubs (rarely trees); only one of which occurs in India, but the literature of that species has been disfigured through its having been described under many names. The genus is named in honour of **Dodonæus** (**Rembert Doddens**), a famous botanist and physician.

[SAPINDACEÆ.

725

Dodonæa viscosa, *Linn.; Fl. Br. Ind., I., 697; Wight, Ic., t. 52;*

Syn.—D. DIOICA, *Roxb.;* D. ANGUSTIFOLIA, *Linn. f.;* D. BURMANNIANA, *DC.;* D. PALLIDA, *Miq.;* D. MICROCARPA, *DC.;* D. WIGHTIANA, *Blume;* D. PENTANDRA, *Griff.;* PTELEA VISCOSA, *Linn.;* DODONEA SPATHULATA, *Sm.;* D. ARABICA, *Hochst.*

Vern.—*Aliár* (Plains of Northern India), HIND.; *Sanatha,* HAZARA; *Sanatta, mendrú. ban mendú, sánthá, mendar,* PB.; *Ghuráske, vera-, vena (shumshad?),* TRANS-INDUS; *Ghuraskai (or ghoráskai), wuraskai,* PUSHTU; *Mírandú,* KANGRA; *Pipalu,* SIMLA; *Banderu,* C. P.; *Bandurgi* (Kanara), BOMB.; *Lutchmi* (according to Dalz. and Gibs.), MAR.; *Dáwa-ka-jhar* (according to Graham), BELGAUM; *Bándári, zakhmi* (according to Dymock), BOMB.; *Virali* (in Ceylon), TAM.; *Bandaru, golla pulleda, bundédu,* TEL.; *Bandurgi, bandrike (bandu, according to* Cameron), KAN.; *Eta-werella* (Trimen), SING.

References.—*Roxb., Fl. Ind., Ed. C.B.C., 324; Voigt., Hort. Sub. Cal., 96; Brandis, For. Fl., 113; Kurz, For. Fl. Burm., I., 287; Gamble, Man. Timb., 101; Thwaites, En. Ceylon Pl., 59; Dalz. & Gibs., Bomb. Fl., 36; Stewart, Pb. Pl., 31; Aitchison, Cat. Pb. and Sind Pl., 34; Fl. Andh. by Sir W. Elliot, 22, 61; Stewart, Bot. Tour in Hazara; Dymock, Mat. Med. W. Ind., 2nd Ed., 191; Baden Powell, Pb. Pr., 578; Atkinson, Him. Dist., 338; Indian Forester, II., 390; V., 13, 32; VI., 238; VIII., 30, 35; IX., 357, 469; XII., 551; Bomb. Gaz., XV., 68; Gazetteer, Dera Ismail Khan, 18; Settlement Rep., Hazara, 95; Gazetteers:—Banu, 23; Shahpur, 69; Hoshiarpur, 12; Peshawar, 27; Rawalpindi, 12.*

Habitat.—An evergreen shrub, met with in the North-West Himálaya, from the plains up to 4,500 feet, in the Panjáb, Sind, and South India (ascending to 8,000 feet, and attaining here the size of a small tree); also in Burma, and planted throughout India as a hedge.

MEDICINE.
Leaves.
726

Wood.
727

Plant.
728

Medicine.—Said to have febrifugal properties. The LEAVES are viscid and have a sour bitter taste, from which fact it is in Jamaica called the "Switch Sorrel." Lindley (*Veg. King., 384*) says the leaves are used in baths and fomentations. The WOOD, he adds, of D. dioica is carminative, and D. Thunbergiana is said to be slightly purgative, febrifugal, and aromatic.

SPECIAL OPINIONS.—§ "This PLANT has been identified for me by Dr. Dymock. It grows about Belgaum. Dr. Graham, in his *Catalogue of Bombay Plants*, mentions that D. Burmaniana is known in Belgaum as *Dáwá-ká-Jhár*. It is believed that the powdered leaves of *Bendugi* applied over a wound will heal it without leaving a *white* scar. It is applied in burns and scalds; said to be useful also in rheumatism. Dr. Dymock gives its Bombay name as *Zakhmi*, from which it may be implied that it is used in the treatment of wounds" (*Surgeon-Major C. T. Peters, M.B., Zandra, South Afghánistan*).

FODDER.
Leaves.
729

Fodder.—Stewart says the LEAVES are hard and dry, and are only eaten by cattle when very hungry. Reported to have nct agreed with the camels at Thal, Afghánistan, during the late campaign.

Products of India. 173

Dogs, Wolves, Jackals, and Foxes.　　(*G. Watt.*)　　DOGS, &c.

Structure of the Wood.—Sap-wood white; heartwood extremely hard and close-grained, dark-brown, with an irregular outline. It is used for engraving, turning, tool-handles, and walking-sticks, and the branches to support the earth of flat roofs. It is likely to be important in reclothing denuded tracts like the Siwálik hills of Hoshiárpur.

Domestic Uses.—The LEAVES and TWIGS are employed to manure fields in Madras. The plant is useful as a hedge. Elliot says the wood is extensively used for fire-wood, and the smaller twigs are formed into faggots. The name *bandedu* in Telegu is said to mean "Touch wood," implying the ease with which it may be ignited.

Dog-rose, see **Rosa canina,** *Linn.*; ROSACEÆ.

[*India, pp. 134—155.*
Dogs, Wolves, Jackals, and Foxes; *Blanford's Fauna of British*

It is not proposed to discuss here the probable history of the domesticated dog or even the forms of it met with in India. The reader is referred to Darwin's *Origin of Domesticated Animals and Plants.* The so-called wild dog of India is, however, more nearly allied to the wolf and the jackal than to the domesticated dog, and is more difficult to tame than either of these animals. This remark is made in order to remove the often-repeated statement (by popular writers) that the Pariah dog of India is the wild dog domesticated, or that the wild dog is the domesticated dog gone wild.

The TRUE WOLF (1) (**Canis lupus**) rarely occurs south of the Himálaya, though specimens have been shot in Sind, and it is fairly common in Baluchistan and Gilgit. The INDIAN WOLF (2) (**C. pallipes**) is common south of the Himálaya in the open country, but is rare in wooded or hilly tracts. It is uncommon in Bengal. The JACKAL (3) (**C. aureus**) is plentiful throughout India and Ceylon, on hills and plains, forest and open country, ascending the Himálaya (for example, at Simla) to an altitude of 8,000 feet. It is rarely found in Lower Burma, but is abundant in Assam and Upper Burma. The INDIAN WILD DOG (4) (**Cyon dukhunensis**) occurs throughout the Himálayan forests, from Baluchistan, Gilgit, and Kashmír, to Assam and Manipur. The MALAY WILD DOG (5) (**C. rutilans**) is said to extend from Borneo, Java, Sumatra, and the Malay Peninsula to Tenasserim, in Burma. The INDIAN FOX (6) (**Vulpes bengalensis**) is common in most open tracts of country, whether cultivated or waste. The HOARY FOX (7) (**V. cana**) occurs in Baluchistan, South Afghánistan, and Sind, while the INDIAN DESERT FOX (8) (**V. leucopus**) inhabits the dry and semi-desert regions of Western India, Sind, Cutch, Rájputana, the Panjáb, and North-West Provinces. The COMMON FOX (9) (**V. alopex**) is met with on the Western Himálaya in brush-woods near cultivation, from about 5,000 feet to the limits of snow; and the SMALL TIBETAN FOX (10) (**V. ferrilatus**) appears to occur chiefly on the northern slopes of the Himálaya as at Lassa. Dr. Stoliczka, however, mentions it in the upper basin of the Sutlej.

Skins.—Most if not all of the above-mentioned animals are killed for their skins, and on that account mainly have they been enumerated in this work. In the *Gazetteers of India* reference is often made to these skins. Thus in Broach the Wolf's skin is said to be soft, handsome, and much valued. The Jackal's skin is made into caps, and the Fox's into fur coats, rugs, &c. Definite information is, however, not available as to the actual extent these skins are utilised nor of their relative merits.

Food.—**Dog's Flesh.**—Being carnivorous most of these animals carry off and devour domesticated animals, the wolf having been often known to

TIMBER.
730
DOMESTIC USES. Manure.
731 Leaves.
732 Twigs.
733
734
True Wolf. 735
Indian Wolf. 736
Jackal. 737
Indian Wild Dog. 738
Malay Wild Dog. 739
Indian Fox. 740
Hoary-Fox. 741
Desert Fox. 742
Common Fox. 743
Tibetan Fox. 744
SKINS. 745
FOOD. 746

| DOLICHANDRONE stipulata. | Dolichandrone Fibre. |

FOOD.

eat even children. The Bengal fox lives largely on fruits, such as those of **Grewia, Zizyphus,** &c.; also field rats, lizards, &c The late Mr. A. de Rœpstorff refers to the fact that the Andaman domesticated dog lives largely on cocoa-nuts, while those of the orange groves of the Khásia hills are fed like pigs on oranges. In the Nága hills and, indeed, throughout India, the dog is mainly fed on rice. But with the Nágas this is so on purpose, as the dog constitutes an important item of human food. Sheep and goats are rare in the Nága country owing to the prefer-

Dog's Flesh.
747

ence paid to dog's flesh. Before being killed, the dog is often made to eat as much rice as possible. Soon after he is killed and cooked, the contents of the stomach being considered a special luxury.

Dog-wood, see **Cornus sanguinea,** *Linn.*; **Vol. II., No. 1975, p. 572.**

DOLICHANDRONE, *Seem.; Gen. Pl., II., 1046.*

748

Dolichandrone falcata, *Seem.; Fl. Br. Ind., IV., 380;* BIGNONIACEÆ.

Syn.—SPATHODEA FALCATA, *Wall.*; BIGNONIA SPATHACEA, *Roxb.*; B. ATROVIRENS, *Roth.*
Vern.—*Háwar,* OUDH ; *Mendal, manehingi,* BANSWARA ; *Kanséri,* MEYWAR ; *Mersingh, bhil,* C. P. ; *Messinge, kanseri, mendal, manchingi,* BOMB. ; *Mersingi,* MAR. ; *Karanjelo,* KURKU ; *Kidatathie,* TAM. ; *Udda, wodi,* TEL. ; *Nir pongilam,* MALAY.
References.—*Roxb., Fl. Ind., Ed. C.B.C., 492 ; Brandis, For. Fl., 350 ; Beddome ; Fl. Sylv., t. 71 ; Gamble, Man. Timb., 276 ; Dalz. & Gibs., Bomb. Fl., 160 ; Indian Forester, III., 204 ; Bomb. Gas., III., 201.*

FIBRE. Bast.
749
MEDICINE. Fruit,
750
TIMBER.
751
DOMESTIC. Fruit.
752
753

Habitat.—A small deciduous tree, met with in Oudh, Rájputana, Central and South India.

Fibre.—A blackish coarse BAST fibre, obtained from this plant, was sent to the Amsterdam Exhibition by the Forest Department of Madras.

Medicine.—A decoction of the FRUIT is used medicinally.

Structure of the Wood.—Whitish, hard, close, and even-grained, seasons well, and becomes shining and glossy ; it has no heartwood. Annual rings indistinct. Is used for building and agricultural purposes.

Domestic Uses.—The FRUIT is placed by the Hindús on a bridegroom's waist.

D. Rheedii, *Seem.; Fl. Br. Ind., IV., 379.*

Syn.—SPATHODEA RHEEDII, *Wall.; Wight, Ic., t. 1339.*
Vern.— *Thakutma,* BURM. ; *Deya-danga (daanga),* SING.
References.—*Kurz, For. Fl. Burm., II., 234 ; Beddome, Fl. Sylv. Man., 168 ; Rheede, Hort. Mal., VI., t. 29 ; Liotard, Dyes, 33.*

FIBRE.
754
TIMBER.
755
756

Habitat. --A small tree of Burma, Malabar, Ceylon, and the Andamans

Fibre.—Yields a fibre similar to that of the preceding.

Structure of the Wood.—White, soft.

D. stipulata, *Benth.; Fl. Br. Ind., IV., 379.*

Syn.—SPATHODEA STIPULATA, *Wall.; BIGNONIA STIPULATA, Roxb.*
Vern.—*Petthan, mahlwa (bet-than* of **Mason**), BURM.
References.—*Roxb., Fl. Ind., Ed. C.B.C., 494 ; Kurz, For. Fl. Burm., II., 234 ; Gamble, Man. Timb., 726 ; Mason's Burma & Its People, app. 411, 543, 794.*

Habitat.—A moderate-sized deciduous tree of Burma and the Andaman Islands.

D. 756

Structure of the Wood.—Heartwood orange-red, beautifully mottled, hard, close-grained; weight 54-58℔ a cubic foot. The wood is used for bows, spear handles, oars, and paddles. Major Ford says it is a durable wood for house-posts, and makes good furniture.

TIMBER.
757

DOLICHOS, *Linn.; Gen. Pl., I., 540.*

A genus of twining herbs containing some 20 species, of which six are natives of India, the others occurring in the tropics of both hemispheres. The generic name **Dolichos** is of Greek origin, but it was more probably originally given to some cultivated species of **Phaseolus** than to any of the plants now designated **Dolichos** by botanists. The word *Dolichos* occurs in **Theophrastus** and *Fasiolos* in **Dioscorides**. The former has now been referred to the scarlet runner (**Phaseolus multiflorus**), and the latter to the dwarf haricot (**Phaseolus vulgaris**), and in modern Greek *fasoulia* survives as the name for the common haricot, a plant once on a time viewed as of Indian origin. (*Conf.* with the remarks at page 185.)

Dolichos biflorus, *Linn.; Fl. Br. Ind., II., 210;* LEGUMINOSÆ.

758

HORSE GRAM or KOOLTEE.

Syn.—D. UNIFLORUS, *Lam.;* GLYCINE UNIFLORUS, *Lam.*

Vern.—*Kúlthi* (or *kúltí*), *gahat,* HIND.; *Kurli-kalai,* BENG.; *Horec,* SANTAL; *Gahat, kalath, kulthi,* KUMAON; *Kalatt, kúlat, kult, kol, barát* (gulatti, the seeds), *roiong, rawan, kúlth, kolth, gágli, bothngt, guar,* PB.; *Kulitha, gagli,* SIND; *Kúdki,* C. P.; *Kulte, kulti, hulga,* BOMB.; *Kulith, kulthi,* DEC., MAR.; *Kalathi,* GUZ.; *Kollú* (vulava in Nellore), TAM.; *Wulawalli,* (or *wúluwúlú*), *ulava* (Elliot), TEL.; *Hurali* (Mysore), *hurlí,* KAN.; *Múthera,* MALAY; *Simbi* (a name for all the **Dolichos**), *kulattha* (according to Dutt), *kolutha* (Birdwood), SANS.

Note.—The name *khúrtí,* or *khúltí,* is in the North-West Provinces also given to **Cyamopsis psoralioides,** *DC.,* which see. **Vol. II., p. 673, No. 2514.**

References.—*Roxb., Fl. Ind., Ed. C.B.C., 563; Dalz. & Gibs., Bomb. Fl. Supp., 23; Stewart, Pb. Pl., 68; Aitchison, Cat. Pb. and Sind Pl., 49; Church, Food Grains of India, 162; Elliot, Fl., Andh., 185; Prof. Wallace, India in 1887, 96, 218; Rev. A. Campbell, Report on the Economic Products of Chutia Nagpur, No. 8147; U. C. Dutt, Mat. Med. Hind., 306, 318; S. Arjun, Bomb. Drugs, 40; Saidapet Exp. Farm, Man. & Guide, 51; Report of Exp. Farm., 1871, 4, 12, 13, 14; 1877, 97; 1879, 25; 1884, 27; Agri. Dept. Report, Madras, 1876, 34 & 35; 1878-79; Baden Powell, Pb. Pr., 241; Atkinson, Him. Dist., 696; Lisboa, U. Pl. Bomb., 153, 217, 277; Birdwood, Bomb. Pr., 119; Jour. Agri-Hort. Soc., 1867-68, Vol. II., 4; (1885) Vol. VII., Proceedings, cxviii; Manual, Coimbatore Dist., 223; Descrip. & Hist. Acc. of Godavery Dist., 68; Man. Trichinopoly, 72; Bombay Manual Rev. Acc., 101; Revenue Settlement Reports for C. P. (Mundlah), 38; (Upper Godavery), 36; (Chanda), 81; for Panjáb (Kumaon), 32d.; (Kangra) 25, 27; (Jhang), 84; (Simla), 58, XL., App.; (Hasara), 88; (Hoshiarpur), 94; for Madras (South Arcot), 109; (Glossary to Nellore); Gazetteers; Mysore & Coorg, I., 60; II., 11; Orissa, II., 15, 133, App.; Bombay, IV., 53; VIII., 182, 189; XIII., 289; XVI., 91; XVII., 269.*

Habitat.—According to the *Flora of British India* there are two forms of this plant: α (**D. uniflorus**), a sub-erect annual, and β (**D. biflorus**), a more or less twining plant. The habitats of these forms are not separately recorded; and Mr. Baker (the author of the LEGUMINOSÆ in the *Flora of British India*), apparently treats of both collectively when he says that it occurs on the "Himálaya to Ceylon and Burma, ascending

DOLICHOS biflorus.

VARIETIES.
759

CULTIVA-
TION.
760

Green
Manure.
761

to 3,000 feet in Sikkim; sometimes cultivated. Distributed everywhere in the tropics of the Old World."

Varieties.—While the writer does not possess the means of testing the accuracy of his opinions by the inspection of specimens obtained from all parts of India, he believes it will be found that a mistake has been made in linking the Himálayan with the plains' plant. **Roxburgh** refers to two forms, one with grey, the other with black seeds, both of which he implies are cultivated in Bengal and Madras. Of the grey-seeded plant (his **D. biflorus**), he remarks that it is erect, with twining branches, and about two to three feet high. He then adds: "I have never found it but in a cultivated state." Again: "This species is much cultivated all over the coast. It requires a dry, light, rich soil. In October and November it is sown either by itself or mixed with **Holcus saccharatus.**" In the writer's opinion there would appear to be considerable room for doubt as to whether the grey and the black-seeded forms of **Roxburgh** are the two forms of modern writers, or whether both of **Roxburgh's** plants constitute but cultivated races of one of these forms. In popular works, on economic products, the Horse-gram of Madras is viewed as **D. uniflorus**, and under either of these names (**D. biflorus** or **D. uniflorus**), a pulse is described as grown, one might almost say, in every district of India, but chiefly in Madras and Bombay. It is somewhat difficult to believe that a pulse of the tropical plains could be the same as that of the Temperate Himaláya, of which **Stewart** wrote that it is "grown at 7,000 feet or more." This will appear the more improbable when it is added that the pulse described as met with in these regions is sown and reaped very nearly during the same periods, though in the one case under tropical, and in the other under temperate, influences.

Cultivation.—It may be said of the plains that the pulse here dealt with is grown for either of two widely different purposes :—*viz.*, as a green manure, or as food and fodder. It has not been found possible to discover the extent to which the former purpose is pursued by the actual cultivators. The reports on the subject are more directly connected with Government experimental farms, although it would appear as if the experiments described had been the outcome of a recognised native practice. **Mr. Robertson**, in several of the Saidapet Farm Reports, deals with the advantages likely to accrue from the use of this pulse as a green manure. He writes: "The action of the green manure is two-fold. *First*, the substance of the plant decaying in the soil leaves behind a large quantity of prepared food, ready for absorption by the roots of the succeeding crop ; *secondly*, when ploughed in, the structures of the green crop add directly to the amount of organic matters in the soil, and thus improve its mechanical condition, increasing its power of absorbing and retaining moisture, and increasing, in the case of stiff soils, their friability." In another place he remarks : "In several fields, crops were ploughed in, during the past season, and although it is not possible to state what actual value the proceeding had, for no experiments were made, yet estimating the value of such a manure at R4 per ton, it was necessary to produce about 4,500lb per acre to cover the cost of growing it." In still another report **Mr. Robertson** says : "The horse-gram (**Dolichos uniflorus**) is well suited for culture on sandy soils, for ploughing in, as a green manure." "Last dry season we raised crops that yielded from 2,000 to 3,000lb of plant per acre in a period of about twelve weeks, during which the rainfall did not amount to one inch. In the neighbourhood around Madras, the 'summer crops,' on dry sandy land, are exceedingly precarious ; on the average we have not more than one year in four, in which crops sown in June or

| Horse Gram or Kooltee (or kúlti). | (*G. Watt.*) | DOLICHOS. biflorus. |

CULTIVA-TION.

July yield returns that repay the expenses of cultivation. I think, there-fore, that instead of attempting to grow 'summer crops,' such as *gingelly*, *cumbú* (**Pennisetum typhoideum**), &c., except on a small scale, on choice land, that the wisest course, after removing the "cold weather crop," would be to clean the soil thoroughly and then to sow it with horse-gram, for ploughing in. These sowings would, in the space of three months or so, yield per acre from 2,000 to 3,000℔ of plant, which, if ploughed in, would prepare the soil admirably, for the succeeding "cold weather crops."

The advantages from growing the crop as a source of FODDER are extolled by various writers. In one report **Mr. Robertson** says : " It pro-duces from 2,000 to 4,000 pounds of fodder in two months, at a cost of about R3 per ton, and thrives with a minimum rainfall in very hot weather. The ease with which it may be cultivated recommends it most highly as a catch crop for forage purposes, either to be grazed on the land or fed in the stalls. The plant may be made to grow at almost any season of the year. It will, in fact, thrive when no other crop can exist. It requires but one shower of rain to start its growth, but if even this be not obtained, the seeds have the power of remaining for months in the soil and of germinating when rain falls. After the removal of the *rabi* crop it is contended that a highly advantageous course is to rapidly dress the soil, sow horse-gram, and in a month's time commence to use the stems and leaves as fodder. By this means the soil is saved from becoming baked with the advancing heat of summer, and the roots left in the soil greatly improve it, even should the cultivator be unable to devote the entire crop as a green manure. **Mr. Robertson** remarks on this point : " The small quantity of moisture present in the land at the time of harvest is generally enough to start the crops, which are found to give a fair outturn of fodder, though there may be no rain whatever during their growth. The advantage of this system is that the land is made use of and kept under tillage during the dry season." He then proceeds to give the results of seven sowings of horse-gram, which took place between the 26th February and the 10th of March. He adds : " No rain fell during the growth of any of these crops. It will be seen that about six tons of green fodder, worth about R48, were obtained without any rain whatever between March and May." "It would be a great boon to the country if the ryots would endeavour to grow horse-gram as far as possible, either for fodder to their cattle, or for green manure to their summer crops, of *gingelly* and *cumbú*, immediately after the harvest of their paddy, instead of allowing their lands to become hard-ened as at present."

FODDER.
762

NATURE OF SOIL, SEASONS OF SOWING AND REAPING, &c., &c., OF HORSE-GRAM.—The earlier writers seem to have been mistaken as to the requirements of this plant, but considerable confusion also exists in the published statements of recent authors, which may to some extent be accounted for by the differences in provincial agriculture and climatic conditions. In the passage quoted above, for example, **Roxburgh** states that this pulse is grown on a "dry, light, rich soil." Every shade of difference of opinion seems, however, to prevail on this and many other features of horse-gram cultivation, many of which (such as yield per acre, cost of cultivation, &c.) have been purposely omitted here, but the follow-ing brief review, province by province, may be found instructive.

NATURE OF SOIL.
763

Madras.—**Mr. Nicholson** (*Manual of Coimbatore*) writes that the ryots were in former days allowed to take up new lands for horse-gram cultivation at a quarter the usual rates. He adds : " It grows on the poorest soils with the least possible trouble and with the minimum of rain-fall. Gram land is seldom manured otherwise than by casual droppings

MADRAS.
764

N

CULTIVA-
TION.

of cattle; they are usually ploughed, sown, and the seed covered by a
second ploughing if there be time, but if not, the seed is simply scattered
broadcast over the natural surface and then ploughed in. As it requires
only one good rain after appearing above ground, it frequently gives a
fair crop when nothing else can live. When the south-west monsoon
rains are too late for *Kambú* it is frequently sown as a substitute in September, but it is also sown largely in November after the first burst of the
north-east monsoons. It is pulled up by the roots, thrown into heaps, and
then trodden out by cattle. The yield is up to 1,200℔." In a recent
report contributed by Mr. H. Sewell, Collector of Cuddapah, there occurs
a similar statement: "It requires no cultivation beyond ploughing,
and grows on any soil." Mr. H. Goodrich, Collector of Bellary, writes:—
"A mixed soil is best suited for the crop. The fields should be
ploughed and harrowed once or twice, but not irrigated nor (generally)
manured." Mr. Robertson's experience of the pulse on the Saidapet
Experimental Farm has been indicated by several passages quoted
above, but with regard to the soil, &c., it may be as well to convey
his meaning still further. He says it is "a valuable fodder-producer for
inferior sandy soils." "The ease with which it may be cultivated recommends it most highly." But several Madras writers give a very different
account of the requirements of this plant. For example, in the Survey
Settlement Report of South Arcot (see *Selections from the Records of the
Madras Govt., 1869, p. 109*), there occurs the following passage regarding "horse-gram (Dolichos uniflorus)": "The land is ploughed four or
five different times after the month of May, and the gram sown between
the latter part of August and the end of September. It is gathered in
the middle of March." In the Manual of the Trichinopoly District (by
Mr. L. Moore), page 72, it is stated that "*Kollu* (Dolichos uniflorus), or
horse-gram, is a four-months crop, being sown in October and reaped in
February. It is a precarious crop, as it requires frequent showers, and is
destroyed equally by excessive drought or moisture. It is grown to a
considerble extent in the Kulittulai Taluk, but not much elsewhere."
Writing of Trichinopoly recently Mr. H. Willock says of *Kollu* that "the
area of this grain under cultivation is about 27,604 acres, of which 1,297
acres are *fasli* lands. It is a four-months crop, sown in October and
reaped in February." "It is cultivated generally in sandy soils and also
in other soils when the season for appropriate crops is over." Of Cuddapah District Mr. H. Sewell gives the extent of cultivation in 1887-88 as
14,755 acres, and the outturn 17,70,600 measures. He adds: "It is sown in
October and reaped in February." Of Bellary District Mr. H. Goodrich
writes of 1887-88, that "the total area under cultivation of this crop is
estimated to be 106,805 acres, of which 90,013 belong to Government and
16,792 are *inam*. The season for sowing is from the 3rd August to 7th
October, and that of harvesting from 20th December to 21st February.
The lowest estimate for the cost of cultivation is given at R1-12, the
highest at R5-8, and the average at R3-2-7 per acre. The profits vary
from 4 annas (lowest) to R4-4 (highest) per acre; the average being
R1-9-2 per acre."

Area.
765

Seed.
766

Yield.
767

The amount of seed per acre and the yield is variously stated, but of
Madras Mr. Robertson wrote in 1871 that in one experiment, 35℔ an
acre was sown in August and yielded in October 5,640℔ of green
fodder. Another experiment with 24℔ an acre, sown in October, gave
in March 450℔ of pulse and 1,800℔ of straw. But reference has
already been made to Mr. Robertson's experiments of cultivating for
fodder or green manure horse-gram sown in February and March. The
present notices regarding the Madras cultivation of horse-gram may there-

Horse Gram or Kooltee (or kúlti.) G. (*Watt.*)	DOLICHOS biflorus

fore be concluded with a passage from the Saidapet Farm Manual and Guide: "It is a hardy plant, thriving in the poorest soils. The soils of this district contain a very small proportion of lime; and this plant like all leguminous plants, requires a good deal of lime before it can mature its seed. It has been ascertained from experiment, that unless the manure applied contains a considerable percentage of lime, the tendency of the plant under better cultivation is to produce leaf rather than seed; this tendency has been utilised, and by deeper cultivation and the application of a moderate dressing of manure we have succeeded in growing good fodder at a very moderate cost."

"Generally, in preparing land for gram, the following method is adopted:—"After ploughing 4 to 6 inches deep, and harrowing, the seed is sown in lines, if the season is unfavourable and the soil poor, close together, if the reverse, far apart, at the rate of from 30 to 40℔ per acre. During growth, the crop should be bullock-hoed, once or twice, as circumstances demand, and hand-hoed at least once. The crop should be cut immediately the flower appears and removed the same day. The cost of growing a ton of fodder is about ℞3. The fodder makes good hay which possesses a pleasant aromatic smell when well made; it, however, loses 75 per cent. of its weight in curing."

"When cut before maturing its seed, the cultivation of gram improves rather than impoverishes the soil. There will always be a slight loss in the mineral constituents; but still, as the plant appropriates such a large amount of atmospheric food, and stores it away in its roots, and as these roots, weighing from 800 to 1,000℔ per acre, are left in the soil, its condition must be greatly improved."

Bombay.—In the *Káthiáwár Gazetteer* (*p. 189*), it is stated that "Horse-gram, *Kulthi*, Dolichos uniflorus, is a crop of small importance, grown to a limited extent in all parts of Káthiáwár. It grows in poor soils, requires ploughing and hoeing, and is sown in July and reaped in October. It is locally used by the poor classes and is given to cattle. Of Ahmadnagar the *Gazetteer* (*p. 269*) says: "Horse-gram, *Kulthi*, or *hulga*, Dolichos uniflorus or biflorus, in 1881-82 had a tillage area of 38,153 acres. It is sown with *bájri* in June and ripens in November. It is eaten boiled whole or split as *dál*, and in soup and porridge, and is also given to horses. The leaves and stalks are good fodder." To contrast with the above, in which the horse-gram is said to be sown in June, the following passage may be given from the *Thána Gazetteer* (*p. 289*): "Horse-gram, *Kulith*, Dolichos uniflorus," "is sown in November after the rice crops have been cut, and ripens about the beginning of March. *Kulith* is eaten in the form of pease-meal, which is called by a number of names. The pease boiled and mixed with gram, make very good food for horses. The stalks are used as fodder." *Kulthi* or *hulga* is referred to in several other volumes of the *Bombay Gazetteer*, in some of which it is said to be sown in June, in others in November. Thus of Sátára (*p. 163*), it is said that, it is "generally sown in June with *bájri* in separate rows and ripens in November." Mr. Lisboa, in his *Useful Plants*, refers to D. biflorus, a twining, and D. uniflorus, a sub-erect, plant, both having trifoliate leaves and yellow flowers. In the figures published in *Church's Food-Grains of India* the twining form has hairy pods, and the erect, glabrous. It would be instructive to know if the June and November sowings of Bombay were of either or both of these forms; in other words, whether the one sowing was the twining plant, and the other the erect.

North-West Provinces.—Very little can be discovered regarding the extent to which this pulse is grown in these Provinces, and the common name *khulti* here more frequently denotes Cyamopsis psoralioides than

Side notes: CULTIVATION. · BOMBAY. 768 · N.-W. PROVINCES. 769

DOLICHOS biflorus.	Horse Gram or Kooltee (or kúlti).

CULTIVA-TION.

Dolichos biflorus. It may be inferred that very little of **Dolichos biflorus** is actually cultivated in the Provinces, from the fact that it is not described in Messrs. Duthie and Fuller's *Field and Garden Crops.* Mr. Atkinson, however, in his *Himálayan Districts, pp. 343, 460, 696,* says: "Horse-gram—*Gahat, kalath,* the *kulthi* of the plains. The horse-gram is occasionally grown in the hills up to 6,000 feet, and in the sub-montane tract. In the Bhábar it ripens in October." A somewhat striking feature of this pulse or bean is the absence of any allusion to it in the *Ain-i-Akbari.* Abul Fuzl, the author of that useful record of Akbar's times, gives particulars of all the grains, pulses, oil-seeds, vegetables, flowers, and fruits known to the Emperor. Among the pulses and vegetables there occurs *Mung, Másh,* and *Moth*—the first two are forms of **Phaseolus Mungo,** and the last is **P. aconitifolius :** then *Adess,* the lentil (**Lens esculentus**) is referred to, and *Nakhúd,* the common gram (**Cicer arietinum**). *Lobiyá* is also mentioned, but whether we are to translate that as **Vigna Catiang** or as **Dolichos Lablab** seems doubtful. At all events, no place is given to *kulthi* and, indeed, it is questionable if that pulse was known to the Persian writers. This fact is difficult to account for, if we admit that the plant of the Himálaya and of Northern India is the horse-gram of Madras, but the absence of any knowledge of it admirably corresponds with the present cultivated area of the plant, *viz.,* in South India and Bombay, the portions of India over which the Emperor Akbar was never able to extend his supremacy. We might, indeed, from this fact be pardoned the assumption that the true habitat of Horse-gram should be looked for in South India rather than on the Himálaya.

History of Kulthi.
770

PANJAB.
771

Panjáb.—Of the Panjáb **Stewart** says : " It is commonly cultivated for its pulse in the Himálaya up to 7,000 feet or more. Occasionally grown outside, near the base of the hills at Ambála *(Edgeworth).*" "*Kulth* (**Dolichos uniflorus**)" is referred to in the *Gazetteer of Hoshiarpur District (page 94)* as a *kharif* crop "sown on the poorest hill slopes, which look as if they could produce nothing but stones." In the *Gazetteer of Simla (page 55)* *kulthi* is alluded to as "the most common pulse, growing freely, even upon high meagre soil. The grain is hard and indigestible, mottled with specks of a dark colour. It is eaten in the form of *dál.*" On a further page it is again alluded to : "*Kulat* or *Kolath* (**Dolichos uniflorus**—horse-gram) is grown in the inferior *bakhíl* lands in the lower villages. Will not grow on the higher lands. Is not sold. Is sown the same time as *Másh* " (=**Phaseolus radiatus,** *viz.,* sown in July and harvested in October), " but ripens 15 days later. To prepare for eating,—it is soaked in water for 12 hours ; then reduced to a *mash* on a stone ; then made into round balls and steamed. Another way is to roast the grains and then boil them, adding rice."

CENTRAL PROVINCES.
772

Central Provinces.—In a recent communication Mr. J. B. Fuller says :— " **Dolichos biflorus** is grown in the southern districts of the Provinces as a cold weather crop. Full details of the area under it are not available, but such statistics as are at hand indicate that its cultivation is of importance only in the Chanda, Bhandara, and Balaghat districts, in each of which it annually covers from 3,000 to 4,000 acres." In the Settlement Reports, referred to under the paragraph of references, mention is made of this pulse, but Mr. **Fuller's** brief note gives the main facts.

BENGAL.
773

Bengal.—Horse-gram is very little cultivated in the Lower Provinces. It is said to be grown to a limited extent in Shahabad as a fodder, but "not grown in lower Bengal." It is, however, "largely cultivated in Chutia Nagpúr Division, on good land. It is usually sown along with *sirguza* in August, and receives the same treatment, and is harvested in November-December. The average quantity of seed sown is ten seers per acre,

D. 773

Products of India. 181

Horse Gram or Kooltee (or kúlti).   (*G. Watt.*)   DOLICHOS biflorus.

and the average produce two maunds, valued at R3. The seed is eaten as *dál*, or ground into *sátu* after being roasted. In Chutia Nagpúr, proper, about 1½ per cent., of the cultivated area is sown under this crop. In Khoorda,* Pooree District, *kulthi* is usually grown as a second crop on paddy lands." The Rev. A. Campbell writes that by the Santals this pulse "is extensively cultivated on good high lands. It is' eaten in the form of *dál* and also as *sátu*. To prepare *sátu*, the pulse is roasted and then ground. It is eaten without being further cooked."

CULTIVA-TION.

AREA UNDER- HORSE-GRAM.—In some of the above passages reference has been made to the extent this pulse is cultivated. With the exception, however, of Madras and Bombay, it is not of such 'mportance as to require being regularly recorded, and a complete statement cannot, therefore, be furnished for all India. The area in Madras, since 1883-84 to present date, has ranged from 1,208,789 acres to 1,498,021 acres. The returns for Bombay may approximately be stated to have indicated between one-third and one-fourth of that area as under the pulse. In 1887-88, the total of these two Provinces was close upon 1,850,000 acres. The Central Provinces have, perhaps, about 10,000 acres, and in the Berars there are usually some 1,500 acres. It is probable that the rest of India would not represent more than 20,000 acres at the outside, so that it may safely be added that if the plant is a native of the Himálaya, its area of cultivation is in Madras and Bombay.

AREA. 774

TRADE IN HORSE-GRAM. —No statistics are available, and it is only necessary to caution intending foreign purchasers that *the gram of Madras* — the *Horse-gram* here discussed — is a perfectly distinct pulse from the *gram* or *Bengal gram* of most writers. (For GRAM see **Cicer arietinum.** *Vol. II., No. 1061, pp. 274 to 284.*) This caution is the more necessary, since every trade journal and agricultural publication is urging the importance of India as a source of pulses and lentils, &c., to be used as cattle food. The importation into Europe of the horse-gram of Madras under the false impression that it was the same as Bengal gram might seriously injure the progress of trade, and the sale of the pulse, **Lathyrus sativa**, as gram, would be attended with such serious consequences (paralysis of the animals so fed) as to prove fatal to the hopes entertained of the expanding pulse and pea trade of India.

TRADE. 775

Another fact of some importance regarding a trade in horse-gram may be here mentioned, *viz.*, that the Madras crop mainly comes into the market in March, April, and May, while the bulk of that of Bombay and Upper India would appear to be available in November and December.

EXTENT TO WHICH USED AS HUMAN FOOD.—It is scarcely necessary to refer to this subject in a separate paragraph, since the most important passages regarding it have already been quoted. Although not deemed a superior pulse, it is largely eaten by the poorer classes, either after being boiled or in the form of a meal variously prepared. Dalzell and Gibson (*Supp. Bombay Flora, p. 23*) say that "when a spur or ergot grows on the seed, it is often very deleterious."

HUMAN FOOD. Seeds. 776

CATTLE-FOOD.—As a fodder for cattle and horses the STEMS and LEAVES of this plant are highly valued all over India, and the BEAN appears to constitute the chief article of diet given to horses in the Madras Presidency. The split husk also is used in Madras as a cattle-food. Numerous experiments have been performed to test the value of *kulthi* both as a fodder and a cattle-food. Mr. Robertson ascertained the merits of boiled as compared with steeped horse-gram on draught cattle. He reports :—" A lot of 16 draught cattle similarly worked were equally

CATTLE FOOD. Stems. 777 Leaves. 778 Beans. 779 Split Husk. 780

* See **Taylor's** Settlement Report on Khoorda Government Estates.

DOLICHOS biflorus.	Horse Gram or Kooltee (or kúlti).

CATTLE-FOOD.

divided. Besides their usual fodder one lot got 12 pounds of boiled gram and 12 pounds of ground nut-cake, and the other lot received daily 12 pounds of steeped gram and 12 pounds of ground nut-cake. The results were as follows : —

Animals on Boiled Gram.

	Pounds.
Weight at the commencement of the experiment	6,339
Do. twenty-seven days afterwards	6,576
Increase .	237

Animals on Steeped Gram.

Weight before the commencement of the experiment . . .	6,310
Do. twenty-seven days afterwards	6,576
Increase .	266

A similar series of experiments were performed with horses, and the verdict arrived at was in favour of steeped gram. **Mr. Robertson** performed a further experiment to test the comparative feeding values of maize and horse-gram. He reports : "for the first few days maize was not readily eaten ; however, at the end of a couple of weeks, the cattle ate it freely and continued to increase in weight, until at the termination of the experiment, they had increased 71 pounds in weight. The other pair ate gram from the first, but they never made the progress observed by the pair fed on maize, and at the termination of the experiment had only increased 3 pounds in weight." This fact might to some extent be accounted for by the beneficial effect of a change, irrespective of the merits or otherwise of the maize diet.

Another series of experiments were conducted in order to determine the value of gram fodder in comparison with grass and cholam fodder as food for sheep. "The animals fed on grass only gave an increase of 8·26 pounds per each 100 pounds of their live weight, whilst thòse fed on gram fodder gave 14·5 pounds, and those on cholam fodder 15·58 pounds. The grass was the inferior stuff usually cut for horses." In the *Khandesh Gazetteer (p. 152)* it is stated that many persons prefer *kulthi* to gram (presumably Bengal gram) in feeding horses. It is much to be regretted that no one appears to have published the results of definite experiments to test the relative merits of Bengal gram (**Cicer arietinum**) and Horse-gram (**Dolichos biflorus**). Such experiments would afford exporters the means of judging whether they should commend most, the Bengal or the Madras staple article of horse food, to European dealers. The chemical analysis taken from *Professor Church's Food-Grains of India* (given below) would, however, justify the preference being shown to Bengal gram :—

CHEMISTRY.
781

CHEMISTRY OF THE HORSE-GRAM.

Professor Church publishes the following table of analysis :—

Composition of Horse-gram.

	In 100 parts, unhusked.	In 1℔	
		oz.	grs.
Water	11·0	1	333
Albuminoids	22·5	3	262
Starch	56·0	8	420
Oil	1·9	0	133
Fibre	5·4	0	378
Ash	3·2	0	224

D. 781

Horse Gram or Kooltee (or kúlti).   (*G. Watt.*)   **DOLICHOS**
                                        Lablab.

The Professor concludes from this result that "the nutrient ratio is 1 : 2·7, and the nutrient coefficient 83. The ash of these beans contains nearly one-third of its weight of phosphoric acid. The long continued use of these beans is regarded as injurious : they are reputed, in some districts, to cause œdematous swellings." The writer is not aware of the source from which **Professor Church** derived the statement that the long continued use of this pulse is injurious. If he alludes to injury done to cattle and horses it would be difficult to account for the fact that it is apparently the chief article of diet given in Madras to cattle and has been so from the very earliest records, but apparently no such opinion of injury done thereby prevails in South India. At the same time, the continued feeding on pulses is by some authors condemned, and one pulse, already alluded to, would seem to have distinctly an injurious effect (**Lathyrus sativa**).

The analysis given above, if compared with that recorded under **Cicer arietinum** (*Vol. II., p. 280*) will be seen to justify the assumption that Bengal gram is superior to that of Madras. In the former a larger percentage of albuminoids and oil exists which manifest a result expressed by **Professor Church** thus :—nutrient ratio of Bengal gram 1 : 3·3 and the nutrient value 84.

CHEMISTRY.

Oil.—The BEANS are said to yield an oil, of which little is known.

              OIL.
          Beans.
            782

Medicine.—**Stewart** says the SEEDS are used medicinally in the Panjáb. **S. Arjun**, in his *Bombay Drugs, p. 40*, has the following remark about **Dolichos uniflorus** : "There are two varieties of this—the red and the white. Both are used for similar purposes. The DECOCTION is used by native females in leucorrhœa and menstrual derangements ; it is also given to parturient females to promote discharge of the lochia."

MEDICINE.
Seeds.
783
Decoction.
784
Pulse.
785

Special Opinion.—§" Sanskrit writers recommend the use of the pulse of this plant as a demulcent in calculus affections, cough, &c. Its employment is said to reduce corpulence. The wild variety is said to be particularly serviceable in eye diseases " (*U. C. Dutt, Civil Medical Officer, Serampore*).

Food.—The PEA is eaten by the poorer classes of natives, and the PODS and PEAS are also eaten by horses and cattle. The STRAW is a much prized fodder.

FOOD.
Peas.
786
Pods.
787
Straw.
788

Dolichos cultratus ; *Syn.* for **Dolichos Lablab.**

D. fabæformis, *L'Herit. ;* see **Cyamopsis psoralioides,** *DC.*

D. Lablab, *Linn. ; Fl. Br. Ind., II., 209.*

789

    Vern.—*Sim* or *sim, makhan-sim, lobia* (or *lóbiyá*), *vál, borboti* (*wall,* according to **Stocks**), HIND.; *Shim, makhan-sim, borboti, gheea-sim, panch-sim, lablab, gurdal-shim, bun-shim, panch-shim, ganchi-shim, &c.,*BENG.; *Malhan,* SANTAL; *Urohi, urshi, uri,* ASSAM ; *Kechu,* NAGA ; *Shimi, chimi, sém, sémbi,* N.-W.P. ; *Katjang* (? **Vigna Catiang**), *kála lobia,* ?PB. ; *Wall* (according to **Birdwood**), SIND. ; *Pauti, valpapri* or *válapápadi,* BOMB. ; *Páote, vál,* MAR. ; *Vál,* GUZ.; *Mochai,* TRI-CHINOPOLY; *Bili manavare,* or *man avare,* MYSORE; *Mutcheh* (according to **Birdwood**), *avarai,* TAM.; *Alsanda, boberlu, tella-chikurkai* (*anumulu, adavi-chikkudu, tella-chikkudu by* **Elliot**), *annapa, anapa chikkudu,* TEL.; *Avare, avre,* KAN.; *Pai,* BURM.; *Simbi* or *shimbi,* a name most frequently assigned to this species (*nespava,* or *nishpáva,* given by some writers, is **Vigna Catiang**), SANS. ; *Lobiyá* (according to **Stocks**), PERS.

    NOTE.—The names *Lobia* and *lóbiyá* given above for this species are in the writer's opinion wrongly so applied, and should be assigned to **Vigna Catiang.**

References.—*Roxb., Fl. Ind., Ed. C.B.C., 560; Dals. & Gibs., Bomb. Fl. Supp., 23; Stewart, Pb. Pl., 67; Aitchison, Cat. Pb. and Sind Pl., 49; Sir Walter Elliot, Fl. Andhr., 10, 15, 16, 175; Rev. A. Campbell, Econ. Prod. Chhutia Nagpur, Nos. 9249 and 8155; Stock, Account of Sind; Church, Food-Grains of India, 161; DeCandolle, Origin Cult. Pl., 346; Murray, Pl. and Drugs, Sind, 127; Mason, Burma and Its People, 466; Atkinson, Him. Dist., 696; Duthie & Fuller, Field and Garden Crops, II., 23; Lisboa, U. Pl. Bomb., 153; Birdwood, Bomb. Pr., 119; Ain-i-Akbari, Blochmann's Transl., 63; Jour. Agri-Hort. Soc., V. (New Series), p. 37; Indian Forester, IX., 203.*

Habitat.—Wild and cultivated throughout India; ascends to 6,000 or 7,000 feet on the Himálaya. This climber may be seen growing along the borders of fields, which contain tall crops being left to twine round the plants near the margin. In some parts of the country the castor oil plant is a favourite support. The *shim* is also grown very commonly round houses, being allowed to climb on the walls and roof.

History.—Some idea of the probable history of **Dolichos biflorus** may be gathered from the series of quotations given above from numerous authors, and from the very extensive collection of vernacular names, most of which seem to be derived from the Sanskrit *Kuluttha*. The remarks made in the paragraph devoted to the cultivation of that species in the N.-W. Provinces may be specially read in this connection. **M. A. DeCandolle** (*Origin of Cultivated Plants*) deals with two (or what the writer regards as only one) species of **Dolichos**, *viz.*, **D. Lablab**, *Linn.*, and **D. Lubia**, *Forskal*. He does not treat of **D. biflorus**, although it is perhaps a more important cultivated plant in India than **D. Lablab**. The line of reasoning urged by **DeCandolle** seems largely to turn on the origin of the word *lubia*. He says: "Oriental scholars should tell us whether *lubia* is an old word in Semitic languages. I do not find a similar name in Hebrew, and it is possible that the Armenians or Arabs took *lubia* from the Greek λοβος, which means any projection, like the lobe of the ear, a fruit of the nature of a pod, and more particularly, according to Galen, **Phaseolus vulgaris**. *Lobion* (λοβιον) in Dioscorides is the fruit of **P. vulgaris**, at least in the opinion of commentators. It remains as *loubion* in modern Greek, with the same meaning." The word *Lobíyá* occurs among the list of autumn crops, known to Akbar. According to some modern writers it is, in Upper India, almost generic for beans, although applied more especially to two plants, *viz.*, **Vigna Catiang** and

Dolichos Lablab. The former comes into season in the autumn (*kharif* crop), while in the N.-W. Provinces and the Panjáb the latter is sown in autumn and reaped in February and March, so that it is a spring (*rabi*) crop. These seasons do not, of course, apply to all parts of India, since, for example, in Assam and some parts of South India, **D. Lablab** ripens in December. The *Ain-i-Akbari* (a work written in Persian) describes the crops grown in Delhi and Agra during the reign of the Emperor Akbar. A pulse, *Lobiyá*, is there spoken of as a *kharif* crop. As at the present day, so in all probability in Akbar's time, this would have been **Vigna Catiang**. This is of importance, since the word *Lobíyá* appears to be of Persian not Sanskrit importation into the languages of India. Persian scholars do not seem to share **M. DeCandolle's** ideas regarding a derivation of *Lobiya, lubiyá, lúbiya,* or *luba,* from λοβος. The word is accepted as of pure Persian origin, and in **Johnson's** Persian, Arabic and English Dictionary is given as "a kind of pulse." It may here be added that *labáb* in Arabic means "green fodder." But even if the Persians borrowed the word from the Greeks, the contention here advanced would still remain in its full force. It came to India through the Persians. Hence the writer is disposed to restrict the word *lobíyá* to **Vigna Catiang**, and if this proves correct it is probable none of the species of **Dolichos** were known to the Persian or Arabic writers of classic times. This conclusion would assign to the species of **Dolichos** an Indian origin, an idea practically confirmed by the almost universality of certain derivative names in the languages of India traceable to the Sanskrit and not to Persian or Arabic, and by the fact that **Dolichos Lablab** exists as a purely wild plant in Bengal and some parts of Madras. No name like *lobíyá* is given to any pulse by the aboriginal races of Indian or by those of Aryan origin. It occurs purely among the people of Upper India, where Persian influence is most pronounced.

In the *Gazetteers of the North-Western Provinces* the name *lobíyá* occurs frequently as that of a pulse, but in Kumáon it is said to be the name for **Vigna Catiang**. In the volume on the Eta District (*p. 27*), it is remarked that " *lobiya*, known as *masína*, is sown with the millets as a rain crop." Again, " *lobiya* is the Persian form of *ramás*, and *ramus* is here usually called *rausa*." *Ramás* and *rausa* are names given throughout these provinces for **Vigna Catiang**. Of the Meerut District it is said that *lobiya* (**Vigna Catiang**) is a *kharif* or rain crop, but that *masína* is applied to linseed. In the *Budaun Gazetteer* " *lobiya*, **Dolichos sinensis** " (a synonym for **Vigna Catiang**), are given as the names of a *kharif* crop, but these names are mentioned in the *Bijnor Gazetteer* as that of a spring crop. This latter statement may be the result of a mistaken identity, or then the plant referred to is not **Vigna Catiang**, but may possibly be **Dolichos Lablab**. But if this be so, in Bijnor alone, of all the districts of the North-West Provinces, is the name *lobíyá* given to a spring pulse, presumably **Dolichos Lablab**. In the *Indian Forester* (*IX., 203*), *lobia*, **Vigna Catiang**, is referred to as one of the most useful of the bean tribe for rainy season cultivation. It is said to continue to yield till the beginning of the cold season.

Of the Panjáb, Stewart says **Dolichos Lablab** is known as *catjang* and *kala-lobia* (the black *lobia*), but he is the only writer who says so. He gives *lobia* itself to **Vigna Catiang**, and it seems probable he was mistaken regarding **Dolichos Lablab**. Mr. Baden Powell, a subsequent writer, speaks of **Dolichos sinensis** as *lobiyan*, but he refers to a black pulse under the name **Dolichos Lablab?** which was obtained from Hushyárpur and Gujrat. This bore the vernacular names of *keo, kaiun*, or *kala múng*. There would seem little doubt but that this is the *kala lobia* of Stewart, and it is probable **Stewart** added the word *lobia* (an Anglo-Indian generic name for beans) much after the same principle as **Baden Powell** gives the paragraph heading *Lobiya* to an account of a Kashmír bean, the botanical name of which he gives as " **Phaseolus vulgaris**, *L.*, and, **P. lunatus**, *L*, red and white haricot beans (mixed)." " These, he adds are exhibited from Srinagar, called in Kashmír *dhakh*." The paragraph heading for the Kashmír bean should therefore have been *dhakh*, but accepting *lobiya* as a better known name it was apparently given instead of the local name.

There is, however, another point of some interest regarding this notice of a Kashmír bean. If correctly referred to **Phaseolus vulgaris** this is the only instance on record of that introduced plant bearing what appears an indigenous vernacular name. Accepting Peddington's Index of the Vernacular Names of the Plants of India as correct, M. DeCandolle discusses the claims of India to a share or otherwise in the production of the haricot bean. Peddington, it would appear, gives that vegetable the names of *loba* and *bakla*, and DeCandolle adds : " This, together with the absence of a Sanskrit name, points to a recent introduction into Southern Asia." The haricot bean, though fairly extensively cultivated in India, is met with only in the gardens of the Europeans, or in the hands of cultivators who

Haricot Bean.
792

HISTORY.

trade in meeting the European demand. It can in no way be said to be a regular article of native cultivation, and the name *loba*, if ever assigned to it, must be viewed as but a modern adaptation of a semi-generic appellation for introduced peas or beans. But to return to the mention of the word *lobiya* in Panjáb recent publications. In the *Gazatteers* of the various districts, **Dolichos Lablab** is practically ignored, while **Vigna Catiang** is frequently mentioned. That pulse is, for example, *rong* in Kangra, *ranyan* in Simla, *rawan* in Montgomery, and *lobia* or *chaula* in Gurgáon. Thus *lobia* appears as a synonym along with other and more distinctly Indian names for **Vigna**. *Chouli* is a very frequently used Hindústani name tor it; *Choulí* in Chanda, *chaunro* in Sind, *Choulí*, *chola*, *safed lobeh* (white *lobeh*), *hurrea lobeh*, and *gat-vál* in Bombay. Thus ever here and there the name *lobiyá* crops up, in connection with **Vigna**, though practically no authentic case is known of its being given to **Dolichos Lablab**. In South India that name scarcely exists, except perhaps with Europeans. **Vigna Catiang** is *alasandi* in Kánarese, *káramanalu*, *alachandalu*, or *bobbarlu* in Telegu, and passing up the east coast to Orissa it becomes *lobiya-chhai* in Uriya.

The final conclusions which the writer has arrived at regarding the word *lobiyá* may be expressed briefly :—(1) It is incorrectly applied to any species of **Dolichos** or **Phaseolus**; (2) it is of Persian origin and may by adaptation have been assigned by the early Persian and Mogul conquerors of India to **Vigna Catiang**; but (3) as used by the Indian market gardeners of the present day it is a generic name for any introduced pulse or bean and is in no way specific. A similar expression exists in the use of **Lablab** for the vegetable or unripe pods of beans such as those of **Dolichos Lablab**. The probable origin of **Vigna Catiang** and its claims to being the true *Lobiya* of Indian (Persian) writers will be dealt with in a further volume of this work.

Having thus in a measure disposed of the confusion caused through the association of *lobiyá* with **Dolichos Lablab**, there remains little to be said regarding the history of **Dolichos Lablab** itself. The existence of it as a wild plant, combined with the extensive series of vernacular names, especially those of Lower, Eastern, and Southern India, leave no room for doubt as to its being a native of India, and more especially of the portion of India indicated as the area of its indigenous habitat. The Sanskrit names given to it are doubtfully correct, and although we may be unable to follow **DeCandolle** in the idea that, according to Sanskrit literature, it has been cultivated in India for 3,000 years, there is everything in favour of the supposition that it was a regularly cultivated crop long anterior to the Aryan invasion of India. It may thus, at an early date, have had assigned to it the Sanskrit names from which some of the vernacular names for the plant are clearly derived. This conclusion would considerably enhance the antiquity of its cultivation in India.

CULTIVA-
TION.
MADRAS.
793

CULTIVATION.

Madras.—In the Trichinopoly Manual **Dolichos Lablab** is said to be a six months crop : sown in July and August, reaped in February and March. In a report, furnished for the present work, the Collector (**Mr. H. Willock**) says the area of cultivation is 3,934 acres. The annual outturn per acre amounts in value to R10, the cost of production being R5. He adds that it is cultivated on all soils along with the staple food-grains. **Mr. H. Goodrich**, Collector of Belláry, writes that the area in his district under this crop is only 350 acres. It is sown from June to August and reaped from October to December. It is usually sown with other pulses in the proportion of 1 to 5. The cost of cultivation and profit cannot therefore be

properly estimated. Black, red, and mixed soils are all adapted for its cul-   CULTIVA-
tivation. The fields should be ploughed and harrowed, and the seeds sown   TION.
with a drill along with other pulses and *cholum*. This grain, he adds, is
eaten by the lower classes in place of *dhál*, and also made into a stew.
The Collector of Cuddapah (Mr. H. Sewell) says he is familiar with three
kinds of this pulse: white, red, and black. The season of sowing and
harvesting, and the cost of production is the same as that of horse-gram
(see above—Dolichos uniflorus). It is also largely grown in Coimbatore
and Salem, and of Acrot it is said to be "sown along with lamp-oil seed."
In the Manual of the Tanjore District repeated mention is made of
various forms of Dolichos. "*Avarei*, Lablab vulgaris," is reaped in December and January; is "cultivated in gardens and supported on poles, often
forming arbours about the doors of native houses. The green pods alone
are cooked, the tender ones being preferred." "*Válavarangáy*, Dolichos
cultratus, is sown and reaped at the same seasons as the above, and is said
to be cultivated "solely for its flat, oblong legumes, which are used in
curries." "*Moccei*, Lablab vulgaris, and *Kárámani*, Dolichos sinensis,
sown in July and August and reaped in January and February on "unirrigated land; often grown as auxiliary crops along with a shorter dry
crop, such as *rági* or *cholum*; more common in the delta." It seems probable that the two last mentioned plants are Vigna Catiang. In the
Madras Manual of Administration (*II., 289*) it is said that Dolichos Lablab is "chiefly used for feeding bullocks."

Mysore and Coorg—In the *Gazetteer* of these provinces repeated refer-   MYSORE AND
ence occurs to this pulse, but definite information is not furnished as to   COORG.
season, soils, method of cultivation, &c.   794

Bombay.—Lisboa (*Useful Plants of Bombay, p. 153*) says: "It is exten-   BOMBAY.
sively cultivated all over India, especially during the cold season, on the   795
sloping lands along the banks of rivers. The seeds are much relished;
they are boiled and eaten." Turning to the Gazetteers and Agricultural
Department Reports for more explicit information as to the cultivation of
this pulse in Bombay, it is said of Thana District that "*vál*, Dolichos
Lablab, an important crop, is, like *udid*, sown in the standing rice in small
holes made between the plants, two seeds being dropped into each hole.
The beans are used as a vegetable and the stalks as fodder for cattle. Of
Káthiáwár it is reported: "The large fruited kidney, *vál*, Dolichos Lablab,
is a crop of small importance, found in the Nagher on the south coast.
There is only one kind of *vál*, which grows in sandy soil, and is sown in the
beginning of the rains and reaped in the middle of the cold weather. The
soil requires ploughing, manuring, and weeding. It is locally used as
human food." Since compiling the above Mr. Muir-Mackenzie has kindly
furnished the following note regarding this pulse in the western Presidency:
"The plant frequently follows rice in the South Marhatta country as a
second crop, and is reaped in February and March, and is sown as a
second crop with the *Kharif* millets (*Bajra*). It is also a favourite crop
in river beds, and is much grown on irrigated plots as a late extra or
catch crop."

Panjáb.—The notices regarding this pulse are so brief that the refer-   PANJAB.
ences already made under the paragraph of history (above) may be   796
accepted as conveying all that is known.

Central Provinces.—A note obtained on this subject from Mr. J. B.   CENTRAL
Fuller conveys the generally accepted opinion that it is a crop of the   PROVINCES.
home-steads—"grown during the rains in the small enclosures which sur-   797
round the village houses."

North-West Provinces.—Messrs. Duthie and Fuller (*Field and*   N.-W.
Garden Crops) give a brief account of this pulse. They say "there are   PROVINCES.
                                                                          798

CULTIVATION· several varieties of this climbing bean, one of the more distinct being that named **D. purpureus,** a separate figure of which is given in *Plate XXXIV. B.* " "Their chief distinguishing characters have reference to the colour of the flowers, the shape and colour of the pods, and the colour' of the seeds." " In these provinces," these authors continue, " *sém* is commonly grown along the borders of tall crops, and allowed to twine itself round the plants standing on the margin. The castor oil plant is a favourite support. It is also occasionally grown in little patches round houses, and allowed to trail over the walls and roof. It is never grown as a field crop by itself, since it would require an artificial support which would add too much to the cost. It is used as a vegetable, its long pods, picked in unripe condition, forming a favourite addition to the daily mess of green food. It is seldom if ever grown for its grain." The reference to its climbing on the castor oil plant may be accepted as showing that it is a *rábi* or spring crop. Mr. Atkinson says of Kumáon that there are six varieties commonly cultivated in gardens and very occasionally as a field crop.

BENGAL. *Bengal.*—The same remark as given under the Central, is applicable to
799 the Lower, Provinces. It is not a regular agricultural crop, though few huts exist without at least one plant trailing over the enclosure. The Director of Land Records and Agriculture says : " Different varieties of *shim* or *lablab,* distinguished from one another by colour, size, form, nature of stripes, &c., of the pod, are cultivated all over Bengal as a garden vegetable. A grass-coloured, small variety, of very indifferent flavour, is found wild in the jungle of Madhupur." The Rev. A. Campbell (a most painstaking observer) has furnished the writer with a complete set of all the cultivated and wild plants of a large portion of Chutia Nagpúr. Of this pulse he says it is largely cultivated, the legumes being eaten, but he does not appear to have found the plant wild. Roxburgh, however, in his *Flora Indica,* says : " Of this species there are known to me five arieties vin a cultivated state, and two wild." Of the two wild plants he calls the one *Ban-shim,* in Bengali, and *Adaví-chikurkai* in Telegu. This he describes as "smooth in every part, and frequently biennial, if not perennial. It is never cultivated, nor any part of it used." Of his other wild form he says: " It is found with the former wild in the hedges, &c., near Samulkota, and differs from it in being very downy; both have red flowers and dark grey mottled seeds. No part of these two varieties is made any use of." Under **Dolichos lignosus,** *Willd.,* he describes some six other cultivated beans. These by the *Flora of British India* have all been reduced to **D. Lablab,** so that, according to Roxburgh, there are some thirteen forms of the plant. Of his cultivated forms under **D. Lablab,** Roxburgh accepts that known as *Annapa* in Telegu as the most typical. He writes of it : "The whole of this plant has a heavy disagreeable smell, something like the green bug. It is much cultivated in the fields during the cold season, and delights in a rich black soil, which cannot be flooded by rains. Like *Bobra,* it requires three months from the sowing, till ripe; yields in a good soil, about forty-fold. These seeds bear a low price, compared to most other sorts of grain. They are much eaten by the poorer classes, particularly when rice is dear. They are not palatable, but reckoned wholesome substantial food. Cattle are also fed with the seeds, and they are remarkably fond of the straw. It is said to make cows yield much milk." Of the other forms of **D. Lablab,** Roxburgh seems to convey the idea that they are garden products and not field crops. Under **D. lignosus** he writes : " I include under the above definition many varieties, some of them hitherto deemed distinct species. All are cultivated during the cold season in the gardens and about the doors of the natives, forming not only cool, shady arbours, but furnishing them with an excellent

pulse for their curries, &c., in the tender legumes. In short, these and the four last mentioned cultivated varieties of *Lablab* may be called the *Kidney Beans of the Asiatics.*"

Assam.—The Director of Land Records and Agriculture furnishes the following note regarding "**Dolichos Lablab,** *urohi*": "The *urshi*, or *uri*," he says, "is a creeper producing beans, and is grown in almost every village. It is nowhere grown as a field crop but is grown on lands adjoining homesteads, which are called *chara* lands. The greater portion is grown for home consumption, and a very small part only finds its way to the markets for sale. There are five kinds of *Urohi*—(1) *Kamtal*, (2) *Dorika*, (3) *Rojala* (purple), (4) *Ranga* (red), and (5) *Boga* (white). Of these the first produces the biggest beans, about 10 inches long and 1½ inches broad, and the fifth kind produces the smallest beans, about 3 inches long and 1 inch broad. The bean of the *dorika urohi* is not flat like the four other kinds but round, about 4 to 5 inches in length and about 1½ inches in diameter. The third, fourth, and fifth kinds have obtained their names on account of the colour of the beans they produce. These creepers are grown only in vegetable gardens in *basti* lands. The seeds are sown in August close to a hedge or large tree. The crop is gathered from November to January, and the plants die in the hot weather. The natives eat the beans either boiled or fried, or use them in curry with fish. About 45 seers of pods are obtained from each plant a year, and the average price is six pice a seer. Not unfrequently the seeds are dried and kept. In this state they last long and are eaten after being ground and cooked like pulses, also in curries. Cattle are never fed on them. The beans have some medicinal properties. The juice is mixed with salt and applied in inflammation of the ear and throat, &c., due to cold. The roots are used for poisoning wild animals."

It may be pointed out that the round podded plant described above, according to the botanical definition, cannot be a form of **Dolichos Lablab,** but is more probably **Vigna Catiang,** and a specimen of *Urohi* sent from Assam to the writer some years ago proved to be **Vigna Catiang.** The pod of **Dolichos Lablab,** as described in the *Flora of British India,* is said to be "flat, linear or oblong recurved, 2—4 seeded and 1½ to 2 inches long, by ⅓ to ¾ inches broad, tipped with the hooked persistent base of the style." The possibility of a mistake may be accepted as a justification for doubting the propriety of dealing with these plants collectively, the more so since Assam, by the above report, would stand by itself in the record of periods of sowing and reaping. The writer may add, however, that he is personally acquainted with **Dolichos Lablab,** as met with in Assam, and he collected a sample of it even in the Naga hills, there known as *kechu*. This latter fact is of very considerable interest, since, till recently, these mountain tracts have been completely closed to visitors from the lower neighbouring tracts. The names given to the plant by the Angami Nagas and Assamese are, therefore, in all probability purely indigenous, and would point to a probable independent cultivation, from the wild stock of the plant, by the hill tribes on the eastern side of India, that is to say, independent of the cultivation in the southern and central table land of India.

Burma.— Mason, in his *Burma and Its People (pp. 466,768),* says : " The Burmese and Karens grow several varieties of one or two species of *lablab,* which occupy the place of kidney beans in Europe." Reverting to this, on a further page, he speaks of "wild **Dolichos,** the *tau bai* : this is **Vigna pilosa** of modern botanists." In a recent official communication on the subject of **Dolichos Lablab** cultivation in Burma, it is stated that "in the Kyaukpyu District it is sown in the latter part of the rains, and bears in the cold weather. It is grown on well-raised manured soil, and

DOLOMÆA macrocephala.	**The Sim or Asiatic Bean.**

when about a foot high it is allowed to twine round bamboo trellis work."

AREA OF CULTIVATION.

It is difficult, if not impossible, to discover the area under a crop which, like the present, exists as a garden climber, each peasant having one or two plants. It is grown all over India, becoming less abundant towards the north than in the southern and western divisions of the country. In Madras and Bombay, however, it is to some extent a field crop. In Madras in 1885-86, there were stated to have been 65,664 acres under the crop, in 1886-87, 78,700 acres, and in 1887-88, 35,724 acres. In Bombay the area appears to be greater. In 1885-86, 72,660 acres, in 1886-87, 91,652, and in 1887-88, 95,188 acres.

The Madras returns for 1887-88 may, however, be incorrect, since ambiguity often exists through the figures of area appearing under different names, such as " beans," " *avari*," " *mochai*," or " *anumulu*," &c.

Chemical Composition.—In his *Food Grains of India*, Professor Church publishes the results of five analyses of this pulse. He accepts the fourth as fully expressing the character of the grain. The following abstract from the Professor's table of analysis may be here given :—

CHEMISTRY.
803

Composition of Lablab Beans.

	In 100 parts.		In 1℔	
	Husked.	With husk.	Oz.	Grs.
	(3)	(4)		
Water	12·1	12·1	1	410
Albuminoids	24·4	22·4	3	255
Starch	57·8	54·2	8	294
Oil	1·5	1·4	0	98
Fibre	1·2	6·5	1	17
Ash	3·0	3·4	0	238

From these figures the Professor concludes that the " nutrient ratio deduced from analysis (4), is 1 : 2·5, the nutrient value is 80. It will be seen, however, on comparing the several analyses given above, that the percentage of albuminoids is rather variable. The extreme range is probably not more than 6 per cent. Of the numerous forms of *Lablab* the majority are eaten as a green vegetable." The concluding sentence is of importance, since, to judge of the value of this plant as a source of human food, the green pods would have to be analysed.

FOOD and
FODDER.
Green Pods.
804
Ripe seeds.
805
Stems.
806

Food and Fodder.—The extensive series of quotations from numerous writers given above will, it is believed, have conveyed the main facts regarding the GREEN PODS as a vegetable, the RIPE SEEDS as a pulse eater by certain classes or employed as cattle food, and of the STEMS as constituting a valued fodder. It is only necessary to repeat these points here in order to establish, in their proper places, the numbers to be assigned to these products.

MEDICINE.
807

Medicine.—The only record of this plant being used for medicinal purposes is that published above in the paragraph of cultivation in Assam.

DOMESTIC.
Roots.
808

Domestic Uses.—The ROOT are said to be used in Assam to poison wild animals. This is a remarkable fact, since the whole plant has hitherto been supposed to be wholesome.

Dolichos sinensis, *Linn.;* see **Vigna Catiang,** *Endl.;* LEGUMINOSÆ.

D. uniflorus, *Lam.;* see **Dolichos biflorus,** *Linn.*

Dolomæa macrocephala, *DC.;* see **Jurinea macrocephala,** *Benth.;* COMPOSITÆ.

Products of India. 191

Domestic and Sacred Products. (*G. Watt.*) DORONICUM Hookeri.

DOMESTIC AND SACRED.

809

Under this heading the reader may have observed, in each article (*e.g.*, Bambusa) a paragraph describing the minor economic objects that could not be treated of as Gums, Dyes, Tans, Fibres, Oils, Medicines, Foods, or Timbers. It is proposed to give, in the Appendix to this work, a collective article for each of these classes of products. The intention is that these collective articles should not serve as keys only to the descriptive accounts (distributed throughout the Dictionary) but prove useful, if possible, in arranging specimens in Museums. Many of the articles dealt with, under the paragraph heading "Domestic and Sacred," have already, to some extent, however, been summarised in the bulk of the work. Thus, for example, lists of timbers suitable for definite purposes have been given in the alphabetical positions of their uses (*e.g.*, Cabinet-work). The article "Beads" enumerates all the animal, vegetable, and mineral substances used *as* beads; and the article "Detergents" gives the materials employed in place of soap, in washing garments, or cleansing the hair and the teeth. But the detailed article on Domestic and Sacred Products would afford the key by which these special lists might be discovered, and at the same time it would indicate the writer's ideas of classification—ideas which have of necessity influenced him in dealing with the minor economic articles which, in the absence of a better title he has designated Domestic and Sacred Products.

DOREMA, *Don.*; *Gen. Pl., I., 918.*

RESIN.
810

Dorema Ammoniacum, *Don.*; UMBELLIFERÆ.

The Eastern Giant Fennel (a native of Persia) is supposed to afford at least some portion of the Gum-resin *Ammoniacum* (the *Ushak* in Persian and Arabic: the *Kandal* in Bokhara), which is largely imported into India. The plant is said by Aitchison to occur in the Harirud valley. He writes of it: "No sooner is the fruit well formed and beginning to ripen than the plant is attacked by some boring insect which causes the milky juice to escape. This dries into hard blocks, frequently enclosing the fruit. The *Kandal, Ushak* or *Ammoniacum* is usually collected from the stem and fruitescence, and often encloses clusters of the fruit."

Dorema Ammoniacum is alluded to by many writers on Indian Economic Products, among whom the following may be mentioned:—*Stewart, Panjáb Pl., 106; R. H. Irvine, Mat. Med. Patna, pp. 80, 84; Dymock, Mat. Med. West India, 2nd Ed., p. 392; Atkinson, Gums and Gum-resins, p. 28; Report on the Gums and Resins of India issued by the P. W. D., 13, 26, 60; Indian Forester, XIII., 91, 93; XIV., 369; Watt's Cat. Econ. Prod. shown at the Calcutta Exhib., Parts I., No. 124; IV., No. 126; V., No. 472.*

DORONICUM, *Linn.*; *Gen. Pl., II., 440.*

Doronicum Falconeri, *Clarke*; *Fl. Br. Ind., III., 333*; COMPOSITÆ.

811

Habitat.—A stout herb 1-1½ feet high and nearly leafless above; found in Kashmír, altitude 13,000, and in Western Tibet 14,000 feet.

D. Hookeri, *Clarke*; *Fl. Br. Ind., III., 332.*

812

Syn.—D. SCORPIOIDES, *Clarke, Compositæ Ind., 169 in part.*

Habitat.—A robust herb 1-2 feet high; found in Sikkim (Lachím and Tungu), altitude 12,000 to 14,000 feet.

DRACOCEPHALUM
moldavicum.   **The Akrabi.**

---

813 | **Doronicum Roylei,** *DC.; Fl. Br. Ind., III., 332.*

Syn.—FULLAROMIA KUMAONENSIS, *DC.*
Vern.—*Darúnaj-akrabí,* PB.; *Darúnaj-i-akrabí,* PERS.
References.—*Dymock, Mat. Med. W. Ind., 2nd Ed., 442; S. Arjun, Bomb. Drugs, 77; Year Book Pharm., 1880, 248; Med. Top. Ajmir, 133; Baden Powell, Pb. Pr., 357; Atkinson, Him. Dist., 312.*

Habitat.—A herbaceous erect plant, 2 to 4 feet high; found on the Western Himálaya from Kashmír to Garhwál, altitude 10,000 feet.

This species is closely allied to the European plant D. **Pardalianches,** *L.*

MEDICINE.
Root.
814

Medicine.—The ROOT is an aromatic tonic, said to be used to prevent giddiness on ascending heights (*Baden Powell*). **Dymock** gives an account of the European drug, D. **Pardalianches,** *Linn.,* and states that there appears to be a demand for it, since it is kept by all the Muhammadan drug-sellers in Bombay. It is described by the author of the *Makhzan-el-Adwiya* as a scorpioid knotted root with greyish exterior and white interior; hard, faintly bitter, and aromatic. Is said to be found in Andulasia and the mountainous parts of Syria, especially about Mount Yabrúrat, where it is known by the name of *Akrabí.* With regard to its medicinal properties, he says that it is a resolvent of phlegm, adust bile, and flatulencies, cardiacal and tonic, useful in nervous depression, melancholy, and impaired digestion, also in pain of the womb, and flatulent dyspepsia.

Dr. Dymock, from whom the facts given above have been compiled, in his Materia Medica, adds: "Besides this it is prescribed for persons who have been bitten by scorpions and other venomous reptiles, and is hung up in houses to keep away the plague; pregnant women wear it round the waist suspended by a silken thread, which must be made by the wearer; it is supposed to act as a charm, protecting the fœtus and procuring a painless delivery. Hung up over the bed it prevents night terrors and ensures pleasant dreams." **Dr. Dymock,** under the heading Chemical Composition, discusses the properties of *Inulin,* the starch equivalent present in the COMPOSITÆ, but gives no special properties to the roots of this plant. It would appear from the virtues attributed to the drug that its reputation depends more on the theory of signatures than to any ascertained properties. Should a greater demand arise for it, it is probable that either of the Indian forms mentioned above might be substituted for the imported root.

DRACOCEPHALUM, *Linn.; Gen. Pl., II., 1199.*

[LABIATÆ.

815 | **Dracocephalum heterophyllum,** *Benth.; Fl. Br. Ind., IV., 665;*

Vern.—*Zanda, shanku, karamm,* N. PB. & LADAK.
Reference.—*Stewart, Pb. Pl., 168.*

FOOD AND
FODDER.
Plant
and Root.
816

Habitat.—A brittle herb with obtusely angled branches. Found in the Panjáb Himálaya and Ladak from 13,000 to 17,000 feet.

Food and Fodder.—The PLANT is browsed by goats and sheep, and its ROOT appears to be used as a vegetable (*Stewart*).

817 | **D. moldavicum,** *Linn.; Fl. Br. Ind., IV., 665.*

Vern.—*Túkhm-ferunjmishk,* HIND.

Habitat.—A glabrous small herb found in the western temperate Himálaya and Kashmír at altitudes of 7,000 to 8,000 feet.

MEDICINE.
Seeds.
818

Medicine.—Irvine (*Mat. Med., Patna, p. 125*) says the "SEEDS are used ground up in fevers and as demulcent: dose ʒii to ʒʒ in infusion."

D. 818

| Dragon's Blood. (G. Watt.) | DREGEA volubilis. |

Dracocephalum Royleanum, *Wall.*; see **Lallemantia Royleana**, *Bth.*;
[LABIATÆ.

DRACÆNA, *Linn.*; *Gen. Pl., III.*, 779.

819

A genus of trees or shrubs belonging to the Natural Order LILIACEÆ. Very little of an economic interest has been recorded regarding the Indian species. Kurz describes eight species as met with in Burma, **D. angustifolia,** *Roxb.*, being there known as *Kwam-lin-nek* (or *kunlinnet*). Roxburgh gives nine species, of which two are natives of Sylhet, *viz.*, **D. ternifolia,** *Roxb.*, the *bun-amtol*, and **D. atropurpurea,** *Roxb.*, the *láll-bun-amtol.* Many Indian writers allude to the species of this genus, more especially the ornamental garden forms now so extensively grown. Baker (*Linnæan Soc. Jour., XIV.,* 525—538) describes 38 species met with in the world, of which only four are natives of India, with one or two forms reduced to varieties, which were formerly treated as separate species. The Indian species are:—

1. **Dracæna angustifolia,** *Roxb.*—A native of the lower Himálaya, ascending to 6,000 feet and distributed to the Khásia Hills, Assam, Sylhet, Burma, &c.
2. **D. atropurpurea,** *Roxb.*—A native of Sylhet, the Khásia Hills, and Chittagong, ascending to 3,000 feet. This has three varieties.
3. **D. elliptica,** *Thunb.*—Met with in Sylhet and the Andaman Islands.
4. **D. spicata,** *Roxb.*—A native of the Himálaya, ascending to 3,000 feet, but distributed to Bombay, the Nilgiri Hills, and Andaman Islands.

The only known economic product obtained from **Dracæna** is the DRAGON'S BLOOD said to be obtained from **D. Draco,** also from **D. schizantha** and **D. Cinnabari.**

See **Calamus Draco,** Vol. II., Nos. 69—73, pp. 17 to 19.

Dragon's
Blood.
820

DRACONTIUM, *Linn.*; *Gen. Pl., III.,* 995.

Dracontium polyphyllum, *Linn.*; *Engler, in DC., Mon. Phaner., Vol. II., 283;* AROIDEÆ.

821

Vern.—*Sévalá,* BOMB.; *Jangli suran,* GUZ.; *Caat karnay kaloung,* TAM.; *Adivie kunda gudda,* TEL.; *Kanana canda,* SANS.

Habitat.—Met with on the Malabar Hills, Bombay, and the Concans. The writer is disposed to regard this as a mistake, some other plant being meant, since **D. polyphyllum** is not a native of India, though frequently met with under cultivation.

Medicine.—The ROOT is large, rugged, and irregular, and supposed to possess antispasmodic virtues and to be a remedy in asthma. It is also used in hœmorrhoids. According to Thunberg, it is highly esteemed in Japan as a powerful emmenagogue, and sometimes used to procure abortion (*Ainslie*).

Special Opinion.—§ " Good medicine for chronic diarrhœa " (*V. Ummegudien, Mettapollian, Madras*).

MEDICINE.
Root.
822

Dragon's Blood, see **Calamus Draco** and **Dracæna** above.

DREGEA, *Meyer; Gen. Pl., II., 775.*

[ASCLEPIADEÆ.

Dregea volubilis, *Benth.; Wight, Ic., t. 586; Fl. Br. Ind., IV., 46;*

823

Syn.—HOVA VIRIDIFLORA, *R. Br.*; ASCLEPIAS VOLUBILIS, *Linn. f.*
Vern.—*Nak-chhikni,* HIND.; *Tit-kunga, tita-kunga,* BENG.; *Marang kongat,* SANTAL; *Dodhi,* BOMB.; *Hirandodi, harandori, khandodi,* MAR.; *Kodic-palay, cúringi-kirai,* TAM.; *Dúdi-palla,* TEL.; *Gway tankpin,* BURM.; *Kiri-anguna,* SING.; *Madhu malati* (according to Ainslie), SANS.

o

References.—*Roxb., Fl. Ind., Ed. C.B.C., 253; Thwaites, En. Ceylon Pl., 199; Dals. & Gibs., Bomb. Fl., 153; Campbell's Econ. Prod., Chhutia Nagpur, No. 9250; Grah., Cat. Bomb. Pl., 119; Griff., Ic. Pl. Asiat., t. 387, 388; Pharm. Ind., 143; Ainslie, Mat. Ind., II., 154; O'Shaughnessy, Beng. Dispens., 454; Moodeen Sheriff, Supp. Pharm. Ind., 155; Dymock, Mat. Med. W. Ind., 2nd Ed., 524; S. Arjun, Bomb. Drugs, 201; Irvine, Mat. Med., Patna, 74; Lisboa, U. Pl. Bomb., 201, 233; Royle, Fib. Pl., 306; Home Dept. Cor. regarding Pharm. of Ind., 239; Indian Forester, III., 237.*

Habitat.—A stout, tall, climbing shrub of Bengal, Assam, the Deccan Peninsula, from the Concan southward to Ceylon.

FIBRE.
824
Rope.
825

Fibre.—Contains an exceedingly strong FIBRE, which is extracted by the natives. The Rev. A. Campbell says that in Chutia Nagpur the Brahmans sometimes make their *poita* or sacred threads from this plant. Lisboa says that in Bombay the creeper is used as a substitute for ROPE to tie up bundles of firewood.

MEDICINE.
Leaves.
826

Roots.
827

Stalks.
828

Medicine.—The LEAVES are much employed as an application to boils and abscesses. The ROOTS and tender STALKS are considered emetic and expectorant. Ainslie tells us that the *Vytians* suppose the root and tender stalks to possess virtues in dropsical cases; "they sicken, and excite expectoration; though I could not obtain much information of a certain nature respecting them, it is to be presumed that they operate in a manner somewhat similar to the root of **Asclepias Curassavica;** which, according to Browne, in his *Natural History of Jamaica*, the Negroes use as a vomit." The *Pharmacopœia of India*, after alluding to the value of the leaves as an external application, adds: "According to native testimony, it has the same emetic and expectorant virtues as **Dæmia extensa.** Irvine (*Mat. Med., Patna*) says this drug is used in colds and eye diseases to cause sneezing; dose gr. i to ½ drachm. Dr. Dymock repeats the above information, but adds that all parts of the FOLLICLES are intensely bitter, and that the brown MEALY SUBSTANCE that covers them is given in Bombay to cattle as a medicine.

Follicles.
829
Mealy
Substance.
830
Juice.
831

Special Opinions.—§ "The tender end of the creeper with its JUICE when touched into the nose causes excessive sneezing. This remedy is commonly used by Hindús to make sick people sneeze" (*V. Ummegudien, Mettapollian, Madras*).

FOOD.
Leaves.
832

Food.—Ainslie, while alluding to the report that the LEAVES are eaten as a green vegetable, doubts the accuracy of this opinion, because of their nauseate reputation. Many subsequent writers, however, affirm that they are regularly eaten. Thus Thwaites says, they are eaten in Ceylon, and Lisboa says of Bombay, the "leaves are used as a vegetable."

833

DREPANOCARPUS, *Mey.; Gen. Pl., I., 546.*

According to the *Genera Plantarum* there are only eight species belonging to this genus, and these are all American. The chief characters, as established by the *Genera Plantarum*, in the separation of this genus from **Dalbergia,** are the versatile anthers, and lunate to reniform pod. These characters, according to **Kurz,** are possessed by three Burmese trees, *viz.,* **Drepanocarpus Cumingii, D. monospermus, D. reniformis,** and **D. spinosus.** Following the usual course pursued in this work, however, of accepting the synonymy of the *Flora of British India*, these have been dealt with under **Dalbergia,** which see.

834

DRIMYCARPUS, *Hook. f.; Gen. Pl., I., 424.*

Drimycarpus racemosus, *Hook. f.; Fl. Br. Ind., II., 36;* ANA-
[CARDIACEÆ.

	DROSERA peltata.

Drosera—Insectivorous herbs. (*G. Watt.*)

Syn.—HOLIGARNA RACEMOSA, *Roxb., Fl. Ind., II., 82.*
Vern.—*Telsur,* BENG. ; *Amdali,* ASSAM ; *Amjour,* SYLHET ; *Kagi,* NEPAL ; *Brong-kúng,* LEPCHA ; *Chengane, sangaipru, sangryn,* MAGH.
References.—*Kurs, For. Fl. Burm., I., 314; Gamble, Man. Timb., 112; Cat. Trees, Shrubs, &c., of Darjeeling, 25.*
Habitat.—A large evergreen tree of the Eastern Himálaya, from 2,000 to 6,000 feet, the Khásia Hills and Sylhet to Chittagong and Pegu.
Structure of the Wood.—Greyish-yellow, hard, close-grained. Used occasionally in Assam for canoes and planking ; in Chittagong for boats, for which it is one of the woods most employed. Major Lewin says that boats 50 feet long and 9 feet in girth are sometimes cut out of logs of this wood.

TIMBER. 835

DROSERA, *Linn.; Gen. Pl., I., 662.*

There are three species of this genus of small annual insectivorous herbs found in India, of which **Drosera Burmanni,** *Vahl.* (found throughout the plains and ascending the hills to 4,000 feet), is the most abundant and resembles closest the European Sun Dew. **D. indica,** *Linn.,* is a very minute species with obovate leaves, met with on Parisnath in Chutia Nagpur, and distributed southwards through the Deccan to Burma and Ceylon, while **D. peltata** is a tall species with peltate leaves arranged along an erect stem. It is found on the Himálaya from 4,000 to 10,000 feet, and also in the Nilgiri Hills.

It seems probable that what little economic information exists, regarding these plants, is fairly applicable to any one or to all the species. Writers on **Drosera** generally allude to **D. peltata** however, but it is, perhaps, safe to relegate the statements made regarding the Gangetic plains to the first species alone and regarding the Himálaya to the last.

Drosera Burmanni, *Vahl.; Fl. Br. Ind., II., 424;* DROSERACEÆ.

836

Vern.—*Mukha-jali,* HIND.
References.—*Stewart, Pb. Pl., 20; Kanara Gazetteer (XV., I.), 433; Indian Forester, II, 24; VIII., 405; Mason's Burma and Its People, 436, 749; Atkinson, Him. Dist., 310, 735; Drury, U. Pl., 118.*
Habitat.—Found throughout India ; plentiful in the Gangetic plains, appearing on the paddy fields in the cold season. It is everywhere seen in Chutia Nagpúr and Orissa, and is common in fields around Burdwan, although not met with in the vicinity of Calcutta. From Behar it passes through the Central Provinces to the Deccan, is very common in Kánara, and extends south to the Madras Presidency, appearing on the lower hills and also in Burma. It prefers a sandy, open soil.

D. peltata, *Sm.; Fl. Br. Ind., II., 424.*

837

Vern.—*Chitra,* PB.
References.—See above.
Habitat.—There are two forms of this plant, the type being found in Moulmein. The form known as **lunata** occurs throughout the Himálaya and on the Nilgiri Hills. It is nowhere, however, met with on the plains.
Dye.—Drury suggests that a dye may be prepared either from **D. Burmanni** or **D. peltata,** as Royle mentions the fact of the paper which contained his dried specimens being saturated with a red tinge.

DYE. 838

Medicine.—It seems probable that both the above species are referred to under the vernacular name of *Mukha-jali.* The LEAVES of this curious and insectivorous plant, bruised and mixed with salt, are used as a blister in Kumáon. This same practice prevails, however, in Kanáwar without the use of salt. All the members of this family have a bitter, acrid, and caustic flavour. If placed in milk they rapidly curdle it.

MEDICINE. Leaves. 839

Fodder.—Cattle will not touch any species of **Drosera.**

FODDER. 840

O 2

Drugs, see Medicines.

DRYOBALANOPS, *Gærtn.; Gen. Pl., I., 191.*

841 **Dryobalanops Camphora,** *Coleb.;* DIPTEROCARPEÆ.
BARUS CAMPHOR.
See Vol. II., No. 259, pp. 84—93.

DUABANGA, *Ham.; Gen. Pl., I., 783.*
[LYTHRACEÆ

84 **Duabanga sonneratioides,** *Ham.; Fl. Br. Ind., II., 579;*

Syn.—LAGERSTRŒMIA GRANDIFLORA, *Roxb.*
Vern.—*Bandorhulla,* BENG.; *Baichuo,* CHITTAGONG, SANTAL; *Kochan, kokan,* ASSAM; *Bondorkella, achúng, bolchim,* GARO; *Jarúl-jhalna,* CACHAR; *Lampatia,* NEPAL; *Dúr,* LEPCHA; *Baichua,* MAGH.; *Myouk-gnau, myan kngo,* BURM.
References.—*Roxb., Fl. Ind., Ed. C.B.C., 404; Kurz, For. Fl. Burm., I. 525; Gamble, Man. Timb., 204; Cat. Trees, Shrubs, &c., Darjeeling, 42; Indian Forester, I., 88, 99; IV., 345; VII., 101; IX., 377; XI., 255, 315; XII., 286, 453.*

Habitat.—A lofty, deciduous tree, with light-brown bark, peeling off in thin flakes; a native of Nepál and Eastern Bengal (ascending to 3,000 feet), Assam, Chittagong, and Burma.

TIMBER. Structure of the Wood.—Grey, often streaked with yellow, soft, seasons
843 well, takes a good polish, and neither warps nor splits. Weight 30℔ per cubic foot. Canoes cut out of it green are at once used, even when liable alternately to wet and the heat of the sun. In Northern Bengal and
Tea-boxes. Assam it is now very extensively used for tea-boxes, for which purpose it is
844 admirably fitted. It is also made into cattle troughs and other ordinary
Cattle. domestic utensils. It came into use for tea-boxes in 1874-75 when *Toon*
troughs. wood became scarce. The seeds are small but germinate freely, so that
845 for planters this is one of the most useful of trees.

846 ## DUCKS, TEALS, GEESE, AND SWANS.

The large and very important assemblage of Indian birds which may be accepted as represented by the Duck, the Goose, and the Swan, constitutes one of the best marked sections of the Order Natatores of Zoologists. They are characterised by a more or less perfect state of web-foot, by having short, compressed tarsi and a flattened bill. In the Goose and the Swan the bill is pointed, has a sharp nail-like hook on the tip, and ascends towards the base. In the Ducks and Teal the bill is nearly of one breadth throughout and quite flat, with well-developed lateral laminations, which are employed in sifting the water in the search for food.

The following are the chief edible birds of the above assemblage, met with in India:—

847 1. **Anas boscas**—The Mallard.

This is universally regarded as the best Indian Duck for the table, being followed in point of merit by the Pintail, and after that the Gadwall. The Mallard is a, comparatively speaking, common species, though less so on the western side of the continent.

848 2. **A. caryophyllacea.**—The Pink-headed Duck.
849 3. **A. pæcilorhyncha.**—The Indian spotted-bill Duck.
850 4. **Anser albifrons.**—The White-fronted or Laughing Goose.

D. 850

Ducks, Teal, Geese, and Swans. (*G. Watt.*)	DUGONG oil.

5. **A. cinereus.**—The Grey Goose or Lag. | 851
6. **A. indicus.**—The Barred-headed Goose. | 852
7. **Casarca rutila.**—The Ruddy Sheldrake or Brahmani Duck. | 853
8. **Chaulelasmus strepera.**—The Gadwall (see note above under No. 1). | 854
9. **C. angustirostris.**—The Marbled Teal. | 855
10. **Clangula glaucion.**—The Golden-eye or Garrot. | 856
11. **Cygnus olor.**—The White or Mute Swan. |
12. **Dafila acuta.**—The Pintail (see note above under No. 1). | 857
13. **Dendrocygna fulva.**—The Large Whistling Teal. | 858
14. **D. javanica.**—The Whistling Teal or Duck. | 859
15. **Fuligula cristata.**—The Tufted Pochard. | 860
16. **F. marila.**—The Scaup Pochard. | 861
17. **F. myroca.**—The White-eyed Pochard or Ferruginous Duck. | 862
18. **F. rufina.**—The Red-crested Pochard. | 863
19. **Mareca penelope.**—The Wigeon. | 864
20. **Mergellus albellus.**—The Smew. |
21. **Mergus castor.**—The Gossander. | 865
22. **M. serrator.**—The Red-breasted Mergauser. | 866
23. **Querquedula circia.**—The Garganey or Blue-winged Teal. | 867
24. **Q. crecca.**—The common Indian Teal. | 868

This is universally eaten and one of the commonest birds offered for sale in the market places of large towns. | 869, 870

25. **Q. formosa.**—The Clucking Teal. |
26. **Sarkidiornis melanonotus.**—The Comb Duck. | 871
27. **Spatula clypeata.**—The Shoveller. | 872
28. **Tadorna cornuta.**—The Shell-drake or Burrow Duck. | 873

Though all of the above birds may be eaten, at most only three or four | 874
can be said to be regular articles of trade. Indeed, after the domesticated duck, the common teal is perhaps the most important. Their feathers are not articles of trade (see **Feathers** on a further page).

In the Gazetteers of India frequent reference occurs to the domesticated Duck and Goose and to the above wild species. The reader is referred to Hume and Marshall's *Game Birds of India* for the wild birds, and to the *Bombay Gazetteers* and other such publications for the domesticated especially: Vols. II., 41; III., 19; IV., 29; V., 36; VI., 17; VII., 45; VIII., 106; XI., 35; XII., 33; XV., Pt. I., 81; XVI., 21; XVII., 39; XXI., 68; XXII., 41. It is perhaps unnecessary to quote the volumes of the other Gazetteers and District Manuals, as the information is of a very similar character to that which will be found in the volumes cited. In some parts of the country special houses (Tealeries) are constructed for the purpose of rearing Teal, but the supply of the wild birds is mainly derived by a wholesale system of trapping. The consumption of the domesticated birds must be very great, since by some classes of the Native population, precluded from eating the barn-door fowl, there exists no injunction against the Duck.

Dugong oil, or the oil of the SEA HOG,—the YUNGAN or MOODA HOORA. | 875

There are two species, each yielding an oil highly valued in medicine and for cookery. One of the species, **Halicore indicus,** is distributed throughout the Indian Ocean, in the Gulf of Manaar, on the west coast of Ceylon, in the Straits Settlements and the Eastern Archipelago. The other species, **H. australis,** is found on the Australian coasts.

Oil.—On boiling down, each animal (weighing from 4 to 6 cwts.) yields from 6 to 14 gallons of oil. The oil has no unpleasant flavour; it is free from odour; when refined it is clear and limpid. It is largely used as a substitute for cod-liver oil (*Spons' Encyclop.*).

DYERA lasiflora.	The Durian or Civet-cat Fruit.

Dulcamara, see Solanum Dulcamara, *Linn.*; SOLANACEÆ.

Dunchi Fibre, see Sesbania aculeata, *Pers.*; LEGUMINOSÆ.

Durian, see Durio Zibethinus, *DC.*

DURIO, *Linn.*; *Gen. Pl., I., 213.*

876   **Durio Zibethinus**, *DC.*; *Fl. Br. Ind., I., 351*; MALVACEÆ.

.DURIAN, or CIVET-CAT FRUIT TREE.

Vern.—*Durian*, MALAY; *Duyin*, BURM.

References.—*Linschoten, Voyage to the East Indies in 1596, Vol. II., pp. 34, 51-53, 68 ; Burma Gazetteer, Vol. I., 429 ; Burma Gazetteer by Major Macneill, p. 230; Mason, Burma and Its People, 447 and 754; Annual Report of the Settlement of Port Blair for 1870-71, pp. 33-40 ; Kew Off. Guide to Bot. Gardens and Arboretum, 67.*

Habitat.—A large tree of the Malay Islands, wild in South Tenasserim, and cultivated as far north as Moulmein. The large flowered form, viewed by many botanists as the wild condition, is, by the *Flora of British India*, treated as a different species, under the name of **D. malaccensis**, *Planch.*

FOOD.
Fruit.
877

Food.—Produces a large FRUIT, 10 inches by 7, called the *Durian*, or civet-cat fruit, of which the cream-coloured fleshy aril or pulp enveloping the seeds, like that of the Jack-fruit, is the part eaten. It is well known and much prized, but eaten by Natives only. It has a strong odour, considered by Europeans as highly offensive, which resembles that of putrid animal matter combined with rotten onions. The fruit is, however, highly prized, even by Europeans, when once the prejudice to the smell is overcome. The Burmans regard it as extremely luscious, and it forms a considerable part of their food. The roasted SEEDS and the boiled unripe fruit are also eaten as vegetables. **John Huyghen van Linschoten's** description of this fruit might be read as if written recently instead of 300 years ago. In his time it was perhaps as extensively cultivated as at the present. The Kings of Burma used to import large supplies of the fruit; indeed, it constituted a by no means unimportant article of traffic from Lower to Upper Burma.

Seeds.
878

Vegetable.
879

"The *Dorian* is regarded with peculiar favour by the natives and also European residents in the country. **Colonel Biggs** writes thus about it : 'It is so rich and highly flavoured, that it resembles marrow rather than fruit, and is subject when ripe to speedy decomposition, when its odour becomes disagreeable, a circumstance which has made it disliked by some who have not been able to eat the fruit fresh from the tree; it is beyond question the finest fruit in the world'" (*Burma Gazetteer, written by Major Macneill*).

DYERA, *Hook. f.*; *Linn. Soc. Jour., XIX.*

880   **Dyera costulata**, *Hook. f.*; *Fl. Br. Ind., III., 644*; APOCYNACEÆ.

881   **D. lasiflora**, *Hook. f.*

Sir J. D. Hooker, in the *Linnæan Society's Journal, Vol. XIX., p. 293,* gives a brief history of these plants, while founding the new genus to which they are referred, a genus named in honour of **Mr. W. T. Thiselton Dyer,** C.M.G., Director of the Royal Botanic Gardens, Kew.

D. costulata was first collected by **Griffith** in Malacca, and has since been re-collected both in Malacca and in Sumatra. **D. lasiflora** seems confined to Singapore.

D. 881

| A useful Timber used for Canoes. (*G. Watt.*) | DYSOXYLUM procerum. |

These interesting trees have been shown to be the source of the *Gutta-jelutong* of commerce. See under **Dichopsis**—GUTTA-PERCHA.

GUTTA-PERCHA. 882

DYES AND TANS.

883

For a detailed account of the Dyes and Tans of India, see the Appendix to this work: also consult the Note under Domestic and Sacred Products above.

DYSOXYLUM, *Bl.; Gen. Pl., I., 332, 994.*

[MELIACEÆ.

Dysoxylum binectariferum, *Hook. f.; Fl. Br. Ind., I., 546;*

884

Syn.—D. MACROCARPUM, *Thwaites;* GUAREA BINECTARIFERA, *Roxb.;* G. GOTADHORA, *Buch.-Ham.*
Vern.—*Rata,* HIND. ; *Borogatodhara,* ASSAM; *Rangirata,* CACHAR; *Katongsu,* LEPCHA; *Yerindi,* BOMB.
References.—*Roxb., Fl. Ind., Ed. C.B.C., 319; Kurs, For. Fl. Burm., I., 215; Beddome, Fl. Sylv., t. 150; Gamble, Man. Timb., 71; Cat. Trees, Shrubs, &c., Darjiling, 16; Grah., Cat. Bomb. Pl., 31; Lisboa, U. Pl. Bomb., 42; Indian Forester, IX., 607.*
Habitat.—An evergreen tree of Sikkim (ascending to 2,000 feet), of Assam, the Khásia Hills, Chittagong, and the Western Ghâts.
Structure of the Wood.—Reddish-grey, rough, and close-grained, hard; weight 44℔ a cubic foot. This timber seems worthy of notice.

TIMBER. 885

D. Hamiltonii, *Heirn; Fl. Br. Ind., I., 548.*

886

Vern.—*Bolashin,* GARO; *Gendelli poma, bosuniya poma (Wall.),* ASSAM; *Bauriphal,* NEPAL.
References.—*Gamble, Man. Timb., 72; Indian Forester, III., 21; IV., 292; VIII., 29.*
Habitat.—A large, evergeen tree of the Darjeeling Terai, Assam, and Sylhet.
Structure of the Wood.—Red, hard, close-grained; weight 40℔ a cubic foot. Used in Assam for boats and planks; said not to be durable. Hamilton mentions that it is used for canoes.

TIMBER. 887
Canoes. 888

D. procerum, *Heirn; Fl. Br. Ind., I., 547.*

889

Vern.—*Dingori, govorpongyota (Wall.),* ASSAM.
References. –*Kurs, For. Fl. Burm., I., 214; Gamble, Man. Timb., 71; Indian Forester, IV., 292.*
Habitat.—An evergreen tree of Assam, the Khásia Hills, and Cachar to Pegu and Tenasserim; also met with in Sikkim and the Western Dúars.
Structure of the Wood.—Bright red, moderately hard; handsome and well deserving of more extensive notice; weight 37 to 40℔ a cubic foot. It is said by Hamilton to be used for canoes.

TIMBER. 890
Canoes. 891

ECHIUM.	(*J. F. Duthie.*) The Gaozabán.

E.

Eagle-wood, see **Aquilaria Agallocha,** *Roxb.*; Vol. I., p. 279.

Earthen-ware, Clays used for, see Vol. II., p. 364.

Earth-nut, see **Arachis hypogæa,** *Linn.*; Vol. I., p. 282.

Earths, see Soils.

Ebony, see **Diospyros Ebenum,** *Kœnig: III., p. 138.*

ECBALLIUM, *Rich.; Gen. Pl., I., 826.*

1

Ecballium Elaterium, *A. Rich.;* Cucurbitaceæ.

THE SQUIRTING CUCUMBER.

MEDICINE.
Fruit.
2

A native of South Europe. The FRUIT yields the **Elaterium** of commerce, which is a very powerful hydragogue cathartic. **Dr. Dymock** says that it does not appear to be known in Hindu medicine, but that the Arabs and Persians are well acquainted with it. The fruit is sold in Bombay under the name of *káteri-indráyan,* and is imported from Persia.

ECHINOCARPUS, *Blume; Gen. Pl., I., 239.*

3

Echinocarpus dasycarpus, *Bth.; Fl. Br. Ind., I., 400;* TILIACEÆ.

Vern.—*Gobria,* NEPAL.

References.—*Gamble, Man. Timb., 56; Ind. For., I., 95.*

Habitat.—A large tree of the Eastern Himálaya, from 5 to 7,000 feet.

TIMBER.
4
Tea-boxes.
5

Structure of the Wood.—Greyish-brown, soft; used for planking, for tea-boxes, and for making charcoal. It is in considerable demand in Darjiling (*Gamble*).

ECHIUM, *Linn.; Gen. Pl., II., 863.*

Echium sp. ? BORAGINEÆ.

MEDICINE.
Leaves.
Flowers.
6

Under the above name **Dr. Moodeen Sheriff,** in *Supp. Pharm. Ind., 133,* and **Dr. Dymock,** in his *Mat. Med. W. India, 2nd Ed., 571,* have described the well-known bazar drugs *Gaozabán* and *Gul-i-gaozabán.* Considerable confusion exists in the literature of this subject, for not only is it probable that the products of entirely different plants are sold in the bazars as *gao-zabán,* but the correct botanical determination of the true *gao-zabán* is still doubtful. **Moodeen Sheriff** sent a specimen, so named, to Kew some years ago, and it was determined as a species of **Echium.** **Stewart** regarded the leaves of **Onosma echioides** as the *gao-zabán* of the Panjáb, and in this opinion he has been followed by **Atkinson, Murray, &c.** **Royle,** in his *Illustrations of the Himalayan Botany, p. 304,* says that, " **Onosma bracteatum** is called *gao-zabán,* or ox-tongue, and has *fúghulus* and *buglúsun* assigned as its Greek names." **Sir W. O'Shaughnessy** (*Beng. Disp., 420, 495*) regarded **Cacalia Kleinia** (COMPOSITÆ)—a synonym for **Notonia grandiflora**—as the true *gao-zabán* of the Indian physicians, and pronounced the drug obtained from **Onosma bracteatum** as useless. But he describes his **Cacalia** as prickly, which it is not, and thus leaves room for a grave doubt as to the accuracy of his determination. He specially mentions that the drug is prized in Bombay, while **Dymock** neither gives **Notonia (Cacalia) grandiflora** the name of *gao-zabán* nor attributes to it the properties of that drug. **Birdwood** wrote: "All Indian authorities refer *gao-zabán* to the above plant (**C. Kleinia**), but the *gao-zabán* of the bazars is also derived from **Anisomeles malabarica,** *R. Br.,* LABIATÆ; **Trichodesma indicum,** *Br.,* **Heliotropium ophioglossum,** *Stocks,* and **Onosma bracteatum,** *Wall.,* BORAGINEÆ." Lastly, **Dr. Aitchison,** in his report on

E 6

| The Gaozabán ; the Kesuri. | (*J. F. Duthie.*) | **ECLIPTA alba.** |

the Botany of the Afghán Delimitation Commission, gives *gao-zeban* as the vernacular name for **Caccinia glauca,** *Savi* (BORAGINEÆ). Turning to Boissier's *Flora Orientalis* for a detailed description of that plant it is found to agree admirably with the flowers and leaves sold in the Indian bazars which will be found fully described by **Dymock** under **Echium.** Boissier gives the synonym **Caccinia Celsii** which, it may be suggested, by a clerical error, might be the origin of **O'Shaughnessy's Cacalia Kleinia.** **Caccinia glauca** is a fairly abundant plant in Quetta and in Gilgit, but neither **Mr. Lace** nor **Dr. Giles** seem to have recorded its vernacular name as *gao-zabán*. On the whole, therefore, it appears tolerably certain that the true *gao-zabán* of Indian bazars is derived from one or more species belonging to the Borage family. See **Onosma bracteatum.** (For the above note on **Echium** the Editor is responsible, and regrets that it was omitted to be described under **Caccinia glauca,** which would appear to be the true source of the *gao-zabún*.)

ECLIPTA, *Linn.; Gen. Pl., II., 361.*

Eclipta alba, *Hassk. ; Fl. Br. Ind., III., 304 ;* COMPOSITÆ.

7

Syn.—E. ERECTA, *Linn.*; E. PROSTRATA, *Linn.*

Vern.—*Moch-kand, bhangra, babri,* HIND.**;** *Kesuti, koysuria, keshwri, kesaraya,* BENG.; *Kesardá,* URIYA; *Lál kesari,* SANTAL; *Tik,* SIND**;** *Máká, bhringurája,* MAR.**;** *Bhángrá, kaluganthi, dodhak,* GUJ.**;** *Karisha-langanni, kaikeshi, kaivishi-ilai,* TAM.; *Galagara, guntakala-gara, gunta-galijeru,* TEL.; *Garagada-sappu, kadigga-garaga,* KAN.; *Kikirindi,* SING.; *Kesarája,* SANS.; *Kadim-el-bint,* ARAB.

Dr. Udoy Chand Dutt, in his *Materia Medica, page 181,* says that the Bengali and Hindi vernacular names *kesaraya, bhánrá,* as also the Sanskrit name, *bhringaraja,* are indiscriminately applied to this plant and to **Wedelia calendulacea,** *Linn.* This was not the case in Roxburgh's time, *kesuri* being **Eclipta alba,** and *bángrá* or *kesaraja* (*pivalá máká, pivalá bhangra,* MAR.) **Wedelia calendulacea,** which see.

References.—*Roxb., Fl. Ind., Ed. C.B.C., 605; Thwaties, En. Ceylon Pl., 164; Dals. & Gibs., Bomb. Fl., 127; Stewart, Pb. Pl., 126; Aitchison, Cat. Pb. and Sind Pl., 75; Rheede, Hort. Mal., X., t., 41; Trimen, Hort. Zeyl., 45; Elliot, Fl. Andhr., 57, 66; Rev. A. Campbell, Econ. Prod. Chutia Nagpur, 9; Pharm. Ind., 128; U. C. Dutt, Mat. Med. Hind., 181; Dymock, Mat. Med. W. Ind., 2nd Ed., 430; S. Arjun, Bomb. Drugs, 77; Murray, Pl. and Drugs, Sind, 181; Bidie, Cat. Raw Pr., Paris Exh., 32; Med. Top. Ajmír, 126; Baden Powell, Pb. Pr., 358; Atkinson, Him. Dist., 735; Drury, U. Pl., 189; Lisboa, U. Pl. Bomb., 162, 260, 292; Balfour, Cyclop., I., 1027; Home Dept. Cor., 221, 238.*

Habitat.—An erect or prostrate weed, abundant throughout India, ascending to 6,000 feet on the Himálaya.

Dye.—There is a popular opinion that the HERB, taken internally and applied externally, will turn the hair black (*Dymock*). In tatooing, the natives, after puncturing the skin, rub the juicy green leaves of this plant over the part, which gives the desired indelible colour, *viz.,* a deep bluish black (*Roxburgh*).

DYE.
Herb.
8

SPECIAL OPINIONS.—§ **Dr. Kanni Lal De** writes : " The practice prevails in Bengal of anointing the heads of infants with the juice of the fresh plant **(Eclipta)** to cause apparent greyish hair to become black. This is repeated once or twice, the hair being shaved. " **Dr. De** does not regard it as having any virtue in permanently changing the colour of the hair. " **Eclipta** is here used for tatooing. I have never seen **Wedelia** used " (*Dr W. Dymock, Bombay*). " **Eclipta prostrata,** *var.* **erecta,** is used on this side of India for imparting a bluish black dye; not the other plant, which is called *pivala* (yellow) *bhangra* " (*Assistant Surgeon Sakharam Arjun Ravat, L. M., Gorgaum, Bombay*).

E. 8

EHRETIA acuminata.	Edgeworthia—Nepal Paper.

MEDICINE.

Medicine.—It is an old-established Hindu medicine, principally used as a tonic and deobstruent in hepatic and splenic enlargements, and in various chronic skin diseases; in the latter case, it is also pounded and applied externally. The YELLOW KIND, *peela bhangra*, described by the author of the *Makhzan-el-Adwiya*, is **Wedelia calendulacea!**; and, according to Dutt, is the kind mostly used in Bengal. **Mr. Wood** considers that the plant will be found eventually of greater service than **Taraxacum** in hepatic derangements. The expressed JUICE is recommended in the Pharmacopœia of India as the best form of administration. In Bombay, the natives use the juice in combination with aromatics, such as *ajowan* seeds, as a tonic and deobstruent, and give two drops of it with eight drops of honey to new-born children suffering from catarrh. It also forms an ingredient of a remedy used in the Concan for tetanus (*Dymock*). The FRESH PLANT, mixed with Sesamum oil, is applied externally in elephantiasis. Murray writes that in Sind the expressed juice of the ROOTS is employed as an emetic. It is also purgative. The **Rev. A. Campbell** states that in Chutia Nagpur the root is applied in conjunctivitis and galled necks in cattle.

SPECIAL OPINIONS.—§ "The juice of the LEAVES is given in one teaspoon-ful doses in jaundice and fevers. The ROOT is given to relieve scalding of the urine in doses of 180 grains mixed with salt" (*C. T. Peters, M.B., Zandra, South Afghánistan*). "It is anodyne and absorbent, and relieves headache when applied with a little oil. It is an excellent substitute for **Taraxacum**" (*Kanni Lal De, Bahadur*).

Yellow kind.
9

Juice.
10

Fresh Plant.
11
Roots.
12

Leaves.
13
Root.
14

Eddoes, see **Colocasia antiquorum,** *Schott.*; Vol. II., p. 509.

EDGEWORTHIA, *Meissn.; Gen. Pl., III., 193.*　[LEACEÆ.

Edgeworthia Gardneri, *Meissn.; Fl. Br. Ind., V., 195;* THYME-
　　Vern.—*Kaghuti, aryili,* NEPAL.
　　References.—*Brandis, For. Fl., 386; Gamble, Man. Timb., 314.*
　　Habitat.—A large elegant bush, almost leafless when covered with its clusters of yellow sweet-scented flowers. Found along the Himálaya from Nepál to Sikkim and Bhután, between 4,000 and 9,000 feet altitude, and recently met with plentifully on the mountains of Manipur, extending to the northern frontier of Burma.

15

FIBRE.
16
Twigs.
17
Nepal Paper
18

　　Fibre.—The strong, tough fibre obtained from the long, straight, sparsely-branched TWIGS of this bush must, sooner or later, become one of the most valuable of Indian fibres. The finest qualities of NEPÁL PAPER are made from this plant, which produces a whiter paper than that obtained from **Daphne cannabina,** *Wall.* The chemistry of **Edgeworthia** fibre, and the probable extent to which it is used in Nepál paper-making, will be found discussed under **Daphne cannabina,** *Wall.*; Vol. III., 20.
　　Structure of the Wood.—Grey, light, soft, with little lustre (*Gamble*).

TIMBER.
19

Edible Birds' nests, see **Collocalia nidifica,** Vol. II., p. 504.

Egg-plant, see **Solanum Melongena,** *Linn.*

EHRETIA, *Linn.; Gen. Pl., II., 840.*

20

Ehretia acuminata, *Br.; Fl. Br. Ind., IV., 141;* BORAGINEÆ.
　　Syn.—E. SERRATA, *Roxb.*
　　Vern.—*Pányan, punjlawái, panden, koda, kurkuna, arjun,* HIND.;
　　Kula-aja, BENG.; *Bual,* ASSAM; *Nalshuna, chillay,* NEPAL; *Puna,*
　　N. W. INDIA; *Narra,* GARHWAL; *Shaursi,* KUMAON; *Punna, pursan,*
　　kalthaun, sum, PB.; *Punra,* PUSHTU; *Rend,* KURKU; *Ridi,* BAIGAS.

| The Ehretia. | (*J. F. Duthie.*) | **EHRETIA lævis.** |

References.—*Roxb., Fl. Ind., Ed. C.B.C., 200 ; Voigt., Hort. Sub. Cal., 445 ; Brandis, For. Fl., 339 ; Kurz, For. Fl. Burm., I.., ?10 ; Gamble, Man. Timb., 272 ; Stewart, Pb. Pl., 154 ; Aitchison, Cat. Pb. and Sind Pl., 93 ; Atkinson, Him. Dist., 314 ; Econ. Prod. N.-W. P., Pt. V., 81 ; Drury, U. Pl., 190 ; Balfour, Cyclop., I., 1034 ; Treasury of Bot., 442.*

Habitat.—A medium-sized tree, found in the Sub-Himálayan tract and outer Himálayan ranges, from the Indus to Sikkim, ascending occasionally to 5,500 feet.

Food.—It yields an insipidly sweet FRUIT, which is eaten ; the unripe fruit is pickled.

Structure of the Wood.—Light brown, with white specks, fairly even and compact, soft, not heavy, easily worked, made into scabbards, sword-hilts, gun-stocks, and employed in building and for agricultural implements. Not durable (*Brandis*).

FOOD.	
Fruit.	**21**
TIMBER.	**22**

Ehretia buxifolia, *Roxb. ; Fl. Br. Ind., IV., 144.* — **23**

Vern.—*Pála*, HIND. ; *Pale*, DEC. ; *Pála*, BOMB. ; *Kuruvingi*, TAM. ; *Bápana-búri, pitta-pisiniki*, TEL. ; *Hin-tambala*, SING.

References.—*Roxb., Fl. Ind., Ed. C.B.C., 201 ; Voigt, Hort. Sub. Cal., 446 ; Beddome, For. Man., 167 ; Gamble, Man. Timb., 272 ; Thwaites, En. Ceylon Pl., 214 ; Dalz. & Gibs., Bomb. Fl. Suppl., 60 ; Trimen, Hort. Zeyl., 54 ; Elliot, Fl. Andhr., 23, 154 ; Dymock, Mat. Med. W. Ind., 2nd Ed., 576 ; Bidie, Cat. Raw Pr., Paris Exh., 36 ; Drury, U. Pl., 190 ; Balfour, Cyclop., I., 1033.*

Habitat.—A shrub, found in the dry jungles of the Deccan Peninsula ; also in the Malaya.

Medicine.—Ainslie describes the ROOT as sweet and slightly pungent when fresh. It is used as an alterative in syphilis. Muhammadans regard it as an antidote to vegetable poisons.

MEDICINE.	
Root.	**24**

E. lævis, *Roxb. ; Fl. Br. Ind., IV., 141 ; Wight, Ic., t. 1382.* — **25**

Vern.—*Chamrár, chamrur, koda, darúr, datranga*, HIND. ; *Tambolli*, BENG. ; *Mosonea*, URIYA ; *Dotti, disti, gilchi*, GOND ; *Tambol* (Banda), BUNDEL. ; *Chumbul*, SIND ; *Tamboli*, BOMB. ; *Datrang*, MAR. ; *Pála dantam, pedda-pulimera, seregad, siragadam, áddabukkudu*, TEL. ; *Kappura, avak*, KAN.

References.—*Roxb., Fl. Ind., Ed. C.B.C., 201 ; Voigt, Hort. Sub. Cal., 445 ; Brandis, For. Fl., 340 ; Kurz, For. Fl. Burm., II., 210 ; Beddome, Fl. Sylv., t. 246 ; Gamble, Man. Timb., 272 ; Thwaites, En. Ceylon Pl., 214 ; Dalz. & Gibs., Bomb. Fl., 170 ; Aitchison, Cat. Pb. and Sind Pl., 93 ; Elliot, Fl. Andhr., 109, 142, 150, 168 ; Baden Powell, Pb. Pr., 578* (**E. aspera**) ; *Atkinson, Him. Dist., 314 ; Econ. Prod. N.-W.P., Part V., 81 ; Lisboa, U. Pl. Bomb., 202 ; Balfour, Cyclop., I., 1034.*

Habitat.—A moderate-sized tree, common throughout India.

Food.—The FRUIT is tasteless, but is eaten, as also the inner BARK, during famine times.

Fodder.—The LEAVES are used as cattle fodder.

Structure of the Wood.—Wood greyish white, hard, tough, and durable, used for building purposes, and for agricultural implements.

In the *Flora of Brit. India* the following varieties are enumerated :—

Var. **floribunda** (*Brand., For. Fl., 340*). *Syn.* E. **floribunda**, *Benth.*, in *Royle, Ill., 306.* Leaves acuminate, softly pubescent and ciliate. It occurs from Behar to the Panjáb, extending into Afghánistan.

Var. **pubescens.** *Syn.* E. **pubescens**, *Benth.*, in *Royle, Ill., 306.* Branchlets hairy as well as the leaves. Throughout India.

Var. **timorensis.** Malaya to Australia.

Var. **canarensis** is distinguished by the symmetric strong-nerved leaves, and is the **Ehretia** usually found on the Nilghiris and other Deccan mountains.

FOOD.	
Fruit.	**26**
Bark.	**27**
FODDER.	
Leaves.	**28**
TIMBER.	**29**
VARIETIES.	**30**
	31
	32
	33

E. 33

ELÆAGNUS hortensis.	**Oleaster or Bohemian Olive.**

VARIETIES.
34

Var. aspera, *Syn.* E. aspera, *Roxb., Fl. Ind., Ed. C.B.C.,* 201 ; *Brandis, For. Fl.,* 340 ; *Beddome, For. Man.,* 166 ; *Kurz, For. Fl. Burm., II.,* 209. This variety appears to be confined to Eastern Bengal, and is distinguished by its small obtuse leaves, which are hairy beneath when mature.

35

Ehretia, obtusifolia *Hochst. ; Fl. Br. Ind., IV.,* 142.

Vern.—*Chamror* (Panjáb Plains), *gin* (Rávi), *chamar* (Bias), *sakkur, dhaman, saggar, ganger, bari kander* (Salt Range), *chambal* (Sind Ságar Doáb), *marag'uune, kharawune, khabarra, tutiri, lor,* Pushtu.

References.--*Brandis, For. Fl.,* 340 ; *Gamble, Man. Timb.,* 272 ; *Stewart, Pb. Pl.,* 153 (E. aspera) ; *Dymock, Mat. Med. W. Ind.,* 2nd Ed., 576.

Habitat.—A small shrub, resembling E. lævis, *var.* aspera, and confined to Sind, Rájputána, and the Panjáb.

MEDICINE.
Root.
36

Medicine.—A decoction of the fresh ROOT is used in venereal diseases (*Dymock*).

TIMBER.
37
38

Structure of the Wood.—Resembles that of E. lævis.

E. Wallichiana, *H. f. & T.T. ; Fl. Br. Ind., IV.,* 143.

Vern.—*Bæri, dowari,* Nepal ; *Kalet,* Lepcha.

Reference.—*Gamble, Man. Timb.,* 272.

Habitat.—A large tree, frequent in Sikkim and Bhután, from 2,000 to 7,000 feet ; also on the Khásia mountains.

TIMBER.
39

Structure of the Wood.—Grey and moderately hard ; it is used for building and for charcoal, and occasionally for tea boxes (*Gamble*).

ELÆAGNUS, *Linn. ; Gen. Pl., III.,* 204.

A genus containing about a dozen species, remarkable for the abundance of delicate silvery or brown scales, with which the leaves and stems are coated. The tint of the foliage and the form of the fruit of some of the species give them a striking resemblance to the olive tree ; hence the generic name.

40

Elæagnus hortensis, *M. Beib., Fl. Br. Ind., V.,* 201 ; Elæagneæ.

Oleaster, Bohemian. Olive, Jerusalem Willow, *Eng.;* Olivier de Boheme, *Fr. ;* Wilde Oelbaume, *Germ.*

Syn.—E. angustifolia, *Linn.,* and E. orientalis, *Linn.*

Vern.—*Sirshing, sirsing,* Tibet ; *Shiúlik,* N.-W. P.; *Sanjít, santij, sanjata,* Afgh.; *Zin-zeid* (fruit), Pers.

References.—*Brandis, For. Fl.,* 389 ; *Irvine, Mat. Med., Patna,* 124 ; *Royle, Ill. Him. Bot.,* 323 ; *Balfour, Cyclop., I.,* 1035.

Habitat.—A small deciduous tree, bearing sweet-scented flowers, found on the Western Himálaya and in Tibet, up to 10,500 feet, and extending westward to Spain.

GUM.
41

Gum.—According to Stocks, a transparent brown and white gum, similar to Gum-arabic, exudes from wounds in the bark.

MEDICINE.
Flowers.
42

Medicine.—The FLOWERS are reported to be medicinal.

FOOD.
Berries.
43

Food.—The acid BERRIES are largely eaten in Tibet, Baluchistan, and Afghánistan, and the tree is cultivated to some extent for that purpose. The dried berries are known under the name of Trebizond dates, and are occasionally made into cakes by the Arabs. In Yarkand a spirit is distilled from these berries.

FODDER.
Leaves.
44

Fodder.—Mr. J. H. Lace states that in the autumn in Baluchistan the LEAVES are given as fodder to sheep and goats.

TIMBER.
45

Structure of the Wood.—Sap-wood narrow ; heart-wood dark brown, porous, soft ; used for fuel.

DOMESTIC.
46
Fuses.
47

Domestic Uses.—Dr. Stewart, in the manuscript copy of his *Forest Flora*, states that in Ladak the roots of this plant are used as fuses for match-locks.

E. 47

| Utrasum Bead Tree. (*J. F. Duthie.*) | ELÆOCARPUS lanceæfolius. |

Elæagnus latifolia, *Linn. ; Fl. Br. Ind., V., 202 ; Wight, Ic., t. 1856.* | 48

Syn.—E. CONFERTA, *Roxb.* ; E. ARBOREA, *Roxb.*
Vern.—*Guara,* BENG.; *Kamboong,* MAGH.; *Sheu-shong* (E. **arborea,** *Roxb.,*) GARO HILLS; *Jarila,* NEPAL; *Ghiwáin, mijhaula,* KUMAON ; *Nagri, ambgul,* BOMB.; *Wel-embilla,* SING.
References.—*Roxb., Fl. Ind., Ed. C.B.C.,* 148 ; *Voigt, Hort. Sub. Cal.,* 304 ; *Brandis, For. Fl.,* 390, *t.* 46 ; *Kurz, For. Fl. Burm.,* II., 331 ; *Beddome, Fl. Sylv., t.* 180 ; *Gamble, Man. Timb.,* 317 ; *Thwaites, En. Ceylon Pl.,* 252 (*Excl. Syn.* parvifolia); *Dalz. & Gibs., Bomb. Fl.,* 224; *Trimen, Hort. Zeyl.* ; *Atkinson, Him. Dist.,* 316 ; *Econ. Prod., N.-W. P., Part V.,* 82 ; *Gaz., Simla District,* 12 ; *Gaz. Bomb., XV.,* 441.
Habitat.—A small evergreen tree or shrub, often scandent, widely distributed throughout the hilly parts of India ; on the Himálaya it occurs westward of Jaunsar, up to 9,000 feet ; also in Burma, Penang, South India, and Ceylon.
Food.—The acid, somewhat astringent, FRUIT is eaten. Dr. Mason says that it makes excellent tarts and jellies, and is a great favourite with the natives in Burma. The Conservator of Forests, Northern Circle, Madras, states that the fruit of this plant, which is very common on the Nilgiri Hills, is eaten chiefly by tenders of cattle ; it does not constitute, however, an article of trade. | FOOD. Fruit. 49
Structure of the Wood.—Resembles that of E. **hortensis.** | TIMBER. 50 51

E. umbellata, *Thunb.; Fl. Br. Ind., V., 201.*

Syn.—E. PARVIFOLIA, *Wall.*
Vern.—*Ghiwáin, ghain, kankoli, kankol mirch, bammewa,* PB.
References.—*Brandis, For. Fl.,* 391 ; *Gamble, Man. Timb.* 318 ; *Baden Powell, Pb. Pr.,* 373 (under E. **orientalis**), 578 under (E. **conferta**); *Atkinson, Him. Dist.,* 736 ; *Royle, Ill. Him. Bot.,* 323, *t.* 81, *f.* 1.
Habitat.—A deciduous-leaved, often thorny, shrub of the temperate Himálaya, extending from Kashmír to Nepál, at 3,000 to 10,000 feet ; also in China and Japan.
Medicine.—The SEEDS and FLOWERS (*gul-i-sanjad*) are said to be used as a stimulant in coughs, and the expressed OIL in pulmonary affections. The flowers are also given as a cardiac and astringent. Baden Powell says that the seeds are used to adulterate black pepper. | MEDICINE. Seeds. 52 Flowers. 53 Oil. 54
Food.—The FRUIT is pickled like olives, or eaten in curries. | FOOD. Fruit. 55
Structure of the Wood.—White, hard, even-grained, but warps on seasoning (*Gamble*).

ELÆOCARPUS, *Linn.; Gen. Pl., I., 239.* [66 ; TILIACEÆ. | TIMBER. 56

Elæocarpus Ganitrus, *Roxb.; Fl. Br. Ind., I., 400 ; Wight, Ic., t.* | 57
UTRASUM BEAD TREE, *Eng.*
Vern.—*Rudrák,* HIND.; *Rudrákhya,* BENG.; *Rudraksh,* MAR.; *Rudra-kai,* TAM.; *Rudra-challu,* TEL.; *Rudráksha,* SANS.
References.—*Roxb., Fl. Ind., Ed. C.B.C.,* 433 ; *Voigt, Hort. Sub. Cal.,* 123 ; *Brandis, For. Fl.,* 43 ; *Kurz, or. Fl. Burm.,* I., 168 ; *Beddome, For. Man.,* 38 ; *Dalz. & Gibs., Bomb. Fl.,* 27 ; *Lisboa, U. Pl. Bomb.,* 286 ; *Balfour, Cyclop., I.,* 1035 ; *Treasury of Bot., I.,* 444.
Habitat.—A large tree found in Nepál, Assam, and the Concan gháts.
Domestic Uses.—The hard tubercled nuts are polished, made into rosaries and bracelets worn by Brahmins (Shívas) and fakirs, and are frequently set in gold. They are mostly imported from Singapore, where the tree is common. See the article **Beads,** Vol. I., p., 431. | DOMESTIC. 58

E. lanceæfolius, *Roxb. ; Fl. Br. Ind., I., 402 ; Wight, Ic., t.* 65. | 59
Syn.—E. LANCEOLATUS, *Wall.*

E. 59

ELÆOCARPUS   The Jalpai and Rudrak.
Varunna.

Vern.—*Sakalang*, ASSAM; *Sufed-pai*, SYLHET; *Bhadras, batrachi,*
NEPAL; *Skepkyew*, LEPCHA.
References.—*Roxb., Fl. Ind., Ed., C.B.C.,* 435; *Voigt, Hort. Sub. Cal.,*
123; Kurz, For. Fl. Burm., I., 169; *Gamble, Man. Timb.,* 57.
Habitat —A large tree of the Eastern Himálaya, from 6,000 to 8,000
feet; the Khásia Hills, Sylhet, and Tenasserim; also in Kánara.

FOOD.   Food.—The FRUIT, which ripens in September and October, is eaten
Fruit.   by the natives.
60   Structure of the Wood.—Light brown and soft; it is used for house-
TIMBER. building, tea-boxes, and charcoal.
61   Domestic Uses.—The seeds of this tree are used for a similar purpose
DOMESTIC. as those of E. Ganitrus. See Beads, Vol. I., p. 431.
62

63   **Elæocarpus oblongus,** *Gærtn.; Fl. Br. Ind., I.,* 403; *Wight, Ic. t.*46.
Vern.—*Bikki,* NILGHIRIS.
References.—*Beddome, For. Man. 38; Gamble, Man. Timb.,* 57; *Dals.*
& Gibs., Bomb. Fl., 27.
Habitat.—A large tree, found in Southern India, and in Burma.
TIMBER. Structure of the Wood.—White, strong, and tough, and adapted for
64   the lathe (*Beddome*).

65   **E. robustus,** *Roxb.; Fl. Br. Ind., I.,* 402; *Wight, Ic., t.* 64.
Vern.—*Chekio,* MAGH; *Jalpai,* SYLHET; *Bepari, batrachi,* NEPAL; *Chekio,*
MAGH.; *Taumagyee,* BURM.
References.—*Roxb., Fl. Ind., Ed. C.B.C.,* 434; *Voigt, Hort. Sub. Cal.,*
123; Kurz, For. Fl. Burm., I., 169; *Gamble, Man. Timb.,* 57.
Habitat.—An evergreen tree of the Eastern Himálaya, ascending to
2,000 feet; the Khásia Hills, Eastern Bengal, Chittagong, Burma, and the
Andaman Islands.
TIMBER. Structure of the Wood.—White, shining, soft, even-grained.
66

67   **E. serratus,** *Linn.; Fl. Br. Ind., I.,* 401.
Syn.—E. PIRINCARA, *Wall.*
Vern.—*Jalpai,* BENG.; *Perinkara,* KAN.; *Weralu,* SING.
References.—*Roxb., Fl. Ind., Ed. C.B.C.,* 434; *Voigt, Hort. Sub. Cal.,*
123; Brandis, For. Fl., 43; *Beddome, For. Man.,* 38; *Gamble, Man.*
Timb., 57 ;*Thwaites, En. Ceylon Pl.,* 32; *Trimen, Hort. Zeyl.,* 12;
Buchanan, Statistics of Dinajpur, 153; *Taylor, Topography of Dacca,* 50.
Habitat.—A tree found in the north-east regions of the Himálaya, in
Bengal, and on the western coast; also in Ceylon.
FOOD.   Food.—The fleshy outer portion of the FRUIT is eaten in curries by the
Fruit. natives, and is also pickled in oil and salt like olives. In Assam the tree
68   is occasionally grown for the sake of the fruit, which is eaten either ripe
or unripe and boiled with vegetables to give them an acid flavour.

69   **E. tuberculatus,** *Roxb.; Fl. Br. Ind., I.,* 404; *Wight, Ic., t.* 62.
Syn.—E. SERRULATUS, *Roxb.*
Vern.—*Rudrak,* HIND.; *Rudrák,* KAN.
References.—*Roxb., Fl. Ind., Ed. C.B.C.,* 433; *Beddome, Fl. Sylv., t.*
113; Dalz. & Gibs, Bomb. Fl., 27; *Lisboa, U. Pl. Bomb.,* 287; *Balfour,*
Cyclop., I., 1037.
Habitat.—A large handsome tree, found in South India, and in Burma.
DOMESTIC. Domestic Use.—The nuts of this tree are used in the same way as
70   those of E. Ganitrus. See Beads, Vol. I., p. 432.

71   **E. Varunna,** *Ham.; Fl., Br. Ind., I.,* 407.
Vern.—*Tuttealy, saul-kuri,* ASSAM.
References.—*Kurz, For. Fl. Burm., I.,* 165; *Gamble, Man. Timb.,* 57.

E. 71

Habitat.—A tree met with in the Himálaya, from Kumaon to Sikkim; also in Assam and Chittagong.

Food.—Like the other species this also produces a FRUIT which is edible. *FOOD. Fruit.* **72**

ELÆODENDRON, *Jacq. f.; Gen. Pl., I., 367.*

Elæodendron glaucum, *Pers.; Fl. Br. Ind., I., 623;* CELASTRINEÆ. **73**

Syn.—E. PANICULATUM, *W. & A.;* E. ROXBURGHII, *W. & A.;* NEERIJA DICHOTOMA, *Roxb.*

Vern.—*Miri, thanki,* KOL.; *Neouri, neuri,* SANTAL; *Chikyeng,* LEPCHA; *Dhakka, nisur,* GOND; *Mamri,* BUNDEL.; *Bakra, jamuwa, chauli, daberi, mámri,* N.-W. P.; *Chauri, metkur,* OUDH; *Shauriya,* KUMAON; *Mirandú, padriún, bakra, jamoá, mir-goo* [Hushiarpur], PB.; *Niru,* MELGHAT; *Jamrasi, mamri,* BANDA; *Bata karas,* BHIL; *Jamrasi, jum, rassi, kala mukha, rohi,* C. P.; *Niru,* KURKU; *Aran, tamruj, bhukas-* BOMB.; *Burkas,* KONKAN; *Aran, tamruj, bhutá-palá,* MAR.; *Bhutrak, shi,* HYDERABAD; *Karkava, irkuli, selupa, siri,* TAM.; *Niriju, bira-nerija-manu, nerasi, nirasi, neradi, botanskam, kanemi , bootigi,* TEL.; *Thá-maroja,* KAN.; *Bíra,* MADRAS; *Bhutápálá, chutoyá, támaruja, nerrelu; pieri,* SING.

References.—*Roxb., Fl. Ind., Ed. C.B.C.,* 214 & 217; *Voigt, Hort. Sub. Cal., 167; Brandis, For. Fl., 82; Beddome, Fl. Sylv., t. 148; For. Man., 67; Gamble, Man. Timb., 87; Thwaites, En. Ceylon Pl., 73; Dals. & Gibs., Bomb. Fl., 48; Grah., Cat. Bomb. Pl., 38; Elliot, Fl. Andhr., 27, 133, 135; Stewart, Pb. Pl., 40; Aitchison, Cat. Pb. and Sind Pl., 32; O'Shaughnessy, Beng. Dispens., 271; Dymock, Mat. Med. W. Ind., 2nd Ed., 179; S. Arjun, Bomb. Drugs, 30; Rev. A. Campbell, Cat. Econ. Prod., Chutia-Nagpur, p. 17; Atkinson, Him. Dist., 736; Drury, U. Pl., 190; Lisboa, U. Pl. Bomb., 49, 264, 274; Cooke, Gums and Gum-resins, 16; Atkinson, Gums and Gum-resins, 15; Balfour, Cyclop., I., 1036; Treasury of Bot., I., 444; For. Adm. Report, Chutia-Nagpur, 1885, 29; Bomb. Gaz., XV., 68.*

Habitat.—A moderate-sized tree, or occasionally only a shrub, occurring throughout the hotter parts of India and in Ceylon. Along the outer Himálaya, it ascends to 6,000 feet.

Gum.—It is supposed to yield the gum called *Jumrasi*, which occurs in roundish tears about ½ inch in diameter, rough or cracked on the surface. It is tasteless, and forms a sherry-coloured solution with water. *GUM.* **74**

Medicine.—The ROOT is a specific against snakebite, and Sir Walter Elliot speaks highly of this property. The BARK is used in native medicine and is said to be a virulent poison. A decoction or cold infusion of the fresh bark of the root is applied to swellings. *MEDICINE. Root.* **75** *Bark.* **70**

Roxburgh states that the fresh bark of the root, rubbed with water is by natives applied externally to remove swellings. According to Sakharam Arjun, the LEAVES (*bhutapála*) dried and powdered act as a sternutatory, and are used as a fumigatory to rouse women from hysterical fits. A snuff of the leaves is also employed to relieve headache. *Leaves.* **77**

Structure of the Wood.—Moderately hard, even and close-grained, works and polishes well, light-brown, often with a red tinge; the outer wood white, but no distinct sap-wood; no annual rings. It is often beautifully curled and flaked. It is used for cabinet work, combs, and picture frames. It is also employed for fuel in the Konkans. *TIMBER.* **78**

Elaterium, see **Ecballium.**

Elderflowers, see **Sambucus nigra,** *Linn.*

Elemi Gum. There is considerable doubt as to the plant or plants from which this substance is obtained. It seems to be a member of the BURSERACEÆ. It is generally supposed to be a species of ICICA, of AMYRIS, or of CANARIUM. (It should not be confounded with *Animi,* for which see **Copal.**) **79**

ELEPHAS indicus.	The Indian Elephant.

Elephant-apple, see **Feronia elephantum,** *Correa,* below.

ELEPHANTOPUS, *Linn.; Gen. Pl., II., 237.*

80

Elephantopus scaber, *Linn.; Fl. Br. Ind., III., 242; Wight, Ic.,*
PRICKLY-LEAVED ELEPHANT'S FOOT, *Eng. [t. 1086; COMPOSITÆ.*

Vern.—*Gobhi, samdulun,* HIND.; *Gojiálata, shamdulun,* BENG.; *Manjur-juti,* SANTAL; *Hastipata,* BOMB.; *Anashovadi,* TAM.; *Eddu-málike-chettu* (bullock's tongue-shaped leaves), *hasti-kasaka, enuga-bira,* TEL.; *Ká-too-pin, ma-too-pin,* BURM.; *At-addeya, et-adi,* SING.; *Gojihbá, go-jihwa,* SANS.

References.—*Roxb., Fl. Ind., Ed. C.B.C., 607; Voigt, Hort. Sub. Cal., 406; Dalz. & Gibs., Bomb. Fl., 122; Rheede, Hort. l'al., X., t. 7; Trimen, Hort. Zeyl., 44; U. C. Dutt, Mat. Med. Hind., 298; Dymock, Mat. Med. W. Ind., 423; Balfour, Cyclop., I., 1041; Treasury of Bot., I., 446.*

Habitat.—A stiff hairy herb, with wrinkled crenate radical leaves, distributed throughout the hotter parts of India.

MEDICINE. Root. 81 Leaves. 82	Medicine.—Rheede says that a decoction of the ROOT and LEAVES is given, on the Malabar coast, in cases of dysuria. In Travancore the natives are reported to boil the bruised leaves with rice, and give them internally for swellings or pains in the stomach. The Rev. A. Campbell states that in Chutia Nagpur, a preparation from the root is given for fever.

Elephant's-foot, see **Elephantopus scaber,** *Linn.*

ELEPHAS.
(George Watt.)

83

Elephas indicus, *Cuv.; Jerdon, Mam. Ind., 229.*
THE INDIAN ELEPHANT: ELEPHANTES, *It.;* FIEL, *Scand.;* ELEPHANTE, *Sp.;* FIL, *Turkish.*

Vern.—*Hati* or *háthi, guj. pil,* HIND.; *Gaj,* BENG.; *Ani* or *anay,* TAM., TEL., KAN., and MAL.; *Yenu,* GOND; *Pil,* PUSHTU; *Hasti, gaja,* SANS.; *Feel,* PERS.; *Allia,* SING.; *Shanh, hsen,* BURM.; *Gadjah,* MALAYAN.

Mukna is a tuskless male elephant; tame females used in hunting are called *kúnkies.*

In the *Rig Veda* the elephant is mentioned once or twice under the name of *Migrohasti* (the beast with a hand), and in the *Atharvan* he is exalted as the mightiest and most magnificent of animals. But there is little in early Sanskrit literature to justify the inference that the elephant was then domesticated. The word Elephant is supposed by some to have been derived from *Pilu* in Sanskrit and *Fel* in Persian, which, with the Arabic article *El,* became *el-fil* and *Elephas* in Greek. The Hindu god of wisdom, Ganesh, has the body of a man with the head of an elephant.

References.—*Natural History of Indian Mammalia by Sterndale, 389; Thirteen Years among the Wild Beasts of India, by G. P. Sanderson, pp. 48 to 242; Through Masai Land by Joseph Thomson, 537; The Natural History of Ceylon, by Sir Emmerson Tennent; The Elephant, by Lieut. Ouchterlony; The Management of Elephants, by Col. Hawkes; Gilchrist—A Practical Treatise on the diseases of Elephants; Slymm, Treatise on the Treatment of Elephants in Health and Disease; Sanderson, The Elephant in Freedom and Captivity—a lecture in the Journal of the United Service Institute in India; Various papers in the Quarterly Journal of Veterinary Science in India; The Elephant, by J. H. Steel, V.S., A.V.D.; The Kuram Field Force, by G. A. Oliphant; Pack Gear of Elephants, by G. P. Sanderson; John Huyghen van Linschoten, Journal of Travels in India, published in 1596; The Ain-i-Akbari by Abul Fazl (Blochmann's Transl.), pp. 117 to 132, and 213, 214, 235, 284, 379, 467, and 618; C. P. Administration Report, 1865-66, p. 64, and 1866-67, p. 91; Bombay Gazetteers, Vols. VIII. (Kathiawar), 97; XII. (Khandesh), 29; XV., Pt. I. (Kánara), 27; Madras Man. Adm., Vol. II.*

The Indian Elephant.	(*G. Watt.*)	**ELEPHAS indicus.**

292; *Ainslie, Mat. Med., II., 479; Mysore and Coorg Gazetteer, I., 148; Falconer and Cautley, Fauna Antiqua Silvalensis; Balfour, Cyclopædia of India, 1037; Encyclopædia Britannica, VIII., 122; Ure, Dictionary, Arts, Manufactures, &c., II., 760; Spons' Encyclopædia.*

Where Found.—Jerdon says : " The elephant is still tolerably common in most of the large forests of India, from the foot of the Himálaya to the extreme south It is found in the Terai from Bhután to Dehra Dún and the Kyarda Dún. It used, not many years ago, to occur in the Rájmahal hills, and it abounds in many parts of Central India from Midnapore to Mandla, and south nearly to the Godavari. On the west coast, it is abundant in many localities, from the extreme south of Travancore to north latitude 17 or 18 degrees, all along the line of the Western Ghâts, more especially on the Anamally hills (named from that circumstance) ; in the Coimbatore hills, Wynaad, the slopes of the Nilghiris, Coorg, and parts of Mysore and Kanara. The Shervroys and Colamallies, and other detached ranges to the east, have occasionally small herds. It is numerous in Ceylon and in Assam, southwards to the Malaya Peninsula."

Sanderson expresses briefly the area over which elephants occur, thus :— " The wild elephant abounds in most of the large forests of India, from the foot of the Himálayas to the extreme south, and throughout the peninsula to the east of the Bay of Bengal, *viz.,* Chittagong, Burma, and Siam ; it is also numerous in Ceylon. There is only one species of elephant throughout these tracts." According to the *Ain-i-Akbari (Blochmann's Translation),* the Emperor Akbar drew his supplies from regions where the elephant rarely if at all now exists, *e.g.,* the Cubah of Agra ; in the jungles of Bayawan and Narwar as far as Barár ; in the Cubah of Iláhábád (Alláhábád) ; in the confines of Punnah ; in the Cubah of Malwah ; in the Cubah of Bihár, &c., &c. Those caught near Punnah in Bundelkhand were regarded as the best.

Varieties and Races of Elephants.—According to most writers there is but one species of elephant met with in Asia. Some authors, however, view the elephant of Ceylon as forming, with that of the Sumatra one, a distinct species (**Elephas sumatranus**). Jerdon says of this form : " The Sumatran Elephant has 20 pairs of ribs " (the Indian has 19 and the African 21) " and the laminæ of the teeth are wider than in the Indian species. It is said to be of a more slender make and to be more remarkable for its intellectual development than the Indian." A belief in the superior intellectual powers of the Ceylon as compared with the Indian elephant seems to have prevailed, at least for the past 300 years. John Huyghen van Linschoten thus wrote of Ceylon : " It hath divers elephants, which are accounted for the best in all India, and it is by daylie experience found to be true, that the elephant of all other places and countries being brought before them, they honour and reverence these." Sanderson, while holding that the Ceylon elephant is the same species as the Indian refers to the fact that the males are in the majority of cases tuskless. He writes: " It is difficult to imagine what can cause the vital difference of tusks and no tusks between the male elephant of Continental India and Ceylon. The climate may be said to be the same, as also their food ; and I have not seen any theory advanced that seems at all well founded to account for their absence in the Ceylon elephant." As an external character, the immensely larger ears of the African elephant distinguish it from the Indian. But even among the Indian elephants, local peculiarities and characteristics have been recorded sufficient to justify the opinion that the elephant of Nepál should be regarded as a different race from that of Mysore, just as the Mysore is different from that of Assam or of the Chittagong hill tracts. The Nepál elephant is reported to be small in

WHERE FOUND.
84

RACES.
85

(a) Ceylon.
(b) Indian.

1. Nepal.
2. Mysore.
3. Bengal.
4. Chittagong.
5. Burma.
6. Shan.
7. Madras.
8. Bombay.
9. Central India.
10. Central Provinces.

P

ELEPHAS indicus.	**The Indian Elephant.**

stature and well adapted for life on the hills. The Shán elephants are tall, massive, and handsome, but, like the Ceylon race, are very frequently tuskless. The Burmese elephant resembles more the Nepál animal in being, as Captain Hood remarks, "more compact than those of Hindústan and superior for hill work, carrying loads over steep places and across swamp, or boggy ground, and they are excellent for draught purposes." Steel remarks of the Chittagong race that they "are good all round and make the best *koonkies*; the Assamese are large, both tall and massive, and excellent for hunting purposes."

DOMESTI-CATED BREEDS.
86
Koomeriah.
Dwasala.
Meerga.

Speaking of the classification of elephants as adopted by the Natives of India from the standpoint of their appearance and utility, **Sanderson** says: "Elephants are divided by Natives into three castes or breeds, distinguished by their physical conformation: these are termed in Bengal *Koomeriah*, *Dwásála*, and *Meerga*, which terms may be considered to signify thoroughbred, half-bred, and third-class. The term *Koomeriah* signifies royal or princely. *Meerga* is probably a corruption of the Sanskrit *Mírga*, a deer; the light build and length of leg of this class of elephants suggesting the comparison. *Dwásála* in Persian means two things or originals, and in reference to the elephant, signifies the blending of the first and third castes into the intermediate one. Only animals possessing extreme divergence, rank as *Koomeriahs* or *Meergas*: and the points of these breeds (if they may be so called) do not amount to per-

Not hereditary characteristics.
87

manent, or even hereditary, variations. Whole herds frequently consist of *Dwásálas*, but never of *Koomeriahs* or *Meergas* alone; these I have found occur respectively in the proportion of from 10 to 15 per cent. amongst ordinary elephants." Sanderson enumerates the characters of the *Koomeriah* as follows, "barrel deep, and of great girth: legs short (especially the hind ones), and colossal, the front pair convex on the front side, from the development of muscles; back straight and flat but sloping from shoulder to tail, as an up-standing elephant must be high in front; head and chest massive; neck thick and short: trunk broad at the base and proportionately heavy throughout: bump between the eyes prominent: cheeks full: the eye full, bright, and kindly: hind quarters square and plump: the skin rumpled, thick, inclining to folds at the root of the tail, and soft. If the face, base of trunk, and ears, be blocked with cream-coloured markings, the animal's value is enhanced thereby. The tail must be long, but not touching the ground, and well feathered."

A pronounced *Meerga* is the opposite of these characters, especially in possessing long legs and an arched back. It is well suited for quick marching on account of its lighter weight and length of legs.

AKBAR'S CLASSIFI-CATION.
88
1. Bhaddar.
Pearl from Elephant.
89
2. Mand.
3. Mirg.
4. Mir.
White Elephant.
90

The *Ain-i Akbari* gives the classification of elephants as recognised in Akbar's time into four classes, viz., (*1*) *Bhaddar*—"It is well proportioned, has an erect head, a broad chest, large ears, a long tail, and is bold and can bear fatigue. They take out of his forehead an excrescence resembling a large pearl, which they call in Hindi *Gaj manik* (Elephant's pearl); (2) *Mand*, a large black form said to have an ungovernable temper; (3) *Mirg*, a lighter coloured animal, and (4) *Mir*, an animal with small head which obeys readily but is easily frightened.

The so-called white elephant, held sacred in Burma, is an albino condition. Steel says: "its very name has become a synonym for something expensive, useless, and extraordinary; yet we are assured that there is no such thing as a white elephant." Archibald Forbes gives, in his *Glimpses Through the Cannon Smoke*, a humorous account of the sacred white elephant of Burma. As a rule the pale-coloured form known as the white elephant is a sickly animal, his legs being swollen at the joints and often covered with tumours. The colour is at most a dirty grey, but the

| The Indian Elephant. | (*G. Watt.*) | **ELEPHAS indicus.** |

skin underneath has often a pinkish colour, seen more especially when the animal goes into the water.

CAPTURE OF WILD ELEPHANTS.

HERDS.—The elephant is a gregarious and polygamous animal, living in herds, the members of which are presumably all related to each other. Each male is specially attentive to a selected number of the females of the herd, but in the question of supremacy, the males often fight amongst themselves, the conquerors expelling their antagonists from the herd. At night the males frequently leave the herd and wander into the fields at a little distance from the favourite haunt of the herd. From both these causes single male elephants are occasionally met with, but according to **Sanderson**, it is incorrect to view all solitary male elephants as "rogues," discontented, vicious, deserters from the herd. Many males, from a liking for solitude, choose to separate themselves from the herd for a time if not completely. A herd consists of from 10 to 50 or more. Herds of 1,000, such as are referred to in some of the older works, do not appear to be known at the present day, if they ever existed. The herds select localities for occupation during fixed seasons of the year, and in grazing in their favourite forests, they have regular runs or paths of communication which they almost invariably follow. These facts have suggested most of the methods of capture which are now, and have for centuries been, in use. In advancing from one locality to another the herd is usually conducted by a female. This, as **Sanderson** explains, appears to be in consequence of the desire to regulate the rate of movement by the weaker not the stronger members of the community. Many writers drawing upon a not unnatural imagination have pictured herds led by powerful tuskers. The author of the article "Elephant" in the *Encyclopædia Britannica* thus alludes to the movements of a herd, which he says marches "under the guidance of a single leader whom they implicitly follow, and whose safety, when menaced they are eager to secure." **Steel** writes—"herds of elephants (which are families, their members presenting family traits) vary much in size, sometimes consisting of even 100 individuals but generally more or less broken up. They make their way through trackless forests preceded by *a female*, generally the largest, and following mostly in Indian file. When fleeing from danger the female assiduously keeps the young *in front of her*. Herds which have been broken up re-collect, and if one herd has been disturbed, even others will leave the place (*Young Shikarry*). The conformation and great weight of the animal specially adapt him for thus making a track through the jungle. The bull rambles much more than the cows, but he always keeps the herd within reach, and will often nobly cover the retreat of his cows." **Sir Victor Brooke** describes the herd from which he bagged the largest Indian tusks on record as follows:—"There were about eighty elephants in the herd. Towards the head of the procession was a noble bull, with a pair of tusks such as are rarely seen now-a-days in India. Following him in direct line came a medley of elephants of lower degree—bulls, cows, and calves of every size, some of the latter frolicking with comic glee, and bundling in amongst the legs of their elders with the utmost confidence. It was truly a splendid sight, and I really believe that, while it lasted, neither **Colonel Hamilton** nor I entertained any feeling but that of intense admiration and wonder. At length the great stream was, we believed, over, and we were commencing to arrange our mode of attack, when that hove in sight which called forth an ejaculation of astonishment from each one of us. Striding thoughtfully along in the rear of the herd, many of the members of which were, doubtless, his children, and his children's children, came a mighty bull,

CAPTURE OF WILD ELEPHANTS. 91

Herds of 1,000. See p. 217.

Female leader. See p. 217.

P 2

ELEPHAS indicus.	The Indian Elephant.

METHODS OF CAPTURE.

(a) Pits.
92
(b) Decoy.
93
(c) Kheddah.
94

the like of which neither my companion, after many years of jungle experience, nor the two Natives who were with us, had ever seen before. But it was not merely the stature of the noble beast which astonished us, for that, though great, could not be considered unrivalled. It was the sight of his enormous tusks, which projected like a long gleam of light into the grass through which he was slowly wending his way, that held us rivetted to the spot."

METHODS OF CAPTURE.—Taking advantage of the fact that these noble animals thus live in herds and frequent definite paths in the forests, they are captured in various ways, *viz.*, by digging pits into which they fall, the mouths of which are covered over with a light frame-work of boughs and leaves : by driving them along one of their most frequented paths into an enclosure. The single elephants occasionally met with are also captured by means of tame females, the riders disguising and screening themselves as much as possible, and after having surrounded their prize, the attendants slip off the tame elephants and secure the feet of their victim.

Sanderson (*Thirteen Years among the Wild Beasts of India, p. 101*), gives a spirited account of his early attempts in capturing herds by driving them into an enclosure (the *Kheddah*). He writes of Mysore in 1873 : "I knew nothing of elephant-catching at the time, nor had I any men at command who did ; but I knew where there were plenty of elephants, and I was well acquainted with their habits. Some of the Maharajah's *mahouts*, who were amongst my following, had been accustomed to catch single elephants with trained females and in pitfalls, but they had never heard of any one attempting the capture of a whole herd. It was said that Hyder had made a trial, a century before, in the Kakankote jungles, but had failed, and had recorded his opinion that no one would ever succeed, and his curse upon any one that attempted to do so, on a stone still standing near the scene of his endeavours. Consequently, all the true Mussulmans who were with me regarded the enterprise as hopless— though they judiciously kept that opinion to themselves." Mr. Sanderson then narrates the features of his system, which may be briefly described as the surrounding of a favourite resort of elephants by certain preliminary works prior to the arrival of the elephants, particularly the construction of a strong *kheddah* protected by a trench. When these preparations have been completed, the arrival of the elephants is awaited, but on their arrival some 300 men are rapidly assembled and the elephants, frightened by the noise made by these beaters, are at first made slowly, and later on with a rush, to advance into the *kheddah*. As soon as the last animal has entered, a man, screened from observation, cuts the rope by which the door of the trap is held, and this, closing by its own weight, the herd is captured. The beaters then surround the *kheddah*, and by drums, guns, and torches frighten any brave animal who may threaten an attack upon the enclosure. After vainly struggling for a time the frightened monsters of the forest crowd together in the centre and offer very little further attempts upon the stockaded trap. Food and water are supplied to them, and after all arrangements have been completed, and the animals have become in a measure accustomed to their captive state, tame female elephants, with one or two attendants, enter the *kheddah*. These singling out the largest victims separate them from the herd, two females, getting one on each side, hustle their prisoner towards a tree. The attendants slip off the tame elephants and secure its hind legs with strong ropes or chains with which they also attach it to the tree. Alarmed at this procedure, when efforts at freedom are now unavailing, it struggles violently, but in time submits. According to Mr. Sanderson the strongest and bravests animals become the most docile when thus convinced that they

E. 94

| The Indian Elephant. | (*G. Watt.*) | **ELEPHAS indicus.** |

have been conquered. As soon as all have been secured, they are each in turn led out of the *kheddah* between tame elephants and picketed in a place previously arranged. Food and water are pressed up to them, and through great kindness in giving them luxuries, such as sugar-cane, they get accustomed to their attendants. In a very few days, owing to the attendants speaking and singing to them and cooking their food hard by, they become so familiar with the presence of human beings, that they allow themselves to be approached and fondled. In many cases so successful is this treatment that the attendants after a few days are enabled to ride them and commence the process of training to a code of signals, gestures, and words. They are then marched off to the Government stables, or are sold locally to traders.

SEASON OF ELEPHANT CAPTURE.—Sanderson gives the season of capture as from the beginning of December, the party being equipped for two or three months. The hunters having previously marked down a good herd, the beaters, a mile or so distant, file off to right and left, two men stopping every 50 yards or so until they meet behind, having thus enclosed the herd within a space of 6 or 8 miles in circumference. Once thus surrounded the elephants can only escape through great carelessness. Within a couple of hours a simple enclosure is constructed along the line taken up by the men, and the elephants finding plenty food make little effort to escape during the day, and at night they are made to retire into the interior of the enclosure by fires, drums, and guns, &c., discharged at them, along the line of capture.

It may suffice in completing this brief review of the capture of herds of elephants to quote here one or two passages from early writers, in order to show how closely the present practice follows that pursued two or three hundred years ago. In the *Ain-i-Akbari* (*Blochmann's Transl.*, 284) it is said of "Elephant hunts":—

"There are several modes of hunting elephants:

"1. *K'heddah.*—The hunters are both on horse-back and on foot. They go during summer to the grazing places of this wonderful animal, and commence to beat drums and blow the pipes, the noise of which makes the elephants quite frightened. They commence to rush about, till from their heaviness and exertions no strength is left in them. They are then sure to run under a tree for shade, when some experienced hunters throw a rope, made of hemp or bark, round their feet or neck and thus tie them to the trees. They are afterwards led off in company with some trained elephants, and gradually get tame. One fourth of the value of an elephant thus caught is given to the hunters as wages.

"2. *Char k'hedah.*—They take a tame female eleplant to the grazing place of wild elephants, the driver stretching himself on the back of the elephant, without moving or giving any other sign of his presence. The elephants then commence to fight, when the driver manages to secure one by throwing a rope round the foot.

"3. *Gad.*—A deep pit is constructed in a place frequented by elephants, which is covered up with grass. As soon as the elephants come near it, the hunters from their ambush commence to make a great noise. The elephants get confused, and losing their habitual cautiousness, they fall rapidly and noisily into the hole. They are then starved and kept without water, when they soon get tame.

"4. *Bár.*—They dig a ditch round the resting place of elephants, leaving only one road open, before which they put up a door, which is fastened with ropes. The door is left open, but closes when the rope is cut. The hunters then put, both inside and outside the door, such food as elephants like. The elephants eat it up greedily; their voraciousness makes them

Marginal notes:
SEASON OF
95

EARLY MODES OF HUNTING.
Kheddah.
96

Char kheddah.
97

Gad.
98

Bar,
99

ELEPHAS indicus.	The Indian Elephant.

MODE OF HUNTING.

forget all cautiousness, and without fear they enter at the door. A fearless hunter, who has been lying concealed, then cuts the rope, and the door closes. The elephants start up, and in their fury try to break the door. They are all in commotion. The hunters then kindle fires and make much noise. The elephants run about till they get tired, and no strength is left in them. Tame females are then brought to the place, by whose means the wild elephants are caught. They soon get tame.

"From times of old, people have enjoyed elephant hunts by any of the above modes; His Majesty has invented a new manner which admits of remarkable *finesse*. In fact, all excellent modes of hunting are inventions of His Majesty. A wild herd of elephants is surrounded on three sides by drivers, one side alone being left open. At it several female elephants are stationed. From all sides male elephants will come to cover the females. The latter then go gradually into an enclosure, whither the males follow. They are now caught as shewn above."

Abul Fazl's description of the construction of an enclosure, the door of which is secured by the cutting of a rope, is practically that pursued by **Sanderson**. The fact that after being frightened for a time by the noises and fires of the men outside the enclosure, the animals, as if in despair, commence to eat the food provided for them, just as described also by Sanderson, shows how accurately the author of the *Ain-i-Akbari* had observed the Elephant-capturing operations pursued in Akbar's time.

Capture in Jahangir's presence.

Mr. **Blochmann** gives, as a footnote to the above, an account of a capture of elephants made in the presence of the Emperor Jahángír, which might be almost read as a scene from **Mr. Sanderson's** most detailed descriptions of his Kheddah operations. The passage is as follows :—"A large number of people had surrounded the whole jungle, outside of which, on a small empty space, a throne made of wood had been put on a tree as a seat for the Emperor (Jahángír), and on the neighbouring trees beams had been put, upon which the courtiers were to sit and enjoy the sight. About two hundred male elephants with strong nooses, and many females, were in readiness. Upon each elephant there sat two men of the Jhairyyah caste, who chiefly occupy themselves in this part of India (Gujrat) with elephant-hunting. The plan was to drive the wild elephants from all parts of the jungle near the place where the Emperor sat, so that he might enjoy the sight of this exciting scene. When the drivers closed up from all sides of the jungle, their ring unfortunately broke on account of the density and impenetrability of the wood, and the arrangements of the drivers partially failed. The wild elephants ran about as if mad; but twelve male and female elephants were caught before the eyes of the Emperor" (*Iqbálnámah, p. 113*). An earlier writer, **Linschoten** (frequently placed under quotation in this work), speaks of herds of a thousand elephants being surrounded, and a selection of a hundred or more made. **Linschoten's** account is historically of interest,

Capture in Burma.

since it shows that the Kheddah system was followed in Burma 300 years ago :—"They are found also, he says, in India, and in Bengala, and in Pegu great numbers, where they (use to) hunt them with great troupes of men, and tame elephantes, and so compasse, and get into a heape a thousand or two (at the least), whereof they choose out a hundreth or more as they néede, and let the other go, that the Countrey may alwaies have great store. Those they (doe) in time (bring up and) learne (them to travel) with (them and to indure) hunger and thirst, (with) other inventions, so long that they beginne to understand men when they speake. Then they annoint them with Oyle, and wash them, and so do them great good, whereby they become as tame and gentle as men so that they want nothing but speech" (*Linschoten, Vol. II., p. 1*). This remarkable

| The Indian Elephant. | (*G. Watt.*) | ELEPHAS indicus. |

MODE OF HUNTING,

observer in another passage alludes to the process of training, to the habit of the rider sitting on the neck with his feet under the ears, and to his using an iron "hook" to direct the action of the animal. His observation as to the elephant giving a rope one turn round his tusk and grasping the end between his teeth is almost in the very words used by Sanderson, so that this clever trick is no modern acquisition :—" Then the keeper getteth upon the necke (of the elephant) and thrusteth his feet under his eares, having a hooke in his hand, which he sticketh on his head, where his stones lye, that is to say, above betweene both his eares, which is the cause, that they are so well able to rule them : and comming to the thing which they are to draw, they binde the fat or packe fast with a rope that he may feele the waight thereof, and then the keeper speaketh unto him : whereupon hee taketh the corde with his snout, and windeth it about his teeth, and thrusteth the end into his mouth, and so draweth it hanging (after him), where they desire to have it. If it be to be put into a boate, then they bring the boate close to the shore of the Key, and the Elephant putteth it into the boate himselfe, and with his snout gathereth stones together, which he laieth under the fat, (pipe, or packe) and with his teeth striketh (and thrusteth the packe or vessel,) to see if it lie fast or not " (*Linschoten, Vol. II., p. 2*). To any person who has seen the elephant piling great logs of timber at Moulmein, this feat of placing stones underneath the pipes of oil, &c., will not appear an overdrawn picture. The Moulmein elephants may be witnessed while at work to carefully examine if the logs lie straight and to tilt them this way or that way until parallel. In both cases, the intelligence may have proceeded, however, from the rider who, by almost imperceptible hints with his heels, knees, hands or words, commands the trained actions of the elephant. But the illustration shows the high state of elephant training that existed in India during Linschoten's time (1596).

TRADE OR SUPPLY AND DEMAND IN ELEPHANTS.

TRADE, 100

Sanderson, while admitting that both the Ceylon and the African elephant may be viewed as threatened with extermination, is fully convinced that the Indian stock is in no way endangered by the present or even a greatly increased demand. The animal is captured purely for the purpose of being utilized as a beast of burden, and is not, as in Africa, ruthlessly destroyed on account of the ivory. Reckless persecution is prohibited, and a vast reform effected by the substitution of the Kheddah system of capturing in place of the cruel method of securing them in pits.

EXTERMINATION.—By the pit process the animal was subjected to the greatest cruelty, being even allowed to starve to death from the apathy of the owners of the pits. By far the largest proportion of the animals so procured also died before or soon after they left the pits. Many were at the same time rendered useless through their limbs being broken by the fall into the pits. In some localities elephants are so numerous that they effect heavy damage on the neighbouring crops, and on this account rewards were at one time (in Madras, for example) offered for their destruction. The greatest enemy to the elephant is human enterprise in reclaiming jungle tracts of country. Sanderson says : "The number annually caught by the Government establishments is comparatively very small ; and there is no doubt that all the forest ground that can be legitimately allowed to the wild elephant is as fully occupied at present as is desirable. I have examined the elephant-catching records of the past forty-five years in Bengal, and the present rate of capture attests the fact that there is no diminution in the numbers now obtainable ; whilst in Southern India, elephants have become so numerous of late years that the rifle will have

Extermination. 101

No diminution observable in India. 102

E. 102

ELEPHAS indicus.	The Indian Elephant.

to be again called into requisition to protect the ryots from their depredations, unless more systematic measures for their capture and utilization
than are at present in vogue be maintained."

PRICES.—According to **Sanderson**, Kabúl merchants are the chief
traders in elephants, and the principal countries which meet the Indian
demands are Ceylon, Burma, Siam, and a few of the forests of continental
India. He adds—"from several causes the number brought into the market
is now smaller than formerly, and prices are rising accordingly." He then
gives a table of statistics of imports from Ceylon from 1863, the highest
number in any one year having been 270, but in the year 1870, the imports
shrank to 30, and in 1876 had still further declined to only 3. The chief
Indian mart where elephants are sold to the public is Sonepoor on the Ganges
opposite Patna, a *mela* being there held some time in October or November. The Government of Bengal obtains its supplies from the Kheddah
Establishment at Dacca in Eastern Bengal. The average annual capture
in connection with that establishment is reported to be about 60, and **Sanderson** adds that an elephant which cost the Government £40 to capture
would be sold in the market for at least £150. In addition to the captures
made direct by Government, licenses are also issued for private traders
to capture, Government reserving the right to purchase a certain class of
animals over and above those stipulated for in payment of license. The
Madras Government is entirely dependent on Burma for its supplies, since
there is no catching-establishment in that Presidency. Elephants are,
however, frequently captured by the Mysore Government. Only recently,
Mr. Sanderson secured, on behalf of that State, a herd of some 80
elephants. According to the published returns, Government possesses on
an average about 1,600 elephants, and by present regulations, only females
are retained for the public service. This is owing to the risk attending
males becoming *must*. It may, in conclusion, be stated that **Mr. Sanderson**
has demonstrated that capturing elephants is actually remunerative to
Government in addition to the fact that continuity of supply at a moderate
charge is secured. It may be said that in the open market a good serviceable elephant costs at present R2,000, but year by year, with the extension of railway communication and the opening up of roads, the necessity
for elephants is becoming less and less. They are of greatest use in
regions where road and rail communication is defective, and chiefly in
carrying large articles, such as tents and other heavy baggage, that cannot
conveniently be broken into smaller portions suitable for cattle and mules.

DOMESTICATION.

In modern times the Indian Elephant has not been bred in captivity,
but this, **Mr. Sanderson** explains, is a matter of economy and convenience,
not of necessity. It is both easier and cheaper to capture full-grown
animals than to be deprived of the usefulness of a female during a certain
period of her pregnancy, and during also the subsequent three or four
months, especially when considered in the light of the expense of rearing
and training the young for a considerable number of years before they
attain the age of maturity. During the Mogul Empire, however, elephants were regularly reared in captivity, and apparently some care was
bestowed on the selection of breeds. In Burma, especially among the
Karens, the female elephants are shackled and left at large in the jungles
(during the non-working months), in order to ensure the attentions of wild
males, and the young obtained by this semi-domesticated system are
regularly reared. But, as **Sanderson** adds, "in Burma fodder is plentiful,
and the young stock cost nothing till taken up for sale." In India
generally fodder is so expensive, and the animals are at the same time so

PRICES.
103

**Capturing
Stations.
104**

**Number
captured.
105**

**DOMESTICA-
TION.
106**

| The Indian Elephant. (*G. Watt.*) | **ELEPHAS indicus.** |

overworked, that the offspring of domestication would be, in the 15 years necessary to rear them, both more expensive and less hardy than the captured wild stock. In the *Ain-i-Akbari* will be found much of great interest both as to the breeds of elephants, their classification, kind of work assigned to each, amount of food given, and the wages of attendants, &c. The following extract with regard to breeding may be here given:—

Breeding.

"In former times, people did not breed elephants, and thought it unlucky; by the command of His Majesty, they now breed a very superior class of elephants, which has removed the old prejudice in the minds of men. A female elephant has generally one young one, but sometimes two. For five years the young ones content themselves with the milk of the mother: after that period they commence to eat herbs. In this state they are called *bál.* When ten years old, they are named *pút*; when 20 years old, *bikka*; when 30 years old, *kalbah.* In fact, the animal changes appearance every year, and then gets a new name. When 60 years old, the elephant is full grown. The skull then looks like two halves of a ball, whilst the ears look like winnowing fans." After the above there follows a careful description of the eyes, teeth, tusks, and trunk. "An elephant is perfect when it is eight *dast* high, nine *dast* long, and ten *dast* round the belly, and along the back." "Some elephants rut in winter, some in summer, some in the rains. They are then very fierce, they pull down houses, throw down stone walls, and will lift up with their trunks a horse and its rider. But elephants differ very much in the amount of fierceness and boldness." "When they are hot, a blackish discharge exudes from the soft parts between the ears and the temples, which has a most offensive smell: it is sometimes whitish mixed with red." "The elephant lives to 120 years." From the above passages it will be seen that the attendants employed by Akbar in his elephant stables knew quite as much about the animal as we do at the present day. Even the habits of the wild elephant were fully understood. Space cannot be afforded for more than a very few other quotations from the *Ain-i-Akbari*, but the following will be of interest to naturalists:—"A herd of elephants is called in Hindí *sahu.* They vary in numbers; sometimes a herd amounts to a thousand: wild elephants are very cautious. In winter and summer, they select a proper place, and break down a whole forest near their sleeping place. For the sake of pleasure, or for food and drink, they often travel over great distances. On the journey one runs far in front of the others, like a sentinel; a young female is generally selected for this purpose. When they go to sleep, they send out to the four sides of the sleeping place pickets of four female elephants, which relieve each other." "The time of gestation of the female is generally 18 lunar months." **Abul Fazl** gives a detailed account of the formation of the fœtus, mentioning the periods at which its parts are formed. "Female elephants have often for 12 days a red discharge, after which gestation commences. During that period, they look startled, sprinkle themselves with water and earth, keep ears and tail upwards, and go rarely away from the male." The Emperor Jahángír (*Memoirs, p. 130*), some time after the date of the *Ain-i-Akbari,* while speaking of the period of gestation in elephants, says: "During this month a female in my stables gave birth before my own eyes. I had often expressed the wish to have the time of gestation by the female elephant correctly determined. It is now certain that a female birth takes place after 16, and a male birth after 18 months, and the process is different from what it is with man, the fœtus being born with the feet foremost."

Gestation.

E. 106

The Indian Elephant.

CHARACTER AND PHYSICAL PECULIARITIES OF ELEPHANTS.

**PECULIARI-
TIES.
107**
    Much has been written regarding the intelligence and sagacity of the elephant. Sanderson contends that, in its wild state, the elephant, in allowing itself to be captured by so many transparent stratagems which it might easily frustrate, manifests far less intelligence than most other animals. Nature appears to have gifted it with a certain conscious security proceeding more from its magnitude and strength than from its intelligence and sagacity. When once captured its timidity appears to make it more docile than almost any other animal. There is, in fact, no other known animal where wild adults can be captured and domesticated with so much ease. By various tricks and contrivances it is readily educated, and to such perfection that the slightest hint from the *mahout* (or conductor) makes it obey his utmost wish. It is the expertness of the *mahout* apparently that has given rise to the numerous tales regarding the intelligence of his pupil. Sanderson ridicules the well known tale of the elephant who revenged itself on the tailor by throwing dirty water over him. The elephant is fond of water, and cannot, he contends, be supposed to reason out that this is not likely to be the case with man also. If fable it be, there would seem to be some ground for the belief that a similar power of remembrance of injury done is fully possessed, however, by the elephant. Linschoten says on this point—" but he that hurteth them, hee must take heede, for they never forget when any man doth them injurie, untill they be revenged." Sanderson, while extolling the obedience, gentleness, and patience of the elephant, says he is decidedly stupid and devoid of originality. This, to a large extent, seems true, but the majority of animals could not be educated, even after centuries of domestication, to perform the useful obediences to man's commands which the adult elephant learns in a month after capture.

**GESTATION.
108**
    GESTATION.—The reason of the elephant not being bred in domestication has already been fully stated, and one or two passages have been quoted in which the period of gestation has been dealt with. It may not be out of place here to revert, however, to this subject. The statement that the male calf is carried longer than the female receives confirmation by modern observers. Sanderson writes : " The period of gestation in the elephant is said by experienced natives to vary as the calf is male or female, being 22 months in the case of the former and 18 in the latter. I cannot of my own observation afford conclusive proof that such is the case, though I believe there is some truth in the statement. I have known elephants to calve 20 months after capture, the young always being males when 18 months were exceeded." According to Corse the duration of pregnancy is 20 months and 18 days, and in the *Asian* (June 5th, 1883) instances of elephants breeding in domestication are given, and the duration of pregnancy stated to have been 583 to 680 days.

**WEIGHT.
109**
    WEIGHT, MEASUREMENTS.—The elephant breeds but once in two and a half years, and only very exceptionally produces twins, though two calves usually suck at the same time. The calf sucks with its mouth not its trunk as has been incorrectly recorded. The calf usually stands three feet high at the shoulder when born, and the trunk is then only two inches long. The average weight at birth is generally 200℔; a large full-grown elephant weighs 6,000 to 7,000℔ ($3-3\frac{1}{2}$ tons).

**AGE.
110**
    AGE.—The medium height of a full-grown elephant is $7\frac{1}{2}$ to 8 feet, but 9 feet 10 inches as the height of the shoulders is often attained. Sanderson points out that the height of an elephant may be obtained by casting a tape twice round the forefoot. Maturity and full growth is attained at from the 20th to the 25th year; but the first calf is generally born when the

E. 110

| The Indian Elephant. | (*G. Watt.*) | **ELEPHAS indicus.** |

cow is 13 to 16 years of age, and this very frequently takes place in September to November. It is believed the full age of an elephant is 120 years. At about 35 years a male obtains the strength to give him command of a herd. Male elephants of mature age are subject to periodical paroxysms, supposed to be of a sexual nature. The animal is then termed *must* or mad. The fits of *must*, Sanderson affirms, differ in duration in different animals : in some they last for a few weeks, in others for even four or five months. "Elephants are not always violent or untractable under their influence, being frequently only drowsy and lethargic. The approach of the period of *must* is indicated by the commencement of a flow of oily matter from the small hole in the temple on each side of the head, which orifice is found in all elephants, male and female. The temples also swell. The elephant frequently acts somewhat strangely, and is dull and not so obedient as usual. In the advanced stages the oily exudation trickles freely down from the temples, which are thus much swollen."

MUST.
III

"On the first indications the elephant is strongly secured. If he becomes dangerous, his food is thrown to him and water supplied in a trough pushed within his reach." Sanderson continues : "The flow of *must* occasionally, but very seldom, occurs in female elephants. I have seen it twice in newly-caught females in the prime of life, and in very full condition. It never occurs, I believe, in tame female elephants."

In the wild state, although the discharge takes place, it does not appear to be often associated with madness. This seems to depend, as Steel expresses it, to some extent on the condition of the domesticated animal, *highly fed and lightly worked.* "It has been supposed that male elephants as well as females 'come into heat,' and although they seem always prepared to pay attentions to females, there are certainly seasons when the sexual instinct in them runs higher than at others, and which may correctly be called 'rutting times.'" The male approaches the female in the attitude common to most quadrupeds, and not in the crouching position assumed by the camel.

PACE.—"The only pace of the elephant is the walk, capable of being increased to a fast shuffle of about fifteen miles an hour for a very short distance. It can neither trot, canter, nor gallop. It does not move with the legs on the same side together, but nearly so. A very good runner might keep out of the elephant's way on a smooth piece of turf ; but on the ground in which they are generally met with any attempt to escape by flight, unless supplemented by concealment, would be unavailing." An elephant cannot jump, can never have all four feet off the ground together. As Sanderson points out a trench seven feet wide is impassable to them, though the step of a full grown animal may be put down at 6½ feet. In a further passage, Sanderson says that four miles an hour is a good pace for an elephant, but long-legged ones will swing along at five or upwards for a moderate distance, say, ten miles. "I have known," he adds, "thirty-nine miles done at a stretch at a moderate pace. Single wild elephants that have been wounded or much frightened will often travel as far as this in a few hours without a halt." The elephant is remarkably sure-footed, being known to charge down hill with as much ease as up. He swims remarkably well, the body being down in the water with the trunk carried erect for breathing. In fording shallow streams he moves cautiously, and may be trained to tramp down materials given him, to ensure a better footing. Should the ground sink underneath him he rolls over on his side to liberate his feet. It is thus recommended to send one elephant over a ford without his load in order to ascertain the nature of the shallow river bed before taking others with loads across.

PACE.
112

Cannot jump.
113

LOADING.—The elephant equipment should be so constructed that the

LOADING.
114

E. 114

ELEPHAS indicus.	The Indian Elephant.

PECULIARI- TIES.

weight of the load rests on the upper part of the ribs not on the spine. Half a ton is considered a good load for an elephant intended for continuous marching. Sanderson says: " I have known a large female carry a pile of thirty bags of rice, weighing 82℔ each, or 1 ton and 2 cwts., from one store room to another, three hundred yards distant, several times in a morning. By the Bengal Commissariat Code elephants are expected to carry 1,640℔, exclusive of attendants and chains, for which 300℔ extra may be added; but this is too great a weight for continued marching."

Load half a ton. 115

Captain Hood gives the following estimate for loads :—elephants 7 feet 6 inches high not to exceed 6 maunds; 8 feet 7 maunds; 9 feet 8 maunds; and 10 feet 9 maunds. This is for hilly country, and for the plains he allows to each of the above animals 2 maunds extra. An excessive load tires the animal too soon, makes the feet sore, and causes it to stumble. An average load is, therefore, equal to that which would be carried by three camels or by seven and a half mules. On the march metalled roads are to be avoided, as these soon injure the feet and render the elephant useless. On this point Steel writes : " No part of the body is more liable to disorder, and complete temporary incapacity results from injury to or disease of these important organs." Although very sure-footed, an

Not suitable for draught purposes. 116

elephant picking his way through rocky dry beds of streams, a trench or precipitous nullah is almost impassable to him owing to his inability to jump. On ascending steep banks of streams, with a load, he is liable to fall on the back, and in such cases is almost invariably killed. He is not suited for draught purposes, but has often proved most useful in extricating guns from awkward positions; in such cases, however, he more frequently shoves than draws the load. It has already been remarked that the small Nepál elephant is more suited for hill work than the Assam or South India animal.

SLEEPING. 117

SLEEPING.—The elephant requires very little sleep, but if disturbed in the few hours that are necessary he soon gets out of working form. There should be strict silence in the elephant camp after 9-30 P.M., and the sleeping ground, as Ouchterlony recommends, should, if possible, be on the incline, the animals being placed with the hind up hill. Unless this precaution be observed, should the animal lie down, he will most probably be unable to rise again without the aid of other two elephants. To raise him it has been recommended to give stimulants, then push him on one side and leave him to rest for a time, thereafter push him on to his legs. In rising the elephant " elevates the forehand first, and in lying he flexes the fore limb at the elbow and the hind limb at the stifle. The fore foot is bent inwards with the sole turned towards the root of the trunk, which organ lies curled upon the ground." (*Steel.*)

DETECTION OF AGE. 118

DETECTION OF AGE.—In detecting the age of elephants no difficulty is experienced with very old or very young animals; with intermediate ages, however, it is very difficult to say within a few years. Up to six or seven years the top of the ear is not turned over (as in man), but with advancing years it laps over,—in old elephants very much so, and with age, also, the margin of the ear gets torn. It is a common saying that no one has seen in the jungles the remains of a dead elephant, from which circumstance the natives believe he never dies. Sanderson and most sportsmen attach little importance to this circumstance, and affirm that it is no more to be wondered at than the rarity of finding the skeletons of other

Dead Elephants. 119

wild animals. The abundance of animals that greedily devour carcasses, when taken into consideration with the powerfully decomposing influences of the climate, are supposed to be sufficient causes for the fact of the rarity with which the bodies of wild animals are found in the forest.

STATELY BEARING.—The elephant is peculiarly suited for the stately

| The Indian Elephant. | (*G. Watt.*) | **ELEPHAS indicus.** |

PECULIARI-
TIES.

Baggage
Animal.
120

Utility in
ancient
warfare.
121

processions, so much beloved by Native Princes. His graceful motion and great size give him a charm which no other animal possesses. To the sportsman he is of exceptional value, since his obedience and courage render it, comparatively speaking, safe to closely pursue the tiger and other large game, until so hard pressed that exposure to the rifle becomes a necessity. The merit of the elephant as a baggage animal, in regions with defective communication, has already been dealt with, and there thus remains only the question of his utility or otherwise in warfare. In the *Ain-i-Akbari* will be found a description of the manner in which very courageous elephants were employed by Akbar on the actual battle-field. Large *howdahs* were constructed to carry a number of soldiers, who discharged their guns, spears, &c., on the elephant charging the enemy. We read also that the African elephant was once upon a time domesticated, and that the Carthaginians employed them as fighting animals in their wars against Rome. On the conquest of Carthage, the Romans for some time after also employed elephants, but more especially in the amphitheatre and in military pageants Thus during the ascendancy of the Roman Empire, elephants were quite common in Europe, but they ultimately disappeared and for centuries were altogether unknown, and what is more remarkable, the African elephant, since the fall of Carthage, has not been again domesticated. We read of the Indian elephants on the battle-field from the date of the wars against Alexander the Great down to modern times, but with the English army in India he is practically purely a baggage animal. In concluding an instructive chapter on the adaptability of the elephant for certain work in modern warfare, Steel summarises his arguments as follows:—

Adaptability
to modern
warfare.

" I.—The elephant, as an actual weight-bearer, is most valuable.

" II.—He is very difficult to feed; therefore but few can be allowed to the front on service.

" III.—But a few are very useful there to assist guns and other heavy draught over awkward places—whether sandy, muddy, or narrow.

" IV.—In siege trains, for slow draught movement of heavy guns, for carriage of scaling ladders, &c., &c., elephant legitimately finds a place.

" V.—At the base, and along the line of communications, where they can easily feed and are not exposed to attack or capture, elephants are a most useful means for the transport of heavy baggage, stores, and munitions of war. In this respect they are an excellent substitute for wheeled transport if roads be impracticable for the latter. But they cannot advantageously replace carts and waggons or traction engines when the roads are fit for draught.

" VI.—The spread of railways and metalled roads lessens the need for elephant transport; but in unopened jungly country the elephant is invaluable for Commissariat purposes. Thus, *wherever there is a want of good roads from the base, the elephant finds his proper place as an animal of transport*; he is there more useful than any other animal, and will, to an important extent, compensate for the impracticability of wheeled transport.

" VII.—To engineers the elephant proves most useful for shifting heavy guns, for moving heavy beams and other weighty articles, in throwing down walls, and in various other ways.

" VIII.—Once the elephant acted the part of artillery in war—breaking up compact masses of Infantry at once by the weight of its charge and by the dread its appearance gave rise to. It is now used at the front for artillery purposes only in carrying small guns, or in drawing those of Heavy Field Batteries.

ELEPHAS indicus.	The Indian Elephant.

DISEASES TO WHICH ELEPHANTS ARE SUBJECT, AND REMEDIAL AGENTS.

DISEASES.
122

WILD AND CAPTIVE ELEPHANTS.—Few travellers appear to have observed the wild elephant suffering from more than the natural infirmity of age. The young are always in good health. In captivity the diseases to which the animal is liable are probably all due to the sudden and complete change of life forced on him. It is often difficult to procure so large a quantity of grass as he requires, and the habit has thus to be learned of feeding on leaves of trees which, in the wild state, the animal rarely eats. In fact, with the exception of a few trees, the leaves and boughs of which are partaken of more as a relish than a regular article of diet, the elephant confines himself to eating grass. His habits are also methodical, and he rarely exposes himself to the scorching influence of the sun. At fixed intervals he drinks and bathes, at others feeds or reclines under deep and grateful shade, while his hours of sleep are equally a matter of rigid habit. All this is to a large extent disturbed by domestication. The mahout finds it easier to procure for his charge a meal of boughs of trees than of grass, and loving himself the midday heat, unless carefully watched, he will invariably start foraging late in the morning, most probably at the very hour he should be returning home with the day's supply.

Yaarba'hd.
123

Sanderson says that there are two diseases to which the recently captured elephant is liable. These are the dropsical *yaarba'hd*—accumulations of water under the skin—and the wasting *yaarba'hd*, in which the animals fall gradually away to mere skin and bone. Freedom, he adds, from restraint, and liberty to graze as the animal likes, is the only cure for both these diseases. Medicine is of little or no avail.

Colds.
124

The elephant is extremely liable to cold, and extremes of climate or too rapid changes should be avoided. Thus, for example, when on the march the elephant should be allowed half an hour's rest to cool down before he is made to swim a river if the water be cold. If this precaution be not observed the animal is very apt to acquire the troublesome disease known

Chowrung.
125

as *chowrung*.

BLOOD.
126

CLASSIFICATION OF DISEASES.—Steel classifies the diseases to which the elephant is subject into—NON-SPECIFIC DISORDERS OF THE BLOOD such as Debility, *e.g.*, *yaarba'hd* (*saarbad*), fever, rheumatism, &c. SPECIFIC DISORDERS OF THE BLOOD, *e.g.*, Pleuropneumonia, doubtfully obtained from the epidemic out-breaks among cattle; Dysentery or Murrain; Anthrax; Rabies, from dog bites; Foot and Mouth disease (*kultá*), Variola Elephanti, or Elephant Small-pox.

DIGESTIVE.
127

But the elephant is also subject to many of the ordinary maladies which affect the DIGESTIVE SYSTEM, such as Simple Colic, Flatulent Colic, Enteritis, Diarrhœa, Dysentery, Parasites in the Alimentary Canal, Fascioliasis, and Hepatitis. Similarly, the RESPIRATORY ORGANS are frequently affected by the usual diseases to which man and animals are alike liable, such as Catarrh, Sore-throat, Inflammation of the lungs, and Bronchitis. Inflammation of the Kidneys, as Gilchrist pointed out, is also of fairly frequent occurrence, and amongst NERVOUS COMPLAINTS may be mentioned simple Phrensy after Anthrax, while Encephalitis or inflammation of the brain and its membrane often occurs, the animal becoming dangerous. Apoplexy, Tetanus, and Paralysis have been observed in certain cases.

RESPIRA-TORY.
128
URINARY.
129
NERVOUS.
130

SKIN.
Ulcerations.
131

THE SKIN, though remarkably thick, is very sensitive, insects often annoying the animal very much, while SKIN DISEASES, Ulcerations, Boils, &c., are frequent and dangerous, the more so since a surface cure is only too frequently effected with serious later consequences. Sanderson remarks that SORE-BACKS from chafing of gear are exceedingly tedious

Sore-backs.
132

| The Indian Elephant. | (*G. Watt.*) | **ELEPHAS indicus.** |

to cure. "A free use of the knife, great care in cleansing the wound, and the application of plenty of turpentine, strongly impregnated with camphor, are, he affirms, the best methods for insuring a speedy cure. The deep, burrowing holes, usually present in sore-backs, should be well packed with tow, steeped in the camphorated turpentine. This stuffing prevents the wounds closing up too quickly; the growth of new flesh should be encouraged from the bottom, not at the surface of the sore. A cloth steeped in *margosa* (*neem*, **Melia Azadiracta**) oil should be tied over the wound to prevent flies approaching it and irritating the elephant." Oliver recommends that the wound should be freely washed, using a Read's enema syringe to pump the water into the wound. Thereafter, a dressing with turpentine will, he affirms, speedily produce healthy granulation. On the march SORE-FEET is one of the most serious disorders, perhaps the most serious after the risk of injury to the back from imperfectly fitting gear. A slipper to fit over the foot is by most authors recommended to be carried in case of need, and a preparation known as *chób* is regarded as most useful in overwearing of the feet. This consists, among other ingredients, of Catechu 3℔, marking nut powder (**Anacardium**) 6 ℔; Gum of *Sal* (**Shorea robusta**) 1½℔; Wax 2 ℔; Jaggery 6℔; Gingili oil (**Sesamum**) 6℔, &c., made into a paste and applied over the surface of the foot. Steel, in concluding his admirable account of the diseases to which the elephant is subject, gives a list of the remedies in most general use. He remarks that the doses may be said to be twice those given to the ox for corresponding maladies. The mahouts rarely prescribe purgatives, but according to Sanderson, the elephant eats earth for that purpose. Emetics, as with the horse, have no action on the elephant. In addition to the ordinary drugs in use for other animals, such as alum, chalk, sulphate of copper, camphor, &c., &c., Steel mentions the following Indian drugs :—the seeds of **Butea frondosa** as a vermifuge; **Calotropis gigantea** (madar, root and flower) as a narcotic; marking-nut (**Anacardium occidentale**) as a stimulant; sweet flag (**Acorus Calamus**) as a tonic and stimulant; thorn apple (**Datura fastuosa**) as a narcotic; Bonduc nut (**Cæsalpinia Bonducella**) as a stimulating tonic, &c., &c.

DISEASES. CURES. 133

Camphorated turpentine. 134

Sore-feet. 135

Chob. 136

Doses of Medicines. 137

FOOD AND FODDER OF ELEPHANTS.

Sanderson urges that if the elephant obtains a sufficient amount of grass no animal is easier kept in a good state of health. He writes : " It is common to see elephants in poor condition, suffering from nothing but partial starvation, being treated with medicines and nostrums for debility, whilst their appetites are good, and only require a sufficiency of fodder to effect a cure. It may truly be said that all ailments to which elephants are subject are directly or indirectly caused by insufficient feeding. Underfed elephants become weak and unable to stand exposure; they cannot perform their work, and are laid open to attack by even such remote maladies as sunstroke and sore-back through poor condition. The elephant, in common with all wild animals, goes to no excess in any of its habits, and there is no reason, except bad feeding, why the rate of mortality should be so high, as it unhappily is, amongst Government elephants in India. The actual work they have to perform is seldom arduous enough to affect elephants in health."

According to Sanderson the elephant should be fed chiefly on grasses, at least where that is procurable. They become accustomed to tree fodder, but in his opinion this is unnatural and has a good deal to say to the liability of the domesticated animal to various diseases. "The amount of fodder," Sanderson says that should be given to an elephant, "is much

FOOD AND FODDER. Chief Causes of Disease. 138

Rate of Mortality. 139

Tree Fodder ; Acquired Habit. 140

E. 140

ELEPHAS indicus.	The Indian Elephant.

FOOD AND FODDER.

greater than is usually supposed. The Government allowance in Bengal and Madras for an elephant of full size is as follows :—

Weight of Fodder necessary. 141

BENGAL.   ℔
Green fodder—*viz.*, grasses, branches of trees, sugar-cane, &c. . 400
Or in lieu of the above, dry fodder, *viz.*, stalks of cut-grain . 203

MADRAS.
Green fodder 250
Or dry fodder 125

But the amount of suitable green fodder which a full-grown elephant will consume in eighteen hours I have found, by numerous experiments, to be much greater than this—*viz.*, between 600 and 700℔. This is what a beast of average appetite will actually *eat*, excluding what it throws aside; and I have seen a large tusker eat 800℔, or 57 stone, in eighteen hours." In another passage **Sanderson** adds, "since representing the inadequacy of the above allowances to Government in official correspondence on the subject, I have been informed that experiments have been made in the Bengal Commissariat Department in continuation of my own, which have proved that an elephant will eat 750℔ of dry sugar-cane, which is more feeding fodder than grass, per diem, and that steps are being taken to remodel the fodder scale." Steel writes : " No doubt grazing when possible is the best method of feeding, but sufficient range is not always procurable, and in the hot season grass runs short; even then, however, the branches of trees can be obtained and the leaves which constitute their hot weather foliage." **Slymm** wrote, "my opinion is that grass should form the principal kind of green fodder all the year round, and that either on the march, or when the good kinds are not obtainable, or as a kind of variation, its use may be substituted either by banian, jack tree, peepul, bamboo, plantain leaves, fresh paddy straw, or sugar-cane. The plantain leaves I would not recommend during cold or chilly weather." **Forsyth** says the elephant will not of choice feed on bamboo, though the young shoots are very acceptable and nutritious. When plenty good grass can be obtained as at the beginning of the rains, the *ratib* (or rations of food) may be reduced and increased when the fodder (*cherrai*) is scanty or of poor quality. The *ratib* consists (as prescribed in the Commissariat Code) of *atta* (coarse flour) or rice of the third quality or of dhán (unhusked rice) in twice the amount of either *atta* or husked rice. This grain is to be cooked by baking on an iron-plate and made into cakes or *chapatis* weighing about 2℔ each. Grain is also often made up with straw or leaves into small packages and placed in the elephant's mouth. He is fond of being thus fed and *is* a slow eater of grain otherwise, as he can only pick it up in small quantities. Much difference of opinion prevails as to whether the grain should be given cooked or uncooked. Salt and oil are also allowed to the elephant attendants, but the latter for external application only. According to the scale of rations 15℔ of grain a day is allowed to each elephant, 2 ounces of salt, and 1 ounce of oil.

Grass should be chief Fodder. 142

GRAIN. 143

Amount of per day. 144

Sanderson is opposed to giving elephants large allowances of grain, and would prefer a better quality and large quantity of fodder. He contends that the grain diet is unnatural. The wild elephant, however, regularly makes depredations on the fields, and, moreover, digs up roots and other farinaceous additions to his fodder diet. Tennent mentions, for example, the destruction of Sago-palms (**Caryota urens**) effected in Ceylon by the elephant. These palms are split open and their farinaceous pith greedily eaten. The chief difficulty appears to be in securing that the ration of grain is actually given to the elephant, since its allowance of fodder is lessened in consideration of its expensive diet of grain.

The Indian Elephant.	(*G. Watt.*)	**ELEPHAS indicus.**

The following enumeration of the fodder plants specially mentioned by authors as given to the elephant has been obligingly furnished by Mr. J. F. Duthie :—

ELEPHANT FODDER.

**FODDER.
145**

TREES AND SHRUBS.

Acacia Catechu, *Willd.*; Vol. I., p. 27.
A. ferruginea, *DC.*; Vol. I., p. 59.
A. lenticularis, *Ham.*; Vol. I., p. 52.
A. Suma, *Kurz*; Vol. I., p. 60.
The above **Acacias** are used as Elephant fodders in the Central Pro-
[vinces.

Ægle Marmelos, *Corr.*; Vol. I., p. 117.
Artocarpus integrifolia, *Linn.* The Jack-fruit Tree; Vol. I., p. 330.
Balanites Roxburghii, *Planch.*; Vol. I., p. 363.
Boswellia serrata, *Roxb.*; Vol. I., p. 515.
Butea frondosa, *Roxb.* This seems doubtful. (See **Vol. I.,** p. 555).
Capparis horrida, *Linn. f.* (See Vol. II, p. 133).
Ficus bengalensis, *Linn.* (*Brandis, 412*).
F. glomerata, *Roxb.* (*Brandis, 422* ; *Gaz. Poona, 53*).
F. infectoria, *Roxb.* (*Brandis, 414; Stewart, Pb. Pl., 214*).
F. nitida, *Thunb.*; eaten in C. P.
F. religiosa, *Linn.* (*Brandis, 415* ; *Gaz. Poona, 51*).
F. Roxburghii, *Wall.* (*Brandis, 422*).
F. tomentosa, *Willd.* ; eaten in C. P.
F. Tsiela, *Roxb.*; eaten in C. P.
Garuga pinnata, *Roxb.*
Musa paradisiaca, *Linn.*
Odina Wodier, *Roxb.* (*Brandis, 123*).
Ougeinia dalbergioides, *Benth.* (according to Mr. A. Smythies, Forest Department, Dehra).
Phœnix acaulis, *Roxb.*; eaten in C. P.
Ricinus communis, *Linn.*
Shorea robusta, *Gaertn*; eaten by wild elephants in dry seasons in C. P.
Tamarindus indica, *Linn.* (in the Baroda State).
Typha elephantina, *Roxb.* (Elephant grass, one of the most extensively-used marshy plants).

GRASSES.

Bambusa arundinacea, *Retz.* (See Vol. I., 391).
Dendrocalamus strictus, *Nees.* (See Vol. III., 77).
Elionurus hirsutus, *Munro.* (*Fodder Grasses of N. India, p. 28*).
Saccharum spontaneum, *Linn.* (*Fodder Grasses of N. India, p. 25*).

It may in conclusion be remarked that the above grasses are only those that are specially mentioned by authors, or which occur in such abundance, as to make them of special merit as Elephant fodders. Any grass eaten by cattle (except perhaps the Lemon grass) may be given to elephants, and the leaves from a few more trees than the above are occasionally collected. In Ceylon, for example, the elephant often destroys the young cocoa-nut palms by eating the central bud or cabbage. **Sir E. Tennent** mentions the thick dark leaves of **Messua ferrea**: the leaves of the wood-apple, **Feronia elephantum**, and those of **Mimusops indica**, and many others, as all eaten. Tennent adds that " the stems of the plantains, the stalks of the sugar-cane, and the feathery tops of the bamboos, are irresistible luxuries. Pine-apples, water-melons, and fruits of every description are voraciously

Q

ELEPHAS **indicus.**	**The Indian Elephant.**

FODDER.

devoured, and a cocoa-nut when found is first rolled under foot to detach it from the husk and fibre, and then raised in the trunk and crushed, almost without an effort, by its ponderous jaws." **Steel** writes : " Practically, most green stuffs, grasses, and leafy branches, are acceptable to the elephant and can be utilized by him as food—much must be left to his judgment in selection on the emergencies of the march, and when the Commissariat stores run short on a campaign."

(For further information see the article **FODDER.**)

ELEPHANT FLESH.

FOOD.
146

Elephant flesh is much relished by certain hill tribes as an article of diet, so that, in addition to its utility as a baggage animal the elephant may be said to be of value as an article of human food. **Sanderson** narrates a remarkable accident where two tame elephants tied to a recently captured one were all three mysteriously drowned while swimming the Kurnafoolie river of Chittagong hill tracts. Next day the Joomas swarmed in their boats over the place where the animals sank. The carcasses soon floated on the surface and were cut to pieces, and every particle of their flesh removed. Amongst the Hindus generally a singular belief prevails as to

MEDICINAL
USES.
147

the medicinal property of elephant flesh boiled in mustard oil. This, probably from the theory of signatures, is viewed as a sovereign remedy for Barbados leg—the *dail-fil* of the Arabs (*Ainslie*).

IVORY.
148

IVORY.

Reference has been made to the fact that the Ceylon elephant frequently has no tusks. In India a tuskless male is called a *múkna*. The tusks of the Asiatic species are considerably less than the African. The largest Indian tusk on record is that obtained by **Sir Victor Brooke.** The animal from which this was obtained had the left tusk diseased, but the right one measured (outside curve) 8 feet ; length of part outside the socket or nasal bones, 5 feet 9 inches ; greatest circumference 1 foot 4'9 inches ; and weight 90 ℔. **Sanderson** states that the largest tusks, of elephants shot by him, measured respectively 4 feet 11 inches and 5 feet in length outside curve : 16½ inches in circumference at the gum : weight 74¼lb the pair. " As a rule tusks show barely one half of their total length outside the jaw of the living animal. The length within and without the nasal bones is generally exact, but the lip or gum hides a few inches of the projecting half. As the sockets or nasal bones of a large elephant are from 1 foot 6 inches to 1 foot 9 inches in length, this admits of an elephant's having a tusk 3½ feet long, of which 1½ foot (the gum hides about 4 inches) is visible " (*Sanderson*). " Tusks if once lost are never renewed, and if, in cutting off the tips, too much be removed, thus endangering the hollow lower portion, the tusk is completely destroyed. One tusk is generally considerably longer than the other from the habit of the animal in

TRADE IN
INDIAN
IVORY.
149

using one more than the other. The Indian elephant is not hunted expressly for its ivory, and consequently the trade in Indian ivory is, comparatively speaking, limited. During the past five years the exports of Indian ivory have averaged in value from R44,635 to R73,315. India, however, imports a large quantity of African ivory, and does a considerable trade in exporting this foreign ivory to other countries During the past five years the imports of foreign ivory have been valued at from R19,01,258 (the lowest annual valuation) to R31,24,861. The re-exports of this foreign ivory during that period have averaged from R9,46,164 to R18,24,670. The traffic in this foreign ivory is mainly concentrated in Bombay, the supply coming from Zanzibar and the East Coast of Africa. The exports of Indian ivory are almost exclusively from Bengal and Burma. The

| The Lesser Cardamom. | (*G. Watt.*) | ELETTARIA Cardamomum. |

above are the figures published by Government of unmanufactured ivory, but India also imports a large amount of ivory goods which, in the trade returns, appear as manufactured ivory. This trade may approximately be put down as valued at a lakh and quarter of rupees. Almost the entire traffic in manufactured ivory passes between the United Kingdom and Bombay.

TRADE IN INDIAN IVORY.

It is said that Indian ivory has an opaque dead-white colour, and manifests a tendency to become discoloured. The Ceylon ivory is distinguished by fine grain, small size, and pearly bluish tint. Siam ivory is in the trade regarded as much superior to the Indian in appearance and density. It has been remarked of Africa that the nearer the equator the smaller the elephants but the larger the tusks. The finest *transparent* ivory is collected along the West Coast, between latitudes 10°N. and 10°S. The best *white* ivory is obtained from the East Coast. African ivory is said to be best when recently cut. It has a mellow warm transparent tint, as if soaked in oil, and has very little appearance of *grain* or texture. It is reported that England alone imports 1,200,000℔ of ivory, to obtain which 30,000 elephants have to be annually killed, and the world's supply must, it has been estimated, necessitate 100,000 being annually slaughtered. It may safely be assumed that if this rate of destruction continues, a comparatively few years will suffice to exterminate the African species of elephant. Should such a calamity be ever brought about it is to be hoped the advances of civilization may have discovered substitutes of sufficient merit to prevent the demand for ivory being diverted into Asia, since, though fairly plentiful at present, a very few years would suffice to exterminate the Asiatic species and thus in time deprive the world of any living representative of the largest terrestrial animal.

Annual slaughter to obtain Ivory.
150

ELETTARIA, *Maton; Gen. Pl., III., 646.*

[*t. 267*, SCITAMINEÆ.

Elettaria Cardamomum, *Maton; Bentley & Trimen, Med. Pl.,* The LESSER CARDAMOM, MALABAR CARDAMOM, *Eng.;* CARDAMOME, *Fr.;* CARDAMOMEN, *Germ.*

151

Syn.—ALPINIA CARDAMOMUM, *Roxb.*

Vern.—*Choti eláchi, iláyechi, chhoti iláyechi,* HIND.; *Eláchi, iláchi, elaich, gujráti eláchi,* BENG.; *Illáchi,* PB.; *Elechi,* KHANDESH; *Iláchi, chhoti. iláchi,* DEC.; *Elchi,* GUJ.; *Ilachi, malabari-elachi, elchi, veldode,* BOMB.; *Velloda,* MAR.; *Ellakay, aila-cheddi, ellaay, ela-ká, ela-kay, elakay-virai,* TAM.; *Ellakay, élaki chettu, sanua élaki, ellaay, ela-káya, elakáya, vittula,* TEL.; *Yálakki, yelaki, yerakki,* KAN.; *Elettari, ailum chedy,* MALABAR; *Panlat, pala,* or *ba-la, phálá, bhálá,* BURM.; *Ensal, enasal,* SING.; *Upakunchika, elá* (according to U. C. Dutt), and the following as given by Roxburgh:—*Prithweeka, chundruvala, ela, nishkooti, bahoola,* SANS.; *Kakilahe-saghir,* and the following given by Moodeen Sheriff:—*Qaqilah, qaqilahe-sighár, hel, hel-bava, kh-air-bava, shoshmir,* ARAB.; *Kakilahe-khurd,* PERS.

References.—*Roxb., Fl. Ind., Ed. C.B.C., 24; Voigt., Hort. Sub. Cal., 568; Thwaites, En. Ceylon Pl., 318; Dalz. & Gibs., Bomb. Fl. Supp., 86; Grah., Cat. Bom. Pl., 206; Stewart, Pb. Pl., 238; Rheede, Hort. Mal., XI., tt. 4 & 5; Elliot, Fl. Andhr., 49, 167; Memor. on Cardamom cultivation in Coorg by E. Ludlow in 1868; Voyage of John Huygen van Linschoten to India published 1596, Vol. II., 86-88; Pharm. Ind., 230; O'Shaughnessy, Beng. Dispens., 651; Moodeen Sheriff, Supp. Pharm. Ind., 88 and 134; U. C. Dutt, Mat. Med. Hind., 257; Dymock, Mat. Med. W. Ind., 2nd Ed., 786; Fleming, Med. Pl. and Drugs, as in As. Res., Vol. XI., 156; Flück. & Hanb., Pharmacog., 643; U. S. Dispens., 15th Ed., 361; Bent. & Trim., Med. Pl., 267; S. Arjun, Bomb. Drugs, 141; Med. Top., Ajmir, 138; K. L. Dey, Drugs of*

ELETTARIA
Cardamomum.   The Lesser Cardamom.

Ind., 51 ; *Baden Powell, Pb. Pr., 300, 301 ; Drury, U. Pl., 191 ; Lisboa,*
U. Pl., Bomb.. 176 ; Spons', Encyclop., II., 1803 ; Balfour, Cyclop., I.,
1042 ; Smith, Dic., 92 ; Treasury of Bot , I., 446 ; Kew Off. Guide to
Bot. Gardens and Arboretum, 62 ; Ind. For., X., 287 ; Mys. & Coorg
Gaz., I., 124-125, II., 411 ; Ind. Agri., IX., 43 ; Mason, Burma and its
People, 496, 804 ; Madras Manual, Vol. II , 135 ; Nicholson, Man. Coim-
batore Dist., 407 ; Special Report· by Collector, Madura; Rail-born Trade
Report of Bombay, 1881-82.

Habitat.—A large perennial herb, with a thick fleshy or woody rhizome, from the upper part of which are given off the horizontally spreading, flowering, and fruiting stems. It is indigenous in West and South India, growing abundantly in the rich moist forests of the hilly tracts of Kánara, Mysore, Coorg, Travancore, and Madura. **Mr. Ludlow** mentions it as "a native of the hilly parts of Cochin China, Travancore, Malabar, Coorg, Munjerabad, and Nugur. It is extensively cultivated in many other parts of South India, at elevations from 2,500 feet to 5,000 feet. It grows wild also in many parts of Burma, and in the Bhamo District is said to be cultivated in sufficient quantity for local consumption.

HISTORY.
152

HISTORIC NOTE.—It is worth mentioning in this place that **Linschoten,** in the Journal of his Indian Travels (*Published in 1596*), describes two forms of Cardamoms as used in South India. These he calls the Lesser and the Greater Cardamom. It would thus seem that 300 years ago, as at the present day, the Nepál Cardamom was carried all over India. Cardamom is in Sanskrit known as *Ela*, and is mentioned by **Susruta,** so that it must have been used by the Hindus from a very remote period. The early Arabian writers were acquainted with it, and the more recent Muhammadan authors speak of the Cardamom under the names of *Kakulah* and *Híl*. **Dr. Dymock,** referring to the first European knowledge of Cardamoms, says—"When they were first introduced into Europe is doubtful, as their identity with the Amomum and Cardamomum of the Greeks and Romans cannot be proved. Garcia thinks that the Amomum of the ancients was the *Hamáma* of the Arabs, a drug still to be found in the Bombay shops, and which appears to be a species of Sphagnum : it is figured by **Clusius.**" **Muhammad Hussain** gives *kátídáús* as the Greek, and *sharfiyún* and *shusma* as the Syrian names for the Cardamom. He describes two forms—the large and the small. Of the Lesser Cardamom **Linschoten** wrote that "it most groweth in Calicut and Cananor, places on the coast of Malabar." Commenting on **Linschoten's** account of this spice, his contemporary, **Dr. Paludanus,** wrote that, according to **Avicenna,** there are two kinds of Cardamoms—the Greater and the Lesser. He then adds that to the ancient Greeks such as "**Galen, Dioscorides,** and others, it was unknown : and although **Galen,** in his *seventh book* of simples, saith that Cardamomum is not so hot as Nasturcium or watercresses, but pleasanter of savour and smell with some small bitternesse, yet those signes or properties doe not agree with the Cardamomum of India. **Dioscorides** in his first booke and fifth Chapter commending the Cardamomum brought out of Armenia and Bosphorus (although hee saith also that such doe growe in India and Arabia) saith that we must choose that which is full, and tough in breaking, sharp and bitter of taste, and the smell thereof causeth a heavinesse in a man's head : yet is the Indian Cardamomum caryed into these places from whence **Dioscorides** affirmeth that his Cardamomum doeth come although it be neither tough in breaking nor annoyeth the head, neyther is bitter of taste nor so sharp as cloves." Thus **Paludanus** held the opinion that has since become current in the literature of the subject that the Amomum and Cardamomum of the ancient Greeks was not the spice of India.

E. 152

The Lesser Cardamom. (*G. Watt.*)	ELETTARIA Cardamomum.

CULTIVATION.

There are two ways of propagating this plant, *viz.*, by bulbs (or rather rhizomes) and by seed. The chief requirements for successful cultivation are a rich loamy soil, and a site sheltered from strong winds and too much direct sunlight. Clearings in forest land, with a few trees left here and there, in order to give the requisite shade and shelter, are found to offer the best conditions for the production of good crops. In the planting of bulbs, young ones of one to two years old should be chosen. Holes one foot deep and 18 inches wide are dug, and into these, after they have been prepared as beds raised a few inches above the surrounding ground, the bulbs are inserted just below the surface of the soil.

The spaces between each plant may be 6 feet to 12 feet, according to the quality of the soil. The ground should be well cleared of weeds, stones, and rubbish, but when the plants have grown to a certain size no further weeding will be necessary, as nothing will grow under their shade. Seeds should be sown in prepared nurseries, care being taken not to sow too deep. The seedlings, when 6 to 8 inches in height, should be transplanted and treated as directed for bulb propagation. Several writers have recommended an artificial germination of the seeds in a closed tin case, the lid of which is kept tight so as to exclude air and light as much as possible. The seeds are placed on a piece of flannel and kept moist from a saturated layer of soil below. On germination the seeds, according to this process, are recommended to be dusted off the flannel on to a prepared nursery bed, by striking the flannel on the reverse side, and thereafter thinly covered with soil.

It may be as well to give here a few passages from the more important authors regarding the various localities where the plant either occurs wild or exists in that state of cultivation which **Mr. Ludlow** very appropriately describes as a singular kind of jungle horticulture. Compiling largely and admittedly from **Mr. Ludlow's** interesting paper, the learned authors of the *Pharmacographia* (*p. 644*) give the following brief abstract of the system as pursued in South India generally :—

" Previous to the commencement of the rains the cultivators ascend the mountain sides and seek in the shady evergreen forests a spot where some cardamom plants are growing. Here they make small clearings, in which the admission of light occasions the plant to develop in abundance. The cardamom plants attain 2 to 3 feet in height during the following monsoon, after which the ground is again cleared of weeds, protected with a fence, and left to itself for a year. About two years after the first clearing the plants begin to flower, and five months later ripen some fruits, but a full crop is not got till at least a year after. The plants continue productive six or seven years. A garden, 484 square yards in area, four of which may be made in an acre of forest, will give on an average an annual crop of 12½℔ of garbled cardamoms. **Ludlow**, an Assistant Conservator of Forests, reckons that not more than 28℔ can be got from an acre of forest. From what he says, it further appears that the plants which come up on clearings of the Coorg forests are mainly seedlings, which make their appearance in the same quasi-spontaneous manner as certain plants do in the clearings of a wood in Europe. He says they commence to bear in about 3½ years after their first appearance. The plan of cultivation above described is that pursued in the forests of Travancore, Coorg, and Wynaad. On the lower range of the Pulney Hills, near Dindigul, at an elevation of about 5,000 feet above the sea, the cardamom plant is cultivated in the shade. The natives burn down the underwood, and clear away the small trees of the dense moist forests called sholas, which are damp all the year round. The cardamoms are then sown, and when a few inches high

Marginal notes:

CULTIVA-TION. 153

Planting.

SOUTH INDIA. 154

Coorg.

WYNAAD.

ELETTARIA
Cardamomum.   **The Lesser Cardamom.**

CULTIVA-
TION.

are planted out, either singly or in twos, under the shade of the large trees. They take five years before they bear fruit; 'in October,' remarks our informant, 'I saw the plants in full flower and also in fruit, the latter not however ripe.' In North Cánara and Western Mysore the cardamom is cultivated in the betel-nut plantations. The plants, which are raised from seed, are planted between the palms, from which and from plantains, they derive a certain amount of shade. They are said to produce fruit in their third year ; cardamoms begin to ripen in October, and the gathering continues during dry weather for two or three months. All the fruits on a scape do not become ripe at the same time, yet too generally the whole scape is gathered at once and dried, to the manifest detriment of the drug. This is done partly to save the fruit from being eaten by snakes, frogs, and squirrels, and partly to avoid the capsules splitting, which they do when quite mature. In some plantations, however, the cardamoms are gathered in a more reasonable fashion. As they are collected the fruits are carried to the houses, laid out for a few days on mats, then stripped from their scapes, and the drying completed by a gentle fire heat. In Coorg the fruit is stripped from the scape before drying, and the drying is

Cochin and
Travancore.

sometimes effected wholly by sun heat. In the Native States of Cochin and Travancore cardamoms are a monopoly of the respective Governments. The Raja of the latter State requires that all the produce shall be sold to his officials, who forward it to the main depôt at Alapalli or Aleppi, a port in Travancore, where his commercial agent resides." "The cardamoms at Aleppi are sold by auction, and bought chiefly by Moplah merchants for transport to different parts of India, and also, through third parties, to England. All the lower qualities are consumed in India, and the finer alone shipped to Europe. In the forests belonging to the British Government cardamoms are mostly reckoned among the miscellaneous items of produce; but in Coorg, the cardamom forests are now let at a rental of £3,000 per annum under a lease which will expire in 1878. Dr. Cleghorn, late Conservator of Forests in the Madras Presidency, observes in a letter to one of us, that the rapid extension of coffee culture along the slopes of the Malabar Mountains has tended to lessen the production of cardamoms and has encroached considerably upon the area of their indigenous growth. A recent writer has shown from his own experience that the cultivation of the cardamom is a branch of industry worthy the attention of Europeans, and has given many valuable details for insuring successful results."

MYSORE.
155

Mysore and Coorg.—Rice's *Gazetteer* (*I., 124*) gives the following description which will be found to amplify the facts narrated in the above passage :—

"Cardamoms are propagated entirely by cuttings of the root, and spread in clumps exactly like the plantain tree. In the month following the autumnal equinox, a cluster of from three to five stems, with the roots adhering, are separated from a bunch, and planted in the same row, one between every two areca nut palms, in the spot from whence a plantain tree has been removed. The ground around the cardamom is manured with *nelli* (Emblica) leaves. In the third year, about the autumnal equinox, it produces fruit. The capsules are gathered as they ripen, and are dried four days on a mat, which, during the day, is supported by four sticks, and exposed to the sun, but at night is taken into the house. They are then fit for sale. Whenever the whole fruit has been removed, the plants are raised, and, all the superfluous stems and roots having been separated, they are set again ; but care is taken never to set a plant in the spot from whence it was raised, a change in this respect being considered as necessary. Next year these plants give no fruit, but in the year following, yield

E. 155

| The Lesser Cardamom. (*G. Watt.*) | ELETTARIA Cardamomum. |

capsules again as at first. After transplantation, the old stems die, and new ones spring from the roots. Each cluster produces from a quarter to one seer weight of cardamoms."

The Collector of the Madura District reports, in a recent communication, that the seeds are there sown from the beginning of July to the end of October, in small plots prepared for the purpose by weeding and hoeing. The young plants, after having attained a height of about four inches, are carefully transplanted into pits. They are again, when about one foot high, removed to pits one foot square, which have been prepared one or two months previously. The plants begin to yield in the fourth year, and the fruit is picked in the months of November and the earlier half of December. The average crop in the first year of fruit is about 10℔, in the second 15℔, and so on, till a maximum of 25℔ is reached.

Speaking of the tradition which prevails in Coorg regarding cardamom cultivation, Mr. Ludlow remarks: "The Coorgs relate that in the olden times, the cardamom plant was seldom met with in their jungles. The seeds being very agreeable to taste, the plant was much sought after. In course of time people noticed that it only grew in places where the ground had been shaken by the fall of some large tree, or of a large branch thrown down by the force of the wind, especially when this had happened a short time previous to the falling of the annual showers in March and April. In imitation of nature, during the months of February and March, they selected in these jungles the largest trees and felled them, previously cutting down all the smaller surrounding trees and brushwood that would otherwise have lessened the shock given to the ground. By these means the plants increased. The people gradually became more and more acquainted with their requirements."

"The Coorgs have many signs by which they are more or less influenced when selecting sites for new gardens. Many know the good jungles by tradition from their ancestors, who had a better knowledge of them than the present generation; for, in the days of the wars with Hyder and Tippoo, they often were obliged to fly for safety into the recesses of their "Males." They will, in a doubtful jungle, in the month of February, here and there fell a few trees, and judge the following year of its capabilities as a cardamom jungle by the presence or absence of young cardamom plants near to the felled trees."

Travancore—In the *Madras Manual* a short notice will be found regarding cardamom cultivation in Travancore State. It is there stated that "in the hills, the cardamom grows spontaneously, in the deep shade of the forest: it resembles somewhat the turmeric or ginger plant, but grows to a height of 6 to 10 feet, and throws out at the roots the long shoots which bear the cardamom pods. The owners of the gardens, early in the season, come up from the low country east of the ghauts, cut the brushwood and burn the creepers and otherwise clear the soil for the growth of the plants as soon as the rains fall. They come back to gather the cardamoms when they ripen, about October or November." It is further said to be an uncertain crop, being greatly dependent on the rains. In the *Madras Mail* there appeared the following particulars regarding cardamom cultivation in Travancore:—

There are two varieties of this crop, caused by difference of rainfall and soil; one crop comes to maturity in October and the other in January. The former grows in a wet climate and a poor soil, while the other flourish in a dry climate and fine rich soil. The writer's experience is confined to the latter variety. This plant will grow only at certain places, and the presence of a few wild plants safely indicates that the soil will suit the cultivation of cardamoms. In April the ground should be cleared

CULTIVA-TION.

MADURA.
156

COORG.
157

TRAVAN-CORE.
158

E. 158

ELETTARIA   The Lesser Cardamom.
Cardamomum.

CULTIVA-
TION.

of all undergrowth and the seed sown before the monsoon. In October, when the young cardamoms sprout up, it is necessary to thin them out where they are too much crowded and where the ground is sparely grown it should be sown with seed. For two years nothing more is to be done. In the third year the plantation should be weeded and the small crop gathered. In the fourth year the garden should be thoroughly weeded, and as it is, by this time, in full bearing, a close attention should be paid to it. "Cardamoms require light showery weather in March and April, when the flowering scapes are ready to blossom, and the absence of this at the proper time almost ruins the crop." Cardamoms ripen in November and are liable to be damaged by rats, snakes, and vermin of every description.

The scapes with the cardamoms are removed from plants; the capsules are then carefully removed from the scapes and dried on the rocks. The fruits soon lose their green colour and are then ready for the market.

The fruit sells at the coast at R4 per ℔ (Dutch), but the grower gets only a third of this.

A little care on the part of the Travancore authorities has brought up the total produce to 1,500 cwt., which was formerly only a few cwt.

"Roughly estimated, about 20,000 acres were under cultivation, and there is land available for extending the cultivation five-fold.

The yield per acre in even favourable time does not exceed 20 to 25℔ of cardamoms."

BOMBAY.
159

Bombay.—The following special report has been furnished, for the present work, by the Officiating Director of Land Records and Agriculture :—

Cardamoms are grown in Kánara only. In 1887-88 that crop occupied 899 acres. It is common in the hill gardens of North Kánara. It requires plenty water. In a new garden, Cardamoms are grown from seed, and in an old one from cuttings. The seed is sown in October after the outer shell is removed. It must be carefully sheltered from the sun, and it takes three months to sprout. When the seedlings are a foot high, they are transplanted, and a year and a half later they are set in shady places among betel-palms, and begin to bear when three years old. In Sirsi about 1,000 seedlings go to an acre, while in Yellapur the number of seedlings required to plant an acre of land is 650. The pods commence ripening in September and October, and are gathered till the end of February or the beginning of March. There are about 17 pickings, more than half the pickings having an interval of a week between them, while the rest from a fortnight to three weeks. The acre yield varies from 7 to 28℔. The pods, after they are dug out of the ground, are dried four days on a mat, which, during the day, is hung in the sun, and at night is taken into the house. The pods are then fit for sale. When the whole crop has been picked, the plant is taken out of the ground, the useless wood and roots are cleared away, and it is again planted in a fresh hole. The year after it has been moved, the plant yields no fruit, but in the following year it again bears. After the plant has been removed, the old stem dies and a new stem springs from the root.

Cost of
cultivation;
yield, &c.

As Cardamom is never grown by itself it is very difficult to ascertain accurately the cost of cultivation. As a rule, it is grown in spice gardens containing betel-nut palms, betel and pepper vines, and plantains. In an experiment conducted in a good specimen of the highest class of spice garden in full bearing, Mr. **J. H. Todd, C.S.**, estimated the cost of cultivation per acre at R90. To this must be added R45, being a moiety of wages for watching, weeding, and taking care of the garden. Thus the amount of charges per acre comes to R135. By the same experiment the value of produce—114℔—comes to R326. Mr. Todd's details of the

E. 159

The Lesser Cardamom.	(*G. Watt.*)	ELETTARIA Cardamomum

cost of cultivation and profit are more reliable than those given in the *Kánara Gazetteer.*

Bleaching of Cardamoms.--Though local taste appears to prefer them unbleached, a good market is found for doctored Cardamoms as far as Bombay and Bangalore, and for this purpose a considerable proportion of the Cardamoms produced in Kánara is taken to Haveri and Dharwar to be bleached with the aid of the water of the well which is supposed to have the virtues of bleaching and improving the flavour of Cardamoms. The well belongs to a Jangam or Ling-ayat priest. He makes no charge for its use, though it is said that he receives occasional voluntary presents from the Cardamom dealers.

With a view to ascertain whether this well had really the virtues ascribed to it, samples of its water were subjected to analysis by **Dr. Lyon**, Chemical Analyser to Government, and **Dr. Cooke**, Principal, College of Science, Poona. Both think that the so-called virtues of the water are totally fanciful. The Chemical Analyser reported—"I have examined a sample of water stated to be a specimen of that used at Haveri for washing Cardamoms. The sample yielded to analysis the results shown below. I was unable to detect in the water the presence of any special constituents such as would account for the reputation stated to be possessed by it of being a water specially suited for washing Cardamoms :—

	ANALYSIS.		Grains per gallon.
Total solids by evaporation	.	.	.	427·00
Chlorine	.	.	.	.	110·60
Sulphuric acid .	.	.	.	.	36·38
Silica .	.	.	.	.	.	2·59
Alumina .	.	.	.	.	.	4·27
Lime .	.	.	.	.	.	60·20
Magnesia	.	.	.	.	.	34·44
			675·48"

Mr. E. C. Ozanne, C.S., who in 1885 saw the whole process of bleaching describes it as follows : —"Water from the well is drawn and taken to a suitable room. A large earthen-ware vessel is filled with the water into which pounded *antalkai* (the fruit of the soap-nut, **Sapindus trifoliatus**) and *sikikai* (**Acacia concinna**), in the proportion of 2℔ of the former to ¼℔ of the latter for about 5 gallons of the water, are placed and well stirred. Another vessel contains a strong solution of common soap in the water of the well. The mixture containing 2℔ of pounded soap-nut and ¼℔ of *sikikai*, supplies for 5 *mans* (1 man = 26℔) of cardamoms.

"Two women seated on tripods place a wide-mouthed earthen-ware vessel between them—the washing tub as it may be styled. Eight *lota-fulls* of the well water (a large supply of which is kept at hand), are poured into the tub, and three *lota-fulls* of the soap-nut *sikikai* mixture. The *lota* holds about one quart of water.

"The tub then receives a basketful of cardamoms weighing 10℔. The two women plunge their hands into the tub and stir vigorously for about one minute and then suddenly rest for about the same length of time, and again stir for another minute. A thick lather results. This completes the first washing. The cardamoms are baled out by hand and transferred, to a basket where they remain a few seconds till the water has drained off. The basketful is received by two other women sitting on tripods with a washing tub between them. This tub contains 7 quarts of the pure water, 1 quart of the soap-nut and *sikikai* mixture, and one of the soap solution. The cardamoms are stirred as in the first washing with the same interval of rest, and are baled out into another basket. When the

ELETTARIA   **The Lesser Cardamom.**
Cardamomum.

CULTIVA-
TION.

water is drained off, the washed cardamoms are thrown on to a mat. The heap becomes large after a few hours' work. A woman is exclusively in charge of it and continually sprinkles the well-water over it. She is relieved at night by another woman, who sprinkles the heap till morning, once every half hour.

Bleaching.

" Next day when the sun has risen, the heap is carried to the flat roof of the house, and the cardamoms are spread on mats for four or five hours to dry. The next operation is to nip off the short stalks. This is done by women sitting in the house. Each woman has a large pair of English scissors. She squats on the floor and rests her right hand which holds the scissors on the floor, and feeds the scissors with her left hand. The pace at which this nipping is done astonished me. The stalk is very small and care must be taken to cut it off without injury to the cardamom itself. I saw an old woman nip 90 cardamoms in one minute.

"This done, the sorting begins. The small ill-shapen cardamoms are separated, and only the well-rounded ones packed for export to distant markets. A woman sorts a *man* per diem.

" I must now return to the first washing. The mixture in the tub, after the first basketful has been baled out, is replenished by two or three quarts of the well-water and a second basketful washed. The tub is then emptied and a fresh mixture made. The mixture for the second washing also does duty for two basketfuls. The women who wash the cardamoms are paid 3 annas per diem. An ordinary wage is 1½ to 2 annas. The night-watcher receives 4 annas. The nipping is paid for by the piece at the rate of ⅓ anna per *padi* (10 padis = 1 *man* = 26℔). It is said that an expert can earn 2½ annas per diem. She must clip 13℔ therefore; all other hands employed are paid by the day at 2 annas."

Starching.

" Besides this bleaching now-a-days cardamoms are starched. Starching was first introduced at Sirsi, where bleachers had recourse to it as they had to compete with the bleachers at Haveri, who were experts in the art of bleaching, and who had established their fame as such. The starched cardamoms look whiter than the ordinary bleached cardamoms of Haveri ; and the bleachers of Haveri have therefore now taken to starching. The starch is prepared by pounding together rice, wheat, and country soap with butter milk. The paste is dissolved in a sufficient quantity of water, and the solution is sprinkled over the cardamoms to be starched as they are being rubbed by the hand."

It may be worth adding in connection with North Kánara, that **Mr. Talbot,** in his interesting paper on the trees and shrubs of that district, makes no mention of the wild cardamom, from which circumstance it may be inferred as not indigenous. In the *Bombay Gazetteers* brief notices are given regarding the cardamom. Of Khandesh it is said to be grown in sufficient quantity to meet local demand, but that there is no export. It is also mentioned as one of the thirteen spices which are grown in Kolhapur.

AREA OF CULTIVATION, PRICES, &c.

AREA.
160

The total area under cardamoms cannot be definitely determined, though it may be affirmed that the crop is chiefly raised in the portion of the mountainous tract of the southern or south-western extremity of India. The chief districts in the Madras Presidency and the areas under the crop during the past three years were Madura (1885-86, 1,200 acres ; 1886-87, 1,000 acres, and 1887-88, 1,800 acres) ; South Canara (1885-86, 1,000 acres, 1886-87, 1,800 acres, and 1887-88, 1,400 acres); and Malabar (1885-86, 1,500 acres, 1886-87, 1,800 acres, and 1887-88, 2,000 acres). In Mysore, cardamoms are mainly grown in the Kadur District, the area under the crop having, in the corresponding years to the above, been 1,600, 2,300, and

The Lesser Cardamom.	(*G. Watt.*)	ELETTARIA Cardamomum.

2,200 acres. In Coorg the crop rarely occupies much over 300 acres. Thus, in Southern India, according to the published statistics, there were 7,700 acres in 1887-88 and 5,590 acres in 1885-86. According to these returns the area under cardamoms has increased, while it will be found the foreign exports have decreased, but the imports greatly increased. There are many other features of the cardamom trade which appear contradictory, so that in compiling from existing literature, it is difficult to decide the course to be pursued. It is hoped, therefore, that this admission may suggest the desirability of another original enquiry, such as that published by Mr. Ludlow in 1868—an enquiry which would place more recent information in the hands of the public. One of Mr. Ludlow's correspondents, while commenting on the rise of prices, accounted for this by saying the demand for coffee land had contracted the area available for cardamoms. He wrote : " Cardamoms come to our market (Cochin), chiefly from the Travancore State, with a small portion from the Cochin hills. That grown in Wynaad very seldom finds its way to our market. When we say that there is scarcely ever any stock on hand, you will understand that purchases are made from immediate shipment—

		Mds.
"Quantity brought for sale at Calicut		1,100
Ditto exported from Madras, January to November 1867 .		708
Ditto ditto 1866 .		988
Ditto ditto 1865 .		1,884
Ditto ditto 1864 .		1,882

" Cardamoms are gradually becoming scarcer as the land is cleared, and consequently dearer. Prices in the country have more than trebled themselves in the last three years. Present quotations are R88 to R100 per maund, at Cochin and Calicut." "The price realized at home, 5s. to 7s. 6d. per ℔. Home charges averaged about 5d. per ℔; last quotation, 5s. 6d. to 7s. 6d. Cardamoms are sorted according to size and colour, but, unlike coffee and colonial produce generally, the small-sized ones, provided they are plump, are considered the best. The large lanky ones form class No. 2. Discoloured empty ones (or nearly so) constitute triage. The quality of the seeds varies very much according to the locality of the plant. Cardamoms are usually distinguished by the places of their growth, and valued accordingly. At present it is not judicious to ship good cardamoms from the eastern coast, but no doubt, when brands become known, the port of shipment will no longer be so much thought of." According to *Spons' Encylopædia* the price of Madras cardamoms ranges from 1s. 6d. to 7s. a pound, while good Malabar fetches from 6s. to 9s. 6d. and inferior 2s. to 7s. 6d., and Ceylon from 2s. 6d. to 5s. 6d. Dr. Trimen, in his *Systematic Catalogue of the Flowering Plants and Ferns of Ceylon*, speaks of the Ceylon Cardamom as **Elettaria Cardamomum,** *Maton,* var. **major**—the *ensál* of the Singhalese. The cardamoms of Ceylon are much larger than those of India, but this fact should not be confused with the statement made above that the Greater Cardamom of Bengal and Nepál is **Amomum subulatum** and the Lesser Cardamom of South India, **Elettaria Cardamomum,** two widely different plants.

TRADE.

The trade in Indian cardamoms seems to have been declining for some years past. In 1880-81 the *exports* to foreign countries were valued at R8,20,257, but the returns for that year were the highest on record. For subsequent years they were as follows :—1883-84, R5,68,334 ; 1885-86, R5,60,012 ; and 1887-88, R2,04,858. In 1883-84, the United Kingdom received of the above, cardamoms to the value of R4,05,649, but last year only R52,658. After the United Kingdom the other receiving

CULTIVA-TION.

Area.

PRICES.
161

TRADE.
162

| ELEUSINE
ægyptiaca. | The Lesser Cardamom: The Makri Millet. |

TRADE.

countries are generally in the following order of importance:—Arabia, Germany, Aden, and Persia. On the other hand, the *imports* of Foreign Cardamoms seem to be on the ascendant. In 1880-81 they were valued at R4,134, and taking the same years as have been given for the exports, these imports were in 1883-84, R18,351; 1885-86, R92,205; and 1887-88, R2,60,450. During the last mentioned year the bulk of the imports (*viz.*, R2,51,211 worth) came from Ceylon, and of the total of these foreign imports, Bombay received R2,16,455 worth. The coast-wise imports and exports (*e.g.*, the inter-provincial trade by sea) were valued at over 10 lakhs of rupees, so that, excluding the trans-frontier trade by land and the railway, road, and river-borne transactions (the exact figures for which cannot be discovered), the total Indian trade in cardamoms was last year valued at R25,11,053. But it must be added that it is not known how much of these figures of Indian trade in cardamoms relate to the Greater or Nepál Cardamom (see **Amomum subulatum**), though, of course, the bulk of the transactions, especially in South India and Ceylon, must be in the Lesser Cardamom, the fruits of the plant presently under consideration.

OIL.
163

Oil.—An essential OIL is extracted by aqueous distillation. It is of a pale yellow colour, about 5 per cent. being generally obtained; it possesses the flavour and odour of Cardamoms, and is said to be distilled to some extent in Madras.

MEDICINE.
Seeds.
164

Medicine.—The SEEDS are agreeably aromatic, but their chief medicinal use is as an ingredient in compound preparations. "They are used as a corrective for foul breath. Finely powdered they are administered as a snuff for headache. The cardamoms, fried and mixed with mastiche and milk, are employed internally in irritation of the bladder. In nausea and vomiting they are used as a *sherbut* with pomegranate, and in cholera they are resorted to as a stimulant" (*Dr. Emerson*). As the seeds rapidly deteriorate on exposure, they should not be removed from the capsules until required for use.

SPECIAL OPINION.—§ "Carminative, employed with other aromatic drugs," (*Assistant Surgeon Shib Chandra Bhuttacharji, Chanda, Central Provinces*).

FOOD.
165

Food.—Cardamoms are used by the natives in flavouring sweetmeats and certain cooked dishes; also as a spice, and are sometimes chewed in *pán* with betel-leaf.

ELEUSINE, *Gærtn.; Gen. Pl., 1172.* [GRAMINEÆ.
(*J. F. Duthie.*)

166

Eleusine ægyptiaca, *Pers.; Duthie, Fodder Grasses, N. Ind., 56;*

Syn.—CYNOSURUS ÆGYPTIACUS, *Linn.*; DACTYLOCTENIUM ÆGYPTIACUM, *Willd.*

Vern.—*Makra, makri,* HIND.; *Kákuriya,* URIYA; *Suntu-bukrui,* SANTAL; *Cavara-pullu,* MAL (S.P.); *Maka-makna, tipakia,* BUNDEL.; *Madana, chimbari, chubrei, bhobra, madhána, kar-madhana,* PB.; *Malicha, maligha, mansa,* RAJ.; *Mathna, chikára, chota mandiya, ute-sirkum, ute-sirla,* C. P.; *Mhar, nachani, natchni, nagli, raj,* BOMB.; *Tamida, sodee,* TAM.; *Muttengapilloo,* TEL.; *Puta-tana,* SING.

References.—*Roxb., Fl. Ind., Ed. C.B.C., 116; Voigt., Hort. Sub. Cal., 712; Thwaites, En. Ceylon Pl., 371; Stewart, Pb. Pl., 254; Aitchison, Cat. Pb. and Sind Pl., 167; Trimen, Hort. Zeyl., 110; Rheede, Hort. Mal., XII., 131, t. 69; Lisboa, U. Pl. Bomb., 208; Royle, Ill. Him. Bot., 421.*

Habitat.—A perennial grass with stems erect, or creeping and rooting at the nodes. It is plentiful all over Northern India, especially on cultivated ground.

MEDICINE.
Seeds.
167

Medicine.—A decoction of the SEEDS is renowned in Africa as an

E. 167

| The Marua Millet. | (*J. F. Duthie.*) | ELEUSINE Coracana. |

alleviator of pains in the region of the kidney, and its herbaceous parts are applied externally for the cure of ulcers (*Le Maout and Decaisne, Descriptive and Analytical Botany, Eng. Trans., 891*).

Food.—The SEEDS are eaten by the poorer classes, especially during times of scarcity.

Fodder.—It is generally considered to be a very nutritious fodder grass for cattle, being both fattening and milk-producing.

MEDICINE.

FOOD. Seeds. 168

FODDER. 169

Eleusine Coracana, *Gærtn. ; Duthie, Fodder Grasses, N. Ind., 57.*

170

Syn.—CYNOSURUS CORACANUS, *Linn.*

Vern.— *Maruá,* BENG.; *Kode,* SANTAL; *Manduá, maruá, makra, rotka,* N.-W. P. & OUDH; *Mandal, chalodra,* PB.; *Kodon, koda, kodra, kutra,* PB. HIM.; *Nangli, nachni,* SIND; *Nangli, nágli,* BOMB.; *Nagli, nachiri,* MAR.; *Bávto nágli,* GUZ.; *Rági,* SOUTHERN INDIA; *Kayur, kelvaragú,* TAM.; *Tamidelu, rágulu,* TEL.; *Ragi,* KAN.; *Kurakkan,* SING.; *Rájika* (according to **Piddington**), *rági*) (according to **U. C.** **Dutt**), SANS.; *Mandwah,* PERS.

References.—*Roxb., Fl. Ind., Ed. C.B.C., 115; Voigt., Hort. Sub. Cal., 712; Thwaites, En. Ceylon Pl., 371; Dals. & Gibs., Bomb. Fl., 97; Stewart, Pb. Pl., 254; Aitchison, Cat. Pb. and Sind Pl., 168; DC., Origin Cult. Pl., 384; Elliot, Fl. Andhr., 44, 162, 173; Trimen, Hort. Zeyl., 110; Atkinson, Him. Dist., 690; Drury, U. Pl., 193; Duthie & Fuller, Field and Garden Crops, II., 10; Lisboa, U. Pl. Bomb., 187; Birdwood, Bomb. Pr., 109; Royle, Ill. Him. Bot., 420; Church's Food Grains of India, 89; Balfour, Cyclop., 1042; Smith, Dic., 285 and 345; General Adm. Report, Bengal, 1882-83, 12; Report, Agri. Hort. Soc., Vol. IV., 54; Bomb. Gaz., XIII., Part I., 288; Bomb. Gaz., XVI., 99; Gaz. Karnál, 172; Gaz., Simla, 57; Gaz., Mysore & Coorg, I., 77; Nicholson, Man. Coimbatore, 220; Special Report by Collector, Madura; Hunter, Orissa, II., App. IV., 133; Set. Rep., Bareilly, 1874, 88; U. C. Dutt, 208, 314.*

Habitat.—A tall annual grass; stems many, erect or decumbent at the base, and somewhat compressed. At the summit of each stem are four to six digitate, and usually incurved spikes. It is largely cultivated as a rainy-season crop, and in many parts of India its grain constitutes the staple food of the poorer classes. It is affirmed that the grain is never attacked by insects, and will accordingly keep for any length of time.

History.—The facts stated by DeCandolle, in his *Origin of Cult. Pl.*, indicate a probable Indian origin for this millet. In Egypt the ancient monuments bear no trace of its cultivation in early times, and Græco-Roman authors, who knew the country, do not speak of it. It is mentioned by Sanskrit writers under the name of *Rájika* or *Rági*; the word **Coracana** comes from *Kurakkan*, its Ceylon name. Its nearest ally in the wild state is **E. ægyptiaca**, an abundant and somewhat variable species, luxuriant states of which sometimes bear a very close resemblance to the cultivated **E. Coracana**.

HISTORY. 171

Varieties.—There are several so-called varieties of this plant, which differ chiefly according to their requirements as to soil and time of sowing. Under the name of **E. stricta**, Roxburgh has described the form which has the spikes quite straight. This kind requires a richer soil, and is often surprisingly productive.

Varieties. 172

CULTIVATION.

As this millet is cultivated over the greater part of India, it will be necessary to describe briefly the mode of growing it in certain typical regions.

CULTIVA-TION. 173

1. *Himálayan Districts*—Mr. **Atkinson** says: "It is the staple autumn crop of the highlands (up to 8,000 feet) between the Tons and the Sárda, and forms the main food-resource of the agricultural classes. It

HIMALAYAN DISTRICTS. 174

| ELEUSINE Coracana. | The Marua or Ragi Millet. |

CULTIVA-TION

gives a larger yield than other crops, and is said to increase in bulk when ground, qualities that have probably led to its more general cultivation, as it is a poor and very coarse grain." "*Mandua* is cultivated both in ordinary agricultural land and in freshly-cleared jungle. In ordinary land, it usually follows a wheat crop, which is gathered in April-May, and the land is at once prepared for the *mandua* in the same manner as for rice. The seed is sown broadcast, and instead of a harrow, the bough of a tree is drawn over the newly-sown land to cover the grain. When the young plants have risen two or three inches, the whole field is harrowed two or three times, and the vacant spaces are filled up from those where the plants are in excess." "Later on the crop is well weeded with the *kútala,* and in October-November the ears of the *mandua* are cut off." It is generally sown as a mixed crop along with pulses, &c., known collectively as *Kán.*

PANJAB.
175

2. *Panjáb.*—"In the Karnál District it is grown in fairly stiff soil, but chiefly in the Khádar, and then only in small quantities. It is sown in seed beds carefully dressed and manured. The seedlings are then planted out in land which has been twice ploughed, and dressed with the *sohágga.* It is watered once, or twice if the rains are late, and weeded once. The heads ripen slowly, and the ripe heads are picked off, and the grain beaten out .   .   .   . In dry seasons its cultivation as a food crop is largely increased, it being put in fields intended for *ziri*, which cannot be planted out owing to the drought" (*Gaz., Karnál Dist., 178*). In the Kangra District it forms an import.

N. W. PROV-
INCES.

3. *North-Western Provinces and Oudh.*—"It is cultivated under two very different circumstances in these provinces. The most important position it fills is that of the chief food-grain of the hill tracts on their northern border, where it is very extensively cultivated. In Jaunsár Báwar it forms the chief article of food of the hill-men, and is grown on the very poorest soil, often yielding a crop from mere stones and shingle. It is, on the other hand, very rarely grown in the hilly country to the south of the Provinces, where its place is taken by *kodon*. But it is grown to a greater or less extent over the whole of these Provinces, and in the more fertile districts its cultivation is often attended with considerable care, and results in a very large weight of produce. It prefers light soils, and is sown at the commencement of the rains, at the rate of 10℔ of seeds to the acre. In the Allahabad and Azamgarh Districts, it is reported to be occasionally sown in seed beds and transplanted like rice. In this case the seed is sown with irrigation in May, and the seedlings are planted out when the rains break. It suffers greatly from heavy rain, and a good year for rice is a bad year for *mandua*, and *vice versá*. It should be weeded two or three times, and when carefully cultivated, often receives a top dressing of manure after the first weeding. The yield is the heaviest of any of the minor millets, since not only is the gross weight of the produce large, but only a small proportion of this weight consists of husk. In this respect *mandua* is the most profitable of the minor millets. With *sawan* and *kodon*, for instance, the husk contributes almost 50 per cent. of the weight, while with *mandua* it only amounts to 4 or 5 per cent. Where carefully cultivated 12 to 14 maunds of grain may be expected to the acre, but in the hills a much smaller produce than this is gathered, and cultivators would be content with 5 or 6 maunds" (*Duthie and Fuller*).

MADRAS.
177

4. *Madras.*—In the Coimbatore District it is sown in nurseries and transplanted, when a few weeks old, to the fields. It is, however, best known as a garden crop, and is sown generally in June or July; in some localities it is a cold weather, in others a hot weather, crop. It is usually

E. 177

The Marua or Ragi Millet. (*J. F. Duthie.*) | ELEUSINE Coracana.

transplanted from the nurseries, but is sometimes sown broadcast in the beds. It is called a four-months' crop and will produce up to 2,520℔ per acre. On dry land *ragi* is rare; it is then grown chiefly near the hills, where rain is more abundant and the soil is better. The land is well prepared by ploughing and manuring, and the seed is sown broadcast with lines of castor, dholl, &c., in furrows at 10 or 12 feet apart; at about a month old it is interploughed and weeded. The *ragi* is harvested about four months after sowing, and the dholl a month or two afterwards Threshing is performed after it has been heaped to sweat, when the grain becomes looser in the husk and is easily trodden out. It is reaped by cutting off the ears as they ripen, leaving the straw standing till it is removed bodily and stacked. (*Extract from Nicholson's Manual of the Coimbatore District.*) The Collector, Madura, reports that the sowings begin in July and end in November, the reaping in November to February. The cost of cultivation is estimated at R16-8, the outturn at R18-12. It is often grown by irrigation, and is suitable for any soil. The millet is used as food, being prepared either as a cake with water or powdered and boiled.

Ragi yields a valuable food-grain under moderate irrigation. It is easily grown and is extensively raised under wells during the hot season, being planted out from seed-beds. The best plan is to ridge up the land, as is done for maize and cotton, and to plant the seedlings on both sides of the ridges. The crop is a difficult and expensive one to harvest, owing to the ears never ripening at one time, and it is also costly to thresh, the grain adhering with great persistency to the panicle (*Saidapet Experimental Farm Manual and Guide*).

In the Trichinopoly District there were 153,614 acres under *ragi* cultivation in 1888. The crop is sown from May to August, and harvested from September to December. In dry lands the annual outturn amounts to the value of R9, the cost of cultivation being R4-8, and the profit R4-8. In wet lands the yield attains to the value of R14, the cost of cultivation and profit being R7 each. This crop is generally cultivated in black clay, black loam, and red soils (*Report of H. Willock, Esq., Collector of Trichinopoly*).

In the South Arcot District the land intended for *ragi* is first ploughed in January, and at different times between the middle of July and the middle of August. Sheep are then penned on the land for manure, and it is ploughed five or six times, till the soil is reduced to a fine consistency. It is sown between the middle of August and the end of October. It is weeded after twenty or thirty days, and a second time after sixty days. The crop is harvested from the latter part of December to the middle of January.

In the Cuddapah District, during 1887-88, there were 115,087 acres under *ragi* cultivation. There are two kinds, the one irrigated and the other unirrigated. The former is planted for the seed-beds in May and June and reaped in September, while the latter is sown in September and reaped in January. The cost of cultivating the former kind is R15, and the profit is R10, and that of the latter is R7 and the profit R10 per *cawny*.

5. *In Mysore, ragi* is by far the most important crop grown on dry fields, and much care is taken in its cultivation. The soil which suits it best is red, next black, then ash-coloured, and the worst is that which contains much sand. A variety called *tota* or *nát ragi*, and which will not thrive on dry lands, is grown in certain parts of Mysore. A brief description of its cultivation is worthy of mention. "Garden *ragi* is always transplanted, and hence it is called *náti*. The following is the process followed in the Kolar District. For the seedling bed, dig the ground in Pushya (December-

ELEUSINE Coracana.	The Marua or Ragi Millet.

**CULTIVA-
TION.**

January) and give it a little dung. Divide it into squares, and let it have some more manure. Then sow the seed very thick ; cover it with dung, and give it water, which must be repeated once in three days. The ground into which it is to be transplanted is, in Pushya, ploughed five times, and must be dunged and divided into squares with proper channels, like a poppy garden. About the end of January, water the seedlings well, and pull them up by the roots, tie them in bundles and put them in water. Then reduce to mud the ground into which they are to be transplanted, and place the young *ragi* in it, with four inches distance between each plant. Next day water, and every third day for a month this must be repeated. Then weed with a small hoe, and water once in four days. It ripens in three months from the time when the seed was sown; and in a middling crop produces twenty-fold. It is only sown on the ground at times when no other crop could be procured, as the expense of cultivation nearly equals the value of the crop " (*Gas. of Mysore ana Coorg, I., 81*).

**BOMBAY.
179**

6. *Bombay.*—It is grown in the hill lands of the Násik District, sometimes under the wood-ash (*dalhi*) system. The seed is sown in burnt beds in the latter part of May, the seedlings are planted out in June or July, and the crop is reaped in October. It is widely grown in the hill forest country of Kánara, and the grain is generally eaten by the poorer classes. It is the principal crop on the hill lands of the Thána District, and is always cultivated as a first crop after a fallow. About twelve varieties are recognized, half of them early-ripening and the rest are late-ripening. The former are ripe in September and the latter in October. The crop is similarly treated and holds an important position amongst the food-grains in many other parts of the Presidency.

AREA UNDER ELEUSINE.

AREA.

The total area for all India cannot be ascertained, but the following are the areas returned as under the crop in Madras and Bombay for 1887-88 :—Madras 1,551,000 acres ; Bombay 802,000 acres.

**CHEMISTRY.
180**

Chemical Analysis of the Grain.—The following is the composition of *ragi* grain according to Professor Ohurch :—

	In 100 parts.		In 1 ℔.	
	Husked.	Whole.		
Water	13·2	12·5	2 oz.	0 grains.
Albuminoids	7·3	5·9	0 ,,	413 ,,
Starch	73·2	74·6	11 ,,	409 ,,
Oil	1·5	0·8	0 ,,	56 ,,
Fibre	2·5	3·6	0 ,,	252 ,,
Ash	2·3	2·6	0 ,,	182 ,,

The nutrient ratio is here 1 : 13, the nutrient value 84. The percentage of phosphoric acid in the whole grain is about 0·4 (*Food Grains of India, p. 89*).

**FOOD,
181**

Food.—Though eaten largely by the labouring and poorer classes of people in many parts of India, it is not considered to be very wholesome, being somewhat difficult of digestion. In Mysore the flour is dressed either in the form of a pudding, or is made into cakes fried in oil.

SPECIAL OPINION.—§ " It forms the food of four-fifths of the people of Mysore, and is largely eaten by the working classes in Southern India. It enters into jail diet. It is a highly nourishing millet, suited to working men. It sometimes produces diarrhœa, but this is due to bad grinding and non-separation of the coarse coating of the grain " (*Surgeon General W. R. Cornish, F.R.C.S., C.I.E., Madras*).

**FODDER.
Straw.
182**

Fodder.—The STRAW is considered excellent fodder for cattle, and is said to improve by keeping. In the Mysore District cattle thrive and

E. 182

work on it alone without requiring gram, which is not the case with respect to paddy straw. Though considered heating it is sometimes given to horses when grass is scarce.

Domestic Use.—A fermented liquor, called *bojah*, or *bojali*, is prepared from the seeds in the Mahratta country, and a similar beverage, either distilled into spirit or consumed as a kind of beer, is manufactured on the Sikkim Himálaya and imbibed through a straw (*Hooker, Him. Jour., I., 175*).

     DOMESTIC. 183

Eleusine flagellifera, *Nees. ; Duthie, Fodder Grasses, N. India, 57.*

Syn. E. ARABICA, *Hochst.*

Vern.—*Chhimbar*, HIND.; *Gurdub*, N.-W. P.; *Chemri, chimbari, chhembar, kharimbar, dubra, gathil, ghantil (chubrei* and *bháru,* Trans. Indus, according to **Stewart**), PB.; *Ganthia, gánth dob,* RAJ.

References.—*Aitchison, Cat. Pb. and Sind Pl., 167 ; Journ. Agri.-Hort., Soc., 1885, Vol. VII., New Series, 237.*

Habitat.—A small, creeping, perennial grass, found in many parts of Northern India, more particularly where the soil is sandy.

Fodder.—Affords very good fodder for cattle and horses; and in parts of the Panjáb it is said to form the special food of donkeys.

     FODDER. 184

     FODDER. 185

E. indica, *Gærtn. ; Duthie, Fodder Grasses, N. India, 57.*

Syn.—CYNOSURUS INDICUS, *Linn.*

Vern.—*Mal-ankuri*, HIND.; *Gurcháwa*, BUNDEL.; *Jhingri, jhinjhor, makraila, gadha, gadha-charwa, gatha-mandwi, lijhar*, N.-W. P. & OUDH; *Mandavi*, KUMAON; *Mandwa*, RAJ.; *Godchabba, gurra-gadi, kakariya, madanya, mandiál, malghi,* C. P.; *Kuror, káru-chodi,* TEL.; *Sin-gno-myet, hsen-gno-myeet,* BURM.; *Wal-kurakkan,* SING.

References.—*Roxb., Fl. Ind., Ed. C.B.C., 116 ; Voigt., Hort. Sub. Cal., 713 ; Thwaites, En. Ceylon Pl., 371 ; Aitchison, Cat. Pb. and Sind Pl., 168 ; Trimen, Hort. Zeyl., 110 ; Elliot, Fl. Andhr., 86 ; Atkinson, Him. Dist., 691 ; Balfour, Cyclop., 1043 ; Mason, Burma and its People, 478, 818.*

Habitat.—A small, rather coarse-looking grass, abundant on waste ground and by road-sides all over India, ascending to moderate elevation on the Himálayas; also in Burma and in Ceylon.

Fodder.—It is eaten by horses and cattle in Northern India, and in some districts is considered to be a good fodder grass, though **Roxburgh** says that cattle are not fond of it, a remak which may, however, apply to the Bengal form, which the nature of the climate would render more rank and less palatable. In Australia and in North America it is highly spoken of as a pasture grass.

     186

     FODDER. 187

E. scindica, *Duthie, Fodder Grasses, N. India, 58.*

Syn.—DACTYLOCTENIUM SCINDICUM, *Boiss.*

Vern.—*Mandjiro*, SIND; *Bhobra, bobriya,* PB.; *Ganthya, ganti ghás, jangli malicha, kharo-makro,* RAJ.

Habitat.—A slender perennial species confined to sandy tracts in Northern India.

Fodder.—It is valued locally as a good fodder grass.

     188

     FODDER. 189

E. verticillata, *Roxb. ; Duthie, Fodder Grasses, N. India, 58.*

Vern.—*Jharna, therna,* PB.; *Chhinke, kuri chinke, kangri,* RAJ.

References.—*Roxb., Fl. Ind., Ed. C.B.C., 116 ; Aitchison, Cat. Pb. and Sind Pl., 168.*

Habitat.—Resembles **E. indica**, but is taller, and has the spikes arranged in verticels.

Fodder.—It is said to be a good fodder grass for cattle, both in the Panjab and in Rajputana.

     190

     FODDER. 191

R

ELIONURUS, *Humb. & Bonpl.; Gen. Pl., III., 1129.*
[GRAMINEÆ.

192   **Elionurus hirsutus,** *Munro; Duthie, Fodder Gr., N. Ind., 28;*
          Vern.—*Bhanjuri,* N.-W. P.; *Sin, sewan, shewar,* PB.; *Shinwan, siwan,
          gawán,* RAJ.
          References.—*Aitchison, Cat. Pb. and Sind Pl., 173; Todd, Rájasthán,
          II., 286.*
          Habitat.—A perennial grass, 1 to 2 feet high, with silvery pubescent
          spikes of florets. It grows in sandy parts of the Panjáb, also in Sind and
          Bundelkhand, and is a characteristic plant of the Rájputána desert tract.

FIBRE.   Fibre.—The ROOTS are said to yield a fibre used for weavers' brushes.
Roots.
193   Food.—Todd mentions that in Bikanir, where this grass is abundant,
FOOD.   the SEED is collected, and, mixed with *bájra* flour, is largely consumed
Seed.   by the people.
194   Fodder.—Nutritious, and when young affords excellent grazing.
FODDER. Coldstream say sit is a good stacking grass and will keep good for ten
195   years.

Elm, see **Ulmus campestris,** *Linn.*

ELSCHOLTZIA, *Willd.; Gen. Pl., II., 1181.*

196   **Elscholtzia polystachya,** *Benth.; Fl. Br. Ind., IV., 643;* LABIATEÆ.
          Vern.—*Bhangria,* KUMAON; *Rangchari, mehndi, dúss, pothi, garudar,
          tappaddar,* PB.
          References.—*Gamble, Man. Timb., 301; Stewart, Pb. Pl., 168; Atkinson,
          Him. Dist., 315.*
          Habitat.—A shrub or under-shrub, common on the Himálaya, from
          Kashmír to Sikkim, up to 9,000 feet; also on the Khásia Hills.

DYE.   Dye.—South of Kashmír it is said to be used as a DYE (*Stewart*).
197   Structure of the Wood.—Grey, moderately hard, splits and cracks, and
TIMBER. in seasoning separates into concentric masses. Annual rings distinctly
198   marked by a belt of numerous and larger pores in the spring wood.

EMBELIA, *Burm.; Gen. Pl., II., 644.*
[MYRSINEÆ.

199   **Embelia Ribes,** *Burm.; Fl. Br. Ind., III., 513; Wight, Ic., t. 1207;*
          Syn.—E. GLANDULIFERA, *Wight.*
          Vern.—*Baberáng, wawrung,* HIND.; *Biranga, bhai-birrung,* BENG.; *Bái-
          bidanga,* URIYA; *Bebrang,* SYLHET; *Himalcheri,* NEPAL; *Vishaul,*
          MAL (S.P.); *Babrung,* PB.; *Bábrang,* PUSHTU; *Baibrang, wonding,*
          C. P.; *Bhringeli,* MELGHAT; *Karkannie, vaivarang,* BOMB.; *Karkan-
          nie, vavadinga* (fruit), MAR.; *Vavading,* GUJ.; *Bebrang,* SYLHET;
          Váyu-vilamgam, vellal, TAM.; *Váyu-vilamgam,* TEL.; *Vayivalanga,*
          KAN.; *Wel-ambilla,* SING.; *Vidanga,* SANS.
          The Conservator of Forests, Panjáb, in a recent report, states that in Hazára
          " the berries called *Bebrang* is the fruit of the *Kokhur* (**Myrsine africana**).
          The fruit of E. Ribes is known as *Baibarang* or *Wai varang.*
          References.—*Roxb., Fl. Ind., Ed. C.B.C., 197; Voigt., Hort. Sub. Cal.,
          337; Brandis, For. Fl., 284; Kurz, For. Fl. Burm., II., 101; Thwaites,
          En. Ceylon Pl., 172; Dals. & Gibs., Bomb. Fl., 137; Elliot, Fl. Andhr.,
          190; U. C. Dutt, Mat. Med. Hind., 187 and 323; Dymock, Mat. Med.
          W. Ind., 471; S. Arjun, Bomb. Drugs, 83; Murray, Pl. and Drugs,
          Sind, 168; Irvine, Mat. Med. Patna, 16; Med. Top. Oudh, 32; Drury,
          U. Pl., 194; Birdwood, Bomb. Pr., 51; Balfour, Cyclop., 1045; Treasury
          of Bot., 448; Kew Off. Guide to the Mus. of Ec. Bot., 90; Mysore, Cat.
          Cal. Exh., 21; Home Dept. Cor., 316.*
          Habitat.—A large climbing shrub, abundant in the hilly parts of India,
          from the Central Himálaya to Ceylon and Singapore; also in Burma.

E. 199

The Báyabirang, a useful Anthelmintic. (*J. F. Duthie.*)

MEDICINE.
Seeds.
200

Medicine. —According to Susruta the SEEDS of the plant have been described as anthelmintic, alternative, and tonic. Later writers (Dr. U. O. Dutt informs us) recommend it as a carminative, stomachic, and anthelmintic medicine. In the special report from Hazara (quoted above), it is stated that "the berries are prescribed by Hakims in affections of the kidney; they are viewed as a perfect anthelmintic. Dose 6¼ drachms of very finely powdered and previously shelled berries being given in a cup full of butter milk taken on an empty stomach, the first thing in the morning." Many authors allude to them as entering into the composition of several applications for ringworm and other skin diseases. **Royle** says that they possess aperient properties, **Dr. Dymock,** that it is a common practice in the neighbourhood of Bombay to put a few berries of the *vaivarang* plant in the milk that is given to young children; they are supposed to prevent flatulence. He also states that the berries are largely collected in the Bombay Presidency, and have lately been exported to Germany.

SPECIAL OPINIONS.—§ "180 grains (a tola) of the powdered seeds administered at bed-time in curdled milk, followed by a dose of castor oil on the following morning, has been found an efficacious remedy in tape worm" (*Assistant Surgeon Sakharam Arjun Ravat, L.M., Gorgaum, Bombay*). "Used in Mysore externally by itself or in combination" (*Surgeon-Major John North, Bangalore*). "Half an ounce in powder mixed with '*dahi*' (curd) taken on empty stomach is a sovereign remedy for tape worm" (*Assistant Surgeon Mokund Lall, Agra*). "The seeds are used as a carminative. For this purpose they are mixed with tobacco and smoked" (*Aligarh*). "An undoubted carminative and stomachic" (*Civil Surgeon S. M. Shircore, Moorshedabad*). "Powdered seeds used in atonic dyspepsia" (*Surgeon-Major J. J. L. Ratton, M.D., M.C., Salem*). [This drug would seem to richly deserve being experimented with in Europe. It is an undoubted anthelmintic, quite devoid of the nauseating property possessed by male fern. The writer has received numerous medical opinions from one end of India to the other in which a singular uniformity prevails. The drug is not referred to in the *Pharmacopœia of India.--Ed.*]

Food.—The SEEDS are said to be extensively employed as an adulterant for black pepper.

FOOD.
Seeds.
201
202

Embeliarobusta, *Roxb.; Fl. Br. Ind., III., 515; Wight, Ic., t. 1209.*

Syn.—E. BASAAL, *A. DC.*

Vern.—*Bayabirang,* HIND.; *Kalay bogoti,* NEPAL; *Kopadalli,* GOND; *Bebrang,* OUDH; *Bharangeli,* KURKU; *Amti, ambat, barbatti,* BOMB.; *Aipmwaynway,* BURM.

References. —*Roxb., Fl. Ind., Ed. C.B.C., 197; Voigt., Hort. Sub. Cal., 338; Brandis, For. Fl., 284; Kurz, For. Fl. Burm., II., 102; Beddome, For. Man., 137; Gamble, Man. Timb., 240; Thwaites, En. Ceylon Pl., 173; Dalz. and Gibs., Bomb. Fl., 136; Rheede, Hort. Mal., V., 23, t. 12; Atkinson, Him. Dist., 736; Treasury of Bot., 448.*

Habitat.—A shrub, or small tree, extending from the Sub-Himálayan tract east of the Jumna to Bengal, Ceylon, and Burma.

Medicine.—The FRUIT of this species, like that of **E. Ribes,** is given as an anthelmintic, and internally for piles. Atkinson remarks that the greater portion of the *bayabirang* exported from Kumáon seems to be the fruit of **Myrsine africana.** In the *Treasury of Botany* it is mentioned that the young LEAVES, in combination with ginger, are used as a gargle in cases of sore-throat; that the dried BARK of the root is a reputed remedy for toothache, and that the BERRIES, mixed with butter, are used as an ointment, which is applied to the forehead as a specific for pleuritis.

SPECIAL OPINION.—§ "Sometimes used as an antispasmodic and carminative" (*Surgeon-Major C. J. McCanna, I.M.D., Cawnpore*).

MEDICINE.
Fruit.
203
Leaves.
204
Bark.
205
Berries.
206

R 2

E. 206

ENHYDRA fluctuans.	Engelhardtia Bark Tan.

FOOD.
Fruit.
207

Food.—In Orissa the FRUIT is eaten by the poorer classes. Like that of E. Ribes it is collected and sold as an adulterant for black pepper. On Parisnath, Behar, this is said to be a regular trade.

Emblic myrobalan, see **Phyllanthus Emblica,** *Linn.*

Emerald, see **Precious Stones and Rubies.**

Endive, see **Cichorium Endivia,** *Linn.*; Vol. II., p. 285.

208

ENGELHARDTIA, *Leschen.; Gen. Pl., III., 399.* [JUGLANDEÆ.

Engelhardtia Colebrookiana, *Lindl.; Fl. Br. Ind., V., 596;*

Vern.—*Khusam,* BUNDEL.; *Mowa, gobar-mowa, bodal-mowa, mao,* KU-MAON; *Timar rákh,* PB.

References.—*Brandis, For. Fl., 499; Kurz, For. Fl. Burm., II., 491; Gamble, Man. Timb., 393; Aitchison, Cat. Pb. and Sind Pl., 140; Atkinson, Him. Dist., 317; Royle, Ill. Him. Bot., 342.*

Habitat.—A small deciduous tree of the outer North-West Himálaya, ascending to 6,500 feet; often gregarious. **Sir D. Brandis** suggests the probability of this being shown to be only a tomentose and small-sized variety of E. spicata, in which opinion **Sir Joseph Hooker** (in *Fl. Br. Ind., l.c.*) is inclined to agree.

TIMBER.
209

Structure of the Wood.—" Grey with a reddish tinge, moderately hard, even-grained, seasons and polishes well, but is not durable (*Gamble*).

210

E. spicata, *Bl.; Fl. Br. Ind., V., 595.*

Syn.—E. ROXBURGHIANA, *Lindl.;* JUGLANS PTEROCOCCA, *Roxb.*

Vern.—*Silapoma,* HIND.; *Bolas,* BENG.; *Rumgach,* ASSAM; *Dinglaba,* KHASIA; *Bor-patta-jam,* CACHAR; *Vakru,* GARO; *Mowa, mahua,* NEPAL; *Suviak,* LEPCHA.

References.—*Roxb., Fl. Ind., Ed. C.B.C., 670; Voigt., Hort. Sub. Cal., 296; Brandis, For. Fl., 500; Kurz, For. Fl. Burm., II., 491; Gamble, Man. Timb., 393; Rumph., Herb. Amb., II., 169; Royle, Ill. Him. Bot., 342; Ind. For., I., 92.*

Habitat.—A large, handsome, sub-deciduous tree, found in the Terai and outer hills of Eastern Himálaya up to 6,000 feet; also in Chittagong and Burma.

TAN.
Bark.
211

Tan.—Roxburgh states that its thick brown BARK possesses much tannin, and is reckoned by the natives as the best material they are acquainted with for tanning purposes.

TIMBER.
212

Structure of the Wood.—Similar to that of **E. Colebrookiana,** showing a beautiful grain on a radial section. It is used in Sikkim for tea-boxes and building; in the Khásia Hills and Cachar for planking, and spoons are made of it. It does not warp.

213

ENHYDRA, *Lour.; Gen. Pl., II., 360.*

Enhydra fluctuans, *Lour.; Fl. Br. Ind., III., 304;* COMPOSITÆ.

Syn.—E. HELONCHA, *DC.;* HINGTSHA REPENS, *Roxb.*

Vern.—*Harhuch,* HIND.; *Hingchá,* BENG.; *Hilamochiká,* SANS.

References.—*Roxb., Fl. Ind., Ed. C.B.C., 609; Voigt., Hort. Sub. Cal., 416; U. C. Dutt, Mat. Med. Hind., 185, 300.*

Habitat.—Found in East Bengal, Assam, and Sylhet, frequenting rich damp soils.

MEDICINE.
Leaves.
214

Medicine.—According to **Dutt** the LEAVES of this aquatic plant are regarded as laxative, antibilious, and useful in diseases of the skin and nervous system. Prescribed as an adjunct to tonic metallic medicines given for neuralgia.

Juice.
215

SPECIAL OPINIONS.—§ " Expressed JUICE of the leaves is used as demulcent in cases of gonorrhœa; it is taken mixed with milk, either of cow or goat. The leaves are pounded and made into a paste which is

E. 215

| The Chota Chiretta; Gilla Nuts. (*J. F. Duthie.*) | ENTADA scandens. |

applied cold over the head as a cooling agent " (*Assistant Surgeon Anund Chunder Mookherji, Noakhally*). " Useful in torpidity of the liver. The infusion should be made the previous evening. It is boiled with rice and used with mustard oil and salt, dose infusion ʒi. (*Mr. Forsyth, F.R.C.S., Ed., Civil Medical Officer, Dinagepore, North Bengal*). Juice of fresh leaves is bitter; and much used in dyspepsia and bilious complaints" (*Shib Chundra Bhattacharji, Chanda, Central Provinces*).

Food.—" The LEAVES of this water plant are eaten by the natives as a vegetable. Being somewhat bitter they are regarded as wholesome and invigorating " (*U. C. Dutt*).

FOOD.
Leaves.
216

ENICOSTEMA, *Blume; Gen. Pl., II., 807.*
[*t. 600 (Adenema) ;* GENTIANACEÆ.

Enicostema littorale, *Blume; Fl. Br. Ind., IV., 101 ; Wight, Ic.,*

217

Syn.—CICENDIA HYSSOPIFOLIA, *W. & A.;* HIPPION ORIENTALE, *Dals. & Gibs.;* GENTIANA VERTICILLATA, *Linn.*
Vern.—*Chota-kiráyata,* HIND.; *Manucha,* SIND; *Kadavinayi,* MAR.; *Mámijwá,* GUZ.; *Vallari,* TAM.; *Nela-guli, nela-gulimidi,* TEL.
References.—*Roxb., Fl. Ind., Ed. C.B.C., 264; Voigt., Hort. Sub. Cal., 520; Thwaites, En. Ceylon Pl., 204; Dals. & Gibs., Bomb. Fl., 157; Aitchison, Cat. Pb. and Sind Pl., 92; Bot. Mag., II., 249; Elliot, Fl. Andhr., 131, 188; Pharm. Ind., 150; O'Shaughnessy, Beng. Dispens., 460; Dymock, Mat. Med. W. Ind., 2nd Ed., 541; S. Arjun, Bomb. Drugs, 193; Bidie, Cat. Raw Pr., Paris Exh., 34; Drury, U. Pl., 133; Lisboa, U. Pl. Bomb., 262; Balfour, Cyclop., I., 727; Home Dept. Cor. regarding Pharm. Ind., 238; Ind. Ann. Med. Sc., Vol. III., 272.*
Habitat.—A small glabrous herb, with whitish flowers in axillary clusters; met with in moist places all over India from the Panjáb and Gangetic plain to Ceylon; more frequent near the sea, but unknown in Bengal.
Medicine.—This is the *chota* (small) *chiretta* of the natives. It possesses marked bitterness, and is much used in Madras as a stomachic. It is also tonic and laxative (*Pharm. Ind.*). **Dr. Dymock** states that it " is brought to Bombay from Guzerat along with other simples ; the plant is collected when in flower and tied up in small bundles which contain a pound or more."

MEDICINE.
218

Ensilage, see **Fodder.**

ENTADA, *Adans.; Gen. Pl., I., 589.*

Entada scandens, *Bth.; Fl. Br. Ind., II., 287 ;* LEGUMINOSÆ.

219

Syn.—E. PURSŒTHA, *DC.;* MIMOSA SCANDENS, *Linn., Roxb.*
Vern.— *Gilla, gila-gach,* BENG.; *Geredi,* URIYA; *Pangra,* NEPAL; *Takto-khyem,* LEPCHA; *Gelha,* OUDH; *Chian,* N.-W. P.; *Kastori-kaman,* PB.; *Gardal, gardul, garbi, ghárbi, garambi, pilpápra* (seeds), BOMB.; *Kongnyin-nway, kung-nyen, gonnyin, gán nyin,* BURM.; *Pus-wel,* SING.
References.—*Roxb., Fl. Ind., Ed. C.B.C., 420; Brandis, For. Fl., 167; Kurz, For. Fl. Burm., I, 416; Gamble, Man. Timb., 145; Thwaites, En. Ceylon Pl., 98; Dals. & Gibs., Bomb. Fl., 83; Rheede, Hort. Mal., VIII., tt. 32-34, IX., t. 77; Elliot, Fl. Andhr., 60, 181; Dymock, Mat. Med. W. Ind., 2nd Ed., 276; S. Arjun, Bomb. Drugs, 50; Med. Top. Ajmir, 197; Baden Powell, Pb. Pr., 343; Drury, U. Pl., 196; Lisboa, U. Pl. Bomb., 154; Birdwood, Bomb. Pr., 344; Royle, Ill. Him. Bot., 183; Spons, Encyclop., 795; Balfour, Cyclop., I., 1050; Smith, Dic., 371; Treasury of Bot., I., 452; Kew Off. Guide to the Mus. of Ec. Bot., 53; Bomb. Gas., XIII., 24, XV., 433; Mason, Burma and its People, 503, 771.*
Habitat.—A large climber of the forests of the Eastern Himálaya (ascending to 4,000 feet in Sikkim); Eastern Bengal, South India, Manipur, Burma, and the Andaman Islands. Cosmopolitan in the tropics.

E. 219

EPHEDRA.	The Gilla Nut ; made into snuff-boxes, &c.

FIBRE.
Bark.
220
OIL.
Seeds.
221
MEDICINE.
Seeds.
222

Fibre —According to Dr. Thwaites the tough BARK of this plant is used in Ceylon for cordage and ropes.

Oil.—An OIL is said to be expressed from the SEEDS, the properties of which are not known.

Medicine.—A preparation from the SEEDS is used in pains of the loins and also in debility. Dr. Dymock remarks that "the properties of the seeds do not appear to have been tested in European practice ; among the natives they have the reputation of being emetic." Dr. Mason says that in Burma they are, in native Materia Medica, used as a febrifuge. Along with the seeds of several other leguminous plants they are often found mixed with Calabar beans in consignments exported from tropical Africa, and all are known to the natives under the name of '*garbee*' beans. An infusion of the spongy FIBRES of the stem is said to be used with advantage for various affections of the skin in the Philippines (*Dalz. & Gibs., Bomb. Fl.*, 84).

Fibres.
223

SPECIAL OPINIONS.—§ "The kernels of the seeds are used by the natives as stomachic, carminative, and anodyne, in cases of recent confinement. The drug is said to excite appetite, check fever, relieve pain, and regulate the functions of the chylopoietic viscera" (*Civil Surgeon J. H. Thornton, B.A., M.B., Monghyr*). "Powdered kernel mixed with some few spices, is commonly taken by native women for some days immediately after delivery, for allaying the bodily pains and warding off cold" (*Assistant Surgeon Anund Chunder Mookherji, Noakhally*).

FOOD,
Pods.
224
DOMESTIC,
225

Food.—The PODS contain large, flat, hard, polished, chestnut-coloured seeds, or rather nuts, which, on being steeped in water and afterwards roasted, are sometimes eaten by the natives.

Domestic Uses.—Birdwood mentions that the pods, which are often as much as 4 feet in length, are used by the police in the West Indies. According to Dr. Thwaites the juice of the leaves is employed in Ceylon for stupefying fish. The large ornamental seeds are frequently made into snuff boxes, match-boxes, &c. ; and Royle alludes to the fact that the Nepálese make use of a preparation from them as a hair-wash. The most general use, however, to which these seeds are applied, is for crimping linen. Dr. Bonavia, writing from Etáwa, contributed the following account of the process of employment, to the Transactions of the Agri-Horticultural Society, Calcutta :—

"Dhobis up here, and probably also down in Bengal, use a curious kind of nut for crimping linen, without using any crimping irons. This nut they call in Oudh '*Gelha*,' and here '*Chian*,' the latter means a seed. They say it is brought from Bengal and sold in Cawnpore . The Dhobis cut one side, and scoop out the kernel ; then they introduce two fingers into the cavity, and quickly stroke the damp linen forwards with its polished surface. This crimps it beautifully crossways."

226

EPHEDRA, *Linn. ; Gen. Pl., III.*, 418.
(*George Watt.*)

A genus of erect or sub-scandent rigid shrubs, comprising some eight or ten species (or, according to certain authors, three times that number); met with in Europe, temperate Asia, and South America. The EPHEDRÆ belong to the natural order GNETACEÆ—a family closely allied to the CONIFERÆ. They have opposite or fascicled, terete, striate, jointed, branches ; also opposite scales at the joints, and in the axils of these occur solitary or fascicled minute cones. The flowers are uni-sexual, and the plants often even diœcious. On this account it is probable the males and females have been described as different species ; and, moreover, they are extremely variable plants, being much influenced by soil and humidity. In India one species only can be said to occur throughout the Himaláya, *viz*, E. vulgaris, *Rich.* (=E. Gerardiana, *Wall.*) ; but this is also distributed to Central and Western Asia and to Europe. The other two Indian

| The Soma and Homa. | (*G. Watt.*) | EPHEDRA. |

species have a more easterly distribution,—the one extending from Gárhwal to Afghánistan and Persia (**E. pachyclada**, *Boiss.*) and the other being met with in the Panjáb, Rájputana, Sind, and distributed to Afghánistan and Syria (**E. peduncularis**, *Boiss.*).

Interest has recently been taken in these curious plants from the observation that the dried twigs of an **Ephedra** imported from Persia into Bombay constitute the sacred *Homa* of the Parsís. A sample of the *Homa* obtained in Bombay was at first determined as **Periploca aphylla**—an erect, leafless perennial, with twigs as thick as a goose-quill or less, and possessing a milky sap. Subsequent examination of other samples, however, revealed the fact that the *Homa* of the Parsís was in reality an **Ephedra**, and this determination has since received support from the information recorded by **Dr. Aitchison** in his botanical report in connection with the Afghán Delimitation Commission, where it is stated **Ephedra pachyclada**, *Boiss.*, bears, in the Hari-rud valley, the names of *Hum*, *huma*, *yehma*. **Dr. Aitchison** states of that plant that it was found "a very common shrub, from Northern Baluchistan, along our whole route, in the Hari-rud Valley, the Badghis District, and Persia, growing in stony gravelly soil." Of **Ephedra foliata**, *Boiss.*, **Dr. Aitchison** further affirms that *it* is known as *Hum-i-bandak*.

The question has thus been suggested, is the *Homa* of the Parsís the *Soma* of the early Sanskrit writers? **Professor Max Müller**, in an article in the *Academy* (1884), writes : " It is well known that both in the Veda and the Avesta a plant is mentioned, called *Soma* (Zend, *Haoma*). This plant," the learned Professor continues, " when properly squeezed, yielded a juice, which was allowed to ferment, and when mixed with milk and honey, produced an exhilarating and intoxicating beverage. This *Soma* juice has the same importance in Veda and Avesta sacrifices as the juice of the grape had in the worship of Bacchus. The question has often been discussed, what kind of plant this *Soma* could have been? When *Soma* sacrifices are performed at present, it is confessed that the real *Soma* can no longer be procured, and that some *Ci-prés*, such as *Pútikás*, &c., must be used instead. Dr. Haug, who was present at one of these sacrifices and was allowed to taste the juice, had to confess that it was extremely nasty and not at all exhilarating. Even in the earliest liturgical works, in the Sûtras and Brâhmanas, the same admission is made, namely, that the true *Soma* is very difficult to be procured, and that substitutes may be used instead. When it was procured, it is said that it was brought by barbarians from the North, and that it had to be bought under very peculiar circumstances." **Professor Max Müller**, in a further passage, furnishes the oldest known description of the *Soma* plant. He writes : " I published, so far back as 1855, in the *Journal of the German Oriental Society* an account of the plant." " After describing the peculiar rules for buying and rebuying the *Soma* from northern barbarians, as given in the *Ápastamba Yagna-páribhastá*, I added a note : ' The only botanical description of the *Soma* plant, which I know at present, is found in an extract from the so-called *Ayur-veda* quoted in the *Dhúrtasvámi-bháshyátíká*.' There we read : ' The creeper, called *Soma*, is dark, sour, without leaves, milky, fleshy on the surface; it destroys phlegm, produces vomiting, and is eaten by goats.' I added that, according to the opinion of **Sir J. D. Hooker**, this description points to a **Sarcostemma**, which alone, of a large family, combines the qualities of sour and milky; but I remarked at the same time that the fact of this **Sarcostemma** growing in the Presidency of Bombay militated against this identification, because the true *Soma* must be a northern plant, which was replaced in India itself by *Pútikás* or similar substitutes.

HISTORY.
227

EPHEDRA.	The Soma.

HISTORY,

I cannot vouch for the exact age of the *Ayur-veda,* but I doubt whether we shall find any scientific description of the *Soma* of an earlier date."

Since, however, it is stated in the Sûtras and Brâhmanas that substitutes at even that early period had to be used, may it not be that the description in the *Ayur-veda* is the description of the best known substitute? **Sarcostemma** would be difficult to procure in most parts of this country; it would, in fact, have to be imported from the Deccan to Upper and Northern India. The description, however, would agree admirably with that of a **Sarcostemma**. Assuming the determination correct the substitutes for it—the *Pûtikâs*—one of which was the *Pui-sak* (**Basella**) would, when deprived of their leaves, closely resemble the twigs of **Sarcostemma**. Added to all this we have the fact that Roxburgh calls **Sarcostemma brevistigma** the *Soma luta* (or *Soma*-climber), and says of it that it has so much milky juice of a mild nature "that native travellers often suck the tender shoots to allay their thirst." **Mr. Duthie** gives the name *Soma* to the grass **Setaria glauca**; and a very large number of other plants in the various dialects of India have names like *Soma* or *Homa.* For example, **Veronia anthelmintica** is, in Hindústani, known as *Soma-raj*; so also is **Pœderia fœtida.** A creeper with fleshy stems and milky sap, however, must of necessity almost, be a member of the Asclepiadeæ or of the Euphorbiaceæ. Some of the species of **Ephedra** are sub-scandent, leafless shrubs, but they have not got a milky sap; and far from being likely to cause vomiting when taken, they are pleasant in flavour and not unlike the hops of Europe. But the twigs of **Sarcostemma** are certainly not dark, but rather of a delicate succulent green colour. They might turn black when removed from the plant in the form ready for export, but would only do so when the whole of the milky sap had been dried up. The word "dark" would, however, be perfectly applicable to the brownish twigs of the leafless shrub **Periploca aphylla.** That plant has a milky sap, and **Dr. Aitchison** informs us that in Northern Baluchistan it is known as *Um* or *Uma.* Of **Periploca hydaspidis,** *Falc.* (which Aitchison collected at Jelamai near Shinak), he wrote—"it is quite impossible to distinguish it, as it grows, from **Ephedra ciliata,** *Fisch. & Mey.*" A wild species of grape vine is in Kashmír known as *Um* or *Umbur,* and in most of the languages of India the imported grapes, brought into this country are known as *Angúr,* a Persian name. Its Sanskrit name is *Draksha.* A grape grown in Europe and Australia is known as "the Kashmír."

Thus it would appear that the evidence derived from modern vernacular names largely breaks down. **Dr. Dymock,** at the writer's suggestion, examined the *Homa* plant used in Bombay by the Parsís, and pronounced it to be **Periploca aphylla.** A sample was afterwards sent to Kew, and **Mr. W. T. Thiselton Dyer** wrote that "the *Homa* of the Parsís is undoubtedly **Ephedra vulgaris.**" Acting on this assurance the writer, through the kindness of **Dr. Dymock,** had a sample of **Ephedra vulgaris** chemically analysed, with the result that the opinion he formerly advanced seemed to be confirmed, *viz.,* that it afforded a bitter principle which might have been employed much after the same manner as hops are used in Europe and **Acacia** bark in India, *e.g.,* as a bitter adjunct in the preparation of an alcoholic beverage similar to beer or to the Angami Naga *Zú* from rice. It would now, however, appear from a renewed study of the facts since brought to light, that **Periploca** may have an even stronger claim to consideration than **Ephedra.** It seems probable that both plants are used by the Parsís, and assuming that the names *Homa* and *Soma* referred to one and the same thing originally, it may be worth while suggesting that a chemical analysis of **Periploca** should be made in order to determine if it affords, like **Ephedra,** a harmless bitter principle. It is a native of Northern and Western India in the drier tracts, and is from thence distributed through Baluchistan and Afghánistan to Persia, Arabia, and Nubia. It is, in fact, of all the ASCLEPIADEÆ the most prevalent Central Asian species, and is a climbing shrub which answers admirably to the description given by **Professor Max Müller** except in the absence of any information as to its being used as an ingredient, still less as the principal constituent of an intoxicating beverage. It is, however, eaten by goats. "The flower buds are sweet, and are eaten, raw or cooked, as a vegetable." The majority of the plants belonging to this family act as emetics, and it is probable that the mature twigs would be found to possess that property, though they are not so mentioned by Indian writers. There is no evidence of a **Sarcostemma** being found in Central Asia, while **Periploca** is abundant. But it is by no means rare in the hotter parts of Upper India also, so that we are confronted with a serious difficulty. If **Periploca** was the *Soma* of the Aryan invaders of Southern Asia, they failed to recognise the plant in India, and it was perhaps only after they had penetrated to the extreme southern and western limits of their new empire (where **Periploca** does not occur) that they first discovered a plant which seemed to deserve the ancient and sacred name *Soma*, the **Sarcostemma** of botanists.

There is, however, another feature of the *Soma* of the *Ayur-veda* that has still to be dealt with, *viz.*, it was imported into India from the North by barbarians, and "when properly squeezed yielded a juice which was allowed to ferment and, when mixed with milk and honey, produced an exhilarating and intoxicating beverage." These are **Professor Max Müller's** words, and it is assumed they express the main ideas conveyed in Sanskrit literature. Now, it may safely be affirmed that we know of no milky plant the severed twigs of which would be found to still possess their sap, on arrival in the plains of India from a Northern trans-Himálayan region. The expression as to their yielding juice when properly squeezed must therefore have some other interpretation assigned to it. But the juice, we are told, was allowed to ferment, and in that state was mixed with milk and honey. May it not, therefore, have been the case that a decoction was made of the dried twigs which was employed as a ferment with the milk and honey? It is enjoined that the juice was to be obtained from the stem of a plant not the fruit,* and that the liquor was not to be prepared by distillation; but all this could have been arrived at by flavouring with the *Soma* decoction (or infusion) a saccharine liquid left until fermentation had set in. The twigs would be softened in the process of preparing the decoction, and the direction to squeeze them might fairly well have reference to this stage of the process. The Angami Nagas pour boiling hot water over rice and leave the infusion for three or four days, by which time the fluid is both refreshing and exhilarating but soon becomes absolutely intoxicating. They are not reported to add any adjunct to their *Zú*, in order to assist fermentation, but doubtless this is unnecessary, since the troughs, in which it is prepared, are not washed out between each fresh brew. In the various parts of India different materials are employed to establish fermentation. This has already been dealt with in Volume II., page 259, of this work. The reader will there find mention of a cyperaceous plant (vulgarly a grass), and among many others a **Terminalia** which might answer to the *Arjuna* specified in certain passages in Sanskrit works as one of the *Soma* substitutes. The Santals use a plant known to them as *Saram lutur* (**Clerodendron serratum**) when they wish

* An expression which might be accounted for by the remarkable similarity of the long round fruits of **Periploca** to portions of the stem.

HISTORY,

to make their liquor specially intoxicating; and it is said that even from the milky juice of **Calotropis gigantea** (the *Ak, ákanda* in Bengal; the *Ushar,* in Arabic; *Khark* in Persian, and the *Arka, alarka* of Sanskrit) an alcoholic beverage may be prepared. Since most writers hold that the long grapes of Afghánistan, which might not inaptly be compared to the joints of the human fingers, cannot be admissible, the final conclusion which the writer has come to regarding the so-called *Soma* plant of the ancient Vedic literature is, that it would be safer to view the references to that plant as indicating *an early discovery of the art of fermentation** than to seek to establish any special and peculiar plant which may have been first so used. The disappearance of all knowledge in any such special plant (the first fermenting agent), might on this hypothesis be attributed to the discovery of better and easier processes both in the original home of the *Soma* and in the country of Aryan adoption, until the practice lost its sacred associations in the prevalent use of the sub-stitutes. The sacerdotal injunctions might have survived for a time, and substitutes which resembled but possessed none of the properties of the original *Soma* might easily be supposed to have been used by the priest-hood, while the art of fermentation became a domestic industry.

Some short time ago, the writer published a few notes on the subject of the *Soma* plant, suggested on reading **Professor R. von Roth's** paper in the German Oriental Society's Journal for 1884. He instituted a cor-respondence on *Soma* with certain eminent scholars, and a few of their replies may appropriately be here reproduced. These will be found to support the main contention advanced above that the *Soma* was an adjunct in the preparation of the beverage of the ancient Aryans, but did not itself afford a sweet exhilarating fluid.

Dr. Dymock wrote: " On looking over the *Zend Avesta,*" &c., &c., " it appears to me that the *Homa* or *Soma* was not used to obtain liquor from its juice, but that only a small portion of it was added to liquor obtained from grain. The Parsí priests say that the *Homa* never decays, and they always keep it for a considerable time before they use it." It may therefore be remarked —if the *Homa* and *Soma* are the same thing, this fact is utterly at variance with **Dr. Roth's** interpretation of the Sanskrit passages regard-ing the *Soma* not keeping.

Dr. Rice, of New York, a distinguished Sanskrit scholar said :—

" For your interesting pamphlet on the *Soma* plant I am much obliged. Of course I had read the papers by **Professor Roth** already in the original German, but the additional remarks now accompanying them are also interesting. I have often tried to reconcile the apparent objections against the *Soma* to be plain and simple sugarcane, but have not been able to overcome the apparently well-authenticated statements as to the altitude over the sea level and other data which positively prohibit such a belief. But the description of the plant, its pleasant juice, &c., &c., aside from other considerations, make one think of sugarcane or some species of **Sorghum.**"

This is certainly a most interesting suggestion; but apart from other difficulties, it seems impossible to suppose that branches of sugarcane could have been carried from Central Asia to India so as to still contain their sweet sap. As a matter of fact, the sugarcane sap, in India, dries up completely in less than a month. Sugarcane **(Saccharum officinarum)** is very likely a native of South-Eastern Asia—from Bengal to Cochin-China. It was probably first systematically cultivated in India. It is therefore highly improbable that any form of sugarcane was cultivated in Central Asia during the Vedic period, or was, perhaps, even known to the Sanskrit-speaking people prior to their invasion of India. Most of the

* In Siberia the ermine-hunters, when their yeast fails, use the inner bark of the pine as a ferment.

Indian and European names for sugar appear to be derived from the

Sanskrit *Sarkara*, but it does not follow that *Sarkara* was, in its original application, the sweet preparation from a species of sugarcane. An ancient name for Bengal is *Gura*, from whence is derived *Gula*, raw sugar, a term which extends from India throughout the Malayan Archipelago. But *Gura* (or *guda*) occurs also in many ancient writers, such as Charaka and Susruta, so that sugar manufacturing was known in Upper India as well as Bengal. May it not have been prepared from some of the palms, such as the date-palm, which to the present day is in Bengal so extensively grown as a source of *gur* or raw sugar? The Sanskrit name of the sugarcane plant, *Ikshu*, as DeCandolle points out, has survived in Bengal as *Ak*, and in Hindústan as *Uk*. But, though perhaps but a coincidence, it is worth while adding that a similar word exists in some of the Southern and Eastern languages of India for the date-palm. Thus it is *Ichan* or *Ishan* in Tamil, and the sugar prepared from the juice, *Ich-cha-vellam*. In Telegu the date-palm is *Ita* and in Malyal *Inte*. The English word candy and the Arabic *kand*, come from the Sanskrit *khanda*, crystallized sugar, and these names recall **Calotropis gigantea**—the *ak*, *ákandá*, which, according to the Arabs and Persians, yields sugar and manna.

In a letter addressed to the Government of India on the subject of the *Soma*, by Raja Rajendra Lala Mitra, LL.D., C.I.E., we are promised to be favoured with a complete series of the passages relating to the *Soma* from Sanskrit authors. Dr. Mitra wrote :—

" In the later Vedas, the juice of the plant appears to have been used like hops in Europe, as an ingredient in the preparation of a kind of beer and not as a beverage by itself. In poetry, of course, they talk of drinking the *Soma* juice, but this, in the Bráhmana period of the Vedas, is looked upon as a figure of speech. The rituals nowhere enjoin the use of the juice by itself as a meat offering. If we may rely on this interpretation of the Bráhmanas and the rituals as the right one, it would be in vain to search for a plant with profuse sweet juice as the *Soma*. The word 'Sweet,' which has so much puzzled the learned Professor von Roth, may be safely, nay appropriately, used in a poem in praise of bitter beer " (*G. Watt, Editor, Dictionary, Economic Products of India*).

Ephedra pachyclada, *Boiss. ; Fl. Br. Ind., V., 641 ;* Gnetaceæ. 228

 Vern.—*Hum, huma, gehma,* Afgh.; *Oman,* Pushtu.
 References.—*Aitchison, Bot. Afghan Del. Comn., in Trans. Linn. Soc.*
 Habitat.—Rather a tall shrub, found in the drier regions of the Western Himálaya and Western Tibet.

 Tan.—Aitchison says :—" The branches are employed in tanning the skins of goats for water-bottles." Dr. Banerji, writing from Duki, in Baluchistan, mentions that this plant is used for tanning leather in that part of the country also.

 Food.—The small red fruit is eaten, according to Aitchison.

 Domestic Use.—The ashes, Aitchison says, are used either mixed with or in lieu of snuff. Griffith also makes mention of an Ephedra near the Khyber as being used for the same purpose.

E. peduncularis, *Boiss. ; Fl. Br. Ind., V., 641 ; Brandis, For. Fl., t. 69.* 232

 Syn.—E. Alte, *Brand.*
 Vern.—*Kuchan, nikki-kurkan, bratta, tandala, lastúk, mangarwal,* Pb.; *Bandukái,* Trans-Ind. ; *Alte,* Arab.
 References.—*Brandis, For. Fl., 501 ; Aitchison, Cat. Pb. and Sind Pl., 142 ; Raj. Gaz., 30 ; Edgew., Journ. Linn. Soc., VI., 194.*
 Habitat.—A tall scandent shrub, often glaucous, with slender branches. common, on stony ground, in Sind, the Panjáb, and Rájputána.

E. 232

ERAGROSTIS **abyssinica.**	**Fodder Grasses.**

DOMESTIC.
233

Domestic Use. —Bunches of the stem and branches sometimes used on the Salt Range for cleaning brass dishes.

234

Ephedra vulgaris, *Rich. ; Fl. Br. Ind., V., 640.*

> **Syn.** —E. GERARDIANA. *Wall.;* E. DISTACHYA and MONOSTACHYA, *Linn.*
> **Vern.** —*Amsánia, butshur, budshur, chewa,* PB. ; *Khanda, khama,* KUNA-WAR; *Tse, tsapatt, trans,* LADAK; *Phok,* SUTLEJ VALLEY.
> **References.** —*Brandis, For. Fl., 501 ; Gamble, Man. Timb., 394 ; Stewart, Pb. Pl., 228 ; Boiss, Fl. Or., V., 713 ; Atkinson, Him. Dist., 318 ; Econ. Prod., N.-W. Prov., Part V., 89; Royle, Ill. Him. Bot., 348 ; Ind. For., January 1885, Vol. XI., 5 ; Jour. Agri.-Hort. Soc. Ind., Vol. IV., Selections, p. 263.*
> **Habitat.** —A small low growing rigid shrub, abundant in the drier regions of the temperate and alpine Himálaya, from Western Tibet to Sikkim, ascending to 16,000 feet. It is abundant on the Shalai hill north of Simla at an altitude close on 10,000 feet.

TAN.
235

> **Tan.** —Specimens of the twigs, &c., collected near Simla, were analysed by **Dr. Dymock.** The yield was only 3 per cent. of tannin, giving a whitish precipitate with gelatine and with acetate of lead, and a greenish precipitate with acetate of iron.

MEDICINE.
236
FOOD.
237
FODDER.
238
TIMBER
239
240

> **Medicine.** —Aitchison remarks that some part of the plant is used medicinally in Lahoul (*Proc. Linn. Soc., X., 77*).
> **Food.** —Dr. Stewart says that the red berries have a not unpleasant, mawkish, sweet taste, and are sometimes eaten by the natives of the Panjáb Himálaya. They are also eaten in Kumaon.
> **Fodder.** —The plant is browsed by goats.
> **Structure of the Wood.** —Whitish-yellow. Occasionally used as fuel.

Epicanta nepalensis, *Moore ;* COLEOPTERA.

> An insect recommended as a substitute for Cantharides; see **Vol. II., 128.**

Epicarpurus orientalis, *Bl.,* see **Streblus asper.**

Epsom salts, or **Epsomite,** see **Magnesia.**

EQUISETUM, *Linn.*

(*J. F. Duthie.*)

241

Equisetum debile, *Roxb. ;* EQUISETACEÆ.

> **Vern.** —*Buru-katkom-charec',* SANTAL ; *Matti, skinung, bandukei, nari, trotak, búki,* PB. ; *Myet-sek,* BURM.
> **References.** —*Roxb., Fl. Ind., Ed. C.B.C., 745 ; Voigt., Hort. Sub. Cal., 560 ; Stewart, Pb. Pl., 267 ; Aitchison, Cat. Pb. and Sind Pl., 178.*
> **Habitat.** —A perennial vascular cryptogam with creeping rhizomes, and weak fluted stems, composed of superposed jointed tubes. Found in wet situations in the Panjáb, North-Western Provinces, Bengal, and Burma.

MEDICINE.
Plant.
242
FODDER.
243
DOMESTIC.
244

> **Medicine.** —"The PLANT is administered as a cooling medicine, and near Jhelum is given for gonorrhœa " (*Stewart*).
> **Fodder.** —According to **Dr. Stewart,** it is at times given to cattle as fodder.
> **Domestic Use.** —Joints of the stem are used by the natives for cleaning the surface of the nails.

Equus, see **" Horses, Mules, and Asses."**

ERAGROSTIS, *Beauv. ; Gen. Pl., III., 1186.*

245

Eragrostis abyssinica, *Link. ; Duthie, Fodder Grasses, N. Ind., 66.*

> An Abyssinian species, largely grown in the mountainous districts of that country for its grain, of which the natives make bread. It is called

E. 245

Fodder Grasses.	(*J. F. Duthie.*)	ERAGROSTIS cynosuroides.

"Teff," "Thaf," or "Thief," and there are two distinct varieties, white and red; the former is sown as a cold season, and the latter as a rainy season, crop. Experiments recently undertaken at Saharanpur with seed received from the Royal Gardens, Kew, indicate the possible utility of the plant in this country for fodder purposes. For further particulars see *Kew Bulletin of Miscellaneous Information,* No. 1 (1887).

Eragrostis bifaria, *W. & A.; Duthie, Fodder Grasses, N. India, 61.*   246

Syn.—Poa BIFARIA, *Vahl.*

Vern.—*Punya-safed, chota-bhánkta* (Ajmere), *moi* (Mt. Abu), RAJ.; *Wooda-tallum,* TEL.

References.—*Roxb., Fl. Ind., Ed. C.B.C., 111; Thwaites, En. Ceylon Pl., 373.*

Habitat.—A perennial grass with wiry stems, about one foot high. Common on dry rocky ground in hilly parts of India. In Ceylon up to 5,000 feet.

Fodder.—At Ajmere it is considered a good fodder grass; it is eaten by cattle on Mount Abu.

     FODDER.
     247

E. Brownei, *Nees; Duthie, Fodder Grasses, N. Ind., 62.*   248

Syn.—Poa BROWNEI, *Kunth.*

Vern.—*Jenkua,* ROHILKHAND; *Khari,* BUNDELKHAND; *Asata, chir* (Seoni), C. P.; *Choti khidi,* BERAR.

References.—*Thwaites, En. Ceylon Pl., 373; Aitchison, Cat. Pb. and Sind Pl., 169.*

Hábitat.—A perennial grass with stems about one foot high, and bearing numerous closely-packed dark-coloured spikelets. It is plentiful in wet places all over India, ascending to moderate elevations on the Himálaya.

Fodder.—No definite information has been obtained regarding the feeding value of this grass in India, though no doubt it is eaten by cattle along with other grasses. In Australia, according to Baron von Mueller, it is looked upon as a good pasture grass, yielding an abundance of food both winter and summer.

     FODDER.
     249

E. ciliaris, *Link; Duthie, Fodder Grasses, N. Ind., 62.*   250

Syn.—Poa CILIARIS, *Linn.*; P. CILIATA, *Roxb.*

Vern.—*Undar-punchha,* JEYPUR; *Tor-chandbol,* SANTAL.

References.—*Roxb., Fl. Ind., Ed. C.B.C., 112; Dals. & Gibs., Bomb. Fl., 298; Aitchison, Cat. Pb. and Sind Pl., 169.*

Habitat.—Annual, with hairy florets in narrow spike-like panicles. Common on sandy ground. A small variety, with the spikelets in short roundish heads, is frequently met with.

Fodder.—Affords good grazing wherever it occurs in sufficient quantity.

     FODDER.
     251

E. cynosuroides, *R. & S.; Duthie, Fodder Grasses, N. Ind., 62.*   252

Syn.—Poa CYNOSUROIDES, *Retz.*; BRIZA BIPINNATA, *Linn.*

Vern.—*Dab, dáb, durva, davoli,* HIND.; *Kusha,* BENG.; *Dabvi,* BUNDEL.; *Dab, dhab, daboi, kush,* N.-W. P.; *Dib, dab, dhab, dráb, drábh, kusa,* PB.; *Kir-thag, drab,* AFG.; *Chir, dabhat, kusha,* C. P.; *Darbh,* BOMB.; *Darbhu,* MAR.; *Darbha, kusa-darbha, dabha, durpa, áswaláyana,* TEL.; *Kusha, kutha, durbha, puvitrung,* SANS.

References.—*Roxb., Fl. Ind., Ed. C.B.C., 112; Voigt., Hort. Sub. Cal., 716; Dals. & Gibs., Bomb. Fl., 298; Stewart, Pb. Pl., 254; Aitchison, Cat. Pb. and Sind Pl., 169; Elliot, Fl. Andhr., 17, 46, 105; Dymock, Mat. Med. W. Ind., 2nd Ed., 854; S. Arjun, Bomb. Drugs, 153; Year Book Pharm., 1878, p. 288; Baden Powell, Pb. Pr., 383; Atkinson, Him. Dist., 736, 807; Lisboa, U. Pl. Bomb., 279, 284, 290; Birdwood, Bomb. Pr., 347; Royle, Ill. Him. Bot., 427; Balfour, Cyclop., III., 237; Taylor, Topography of Dacca, 60.*

     FODDER.
     251
     252

Habitat.—A strong coarse, perennial grass, with thick far-creeping rhizomes, common in barren ground and sandy soil on the plains of the North-Western Provinces, the Panjáb, and Sind; it grows luxuriantly also on the low lying portions of the *usár* lands in the North-Western Provinces.

FIBRE.
253

Fibre.—It produces a fairly strong fibre, which is used for making ropes. In the Karnál Settlement Report it is stated that the fibre is used for the ropes of Persian wheels, and they are said to last for three months or more. Stewart remarks that the upper part of the stem is in some places used for making the sieves employed in paper manufacture.

MEDICINE.
Culms.
254

Medicine.—The stout CULMS are said to possess diuretic and stimulant properties, with a bitter taste. Dr. Dymock writes: "It is the *Gramina* of the Portuguese at Goa. The *Gramen* of the Romans and αγρωστις of the Greeks was **Triticum repens**, still much used as a diuretic in Europe." The same author states that in the Concan it is prescribed in compound decoctions with more active drugs for the cure of dysentery, menorrhagia, &c.

FODDER.
255

Fodder.—Cattle do not eat it as a rule, though it is liked by buffalos. Captain Wingate, however, says that it is the principal fodder grass on both sides of the Indus in the Deraját tract. According to Dr. Aitchison it is considered by the Afghans to be a good fodder grass, and was largely used as such for the animals belonging to the Delimitation Commission along portions of their route.

SACRED AND
DOMESTIC.
256

Sacred and Domestic Uses.—Dr. Dymock says that it is in constant requisition at the funeral ceremonies of the Hindús, and that the chief mourner wears a ring of the grass upon his finger; it is also placed beneath the *pindás*. Dr. Lisboa, in the Botanical Volume of *Bombay Gazetteer*, states that it is mentioned in Chapter XX. of *Chaturmas Máhátma*, that this plant is a transformation of *Ketu*, and that Chapter XXVI. of *Shrávan Purán* orders that these *darbhs* should be pulled out of the ground on *Pithori Amváshya*, and that unless this is done the plants are not considered fit for use in sacred ceremonies.

The following account is given by Balfour:—

"Some Hindú legends make Garuda the offspring of Kasyapa and Diti. This dame laid an egg, which it was predicted would produce her a deliverer from some great affliction. After a lapse of five hundred years Garuda sprang from the egg, flew to the abode of Indra, extinguished the fire that surrounded it, conquered its guards, the *devata*, and bore off the *amrita* (ambrosia), which enabled him to liberate his captive mother. A few drops of this immortal beverage falling on the *Kusa*, it became eternally consecrated; and the serpents, greedily licking it up, so lacerated their tongues with the sharp grass, that they have ever since remained forked; but the boon of eternity was ensured to them by their thus partaking of the imperishable fluid. This cause of snakes having forked tongues is still, in the popular tales of India, attributed to the above greediness. At the Ganges bathing places for pilgrims, the Brahman guides usually present the pilgrim with blades of this grass."

This grass is frequently used for thatching, and sometimes for the doors and walls of huts. (*Conf.* with **Cynodon Dactylon**, Vol. II., p. 679.)

257

Eragrostis megastachya, *Link.; Duthie, Fodder Grasses, N. Ind., 63.*

Syn.—E. MAJOR, *Host.*

Vern.—*Chiriya-ke-chaolai*, N.-W. P.

References.—*Thwaites, En. Ceylon Pl., 373; Aitchison, Cat. Pb. and Sind Pl., 169.*

Habitat.—This and **E. poæoides**, *Beauv.*, regarded by some writers as varieties of one species, are commonly met with in most parts of India, ascending to 7,000 feet on the Himálaya. Both are annual grasses with spreading many-flowered panicles.

Fodder.—Used more or less as fodder for cattle and horses.

FODDER.
258
259

Eragrostis nutans, *Nees.; Duthie, Fodder Grasses, N. Ind., 63.*

Syn.—Poa NUTANS, *Retz.*; P. INTERRUPTA, *Roxb.*
Vern.—*Lál-báli, asaunra, mumkára,* BUNDEL. ; *Lamcha, rasaurah, ghui,* N.-W. P.; *Kutti-pushli, sur, lumra,* PB. ; *Ghodila, ghorila, khajuria,* C. P. ; *Nakurmaral, naka-náru, urenka, uranké,* TEL.
References.—*Roxb., Fl. Ind., Ed. C.B.C., 112; Voigt., Hort. Sub. Cal., 715 ; Thwaites, En. Ceylon Pl., 373 ; Dalz. & Gibs., Bomb. Fl., 298; Aitchison, Cat. Pb. and Sind Pl., 169; Elliot, Fl. Andhr., 123, 187.*
Habitat.—A tall annual grass, having long narrow spikes, which often assume a pinkish red tinge when mature. It is usually met with in heavy retentive soils and along the banks of water-courses and borders of rice-fields.

Fodder.—Though not a first-class fodder grass, cattle eat it readily when other better kinds have failed.

FODDER.
260

E. pilosa, *Beauv.; Duthie, Fodder Grasses, N. Ind., 64.*

Syn.—E. VERTICILLATA, *R. & S.*; POA PILOSA, *Linn.*
Vern.—*Nika-sanwak, gádar punch,* PB.; *Palichhi,* RAJ. ; *Kutaki,* C. P.
References.—*Thwaites, En. Ceylon Pl., 373 ; Aitchison, Cat. Pb. and Sind Pl., 170.*
Habitat.—An annual species with slender stems and numerous minute spikelets borne on spreading panicles, common in India, and usually found in damp localities.

Fodder.—Buffalos are said to be fond of this grass. Mr. Symonds remarks that cattle eat it readily, and that it would make good hay. According to Mr. Lowrie it is considered to be a good fodder grass at Ajmere.

FODDER.
262

E. plumosa, *Link.; Duthie, Fodder Grasses, N. Ind., 64.*

Syn.—POA PLUMOSA, *Retz.*
Vern.—*Phularwa,* BUNDEL.; *Bara bhurbhura, bholoni, galgala, jhusa,* N.-W. P.; *Budhan, palinji,* PB.; *Chiri-ka-khet, chiri-ko-bajro,* RAJ.; *Sipar, bharbusi, pithi, safed bhurki, chikti, chippal,* C. P.
References.—*Roxb., Fl. Ind., Ed. C.B.C., 113 ;Voigt., Hort. Sub. Cal., 715 ; Thwaites, En. Ceylon Pl., 373 ; Aitchison, Cat. Pb. and Sind Pl. 170; Rheede, Hort. Mal., XII., 75, t. 41 ; Rumph., Amb., VI., 10, t. 4, f. 3 ; Atkinson, Him. Dist., 320.*
Habitat.—A slender annual species, very common, especially on sandy soils. Variable both as to size and habit. E. viscosa, *Trim.,* is probably only a variety with sticky inflorescence. Another variety (*var.* densiflora, *Hack*), with congested spike-like panicles, and resembling forms of E. ciliaris, is common on *usár* soils.

Fodder.—Mixed with *dub* it has been found to produce excellent hay at Allahabad. In Rájputana it is valued as a fodder grass.

FODDER.
264

263

E. tenella, *Beauv.; Duthie, Fodder Grasses, N. Ind., 65.*

265

Syn.—POA TENELLA, *Linn.*
Vern.—*Ich koic',* SANTAL ; *Bharburi,* N. W. P.; *Mondiajori,* C. P.
References.—*Roxb., Fl. Ind., Ed. C.B.C., 113 ; Voigt,\Hort. Sub.Cal.,716.*
Habitat.—An annual with stiff, rather brittle, flowering stems, bearing minute spikelets, which are often tinged with red when mature. Common on cultivated ground and frequently associated with rainy season crops.

Fodder.—Eaten by cattle, both green and as hay, and the grain is said to be nutritious.

FODDER.
266

E. 266

267 | **Eragrostis rachitricha**, *Hochst.; Duthie, Fodder Grasses, N. Ind.,*
65, 89.

Syn.—POA MULTIFLORA, *Roxb.*; E. TREMULA, *Hochst.*
Vern.—*Kalunji, bhamiri, bánsa,* N.-W. P.; *Chankan buti, laki,* PB.; *Chiri-ka-khet, chiri-ka-chunwalia,* RAJ.
References.—*Roxb., Fl. Ind., Ed. C.B.C.,* 114; *Voigt, Hort. Sub. Cal.,* 716; *Dals. & Gibs., Bomb. Fl., 298; Aitchison, Cat. Pb. & Sind Pl., 169.*
Habitat.—An annual, with stems 1—1½ feet. The extremely slender pedicels, which support the long many-flowered spikelets, give rise to the constant tremulous motion exhibited by this species when in flower. It is a characteristic grass of sandy soils in North India.

FOOD.
268

Food.—The grain is said to have saved many lives during the severe famine of 1813, and which is now alluded to as the *lakiáwála sál.*

FODDER.
269

Fodder.—Regarded as a good fodder grass at Ajmere.

EREMOSTACHYS, *Bunge; Gen. Pl., II., 1215.*

270 | **Eremostachys Vicaryi**, *Benth.; Fl. Br. Ind., IV., 695;* LABIATÆ.
Vern.—*Gurgunna, khalátrá, rewand chini,* PB.
References.—*Stewart, Pb. Pl., 168; Aitchison, Cat. Pb. & Sind Pl., 119.*
Habitat.—A beautiful, yellow-flowered plant, common on the Salt Range, ascending to 2,500 feet; also met with at Peshávar.

MEDICINE.
Seeds.
271

Medicine.—The SEEDS are given as a cooling medicine.

DOMESTIC.
272

Domestic Use.—The plant is said to be used in the Eusufzai near Peshá-war for poisoning fish.

EREMURUS, *Bieb.; Gen. Pl. III., 787.* [*280;* LILIACEÆ.

273. | **Eremurus spectabilis**, *M. Bieb.; Baker in Linn. Soc. Journ. XV.,*
Vern.—*Shili, bre, prau,* ᴸ B.
References.—*Stewart, Pb. Pl., 234; Balfour, Cyclop., I., 1052.*
Habitat.—A handsome herbaceous plant, with close spikes of white flowers and linear radial leaves, found on the Panjáb Himálaya between 6,000 and 9,000 feet.

FOOD.
274

Food.—"The leaves when young are much eaten, both fresh and dry, cooked as vegetables" (*Dr. Stewart*).

DOMESTIC.
275

Domestic Use.—Dr. Aitchison, in his Report on the Botany of the Afghán Delimitation Commission, draws attention to an interesting econo-mic product derived from **Eremurus Aucherianus**, *Boiss., var.* **Korolkowi.** Its long fleshy roots are dried and ground into powder, which forms into a jelly with boiling water. This jelly is then hardened into variously-shaped vessels called *dabba,* used for holding oil and clarified butter. There is a large trade in this material in Khorásan, and Dr. Aitchison believes that the introduction of these vessels into India would be much appreciated by the Hindú community as a substitute for the animal skins at present employed in the oil and *ghí* trade. It is not known if any of the Indian species could be similarly used.

Ergot or **Ergota ;** see **Claviceps purpurea**, Vol. II., 359.

Eria, see **Silk.**

ERIGERON, *Linn.; Gen. Pl., II., 279.*

276 | **Erigeron asteroides**, *Roxb.; Fl. Br. Ind., III., 254;* COMPOSITÆ.
Vern.—*Maredi, sonsali,* BOMB.
References.—*Roxb., Fl. Ind., Ed. C.B.C., 603; Dymock, Mat. Med. W. Ind., 429.*
Habitat.—A coarse, hairy annual, 1-2 feet high, found in Bengal and the Western Peninsula, and up to 4,000 feet on the Eastern Himálaya.

The Loquat Fruit.	(*J. F. Duthie.*)	ERIOCHLOA annulata.

Medicine.—Dr. Dymock states that this HERB, together with other simples, is brought for sale into the Bombay bazar from Guzerat as a stimulating and diuretic medicine.

MEDICINE.
277

ERINOCARPUS, *Nimmo; Gen. Pl., I., 234.*

Erinocarpus Nimmoanus, *Grah.; Fl. Br. Ind., I. 394;* TILIACEÆ.

278

    Vern.—*Chera, chira,* BOMB.; *Chowra, jangli-bhendi, haladi, adavi,* KAN.

    References.—*Beddome, Fl. Sylv., t. 110; Gamble, Man. Timb., 52; Grah., Cat. Bomb. Pl., 21; Dals. & Gibs., Bomb. Fl., 27; Lisboa, U. Pl. Bomb., 28; Balfour, Cyclop., I., 1052.*

    Habitat.—A tree, with large yellow flowers, found in the Deccan and parts of the Bombay Presidency.

    Fibre.—The BARK is said to yield an excellent fibre for ropes.

FIBRE.
Bark.
279

    Structure of the Wood.—Soft, used for yokes and rafters.

TIMBER.
280

ERIOBOTRYA, *Lindl.; Gen. Pl., I., 627* (under **Photinia**).

281

Eriobotrya bengalensis, *Hook. f.; Fl. Br. Ind., II., 371;* ROSACEÆ.

    Syn.—MESPILUS BENGALENSIS, *Roxb.*

    Vern.—*Berkung,* LEPCHA.

    References.—*Roxb., Fl. Ind., Ed. C.B.C., 406; Voigt., Hort. Sub. Cal., 198; Kurz, For. Fl. Burm., I., 443; Gamble, Man. Timb., 167; Balfour, Cyclop., III., 206.*

    Habitat.—A large tree found in the Eastern Himálaya and the Khásia Hills up to 4,000 feet; also in Chittagong and Burma.

    Dye.—The BARK is said to be used in Nepál for dyeing scarlet.

DYE.
Bark.
282

E. elliptica, *Lindl.; Fl. Br. Ind., II., 372.*

283

    Syn.—MESPILUS CUILA, *Ham.*

    Vern.—*Mihul, mya,* NEPAL; *Yelnyo,* LEPCHA.

    References.—*Gamble, Man. Timb., 167; Don, Prod. Nep., 238.*

    Habitat.—A moderate-sized evergreen tree of the Eastern Himálaya, from Nepál to the Mishmi Hills; altitude 6,500 to 8,000 feet.

    Structure of the Wood.—Reddish-brown, compact, hard, apt to warp slightly; it is good but not used. Weight 58lb per cubic foot (*Gamble*).

TIMBER.
284

E. japonica, *Lindl.; Fl. Br. Ind., II., 372; Wight, Ic., t. 226.*

285

    THE LOQUAT, or JAPAN MEDLAR.

    Vern.—*Lakote,* KAN.

    References.—*Roxb., Fl. Ind., Ed. C.B.C., 406; Voigt., Hort. Sub. Cal., 198; Brandis, For. Fl., 575; Kurz, For. Fl. Burm., I., 443; Gamble, Man. Timb., 167; Dals. & Gibs., Bomb. Fl., Suppl., 32; Aitchison, Cat. Pb. and Sind Pl., 58; Econ. Prod., N.-W. P., Part V., 69; Lisboa, U. Pl. Bomb., 155; Birdwood, Bomb. Pr., 150; Balfour, Cyclop., I., 1052; Smith, Dic., 251; Treasury of Bot., I., 462; Mueller, Sel. Ext. Trop. Pl., 293.*

    Habitat.—A handsome evergreen fruit tree, introduced from Japan. Extensively cultivated for its fruit.

    Food.—The *Loquat* tree is well known in gardens, especially in Northern India. By careful cultivation, fruit of excellent quality can be obtained. It is grown easily, either from seed or by grafts, the latter method being preferred. The fruit ripens towards the end of the cold season. There are two distinct varieties, one pear-shaped and of a deep apricot colour, the other roundish and white; the latter kind ripens a few days later, but is less sweet.

FOOD.
286

ERIOCHLOA, *H. B. & K.; Gen. Pl., III., 1099.*

[GRAMINEÆ.

Eriochloa annulata, *Kunth; Duthie, Fodder Grasses N. India, 2.;*

287

S

ERIODENDRON anfractuosum.	**The White Cotton Tree.**

FODDER.
288

Syn.—E. POLYSTACHYA, *H. B. & K.*; PASPALUM ANN DLATUM, *Flügge.*
Habitat.—A quick-growing perennial grass, found on wet ground in many parts of the plains.

Fodder. Eaten by buffalos. In Australia it is said to afford fodder all the year round, and to be highly relished by stock.

289

ERIODENDRON, *DC.; Gen. Pl., I., 210.*
(*George Watt.*)

Eriodendron anfractuosum, *DC.; Fl. Br. Ind., I., 350;*
THE WHITE COTTON TREE; KAPOK FLOSS.   [MALVACEÆ.

Syn.—BOMBAX PENTANDRUM, *Linn.*; B. ORIENTALE, *Spreng.*; CEIBA PENTANDRA, *Gœrtn*; ERIODENDRON ORIENTALE, *Steud.*; E. RHEEDII, *Planch.*

Vern.—*Hattian, katan, safed-semal,* HIND.; *Shwet-simul,* BENG.; *Ilavam,* TAM.; *Búruga, púr, buraga-sánna,* TEL.; *Paniá, paniala,* MAL.; *Khatyán, suféd-khatyán,* DUK.; *Katsávar,* KHANDESH; *Shamieula, saphetasávara, sálmali, pandhari, savar,* MAR.; *Bili burga, bili-barlu,* KAN.; *Imbul,* SING. (*elavum, illaku,* TAM.; in CEYLON); *Thinbawle,* BURM.

References.—*Roxb., Fl. Ind., Ed. C.B.C., 513; Voigt., Hort. Sub. Cal., 105; Grah., Cat. Bomb. Pl., 17; Dals. & Gibs., Bomb. Fl, 22; Wight & Arn., Prod., I., 61; Wight, Ic., t. 400; Griff., Not., IV., 533; Beddome, Flor. Sylv., XXX., and Anal. Gen., t. 4; Hamilton (Gossypinus) in Trans. Linn. Soc., XV., 126; Thwaites, Enum. Ceylon Pl., 28; Kurz, For. Fl. Burm., I., 131; Moodeen Sheriff, Supp. Pharm., 135; also (new work, proof sent to author) Mat. Med. South India; Gamble, Man. Timb., 42; Report on Ind. Fibres, Col. & Ind. Exhib. (1886), 63; Lisboa, U. Pl. Bomb., 195 and 229; Gray, Botany of Bombay in Gazetteer, XXV., 322; Baden Powell, Pb. Prod., 333; Murray, Pl. and Drugs of Sind, 56; Drury, U. Pl. Ind., 197; also Handbook Fl. Ind., I., 86; Cooke, Gums and Gum-resins, 34; Ainslie, Mat. Ind., II., 96; Balfour, Cyclopœdia of India, I., 1053; O'Shaughnessy, Beng. Disp., 227; Dymock, Mat. Med. West Ind., 2nd Ed., 106; Rheede, Hort. Mal., III., t. 49, 50; Rumph., Amb., I., t. 80; Sir W. Elliot, Fl. Andh., 32.*

Habitat.—A tall tree with straight trunk, prickly when young; branches horizontal and whorled. Flowers dirty white and much smaller than those of **Bombax**, with staminal tube splitting into five portions, each with two anthers, instead of into many divisions, each with one anther, as in **Bombax.** According to the *Flora of British India* this tree occurs " in the forests throughout the hotter parts of India and Ceylon : distributed to South America, the West Indies, and Tropical Africa."

LOCALITIES where met with.
290

Although occasionally met with in most districts of India, in only a few localities is it reported to be fairly abundant. With the view of affording information as to the localities where an effort might be made to develop a trade in kapok fibre—the floss from the seeds of this plant—the following review of the official correspondence and writings of Indian authors may be given :—

BENGAL.
291

Of Bengal Roxburgh says : " On the Coromandel Coast, the Tamils plant the tree about their temples. In Bengal, where the winters are colder, the leaves drop off during the hot season. In February, when destitute of foliage, the blossoms appear, and soon afterwards the leaves ; the seed ripens in May." The writer is not aware of having seen the tree in Bengal except as planted along road-sides and in gardens. Mr. Gamble does not mention any special locality, but remarks that it is " often planted."

N. W. P.
292

Dr. King, in a list of the plants of the North-West Provinces (printed in the *Gazetteer, Vol. IV., p. LXVIII.*), simply mentions it by name. In his *Forest Flora of North-West and Central India,* Sir D. Brandis makes no mention of the tree, and Dr. Stewart is also silent as to its occurrence in the Panjáb; but **Mr. Baden Powell,** in his *Panjáb Products,* mentions it

PANJAB.
293

Kapok Floss.	(*G. Watt.*)	ERIODENDRON anfractuosum.

briefly without stating where it is met with. It would thus appear that as far as these Provinces are concerned, while the plant occurs occasionally under cultivation, there is little or no prospect of a trade being done in the fibre from the wild or naturalised plant. Of Coromandel it may be otherwise. **Roxburgh** appears to have found it fairly plentiful. **Sir Walter Elliot** speaks of it as met with in that region, and gives it the Telegu name of *Búruga.* Turning to Burma, Kurz remarks that it is "here and there cultivated in Pegu and Tenasserim; a single tree was observed wild in the coast forests of South Andaman." **Mr. Baden Powell**, in a re-cast of Brandis's classification of the Burmese Teak forests, mentions **Eriodendron** as occurring on the lower undulating hills along with bamboo, **Xylia, Pterocarpus, Albizzia, Terminalia, Dillenia, Hibiscus**, and **Bombax** (*Indian Forester, VIII., 415*).

In the Madras Presidency generally, the tree would appear to be by no means of unfrequent occurrence. The Conservator of Forests, Northern Circle, says that, except as a cultivated tree, and in a few isolated cases in the South East Wynaad (Nilghiri District), he has not met with it in his circle. In the Salem District (Southern Circle), it is reported to be prevalent to a small extent on north-west slopes of the Sheveroy Range, and scantily in other parts of the district. In the Hosur taluk scattered trees are met with towards the middle and low lands. From other districts in the South Circle it is reported that the tree is found chiefly in a cultivated state, especially near temples. In Tinnevelly it is found scattered about in the Ghát forests, and it is estimated that about three tons of cotton could be gathered yearly. In North Malabar the tree is found chiefly on the lower slopes of the Chenat Nair forests, but there only at scattered intervals, and it disappears further west and north where the rainfall is heavier. In Southern Malabar there is little trade in the silk cotton, such trade as there is being more often in the cotton of the **Bombax malabaricum.** Dr. Shortt (*Indian Forester, III., 236*) alludes to it in a list of plants parts of which are eaten in times of famine. He gives it the following names: *Elevam,* Tam.; *Pur,* Tel.; and he remarks that it is "found in gardens, the seeds being roasted and eaten." **Dr. Moodeen Sheriff** (in his forthcoming work on the *Materia Medica of South India*) gives a detailed account of the plant, distinguishing it from **Bombax malabaricum**; but while he states that "the cotton is always found in the bazars and is much cheaper than the common cotton," he does not mention from what source it is obtained. In the Manual of Trichinopoly the tree is referred to in a list of the "More important fruit and timber trees found in the district." It is said to be the *ilavam* or *ilava* of Tamil, and the remark is made: "The seeds are embedded in silky cotton, which is used for stuffing beds, cushions, &c." In the Nellore Manual **Eriodendron** is given in a "List of the Principal Trees of the District" and receives the Telegu name *Buraga.* In the *Mysore Gazetteer* (Volume I., 58) the tree is alluded to as grown in the Bangalore gardens, but in a List of the trees of Mysore, **Mr. Cameron** gives it the Kanarese name of *Bili burga.* Dr. George Bidie, C.I.E., in a Catalogue of the drugs of the Madras Presidency, refers to the unripe fruits of this plant as being "demulcent and astringent, and used in medicine as well as cookery." He gives the drug the following names—*Khatyan-kakalli,* Duk.; *Maratimoggu,* Tam.; and *Buraga-pintha,* Tel. At the Colonial and Indian Exhibition an interesting series of the products of this plant was shown contributed by **Mr. J. W. Cherry**, of the Forest Department, from Salem.

Bombay.—Sir George Birdwood, in his *Bombay Products*, mentions the plant as met with in "Khandesh, Travancore, and Coromandel" It has been customary to read of the plant being, as far as India is

s 2

ERIODENDRON
anfractuosum.   The White Cotton Tree.

concerned, most abundant in the Deccan. A recent correspondence would, however, appear to throw doubt on this prevalent opinion. **Mr. McGregor**, Conservator of Forests, Southern Circle, Bombay, in a letter on the subject, wrote that **Eriodendron anfractuosum** "is said to occur in Kánara, but its occurrence is doubtful." He gave it the names of *Pandhari, savar*, MAR., and *Bili barlu*, KAN. In the same correspondence the Conservator of the Northern Circle, Bombay, was asked for information regarding the tree, and **Mr. A. T. Shuttleworth** replied that "**Eriodendron anfractuosum**, though stated in some botanical publications to be a common tree in the forests of the northern circle, is exceedingly rare, and·in Khandesh, where it is supposed to grow in large numbers, there is scarcely a tree of the kind to be seen in the forests. Authors of works on the Forest Flora of Western India have evidently mistaken some other tree— probably **Bombax malabaricum**—for the Kapok or white cotton tree."

It would thus appear that there is room for doubt as to the existence in Bombay of **Eriodendron** as an abundant tree, and much confusion appears still to prevail in the identification of the kapok tree; popular writers are apparently unable to recognise it from **Bombax**. The vernacular names now given to the kapok tree might easily enough be adaptations from the names given to **Bombax**. The Sanskrit word *sálmali* is by some writers given to the one, by others to the other. *Sálmali*-wood is prescribed in the Institutes of Manu as that on which washermen should wash clothes. No writer definitely affirms that **Eriodendron** is wild; nearly all speak of it as cultivated, and it may be the case that India can only hope to take part in the growing kapok trade after some years when the tree has been still further cultivated, in some of the regions where it is now successfully grown. It would, however, be undesirable to accept as final the present information; and as in a measure opposed to the opinions expressed by **Mr. McGregor** and **Mr. Shuttleworth** it may be as well to complete this brief review by quoting some of the passages in which it is affirmed that **Eriodendron** is a fairly common tree in both the southern and northern divisions of Bombay. **Lisboa** (*Useful Plants of Bombay, p. 195*) says: It is "a very common prickly tree with palmate leaves and dingy white flowers." There can be no mistake as to the plant there meant; it is **Eriodendron** and not **Bombax**. **Dalzell and Gibson** also describe the tree in language which cannot be mistaken, and they add: "It grows in Khandesh;" its native name being "*Shameula*." **Dr. Gray**, in his essay on the "Botany of the Bombay Presidency" (*Gazetteer, XXV., 322*), says that "**Eriodendron** is another large tree" (he has just been speaking of **Bombax**) "similarly distributed in this country." "It is known as the white silk-cotton tree." Of **Bombax** he says it "is common in all the forests of the Presidency from Gujarat and Khandesh to Kánara." Turning now to the *Bombay Gazetteers* **Mr. Talbot**, in the Kánara volume, says that **Eriodendron anfractuosum** is the *Bile burlu* of the Konkan, and *pándhari sávar* of MAR.; he remarks that "the white-cotton, though fairly large, does not grow to the same size as **Bombax malabaricum**. The pods are gathered for their cotton." Of the Panch Maháls it is stated that "the *Shamla*, **Eriodendron anfractuosum**," is similar in appearance to **Bombax malabricum**, the *Shimal* or *Shimar*, "but differs in the flower," those of **Bombax** "a dull crimson and those of **Eriodendron** a dirty white." The writer of the chapter on the Panch Mahál forests thus made no mistake, and **Mr. Talbot's** reputation as a botanist warrants the most complete confidence being placed on his statement that the tree occurs in Kánara. Of the Poona District it is stated: "*Hattian*, **Eriodendron anfractuosum**, though not plentiful, is found in the thicker forests on the western hills. The light and soft wood is

Kapok Floss.	(*G. Watt.*)	ERIODENDRON anfractuosum.

used in tanning leather and for making toys. The fine soft silky wool which surrounds the seeds is used for making cushions. It yields a gum called *hattian-ke-gond*, which is valued in bowel complaints." Of the Khandesh District it is stated : "*Katsávar*, Eriodendron anfractuosum, sometimes called a Bombax and confounded with the *simal*, has a white soft wood of no use, save for making toys or fancy articles. The down round its seeds is used for stuffing pillows. It is not common anywhere in Khandesh"

Gum.—This gum is of a dark-red colour and almost opaque. It is generally known as *hattian-ke-gond*, and by European writers is said to be one of the forms of the *Katéra* or hog-gums, *e.g.*, the pseudo-gums, or those which are insoluble in water but swell and form a pasty mass. Accordingly **Dr. Cooke**, in his Report on the Gums of India, places it along with the gums from **Cochlospermum**, **Gossypium**, **Sterculia urens**, and **Uvaria tomentosa**, but these being pale-coloured, it is assigned its more immediate position in the sub-series—the dark-coloured pseudo-gums— such as **Moringa pterygosperma**, **Stereospermum suaveolens**, **Ailanthus excelsa**, **Macaranga tomentosa**, and **Bombax malabaricum**. This gum is, however, said to be astringent and to be employed medicinally in bowel complaints. **Ainslie**, who wrote in Madras at the beginning of the present century, says : " A solution of this gum is given in conjunction with spices in certain stages of bowel complaints. We are told by **Rumphius** (*Amb. I., p. 194, t. 80*), who speaks of the tree under the name of **Eriophorus javana**, that the inhabitants of the island of Celebes eat the seeds of it. It is the *Capock* of the Malays." Then follows a botanical description which shows that **Ainslie** clearly distinguished this plant from **Bombax**. This fact is of considerable importance, as it confirms the suggestion, already thrown out, that the true Indian habitat of the plant may be Southern India. He gives the tree the Sanskrit name of "*mullie,*" and adds that it is the *pania-paniala* of the *Hort. Mal.* (*III., p. 59, t. 49, 50, & 51.*) It is interesting to note that the name *kapok* (or *capock*) was known a hundred years ago, and that it is a Malayan and not a Dutch name as some writers have stated. Only the other day a great advance was supposed to have been made by the discovery of the plant from which the Dutch fibre *kapok* was obtained. This fibre was well known to **Ainslie** nearly a century ago, and it is worthy of remark that the word *kapok* bears a close resemblance to the *kárpási* of the Sanskrit writers, and that the most general modern names for the plant *Hattian* and *Kattan* seem to be directly derived from the Arabic *Kattan*; both these classical names are, however, now stated to be synonyms for cotton. (*Conf.* with the remarks at page 324, Vol. I. of the Selections from the Records of the Government of India.) **Sir William O'Shaughnessy** (*Beng. Dispens., 227*) also alludes to the gum as being medicinal.

Tan.—The WOOD is said to be used in tanning leather.

Fibre.—An inferior reddish fibre is sometimes prepared from the BARK which is used locally for making ropes and paper. This was analysed by **Messrs. Cross, Bevan and King**, and their results, published in the recent report on Indian fibres, are as follows :—moisture 12·4 per cent.; ash 95 per cent.; loss by hydrolysis (one hour's boiling in solution 1 per cent. Na₂O) 50·5 per cent. ; cellulose 33·6 per cent. ; loss by mercerising 7·5 per cent. ; and by acid purification 6·1 per cent. The ultimate fibres were only 1—2 mm. in length. These figures may be accepted as fully disposing of the dark fibre of this plant from all further consideration. The barking of the trees should, if possible, be prohibited, since the proceeds from the fibre thus obtained would by no means compensate for the injury done to the tree as a source of floss. The *Kapok* or FLOSS from the

---

ERIODENDRON anfractuosum.   **The White Cotton Tree.**

FIBRE.

seeds is, however, according to the present demand, a fibre of great merit. The modern trade in it was created by the Dutch merchants, their supply being drawn from Java. It is used in upholstery, being too short a staple to be spun, and, indeed too brittle and elastic. But these are the very properties that commend the floss to the upholsterer. In cushions, mattresses, &c., its elasticity and harshness prevent its becoming matted as is the case with *simal* floss, and it is, therefore, considerably superior to that fibre. Indeed, it is probable that the even still shorter staple of **Cochlospermum** would in time command a better price than that of the *simal*. Like *Kapok* it is very elastic, the fibre springing up to its former position the moment the weight is removed from the cushion. With *simal*, on the other hand, a very short time suffices to make a mattress assume permanently a compressed condition, in which it occupies perhaps less than half its original bulk and at the same time becomes knotted. This necessitates the removal of the stuffing to be teased or rudely carded.

It will thus be seen that if future extended usage of *Kapok* confirms the properties attributed to it, the demand for the fibre will year by year increase. But while endeavouring to participate in this trade it becomes essentially necessary that an error made by many writers be guarded against, namely, that of viewing *Kapok* as a generic trade-name for all the silk-cottons—including that of the *simal*—the floss of **Bombax malabaricum**. When the demand for *Kapok* first started, Indian exporters placed in the market a quantity of very dirty *simal*, having a large percentage of dust as well as seed. This was at once condemned, and fetched a price that would not even cover the transport charges. India thus fell into an inferior position which it is possible might never have been the case had carefully-cleaned *simal* been sent to Europe. A low-priced fibre like that of either *Simal* or *Kapok* cannot bear the extra freight of a large percentage of dust. It becomes essentially necessary that the floss be cleaned, freed of seed, and carefully baled. At the Colonial and Indian Exhibition, a large assortment of *simal* floss was shown, and the writer had the opportunity of conversing with several Dutch and English dealers in *Kapok*. These gentlemen assorted the *simal* samples and pointed to the fact that even among these there were inferior and superior qualities. Some had twice as long a staple as others, while the *Kapok* property of elasticity was possessed by but few. After this had been done **Mr. Cherry's** true *Kapok* floss was shown, when in every case the experts recognised it as *Kapok* and were eager to know the price, amount available annually, and the names of merchants with whom they might open up dealings. Unfortunately these were points regarding which no information could be furnished.

The necessity for care in future efforts may be apparent when it is here stated that nearly every trade journal which has published notices regarding *Kapok* has viewed it as one and the same thing with *simal*. Thus the *Indian Agriculturist* (October 16th, 1886) says that every person in India is familiar with "the value of the 'tree-cotton,' as stuffing for pillows and bedding, and *Kapok*, which is really the Malayan name for it, is the designation by which it is known in the Dutch and Australian markets, &c., &c." The writer of that article was apparently very liberally compiling from a paper which appeared in *Buchanan's Monthly Register* (Melbourne, June 21st, 1886), in which the tree-cotton—**Bombax malabaricum**—is incidentally mentioned along with **Cochlospermum Gossypium** and the *Baobab* tree of Africa as "in their growth and products" possessing "very little difference." Indeed, it seems probable that, as far as the Australian trade in Kapok from India is concerned, the floss of **Bombax**

FIBRE.

malabaricum is that which is so designated. While that may be so, the necessity for distinguishing the two fibres none the less remains in its full force, and the above reference to the *Indian Agriculturist* has been made in the hope of guarding against any ignorant or mistaken continuance of the error here indicated. A reference was recently made by the Government of India to Her Majesty's Consul at Batavia, Java, asking the name of the plant from which the Kapok fibre was obtained, and " also whether the exports of the tree cotton obtained from the **Bombax** or **Eriodendron** trees are the larger." The reply may be here published, as it is highly instructive :—

" BRITISH CONSULATE, " BATAVIA ; " *10th November 1887.*

" I have the honour to acknowledge receipt of your letter No. 214—24-1 F. & S. of the 29th September last, and, in reply thereto, I beg to inform you that the scientific name of the tree from which Kapok is chiefly obtained in Java is **Eriodendron anfractuosum.** The exports of Kapok from the Netherlands India have been as follows :—

Kilos.		Kilos.
1882 . . . 302,201	1884 426,061
1883 . . . 341,136	1885 . . . . 600,269 "

Thus there can no longer remain any doubt as to the Kapok of Java, and it is instructive to observe how the exports of that fibre have steadily increased, having been in 1885 twice those of 1882. It is worth adding also that **Bombax malabaricum** is a native of Java, and apparently, a more abundant tree than **Eriodendron anfractuosum.**

OIL. Seeds. 301

Oil of Eriodendron.—The SEEDS are said to yield a bright red or dark-brown clear oil, Dr. Cooke, in his *Report on the Oils and Oil-seeds of India* (p. 43), mentions a sample as in the possession of the Indian Museum (obtained from Chingleput), and adds that the oil was first made known at the Madras Exhibition of 1857. The peculiar properties of this oil are unknown, but from the fact that the seeds are often eaten it may be inferred that the oil is edible.

MEDICINE. Gum. 302 Floss. 303

Fruits. 304

Medicine.—It has already been stated that the GUM obtained from this plant is used medicinally in bowel complaints, having attributed to it a useful astringent property. **Dr. Moodeen Sheriff** recommends the FLOSS or cotton for medicinal use, as it is cheaper than common cotton. It is also cooler and more elastic and on that account might be recommended for cushions and pillows used in hospitals and also for stuffing to bandages and other such surgical dressings. The DRY YOUNG FRUITS have also been alluded to as used medicinally, and **Dr. Moodeen Sheriff** explains that the best mode of procuring them is to have them collected from the ground underneath the trees. A large number of things are often sold as the fruits of this tree, some even poisonous, such as the unripe fruits of **Datura.** By collecting them from the ground below the trees this is prevented, but at the same time immature or rather unfertilized fruits would be so collected, since if fertilized they would not fall to the ground. These fruits are similar in their properties though inferior to those of **Bombax.** " The dry young fruits of **Eriodendron anfractuosum** are sometimes sold in the bazars under the same name," *viz.*, *Maráti-Moggu,* " and used for the purpose of adulteration and substitution of those of **Bombax malabaricum.** Although the similarity between the fruits of both plants is very great, yet the difference between their stalks, which are almost always attached to them, is so distinct that they can be very easily distinguished from each other. The fruits of **Eriodendron** are always round, not angled, and somewhat larger and of a darker colour; the fruit-stalk of the **Eriodendron,**

ERIOGLOSSUM edule.	The White Cotton Tree ; the Ritha.

MEDICINE.

Roots.
304 a.

however, is round, about the thickness of a pin, and two or three times longer than the fruit." These unripe fruits are regarded as demulcent and astringent. The exact original use of the expression *Maráti-Moggu* is not quite clear—*Moggu* means buds. The ROOTS are also used medicinally, being one of the forms of *Músla* or *Músli-sémul* (described under **Bombax malabaricum**, Vol. I., No. B. 653). **Dr. Dymock** explains that in the Concan the young roots of **Eriodendron** are preferred to those of **Bombax**. "They are dried in the shade, powdered and mixed with the juice of the fresh bark and sugar. This tree is called *Pándhra Saur* in Marathi and *Dolo Shamlo* in Guzerathi" (*Mat. Med. West. India, 2nd Ed., 106*). The LEAVES and also the SEEDS have medicinal virtues attributed to them, but they do not seem of sufficient merit to deserve separate description.

Leaves
304 b.
Seeds.
304 c.

SPECIAL OPINIONS.—§ "A handful of the tender leaves of this plant is ground into a paste and is administered to a patient newly attacked with gonorrhœa. One dose at 6 A.M. is given daily for three or four days, and a little butter-milk is taken with it" (*Surgeon W. F. Thomas, Madras Army, Mangalore*). "The gum is also used in the incontinence of urine of children" (*Surgeon-Major J. J. L. Rutton, M.D., M.C., Salem*). "The root of the young plant is used in the form of decoction in cases of chronic dysentery and diarrhœa, also in cases of ascites and anasarca when it acts as a diuretic" (*Civil Surgeon J. H. Thornton, B.A., M.B., Monghyr*).

FOOD.
Seeds.
305
Fruits.
306
FODDER.
Seed-cake.
307
CHEMISTRY.
308

Food.—The SEEDS are said to be eaten, and the young or unripe FRUITS are also stated to be used in cookery. The seed-cake is sometimes given as FODDER. **Dr. Warden** has kindly furnished the following note on this subject showing the comparative composition of Kapok to Cotton-seeds :—

§ "The seeds of the Kapok tree have been made into cakes, and the comparative value of these cakes and ordinary cotton-seed cake for cattle-feeding purposes has formed the subject of an enquiry by **Mr. G. Reinders**. The following analytical results were obtained :—

	Kapok cake.	Cotton cake.		Kopok cake.	Cotton cake.
Water . . .	13·28	12·0	Non-nitrogenous extraction .	19·92	35·42
Nitrogenous matter; albuminous compounds. . .	26·34	20·62	Woody fibre . .	28·12	20·36
Fat . . .	5·82	6·36	Ash . . .	6·52	5·64

"The ash of the Kapok-tree seed contains 28·5 % of phosphoric acid and 24·6 % of potash ; it ought, therefore, to be of value as a manure."

TIMBER.
309

Structure of the Wood.—Soft, very light ; 30℔ per cubic foot. According to some writers this is the *Salmali* of Sanskrit writers. It is used for toys and other such purposes, and is sometimes hollowed out into canoes.　　　　　　　(*J. F. Duthie.*)

ERIOGLOSSUM, *Blume; Gen. Pl., I., 396.*

[*t. 73 ;* SAPINDACEÆ.

310

Erioglossum edule, *Bl. ; Fl. Br. Ind., I., 672 ; Beddome, Fl. Sylv.,*
Syn.—E. RUBIGINOSUM, *Bl.* ; SAPINDUS RUBIGINOSA, *Bl.*

Vern.—*Ritha,* HIND. ; *Mukta-moya,* URIYA ; *Manipangam,* TAM. ; *Isakarási, undurugu,* TEL. ; *Tseikchay,* BURM.

References.—*Roxb., Fl. Ind., Ed. C.B.C., 332 ; Voigt, Hort. Sub. Cal., 94 ; Brandis, For. Fl., 108 ; Kurz, For. Fl. Burm., I. 296 ; Gamble, Man. Timb., 94 ; Grah., Cat. Bomb. Pl., 29 ; Dals. & Gibs., Bomb. Fl. Suppl., 14 ; Elliot, Fl. Andhr., 71, 186 ; Drury, U. Pl., 385 ; Lisboa, U. Pl. Bomb., 52 ; Royle, Ill. Him. Bot., 138 ; Balfour, Cyclop., III., 531 ; Treasury of Bot., 463.*

Habitat.—A large tree of Sikkim, Assam, South India, and Burma.

	ERIOLÆNA Wallichii.
Eriolena Fibre. (*J. F. Duthie.*)	

Structure of the Wood.—Strong and durable, with chocolate-coloured heartwood (*Roxburgh*).

TIMBER.
311

ERIOLÆNA, *DC.; Gen. Pl., 220.*

Eriolæna Candollei, *Wall.; Fl. Br. Ind., I., 370;* STERCULIACEÆ.

Vern.—*Búte,* BOMB. ; *Dwani,* BURM.
References.—*Voigt., Hort. Sub. Cal., 108 ; Kurz, For. Fl. Burm., I., 148 ; Gamble, Man. Timb., 51 ; Dals. & Gibs., Bomb. Fl., 24 ; Lisboa, U. Pl. Bomb., 24 ; Burm. Gaz., 127.*
Habitat.—A deciduous tree, found in the Western Peninsula; in Bhutan and in Burma.
Structure of the Wood.—Heartwood brick-red, with orange and brown streaks, old pieces, however, losing their bright colour ; hard, close-grained, shining, takes a beautiful polish, seasons well. Weight about 50℔ per cubic foot.
It is used for gunstocks, carpentry, paddles, and rice-pounders ; is very handsomely marked, and is well worthy of greater attention.

312

TIMBER.
313

E. Hookeriana, *W. & A.; Fl. Br. Ind., I., 370.*

Vern.—*Búndún, oit bulung,* KOL. ; *Gua-goli,* SANTAL ; *Gua-kasi,* MAL. (S.P.) ; *Kutki, bhonder,* GOND ; *Arang,* BERAR ; *Bute, bother, botku, arang,* BOMB. ; *Ponra,* ORAON ; *Nar-botku,* TEL. ; *Hadang,* KAN.
References.—*Brandis, For. Fl., 36 ; Beddome, For. Man., 35 ; Gamble, Man. Timb., 50 ; Elliot, Fl. Andhr., 129 ; Lisboa, U. Pl. Bomb., 24 ; Kew Reports, 1879, 34 ; Forest Admn. Report, Ch. Nagpore, 1885, 28 ; Bomb. Gaz., XV., 68 ; XII., 25.*
Habitat.—A small tree of Central and South India ; Behar and the Western Peninsula.
Fibre.—The BARK yields a good fibre, of which fine specimens were sent to the Paris Exhibition of 1878, and by the Rev. A. Campbell, to the Colonial and Indian Exhibition of 1886.
Structure of the Wood.—Light-red, tough. Annual rings marked by an almost continuous line of pores. Said to be commonly used in the Kánara District for axe handles.

314

FIBFE.
Bark.
315
TIMBER.
316

E. quinquelocularis, *Wight; Fl. Br. Ind., I., 371 ; Wight, Ic., t. 882.*

Vern.—*Budjari-dha-mun,* BOMB.
References.—*Beddome, For. Man., 35 ; Gamble, Man. Timb., 50 ; Lisboa, U. Pl. Bomb., 25.*
Habitat.—A small tree found in Behar, the Bombay Gháts, and, according to Beddome, very common on the Nilghiris and in the Wynaad ; widely distributed in the western forests of the Madras Presidency, and in Mysore.
Structure of the Wood.—Said to be strong, and to be used by the natives for various purposes.

317

TIMBER.
318

E. spectabilis, *Planch.; Fl. Br. Ind., I., 371.*

References.—*Beddome, Fl. Syl. An. Gen., t. 5 ; Gamble, Man. Timb., 50.*
Habitat.—A small tree of the Central Himálaya to Nepál. It is also plentiful everywhere on the dry, red clay hills in the arid districts of Manipur.
Fibre.—The BARK yields a good fibre.
Structure of the Wood.—Heartwood hard and close-grained, reddish, mottled.

FIBRE.
Bark,
319
TIMBER.

E. Wallichii, *DC.; Fl. Br. Ind., I., 370.*

Vern.—*Kubindé,* NEPAL.
References.—*Voigt., Hort. Sub. Cal., 108 ; Gamble, Man. Timb., 50.*

320
321

E. 321

ERUCA sativa.	The Bhabar grass.

TIMBER.
322

Habitat.—A small tree of Nepál and the Sikkim Himálaya.

Structure of the Wood.—Sapwood grey; heartwood reddish-brown, hard, mottled; much esteemed by Nepalese.

ERIOPHORUM, *Linn. ; Gen. Pl., III., 1052.*

323

Eriophorum comosum, *Wall.*; CYPERACEÆ.

Syn.—ERIOPHORUM CANNABINUM, *Royle ;* SCRIPUS COMOSUS, *Roxb.*

Vern.—*Bábar, bab, babila, bhabhur, bhabhuri,* N.-W. P.; *Pan-babiyo* (Almora), KUMAON.

References.—*Atkinson, Him. Dist., 808 ; Royle, Ill. Him. Bot., 415 ; Royle, Fib. Pl., 34 ; Huddleston, Trans. Agri. Hort. Soc. Ind., VII., 272 ; Balfour, Cyclop., I., 1053 ; Ind. For., IV., 168 ; IX., 569 ; Linn. Soc. Jour., XX., 409.*

Habitat.—A coarse sedge-like perennial herb, the heads of flowers clothed with long silky hairs. Common in the Siwaliks and outer Himálayan ranges. Allied to the *Cotton grasses* of Europe.

FIBRE.
324

Fibre.—The fibre yielded by this plant forms a very small portion of what is exported to the plains under the name of *bhábar.* This latter is the produce of a grass named **Ischœmum angustifolium.** The **Eriophorum** fibre is utilised locally, but it is often difficult to discover whether it is pure or mixed with **Ischœmum.** Former writers are in error who have attributed *Bhábar* entirely to **Eriophorum.**

Captain Huddleston, in *Trans. Agri.-Hort. Soc. Ind., l.c.,* mentions that " All the *jhoolas* or rope bridges, which are erected over the large rivers, where *sanghas* or wooden planked bridges cannot be made, on all the principal thoroughfares of this district, are constructed of this silky species of grass, the cables of which are of a considerable thickness. This grass grows abundantly in all the ravines up the sides of the mountains, and is to be had only for the cutting, but it is not of a very durable nature, though pretty strong when fresh made into ropes. It lasts about a twelvemonth only, or a little more, and the people in charge of the rope bridges are constantly employed in repairing and annually renewing the ropes and stays. The '*chinkas*' or temporary bridges of a single cable, upon which traverses a seat in the shape of an ox-yoke, are also sometimes made of this grass." For further information regarding *bábar* grass, see **Ischœmum angustifolium.**

ERIOSEMA, *DC. ; Gen. Pl., I., 543.*

325

Eriosema chinense, *Vogel ; Fl. Br. Ind., II., 219 ;* LEGUMINOSÆ.

Vern.—*Konden,* SANTAL,

Reference.—*Rev. A. Campbell, Cat. Econ. Pl. of Chutia Nagpur, 64.*

Habitat.—A perennial herb with tuberous root, common on the Central and Eastern Himálaya, ascending to 6,000 feet. Recorded as occurring also in Chutia Nagpur, Burma, and Ceylon.

FOOD.
326

Food.—The Rev. A. Campbell states that the root is about the size of a marble, and is eaten by the Santáls.

ERUCA, *Tourn. ; Gen. Pl., I., 84.*

327

Eruca sativa, *Lam. ; Fl. Br. Ind., I., 158 ;* CRUCIFERÆ.

Syn.—BRASSICA ERUCA, *Linn ;* B. ERUCOIDES, *Roxb.*

Vern.—*Taramira,* HIND.; *Suffed-shorshi, shwet-sursha,* BENG.; *Duan, sahwan, tira, tara, taramira, lalu,* N.-W. P. & OUDH; *Dua, chara,* KUMAON ; *Tara, assu, usan, jamnia,* PB.; *Mandao,* AFG. ; *Jambho,* SIND; *Siddartha,* SANS.; *Jambeh,* PERS.

References.—*Roxb., Fl. Ind., Ed. C.B.C., 497 ; Voigt, Hort. Sub. Cal., 72; Stewart, Pb. Pl., 11 ; Aitchison, Cat. Pb. and Sind Pl., 7 ; Murray, Pl.*

E. 327

	ERVUM
Taramira—Eruca Sativa. (*J. F.Duthie.*)	**Lens.**

and Drugs, Sind, 50; Atkinson, Him. Dist., 708; Econ. Prod., N.-W. Prov., Part V., 39; Duthie & Fuller, Field and Garden Crops, Part II., 26; Baden Powell, Pb. Prod., 419; Balfour, Cyclop., I., 441; Oudh Gaz., I., 498.

Habitat.—An erect herb, closely allied to the mustards, said to be a native of South Europe and North Africa. It is extensively cultivated as a cold weather crop in N. W. India; and, according to the *Flora of British India*, it is met with up to 10,000 feet on the Western Himálaya.

CULTIVATION.

CULTIVATION. N. W. P. **328**

North-Western Provinces.—"Its cultivation is most general in the western portions of the Provinces. It is most commonly grown mixed with gram or barley, or the combination of gram and barley known as *bejhar*, taking with these crops the place which rape fills in wheat fields. It is occasionally grown alone on land which has become too dry for the germination of any of the cold-weather cereals, and it is very frequently sown in cotton fields, its seed being scattered over the ground before the cotton receives its first weeding, in which process they are buried. No returns are available of the area on which *Dúan* is grown mixed with rabi crops, although it is known to be very large, especially in the western districts. Taking into account only the land on which it is grown by itself or in company with cotton, it is reported to occupy some 14,000 acres in the Meerut, 17,500 in the Agra, and 8,500 acres in the Rohilkhand Divisions. In the Allahabad Division it is only grown alone or with cotton on between 300 to 400 acres, and in the Jhánsi and Benares Divisions its cultivation seems to be almost unknown. *Dúan* may be sown at any time between the beginning of September to the end of November, and ripens about the same time as the rabi cereal harvest commences. * * * * When grown alone or with cotton its produce of seed per acre varies from 4 to 12 maunds (*Duthie & Fuller, Field and Garden Crops, II., 26*). Mr. E. T. Atkinson says that about Almora it comes up accidentally with the other species of mustard, but is also sparsely cultivated both in the hills and plains along the edges of corn fields.

PANJAB. **329**

Panjáb.—In 1882-83 the total area under this crop was given as 210,000 acres, in 1883-84 it was 253,000 acres, and in 1884-85 it increased to 256,000 acres. When grown with peas or gram it is intended for fodder. In the Jhelum District it is not unfrequently sown into a poor *bajra* crop.

OIL. Seeds. **330**

Oil.—The oil expressed from the SEEDS of this plant is used chiefly for burning, and "resembles," Roxburgh says, "Colza oil in all respects but in colour." It is sometimes used by the natives as a hair oil, and to a certain extent as food. "The cost is from 3 to 10 seers per rupee" (*Balfour Cycl.*). In Southern Europe it is said to be used as a salad oil.

FOOD. **331**

Food.—Stewart remarks that "the young plant is used as greens, as in France." The oil is sometimes employed in the preparation of sweet-meats.

FODDER. **332**

Fodder.—*Usan* is largely grown in the Panjáb, to be used as green fodder for cattle, camels, and goats. In some districts it is cultivated during the hot weather, and given, mixed with bruised barley, as a cooling food to buffalos. According to Dr. Stocks "the oilcake is universally used for oxen, camels, goats, and sheep."

Conf. with account of **Brassica**, Vol. I., pp. 520-534.

Ervalenta, see **Lens esculenta,** *Mœnch.*

Ervum Lens, *Linn.;* see **Lens esculenta,** *Mœnch.*

E. 332

ERYCIBE, *Roxb.; Gen. Pl., II., 868.*

333   Erycibe paniculata, *Roxb.; Fl. Br. Ind., IV., 180;* CONVOLVULACEÆ.
Vern.—*Urumin,* KOL; *Kari,* SANTAL; *Atta-meeriya,* SING.
References.—*Roxb., Fl. Ind., Ed. C.B.C., 197; Voigt., Hort. Sub. Cal.,
441; Brandis, For. Fl., 344; Kurz, For. Fl. Burm., II., 214; Gamble,
Man. Timb., 273; Thwaites, En. Ceylon Pl., 213; Dals. & Gibs., Bomb.
Fl., 169; Rheede, Hort. Mal., VII., 73, t. 39; Journ. As. Soc., Pt. 2,
No. 2, 1867, 80; For. Adm. Report, Ch. Nagpur, 1885, 32.*
Habitat.—A diffuse or sub-scandent shrub, or an erect tree, 40 feet
high, found throughout India from Oudh eastward, and southward to
Ceylon, Tenasserim, and the Nicobars.

MEDICINE   Medicine.—The Rev. A. Campbell mentions that in Chutia Nagpur the
Bark.   BARK is given for cholera.

334   ## ERYNGIUM, *Linn.; Gen. Pl., I., 878.*

335   Eryngium cœruleum, *Bieb.; Fl. Br. Ind., II., 669;* UMBELLIFERÆ.
Syn.—ERYNGIUM PLANUM, *Lindl.,* in *Royle Ill., 232* (not of *Linn.*)
Vern.—*Dhudhali,* HIND.; *Poli, mittúa, kandú, pahari gájar, núrálam,*
PB.; *Shakakul-misri,* ARAB.; *Gurs-dusti,* PERS.
References.—*Stewart, Pb. Pl., 105; Aitchison, Cat. Pb. and Sind Pl., 67;
Royle, Ill. Him. Bot., 232.*
Habitat.—A glabrous, perennial herb, with spinescent glaucous leaves,
found wild in Kashmír up to 6,000 feet.

MEDICINE.   Medicine.—The ROOT is considered to be aphrodisiac and to act as a
Root.   nervine tonic. In Kandahar the SEEDS are said to be officinal.
336
Seeds.   SPECIAL OPINION.—§ "The root is much used on account of its supposed
337   aphrodisiac properties" *(Civil Surgeon J. Anderson, M.B., Bijnor).*

ERYTHRÆA, *L. C. Rich.; Gen. Pl., II., 809.*
[*Ic., t. 1325;* GENTIANACEÆ.

338   Erythræa Roxburghii, *G. Don.; Fl. Br. Ind., IV., 102; Wight,*
Syn.—CHIRONIA CENTAURIOIDES, *Roxb.*
Vern.—*Charáyatah,* HIND.; *Girmi, gima,* BENG.; *Gada-sigrik,* SANTAL;
Luntak, kurúnai, kadavi-nai, BOMB.; *Jangli kariátu,* GUZ.
References.—*Roxb., Fl. Ind., Ed. C.B.C., 196; Dals. & Gibs., Bomb. Fl.,
157; Pharm. Ind., 150; O'Shaughnessy, Beng. Dispens., 461; Moodeen
Sheriff, Supp. Pharm. Ind., 99; Dymock, Mat. Med. W. Ind., 2nd Ed.,
541; S. Arjun, Bomb. Drugs, 90; Drury U. Pl., 198; Lisboa, U. Pl.
Bomb., 262.*
Habitat.—A slender annual with rose-coloured flowers, found through-
out India, ascending to 2,000 feet, from the Panjáb and Bengal to
Travancore.

MEDICINE.   Medicine.—The whole plant is powerfully bitter, and may be substitut-
339   ed for chiretta, when the latter is not available. According to Rev. A.
Campbell it is used by the Santáls in fever.

ERYTHRINA, *Linn.; Gen. Pl., 531.*
The Erythrinas are mostly trees or shrubs, rarely herbs. They are
chiefly remarkable for their brilliantly coloured red flowers, which are
usually produced before the new leaves are developed.
[LEGUMINOSÆ.

340   Erythrina arborescens, *Roxb.; Fl. Br. Ind., II., 190;*
Vern.—*Dingsong,* KHASIA; *Rodinga, fullidha,* NEPAL; *Gyesa,* LEPCHA;
Rungara, KUMAON.
References.—*Roxb., Fl. Ind., Ed. C.B.C., 544; Voigt., Hort. Sub. Cal.,
237; Brandis, For. Fl., 140; Gamble, Man. Timb., 122; Balfour,
Cyclop., I., 1054.*

		ERYTHRINA
The Indian Coral Tree.	(*J. F. Duthie.*)	**indica.**

Habitat.—A small or moderate-sized tree, found in the outer Himálaya from the Ganges to Bhután up to 7,000 feet, and in the Khásia Hills.

Structure of the Wood.—Similar to that of **E. suberosa** and **indica**, but is more compact, less spongy, and has more numerous concentric bands of soft texture.

TIMBER.
341

Erythrina indica, *Lam. ; Fl. Br. Ind., II., 188 ; Wight, Ic., t. 58.*
INDIAN CORAL-TREE ; MOCHI WOOD.

342

Vern.—*Pangra, panjira, pangara, pharad, pángrá, mandára,* HIND.; *Palita mandar, palitá-mádár,* BENG.; *Birsing,* KOL.; *Katheik,* MAGH; *Marur-baha,* SANTAL; *Chaldua, paldua,* URIYA; *Madar,* CACHAR; *Pángra,* BERAR; *Pángárá, phandra, pangaru,* MAR.; *Panaraweo, panarvo,* GUZ.; *Muruká, kalyáná-murukku, kalayána-murukku, muráka,* TAM.; *Barijamu, bádapu, modugu, badidapu, barjapu, bádise, mahá-meda,* TEL.; *Háliwára, halivana, páliwára, paravala-damara,* KAN.; *Dudap,* MALAY; *Penlaykathit, kathit,* BURM.; *Errabadu,* SING.; *Palitmandár,* SANS.

References.—*Roxb., Fl. Ind., Ed. C.B.C., 541 ; Voigt., Hort. Sub. Cal., 237 ; Brandis, For. Fl., 139 ; Kurz, For. Fl. Burm., I., 368 ; Beddome, Fl. Sylv., 87 ; Gamble, Man. Timb., 122 ; Dals. & Gibs., Bomb. Fl., 70 ; Rheede, Hort. Mal., VI., t. 7 ; Elliot, Fl. Andhr., 19, 20, 23, 110 ; U. C. Dutt, Mat. Med. Hind., 308 ; Bidie, Cat. Raw Pr., Paris Exh., 52 ; Irvine, Mat. Med., Patna, 89 ; Lisboa, U. Pl. Bomb., 59 ; Bird-wood, Bomb. Pr., 329 ; Cooke, Gums and Gum-resins, 17 ; McCann, Dyes and Tans, Beng., 66 ; Liotard, Dyes, 33 ; Liotard, Paper-making Mat., 11 ; Watson's Report on Gums, 18, 34 ; Gums & Resinous Prod. P. W. D. (1871), 14, 59 ; Balfour, Cyclop., I., 1055 ; Smith, Dic., 132 ; Treasury of Bot., I., 468 ; Kew Off. Guide to the Mus. of Ec. Bot., 43 ; Bomb. Gaz., XV., Pt. I., 68 ; XIII., Pt. I., 26.*

Habitat.—A moderate-sized, quick-growing tree with straight trunk, which is usually armed with prickles when young. It occurs throughout India from the foot of the Himálayas and in Burma. Often grown in gardens.

Gum.—It yields a dark-brown gum of little importance.

GUM.
343

Dye and Tan.—The dried red FLOWERS on being boiled yield a red dye. The BARK is also said to be used in dyeing and tanning.

DYE & TAN.
Flowers.
344
Bark.
345

Fibre.—The Rev. A. Campbell (*Chutia Nagpur*) states that the BARK yields an excellent cordage fibre of a pale straw colour.

FIBRE.
Bark.
346

Medicine.—The BARK is used medicinally, being antibilious and a febrifuge. It is also useful as a collyrium in ophthalmia. The JUICE of the leaves taken in a dose of two ounces is considered as a good vermifuge and cathartic. **Dr. Kani Lal De, C.I.E.**, says that the LEAVES are applied externally to disperse venereal buboes and to relieve pain on the joints.

MEDICINE.
Bark.
347
Juice.
348
Leaves.
349

SPECIAL OPINIONS.—§"Inner side of the bark is smeared with *ghí* and held over the flame of a lamp, the soot thus deposited is used in watery eye, being applied to the inner side and edges of the lower lid" (*Assistant Surgeon Anund Chunder Mookerji, Noakhally*). "Used as an anthelmintic. The fresh juice of the leaves is used in conjunctivitis. Soot deposited on the raw surface of a fresh piece of the bark is an useful application in tineatarsi and purulent ophthalmia. The fresh juice of the leaves is used as an injection into the ear for the relief of earache, and as an anodyne in toothache" (*J. H. Thornton, B A., M.B., Monghyr*).

Food.—The tender LEAVES are eaten in curry.

FOOD.
Leaves.
350

Fodder.—The LEAVES are used as cattle fodder in the Trichinopoly District.

FODDER.
Leaves-
351

Structure of the Wood.—Rather durable, though light and open-grained; it does not warp or split, and takes a good varnish. Structure the same as that of **E. suberosa.**

TIMBER.
352

It is used for light boxes, toys, scabbards, trays, as well as for fire-

E. 352

ERYTHROXYLON	
Coca.	**The Coca Plant.**

wood. Carpenters prefer it to all others for the poles of palanquins. According to Brandis it is used for much of the lacquered-ware of different parts of India. In Madras it is known as *mochi* wood, and according to Wight is generally employed for constructing catamarans.

DOMESTIC.
353

Domestic Uses.—It is said to be largely planted in Bengal and South India to support and shelter the betel and black pepper vine. It is also used for hedges.

354

Erythrina stricta, *Roxb.; Fl. Br. Ind., II., 189.*

Vern.—*Falleto, fullidha,* NEPAL; *Mouricon, kichige,* KAN.; *Taung kathit,* BURM.

References.—*Roxb., Fl. Ind., Ed. C.B.C., 542; Voigt., Hort. Sub. Cal., 237; Kurz, For. Fl. Burm., I., 369; Beddome, Fl. Sylv., t. 175; Gamble, Man. Timb., 122; Dalz. & Gibs., Bomb. Fl., 70; Balfour, Cyclop., I., 1055; Ind. For., XIV., 391.*

Habitat.—A large tree with pale-coloured prickles when young; is found in Burma and the western half of the Peninsula.

TIMBER.
355

Structure of the Wood.—Soft, resembling that of **E. suberosa**; it is sometimes used for planks.

356

E. suberosa, *Roxb.; Fl. Br. Ind., II., 189.*

Vern.—*Pangra, dauldhák, rúngra, rowanra, nasút, madára,* HIND.; *Farhud,* KHARWAR; *Mandal,* GARO; *Fullidha,* NEPAL; *Katiang,* LEPCHA; *Phangera,* GOND; *Gúlnashtar, pariára, thab,* PB.; *Gadaphassa,* KURKU; *Nangtháda,* MELGHAT; *Pangra,* KON.; *Pángára,* DEC.; *Mandal,* GARO.; *Muni, maduga,* TAM.; *Mulu modugu, badadam* (*var.* **sublobata**), TEL.

References.—*Roxb., Fl. Ind., Ed. C.B.C., 543; Voigt., Hort. Sub. Cal., 237; Brandis, For. Fl. 140; Kurz, For. Fl. Burm., I., 369; Beddome, For. Man., 87; Gamble, Man. Timb., 121; Grah., Cat. Bomb. Pl., 54; Dalz. & Gibs., Bomb. Fl., 70; Aitchison, Cat. Pb. and Sind Pl., 47; Elliot, Fl. Andhr., 19, 119 (var. sublobata); Atkinson, Him. Dist., 309; Balfour, Cyclop., I., 1055; Raj. Gaz., 35; Bomb. Gaz., XV., 68.*

Habitat.—A moderate-sized deciduous tree of the Himálaya from the Ravi to Bhután, up to 3,000 feet, and extending to Central and South India and Burma. **E. sublobata,** *Roxb.,* is only a variety with larger and lobed leaflets.

TIMBER.
357

Structure of the Wood.—Very soft, spongy, white, fibrous but tough; darker-coloured near the centre, but no regular heartwood. It is used for scabbards, sieve-frames, and occasionally for planking.

(*J. Murray.*)

ERYTHROXYLON, *Linn.; Gen. Pl., I., 244.*

A genus of shrubs or trees containing about 50 species, natives of warm countries—10 in Africa, 6 in India and Ceylon, 1 in Australia, and the rest in America. The generic name has been given in allusion to the red-sandal-like wood which the majority possess.

358

Erythroxylon Coca, *Lam., Bent. & Trim.; Med. Pl., t. 40;* LINEÆ.

References.—*DC. Origin Cult. Pl., 135; U. S. Dispens., 15th Ed., 563; Bent. & Trim., Med. Pl., 40; Warden, Prof., Chemistry, Calcutta,—Note on Erythroxylon Coca as grown in India; Agri.-Hort. Soc. Jour., VIII. Pt. II. (new series), 1888, pp. 127–170; Kew Bulletin, January 1889; Christy, Com. Pl. and Drugs, No. 3, 24, No. 4, 43, No. 5, 55, No. 6, 85, No. 7, 45, No. 8, 47, No. 9, 62; Spons, Encyclop., 1307; Balfour, Cyclop., 1055; Treasury of Bot., 469; Weddel, Voyage dans le Nord de Bolivie (Paris, 1853); Johnstone, Chem. of Common Life, Ed. Church, 357; Watts, Dic. of Chem. I., 1059; Gosse, Monographie de l' Erythroxylon Coca (1861); Christison, Brit. Med. Journ., April 29, 1876; Crockman in Journal Pharm. Society, April 23rd, 1887; Dowdeswell, in Lancet, April 29, 1876, and May 6, p. 664; Bidie, Pamphlet on Erythroxylon Coca,*

| The Coca Plant. (*J. Murray.*) | ERYTHROXYLON Coca. |

Madras, March 1885, Rusby. "The Cultivation of Coca," Therapeutic Gazette, January 1886.

Habitat.—A shrub, 2 to 6 feet high, much branched, somewhat resembling the black thorn.

It is found in Peru, Bolivia, Brazil, the Argentine Republic, and other parts of South America, growing from 2,000 to 8,000 feet above the sea-level, but according to **De Candolle** the plant is indigenous only to the two former countries. It is an escape from cultivation as generally met with in other parts of America, and is cultivated in various parts of India and Ceylon. In a recent pamphlet on Coca (*Kew Bulletin* of January, 1889) two distinct varieties are described :—

(1)—The typical **E. Coca,** *Lam.,*
(2)—**E. Coca,** *var.* **novo-granatense.**

An intermediate form is provisionally adopted as exhibiting the general characteristics of Bolivian Coca. According to the *Bulletin* the second variety is the plant figured by **Bentley and Trimen**; leaves received from Ceylon corresponded with those of the typical **E. Coca,** while those from India exhibited the characters of the variety **novo-granatense,** or of the intermediate form, the Bolivian Coca.

History.—The name *Coca,* sometimes called *Cuca,* is a corruption of the Aymara Indian word *Khoka,* signifying "plant," *the* plant *par excellence.*

HISTORY, 359

The natives of Peru have utilized this bush from the earliest period, and its employment was general at the time of the conquest of that country by Spain. From Peru (and, according to **De Candolle** from Bolivia also,) it seems to have spread over the other parts of South America, where the cultivated plant is now to be found in all localities, the natural conditions of which allow of its growth. The exact date of the introduction of **Coca** into England is not known, but it was probably not much before the year 1870.

Introduction into India.—Coca in 1870 was introduced into Ceylon from Kew, and from the Peradeniya stock have been derived the plants now in that island. It seems probable that from this same source came also the plants originally grown in the gardens of the Agri.-Horticultural Society of Madras, and these furnished some of the first plants cultivated in India.

INTRODUC-
TION.
360

At a meeting of a Committee of that Society held in May 1876, a letter was read from **Mr. Joseph Stevenson,** suggesting that the propagation of the plant might be attempted, as it then had become evident that **Coca,** the wonderful *sustaining* effects of which were beginning to be recognised in Europe, would rapidly become an important article of commerce. No steps of any importance, however, seem to have been taken till 1885, when, owing to the discovery of the value of *Cocaine* as an *anæsthetic,* the demand in Europe for the **Coca** leaf was rapidly increased. As a consequence, applications for the plant became very numerous, and, as far as the limited supply from a single specimen in the Madras Gardens allowed, seedlings were distributed amongst planters and others in various parts of the country. In 1885-86, the Agri.-Horticultural Society of India distributed young plants from the Calcutta Gardens to the tea-growing districts, *i.e.,* Assam, Cachar, the Duars, Darjeeling Terai, and Jaunpore. Certain cultivators at Ranchi obtained seeds direct from Paris.

In 1885 the Government of India addressed Her Majesty's Secretary of State for India with a view of ascertaining the method of preparation of the leaf as pursued in South America. This resulted in accounts of the methods pursued, from **Surgeon General Balfour** and **Mr. W. T. Thiselton Dyer,** the latter reporting that the method described by **Deputy Surgeon General G. Bidie, C.I.E.,** in his lecture at Madras in March 1885, left nothing to be added.

Owing most probably to the great increase in exportation of the plant

E. 360

ERYTHROXYLON
Coca.   The Coca Plant.

HISTORY.

from South America, and its consequent cheapening in the European markets, **Coca** cultivation in India has not materially developed since its introduction. Indeed, of the tea districts of Ceylon, Madras, Mysore, and Bengal, it may practically be said that it is now, as it was three years ago, grown only experimentally.

CULTIVATION.

CULTIVATION
361

Method followed at the source of supply.—The *Tropical Agriculturist* of November 1885 publishes an account taken from *The Ephemeris*, of which the following may be given as the substance :—

Coca is grown on terraces on the sides of deep narrow valleys, between the heights of 3,000 and 6,000 feet. In August the seeds are sown in boxes or beds, and in June they are transplanted to the hill terraces and deposited about three feet apart. The soil must be rich in vegetable manure, and free from weeds ; the crop, in other words, is an exhausting one, necessitating virgin soil. In consequence a forest clearing is generally chosen, the ground being already rich in decayed vegetable matter.

Dr. Warden, in his note on "*Erythroxylon Coca grown in India*" writes :—

"From a high altitude, the best results as to total alkaloids have been yielded by plants grown on a hill-side soil rich in vegetable manure. But a rivalry exists between this variety of soil and a yellow clay. The author is inclined to think that those who prefer the latter soil do so because it yields a somewhat larger crop.

"The ground for the nursery beds is prepared during the latter part of the dry season, by breaking it up very thoroughly to the depth of a foot or more. The plan of sowing the seeds broadcast as soon as gathered, and covering with a little earth, or better, a layer of banana leaves or decaying vegetable matter, has been found to answer. Germination requires from eight to twelve days longer than by adopting the native method. which consists in depositing the seeds, as soon as gathered, in a shaded place in layers an inch or more deep, and covered with a thin layer of decaying leaves. The heat generated by the decomposition of the fleshy pericarps seems to induce germination, and the embryo bursts its bony covering. This growth unites them in from eight to fourteen days into a solid mass which is broken up into small pieces and planted in furrows in the nursery. In this process very many of the sprouts are broken off and the plants destroyed. A covering of brush or straw must be placed over the nursery, at first only three or four inches above the surface, and elevated as the plants grow.

"On the manner in which the ground is prepared for the plantation much of the future well-being of the plant depends. The ground should be thoroughly powdered to the depth of two, or, if possible, three feet, and all roots and large stones removed. It is generally believed that shade tends to the production of the best quality of leaves, and the cocales are therefore planted thickly with a small broad-topped leguminous tree related to the St. John's head plant. The custom appears to have arisen from two considerations. There is a period, already referred to, of two or three months during which no rain falls ; and then these trees afford protection from the sun. Secondly, because shade conduces to the production of a large smooth leaf of elegant colour and thus adds to the appearance of the product. From repeated comparative assays made by Rusby of shade and sun-grown leaves from adjoining plants, the sun-grown leaves were invariably much richer in total alkaloids.

"The plants are transplanted from the nursery at the advent of the permanent rains, and are set out from half an inch to three feet apart. They grow to a height of two to six feet, but the largest plants do not yield the

E 361

| The Coca Plant. (*J. Murray.*) | ERYTHROXYLON Coca. |

best leaves. Great care must be taken to keep the soil thoroughly stirred and free from weeds. (*Jour. Agri.-Hort. Soc., Vol. VIII., Pt. II., New Series, p. 149.*)"

Most American writers appear to hold that the plant is better cultivated in the open than in the shade, an opinion which the above chemical analysis would seem to corroborate. On the soil becoming exhausted, fresh plantations are opened out in the forest, in preference to resorting to manure.

In India.—The following interesting facts regarding the effects of manure are given by Dr. Warden in the paper already quoted :—

" As regards the effects of cultivation and manure on the yield of alkaloid, it would appear from the reports I have received, that in only two of the districts was the soil specially manured. At Arcuttipore, the manure consisted of old cow-dung with a top dressing of soot. In the Jaunpore district the soil is stated to have been highly manured, but no particulars as to the precise nature of the fertiliser are afforded. The leaf grown at Arcuttipore yielded very considerably more alkaloid than any of the other samples examined ; while that grown in the Jaunpore district contained only ·571 per cent. of alkaloid. I have no information whether the Arcuttipore plants were grown in the shade or open. On the Jaunpore Tea Estate there appear to be four plants, two in full sun-shine and measuring 5½ feet and 5 feet 2 inches in height respectively, one in partial shade 3 feet high, and one in shade 5 feet high, and I gather from the Manager's letter that the leaf sent me was collected from the plant which grew in the shade.

" Taking into consideration the amount of potash contained in the leaves and the rapid exhaustion of the soil which would necessarily ensue from repeated plucking of the leaves, it appears to me that though at first a nitrogenous fertiliser would be beneficial, yet after a time the addition of a fertiliser containing potash in some form in addition to nitrogenous matter would be necessary. The amount of nitrogen in the soot or cow-dung might possibly suffice, but whether the amount of potash in the cow-dung would be sufficient to supply the place of that removed from the soil by the leaves is an open question " (*l. c. p. 153*).

From reports furnished at various times to the agricultural journals, it appears that the slightest degree of frost is fatal to the plant—at least during its infancy. For this reason experiments in the tea plantations on the higher Himálaya have been unsuccessful, but more encouraging reports exist of its cultivation at lower altitudes in India, as, for example, from about 100 feet to 2,000 feet above the sea level.

The essential conditions seem to be :—

(1) a rich soil, preferably of virgin forest ; (2) a considerable rainfall ; (3) a complete absence of frost ; (4) a careful system of cultivation, paying special attention to weeding.

SPECIAL REPORTS.—The Conservator of Forests, Southern Circle, Madras, writes :—"This Circle has not got beyond the experimental stage. No regular areas have been planted. In Wynaad the planters have a few plants here and there, but apparently more as curiosities than anything else, and the Forest Department there has about one hundred plants. The District Forest Officer has observed that it seeds less freely in Wynaad than on the coast." The Deputy Conservator of Forests, Coorg, reports :—"Coca has only been cultivated in gardens in Coorg. Flowers and fruits in Mercara. It seems doubtful if its cultivation would 'pay."

COLLECTION AND MANUFACTURE.

In Peru and Bolivia two crops are gathered, the first, the " March

ERYTHROXYLON
Coca.

The Coca Plant.

crop," commences in January, the second, the " St. John's " crop, begins in May. The first picking of leaves is made one year and a half from the date of transplanting. During the first five years the percentage of the alkaloid, cocaine, yielded by the leaves increases rapidly, reaching its maximum about the tenth year. The plant retains its full productive power till about the twentieth, after which it slowly declines till about the fortieth year, probably owing to exhaustion of the soil.

The women and children collect the mature leaves, which are known by being bright green on the lower, and yellowish on the upper, surface. Each leaf is picked separately, and very carefully, and every precaution is taken not to touch the top of the bush. The leaves are then conveyed to the place of preparation, where they are laid out in a single layer on a pavement, kept scrupulously flat and clean, which has been previously heated by the sun. The necessity of the pavement being already hot is greatly insisted on. The leaves are stirred occasionally until dry, which they become in about three hours. They are then either placed for a short time in storage houses, where they undergo a slight sweating process, or are at once packed. The slightest amount of moisture is fatal to the leaf after being dried. The leaf should therefore always be packed in zinc, or tin-lined, air-tight boxes.

In India several methods of drying artificially, in tea driers or charcoal *chulas,* have been experimentally tried. According to Dr. Warden the results by all are equally good.

He writes:—" The object which should, I think, be kept prominently in view, is to dry the leaves as thoroughly and quickly as possible at the lowest temperature. The plan adopted at Arcuttipore of first withering the leaves in the shade, and then drying them in a tea drier at 150° Fh. for 10 minutes, appears to me as good as any. I do not think any advantage is to be gained by employing a higher temperature. In whatever way dried, the leaves should be at once packed in air-tight boxes directly they are cold."

Medicine.—From the earliest dates the Indians of Bolivia, Peru, and Brazil have ascribed marvellous properties to the LEAF of the Coca plant. Chewed, either alone or mixed with lime, or taken in various forms of syrup and decoction, the consumer is enabled to sustain the greatest fatigue and hardship, without either food or sleep, for a lengthened period. The drug is also said, especially when taken as an infusion or decoction, to prevent difficulty of respiration in ascending hills. Consumed in any form it produces a peculiar excitement, slow and sustained, and diffused generally over the nervous system, accompanied by a general feeling of well-being. When eaten along with tobacco it is reported to produce a condition of intoxication very similar to that caused by alcohol or Indian hemp. Prolonged or excessive use of the drug is followed by much the same results as over-indulgence in opium. The Coca-eater loses his appetite, suffers from impaired digestion, and, when not under its influence, becomes phlegmatic and apathetic. According to Johnston in his *Chemistry of Common Life* quoting Von Tschudi, " The inveterate Coquard (or Coca-eater) is known at the first glance. His unsteady gait, his yellow skin, his dim and sunken eyes, encircled by a purple ring, his quivering lips, and his general apathy, all bear evidence of the baneful effects of the Coca juice when taken in excess." Von Tschudi, however, states as the result of his inquiries, that " the moderate use of Coca is not merely innocuous, but that it may even be very conducive to health."

Dependent on these properties, the infusion of Coca is viewed as a valuable remedy in asthma and colic, and that the leaves applied externally as a plaster to cure boils and ulcers.

E. 363

| The Coca Plant. (*J. Murray.*) | ERYTHROXYLON Ccca. |

The Indians of Peru, probably influenced by their experience of the wonderful properties of the leaf, are said to regard it as sacred. Its use is much intermingled with their religious rites, and to the plant itself worship is rendered.

Since the introduction of the leaf into Europe, many writers have extolled the advantages to be derived from the drug, and actual experiments by **Sir Robert Christison** and others have shewn that it possesses nearly, if not quite, all the qualities ascribed to it by the Indians. In 1860 Neimann separated the now very important alkaloid *Cocaine* from the leaf, and described it as producing insensibility to the tongue when applied to it. This important fact seems to have lain dormant till 1884, when **Herr Koller**, a medical student in Vienna, rediscovered this valuable anæsthetic action of the alkaloid. It is now most extensively employed as a local anæsthetic in many minor operations, and is specially valuable in ophthalmic surgery, since it produces complete insensibility to pain in the superficial structures of the eye. It is also mydriatic, and paralyses the accommodation. *Cocaine* seems to act by paralysing the termination of the sensory nerves in any structure to which it is applied, but this paralysis remains purely local and does not last long. Indeed, this limitation of its action to the tissues to which it is directly applied, is the most valuable property of the drug; as an external remedy for painful diseases of the skin or mucous membrane it is therefore most useful.

Chemistry.—" The **Coca** leaf is said to contain the following principles: the *alkaloid Cocaine, Hygrine, amorphous Cocaine, Ecgonin, coca-tannin,* and a *peculiar wax.*

From recent researches, however, it would appear that the amorphous cocaine, formerly described, is in reality a solution of cocaine in the volatile oily body *hygrine* (*Stockman, Journal Pharm. Society, April 23rd, 1887*). Regarding *ecgonin,* it appears that it also does not exist ready formed in the leaves, but is a product of the decomposition of Cocaine (*Dr. Warden's note on Erythroxylon Coca grown in India—March 1888*). The further elucidation of this question is to be hoped for, as the cocaine of commerce at present seems to vary much in character, and a more exact knowledge of its true chemical nature is required to determine whether the amorphous substance often connected with the alkaloid and its salts, may not be the cause of the objectionable effects which sometimes follow its use. Excluding these doubtful substances, therefore, there remain to be considered the alkaloid cocaine, hygrine, coca-tannin, and the wax:—

1. Cocaine.—$C_{17}H_{21}NO_4$ (Zorsen). Has been generally described as possessing all the properties of an alkaloid and as crystalline. **Dr. Warden's** recent analyses, however, show that the alkaloid obtained from the leaves of **E. Coca** grown in India possess the marked peculiarity of in no single instance shewing any tendency to spontaneous crystallization. But this result is at variance with the analyses of **Mr. Alfred G. Howard, F.C.S.**, given in the *Kew Bulletin*, already referred to. That chemist found that the leaves received from Darjiling, Bogracote, Alipore, and Ranchi yielded from '23 to '45 per cent. of crystallisable cocaine, and from '17 to '35 per cent. of the uncrystallisable alkaloid. The leaves from Ceylon, on the other hand, which belonged to the typical **E. Coca**, were found to contain from '47 to '60 per cent. of crystallisable and no uncrystallisable Cocaine. The alkaloid forms salts, of which the citrate, salicylate, and hydrochlorate are used in medicine. It is very sparingly soluble in water (1 in 700 parts), more so in alcohol, and freely so in ether and volatile oils. It is also soluble in fats. The fact of its being soluble in ether, while its salts are not, is taken advantage of in the preparation of the pure alkaloid.

MEDICINE.

CHEMISTRY.
364

Cocaine.
365

E. 365

T 2

ERYTHROXYLON
Coca.
The Coca Plant.

CHEMISTRY.

It has a bitterish taste, and crystallizes in small shining monoclinic prisms.

The pure alkaloid is much used in medicine, especially in the manufacture of oleates and ointments, for which it is more suitable than its salts, owing to its solubility in fats and oils. The amount of the alkaloid obtainable from the leaf of commerce is variously stated as from ·2 to ·5 per cent.

Hygrine.
366

2. HYGRINE—is described by **Wholer** and **Lossen** as "a pale yellow, volatile, oily body, giving the ordinary reactions of alkaloids, hygroscopic, and forming hygroscopic salts which crystallize with great difficulty."

Coca-tannin.
367

3. COCA-TANNIN—resembles the tannin of tea in lgiving a deep brownish green colour with the persalts of iron. It has been found to vary much in quantity in the different leaves examined in this country. Dr. Warden writes :—" It is of interest to note that the largest deposits of *coca-tannin* occurred in those samples which yielded the highest percentage of alkaloid. It appears to me, therefore, as not improbable that in the leaves the *cocaine* is in combination with the acid to which this term of *coca-tannin* has been applied."

4. The WAX is unimportant.

Wax.
368

Dr. Warden, in his paper above quoted, gives a number of very interesting analyses of leaves grown in different parts of India, from which it would appear, that the percentage of *cocaine* is higher than that recorded in any previously published assay. Those which yielded the best results were leaves from Ranchi, Arcuttipore, and from the Central Terai Tea Co. They contained an average percentage, on the anhydrous leaf, of over 1, the highest being 1·671. Though, as above noted, the physical character of the alkaloid obtained by Dr. **Warden** differs from that of the American leaf, it has been proved that it is equally efficacious as a local anæsthetic. Dr. **Saunders**, Professor of Ophthalmic Surgery at the Medical College, Calcutta, used a 4°/₀ solution in thirteen cases of operation for cataract and many minor operations, and found that it differed in no way from other *cocaine*, except that it appeared to have a quicker and slightly stronger action. Should **Warden's** analysis be confirmed that the cocaine of the Indian plant neither spontaneously crystallizes itself nor possesses spontaneously crystallizable salts, it might be objected to on purely pharmaceutical grounds; but it is to be remembered that the salts of the alkaloid are mainly used in solution.

TRADE.
369

Trade.—It has been clearly established that the climate and physical conditions of many parts of India are in every way suitable for the growth of **Coca**; but whether it will pay to cultivate the plant is another question.

According to **Squibb** in the *Ephemeris*, May 1887, the Peruvian Government records and taxes a production of 15,000,000℔ per annum, and the Bolivian Government 7,500,000℔. Of the latter quantity 5°/₀ or 375,000℔ is exported to the United States and Europe. Assuming that from the doubly great produce of Peru, twice the quantity above mentioned reaches the United States and Europe, an aggregate export of 1,125,000℔ annually is arrived at. This amount of leaf, if manufactured, would yield from 2,000 to 3,000℔ of cocaine. When it is remembered that the uses of **Coca** are very limited in Europe, that it is employed almost entirely as a medicine, and that there are no indications of **Coca** preparations coming into general use as a beverage, it seems very improbable that cultivation of the plant to any great extent would pay. Still, the Indian plant seems to be peculiarly rich in the alkaloid, and small quantities carefully prepared and packed would probably find a ready sale in Europe.

E. 369

The Bastard Sandal.	(*J. Murray.*)	**ERYTHROXYLON monogynum.**

Recent returns give the price of the dried leaf at from 10*d.* to 1*s.* 6*d.* a pound.

Erythroxylon monogynum, *Roxb.* ; *Fl. Br. Ind., I., 414 ; Beddome,*   **370**
BASTARD SANDAL OR RED CEDAR.   [*Fl. Sylv., t. 81.*

Syn.—E. INDICUM, *Beddome ;* SETIHA INDICA, *DC.*

Vern.— *Dévadáram,* or *dévadári* (in Arcot, Salem, and Coimbatore) *Nát-ká-deodár, simpuliccai,* or *simpulicham* (in Madras, South Arcot, Trichinopoly, &c.), *sammanathi* (in Madras, Tanjore, Madura, and Tinnevelly), *Kat santhanam,* (in Salem), *thasadaram* (in Madras), TAM. ; *Benáde,* NILGHIRIS ; *Huli,* BADAGA ; *Kuruvakumara,* KAN. ; *Dévadáru* (the name in most Telegu-speaking districts), *adavigóránta* (in Anantapur), *gathiri* (Cuddapah), TEL.

References.—*Roxb., Fl. Ind., Ed. C.B.C., 386 ; Voigt., Hort. Sub. Cal., 172 ; Kurs, For. Fl. Burm., I., 171 ; Gamble, Man. Timb., 58 ; Thwaites, En. Ceylon Pl., 53 ; Moodeen Sheriff, Mat. Med., Madras, 70 ; Pharmacog. Ind., 242 ; Dymock, Mat. Med. W. Ind., 892 ; Drury, U. Pl., 391 ; Lisboa, U. Pl. Bomb., 195 ; Cooke, Gums and Gum-resins, 129 ; Spons, Encyclop., 1684 ; Balfour, Cyclop., 1055 ; Kew Off. Guide to the Mus. of Ec. Bot., 22 ; Paxton, Bot. Dic., 516 (under Sethia); Mysore and Coorg Gaz., II., 87 ; Official Correspondence (Proceedings, Board of Revenue) Madras, No. 165, 1889; Special Report by J. Cameron, Esq., Bangalore.*

Habitat.—A shrub or small tree of the hilly tracts of the Western Peninsula ; also met with in the Kurnool, Bellary, Cuddapah, Nellore, Chingleput, and North Arcot Districts of South India. It occurs plentifully in Tanjore, Tinnevelly (ascending the Gháts to an elevation of 2,500 feet), throughout the Sígúr range, and in the forest reserves of the Nilghiri hills. In Ceylon it is said to be found in the hot dry parts of the Island.

In a recent official report regarding this plant as a source of Madras fodder, it is stated that the belief prevails that the plant is " well able to withstand drought, and evidently flourishes on the driest soils in the very hottest climates "

Oil.—The WOOD is reported to yield an oil used as a preservative for native boats. This oily substance, resembling tar, is known in Ceylon under the name of *dummele ;* it is extracted by packing pieces of the wood in an earthen pot, inverted over a similar empty one and surrounded by fire. The tar thus distilled is soluble in ether, alcohol, and turpentine, and is an excellent preservative of timber. It is not a commercial article, but might become so. This information was first published by Mr. W. C. Oudnatje, and his account of its preparation and uses has been reproduced in various works, such as *Cooke's Gums and Resins ; Spons' Encylcop., &c.,* &c., without anything new being added to our knowledge of the substance.   **OIL, Wood. 371**

Medicine.—Dr. Bidie says that "during the Madras famine of 1877 the LEAVES were largely eaten by the starving poor, and as there is nothing in them structurally likely to satisfy the pangs of hunger, it seems probable that they contain some principle like that of E. Coca." Specimens analysed by the Government Quinologist at Madras were found, however, to have no anæsthetic property analogous to that met with in E. Coca, but to possess a bitter and tonic principle, which might mitigate the pangs of hunger. This same result was obtained by Dr. L. A. Waddell, in his chemical examination of a large quantity of the leaves of this plant furnished by Dr. King, Superintendent, Royal Botanic Gardens, Calcutta (see *Indian Medical Gazette for September 1884, p. 281*). Dr. Waddell found that it contained no alkaloid whatsoever, and he accordingly arrived at the conclusion that had any such alkaloid as that met with in E. Coca existed in this species, the famine-stricken people of Madras would not have continued to eat the leaves. Dr. Moodeen Sheriff describes the plant as possessing stomachic, diaphoretic, and stimulant-diuretic properties, the   **MEDICINE, Leaves. 372**

EUCALYPTUS.   Eucalyptus—Gum Trees.

MEDICINE.
Wood.
373
Bark.
374

WOOD being sold in Madras and used in slight cases of dyspepsia and continued fever, and also in some cases of dropsy. He says the wood has a strong aromatic and agreeable smell. Dr. Bidie mentions the powder as used medicinally as a substitute for sandal wood. The BARK is said by Dr. Shortt to be employed as a tonic in fever, being prepared as an infusion. The leaves when eaten as a vegetable are believed to possess refrigerant properties, and the pulp beaten into a LINIMENT with gingelly oil is used, as an external application to the head. (*Madras Agri.-Hort. Soc. Journal, IV., 41*). This statement, regarding the preparation of a liniment, was apparently first made by Ainslie (*Mat. Med. II., 421*) regarding the plant he calls **E. areolatum,** *Willd.*

Liniment.
375

FOOD.
Leaves,
376
Fruit.
377
Famine
Food.
378
FODDER.
Leaves.
379

Food.—Both in the various Madras accounts of this plant and in **Mr. Lisboa's** *Useful Plants of Bombay*, the LEAVES are said to be regularly eaten as a green vegetable. Of Madras it is reported that they are used in curries, and that in famine times they are boiled with rice, *rági*, &c., to increase the volume of food. **Mr. Lodge** (*District Forest Officer, Cuddapah*) writes that, the plant yields a "small red juicy FRUIT with a refreshing taste, and a flavour somewhat resembling that of a cherry."

Fodder.—The Government of Madras recently called for information as to the extent the LEAVES of this plant were used as fodder. The replies showed that they were "sometimes but rarely used in the Godavari, Cuddapah, and Anantapur districts." The Collector of Salem, however, reported that "no one recognises it as a fodder-plant, and that cattle have been seen to pass close to young succulent coppice shoots without touching them." The Madras report concludes, however, by saying that elsewhere, when other supplies fail, cattle, sheep, and goats eat the plant.

TIMBER.
380

Structure of the Wood.—Sapwood white; heartwood dark-brown, with a pleasant resinous smell; it is very hard, takes a beautiful polish, and is sometimes used as a substitute for sandalwood (**Santalum album**).

Esparto Grass. *See* **Lygeum Spartum** and **Stipa tenacissima.**

(*J. F. Duthie.*)

EUCALYPTUS, *L'Her.; Gen. Pl., I., 707.*

HISTORY.
381

The majority of the species, of which about 140 have been described, are confined to Australia and Tasmania, where they afford characteristic features in the scenery of those countries. A few occur in New Zealand, and in some of the islands of the Indian Archipelago.

Popularly known under the general name of 'gum trees,' they are locally distinguished in Australia by characters observable in the bark; which, in some of the species, is fibrous or stringy, in others hard and fissured, whilst sometimes it presents a smooth and polished surface, and occasionally it scales off in flakes. The botanical determination of the species is often difficult, owing to the close similarity of their floral structure, as well as to the various forms sometimes assumed by the foliage on different portions of the same tree, and at different periods of its life. This task has, however, been greatly lessened by the researches of the eminent local botanist, **Baron von Mueller**, brought to light in his very valuable illustrated monograph entitled "Eucalyptographia."

As trees they are chiefly remarkable for their rapid growth, and the enormous height to which some of the kinds attain; one specimen in Victoria, a fallen one, having been found to measure 480 feet in length; and specimens of **E. obliqua** (the String-bark) have been felled in Tasmania, the trunks of which measured 300 feet high and 100 feet in circumference.

The timber yielded by some kinds, notably that of **E. Globulus** (Blue gum), **E. marginata** (Jarrah or Mahagony of South-West Australia), and

E. 381

Eucalyptus—Gum Trees. (*J. F. Duthie.*) EUCALYPTUS.

E. robusta (Red gum of South Australia), is extremely valuable on account of its strength and durability under water, and its immunity from attacks by white-ants.

An astringent substance resembling *kino* (a product of **Pterocarpus Marsupium**) s yielded by several of the species, and is used medicinally, as well as for tanning and dyeing.

A still more important product is Eucalyptus oil, which, through the exertions of **Mr. J. Bosisto** of Melbourne, has recently been extensively brought into commerce, and is now being employed for various purposes. The existence of this oil, which can be distilled in greater or less quantities from the different species of **Eucalyptus**, has no doubt some influence in improving the climate of districts where malarious fever prevails, though the beneficial results are in all probability mainly due to thorough drainage of the soil effected by the rapid growth of these trees. The following list, taken from **Mueller's** "Select Extra-Tropical Plants," p. 146, gives the percentage of oil yielded by certain species :—

E. amygdalina .	.	.	.	. 3·313	per cent. of volatile oil.
E. oleosa	.	.	.	. 1·250	„ „
E. Leucoxylon	.	.	.	. 1·060	per eent. of volatile oil.
E. goniocalyx .	.	.	.	. 0·914	„ „
E. Globulus	.	.	.	. 0·719	„ „
E. obliqua	.	.	.	. 0·500	„ „

Baron von Mueller then goes on to explain that the lesser quantity of oil of **E. Globulus** is compensated for by the vigour of its growth and the early copiousness of its foliage, and that the proportion of oil varies somewhat according to locality and season. "**E. rostrata**," he says, "though one of the poorest in oil, is nevertheless important for malaria-regions, as it will grow well on periodically-inundated places and even in stagnant waters not saline."

Though confined to the Australian continent and its neighbourhood, the various species of **Eucalyptus** are found to thrive under very different influences as regards climate and soil. Some occur at elevations where snow remains on the ground for several months of the year ; others flourish best in the northern and warmer parts of the continent ; others again are more at home in swampy ground, whilst some seem to prefer sandy or calcareous soils. The experimental cultivation of gum trees in other countries must therefore be regulated by a consideration of these facts. As regards Eucalyptus cultivation in India, the most successful results have been obtained on the Nilghiris, where, according to the latest report received from the Conservator of Forests, South Circle, Madras, it is stated that " there are several extensive plantations, both Government and private, and several species, but chiefly **E. Globulus**, are cultivated on most of the hills in Southern India at from 4,000 to 8,000 feet. It is quite impossible to estimate the area. In Wynaad, too, several varieties have been introduced from Queensland and are growing vigorously. Some trees planted in 1884 are now over 60 feet high and 42 inches girth at 4 feet from the ground. The species to which these Wynaad trees belong have yet to be determined. Eucalyptus oil is extracted in a small way on the Nilghiris."

In Northern India extensive trials were made in 1876 with seeds of various kinds of **Eucalyptus**, and it was then ascertained that of these **E. resinifera**, *Smith*, and **E. rostrata**, *Schlech.*, were the most promising for cultivation in the plains. These two species have since maintained their character, and there are now several vigorous specimens both at Saharanpur and Lucknow, which yield seed abundantly. The localities in Northern India best suited for the blue gum (**E. Globulus**) are

EUCALYPTUS Globulus.	**The Blue Gum Tree.**

HISTORY.

Ránikhet and Abbottabad. **E. citriodora,** *Hook.*, and **E. melliodora,** *A. Cunn*, both having deliciously-scented foliage, are thriving well in many places in the plains of North India.

The following communication was received from the Conservator of Forests, Panjáb, in August 1889 :—" A considerable number of different species of Eucalyptus have been tried in various parts of the Province, but on the whole the results have not been satisfactory; it has been found, however, that planting in groves gives a better chance of success than when the trees are grown singly along roads, &c. In Kangra, in the Kothala estate, and in Kulu, a few specimens of the "blue gum" and other unknown kinds, have done well, and experiments are now being made in the Dera Tahsil.

" The species has been introduced into Bashahr, but has not yet established itself; but in Hazara, the experiments have been successful, and there are now a number of trees round Abbottabad 80 feet high. In Chamba attempts were made at Kalatop, Chamba, and Bakloh; at the two former places they failed, but there are about 100 trees flourishing at Bakloh. The most extensive experiments that have been made were in the Lahore District at Changa Manga and in the carob plantation at Lahore. In all twenty-five species have been tried; but out of these only three, **E. rostrata, E. citrioides,** and **E. resinifera,** have had any real success.

" The cause of this failure may be mainly attributed to three sources: *1st*, failure in the rains; *2nd*, injury to the young stems by sunburn; *3rd*, the worst of all, the white ants which attacked the tree by eating away the supporting roots. From these causes, but mainly from the last, only some 300 Eucalypti have succeeded in Changa Manga out of the several lakhs that have been planted out.

382

Eucalyptus Globulus, *Labill.* ; MYRTACEÆ.

BLUE GUM-TREE OF VICTORIA AND TASMANIA.

Vern.—*Kurpúra maram,* MADRAS.

References.—*Brandis, For. Fl., 231 ; Gamble, Man. Timb., 188 ; Flück. & Hanb., Pharmacog., 280; U. S. Dispens., 15th Ed., 565 ; Bent. & Trim., Med. Pl., 109 ; Year-Book, Pharm., 1874, 25, 113, 221; 1875, 5 ; Christy, Comm. Pl., V., 45 ; Drury, U. Pl., 199 ; Kew Reports, 1877, 29 ; 1879, 16 ; 1881, 12 ; 1882, 20 ; Kew Off. Guide to the Mus. of Ec. Bot., 65 ; Kew Off. Guide to Bot. Gardens and Arboretum, 116, 117 ; Journ. Agri.- Hort. Soc., 1885, Vol. VII., pt. iii.; Procgs., xcviii.; Ind. For., 1885, Vol. XI., No. 2, 51; Journ. Agri.-Hort. Soc., 1875-78, Vol. V., 1 ; Madras Man. of the Administration, II., 110; Müeller, Select Extra-Trop. Pl., 150.; Report, Horticultural Gardens, Lucknow, 1888-89, 7.*

Habitat.—A lofty tree, gregarious in Victoria and the south of Tasmania. Its introduction into India has met with complete success on the Nilghiris, where the plantations, which were started in 1863, are well established. It has also been successfully cultivated at Abbottabad and Ranikhet. It does not thrive in the plains, nor on the outer Himálayan ranges.

CULTIVATION.
383

Cultivation.—The seeds of the 'Blue gum' are unsually large for the genus; they germinate freely, and the seedlings at once begin to shoot up with marvellous rapidity. Great care, however, is required in transplanting them.

GUM.
Bark.
384

Gum.—The BARK of this tree exudes an astringent gum resembling, both in appearance and properties, that which, under the name of *kino*, is yielded by **Pterocarpus Marsupium.** It is known in trade as ' Australian,' ' Botany Bay,' or, ' Eucalyptus kino.' A kino of better quality is obtainable from other species of **Eucalyptus,** such as **E. rostrata, E. corymbosa,** and **E. citriodora,** and, according to the authors of the *Pharmacographia,* might with no disadvantage be substituted for that of true kino.

E. 384

The Blue Gum Tree. (*J. F. Duthie.*)	**EUCALYPTUS** Globulus.

Tan and Dye.—The gum above mentioned is used for 'tanning and dyeing.'

Fibre.—The BARK of this tree yields a material which has been found suitable for making paper.

Oil.—The LEAVES and young SHOOTS yield an essential oil used in the preparation of the much-advertised ' Eucalyptus Soap.' It is also said to be employed as a substitute for ' Cajeput Oil.' The chemical properties of Eucalyptus oil, as determined by M. Cloez, are reviewed in the *United States Dispensatory*, as follows : "Of this oil the fresh leaves afforded 2·75 parts per hundred, the recently-dried parts 6 parts." " M. Cloez believes the oil to be composed of two camphors, differing in their volatility. The bulk of the oil yielded is the portion first distilled ; to this Cloez has given the name of *Eucalyptol*. To obtain it pure a redistillation from caustic potash or chloride of calcium is necessary. It is very liquid, nearly colourless, with a strong aromatic camphoraceous odour, polarises to the right, is slightly soluble in water, but very soluble in alcohol, and has the formula $C_{12}H_{20}O$. Nitric acid produces with it a crystallizable acid ; by the action of phosphoric acid it is converted into *eucalyptene*, a substance allied to *cymene*, and *eucalyptolen*." (*15th Ed.*, *566*.)

Medicine.—The leaves yield an essential OIL used in medicine, and sometimes as a substitute for Cajeput oil.

" Eucalyptus was originally recommended as a remedy in intermittent fever, but experience has failed to establish its value as an antiperiodic. Whatever medical virtues it possesses beyond astringency reside in the volatile oil. This, when applied locally, acts as a powerful irritant. . . . As a stimulating narcotic, the oil of Eucalyptus has been used with asserted success in migraine and other forms of neuralgia. As an antispasmodic it has been highly lauded in asthma. In chronic or subacute bronchitis it may often be employed with advantage, especially when there is a tendency to spasm." (*U. S. Dispens.*, *566*.)

SPECIAL OPINIONS.—" I have used ℨss doses of the leaves infused in an inhaler in cases of chronic thickening of the mucus membrane of fauces and throat with marked good results : one case of over 3 years' standing quite recovered under its use." (*Honorary Surgeon Easton Alfred Morris, in Medical charge, Tranquibar.*) " Prof. Lister has lately made use of the oil as an antiseptic dressing in place of carbolic acid. It is used undiluted. It is largely employed in the form of ointment, and as antiseptic gauze. The oil, with hot water, as an inhalation has been used with the best effects in diphtheria, in America." (*E. G. Russell, Superintendent, Asylums at Presidency General Hospital, Calcutta.*) " Dose of the oil from 10 to 30 mimims for true leprosy with good effect." (*Apothecary Thomas Ward, Madanapalli, Cuddapah.*) " A powerful antiseptic, and used by Prof. Lister in preparation of antiseptic gauze." (*S. Westcott, A.M.D.*) " Much used in antiseptic surgery as a dressing. Also in diphtheria in the form of blue-gum stem,—*vide* Gibbes in the *Lancet*, February 24th and March 31st, 1883. The tincture is much lauded by some for ague." (*G. B.*) " The inhalation of the essential oil is useful in bronchial and phthisical cases. The oil can be supplied from the Nilghiri plantations." (*Surgeon-General William Robert Cornish, F.R.C.S., C.I.E., Madras.*) " Used as an antiseptic." (*Brigade Surgeon G. A. Watson, Allahabad.*) " Also employed in intermittent fever on account of its antiperiodic properties." (*Civil Surgeon J. Anderson, M.B., Bijnor.*) " An infusion of the leaves, or ten to twenty drops of the oil in a pint of boiling water, excellent for steaming the throat when ulcerated." (*Surgeon-Major W. Farquhar, M.D., I. M. D., Ootacamund.*)

Margin notes:
TAN AND DYE.
385
FIBRE.
Bark.
386
OIL.
Leaves.
387
Shoots.
388

MEDICINE.
Oil.
389

E. 389

EUGENIA.	The Blue Gum Tree ; Teosinte Grass.

FOOD.
390

Food.—A liquor is made from **Eucalyptus** that has attained some reputation in Australia.

TIMBER.
391

Structure of the Wood.—Strong, tough, and durable, and extensively used in Australia for ship-building, house-building, sleepers, telegraph poles, &c. It has been found by experiments to rival in strength the best English oak.

INDUSTRY.
392

INDUSTRIAL USE.

In a recently published report on the Lucknow Horticultural Gardens, it is mentioned that a new demand for the leaves of the tree has arisen, owing to the discovery having been made that a decoction has the power of removing the scale or incrustation which forms in locomotive boilers, as a deposit from the water.

The matter is now engaging the attention of the Locomotive Department of the Bengal and North-Western Railway at Gorakhpur, and it is reported that the trials there made have had good results.

The following extract from a letter regarding the method of use is published : " We have a large tank, which we fill with leaves and small branches. the water is then put in, and boiled or made warm with waste steam. This continues till the fluid has a dark colour, when it is used ; say, two or three gallons of the decoction is put into the tender, and so mixes with the water or enters the boiler with the feed.

" I learn excellent results are being obtained, as the scale tumbles off the plates when the boilers are being washed out."

EUCHLÆNA, *Schrad. ; Gen. Pl., III., 1114.*

393

Euchlæna luxurians, *Ascheron ; Duthie, Fodder Grasses, N. Ind.,*
TEOSINTE, *Fr.*   [*19 ;* GRAMINEÆ.

Syn.—REANA LUXURIANS.

References.—*Christy, Comm. Pl., III., 5 : Smith, Dic., 409 ; Kew Reports, 1879, 17 ; 1880, 80 ; Journ. Agri.-Hort. Soc., 1885, Vol. VII., pt. 3, New Series ; Procgs., Soc., CVII. ; Ind. For., X., III., 111 ; Journ. Agri.-Hort. Soc., VI., 117 ; Müeller, Select Extra-Trop. Pl., 165.*

Habitat.—A native of Guatemala. It is a quick-growing succulent grass resembling maize. It requires 9 or 10 months from sowing to the ripening of its seed, and within that period single cultivated specimens have been known, under generous treatment, to produce as many as 90 stems, and to attain 18 feet in height. It is a prolific seed-bearer. **Dr.** Schweinfurth is reported to have secured from three seeds about 12,000 grains.

FODDER.
394

Fodder.—The grass is described as a most excellent fodder for cattle. The attempts hitherto made to introduce it into India have not had any definite results, for while in some places it has been favourably reported on, in others it has failed, and the general opinion is that it could never compete with the existing fodder plants of India, such as *juar*, &c., as its cultivation on a large scale would be too expensive owing to its requiring rich soil and constant irrigation.

395

EUGENIA, *Linn. ; Gen. Pl., I., 718.*

A large genus, containing over 700 species, of which about one fifth are represented in British India. They consist of trees or shrubs with evergreen, smooth foliage ; and many of them are very handsome when in flower. They are found most abundantly in the humid regions of North, East, and South India, also in Burma, Malaya, and Ceylon. A few only of the Indian species are of economic importance. The three sections, Jambosa, Syzygium, and Eugenia, have by many writers been treated as separate genera. In Syzygium the petals are combined, and usually fall off in one piece ; many of the species are fine large timber trees. In Jambosa and Eugenia the petals are free and spreading. **Linnæus** is said to have named the genus after Prince Eugene of Savoy.

Eugenia alba, *Roxb.*; MYRTACEÆ, see **E. javanica**.

E. aquea, *Burm.*; *Fl. Br. Ind., II., 473; Wight, Ic., tt. 216, 550.* | 396

Syn.—JAMBOSA AQUEA, *DC.*
Vern.—*Jambo*, BENG.; *Wat jambu*, SING.
References.—*Roxb., Fl. Ind., Ed. C.B.C., 400; Voigt, Hort. Sub. Cal., 47; Kurz, For. Fl. Burm., I., 494; Gamble, Man. Timb., 193; Thwaites, En. Ceylon Pl., 115; Trimen, Hort. Zeyl., 32; Rumph., Herb. Amb., I., 126; Balfour, Cyclop., II., 411.*
Habitat.—A medium-sized tree, with large white flowers. It is a native of the Moluccas, and is wild also in Ceylon. It has been planted extensively in Bengal and Burma.
Food.—The fruit, which is of about the size of a loquat and flattened at the end, is either pale rose-coloured or white; the former has an aromatic taste; the latter is the *jambo-ayer* of Rumphius. | FOOD. 397

E. Arnottiana, *Wight*; *Fl. Br. Ind., II., 483; Wight, Ic., t. 999.* | 398

Vern.—*Nawal*, S. INDIA.
References.—*Beddome, For. Man., 107; Ind. For., X., 552.*
Habitat.—A large spreading tree, common in the moist woods on the Nilghiri, Pulney, and Anamallay hills of South India.
Food.—Fruit dark purple; Beddome says that it is eaten, but is very astringent. | FOOD. 399
Structure of the Wood.—The timber is said to be valuable. | TIMBER. 400

E. calophyllifolia, *Wight*; *Fl. Br. Ind., II., 494; Wight, Ic., t. 1000.* | 401

References.—*Beddome, For. Man., 107; Thwaites, Enum. Ceylon Pl., 118; Ind. For., X., 552.*
Habitat.—A large and beautiful tree found on the Nilghiri range, and on Adam's Peak in Ceylon.
Food.—The purple oblong FRUIT is edible. | FOOD. 402
Structure of the Wood.—"Its timber is valuable and used for building and other purposes." (*Beddome.*) | TIMBER. 403

E. caryophyllæa, *Wight*; *Fl. Br. Ind., II., 490; Wight, Ic., t. 540.* | 404

Syn.—SYZYGIUM CARYOPHYLLÆUM, *Gærtn.*
Vern.—*Jáman*, HIND.; *Chota jám*, SANTAL; *Dan, dáng*, SING.
References.—*Thwaites, En. Ceylon Pl., 117; Dalz. & Gibs., Bomb. Fl., 93; Trimen, Hort. Zeyl., 33; Rheede, Hort. Mal., V., t. 27; Lisboa, U. Pl. Bomb., 78, 156, 339; Bomb. Gaz., XV., 434.*
Habitat.—A small tree found in the Western Ghâts, South India, and in Ceylon.
Gum.—The tree is said to yield a gum somewhat resembling *kino*. | GUM. 405
Food.—The round, black, pea-sized berries are eaten in the Bombay Presidency, and also by the Singhalese. | FOOD. 406

E. caryophyllata, *Thunb.*; see **Caryophyllus aromaticus**, Vol. II., 202.

E. caryophyllifolia, *Lam.*, a variety of **E. Jambolana**.

E. cerasiflora, *Kurz.*; see **E. Kurzii**.

E. cerasoides, *Roxb.*; see **E. operculata**.

E. claviflora, *Roxb.*; *Fl. Br. Ind., II., 484; Wight, Ic., t. 606.* | 407

Syn.—SYZYGIUM CLAVIFLORUM, *Wall.*
Vern.—*Lumba-nuli-jamb*, CHITTAGONG.
References.—*Roxb., Fl. Ind., Ed. C.B.C., 399; Voigt., Hort. Sub. Cal., 48; Kurz, For. Fl. Burm., I., 480.*

E. 407

**FOOD.
408**

Habitat.—A large tree found on mountains in Sikkim and Khasia, altitude 2,000 to 4,000 feet ; also in Sylhet, Chittagong, Pegu, Nicobar and Andaman Islands, Tenasserim, Singapore and Penang.

Food.—The fruit, which ripens in May, is eaten by the natives.

409

Eugenia cymosa, *Roxb.;* see E. grandis.

E. formosa, *Wall.; Fl. Br. Ind., II., 471 ; Wight, Ic., t. 611.*

Syn.—E. TERNIFOLIA, *Roxb.*

Vern.—*Bolsobak, panchidung,* GARO; *Phul-jamb, lálphul-jamb* (*Roxb.*), CHITTAGONG; *Bara-jáman,* NEPAL; *Famsikol,* LEPCHA; *Bunkonkri,* MICHI.

References.—*Roxb., Fl. Ind., Ed. C.B.C., 399; Voigt., Hort. Sub. Cal., 48; Kurz, For. Fl. Burm., I., 492; Gamble, Man. Timb., 193; Balfour, Cyclop., 1059.*

Habitat.—A handsome moderate-sized tree with very large leaves, met with near streams on the Eastern Himalaya and in Burma. There are two forms, one with white and the other with red flowers.

**FOOD.
Fruit.
410
TIMBER.
411**

Food.—The FRUIT, of about the size of a walnut, is eaten by the natives.

Structure of the Wood.—" Heavy, uniformly brown, close-grained, takes a fine polish." (*Kurz.*)

412

E. grandis, *Wight; Fl. Br. Ind., II., 475 ; Wight, Ic., t. 614.*

Syn.—E. CYMOSA, *Roxb.*

Vern.—*Zebri,* MAGH; *Jam,* BENG.; *Battijamb,* SYLHET; *Taung-thabye, thabyegyi,* BURM.

References.—*Roxb., Fl. Ind., Ed. C.B.C., 400; Voigt., Hort. Sub. Cal., 49; Kurz, For. Fl. Burm., I., 489; For. Man., 1071; Gamble, Man. Timb., 193; Thwaites, En. Ceylon Pl., 116 & 417; Trimen, Hort. Zeyl., 33.*

Habitat.—An evergreen tree of Eastern Bengal, Burma, and the Andaman Islands.

**TIMBER.
413**

Structure of the Wood.—" Red, rough, and hard" (*Gamble*). "The wood is used for various economical purposes." (*Roxburgh.*)

414

E. hemispherica, *Wight; Fl. Br. Ind., II., 477 ; Wight, Ic., t. 525.*

References.—*For. Man., 203; Thwaites, En. Ceylon Pl., 116; Trimen, Hort. Zeyl., 33; Balfour, Cyclop., 1059.*

Habitat.—A large handsome tree, common in mountain forests in Southern India and in Ceylon.

**TIMBER.
415**

Structure of the Wood.—The timber is said to be useful for various purposes.

416

E. Heyneana, *Wall; Fl. Br. Ind., II., 500; Wight, Ic., 539.*

Syn.—E. SALICIFOLIA, *Wight;* SYZYGIUM SALICIFOLIUM, *Grah.*

Vern.—*Gara-kuda,* KÓL.; *Gara-kud,* SANTAL; *Jamti,* KHARWAR; *Hend,* GOND; *Gambu,* KURKU; *Jánbu, jámun,* C. P.; *Panjam-but,* MAR.

References.—*Brandis, For. Fl., 234; Grah. Cat. Bomb. Pl., 73; Elliot, Fl. Andhr., 40; Gamble, Man. Timb., 195; Dalz. & Gibs., Bomb. Fl., 94; For. Man., 109; Lisboa, U. Pl. Bomb., 339; Balfour, Cyclop., II., 411; For. Adm. Report, Ch. Nagpur, 1885, 31.*

Habitat.—A shrub or small tree found in the Bombay Gháts, and in the beds of rivers in Berar and the Central Provinces.

**FOOD
417
TIMBER.
418**

Food.—The FRUIT is eaten by the natives in the Central Provinces.

Structure of the Wood.—Similar to that of E. Jambolana, but pores smaller. (*Gamble.*)

419

E. Jambolana, *Lam.; Fl. Br. Ind., II., 499; Wight, Ic., t. 535.*

BLACK PLUM, *Eng.*

Syn.—SYZYGIUM JAMBOLANUM, *DC.*

E. 419

The Black Plum.	(*J. F. Duthie.*)	EUGENIA Jambolana.

Vern.—*Jáman, jám, jamun, phalinda, phalanda, jamni phaláni, pharenda, phaunda, paiman,* HIND. ; *Jám, kála-jám,* BENG. ; *Zebri, chakukau,* MAGH ; *Kuda,* KÓL. ; *Kudu, kud,* (*so-kod, chuduk'-bad,* Rev. A. Campbell), SANTAL ; *Jamo, jámkuli,* URIYA ; *Jamu,* ASSAM ; *Chambu,* GARO ; *Phoberkúng,* LEPCHA ; *Kor-jam,* MICHI ; *Jam,* MAL. (S. P.) ; *Naindi,* GOND ; *Jambun,* ORAON ; *Jamun,* RAJ. ; *Jamun, jamin, jamul,* C. P. ; *Jambul, jámbu, jámbhul, jambudo,* BOMB. ; *Jambúl,* MAR. ; *Jambu, jambura, jámbudi,* GUZ. ; *Nával, narvel, nawal, nawar, naga,* TAM. ; *Nerale,* MYSORE ; *Naredu, rácha-neredu, pedda-neredu* (large fruited var.), *nairuri, nareyr, nasodu, nasedu,* TEL. ; *Narala, nerlu, nerale,* KAN. ; *Thabyebyu,* BURM. ; *Mahadan, madan-naval, mudang,* SING. ; *Jambu, jambula,* SANS.

References.—*Roxb., Fl. Ind., Ed. C.B.C.,* 398 ; *Voigt, Hort. Sub. Cal.,* 49 ; *Brandis, For. Fl.,* 233 ; *Kurz, For. Fl. Burm., I.,* 485 ; *Gamble, Man. Timb.,* 194 ; *Dals. & Gibs., Bomb. Fl.,* 93 ; *Aitchison, Cat. Pb. and Sind Pl.,* 60 ; *Elliot, Fl. Andhr.,* 72, 133, 162 ; *Rheede, Hort. Mal.,* 5., t. 29 ; *U. C. Dutt, Mat. Med. Hind.,* 164 ; *Dymock, Mat. Med. W. Ind.,* 333 ; *S. Arjun, Bomb. Drugs,* 57 ; *Ind. Agri.,* (*Oct. 9th,* 1886), *p.* 497 ; *Atkinson, Econ. Prod. N.-W. Prov.,* 74 ; *Him. Dist.,* 736 ; *Drury, U. Pl.,* 409 ; *Lisboa, U. Pl. Bomb.,* 77, 156, 211, 245, 259, 279, 284, 291 ; *Christy, Com. Pl. and Drugs, No. 8, p.* 77, *and No. 10, p.* 63 ; *Cooke, Gums and Gum-resins,* 11, 39 ; *Atkinson, Gums and Gum-resins,* 12 ; *McCann, Dyes and Tans, Beng.,* 49, 135, 144, 159, 160, 168 ; *Baron F. Muell., Sel. Extra-Trop. Pl.,* 167 ; *Balfour, Cyclop.,* 1059 ; *Smith, Dic.,* 227 ; *Home Dept. Cor.,* 238 ; *Journ. As. Soc.,* 1867, 80 ; *Bomb. Gaz., XIII., i.* 24 ; *XV,* 68 ; *Mason, Burma and Its People, p.* 45 ; *Special Report of Collector of Madura ; Ind. Agri., Oct. 9th, 1886, p.* 497.

Habitat.—A moderate-sized tree, found wild or cultivated over the greater part of India from the Indus eastward and to the extreme south of the Madras Presidency. It ascends to 3,000 feet on the Panjáb Himalaya, and to 5,000 feet in Kumáon.

Gum.—Yields a gum somewhat resembling *kino.*

GUM. 420

Dye and Tan.—The BARK is used for dyeing and tanning. In Assam it is employed along with the red *Munjit* dye to impart brilliancy to colour ; it is also used to colour fishing-nets. It is mentioned as one of the ingredients (in Lohárdaga, Chutia Nagpur) in a preparation of a permanent black (*McCann*). In tanning it is often combined with *Garán* bark (**Ceriops Roxburghiana, Vol II.,** 261). [A decoction of the bark is very generally employed to precipitate indigo from the infusion obtained from the plant. See **Indigo,** *Ed.*]

DYE and TAN. Bark. 421

Medicine.—The BARK is astringent and used in cases of dysentery, and the decoction as a tooth gargle. A vinegar, prepared from the JUICE of the unripe fruit, is an agreeable stomachic and carminative; it is also used as a diuretic. The fresh juice of the bark is given with goat's milk in the diarrhœa of children. The expressed juice of the leaves is used alone or in combination with other astringents in dysentery. (*U. C. Dutt.*) The powdered SEEDS have had the reputation in recent years of being useful in the treatment of diabetes.

MEDICINE. Bark. 422

Juice. 423

Seeds. 424

SPECIAL OPINIONS.—" The powder of the dried stone of the fruit is used in cases of diabetes. It certainly does diminish the quantity of sugar in the urine very quickly, and, in some cases, even permanently." (*Surgeon D. N. Parakh, Indian Medical Department, Bombay.*) " The dried seeds, in combination with those of **Mangifera indica,** are administered with very good effect, in the form of powder, in cases of diarrhœa and dysentery." (*Sakharam Arjun Ravat, L.M., Bombay.*) " Decoction of the bark used as an astringent gargle in sore-throat, and juice of the fresh tender leaves is given with goat's milk in cases of dysentery." (*Bolly Chund Sen, Teacher of Medicine, Campbell Medical School, Sealdah, Calcutta.*) " A decoction of the bark is used largely for diarrhœa and dysentery in combination with carminatives such as cardamoms and cinnamon." (*Dr. Bensley, Civil*

EUGENIA
Jambolana.   The Black Plum.

MEDICINE.
Surgeon, Rajshahye.) " 160 grains of the pulverized seed is taken as an anti-
dote in cases of Nux vomica poisoning." (*Surgeon W. F. Thomas, Madras
Army, Mangalore.*) "Used in diabetes and in enlargement of spleen. Dose of
extracted juice, about 4 drachms." (*Civil Surgeon John McConaghey, M.D.,
Shahjahanpore.*) " The syrup of the fruits is used in diarrhœa." (*Civil
Surgeon, R. Gray, Lahore.*) " The decoction of the bark is used as a gargle in
salivation, whether brought on by prolonged use of mercury or other causes."
(*Civil Surgeon Bankabehari Gupta, Poor e.*) " The ripe fruit is considered
curative for calculous affections. The leaves are used as a poultice for scorpion bite." (*Surgeon-Major Robb, Civil Surgeon, Ahmedabad.*) " The vinegar manufactured from the ripe fruit is much used as a stomachic by the
natives and is useful in cases of enlargement of the spleen. The doses
used are one to two drachms. The fruit is useful in diarrhœa." (*Narain
Misser, Kothe Bazar Dispensary, Hazaribagh.*) "The vinegar of ripe
fruit is cooling and used in indigestion. The juice of fresh leaves is used
in spongy and painful gums." (*Shib Chundra Bhattacharji, Chanda,
Central Provinces.*) " Grows very commonly and is extensively used as
an astringent in Mysore." (*Surgeon-Major John North, Bangalore.*)

FOOD.
425
Food.—The fruit, which is sometimes as large as a pigeon's egg and
of a purple colour, is eaten by all classes of people : it is sub-acid and
rather astringent, and is improved in taste by being pricked and rubbed
with a little salt, and allowed to stand an hour. In Goa, a wine, faintly
resembling port, is prepared from the ripe fruit.

" A sort of spirituous liquor called *Jámbava* is described in recent
Sanskrit works as prepared by distillation from the juice of the ripe fruits."
(*U. C. Dutt, Mat. Med. Hind., 164.*) The Collector of Madura reports
that the fruit should not be extensively eaten as it is apt to bring on fever.
Paludanus in a note appended to *Van Linschoten's Voyage,* says :—
" This fruit is little used by Physitions, but is much kept in pickle, eaten
with Sodden Ryce, for they procure an appetite to meat."

TIMBER.
426
Structure of the Wood.—Reddish-grey, rough, moderately hard,
darker near the centre ; no distinct heartwood. It is fairly durable. Five
sleepers of it were laid down in 1870 on the Oudh and Rohilkhand Railway, and taken up in 1875, when they were reported to be fairly sound
and not touched by white-ants. It is used for building, agricultural implements and carts, also for well-work, as it resists the action of water.

DOMESTIC
and SACRED.
427
Domestic and Sacred Uses.—It is often planted as a shelter tree for
groves, and, as such, is known under the name of *jamoa* in the Saharanpur
and Karnál districts. In habit it is very different from the type, and
should perhaps be considered as a distinct variety.

" The god *Megh* is said to have been transformed into a jambul tree.
The colour of the fruit being dark like that of Krishna, this plant is very
dear to him ; it is, therefore, worshipped, and Brahmins are fed under it.
The leaves are used as platters or *panch pallows* and for pouring libations." (*Lisboa, U. Pl. Bomb., 284.*)

428
Var. caryophyllifolia, *Fl. Br. Ind., II., 499 ; Wight ; Ic., t. 553.*

Syn.—E. CARYOPHYLLIFOLIA, *Lam.* ; SYZYGIUM CARYOPHYLLIFOLIUM,
DC. ; S. LATERIFLORUM, *Royle.*

Vern.—*Chota-jamb,* BENG.; *Jamun,* KOL.; *Bir-kod,* SANTAL ; *Bata-jania,*
TEL.; *Nairla,* KAN.

References. — *Roxb., Fl. Ind., Ed. C.B.C., 398 ; Voigt., Hort. Sub. Cal.,
49 ; Brandis, For. Fl., 234 ; Thwaites, En. Ceylon Pl., 417 ; Drury, U.,
Pl., 410 ; Lisboa, U. Pl. Bomb., 77 ; Cooke, Gums and Gum-resins, 11 ;
Gums and Resinous Prod. of Ind. (P. W. D., 1871), 68 ; Balfour,
Cyclop., I., 1059.*

Habitat.—Found in most parts of India. Its lanceolate acuminate
leaves, and small pea-shaped fruit, distinguish it from the type.

| The Rose Apple. | (*J. F. Duthie.*) | **EUGENIA Jambos.** |

Gum.—" Yields a very good gum, grows in the Mysore district."
(*Gums and Resinous Prod. of India, P. W. D., 1871.*)

GUM. 429

Medicine.—The Rev. A. Campbell states that in Chutia Nagpur the
LEAVES are used medicinally.

MEDICINE. Leaves. 430

Structure of the Wood.—" Whitish, very strong, close-grained, hard
and durable." (*Roxb.*)

TIMBER. 431

Eugenia Jambos, *Linn. ; Fl. Br. Ind., II., 474 ; Wight, Ic., 435.*
ROSE-APPLE.

432

Syn.—JAMBOSA VULGARIS, *DC.*

Vern.—*Guláb-jáman,* HIND.; *Guláb-jamb,* BENG. ; *Golápjám,* URIYA ;
Jamu, SIND. ; *Jámb,* DECCAN ; *Malle-nerale, pan-nerale,* COORG ;
Pannerali, KAN. ; *Jambu,* SING.; *Jamba* (*Roxb.*), *jambu,* SANS.; *Toffah,*
ARAB.

References.—*Roxb., Fl. Ind., Ed. C.B.C., 401 ; Voigt., Hort. Sub. Cal.,
47 ; Brandis, For. Fl., 233; Kurz, For. Fl. Burm., I., 495; Gamble, Mán.
Timb., 193; DC. Origin Cult. Pl., 240 ; Trimen, Hort. Zeyl., 33 ; Rheede,
Hort. Mal., I., 27, f. 17 ; Atkinson, Econ. Prod., N.-W. Prov., 74;
Drury, U. Fl., 265 ; Lisboa, U. Pl., Bomb., 156 ; Smith, Dic., 355 ;
Mason, Burma, 450 ; Gaz. Bomb., XVIII., Pt. I., 46.*

Habitat.—A small handsome tree, a native of the East Indies. Largely
cultivated in India and in other tropical countries. Kurz says that it is
frequently cultivated in native gardens all over Burma. The beauty of
its flowers, fruit, and foliage renders it a fit ornament in any garden.

DESCRIPTION. 433

History.—Linschoten in his *Voyage to the East Indies* (*1598*), gives the
following description of this tree and its fruit :—" The trees whereon the
Jambos do grow are as great as Plum trees, and verie like unto them : it is
an excellent and a (verie) pleasant fruite to looke on, as big as an apple ; it
hath a red colour and somewhat whitish, so cleare and pure that it
seemeth to be painted or made of waxe : it is very pleasant to eate, and
smelleth like rose water ; it is white within, and in eating moyst and
waterish, it is a most daintie fruite, as well for bewtie to the sight, so for
the sweet savour and taste ; it is a fruite that is never forbidden to any
sicke person, as other fruites are, but are freelie given unto sicke men to
eate, that have a desire thereunto, for it can doe no hurt. The blossoms
are likewise very faire to the sight, and have a sweet smell : they are red and
somewhat whitish (of colour). This tree beareth fruite three or foure
tymes every yeare, and which is (more) wonderfull, it hath commonly on the
one side or halfe of the tree ripe Jambos and the leaves fallen off, and on
the other side or halfe it hath all the leaves, and beginneth (againe) to
blossome, and when that side hath fruite, and that the leaves fall off,
then the other side beginneth again to have leaves, and to blossome, and
so it continueth all the yeare long : within they have a stone as great (and
very neere of the same fashion) as the fruite of the cypres tree."

HISTORY. 434

The note by Paludanus appended to Linschoten's account of the Rose-
apple tree probably refers to **E. malaccensis,** *Linn.*, as suggested by Col.
Sir H. Yule. In a foot-note to the English edition (1885) of *Linschoten's
Voyage to the East Indies,* Yule says :—" The name of the tree and
fruit, *jambu, jambû,* is Sanskrit; one of the ancient names of India, *e.g.,*
in the oldest writings of the Buddhists and in inscriptions from the third
century B.C., was *jambu-doipa.*"

The following is from **Mason's** work on Burma (1860), p. 451 :—
" According to Burman geography there is a **Eugenia** tree on the great
island or continent which we inhabit,—that is, twelve hundred miles high,
one hundred and eighty-six in circumference, with five principal branches,
each six hundred miles long. From this tree the island derives its name
Jambudeba, Eugenia island." Buchanan, in his " Statistics of Dinaj-

pur," p. 156, referring to this tree, also says :—"The Indians indeed are said to have given its name to their position of the world, *Jambudwip* or the Island of the Jumbu tree." It may be here added that the Rose-apple is wrongly referred by Yule, Mason, and others to E. Jambolana.

**MEDICINE.
Leaves.
435**

Medicine.—In Bhamo, Upper Burma, the LEAVES are boiled and used as a medicine for sore eyes.

**FOOD.
Fruits.
436**

Food.—The FRUIT, which is usually produced during the rainy season, is about the size of a small apple. By many persons it is highly esteemed on account of its delicate flavour which resembles rose-water, but there is a want of juice which renders it unpalatable. In the neighbourhood of Calcutta the fruiting branches are covered with pieces of cloth, and this is believed to increase the size as well as the flavour of the fruit. A preserve is sometimes made of the fruit.

**TIMBER.
437
DOMESTIC.
438
439**

Structure of the Wood.—"Reddish brown" (*Brandis*).

Domestic Use.—In Burma the leaves are said to be much prized for ornamental purposes.

Eugenia javanica, *Lamk. ; Fl. Br. Ind., II., 474; Wight, Ic., t. 548.*

Syn.—E. ALBA, *Roxb.*

Vern.—*Jamrool, amrool,* HIND.

References.—*Roxb., Fl. Ind., Ed. C.B.C., 400; Voigt., Hort. Sub. Cal., 48; Kurz, For. Fl. Burm., I., 494; Trimen, Hort. Zeyl., 33.*

Habitat.—A tree of Malacca, Andaman and Nicobar Islands. Introduced into Bengal, where it is now common, chiefly in gardens.

**FOOD.
440**

Food.—Produces abundantly during the hot and rainy seasons a fruit which when ripe is pure white and shining ; though juicy and refreshing it is almost tasteless ; it is eaten, however, by all classes of people.

**TIMBER.
441
442**

Structure of the Wood.—"Red, rough, and hard" (*Gamble*).

E. Kurzii, *Duthie ; Fl. Br. Ind., II., 478.*

Syn.—E. CERASIFLORA, *Kurz.*

Vern.—*Jámun,* NEPAL ; *Sunom,* LEPCHA.

References.—*Kurz, For. Fl. Burm., I., 491; Gamble, Man. Timb., 193; Journ. As. Soc. Beng., XLVI. (1877), ii., 68.*

Habitat.—A large evergreen tree, met with in the hills of Bengal and Burma, from 3,000 to 6,000 feet.

**TIMBER.
443**

Structure of the Wood.—"Reddish-grey, moderately hard, rough" (*Gamble*).

444

E. malaccensis, *Linn.; Fl. Br. Ind., II., 471; Wight, Ic., t. 98.*

MALAY APPLE or the KAVIKA TREE.

Syn.—JAMBOSA MALACCENSIS, *DC.*

Vern.—*Maláka jamrul,* BENG. ; *Nati-shambu* (Rheede),I MALAY ; *Thabyoo-thabyay,* BURM.

References.—*Roxb., Fl. Ind., Ed. C.B.C., 397 ; Voigt., Hort. Sub. Cal., 47 ; Kurz, For. Fl. Burm., I., 493 ; Gamble, Man. Timb., 193 ; DC., Origin Cult. Pl., 241 ; Trimen, Hort. Zeyl., 33 ; Rheede, Hort. Mal., I., 29, t. 18 ; Lisboa, U. Pl. Bomb., 155 ; Baron F. von Müell., Sel. Extra-Trop. Pl., 167 ; Smith, Dic., 260 ; Mason, Burma, 450.*

Habitat.—A handsome tree, with a profusion of either white or scarlet flowers, followed by an abundance of fruit of the size of a small apple. It is a native of the Malay Islands, and is now cultivated in Bengal and Burma, chiefly in gardens. The Malay looks upon the ' *Kavika* ' tree as representing all that is lovely and beautiful.

445

The note by Paludanus appended to Linschoten's description of the Rose-apple tree evidently refers to E. malaccensis. He mentions the fact of its having been "first brought out of Malacca into India," and he describes the flowers as "of a reddish purple colour," and the fruit "as bigge as a Peare." He also says :—" There are two sorts of this fruit, one a browne red, seeming as though it was black, most part without stones and more savory, than the other which is a pale red or a pale purple colour, with a lively smell of roses."

Food.—Produces at different periods of the year a large, juicy fruit, which is very commonly eaten, though rather insipid (*Roxb.*) The pulp of the fruit is said to be wholesome and agreeable. Paludanus (*l. c.*) says :—"[This fruite is ordinarily eaten before other meate be set upon the table, and also at all times of the day."

Structure of the Wood.—" Reddish-grey, rough, soft. Weight, **Wallich** gives 30, our specimen 38℔ per cubic foot " (*Gamble*).

Eugenia montana, *Wight, Fl. Br. Ind., II., 488 ; Wight, Ic., t. 1060.*
References.—*Beddome, For. Man., 107 ; Ind. For., X., 552.*
Habitat.—A large tree, common on the higher ranges of the Nilghiris.,
Structure of the Wood.—"Is in use for building purposes, &c ' (*Beddome*).

E. oblata, *Roxb. ; Fl. Br. Ind., II., 492 ; Wight, Ic., t. 622.*
Vern.—*Goolum* (Roxb.), CHITTAGONG; *Thabyay-nee,* BURMA.
References.—*Roxb., Fl. Ind., Ed. C.B.C., 400 ; Voigt., Hort. Sub. Cal., 48 ; Kurz, For. Fl. Burm., II., 488.*
Habitat.—A tree found in Eastern Bengal, Burma, Penang, and Singapur. In Chittagong it is cultivated for its fruit.
Food.—The fruit, about the size of a cherry, is, according to **Roxburgh,** edible.
Structure of the Wood.—Roxburgh also states that the wood is in some estimation.

E. obovata, *Kurz,* a variety of **E. operculata.**

E. operculata, *Roxb. ; Fl. Br. Ind., II., 498 ; Wight, Ic., tt. 552 & [615.*
Syn.—E. CERASOIDES, *Roxb.*
Vern.—*Rai-jáman, paiman, jamawa, dugdugia,* HIND. ; *Topa,* KOL.; *Totonopak',* SANTAL ; *Botee-jam* (Roxb.), CHITTAGONG; *Teathaby-ay* (*yethabyay*), *thabyay-chin,* BURM. ; *Batatdomba,* SING.
References.—*Roxb., Fl. Ind., Ed. C.B.C., 398 ; Voigt., Hort. Sub. Cal., 49 ; Brandis, For. Fl., 234; Kurz, For. Fl. Burm., I., 483 & 484; For. Man., 106 ; Gamble, Man. Timb., 194: Thwaites, En. Ceylon Pl. 417 ; Trimen, Hort. Zeyl., 33 ; Atkinson, Econom. Prod., N.-W. Prov., 74; For. Adm. Report, Ch. Nagpur, 1885, 31.*
Habitat.—A moderate-sized or even large evergreen tree, met with in the sub-Himalayan tract from the Jumna to Assam up to 2,000 feet, in the forests of Chittagong, Burma, the Western Gháts, and in Ceylon up to 3,000 feet. Brandis says that under favourable conditions it grows to be one of the largest and most handsome trees of the genus. The leaves turn bright-red before falling.
Medicine.—" The FRUIT is eaten for rheumatism ; the ROOT, boiled down to the consistence of *gúr*, is applied to the joints by rubbing ; the LEAVES are much used in dry fomentation ; the BARK is also employed medicinally " (*Rev. A. Campbell, Chutia Nagpur*).
Food.—It yields an edible FRUIT which ripens towards the end of the hot weather.
Structure of the Wood.—" Reddish-grey, hard ; used for building and agricultural implements " (*Gamble*).
Var. obovata, *Kurz, Fl Br. N. I., II., 498; Sp. (Wall., Gamble 194).*
Syn.—SYZYGIUM OBOVATUM, *Wall.*
Vern.—*Kiamoni,* NEPAL ; *Jung-song,* LEPCHA ; *Boda-jam,* MICHI.
References.—*Kurz, For. Fl. Burm., I., 482 ; Gamble, Man. Timb., 194; For. Adm. Report, Ch. Nagpur, 1885, 31.*

DESCRIPTION.

FOOD.
446

TIMBER.
447

448

TIMBER.
449

450

FOOD.
451
TIMBER.
452

453

MEDICINE.
Fruit.
454
Bark.
455
FOOD.
Fruit.
456
TIMBER.
457
458

U

E. 458

EULOPHIA.	The Eugenias ; Salep.

TIMBER.
459
460

Habitat. —Found in the savannah forests of Bengal and Burma.
Structure of the Wood.—Grey, rough, moderately hard.

Var. **Paniala** (*Roxb., sp.*) ; *Fl. Br. Ind., II., 498.*
  Vern.—*Paniala-jamb,* BENG.
  Reference.—*Roxb., Fl. Ind., Ed. C.B.C., 399.*
  Habitat.—Found in Chittagong, Sylhet, and Burma. Roxburgh describes it as one of the largest and most robust trees of the genus.

FOOD.
Fruit.
461

  Food.—The FRUIT ripens in June, and is "about the size of a small gooseberry and very juicy " (*Roxburgh*).

Eugenia Pimenta, *DC. ;* see **Pimenta officinalis.**

E. salicifolia, *Wight ;* see **E. Heyneana.**

E. ternifolia, *Roxb. ;* see **E. formosa.**

462

E. tetragona, *Wight; Fl. Br. Ind., 497.*
  Vern.—*Kemma, chamlani,* NEPAL ; *Sunóm,* LEPCHA.
  References.--*Voigt, Hort. Sub. Cal., 49 ; Kurz, For. Fl. Burm., I., 484 ; Gamble, Man. Timb., 194.*
  Habitat.—A large evergreen tree, found in the hills of Northern Bengal up to 6,000 feet, and in Chittagong.

TIMBER.
463

  Structure of the Wood.—Brownish or olive-grey, shining, hard; it is used occasionally for building, for the handles of tools, and for charcoal.

464

E. zeylanica, *Wight; Fl. Br. Ind., II., 485 ; Wight, Ic., 73.*
  Vern.—*Sagarabatna,* URIYA ; *Bhedas,* MAR. ; *Nerkal,* KAN. ; *Thabyepauk,* BURM.
  References.—*Roxb., Fl. Ind., Ed. C.B.C., 402 ; Kur-, For. Fl. Burm., I., 481 ; Beddome, Fl. Sylv., ccii ; Thwaites, En. Ceylon Pl., 118 ; Dals. & Gibs., Bomb. Fl., 94 ; Rheede, Hort. Mal., V., t. 20 ; Gas. Bomb., XV., Pt. I., 68.*
  Habitat.—A small myrtle-like shrub of the scrubby forests of Orissa ; a shrub or small tree in the Concan and southwards, also in Sylhet, the Malay Peninsula, and the Andamans.

TIMBER.
465

  Structure of the Wood.—In Kánara it is used for building purposes and for field tools.

466

EULOPHIA, *R. Br. ; Gen. Pl., III., 535.*

  The *salep* obtainable in Indian bazars has been ascertained to be the product of two species of **Eulophia,** *viz.,* **E. campestris** and **E. herbacea,** and possibly of others. *Salep* or *sálib misri* consists of the dried tubers of the abovementioned orchids and of several species of **Orchis,** which latter constitute the bulk of the *salep* of European commerce. Its oriental reputation as an aphrodisiac was founded merely on superstition in connection with the so-called doctrine of signatures. It possesses no medicinal properties whatsoever. A decoction prepared from powdered *salep,* and flavoured with wine and spice, is considered a more or less nutritious and agreeable drink for invalids. **Mr. J. G. Baker** of Kew, in the discussion which followed the reading of **Dr. Aitchison's** paper before the British Pharmaceutical Society (December 8th, 1886), said that **Dr. Aitchison** had practically disposed of the much-debated question as to the source of the *Royal salep* or *badjah.* **Mr. Baker,** acting on a suggestion made by **Hanbury,** said that this form of *salep* resembled a bulb more than a tuber, and that he had succeeded in tracing out what appears to be the source of that drug. **Dr. Aitchison** had brought fresh specimens of these bulbs, and they proved to be **Ungernia trisphœra,** a plant belonging to the AMARYLLIDACEÆ. **Dr. Dymock** writes that the *salep* of Bombay commerce is imported from Persia, Cabul, and Northern India, and is probably obtained from various species of **Orchis,** under which genus further information on this product will be found. (For further information see **Curculigo Orchioides,** Vol. II., 650).

E. 466

Eulophia campestris, *Lindl.* ; ORCHIDEÆ.   **467**

Vern.—*Sung-misrie* (Irvine), BENG. ; *Bonga taini,* SANTAL ; *Hatti-paila,*
NEPAL ; *Sálib-misri,* PB. ; *Sálum,* GUZ.

References.—*Stewart, Pb. Pl., 236 ; Dymock, Mat. Med. W. Ind., 789 ;*
S. Arjun, Bomb. Drugs, 137 ; Murray, Pl. and Drugs, Sind, 22 ; Irvine,
Mat. Med. of Patna, 101 ; Baden Powell, Pb. Pr., 262 ; Atkinson, Him.
Dist., 318 ; Royle, Ill. Him. Bot., 370 ; Balfour, Cyclop., I., 1060 ; Rev.
A. Campbell's Report on Econ. Prod., Chutia Nagpur, Eulophia, sp. ?

Habitat.—An orchid found in Oudh and Rohilkhand, and in the
Gangetic Doab ; also, according to Aitchison, in the Panjáb, on the
islands formed by the recurving of the rivers. Dr. Stewart records
having gathered the tubers near Lahore in the Rávi. Dr. Royle mentions
that the plant was of common occurrence in and near the Kheree Pass.
Admirable samples of what appear to be the saleep of this orchid were
recently sent to the writer from the Sirohi State, Western Rajputana.

Medicine.—By the natives the *salep* is chiefly esteemed as a tonic and   **MEDICINE.**
aphrodisiac.   **468**

SPECIAL OPINIONS.—§ " Saleep is considered as nutritious and is largely consumed by persons suffering from phthisis and other exhausting diseases." (*Surgeon-Major A. S. G. Jayakar, I.M.D., Muskat, Arabia.*)
" Useful form of conjee for nursing mothers." (*Surgeon-Major G. Y.*
Hunter, Karachi.) " Is extensively used in cases of impotence and when
lithates of a pink colour are passed in the urine, a condition which the
native hakims almost always confuse with spermatorrhœa." (*Surgeon-*
Major C. W. Calthrop, M.D., Morar.) " Salip misri is very useful as a diet
in dysentery ; the tubers should be grated and boiled down in milk."
(*Civil Surgeon George Cumberland Ross, Delhi, Panjáb.*) " Used in spermatorrhœa and impotence. Infusion made from pounded tuber." (*Civil*
Medical Officer Mr. Forsyth, F.R.C.S., Ed., Dinagepore, North Bengal.)

Food.—The Europeans in Northern India and at some of the Himá-   **FOOD.**
layan and Nilghiri hill stations collect the tubers of this and other allied   **469**
species and use them for family consumption as salep ; they regard them
as an easily digestible kind of farinaceous food.

E. herbacea, *Lindl.*   **470**

Syn.—E. VERA, *Royle* (?)

References.—*Stewart, Pb. Pl., 236 ; Lindl., Gen. and Sp. Orchid, 182 ;*
Royle, Ill. Him. Bot., 366 and 370 ; Balfour, Cyclop. I., 1060.

Habitat.—Royle's specimens, named by him **E. vera,** were gathered
near the banks of the Jhelam river in the Panjáb Himalaya. This he
believed to be the source of the true *salep misri* of commerce, and distinct
from that of **E. herbacea.** According to other writers this species occurs
on the mountains of South India.

EUONYMUS, *Linn. ; Gen. Pl., I. 360.*

Euonymus crenulatus, *Wall. ; Fl. Br. Ind., I., 608 ; Wight. Ic.,*   **471**
[*t., 973 ;* CELASTRINEÆ.

References.—*Beddome, Fl. Sylv. t., 144; Gamble, Man. Timb., 84;*
Drury, U. Pl., 203 ; Balfour, Cyclop., I., 1060.

Habitat.—A small tree, common in hilly parts of South India.

Structure of the Wood.—White, very hard and close-grained ; answers   **TIMBER.**
for wood engraving, and is about the best substitute for boxwood in the   **472**
Madras Presidency (*Beddome*).

E. glaber, *Roxb. ; Fl. Br, Ind., I., 609.*   **473**

References.—*Roxb., Fl. Ind., Ed. C.B.C., 211 ; Voigt., Hort. Sub. Cal.,*
165 ; Kurz, For. Fl. Burm., I., 249.

U 2   **E. 473**

EUONYMUS pendulus.	**The Euonymus.**

TIMBER.
474

Habitat.—A small tree found in East Bengal and in Burma.
Structure of the Wood.—Brownish yellow, turning brown; heavy, rather close-grained and hard, but soon attacked by xylophages. Fine wood for furniture (*Kurz*).

475

Euonymus grandiflorus, *Wall.; Fl. Br. Ind., I., 608.*

Syn.—E. LACERUS, *Ham.*
Vern.—*Siki, pattali, papar, banchir, dudhapár, hanchu, pásh, mara, chikan, rangchúl, kioch,* PB.; *Gule, grui,* SIMLA.
References.—*Voigt., Hort. Sub. Cal., 166; Brandis, For. Fl., 78; Gamble, Man. Timb., 84; Wall., Pl. As. Rar., III., 35, t. 254; Atkinson, Him. Dist., 307.*
Habitat.—A small deciduous tree of the Himálaya, from the Indus to Sikkim, between 6,000 and 11,000 feet.

FODDER.
476
TIMBER.
477
DOMESTIC.
478
479

Fodder.—The young shoots and leaves are lopped to feed goats.
Structure of the Wood.—White, moderately hard, exceedingly compact, close and even-grained. It is used for carving (*Gamble*).
Domestic Use.—According to Brandis, the seeds with their bright red arils are strung up and used as ornaments in Bussahir.

E. Hamiltonianus, *Wall.; Fl. Br. Ind., I., 612.*

Syn.—E. ATROPURPUREUS, *Roxb.*
Vern.—*Agniun, agnu,* KUMAON; *Brahmáni,* KASHMIR; *Siki, singi, chual, watal, papar, rithu, ranái, banchor, karún, skioch, sidhera, naga,* PB.
References.—*Roxb., Fl. Ind., Ed. C.B.C., 211; Voigt., Hort. Sub. Cal., 165; Brandis, For. Fl., 78; Gamble, Man. Timb., 84; Stewart, Pb. Pl., 41; U. S. Dispens., 15th Ed., 567.*
Habitat.—A large deciduous shrub, or small or occasionally moderate-sized tree of the outer Himalaya, from the Indus to Bhután, and of the Khásia Hills from 4,000 to 8,000 feet (*Gamble*).

FODDER.
480
TIMBER.
481

Fodder.—The young leaves and shoots are lopped for fodder (*Brandis*).
Structure of the Wood.—White, with a slightly yellow tinge, soft, close-grained. It is used for carving into spoons (*Gamble*).

E. japonicus, *Wall.;* see E. pendulus, *Wall.*

482

E. pendulus, *Wall.; Fl. Br. Ind., I., 612.*

Syn.—E. JAPONICUS, *Wall.* (not of *Thunb.*)
Vern.—*Chopra, pincho, garúr, kúnku,* N.-W.|P.
References.—*Brandis, For. Fl., 79; Gamble, Man. Timb., 84; Atkinson, Him. Dist., 307.*
Habitat.—A moderate-sized evergreen tree, found in the Himáláya, from the Jhelum to Nepal, between 2,500 and 7,500 feet (*Gamble*).

CHEMISTRY.
483

Chemistry.—Dr. Dymock writes the following as the result of his analysis of a specimen of the bark furnished by **Dr. G. Watt** from Simla:—
"The young branches give a green tincture with spirit and the older bark a red tincture; in each case, on dissipating most of the alcohol and treating with water, a greenish yellow resinous substance falls, and a bright red liquid remains. The resins are soluble in ether and partly in alkalies, and the red astringent supernatant liquor contains tannin, giving a murky green colour with a ferric chloride, and a quantity of saccharine matter. No bitterness was perceived in the extract, and nothing alkaloidal was detected. The aqueous extract of the bark, after exhaustion with spirit, contained a large quantity of a white, neutral, crystalline body, which was dissolved by hot alcohol and crystallized out on cooling. The bark had no marked odour or taste, and afforded a light buff-coloured powder. The powder, treated directly with rectified spirit, gave 45·5 per

E. 483

cent. of extract, and when burnt left 12·8 per cent. of carbonated ash. The crystalline body appears to be mannite. Mr. Hooper does not think that this bark, or that of **E. crenulatus**, are likely to replace that of **E. atropurpureus**. If we could find an Euonymys with a bitter bark, a better result might be obtained."

Structure of the Wood.—White, moderately hard, compact, with a light-red tinge, very close and even-grained. TIMBER. 484

Enonymus tingens, *Wall.; Fl. Br. Ind., I., 610.* 485

Vern.—*Newar, kasúri*, NEPAL; *Kungku*, N.-W. P.; *Chopra, mer mahaul*, SIMLA.

References.—*Brandis, For. Fl., 79; Gamble, Man. Timb., 85; O'Shaughnessy, Beng. Dispens., 272; Royle, Ill. Him. Bot., 167.*

Habitat.—A small evergreen tree of the Himálaya, from the Sutlej to Nepal, between 6,500 and 10,000 feet (*Gamble*).

Dye.—The inner BARK is said to yield a beautiful yellow dye. DYE. Bark. 486

Medicine.—Royle was informed of the PLANT being used in diseases of the eye. MEDICINE. Plant. 487

Structure of the Wood.—Similar to **E. grandiflorus**, except that the wood of this species has a slightly reddish tinge. TIMBER. 488

Domestic Use.—The dye is said to be used in Nepal for marking the *tika* on the foreheads of Hindus. DOMESTIC. 489

EUPATORIUM, *Linn.; Gen. Pl., II., 245.*

Eupatorium Ayapana, *Vent.; Fl. Br. Ind., III., 244;* COMPOSITÆ. 490

References.—*Pharm. Ind, 127; O'Shaughnessy, Beng. Dispens., 422; Dymock, Mat Med. W. Ind., 424; Fleming, Med. Pl. and Drugs, as in As. Res., Vol. XI., 166; U. S. Dispens., 15th Ed., 560; S. Arjun, Bomb. Drugs, 78; Fleming, New Pl. and Drugs, in As. Res., XI., 167; K. L. Dey, Indig. Drugs of Ind., 53; Drury, U. Pl., 203; Balfour, Cyclop., 1061; Mueller, Select Extra-Trop. Pl., 168.*

Medicine.—A small aromatic shrub naturalised in many parts of India, and known under its Brazilian name, *Aya-pana*. For long it held a high position as a medicinal plant, but the exaggerated ideas of its virtues have now exploded. It is a good simple stimulant, tonic, and diaphoretic. In cholera it has been used to restore warmth to the body, and it is said also to be used internally and externally in the treatment of snake-bite. Fleming (in *Asiat. Res., l.c.*) says that "instances are not unfrequent of medicines which had been at first too highly extolled having afterwards met with unmerited neglect; and such may, perhaps, be the case in respect to the plant in question." Dymock says that it is not uncommon in gardens in Bombay, and, though not generally known, is held in considerable esteem by those who are acquainted with it. MEDICINE. 491

E. cannabinum, *Linn.; Fl. Br. Ind., III., 243.* 492

THE HEMP AGRIMONY.

References.—*Voigt., Hort. Sub. Cal., 407; Fleming, Med. Pl. and Drugs, in As. Res., Vol. XI., 167.*

Habitat.—Exceedingly plentiful, tall, erect plant, with downy leaves and terminal crowded head of dull purple flowers, inhabiting damp, watery places on the temperate Himálaya, Khásia Hills, and Burma, between 3,000 and 6,000 or even up to 10,000 feet in altitude.

Medicine.—"Was strongly recommended by Tournefort as a deobstruent in visceral obstructions consequent to intermittent fevers, and externally as a discutient in hydropic swellings of the legs and scrotum." (*Fleming, in Asiat. Res., l.c.*) MEDICINE. 493

494
EUPHORBIA, *Linn.; Gen. Pl., III., 258.*

A large genus, containing more than 600 species, which are widely distributed over the greater part of the world. They are popularly known as Spurgeworts, a name which is sometimes applied to the whole family. Linnæus is said to have named the genus after Euphorbus, a physician to Juba, King of Mauritania. The species consist of herbs or shrubs, but in some instances they assume the form of small cactus-like trees, with thick, soft-wooded jointed branches. Though differing so widely in general appearance, they can generically easily be recognized by the structure of their flowers. The monœcious flowers are arranged in clusters, and each cluster, consisting of several jointed stamens (male flowers) surrounding a single female flower, is enclosed within a common involucre. All the species abound in a more or less acrid milky juice, which contains active medicinal properties. The most important extract, known under the name of Euphorbium, is obtained chiefly from E. resinifera, one of the fleshy-stemmed species, indigenous to Morocco. This resinous substance used to be given as a purgative and emetic, but owing to its extremely powerful action, it is now never used as an internal remedy. Its anticorrosive properties have recently created a demand for it as an ingredient of paint for ships' bottoms. Euphorbium occurs in small roundish masses resembling tragacanth ; it is of a light yellow or reddish colour, it has no smell, and its taste,at first slight, becomes painfully acrid and burning.

CHEMISTRY.
495
Its chemical composition, according to Flückiger (1868), is as follows :—

Amorphous resin, $C^{10} H^{16} O^2$	38
Euphorbon, $C^{13} H^{32} O$	22
Mucilage	18
Malates, chiefly of calcium and sodium	12
Mineral compounds	10
	100

" The amorphous resin is readily soluble in cold spirit of wine containing about 70 per cent. of alcohol. The solution has no acid reaction, but an extremely burning acrid taste : in fact it is to the amorphous indifferent resin that Euphorbium owes its intense acridity." (*Flück. and Hanb., Pharmacog., 560.*) See also *Spons' Encyclop., II., 1649; U. S. Dispens., 15th Ed., 1641; Ainslie, Mat. Med., I., 120.*

496
Euphorbia antiquorum, *Linn. ; Fl. Br. Ind., V., 255 ; Wight. Ic.,*
[*t. 897 ;* EUPHORBIACEÆ.

Vern.—*Tindhára sehund, tidhára-sehnr, tidhára-sehnd,* HIND. ; *Nara sij, tekáta sij, bajvaran, lariya-dáona,* BENG. ; *Etkec',* SANTAL ; *Dokánásiju,* URIYA ; *Shidu,* MICHI ; *Tidhári-send, tin-dhári-send,* DECCAN ; *Naraseja,* MAR. ; *Tandhári-send,* GUZ. ; *Shadhurak-kalli, shadray kullie* (Ainslie), *tirikalli,* TAM. ; *Bomma jemudu, bonta-chemudu,* TEL. ; *Buma-chumadoo* (Roxb.), *mudu, mula-jemudu,* KAN. ; *Katak-kalli, chatirak-kalli, sudusudu,* MALAY ; *Shasoung-pya-thal, shasánv-ji,* BURM. ; *Daluk,* SING. ; *Situnda, vajri, seehoondee* (Roxb.), *vajrakantaka,* SANS. ; *Zaqqume-hindi,* ARAB. ; *Zaqunniyæ-hindi, saqqume-hindi,* PERS.

References.—*Roxb., Fl. Ind., Ed. C.B.C., 392 ; Voigt., Hort. Sub. Cal., 162 ; Brandis, For. Fl., 438 ; Kurs, For. Fl. Burm., II., 416 ; Beddome, For. Man. 216 ; Gamble, Man. Timb., 368 ; Dals. & Gibs., Bomb. Fl., 226 ; Elliot, Fl. Andhrica, 29 ; Trimen, Hort. Zeyl., 70 ; Rheede, Hort. Mal. ii., t. 42 ; Pharm. Ind., 204 ; Ainslie, Mat. Ind., I., 121 ; O'Shaughnessy, Beng. Dispens.. 564 ; Moodeen Sheriff, Supp. Pharm. Ind., 136 ; U. C. Dutt, Mat. Med. Hind., 322 ; S. Arjun, Bomb. Drugs, 198 ; Drury, U. Pl., 203 ; Lisboa, U. Pl. Bomb., 114 ; Atkinson, Gums and Gum-resins, 29 ; Gums and Resinous Prod. of India (P. W. D., 1871), 28 ; Balfour, Cyclop., I., 1061 ; Treasury of Bot. 477 ; Kew Off. Guide to the Mus. of Ec. Bot., 115 ; Home Dept. Cor. regarding Pharm. Ind., 240.*

E. 496

The Euphorbias.	(*J. F. Duthie*)	**EUPHORBIA granulata.**

Habitat.—A shrub or small tree, with three- or five- angled branches, common on the dry hills of Bengal and the Peninsula generally. Mr. J. O. Hardinge states that it is common all over Burma, being often cultivated for hedges.

Gum.—This species was supposed for a long time to be capable of yielding the commercial **Euphorbium** resin. Buchanan Hamilton (*Linn. Trans., Vol. XIV.*) and Royle (*Ill., i., 328*) have, however, clearly demonstrated that the true **Euphorbium** is not a product of India.

GUM.
497

Medicine.—The JUICE which flows from the branches is used as a purgative to relieve pain in the loins. It is an acrid irritant in rheumatism and toothache. When taken internally it acts as a drastic purgative. It is also employed in nervine diseases, dropsy, palsy, deafness, and amaurosis (*Baden Powell*). A plaster prepared from the ROOTS and mixed with asafœtida is applied externally to the stomachs of children suffering from worms. The BARK of the root is purgative, and the STEM is given in decoction in gout (*Wight; Rheede*). The Rev. A. Campbell states that a preparation from the plant is in Chutia Nagpur given as a cure for cough.

MEDICINE.
Juice.
498

Roots.
499
Bark.
500
Stem,
501

SPECIAL OPINIONS.—§" The fresh juice of cut branches is irritant; it is applied to painful joints." (*Shib Chunder Bhuttacharji, Chánda, Central Provinces.*) "The juice mixed with burnt borax and common salt is used as an application in painful joints and swellings. The fresh milky juice is a direct irritant both when taken internally and applied externally. Taken in very small quantities it is a drastic purgative. It is also used as an antidote in cases of snake-bite." (*Civil Surgeon J. H. Thornton, B.A., M.B., Monghyr.*)

Fodder.—The Rev. A. Campbell states that in Chutia Nagpur goats and sheep feed on this plant.

FODDER.
502

Structure of the Wood.—"White, light, soft, but even-grained" (*Brandis*).

TIMBER.
503

Domestic Uses.—"This plant is supposed to ward off lightning strokes, and is generally kept in tubs or pots on the roofs or other exposed parts of native houses" (*U. C. Dutt*). This fact is also corroborated by Shib Chunder Bhuttacharji, Chánda, Central Provinces.

DOMESTIC.
504

Euphorbia dracunculoides, *Lam. ; Fl. Br. Ind., V., 262.*

505

Syn.—E. LANCEOLATA, *Heyne* ; E. UNIFLORA, *Wall.*

Vern.—*Jy-chee, chhagul-puputi,* BENG. ; *Parwa,* SANTAL ; *Ric'ini, sudáb* (the fruit), *kangi* (the plant), PB. ; *Tilla káda,* TEL.

References.—*Roxb., Fl. Ind., Ed. C.B.C., 394; Voigt, Hort. Sub. Cal., 164; Stewart, Pb. Pl., 193; Aitchison, Cat. Pb. and Sind Pl., 131; Fl. Andhrica, 182; Spons, Encyclop., II., 1414.*

Habitat.—A much-branched annual, met with in the Panjáb, Bengal, Madras (Coromandel), and Konkan.

Oil.—It yields a limpid, clear oil, of a yellowish or greenish-yellow colour, used as a drying oil and for burning. In 1843 it was submitted to London brokers, who pronounced it more valuable than linseed oil. The *Agri-Horticultural Society of India Journ., 1843, ii., p. 52,* draws attention to this oil.

OIL.
506

Medicine.—The FRUIT is official and is said to be used to remove warts.

MEDICINE.
Fruit.
507
508

E. granulata, *Forsk. ; Fl. Br. Ind., V., 252.*

Syn.—E. ARILLATA, *Edgew.*

Vern.—*Kantha arak',* SANTAL.

References.—*Edgew., in Journ. As. Soc. Beng., XVI., 1218.*

Habitat.—A hispid, perennial herb, with prostrate stems, inhabiting the plains of Northern and Central India from Rohilkhand to Sind.

E. 508

EUPHORBIA microphylla.	The Euphorbias.

FOOD.
Leaves.
509

Food.—"The LEAVES are eaten as a pot-herb by the Santals" (*Rev. A. Campbell*).

Euphorbia helioscopia, *Linn.; Fl. Br. Ind., V., 262.*

SUN SPURGE.

Vern.—*Hirruseeah, mahabi*, HIND.; *Ganda búte, dúdal, kulfa-dodak, chatriwal*, PB.

References.—*Stewart, Pb. Pl., 193; Aitchison, Cat. Pb. and Sind Pl., 132; Murray, Pl. and Drugs, Sind, 32.*

Habitat.—A common field-weed in spring throughout the Panjáb plains and the Siwalik tract, ascending to 7,000 feet in the outer Himálaya. Introduced into the Nilghiri hills.

MEDICINE.
Juice and
Seeds.
510
Roots.
511

Medicine.—The milky JUICE is applied to eruptions and the SEEDS are given with roasted pepper in cholera. The juice is also used in the form of a liniment in neuralgia and rheumatism, and the ROOT is employed as an anthelmintic (*Murray*).

512

E. hypericifolia, *Linn.; Fl. Br. Ind., V., 249.*

Syn.—E. INDICA, *Lamk.*; E. PARVIFLORA, *Linn.*

Vern.—*Hasárdána* (seeds and leaves), PB.; *Nayeti*, BOMB.; *Dhákti-dudhi*, MAR.; *Ela-dada-kiriya*, SING.

References.—*Roxb., Fl. Ind., Ed. C.B.C., 394; Voigt, Hort. Sub. Cal., 163; Thwaites, En. Ceylon Pl., 268; Dals. & Gibs., Bomb. Fl., 227; Stewart, Pb. Pl., 194; Aitchison, Cat. Pb. and Sind Pl., 132; Trimen, Hort. Zeyl., 71; Rheede, Hort. Mal., X., t. 51; Dymock, Mat. Med. W. Ind., 2nd Ed., 694; U.S. Dispens., 15th Ed., 1640; S. Arjun, Bomb. Drugs, 124; Atkinson, Him. Dist., 317; Treasury of Bot., 477.*

Habitat.—A small, slender annual, common throughout the hotter parts of India (from the Panjáb to the Southern Deccan), and occurring up to 4,000 feet on the Himálaya.

MEDICINE.

Medicine.—Stewart mentions that in some parts of the Panjáb it is given with milk to children suffering from colic. S. Arjun remarks that it possesses properties similar to those of E. pilulifera and E. thymifolia. Dr. W. Zollickoffer (in *Am. Journ. of Med. Soc., XI., 22*) recommends

Leaves.
513

an infusion of the dried LEAVES of this herb as a remedy in dysentery, diarrhœa, menorrhagia, and leucorrhœa, and finds that it affects the system as an astringent and feeble narcotic.

514

E. Lathyris, *Linn.*

CAPER SPURGE, *Eng.*

Vern.—*Burg-sadab* (Irvine), BENG.; *Sudab*, PB.

References.—*Ainslie, Mat. Ind., I., 599; O'Shaughnessy, Beng. Dispens., 565; U.S. Dispens., 15th Ed., 1713; Irvine, Mat. Med., Patna, 18; Am. Journ. Pharm., XXVI, 305; Spons, Encyclop., II., 1414.*

Habitat.—A perennial herb with narrow glaucous leaves, a native of Central and South Europe.

OIL.
Seeds.
515
MEDICINE.
Oil.
516

Oil.—The SEEDS yield by expression, or by the agency of alcohol or ether, a colourless tasteless OIL.

Medicine.—The OIL formerly found much favour with certain French and Italian physicians as a purgative owing to its tastelessness (when fresh), and because of the small amount required for a dose. In this country the seeds are said to be used in dropsy, and also to procure abortion. According to Irvine (*Mat. Med. of Patna*), the imported dried leaves, fruits, and stalks are used as a carminative in dyspepsia, and as a deobstruent.

DOMESTIC.
517
518

Domestic Use.—The capsules are said to intoxicate fish.

E. microphylla, *Heyne; Fl. Br. Ind., V., 252.*

Syn.—E. UNIFLORA, *Dals. & Gibs.;? E. CHAMÆSYCE, Roxb.*

E. 518

| The Euphorbias. | (*J. F. Duthie.*) | EUPHORBIA neriifoia. |

Vern.—*Choto-keruee* (**V**oigt), BENG.; *Dudhia-phul*, SANTAL.
References.—*Roxb., Fl. Ind., Ed. C.B.C., 394; Voigt, Hort. Sub. Cal., 163; Dals. & Gibs., Bomb. Fl., 227 : Aitchison, Cat. Pb. and Sind Pl., 131 ; Rev. A. Campbell, Report Econ. Prod., Chutia Nagpur, No. 7921.*
Habitat.—A slender, prostrate, much-branched annual, found in Bengal, Bundelkhand, Southern India, and Burma.
Medicine.—The Rev. A Campbell mentions that, in Chutia Nagpur, a preparation of this plant, along with that of **Cryptolepis Buchanani**, (Vol. II., 624), is given to nursing mothers when the supply of milk fails or is deficient.

MEDICINE.
519

Euphorbia neriifolia, *Linn.; Fl. Br. Ind., V., 255.*

520

Syn.—E. LIGULARIA, *Roxb.*
Vern.—*Sehund, thohar, sij, patton-ki-send*, HIND.; *Mansa-sij, páta-shij, hij-dáona*, BENG.; *Gángichu*, PB.; *Nivadunga, minaguta, thohur*, SIND; *Kutte-ki-jibh-ki-send, kutte-ki-jibh-ka-patta*, DECCAN; *Minguta, thor, newarang*, BOMB.; *Nevadunga, mingut*, MAR.; *Thor*, GUZ.; *Nevul-kanta*, GOA; *Ilaik-kalli*, TAM.; *Aku-jemudu*, TEL.; *Yalekalli*, KAN.; *Elakalli*, MALAY; *Shasaung, sha-soung, shazávn-mina*, BURM.; *Patuk*, SING.; *Snuhi* (U. C. Dutt), *vujri, sehunda*, SANS.
References.—*Roxb., Fl. Ind., Ed. C.B.C., 392; Voigt, Hort. Sub. Cal., 161; Brandis, For. Fl., 439; Kurz, For. Fl. Burm., II., 416 ; Beddome, For. Man., 216; Gamble, Man. Timb., 368 ; Dals. & Gibs., Bomb. Fl., 226; Stewart, Pb. Pl., 195 ; Aitchison, Cat. Pb. and Sind Pl., 132 ; Trimen, Hort. Zeyl., 71; Pharm. Ind., 204; O'Shaughnessy, Beng. Dispens., 564; Moodeen Sheriff, Supp. Pharm. Ind., 137 ; U. C. Dutt, Mat. Med. Hind. 233 and 318; Dymock, Mat. Med. W. Ind., 2nd Ed., 689 ; S. Arjun, Bomb. Drugs, 198 ; Irvine, Mat. Med., Patna, 65 ; Atkinson, Him. Dist., 317 ; Drury, U. Pl., 205; Lisboa, U. Pl. Bomb., 114, 275 ; Balfour, Cyclop., I., 1061 ; Treasury of Bot., 477 ; Bomb. Gaz., VI., 14; XV., 68 ; Journ. Agri.-Hort. Soc., VIII., pp., 223-226.*
Habitat.—A small, erect, glabrous tree, with fleshy, cylindrical stems, spirally twisted, 5-angled branches, and sharp stipular thorns at the bases of the subterminal fleshy leaves. It is found wild on rocky ground in Central India, and is extensively cultivated in the neighbourhood of villages in Bengal and elsewhere. It is cultivated, and according to Kurz, is also found wild, in Burma. Distribution to Baluchistan, Malay Islands, &c.
Gum.—It yields a gum, or GUTTA-PERCHA-LIKE substance, on boiling.
Medicine.—"The milky JUICE is considered purgative and rubefacient. As a purgative it is generally used in combination with other medicines which are steeped in it. Chebulic myrobalan, long pepper, *trivrit* root, &c., are thus treated and administered as drastic purgatives, in ascites, anasarca, and tympanitis. It enters into the composition of several compound prescriptions of a drastic character" (*U. C. Dutt, Mat. Med. Hind., 233*). "The ROOT, mixed with black pepper, is employed in cases of snake-bites, both internally and externally. Every part abounds with an acrid milky juice, employed to remove warts and cutaneous eruptions, &c. The PULP of the stem mixed with green ginger is given to persons who have been bitten by mad dogs before the accession of hydrophobia" (*Taylor, Topog. of Dacca, 57*).
SPECIAL OPINIONS.—§ "The juice is employed in ear-ache and mixed with soot is employed in ophthalmia as an *anjan*." (*Assistant Surgeon T. N. Ghose, Meerut.*) "The tender terminal portions of the branches are slightly roasted, and the juice is then squeezed out and given with molasses for producing vomiting and purging in bronchitis of children." (*Surgeon-Major Robb, Civil Surgeon, Ahmedabad.*) "Is largely used with clarified or fresh butter as an application to unhealthy ulcers and in scabies. Is also employed as an antidote in snake-poisoning." (*Civil Sur-*

GUM.
Gutta percha.
521
MEDICINE.
Juice.
522

Root.
523

Pulp.
524

E. 524

| EUPHORBIA | The Euphorbias. |
| pilulifera. | |

geon J. H. Thornton, B.A., M.B., Monghyr) "The milky juice is applied to glandular swellings to prevent suppuration." (*Shib Chundra Bhatta-charji, Chanda, Central Provinces.*)

TIMBER.
525

Structure of the Wood.—Attains 20 feet, stem often 12 inches in diameter.

SACRED and DOMESTIC.
526

Sacred and Domestic Uses.—"This shrub is sacred to Mansá, the goddess of serpents. On the fifth day after full moon of the month Sravana (July-August) it is planted in the courtyard of Hindu houses and worshipped as the representative of Mansá." (*U. C. Dutt, Mat. Med. Hind., 233.*)

[*1862.*

527

Euphorbias Nivulia, *Ham.; Fl. Br. Ind., V., 255; Wight, Ic., t.*

Syn.—E. NERIIFOLIA, *Roxb.*

Vern.—*Sij,* BENG.; *Tor. raj,* RAJ.; *Patteoon* (O'Shaughnessy), DECCAN; *Newrang,* MAR.; *Aku jemudu* or *chemudu,* TEL.; *Ela-calli* (Roxb.), MALAY; *Sha-soung,* BURM.; *Pattakarie,* SANS.

References.—*Roxb., Fl. Ind.. Ed. C.B.C., 392; Voigt, Hort. Sub. Cal., 162; Brandis, For. Fl., 439; Kurz, For. Fl. Burm., II., 417; For. Man., 216; Gamble, Man. Timb., 368; Dalz. & Gibs., Bomb. Fl., 225; Elliot, Fl. Andhr., 13; Rheede, Hort. Mal., II., t. 43; Pharm. Ind., 204; O'Shaughnessy, Beng. Dispens., 555; Moodeen Sheriff, Supp. Pharm. Ind., 137; Murray, Pl. & Drags, Sind, 32; Atkinson, Him. Dist., 317; Drury, U. Pl., 206; Balfour, Cyclop., I., 1061; Treasury of Bot., 477.*

Habitat.—A large, fleshy-stemmed shrub, or small tree, with smooth roundish, whorled, branches, found in dry rocky places in Northern and Central India; also in Burma. Often planted for hedges.

MEDICINE.
Milk.
528

Medicine.—The MILK has properties similar to those of **E. neriifolia.**

E. parviflora, see **E. hypericifolia.**

529

E. pilosa, *Linn.; Fl. Br. Ind., V., 260.*

Habitat.—A tall, erect, perennial herb, found on the Himálaya from Garhwal westward.

MEDICINE.
Root.
530

Medicine.—This is, no doubt, the plant referred to by **Stewart** under **E. longifolia,** *Don.,* and the ROOT of which **Honigberger** mentions as being used for the cure of fistulous sores.

531

E. pilulifera, *Linn.; Fl. Br. Ind., V., 250.*

Syn.—E. HIRTA, *Linn.*

Vern.—*Dudhi,* HIND.; *Bura keru* (Roxb.), *buro-keruee* (Voigt), BENG.; *Pusi-toa,* SANTAL; *Gordon, C. P.; Nayeti,* BOMB.; *Dudhi,* or *mo-thidudhi,* MAR.; *Dudeli,* GUZ.; *Amumpatchay-arissi,* TAM.; *Bidarie, nanabeeam, nánabála,* TEL.; *Bú-dada-kiriya,* SING.

References.—*Roxb., Fl. Ind., Ed. C. B. C., 394; Voigt, Hort. Sub. Cal., 163; Dalz. & Gibs., Bomb. Fl., 227; Aitchison, Cat. Pb. and Sind. Pl., 132; Elliot, Fl. Andhr., 129; Trimen, Hort. Zeyl., 71; Dymock, Mat. Med. W. Ind., 2nd Ed., 693; S. Arjun, Bomb. Drugs, 123; Atkinson, Him. Dist., 317; Christy, Com. Pl. and Drugs, No. 5, p. 64; No. 7, p. 47; No. 8, p. 55; No. 9, p. 35.*

Habitat.—A small erect or ascending herb, with acute, hispid leaves (having copious crisped hairs), and small fruits. Found throughout the hotter parts of India from the Panjáb eastward, and southward to Ceylon and Singapore.

MEDICINE.

Medicine.—[Indian writers have very little to say as to the properties of this plant. They regard it as equal with **E. thymifolia,** but appear never to have learned that either had a special virtue in the treatment of asthma. The following are the only Indian passages the writer can discover that deal with the properties of these plants: "The PLANT is chiefly used in the affections of childhood, in worms, bowel complaints, and cough. Sometimes prescribed also in gonorrhœa" (*S. Arjun, Bomb. Drugs*).

Plant.
532

E. 532

The Rev. A. Campbell states that the ROOT is given to allay vomiting, and the plant to nursing-mothers when the supply of milk is deficient or fails. Dr. Dymock speaks of this species conjointly with **E. thymifolia**, and says they have a reputation as vermifuges. Though Baron F. von Mueller is apparently silent as to the merits of **E. pilulifera**, certain popular writers, especially in Australia, have extolled "the weed" as a "most valuable remedy" in the treatment of asthma. Mr. Thomas Christy has republished, from the Australian newspapers, the various letters which have appeared on this subject, but Mr. O. G. Levison (in the *Therapeutic Gazette*) has furnished us with a chemical analysis which goes a long way towards destroying the claims of the drug to the consideration Mr. Christy has urged. He says the analysis did not demonstrate the presence of "anything of importance besides the usual constituents found in most drugs," not the least trace of an alkaloid, and although it gave off a characteristic odour, no volatile substance could be discovered nor any fixed oil. "When subjected to destructive distillation, a distillate was obtained, which had a very powerful and empyreumatic odour, somewhat resembling nicotine. Possibly, this may be a principle of some importance, which, later on, I will investigate." Experimenting upon its physiological effects, Mr. Levison found it to act "slightly as a stimulant and narcotic; but as far as being a specific for asthma, did not find it to act as such, sometimes increasing the sufferings of the patient by producing a more marked dyspnœa." (*Ed. Dict. Ec. Prod.*)

MEDICINE.
Root.
533

Food.—The LEAVES and tender SHOOTS have, according to Dr. Shortt, been eaten in the Madras Presidency in times of famine.

FOOD.
Leaves.
534
Shoots.
535
536

Euphorbia pulcherrima, *Willd.; Fl. Br. Ind., V., 239.*

POINSETTIA, *Eng.;* FLOR DE PASQUA, *Span.*

Syn.—POINSETTIA PULCHERRIMA, *Grahm.*

References.—*Voigt, Hort. Sub. Cal., 164; Brandis, For. Fl., 439; Kurz, For. Fl. Burm., II., 418; Gamble, Man. Timb., 368.*

Habitat.—An ornamental shrub, discovered in Mexico by Graham in 1828. It is cultivated in most Indian gardens; its bright crimson floral leaves appearing about Christmas time.

Gum.—It yields freely a milky sap, which hardens into a black gum, or may be boiled down to a sort of gutta-percha.

GUM.
537
538

E. Royleana, *Boiss.; Fl. Br. Ind., V., 257.*

Syn.—E. PENTAGONA, *Royle, Ill., t. 82.*

Vern.—*Shakar-pitan (Balf., Cycl.),* HIND.; *Sohund,* KUMAON; *Shakar pitan, thar, thor, tordanda* (Salt Range), PB.; *Suli* (J.), *chula* (C.), *chun* (R.), *chu, chunga & surs* (B.), *suro & tsui* (S.), PB. HIM.; *Thor,* RAJ.

References.—*Brandis, For. Fl., 438; Gamble, Man. Timb., 368; Stewart, Pb. Pl., 194; Aitchison, Cat. Pb. and Sind Pl., 132; Atkinson, Him. Dist., 736; Balfour, Cyclop., I., 1062.*

Habitat.—A large fleshy shrub, common on dry rocky hillsides of the outer Himálaya, from Kumáon westwards, ascending to 6,000 feet. It occurs also on the Salt Range.

Guttapercha.—§ ["The milky sap of this plant contains a large amount of superior guttapercha. The sap has, when fresh, a rich sweet odour, and does not blister the fingers even when handled and worked with for hours." It is, however, very injurious to the eyes and flavours anything handled for days after all trace has been removed from the fingers. *Ed. Dict. Ec. Prod.*]

GUM.
Guttapercha.
539

Medicine.—The acrid, milky JUICE of this plant possesses cathartic and anthelmintic properties.

MEDICINE.
540

Structure of the Wood.—"This is probably the species on the dry hills near Jeypur, which furnishes a great part of the fuel for that city.

TIMBER.
541

E. 541

| EUPHORBIA thymifolia. | The Euphorbias. |

Attains 15-16 feet; the stems have generally a girth of 2-3 feet, but sometimes of 5-6 feet. The wood is soft and useless." (*Brandis.*) Near Simla the dry white wood is largely used by the poor classes as fuel.

542

Euphorbia, sp.

The dried roots of an undetermined **Euphorbia** are used in Kuram as a purgative. In large doses it causes vomiting, hence it is called the "vomit-weed." The fresh milk of the leaves causes blisters on the hands when collecting the root (*Aitchison, Kuram Valley Flora, in Journal Linnæan Society, XVIII., page 25*). May this not be **E. Thomsoniana** referred to by the author in Vol. XIX., *l. c.*, page 147?

543

E. Thomsoniana, *Boiss. Fl. Br. Ind., V., 260.*

Vern.—*Hirtis* (Aitchison), KASHMIR.

Habitat.—A very distinct plant, with glabrous, simple stems, a foot high, rising from a perennial root-stock. It occurs in Western Thibet, Gilgit, &c., at altitudes of 110,000 to 12,000 feet above the sea. (*Fl. Br. Ind.*)

Medicine.—The crushed ROOT-STOCKS are employed by the natives of Kuram as detergents for washing the hair, and when boiled are given as purgatives (*Aitchison, Kuram Valley Flora; Linnæan Journal, XXIX., page 147*). In Kashmír the root-stock is employed to adulterate "*kut*" (Saussurea Lappa), and is called by the Kashmíris "*Hirtiz.*" The STEM, ROOT, and LEAVES are said to be used medicinally. (*Aitchison.*)

Domestic Uses.—The root as a detergent

E. thymifolia, *Burm.; Fl. Br. Ind., V., 252.*

Syn.—E. FOLIATA, *Ham.*; E. PROSTRATA, *Grah. (not of Aiton)*; E. RUBICUNDA, *Bl.*

Vern.—*Chotka dudhi,* HIND.; *Swet-kerua* (Roxburgh), *Shwet-keruee* (Voigt), *Dudiya* (Irvine), BENG.; *Nanha-pusi-toa,* SANTAL; *Bara dodak, hasárdána,* PB.; *Nayeti, nayata,* BOMB.; *Mathi-dudhi,* MAR.; *Sittrapaladi, chin-amam, patcha arise* (Balf., Enc.), TAM.; *Reddi vári mánu bála, biduru nána biyyam,*[*] TEL.; *Bin-dada kuriya,* SING.; *Racta-vinda-chada* (O'Shaughnessy) SANS.; *Hasárdánah,* PERS.

References.—*Roxb., Fl. Ind., Ed. C.B.C., 394; Voigt., Hort Sub. Cal., 163; Dalz. & Gibs., Bomb. Fl., 227; Stewart, Pb. Pl., 195; Aitchison, Cat. Pb. and Sind Pl., 131; Elliot, Fl. Andhr., 27 & 164; Rheede, Hort. Mal., X., t. 33; O'Shaughnessy, Beng. Dispens., 565; Dymock, Mat, Med. W. Ind., 693; S. Arjun, Bomb., Drugs., 123; Murray, Pl., and Drugs, Sind, 33; Irvine, Mat. Med. Patna, 27; Drury, U. Pl., 206; Balfour, Cyclop., I., 1062; Treasury of Bot., 477.*

Habitat.—A prostrate hairy annual, common throughout the greater part of India and Ceylon; ascending in Kashmír up to 5,500 feet; often a conspicuous object as a weed on gravel walks.

Medicine.—The JUICE of this plant is known to be a violent purgative. The dried LEAVES and SEEDS are aromatic and astringent, and used in native practice in diarrhœa and dysentery of children along with butter milk (*Murray*). Irvine (*Mat. Med., l.c.*) says that it is common everywhere, and is used as a stimulant and laxative. In the Southern Concan, according to Dymock, the juice is used for the cure of ringworm (hence the name *nayeti*); and mixed with chloride of ammonium, to cure dandriff. O'Shaughnessy says that the juice of the stalks and flowers is a violent purgative; that the fresh plant is, by the Arabs, applied to wounds, and the leaves and seeds given by the Tamúls in cases of worms and in the

---

* Elliot remarks :—"This is a very doubtful name. It is, however, a Telegu word, and has the signification of 'green or raw rice of *Biduru*.' It may, however. be merely a misprint of *Reddi-vári nánu-pála*. But on the other hand the term 'raw rice,' or *pachchi arise*, is applied to several of the smaller species of **Euphorbia** in the Tamil tongue" (*Fl. Andhrica, p. 27*).

E. 552

bowel affections of children. **Rev. A. Campbell** mentions that the root is used by the Santals in Amenorrhœa.

Fodder.—Eaten by camels and goats in the Multán District.

E. Tirucalli, *Linn. ; Fl. Br. Ind., V., 254.*

> MILK-HEDGE, MILK-BUSH, OR INDIAN TREE—SPURGE, *Eng.*
>
> **Vern.**—*Sehnd, sendh, konpal-sehnd, shir thohar, sehund,* (BOMB., **Dr. Dymock**), *Sehud, sehnr,* HIND. ; *Lanka sij, láta-dáona,* BENG. ; *Siju,* SANTAL ; *Seju, lodhoka-sijhu, ksharisiju, lanka,* URIYA ; *Thora,* SIND ; *Send, hári-ki-send, bár-ki-send,* DECCAN ; *Shera, thora, thor, seyr, tej, niwal,* BOMB. ; *Nival,* GOA ; *Shera, seyr-teg, vajraduhu,* MAR. ; *Thordandalio,* GUZ. ; *Tirukali, kalli, kombu-kalli,* TAM. ; *Jemudu, jemudu-kádalu, káda-jemudu, kalli-chemudu, manche, koyya-jemudu kad-jemudu,* TEL. ; *Bonta-kalli, newli,* KAN. ; *Tirukalli, Kol-kalli,* MALAY ; *Sha-shoung-leknyo, sha-soung-lek-hnyo,* BURM. ; *Nawa-handi, navahandi, thovar,* SING. ; *Zaqqume-hindi, asfur sukkum (Balf. Enc.),* ARAB. ; *Zaquniyæ, hindi, shir tothar (Balf. Enc.),* PERS.
>
> **References.**—*Roxb., Fl. Ind., Ed. C.B.C., 393 ; Voigt, Hort. Sub. Cal., 162 ; Brandis, For. Fl., 439 ; Kurz, For. Fl. Burm., II., 417 ; Beddome, For. Man., 217 ; Gamble, Man. Timb., 368 ; Aitchison, Cat. Pb. and Sind Pl., 133 ; Elliot, Fl. Andhr., 36, 73 ; Rheede, Hort. Mal., t. 44 ; Trimen, Hort. Zeyl., 71 ; Pharm. Ind., 204 ; O'Shaughnessy, Beng. Dispens., 563 ; Moodeen Sheriff, Supp. Pharm. Ind., 137 ; Dymock, Mat. Med. W.Ind., 694 ; S. Arjun, Bomb. Drugs, 124 ; Murray, Pl. and Drugs Sind, 33 ; Irvine, Mat. Med. Patna, 62 ; Drury, U. Pl., 206 ; Lisboa, U. Pl. Bomb., 2, 114, 268, 273 ; Birdwood, Bomb. Pr., 271, 336 ; Liotard, Dyes, 11 ; Watson, Report on Gums, 28 ; Gums and Resinous Prod. of India, P. W. D., (1871) 28 ; Balfour, Cyclop., I., 1052 ; Treasury of Bot., 478 ; Kew Off. Guide to Bot. Gardens and Arboretum, 115 ; Home Dept. Cor. regarding Pharm. Ind., 240 : Bomb. Gaz., XX., 69.*

Habitat.—A small tree, with round stems and smooth branches. A native of Africa, but has become naturalized in the drier parts of Bengal, the Deccan, South India, and Ceylon ; elsewhere, it is largely cultivated for hedges, and in Berar is much grown to shelter young mango plants from direct sunlight.

Gum.—Dr. Riddell, writing of this plant, says that the milk, when it "hardens after boiling, becomes brittle ; whilst warm it is as ductile as mudar gutta-percha." The juice is, however, very difficult to deal with, as it causes excruciating pain if it gets into a cut in the skin or into the eye. On this account it is said to be used criminally to destroy the eyes of certain domesticated animals.

Dye.—The ASHES are employed in Southern India as a mordant. Dr. Bidie, however, states (quoted by Liotard) that it is not, properly speaking, a dye-yielding plant, but that it is burnt, and the ashes form an ingredient of the red dye with *chay* root. (see **Oldenlandia**.)

Medicine.—The JUICE of this plant is used as a warm remedy in rheumatism, tooth-ache, and debility. The MILK is said to cure affections of the spleen, and to act as a purgative in colic. Externally it is a vesicatory. It is also cathartic, emetic, and antisyphilitic. According to Irvine (*Mat. Med.*) the acrid juice is applied externally to ulcers.

SPECIAL OPINIONS.—§ "A good application in neuralgia" (*Surgeon-Major G. Y Hunter, Karachi*). "Fluckiger has separated *Euphorbon* from **E. Tirucalli** and **E. Cattimandoo** ; it is probably present in the other Indian **Euphorbias**" (*W. Dymock, Bombay*).

Fodder.—Goats and camels eat both the leaves and the bark.

Structure of the Wood.—Attains a height of 20 feet ; the wood is white, close-grained, and strong ; is used for rafters, &c. Also used for veneering purposes and for toys. "Its wood produces a good charcoal for the manufacture of blasting powder" (*Shuttleworth, Consvr. Forests, Bombay*).

Margin notes:

FODDER.
553

GUM.
554

DYE.
Ashes.
555

MEDICINE.
Juice.
556

Milk.
557

FODDER.
558
TIMBER.
559

EURYA symplocina.	Catimandoo Cement.

DOMESTIC.
560

Domestic Use.—Extensively employed as a hedge plant. Dr. Lisboa states that in the Southern Marátha country and in Goa, the milk is made use of for poisoning fish. The Conservator of Forests, Southern Circle, Madras, says that the bark yields a good charcoal, which is in great demand amongst the blacksmiths and chunam-burners of the Coimbatore District.

561

Euphorbia trigona, *Haworth ; Fl. Br. Ind., V., 256 ; Wight, Ic.,*
    Syn.—E. CATTIMANDOO, *W. Elliot.*   [*1863.*
    Vern.—*Katti-mandu* (Knife medicine), TEL.
    References.—*Roxb., Fl. Ind., Ed. C.B.C., 393 ; Voigt, Hort. Sub. Cal., 162 ; Brandis, For. Fl., 438 ; Beddome, For. Man., 216 ; Gamble, Man. Timb., 368 ; Elliot, Fl. Andhr., 89 ; Moodeen Sheriff, Supp. Pharm. Ind., 137 ; Drury, U. Pl., 204 ; Balfour, Cyclop., I., 1061 ; Smith, Dic., 98 ; Treasury of Bot., 477.*
    Habitat.—An erect, glabrous shrub, with branches acutely 3-5 winged. It inhabits dry rocky hills in the Deccan, and probably other parts of India.

GUM.
Milk.
562

    Gum.—The MILK yields a cement, which is largely used by the country people for fixing knives, &c., into handles, and for similar purposes.
    "Fluckiger has obtained from this plant, as also from E. Tirucalli, *Euphorbon,* the active principle of the officinal *Euphorbium,* and it is probable that most of the Indian species will yield a gum of the same properties as commercial *Euphorbium*" (*Dr. Dymock*).

EURYA, *Thunb.; Gen. Pl., I., 183.*

563

Eurya acuminata, *DC. ; Fl. Br. Ind., I., 285 ;* TERNSTRŒMIACEÆ.
    Vern.—*Sanujhingni,* NEPAL ; *Flotungchoug,* LEPCHA.
    References.—*Brandis, For. Fl., 24 ; Kurz, For. Fl. Burm., I., 101 ; Gamble, Man. Timb., 28 ; Thwaites, En. Ceylon Pl., 41 ; Royle, Ill., Him. Bot., 127, t. 25.*
    Habitat.—A small evergreen tree or shrub of the hills of the North-Eastern Himálaya, from Kumaon to Bhotan and Martaban, on altitudes of from 3,000 to 8,000 feet.

TIMBER.
564

    Structure of the Wood.—Differs from that of **E. symplocina** in having the larger medullary rays less broad and less prominent. Weight, 32 to 47℔ per cubic foot.

565

E. japonica, *Thunb. ; Fl. Br. Ind., I., 284.*
    Vern.—*Baunra, gonta, deura,* HIND.; *Jhingni,* NEPAL ; *Tungchung,* LEPCHA ; *Baunra, gonta, deura,* BOMB.; *Hoolooni,* NILGIRIS ; *Tounglet-pet,* BURM. ; *Neya-dasse,* SING.
    References.—*Voigt, Hort. Sub. Cal., 91 ; Brandis, For. Fl., 24 ; Kurz, For. Fl. Burm., I., 101 ; Beddome, Fl. Sylv., t. 92 ; Gamble, Man. Timb., 28 ; Thwaites, En. Ceylon Pl., 41 ; Trimen, Hort. Zeyl., 7 ; Atkinson, Him. Dist., 306 ; Lisboa, U. Pl. Bomb., 13 ; Balfour, Cyclop., I., 1064.*
    Habitat.—A shrub or small tree, found on the Himálaya from the Jumna eastwards, above 3,000 feet in altitude ; it also occurs in the Western Gháts, and in Burma.

TIMBER.
566

    Structure of the Wood.—Brown, soft, close-grained. It is sometimes used for fuel.

567

E. symplocina, *Blume ; Fl. Br. Ind., I., 284.*
    Vern.—*Barajhingni, kisi,* NEPAL ; *Flotungchoug,* LEPCHA.
    References.—*Kurz, For. Fl. Burm., I., 102 ; Gamble, Man. Timb., 28.*
    Habitat.—A small evergreen tree of the hills of the North-Eastern Himálaya, from 5,000 to 7,000 feet ; also found in Burma.

TIMBER.
568

    Structure of the Wood.—Reddish-white, soft, close-grained. Used only for firewood.

E. 568

EURYALE, *Salisb.; Gen. Pl., I., 47.*

Euryale ferox, *Salisb.; Fl. Br. Ind., I., 115; Bot. Mag., 1447;*   **569**
    The Gorgon Fruit. [Nymphæaceæ.

    Syn.—Anneslea spinosa, *Roxb.;* Euryale indica, *Planch.*
    Vern.—*Makhana,* Hind.; *Makhana,* Beng.; *Kunta pudena,* Uriya; *Jewar,* Pb.; *Mallani padman,* Tel.; *Padma* (? Nelumbium) U. C. Dutt gives *Makhánna,* Sans.
    References.—*Roxb., Fl. Ind. Ed., C.B.C., 427; Voigt, Hort. Sub. Cal., 8; Stewart, Pb. Pl., 8; Le Maout and Dexcaisne, Dscrip. & Ajnal. Bot., 212 (Eng. Ed.); Elliot, Fl. Andhr., 126; O'Shaughnessy, Beng. Dispens., 622; U. C. Dutt, Mat. Med. Hind., 110, 308; Dymock, Mat. Med. W. Ind., 2nd Ed., 38; S. Arjun, Bomb. Drugs, 7; Drury, U. Pl., 207; Balfour, Cyclop., I., 1064; Smith, Dic., 196; Treasury of Bot., 479; Kew Off. Guide to Bot. Gardens and Arboretum, 24, 57.*
    Habitat.—A stemless, aquatic plant of the sweet-water lakes and ponds of East Bengal, Assam, Manipur, Oudh, and Kashmír. It is said to have been cultivated in China for upwards of 3,000 years. It has circular prickly leaves, 2 to 3 feet in diameter, which float on the surface of the water. The flowers are blue, violet or bright red and green on the outside. The fruits are round and prickly, of the size of an orange, and on ripening, they swell out in various places by the growth of the seeds within.

    Medicine.—Roxburgh says that the Hindus "consider the seed as   **MEDICINE.**
possessed of powerful medicinal virtues, such as restraining seminal gleet,   Seed
invigorating the system, &c." A light and invigorating food suited for   **570**
the sick (*Dutt*).

    Food.—The seeds, which are black in colour and of the size of peas,   **FOOD.**
are farinaceous. They are sold in the bazars of Eastern Bengal and eaten   Seeds.
by the natives who consider them light and easily digestible. They are   **571**
largely used in Manipur as an article of food; the women sitting on the roadsides sell the spiny fruits along with betel nuts, *singara* nuts, &c. (*Watt*) Roxburgh describes the process of cooking the seeds, which consists in roasting them in hot sand. They swell and burst, when the seed-coat is easily removeable.

EURYCOMA, *Jack.; Gen. Pl., I., 312.*

Eurycoma longifolia, *Jack.; Fl. Br. Ind., I., 521;* Simarubeæ.   **572**
    Vern.—*Penvar-pet,* Malay.
    References.—*Kurz, For. Fl. Burm., I., 202; Gamble, Man. Timb., 63; Pharm. Ind., 50; Moodeen Sheriff, Supp. Pharm. Ind., 138; Treasury of Bot., 479.*
    Habitat.—A small tree of the Malayan Peninsula and Archipelago.
    Medicine.—The bark and root of this tree possess bitter properties.   **MEDICINE.**
A decoction of the root is a remedy in intermittent fevers, and as a febri-   Bark & Root
fuge stands in the opinion of Mr. Oxley (1850) next to quinine (*Pharm.*   **573**
Ind.).

EUXOLUS, *Raf.*—an old generic name for the species of **Amarantus,** see
Vol. I., p. 208.   **574**
    The classification of the species under **Amarantus** in Vol. I. having been prepared before the publication of Sir Joseph Hooker's monograph of Amarantaceæ in Vol. IV., 718 to 722, of the *Flora of British India,* a revised list of the Indian species, with some of the more important synonyms and references, may be found useful.
        * *Bracts setaceous or awned, exceeding the five sepals; Stamens five; utricle circumsciss; top 2 to 3-fid.*

1. **Amarantus spinosus,** *Linn.; Roxb., Fl., Ind., Ed. C.B C., 663; Grah., Cat. Bomb. Pl., 169; Dalz. & Gibs., Bomb. Fl., 216; Wight, Ic, t. 513.* Waste ground throughout India and Ceylon. *(Sir J. D. Hooker, in Fl. Br. Ind., IV., 718).*

2. **A. paniculatus,** *Linn.; Dalz. & Gibs., Bomb. Fl., 215.* **A. frumentaceus,** *Ham.; Roxb., Fl. Ind, Ed. C.B.C., 663; Wight, Ic., t. 720.* **A. Anacardana,** *Ham.* (A. Anardana). Cultivated throughout India and Ceylon, and up to 9,000 feet on the Himálaya. " Like the following, of which it may be a form, the seeds vary extraordinarily in size, form, and colour " *(Sir J. D. Hooker, l.c., 719.)*

3. **A. caudatus,** *Linn.;* **A. cruentus,** *Willd.; Roxb., Fl. Ind., Ed. C.B.C., 663.* Cultivated in various parts of India. " I find it very difficult to distinguish some states of this from **A. paniculatus.** In its typical state it is a smaller plant with the leaves obtuse at the tip, more globose softer masses of smaller red green or white flowers on the thyrse, the terminal spike of which is very long, thick, and drooping." *(Sir J. D. Hooker, l.c., 719.)*

    * * *Bracts subulate, equalling or exceeding the three lanceolate sepals and utricle. Stamens three; utricle circumsciss.*

4. **A. gangeticus,** *Linn.; Roxb., Fl. Ind., Ed. C.B.C., 662.* **A. tricolor,** *Linn.; Roxb., l.c., 663.* **A. lanceolatus,** *Roxb., l.c., 662.* **A. tristis,** *Linn.; Roxb., l.c., 661; Grah., Cat Bomb. Pl., 169; Wight, Ic., t. 713; Dalz. & Gibs., Bomb. Fl., 215.* **A. oleraceus,** *Roxb., l.c., 662; Wight, Ic., t. 715; Thwaites, En. Ceylon Pl., 247 (not of Linnæus),* A. polygamus, *Roxb., l.c., 661; Wight, Ic., t. 714.* **A. lividus,** *Roxb., l.c., 662.* **A. melancholicus,** *Linn.; Roxb., l.c., 663.* Throughout India and Ceylon. Cultivated or found on cultivated ground. " This is **Roxburgh's A. tristis,** and possibly that of **Linnæus,** but the latter describes the leaves as ovate-cordate, which these are not. **Roxburgh** says that his **gangeticus** and **oleraceus** differ from his **polygamus** and **tristis** and their varieties, in not admitting of being cut for successive crops, but being hence unrooted for market " *(Sir J. D. Hooker, l.c., 720).*

5. **A. mangostanus,** *Linn.* A. polygamus, *Thw. En. Ceylon Pl., 247.* Throughout India and Ceylon, in cultivated ground.

6. **A. Caturus,** *Heyne.* Deccan Peninsula. *(Sir J. D. Hooker, l. c., 720.)*

    * * * *Bracts usually shorter than the two or three sepals and utricle; stamens two or three; utricle indehiscent or circumsciss.*

7. **A. viridis,** *Linn; Roxb., Fl. Ind., Ed. C.B C., 661; Grah., Cat. Bomb. Pl., 169.* **A. fasciatus,** *Roxb., l.c., 663; Wight, Ic., t. 717.* **Euxolus caudatus,** *Moq.; Wight, Ic., t. 1773.* Waste places throughout India. " **A. fasciatus,** *Roxb.,* is a sport, with a pale crescentic band across the leaf." *(Sir J. D. Hooker, l.c., 721)*

8. **A. Blitum,** *Linn.* *(Sir J. D. Hooker, l. c., 721.)*

Var.—A. oleraceus, *Linn.;* E. oleraceus, *Dalz. & Gibs., Bomb. Fl., 216.* Cultivated in India and elsewhere.

Var.—A. sylvestris, *Desf.* Kashmír, 4,000 to 6,000 feet *(Thomson).*

9. **A. polygamus,** *Linn (not of Roxburgh); Thwaites, En. Ceylon Pl., 247.* **A. polygonoides,** *Roxb., Fl. Ind., Ed. C.B.C., 661; Wight, Ic., tt. 512, 719.* **Amblogyna polygonoides,** *Dalz. & Gibs, Bomb. Fl., 219.* **Euxolus polygamus,** *Moq.; Thwaites, En. Ceylon Pl., 248.* Throughout India and Ceylon. " I believe that this can only be ranked as a form of **A. Blitum,** with small usually abovate apiculate leaves, fewer flowers in a cluster, often larger, more subulate sepals and smaller, more acute utricles."

Var.—angustifolia. Occurs in the Panjáb and Carnatic.

10. A. tenuifolius, *Willd.; Roxb. Fl. Ind., Ed. C.B.C.,* 660 *; Wight, Ic., t. 718.* **Mengea tenuifolia,** *Moq.; Dals. & Gibs., Bomb. Fl.,* 218. Bengal, Gangetic Valley, and Panjab. (*Sir J. D. Hooker, l. c., 722.*)

EVODIA, *Forst.; Gen. Pl., I., 296.*

Evodia fraxinifolia, *Hook. f.; Fl. Br. Ind., I., 490;* RUTACEÆ.   575

Vern.—*Kan·koa,* NEPAL; *Kanú,* LEPCHA.
Reference.—*Gamble, Man. Timb.,* 60.
Habitat.—A small tree of the Eastern Himálaya in Sikkim, between 4,000 and 7,000 feet, and of the Khásia Hills from 3,000 to 5,000 feet. It is said to emit a strong scent of caraway when bruised.
Structure of the Wood.—White, soft; used only for posts of huts.   TIMBER. 576

E. Roxburghiana, *Benth.; Fl. Br. Ind., I., 487; Wight, Ic., t. 204.*   577

Syn.—E. TRIPHYLLA, *Beddome;* FAGARA TRIPHYLLA, *Roxb.;* ZANTHOXYLUM TRIPHYLLUM, *Thwaites.*
Vern.—*Nebede, lunu, ankenda,* SING.
References.—*Roxb., Fl. Ind., Ed. C.B.C.,* 139 *; Kurz, For. Fl. Burm., I.,* 180 *; Gamble, Man. Timb.,* 60 *; Thwaites, En. Ceylon Pl.,* 69 & 409 *; Dals. and Gibs., Bomb. Fl.,* 45 *; Grah., Cat. Bomb. Pl.,* 36 *; Lisboa, U. Pl. Bomb.,* 30.
Habitat.—A small tree found in the Khásia Hills, South India, Tenasserim, and the Andaman Islands; also met with in Ceylon.
Structure of the Wood.—Greyish-brown, moderately hard.   TIMBER. 578

E. triphylla, *DC.; Fl. Br. Ind., I., 488.*   579

References.—*Kurz, For. Fl. Burm., I.,* 180 *; Gamble, Man. Timb.,* 60.
Habitat.—A small tree much resembling **E. Roxburghiana.** It inhabits damp localities in Burma and the Andaman Islands, Japan, China, and Borneo.
Structure of the Wood.—"Light, soft, pale-pinkish, close-grained, straight, fibrous, with silvery lustre" (*Gamble*).   TIMBER. 580

EVOLVULUS, *Linn.; Gen. Pl., II., 875.*   581

Evolvulus alsinoides, *Linn.; Fl. Br. Ind., IV., 220;* CONVOLVU-
[LACEÆ.

Syn.—E. HIRSUTUS, *Lam.;* E. ANGUSTIFOLIUS, *Roxb.*
Vern.—*Tandi-kode-baha,* SANTAL; *Sankhpushpi,* PB.; *Shankhávalli,* BOMB.; *Vishnu-karandi,* TAM.; *Vishnu-kranta* TEL.; *Vishnukranti,* KAN.; *Visnú-kraanta.* SING.; *Vishnugandhi,* SANS.
References.—*Roxb., Fl. Ind., Ed. C.B.C.,* 276 *; Voigt, Hort. Sub. Cal.,* 363 *; Thwaites, En. Ceylon Pl.,* 213 *; Dals. & Gibs., Bomb. Fl.,* 162 *; Stewart, Pb. Pl.,* 150 *; Elliot, Fl. Andhr.,* 128, 193 *; Rheede, Hort. Mal., XII., t.* 64 *; Dymock, Mat. Med. W. Ind., 2nd Ed.,* 564 *; S. Arjun, Bomb. Drugs,* 94 *; Bidie, Cat. Raw Pr., Paris Exh.,* 56 *; Atkinson, Him. Dist.,* 314 *; Balfour, Cyclop., I.,* 1057.
Habitat.—A prostrate perennial herb, with wiry stems, and blue or white flowers; found nearly all over India on dryish ground.
Medicine.—Muhammadan physicians believe that this PLANT has the power of strengthening the brain and memory. It is also extensively used as a febrifuge and tonic. Ainslie says that it is given in bowel complaints. In the Vedic period it was believed to possess the power of promoting conception (*Dymock*).   MEDICINE. Plant. 582

SPECIAL OPINIONS.—§ "The ROOTS used in intermittent fever of children" (*Rev. A. Campbell*). "The LEAVES are made into cigarettes and smoked in chronic bronchitis and asthma. The plant is astringent, useful in internal hœmorrhages" (*Surgeon-Major J. M. Hunston, Travancore, and John Gomes, Medical Storekeeper, Travandrum*). "The blue-coloured   Roots. 583 Leaves. 584

X

E. 584

flowered form is called *Vishnugrandie.* The other kind has white flowers and is called *Sivagrandie*" (*V. Ummegudiem, Mettapollian, Madras*).

EXACUM, *Linn.; Gen. Pl., II., 803.*

585   **Exacum bicolor,** *Roxb.; Fl. Br. Ind., IV., 96; Wight, Ic., t. 1321;*
   Vern.—*Bará-charáyatah,* HIND.   [GENTIANACEÆ.
   References.—*Roxb., Cat. Pl.* (*1813*) *; Dalz. & Gibs., Bomb. Fl., 156,*
   (*Syn., excl.*) *; Dymock, Mat. Med. W. Ind., 2nd Ed., 540 ; Drury, U. Pl.,*
   208 ; Lisboa, U. Pl. Bomb., 262 ; Balfour, Cyclop., I., 1067 ; Clarke, in
   Journ. Linn. Soc., XIV., 425.
   Habitat.—An erect herbaceous plant, 1 to 2 feet high, frequent in the
Deccan Peninsula. Flowers large, white tipped with blue.

MEDICINE.   Medicine.—The dried STALKS are sold in South India under the name
Stalks. *Country Kariyát.* The plant possesses tonic and stomachic properties,
586 and may well be substituted for *Gentian* (*Pharm. Ind.*).

587   **E. pedunculatum,** *Linn.; Fl. Br. Ind., IV., 97 ; Wight, Ic., t. 336.*
   Syn.—E. SULCATUM, *Roxb.*
   References.—*Roxb., Fl. Ind., Ed. C.B.C., 134; Voigt, Hort. Sub. Cal.,*
   520 ; Thwaites, Enum. Ceylon Pl., 203 ; Pharm. Ind., 150; Drury,
   U. Pl., 209 ; Clarke, in Journ. Linn. Soc., XIV., 427 ; Edgeworth, Cat.
   Pl. Banda, 51 (**E. rivulare**).
   Habitat.—A small herb, usually under a foot in height, found through-
out India, ascending to 3,000 feet from Oudh and Bengal to Ceylon.
MEDICINE.   Medicine.—The PLANT is less bitter than *Chiretta,* and more so than
Plant. *Gentian,* for which it may be substituted.
588
589   **E. tetragonum,** *Roxb.; Fl. Br. Ind., IV., 95.*
   Vern.—*Titakhana, ava* (purple) *chiretta,* HIND.; *Koochuri,* BENG.
   References.—*Roxb., Fl. Ind., Ed. C.B.C., 133 ; Voigt, Hort. Sub. Cal.,*
   520 ; Grah., Cat. Bomb. Pl., 123 ; Royle, Ill. Him. Bot., I., 277 ; Pharm.
   Ind., 149 ; O'Shaughnessy, Beng. Dispens., 460 ; Irvine, Mat. Med.,
   Patna, 81 ; Balfour, Cyclop., I., 1067 ; Clarke, in Journ. Linn. Soc.,
   XIV., 424.
   Habitat.—An erect herbaceous plant, 1 to 4 feet high, with deep-blue
flowers, found in North India, ascending to 5,000 feet, common from
Garhwál to Central India, Bhután, and the Khásia Mountains; also in
Bombay, Salsette, Khandalla, Morung, Wurgaum, and Bengal.
MEDICINE.   Medicine.—The plant is used as a tonic in fevers and a stomachic bitter
590 (*Pharm. Ind.*).

EXCÆCARIA, *Linn.; Gen. Pl., III., 337.*

   The name is said to be derived from *Exccæco,* because of the powerfully
acrid juice, especially that of **E. Agallocha,** which causes blindness if applied
to the eyes.

591   **Excæcaria acerifolia,** *Didrichs.; Fl. Br. Ind., V., 473;* EUPHORBIACEÆ.
   Vern.—*Básingh,* KUMAON.
   References.—*Brandis For. Fl., 441 ; Ind. For., XI., 5.*
   Habitat.—An evergreen shrub, or small tree found up to 6,000 feet in
Kumáon, Nepál, and on the Khásia Hills.
MEDICINE.   Medicine.—The Bhutias inhabiting East Kumáon use the LEAVES of this
Leaves. plant as a remedy for rheumatism.
592
593   **E. Agallocha,** *Linn.; Fl. Br. Ind., V., 472 ; Wight, Ic., t. 1865 B.*
   THE BLINDING TREE.
    Vern.—*Gangwa, geor, uguru, geria, goria,* BENG.; *Gnua,* URIYA; *Geva,*
    BOMB.; *Chilla, tella-chettu,* TEL.; *Haro,* KAN.; *Tayau, kayau,* BURM.;
    Yekin, ANDAMANS; *Tella kwiya,* SING.

  E. 593

| The Blinding Tree. | (*J. F. Duthie.*) | EXOGONIUM Purga. |

References.—*Roxb., Fl. Ind., Ed. C.B.C.,713; Voigt, Hort. Sub. Cal., 161; Brandis, For. Fl., 442; Kurz, For. Fl. Burm., II., 414; Beddome, For. Man., 255; Gamble, Man. Timb., 368; Dals. & Gibs., Bomb. Fl., 227; Rheede, Hort. Mal., V., t., 45; Elliot, Fl. Andhr., 175; Rumph., Amb., II., t. 79, 80* (Arbor excœcans); *Ainslie, Mat. Ind., II., 438; O'Shaughnessy, Beng. Dispens., 563; Dymock, Mat. Med. W. Ind., 2nd Ed., 676; Drury, U. Pl., 209; Lisboa, U. Pl. Bomb., 125; Birdwood, Bomb. Pr., 345; Balfour, Cyclop., I., 1067; Smith, Dic., 5; Treasury of Bot., 483; Kew Reports (1877), 42; Mason, Burma and Its People, 762; Bomb. Gaz., XV., 443.*

Habitat.—A small evergreen tree of the Coast and tidal forests of India, Burma, and the Andaman Islands. The famous *Agallochum* or *Aloes* of the Old and New Testament, formerly supposed to be the product of this tree, is yielded by **Aquilaria Agallocha,** which belongs to an entirely different family, the THYMELACEÆ. See Vol. I., p. 279.

Gum.—The wood contains a poisonous sap, which hardens into a black caoutchouc-like substance. The fresh sap is extremely acrid, and causes intolerable pain if it accidentally gets into the eyes, and which sometimes happens to the woodcutters when the tree is cut for fuel; "hence," says Balfour, "Rumphins' name 'Excœcans'." The Conservator of Forests, Southern Circle, Madras, writes that a species, supposed to be this one, is known in Travancore as the Tiger's milk-tree; it blisters the skin, and the juice coagulates when stirred.

GUM. 594

Medicine.—In Fiji it is employed for the cure of leprosy, its mode of application being very singular. The body of the patient is first rubbed with green leaves; he is then placed in a small room and bound hand and foot, when a small fire is made of pieces of the wood of this tree from which rises a thick smoke; the patient is suspended over this fire, and remains for some hours in the midst of the poisonous smoke and under the most agonizing torture, often fainting. When throroughly smoked, he is removed, and the slime is scraped from his body; he is then scarified and left to await the result. In some cases he is cured, but frequently the patient dies under the ordeal. (*Smith, Econ. Dic., 5.*)

MEDICINE. 595

Structure of the Wood.—White, very soft, and spongy. Grows occasionally to 5 feet in girth and 40 feet in height, though generally cut for posts when of small girth. It is a useful wood for general carpentry purposes, such as toys, bedsteads, tables, &c. Roxburgh remarks that it is only used for charcoal and firewood.

TIMBER. 596

Domestic Use.—Fishing floats are made from the roots of the tree.

DOMESTI 597

E. baccata, *Müll.-Arg.,* see Sapium baccatum.

E. indica, *Müll.-Arg.,* see Sapium indicum.

E. insignis, *Müll.-Arg.,* see Sapium insigne.

E. sebifera, *Müll.-Arg.,* see Sapium sebiferum.

Exogonium Purga, *Benth.,* see Ipomæa Purga.

| FAGONIA arabica. | The Field or Broad Bean: Fagonia. |

(G. Watt.)

FABA, *Tourn.; Gen. Pl., I., 525.*

1

Faba vulgaris, *Mœnch.;* LEGUMINOSÆ.

THE BROAD BEAN.

Vern.—*Káiún,* KASHMIR; *Chástang,* SUTLEJ; *Nakshan,* LADAK. *Bákla,* a name given to it in the plains and lower hills of India.

Habitat and Area of Indian Cultivation.—The *Flora of British India* does not allude to this plant, from which fact the inference is unavoidable that it is not regarded as a native of India. But introduced cultivated plants are usually described in the *Flora,* and the absence of any notice of the Field or Broad Bean may be assumed as an indication that it is supposed to be scarcely, if at all, cultivated in this country. It is, however, to a considerable extent, cultivated on the Himálaya, and in Kashmír and Ladak may be regarded as a regular crop. De Candolle says it has no Sanskrit name nor any modern Indian name. From this circumstance he infers that it is of modern introduction into India. The vernacular names given above would, however, seem opposed to this opinion. Mr. Atkinson states that it is cultivated in Kumáon up to 8,000 feet, and that there are two or three varieties raised from "introduced and native seed." Mr. Baden Powell refers to its cultivation in Kashmír and Peshawar. Balfour goes even still further, and affirms that it "is found wild in the Sutlej valley, between Rampur and Sungnam, at an elevation of 8,000 to 14,000 feet." Stewart, while not supporting the verdict that it is a native of the Sutlej valley, speaks of it as a regular crop, adding that "beans are ground into flour for food, and are, on the Sutlej, given to cattle." In the Settlement Report of the Kángra District it is alluded to as a regular spring crop. The Director of Land Records and Agriculture in Bengal, replying to an enquiry regarding this plant, reports that it is "not yet grown as a field crop in the Lower Provinces." The Director in Burma, on the other hand, states that "in Pegu District it is cultivated by the Chinese and Shan gardeners in moderate quantities, but has not been taken up as a field crop. This vegetable finds a ready sale in the market. The plant is said to thrive on any land which can be cultivated during the dry season." In the *Indian Forester (Vol. IX., p. 452)* will be found an interesting note on its cultivation in the North-West Provinces. See also in the *Jours. Agri.-Hort. Soc., IV., 7; V., 37.*

For further information consult the article **Vicia.**

FAGONIA, *Linn.; Gen. Pl., I., 267.*

A genus of branching woody herbs, of so variable a nature that it is difficult to fix the number of species. Two occur in India—one in the North-West, to Peshawar, distributed to Algeria; the other also occurs over Northern India, but shows in India a more westerly tendency, being dispersed through the Panjáb and Sind to Bombay. It is often difficult to determine to which of these species writers on Economic Botany allude, and the statements made below may therefore have to be rearranged in the future.

2

Fagonia arabica, *Linn.; Fl. Br. Ind., I., 425;* ZYGOPHYLLEÆ.

Syn.—FAGONIA MYSORENSIS, *Roth.; Wall., Cat., 6853;* F. CRETICA, var. ARABICA, *Dals. & Gibs.*

Vern.—*Usturgar, ústarkhár,* HIND.; *Jowasa* (Ajmere), RAJ.; *Drummahú* (or *drammaho*), SIND.; *Dhamásá, dumaso,* MAR.; *Dhamaso,* GUZ.; *Dusparsha,* SANS.; *Báddvard,* PERS.

References. *Dals. & Gibs, Bomb. Fl., 45; Aitchison, Cat. Pb. and Sind Pl., 27; Aitchison's Report Del. Cor. Afg., 44; Pharmacog. Ind., I., 246; Dymock, Mat. Med. W. Ind., 2nd Ed., 120; S. Arjun, Bomb. Drugs, 27; Murray, Pl. and Drugs, Sind, 91; Baden Powell, Pb. Pr., 335; Stocks, Account of Sind; List of Drugs exhibited by Baroda Durbar*

F. 2

at·Cal. Inter. Exh.; Gasetteers, Mysore and Coorg, I., 56; Agra, IV., LXIX; Ina. For., XII. (App.), 2, 8.

Habitat.—Throughout North-West India, Sind, the Panjáb, and the southern provinces of the Western Peninsula. Spines shorter than the linear leaflets.

Medicine.—Dr. Stocks was the first writer apparently who made the medicinal properties of this plant known to Europeans. He says: "The LEAVES and TWIGS are supposed to have cooling properties, and, according to the Arabian system of medicine, must be good against all disorders arising from heat (external and internal). They are much used as preventatives in the hot weather to keep the system cool, and ward off disorders incident to that season." The authors of the *Pharmacographia Indica* write that it is used in Sind and Afghánistan as a popular remedy for fever among the hill people. Many writers allude to the reputation which the leaves possess as an external application to abscess from thorns. Dalzell and Gibson believed this to be "fanciful," but in the report on the Baroda drugs, shown at the Calcutta International Exhibition, the plant is said to have " a great reputation as a suppurative in cases of abscesses from thorns. An infusion is used as a gargle in sore-mouth." Dr. Dymock refers to the property of a suppurative in equally strong terms, adding, however, that "it is also used for cooling the mouth in stomatitis, the JUICE being boiled with sugar-candy until quite thick, and a small quantity allowed to dissolve in the mouth frequently; the juice is thought to prevent suppuration when applied to open wounds." Mr. Sakharam Arjun remarks that "Mr. Rahim Khan only mentions that this drug 'purifies the blood and acts as a deobstruent.'" Mr. Arjun adds (of F. **mysorensis**) that "it is largely used by the native practitioners as a bitter and astringent tonic."

MEDICINE.
Leaves.
3
Twigs.
4

Juice.
5

Fagonia Bruguieri, *DC.; Fl. Br. Ind., I., 425.*

6

Syn.—F. CRETICA, *var. ?; T. Anders.*

Vern.—*Damáhán* (or *dam-áhár*=carried by the wind), HIND.; *Spalaghsái, aghsái,* TRANS-INDUS; *Dhamá* (or *dhámáh), damá, damiyá, dramah, dhamánh,* PB. and SIND; *Dhamaso,* GUZ.; *Buduwurd* (=carried by the wind), PERS.

References.—*Stewart, Pb. Pl., 37; Baden Powell, Pb., Pr., 335; Settl. Rep., Montgomery, 20; Gas., Musaffugarh, 27; Gas., Agra (IV), LXIX.*

Habitat.—Found in North-West India to Peshawar, and distributed westward to Algeria. Spines exceeding the ovate leaflets.

Medicine.—"The PLANT is given as a febrifuge and tonic, and Bellew states that in the Peshawar valley it is administered to children as a prophylactic against small-pox" (*J. L. Stewart*). Baden Powell writes that it is "useful as an application to tumours: also in chronic fever, dropsy, and delirium, and in any disorder which arises from poisoning."

MEDICINE.
Plant.
7

[*A. No. 1665;* SALVADORACEÆ.

F. montana, *Miq.,* see **Azima tetracantha,** *Lam.; Dic. Econ. Prod., Vol. I.,*

FAGOPYRUM, *Gærtn.; Gen. Pl., III., 99.*

Fagopyrum cymosum, *Meissn.; Fl. Br. Ind., V., 55;* POLYGONACEÆ.

8

Syn.—FAGOPYRUM TRIANGULARE, *Meissn.;* F. EMARGINATUM, var. KUNAWARENSE, *Meissn.;* POLYGONUM CYMOSUM, *Treviran;* P. *not* TRIANGULARE, *Wall.;* P. EMARGINATUM, *Wall.;* P. DIBOTRYS, *Don.;* P. VOLUBILE, *Turcz.;* P. RUGOSUM, *Ham.*

Vern.—*Banogal* (Sutlej Valley), PB.

References.—*Stewart, Pb. Pl., 183; Atkinson, Him. Dist., 316.*

Habitat.—A tall, delicately-branched annual, growing on perennial roots. This appears to be the wild plant from which, perhaps both, or at least one

of the species of BUCKWHEAT has been derived. It occurs on the temperate
Himálaya, frequenting glades between 5,000 and 11,000 feet in altitude. It is
distributed from Kashmír to Sikkim and the Khásia Hills. Mr. Atkinson
calls this the *ban* (wild) *-ogal*, and adds that in Kumáon it occurs wild
on the lower hills.

FOOD.
9

Food and Fodder.—Although eaten as fodder by cattle, it is commonly
reported that this species is not used for any economic purpose. It is, however, so much like **F. esculentum** that it is often doubtful, when in flower,
whether the plants met with in glades, near fields, are truly wild, or only
escapes from cultivation.

10

Fagopyrum esculentum, *Mœnch.; Fl. Br. Ind., V., 55.*

THE BUCKWHEAT OR BRANK.

Syn.—FAGOPYRUM EMARGINATUM, *Meissn.*; POLYGONUM FAGOPYRUM,
Linn.; P. DIOICUM, *Ham. MS.*; P. EMARGINATUM, *Roth.*

Vern.—*Phaphra, kotu, kúltú,* HIND.; *Doron,* ASSAM; *Titaphapur* [Darjíl-
ing], NEPAL; *Bhe, pálti,* BHUTIA; *Kotu,* GARHWAL; *Pháphar, ogul,*
KUMAON; *Daráu, obal, phúlan, ogal, pháphar,* PB.; *Phaphra, úgla,
pagua, kathu, dhanphari,* SIMLA; *Bares katú,* KANGRA; *Káthu, brés,*
KULLU; *Tramba shirin,* KASHMIR.
See the note on the vernacular names of F. TATARICUM.

References.—*O'Shaughnessy, Beng. Dispens., 523; Church on Food-
grains of Ind., 114; Baden Powell, Pb. Pr., 244; Atkinson, Him. Dist.,
698; McCann, Dyes and Tans, Beng., 143; Crookes, Handbook, Dyeing
and Calico Printing, 412; Report, Nilghiri Hills, by W. R. Robertson,
22; Smith, Dic., 67; Settle. Report of Simla [App.], XLI; Settle. Report,
Kumaon [App.], 32nd; Settle. Report, Kangra, 25; Assam—Note on
Condition of the People of; W. R. Robertson in Report, Agri. Dept., Madras, 1878, pp. 136-137; Gazetteers: Kangra, I., 153; II., 57; Mysore and
Coorg, I., 65.*

Habitat.—Extensively cultivated on the Himálaya from Western
Tibet to Sikkim, the Khásia Hills, Manipur, and the Nilghiri Hills.

There would appear to be many very distinct varieties, some with white,
others with pink, flowers. All are more robust and stunted than **F. cymosum**;
but it seems probable that every intermediate condition exists between these
two species. A form occurs which seems to correspond to the **F. emarginatum** as described by Stewart, but the writer, not having the opportunity of
studying specimens of the various cultivated plants, can do no more
than suggest the necessity for such a study. When finally determined,
the vernacular synonyms will have to be rearranged. Indeed, so confused
are the names given to the forms of Buckwheat, that it is impossible to
assign distinctive vernacular terms for two so widely different plants as
F. esculentum and **F. tataricum.** The latter is a much coarser plant,
grows at higher altitudes, and the nut has the angles rounded off instead
of being sharp.

**CULTIVA-
TION.**
II

Cultivation.—On the Himálaya between 4,000 and 10,000 feet
F. esculentum is a rainy season crop, being sown in July and reaped in
October. The forms met with at lower elevations are stunted, and have
thick swollen stems of a red shining colour, with pink flowers. In experimental cultivation at the Saidapet Farm, Madras, Buckwheat from
Australian seed was sown on the 9th November: it was irrigated several
times, and yielded on the 21st January 167℔ of grain and 1,138℔ of
straw per acre. But Mr. Robertson did not apparently form a favourable opinion of Buckwheat as an auxiliary corn-crop. "We have,"
he adds, "several indigenous grain and pulse crops equal, for ordinary cultivation, to the Buckwheat, if only the ryots could be induced to manure
and cultivate better." Mr. Atkinson, speaking of Kumáon and Garhwál,
says that Buckwheat "is grown chiefly as a vegetable in the hills and is

The Buckwheats—Kotu.	(G. Watt.)	FAGOPYRUM tataricum.

recognisable by its red flowers. It is frequently sown in newly-cleared forest lands and ripens in September. The grain is exported to the plains under the name *Kotu*, and is eaten by the Hindus during their fasts (*bart*), being one of the *phaláhas* or food-grains lawful for fast-days. It is said to be heating but palatable, and is sold by the *pansári* or druggists, and not by the general grain-dealers." Stewart remarks under F. emarginatum that he thinks there are at least three cultivated species in the Panjáb Himálaya; "this with reddish flowers is generally said to grow lower than the other, but I have seen both at the same level about 8,500 feet on the Sutlej. The leaves of this are used as a pot-herb." Speaking of the Nilghiri Hills, Mr. Robertson says: "I did not see any crops of this plant; but I was informed that Buckwheat grows readily, and produces heavily, even on exposed parts of the higher portions of the plateau near Ootacamund." "Its flour from decorticated seeds is white and wholesome."

CULTIVATION.

Phalahas. 12

Pot-herb. 13

Dye.—Dr. McOann mentions having received from Darjíling a "sample of woollen yarn dyed a light purple by *tilaphapur* (Buckwheat) and *manjistha*." A specimen of the plant *tilaphapur* alluded to was identified at the Royal Botanic Gardens as F. esculentum. Crookes gives an abstract of Schunck's results obtained on chemically examining Buckwheat. A yellow crystalline colouring matter may be extracted from the leaves identical with *rutin*, and also with *ilixanthin*. This dye yields on mordanted cotton bright yellow shades. It may be obtained by adding acetate of lead to a decoction of the leaves; filtering while hot and adding acetic acid, when the yellow crystals will be precipitated.

DYE. 14

Food.—The LEAVES and tender SHOOTS are boiled as a spinach, and the NUTS are husked and ground into flour which is eaten as bread. The unhusked nuts are regarded as a superior food for poultry. As an article of human food Buckwheat does not hold a high place. About 20 per cent. of the weight is lost in the process of decortication. Professor Church publishes the analysis of what would appear to have been an ordinary sample of Buckwheat, but not of Indian origin. The table given by the Professor may be here reproduced, but it would seem desirable to have authentic samples of the Indian grain subjected to chemical examination:—

FOOD. Leaves. 15 Shoots. 16 Nuts. 17

	In 100 parts.	In 1 ℔.
Water	13·4	2 oz. 63 grains.
Albuminoids	15·2	2 „ 189 „
Starch	63·6	10 „ 77 „
Oil	3·4	0 „ 238 „
Fibre	2·1	0 „ 147 „
Ash	2·3	0 „ 161 „

From this result Professor Church concludes that the nutrient-ratio is 1 : 4·7 and the nutrient-value 86.

Mr. Baden-Powell says: "The seeds yield a hard, bitter, and unpalatable BREAD, which is said to be heating: it is only eaten in the plains during the *bart* or fast-days."

Bread. 18

19

Fagopyrum tataricum, *Gærtn.; Fl. Br. Ind., V, 55.*

Syn.—F. ROTUNDATUM, *Bab.*; POLYGONUM TATARICUM, *Linn.*

Vern.—*Kaspat* [bazar name], HIND.; *Kála trúmba, chin, karma bres, kdtú, brapú, drawo, phaphra, ulgo, ugal, tsábri, káthú*, PB.; *Tráo, rjao,* LADAK.

Note.—On the lower Himálaya it would appear the name *Ogal* or *Ugal* is practically restricted to this species, and *phaphra* given to F. esculentum.

References.—*Stewart, Pb. Pl., 184; Atkinson, Him. Dist., 316, 698; Church, Food-Grains of India, 114.*

Habitat.—Cultivated throughout the higher Himálaya, but more

FAGRÆA **obovata.**	**The Fagræas.**

especially on the western extremity and at altitudes from 8,000 to 14,000 feet. It is a taller, much coarser plant than F. esculentum, and the nuts, which are long and not triangular, have the angles rounded off and keeled towards the top It seems probable that there are several varieties, the nut in some being less than half the size in others.

CULTIVA-
TION.
20

Cultivation.—This seems to be the form grown in Ladak, Zanskar, and Western Tibet. In the Simla neighbourhood it is never seen below 9,000 feet.

FOOD.
Nuts.
21

Food.—There seems to be little or no difference in taste between this and the previously described species. Stewart says, however, that, if anything, this is inferior in point of quality. Bears are said to be more fond of it than almost of any other food, and they commit much damage to the standing crop. In Lahoul, Aitchison states that "the LEAVES are much used as a pot-herb in summer, when other greens are not easily got."

Leaves.
22

Professor Church writes : " An imperfect chemical analysis of the fruits or unhusked seeds of the present species shows it to resemble very closely the common kind cultivated in Europe, the albumenoids being 10·9 per cent., the oil 2·4, and the ash 7 :" he adds, "the percentage of albumenoids and oil would be considerably raised by the removal of the husk."

FAGRÆA, *Thunb. ; Gen. Pl., II.,* 794.

23

Fagræa fragrans, *Roxb. ; Fl. Br. Ind., IV., 85 ;* LOGANIACEÆ.

Vern.—*Anan* (or *a-nan*), BURM.

References.—*Roxb., Fl. Ind., Ed. Carey & Wall., II., 32 ; Kurz, For. Fl. Burm., II., 205 ; Gamble, Man. Timb., 2'7 ; Mason, Burma and Its People, 543, 802 ; Pharm. Ind., 146 ; Moodeen Sheriff, Supp. Pharm., Ind., 138.*

Habitat.—A small evergreen tree of Burma and the Andaman Islands to China.

MEDICINE.
Bark.
24

Medicine.—The BARK of this plant is said to be a remedy for malarious fever. In experiments made by Dr. Kanny Lall De, C.I.E., it was found to contain strychnia. The *Pharmacopœia of India* remarks "the remedy appears worthy of further investigation."

TIMBER.
25

Structure of the Wood.—Hard, brown, close-grained, beautifully mottled. It is very durable, and is not liable to the attacks of the "Teredo." It is one of the most important of the reserved trees of Burma, especially in Tavoy; and is used for house-building, bridge and wharf piles, boat anchors, and other purposes. Weight from 53 to 70℔ a cubic foot.

SACRED
USES.
26

Sacred Uses.—The Burmese regard the wood of this tree as too good for the laity, and hold that it should be reserved for sacerdotal purposes. At Tavoy it is employed principally for the posts of Buddhist edifices (*Mason*).

27

F. obovata, *Wall. ; Wight, Ic., t. 1316, 1317 ; Fl. Br. Ind., IV., 83.*

Vern.—*Sunakhari,* NEPAL ; *Longsoma,* MAGH. ; *Nyoungkyap (nyaung-gyat),* BURM.

References.—*Kurz, For. Fl. Burm., II., 205 ; Gamble, Man. Timb., 267 ; Thwaites, En. Ceylon Fl., 200 ; Rheede, Hort. Mal., 4, tab. 58 ; Indian Forester, II., 25 ; X., 34 ; Bombay Gazetteer (Kánara), XV., Pt. I., 438.*

Habitat.—An evergreen tree, often scandent or stem-clasping, found in the forests of the Deccan Peninsula and in Northern and Eastern Bengal, the Khásia Hills, Chittagong, and Burma.

In Burma it is said to be characteristic of the lower hills, and it is also reported to be one of the most beautiful plants found on the lower slopes of the Nilghiris. It is common in the forests of North Kánara, flowering during the rainy season. In Burma the fruit ripens in the cold season.

TIMBER.
28

Structure of the Wood.—Hard and durable. Weight 56℔.

F. 28

Famine Foods. (*G. Watt.*)	**FAMINE Foods.**

Fagræa racemosa, *Jack.; Fl. Br. Ind., IV., 84.*   29
    Vern.—*Thit-hpaloo,* BURM.
    References.—*Kurs, For. Fl. Burm., II., 205; Gamble, Man. Timb., 268.*
    Habitat. — A moderate-sized evergreen tree, frequent in the forests of the
Andaman Islands, and distributed to Penang and Malacca. It flowers and
fruits from February to May.

    Medicine.—Major Ford says that the ROOT-BARK is used as a cure for   **MEDICINE.**
fever (*Gamble*).   **Root-Bark.**
    Structure of the Wood.—Moderately hard, greasy to the touch, and   ₃₀
with a scent like that of india-rubber. Weight 50℔ per cubic foot.   **TIMBER.**
Major Ford remarks that it is strong and durable, and that the wood is   3¹
used for house-posts.

Fagus sylvatica, the BEECH; not indigenous to India.

FAMINE FOODS.   32

    The following are some of the more important articles reported to have
been eaten in times of SCARCITY AND FAMINE. Those marked *Dec. Fam.*
(=Deccan Famine of 1877-78) are taken from Dr. Dymock's list, appended to his *Mat. Med. of Western India.* But the Famine Commission's
Report, Dr. Shortt's special list of Madras Famine Foods, and numerous
other works have also been drawn upon in compiling the enumeration here
given. The literature of famine food materials appears to have been more
carefully investigated in Bombay than in any other part of India, and it
seems probable that future enquiry may more than double the number of
plants which have been eaten, or which might with safety be recommended
to be eaten, in times of scarcity or famine The reader is referred to their
respective alphabetical places for full particulars regarding these famine
foods. It is commonly stated that the low-caste people have a super-
abundance of food during famines, since they eat the animals that have
died of starvation. The higher-caste Hindus will not do so but prefer
rather to die.

Abrus precatorius.—The *Rati* seeds. These are poisonous if a powder
    prepared from them be injected under the skin, but boiled as a pulse
    they are wholesome, and in Egypt are regularly cultivated as an article
    of diet.
Abutilon indicum.—*Behar Famine.*
A. muticum.—Seeds. *Dec. Fam;* see also *Lisboa, U. P. B., p. 194.*
Acacia arabica.—Seeds. *Dec. Fam;* see *Lisboa, U. P. B., p. 199.*
    The gum and powdered bark are also largely eaten in famine.
A. leucophlœa.—Bark ground into flour and young pods. *Brandis.*
Acalypha indica.—Leaves. *Lisboa, U. P. B., p. 204; Shortt, Ind. For.,
    III., 235.*
Achyranthes aspera. Leaves and seeds. *Dec. Fam.* See also *Lisboa,
    U. P. B., p. 203; Shortt, Ind. For., III., 235.*
Adansonia digitata.—Bark and leaves eaten in Senegal.
Adenanthera pavonina.—Leaves eaten in Orissa Famine.
Ærua lanata.—Seeds.—*Dec. Fam.* *Shortt, Ind. For., III., 235.*
Ægle Marmelos.—Fruit. *Dec. Fam.*
Æschynomene aspera. Leaflets. *Lisboa, U. P. B., 198.* "(Not found
    on the Bombay side, but grows in Bengal.)" *Shortt, Ind. For., III.; 235*
Æsculus indica.—Nuts. *Drury, U. P., 334.*
Agave americana.—⎱
A. vivipara.— ⎰ The flowering stalks. *Lisboa, U. P. B., p. 205.*
Alangium decapetalum.—Fruit. *Dec. Fam.*
Allophyllus Cobbe (Schmidelia).—Fruit. *Shortt, Ind. For., III., 238.*

F. 32

| FAMINE Foods. | Famine Foods. |

Albizzia procera.—*Lisboa, U. P. B., p. 199.*
Aloe vera, *var.* officinalis.—Leaves. *Shortt, Ind. For., III., 235.*
A. indica.—
A. litoralis.— } The leaf—bud or cabbage. *Lisboa, U. P. B., p. 206.*
Alpinia Galanga.—Tubers. *Dec. Fam.*
Altenanthera sessilis.—Leaves. *Shortt, Ind. For., III., 235.*
Alysicarpus rugosus.—Seeds. *Lisboa, U. P. B., p. 198.*
A. vaginalis.—Herb. *Dec. Fam.*
Amarantus gangeticus (A. tristis).—Herb. *Shortt, Ind. For., III., 235.*
A. oleraceus.—Herb. *Dec. Fam.*
A. paniculatus.—Herb. *Shortt, Ind. For., III., 235.*
A. spinosus.—Leaves. *Bengal Famine. Shortt, Ind. For., III., 235.*
Amorphophallus campanulatus.—Tuber. *Lisboa, U. P. B., p. 207 ; Shortt, Ind. For., III., 235.*
A. sylvaticus.—Tuber and leaf. *Dec. Fam.*
Andropogon pertusus.—One of the best grasses to withstand long droughts, hence a cattle-famine fodder, though largely eaten at other times.
Anthocephalus Cadamba.—Fruit. *Dec. Fam.*
Arisœma curvatum.—Roots. *Lisboa, U. P. B., p. 207.*
Arthrocnemum indicum.—Herb, pickled. *Dec. Fam. Shortt, Ind. For., III., 238.*
Arundinaria Wightiana. *Lisboa, U. P. B., p. 209.* Rice, from the flowering stem, formed the principal food of the poor during the famine of Orissa in 1812; of Kánara in 1864; and of Malda, 1866.
Asparagus sarmentosus.—Roots. *Dec. Fam.*
Asphodelus fistulosus.—The Piazi, the tubers of which are, in the Panjáb, eaten in times of scarcity. Stewart says this appears to have been the plant alluded to by Griffith as eaten by the camp followers of the Kandahar Force when provisions ran scarce.
Asterocantha longifolia.—Herb. *Dec. Fam.*
Asystacia gangetica.—Vegetable. *Lisboa, U. P. B., p. 202.*
Atriplex hortensis.—Herb. *Shortt, Ind. For., III , 235.*
Bamboo seeds.—Saved thousands in the Orissa Famine of 1812; Kánara of 1864, when 50,000 people went to Dharwar and Belgaum to collect the seeds, Malda, of 1866; &c., &c.
Bambusa arundinacea.—Seeds. *Dict. Ec. Prod., Vol. I., p. 391.*
B. vulgaris.—*Lisboa, U. P. B., p. 209.*
Bassia latifolia.—Fruit and also flowers when dried in the sun are eaten normally by the hill tribes, but in times of scarcity by all classes. *Shortt, Ind. For., III., 235.* Lisboa writes :—" During the famine of 1873-74 in Behar this is said to have kept thousands of people from starvation."
B. longifolia.—Seeds and flowers. *Lisboa, U. P. B., p. 201.*
Bauhinia malabarica.—Leaves. *Dec. Fam.* Largely eaten as a vegetable by the hill tribes.
B. racemosa.—Flowers. *Dec. Fam.*
Betula acuminata.—The inner bark is eaten by the Lahupás of Manipur. *Dic. Ec. Prod., Vol. I., 451.*
Bœrhaavia diffusa.—Herb. *Dec. Fam.* Revd. A. Campbell says the Santals grow the plant. *See Vol. I., 485.*
B. repanda.—Leaves. *Lisboa, U. P. B., p. 203 ; Shortt, Ind. For., III., 235.*
Borassus flabelliformis.—Roots. *Vol. I., 502 ; also Lisboa, U. P. B., p. 207 ; Shortt, Ind. For., III., 235.*
Boswellia serrata.—Flowers and seeds eaten by the Bhíls. *Vol. I., 516.* Drury says the Uriyas make a soup from the fruits in times of famine.
Brassica.—Mustard, Rape, &c. The leaves of these plants are eaten in times of famine. *Fam. Com. Rept.*

F. 32

Bryonia laciniosa.—Leaves boiled and eaten. *Dec. Fam.*
Buchanania latifolia.—Fruit. *Dec. Fam.*
Buettneria herbacea.—Leaves. *Shortt, Ind. For., III., 236.*
Bupleurum falcatum.—Root eaten by the Himálayan tribes.
Butea frondosa.—Roots. *Dec. Fam.*
Caladium ovatum.—Herb. *Dec. Fam.*
Canna indica.—Roots yield a useful arrowroot. *Vol. II., 102.*
Canthium parviflorum.—Leaves. *Dec. Fam. Shortt, Ind. For., III.,* *236;* eaten also in normal seasons; *Vol. II., 129.*
Carallum adscendens.—Shoots cooked. *Shortt, Ind. For., III., 236.*
C. fimbriata.—Green follicles. *Dec. Fam. Vol. II., 141.*
Cardiospermum Halicacabum.—Herb. *Dec. Fam. Shortt, Ind. For., III.,* *236; Lisboa, U. P. B., p. 197; Vol. II., 156.*
Carissa Carandas & C. spinarum.—Fruits. *Shortt, Ind. For., III., 236.*
Carthamus tinctorius.—Leaves and seeds. *Dec. Fam.* The rich ate the seeds during the famine at Sholapur. *Vol. II., 195.*
Caryota urens, *Willd.* The farinaceous part of the trunk was largely used in the famine of 1830. (*Roxb., Ed. C.B.C., 668.*) *Vol. II., 208.*
Cassia auriculata.—Leaves. *Dec. Fam.;* also *Lisboa, U. P. B., p. 198, Vol. II., 216.*
C. Fistula.—Flowers largely eaten by the Santals (*Rev. A. Campbell*). *Shortt, Ind. For., III., 236.*
C. occidentalis.—Leaves. *Lisboa, U.P.B., p. 198.*
C. pumila.—Herb. *Dec. Fam.*
C. Sophora.—Leaves. *Lisboa, U. P. B., p. 198; Shortt, Ind. For., III.,* *236.* The disagreeable smell and flavour is removed by boiling.
C. siamea.—Leaves. *Dec. Fam.*
C. Tora.—Leaves. *Dec. Fam.; Stewart, Pb. Pl., 62;* also *Lisboa, U. P. B., 198.* "Largely used during famine, but eaten also at all seasons, especially during the month of *Shráwan.* The seeds afford a good substitute for coffee." *Vol. II., 226.*
Celosia argentea.—Herb. *Dec. Fam.;* Stewart says that in the Panjáb it is used as a pot-herb in times of scarcity.
C. cristata.—Leaves and shoots. *Shortt, Ind. For., III., 236; Vol. II.,* *241.*
Cenchrus echinatus.—Seeds. *Vol. II., 246.*
Cephalandra Indica—Cephalostachyum capitatum, *Munro:* GRAMINEÆ. *Vol. II., 253.*
Ceropegia bulbosa.—Root. *Dec. Fam. Vol. II., 262.*
Chenopodium album.—Herb. *Dec. Fam.;* also regularly cultivated. *Vol. II., 265.*
Chlorophytum parviflorum.—Leaves. *Dec. Fam. Comp. with Vol. II., 269-270.*
Chrysopogon montanus.—The seeds of this grass are eaten in Rájputana. *Vol. II., 274.*
Cicer arietinum.—Gram. The leaves and stalks are eaten in times of famine. *Fam. Com. Rep.*
Clerodendron serratum.—Herb. *Dec. Fam. Vol. II., 375.*
Cleome viscosa.—*Shortt, Ind. For., III., 236.*
Cocculus villosus.—Leaves. *Dec. Fam. Vol. II., 398.*
Coffee pulp.—See *Vol. II., 489.*
Coix lachryma.—Seeds. *Dec. Fam.* The *Kew Bulletin for 1888, p. 267,* says the cultivated edible **Coix** is **C. gigantea :** the writer's specimens obtained in Manipur were cultivated **Coix,** but by the Kew authorities these were some time ago named as **C. lachryma.** It seems probable that there is no specific difference, the one being the more readily

Famine Foods.

recognisable cultivated state of the other (*Comp. with Vol. II., pp. 491—500*).

Commelina bengalensis.—Leaves. *Stewart, Pb. Pl., 236.*

C. communis.—Seeds. *Lisboa, U. P. B., p. 206.*

C. obliqua.—Leaves eaten in famine. (*Atkinson.*)

Corchorus trilocularis.—Herb and seeds. *Dec. Fam.;* also given by *Lisboa, U. P., p. 195.*

C. olitorius.—Herb. *Lisboa, U. P. B., p. 195.*

Cordia obliqua.—Flowers and fruit. *Dec. Fam.*

C. Myxa. - Fruit. *Dec. Fam Shortt, Ind. For., III., 236.*

Corypha umbraculifera. Yields starch from the pith. *Vol. II., 575.*

Cressa cretica. - Herb. *Dec. Fam. Vol. II., 588.*

Crinum defixum. —The bulbous root. *Lisboa, U. P. B., p. 204; Vol. II., 590.*

Crotalaria juncea.—Leaves and pods. *Dec. Fam. Vol. II., 613.*

Curcuma caulina.—Tubers. *Dec. Fam. Vol. II., 658.*

C. pseudomontana.—Tubers. *Dec. Fam. Vol. II , 669.*

Cyanotis axillaris.—Seeds. *Dec. Fam.;* also *Lisboa, U. P. B., p. 206.*

Cycas circinalis, pectinata, & Rumphii.—Yield starch from the interior of the stem.

Cynanchum pauciflorum.—The leaves eaten in Ceylon : this does not appear to be known in India. *Vol. II., 678.*

Cynodon Dactylon.—Leaves and culms. *Lisboa, U. P. B., p. 208; Shortt, Ind. For., III., 236.*

Cyperus jeminicus.—Tuber and leaf. The former are ground into flour and eaten (*Vol. II., 685*). (*Roxb., Fl. Ind., Ed. C.B.C., 65.*)

Dalbergia paniculata.—Leaves. *Dec. Fam.*

Daucus Carota.—Recommended as an emergent crop in times of threatened famine. *Fam. Com. Rep., Vol. II., 151; Conf. also with Vol. III. of this work, pp. 48—52.*

Dendrocalamus strictus.—Male bamboo. The seeds and shoots. *Vol. III., 77.*

Digera arvensis.—Herb. *Dec. Fam. Vol. III., 112.*

Dillenia indica.—Calyx. *Dec. Fam. Vol. III., 113.*

Dioscorea anguina. —This, according to Roxburgh, yields a tuber which is eaten in times of famine.

D. oppositifolia.—Tubers. *Dec. Fam.*

D. pentaphylla.—Leaves, tubers, and flowers. *Dec. Fam.*

D. triphylla.—Tubers. *Dec. Fam.*

Diospyros Embryopteris.—Fruit. *Dec. Fam.*

Dolichos biflorus—Is spoken of by Roxburgh as a crop that requires little rain, and may, therefore, be grown when rice fails.

Dracontium polyphyllum (see *Vol. II., p. 192*) – Is said by Drury (*U. P., 187*) to afford a tuber which is eaten in times of famine. *Shortt, Ind. For., III., 236.*

Dregea volubilis. —Leaves. *Shortt, Ind. For., III., 237.*

Ehretia lævis, *Roxb.* Fruit and inner bark. *Stewart, Pb. Pl., 153; Lisboa, U. P. B., p. 202.*

Elæagnus latifolia.—Fruit. *Dec. Fam.*

Eleusine ægyptiaca. —Seed-grains. *Lisboa, U. P. B., p. 268.*

Embelia robusta. —Leaves. *Dec. Fam.*

Erinocarpus Nimmoanus.—Fruit. *Dec. Fam.*

Eriodendron anfractuosum.—Seeds. *Lisboa, U. P. B., p. 195; Shortt, Ind. For., III., 236.*

Erythroxylon monogynum.—Leaves and young shoots. *Lisboa, U. P. B., p. 195.* Said to have afforded food to many thousand people during the famine in Madras of 1877. *Shortt, Ind. For., III., 236-238.*

Eugenia Jambolana.—Kernels. *Shortt, Ind. For., III., 238.*
Euphorbia pilulifera. (*Hirta, Dals. & Gibs.*)—Leaves. *Lisboa, U. P. B., p. 203; Shortt, Ind. For., III., 236.*
E. thymifolia.—Herb. *Dec. Fam.*
Feronia elephantum. – Fruit. *Dec. Fam.*
Ficus bengalensis.—Fruit. *Lisboa, U. P B., 204; Shortt, Ind. For., III., 236.*
F. glomerata.—Fruit. *Dec. Fam. Lisboa, U. P. B., 204; Shortt, Ind. For., 236.*
F. indica. –*Fam. Com. Rep., Vol II., p. 154, C. P.*
F. religiosa.—Fruit. *Dec. Fam. Lisboa, U. P. B., 204; Shortt, Ind. For., III., 236.*
Fimbristylis Kysoor, *Roxb.; Dals. & Gibs., Bomb. Fl., 288.* The tuberous root. *Lisboa, U. P. B., p. 208.*
Fungi. – Nearly all the species are eaten in famine.
Garcinia xanthochymus.– Fruit. *Shortt, Ind. For., III., 238.*
Gisekia pharnaceoides.—Herb. *Lisboa, U. P. B., p. 200.*
Glossocardia linearifolia.—Leaves. Lisboa (*U. P. B., p. 200*) thinks the identification is not correct, and that the plant may be **Cyathoclyne lyrata.**
Grasses.—Seeds of wild species are collected and eaten in times of famine.
Grewia Microcos.—Fruit. *Dec. Fam.*
Guatteria longifolia. Fruit. *Dec. Fam.*
Guazuma tomentosa.—Capsules. *Lisboa, U. P. B., p. 195; Shortt, Ind. For., III., 236.*
Gynandropsis pentaphylla.– Leaves. *Shortt, Ind. For., III., 236, 237.*
Hedychium coronarium. (also H. scaposum, *Nimmo; Dals. & Gibs. Bomb. Fl., 273*)– Tubers. *Dec. Fam.*
Helmia, see Dioscorea.—Tubers. *Dec. Fam.*
Hibiscus tiliaceus. Bark. *Lisboa, U. P B., p. 194.* Drury, quoting Forster, states that the stalks are sucked in times of scarcity.
Holostemma Rheedii.– Flowers. *Dec. Fam. Shortt, Ind. For., III., 237.*
Hoya viridiflora,– Dregea volubilis. Leaves. *Dec. Fam.; also mentioned by Lisboa, U. P. B., p. 201.*
Indigofera cordifolia. – Seeds. *Dec. Fam.;* also mentioned by *Lisboa, U. P. B., p. 197.* "A highly nitrogenous pulse."
I. enneaphylla.—Seeds. *Dec. Fam.*
I. glandulosa.—Seeds. *Dec. Fam.;* also mentioned by *Lisboa, U. P. B., p. 197.* "Rich in nitrogen." According to Roxburgh the seeds of this species are made into bread in times of scarcity (*Roxb., Fl. Ind., Ed. C.B.C., 583*).
I. linifolia.—Seeds. *Dec. Fam.;* also mentioned by *Lisboa, U. P. B., p. 197.* "Seeds largely consumed by the people of Kaládgi, Dharwar, Sholapur, Ahmednagar, &c.; pounded and made into cakes either alone or with some cereals. Rich in nitrogen."
Ipomæa aquatica.—Herb. *Dec. Fam.; Bengal Famine. Shortt, Ind. For., III., 237.*
I. eriocarpa, *Br.*—The plant.
I. muricata.—Peduncles. *Dec Fam.*
I. reniformis.—Herb. *Lisboa, U. P. B., p. 202; Shortt, Ind. For., III., 237.*
I. sepiaria.—Herb. *Lisboa, U. P. B., p. 202; Shortt, Ind. For., III., 237.*
Jasminium arborescens, *var.* latifolia.—Seeds. *Dec. Fam.*
Launæa pinnatifida.–-Herb. *Dec Fam.*
Leea macrophylla.—Leaves. *Dec Fam.*
Leptadenia reticulata.—Leaves. *Lisboa, U. P. B, 201; Shortt, Ind. For., III., 238.*
Leucas aspera.—Herb. *Dec. Fam. Lisboa, U. P. B., 203; Shortt, Ind. For., III., 237.*

F. 32

FAMINE Foods.	Famine Foods.

Leucas cephalotes.—Herb. *Dec. Fam. ; Behar Famine.*

Limnanthemum cristatum.—Stems and fruit. *Shortt, Ind. For., III., 238 ; Lisboa, U. P. B., p. 202.*

Linum usitatissimum.—Green pods. *Dec. Fam.*

Maba buxifolia.—Fruit. *Shortt, Ind. For., III., 237.*

Macaranga Roxburghii.—Fruit. *Dec. Fam.*

Malva parviflora, *Linn.*—A pot-herb eaten largely in famine.

Mangifera indica.—Kernels used in times of scarcity and famine. *Roxb., Fl. Ind., Ed. C.B.C., 216 ; Shortt, Ind. For., III., 237.*

Melia Azaderachta.—Fruit. *Lisboa, U. P. B., p. 196 ; Shortt, Ind. For., III., 237.*

Mengea (Amarantus) tenuifolia.—Herb. *Dec. Fam.*

Mimusops Elengi.—Fruit. *Shortt, Ind. For., III., 237.*

M. hexandra.—Fruit. *Dec. Fam.*　　　　　　　　　　[*III., 237.*

Mirabilis Jalapa.—Leaves. *Lisboa, U. P. B., p. 203 ; Shortt, Ind. For.,*

Mollugo stricta.—Herb. *Dec. Fam.*

Momordica Charantia.—Leaves. *Dec. Fam.*

Morinda citrifolia.—Green fruit. *Shortt, Ind. For., III., 237.*

M. umbellata.—Fruit. *Lisboa, U. P. B., p. 200 ; Shortt, Ind. For., III., 237.*

Mucuna pruriens.—Seeds. *Dec. Fam.*

Murraya Kœnigii.—Fruits. *Shortt, Ind. For., III., 235.*

Musa ornata.— ⎰ Root. *Dec. Fam.* Also mentioned by *Lisboa, U. P. B.,*
M. superba.— ⎱ *204.* "The scape and the convolute leaf-sheaths of both these plants."

Mussænda frondosa.—Flowers. *Dec. Fam.*

Nelumbium speciosum.—Root. *Dec. Fam.*

Neptunia (Desmantus) oleracea.—Herb and pods. *Lisboa, U. P. B., p. 199 ; Shortt, Ind. For., III., 236.*

Nymphœa lotus.—Roots and seeds. *Shortt, Ind. For., III., 237.*

N. stellata, *Willd.*—Roots and seeds.

Olea dioica.—Fruit. *Dec. Fam.*

Opuntia Dillenii.—Fruit. *Lisboa, U. P. B., p. 199.*　　　[*III., 233—237.*

Orygia decumbens.—Leaves. *Lisboa, U. P. B., p. 200 ; Shortt, Ind. For.,*

Oxalis corniculata.—Seeds. *Dec. Fam.* Leaves. *Lisboa, U. P. B., p. 196 ; Shortt, Ind. For., III., 237.*

Oxystelma esculentum.—Follicle. *Dec. Fam.*

Pachyrhizus angulatus.—The tuberous root.

Pandanus odoratissimus.—Pulpy part of drupes (*Roxb., Fl. Ind., Ed. C.B.C., 707*), eaten in times of famine; *Shortt, Ind. For., III., 237.*

Panicum colonum.—Seeds. *Dec. Fam.*

P. frumentaceum. Should be extensively cultivated in seasons of drought, as with little irrigation, on any light soil, it will afford a harvest within six weeks of the date of sowing. *Fam. Com. Report, II., 151.*

Penicillaria spicata (Holcus spicatus, *Dals. & Gibs.*). *Lisboa, U. P. B., 208.*

Phaseolus adenanthus.—The tuberous roots. *Shortt, Ind. For., III., 237.*

P. Mungo—Is, by Roxburgh, spoken of as a crop that will grow in times of threatened famine, when rice fails.

P. trilobus.—Seeds. *Dec. Fam.*　　　　　　　　　　[principle."

P. trinervius.—*Lisboa, U. P. B., p. 198.* "Seeds rich in nitrogenous

Phœnix farinifera.—The farinaceous substance in the trunk (*Roxb., Fl. Ind., Ed. C.B.C., 723*) (*Drury, U. P., 339*). Leaf-bud. *Shortt, Ind. For., III., 237.*

P. sylvestris.—Fruit. *Dec. Fam.* Also leaf-bud or cabbage. *Lisboa, U. P. B., p. 206 ; Shortt, Ind. For., III., 237.*

F. 32

Pistia stratiotes.—Herb. *Dec. Fam.*

Pithecolobium dulce.—Fruit. *Shortt, Ind. For., III., 237.*

Pogostemon parviflorus.—Leaves. *Dec. Fam.*

Polygala chinensis.—Leaves. *Lisboa, U. P. B., p. 194.*

Porana racemosa.—Peduncles. *Dec. Fam.*

Portulaca oleracea.—Shoots. *Shortt, Ind. For., III., 237.*

Pouzolzia tuberosa.—The tuberous roots. *Lisboa, U. P. B., 204.*

Premna latifolia.—Leaves. *Lisboa, U. P. B., p. 202; Shortt, Ind. For., III., 237.* [*III., 237.*

P. integrifolia.—Leaves. *Lisboa, U. P. B., 203; Shortt, Ind. For.,*

Prosopis spicigera, *Linn.*—Pods. *Dec. Fam. Shortt, Ind. For., III., 237.*

Pteris aquilina.—The underground stems.

Pterocarpus Marsupium.—Seeds and flowers. *Dec. Fam.*

Randia uliginosa.—Green fruit. *Dec. Fam.*

Ranunculus sceleratus.—This is eaten by the inhabitants of Wallachia when cooked. It is a powerful poison when not cooked.

Rhynchocarpa fœtida.—Fruit and leaves. *Lisboa, U. P. B., p. 200; Shortt, Ind. For., III., 235.*

Rivea hypocrateriformis.—Leaves. *Lisboa, U. P. B., p. 202; Shortt, Ind. For., III., 237.*

Rothia trifoliata.—Leaves and pods. *Lisboa, U. P. B., p. 197; Shortt, Ind. For., III., 237.*

Sagittaria & Alisma.—Yield edible tubers, the former being cultivated for this reason in North America. There are several species in India, but no record exists of their being eaten.

Salicornia brachiata.—Leaves and shoots. *Shortt, Ind. For., III., 238.*

Salsola fœtida.—Herb. *Dec. Fam.*

Santalum album.—Seeds. *Lisboa, U. P. B., p. 204.*

Schleichera trijuga.—Fruit. *Dec. Fam. Shortt, Ind. For., III., 238.*

Schrebera swietenioides.—Leaves. *Dec. Fam.*

Semecarpus Anacardium.—Green fruit. *Dec. Fam.*

Sesamum indicum.—Seeds made into oil cake.

Sesbania aculeata.—Seeds. *Dec. Fam.*

S. ægyptiaca.—Seeds highly nitrogenous. *Lisboa, U. P. B., p. 197.*

S. procumbens.—Seeds. *Dec. Fam.*

S. grandiflora.—*Shortt, Ind. For., III., 235.*

Sesuvium Portulacastrum.—Seeds and herb. *Dec. Fam.*

Shorea robusta.—Seeds roasted and mixed with the flowers of the Mahua tree.

Sida cordifolia.—Herb. *Dec. Fam.*

Smilax ovalifolia.—Leaves and root. *Dec. Fam.*

Smithia sensitiva.—Herb. *Dec. Fam.*

Solanum Jacquinii.—Unripe fruit curried. *Lisboa, U. P. B., p. 202.*

S. nigrum & xanthocarpum.—Herb. *Dec. Fam. Shortt, Ind. For., III., 238; Lisboa, U. P. B., 202.*

S. torvam.—Curried. *Lisboa, U. P. B., 202; Shortt, Ind. For., III., 238.*

Sorghum vulgare (Holcus saccharatus, *Dalz. & Gibs.*)—*Lisboa, U. P. B., p. 208.*

Spathium chinense (Aponageton monostachyon).—Tubers are boiled and eaten. *Shortt, Ind. For., III., 235.*

Spermacoce hispida.—Seeds. *Dec. Fam.* Rev. A. Campbell mentions this as eaten by the Santals in times of great distress.

Spondias acuminata.—Green fruit. *Dec. Fam.*

S. mangifera.—Leaves and fruit. *Shortt, Ind. For., III., 238.*

Sterculia fœtida.—Seeds. *Dec. Fam. Shortt, Ind. For., III., 238.*

S. guttata.—Seeds. *Dec. Fam.*

Strychnos potatorum.—Fruit. *Shortt, Ind. For., III., 238.*
Suœda maritima & nudiflora.—Leaves. "The leaves of this plant alone,
the natives say, saved many thousand lives during the famine of
1791, 1792, and 1793." *Roxb., Fl. Ind., Ed. C.B.C., 262; Shortt, Ind.
For., III., 238.*
Synantherias sylvatica.—Root, petioles, and leaves. *Lisboa, U. P. B., 208.*
Syzygium Gibsonii. (Eugenia *sp.?*)—Fruit. *Dec. Fam.*
Tacca pennatifida.—Root. *Dec. Fam.*
Tamarindus indica.—Leaves and seeds. *Dec. Fam. Roxb., Fl. Ind.,
Ed. C.B.C., 531; Shortt, Ind. For., III., 233—238.*
Tephrosia purpurea.—Seeds. *Dec. Fam.*
Terminalia belerica.—Seeds. *Dec. Fam.* Gum eaten by the Santals.
Theriophonium Dalzelii.—Leaves and petioles. *Lisboa, U. P. B., p. 208.*
Toddalia aculeata.—Leaves. *Shortt, Ind. For., III., 238.*
Trapa bispinosa.—Seeds. *Shortt, Ind. For., III., 238.*
Trianthema crystallina.—Seeds.
T. monogyna.—Leaves and shoots. *Shortt, Ind. For., III., 238.*
T. pentandra.—Leaves and shoots. *Lisboa, U. P. B., p. 200.*
Tribulus alatus, *Delile.*—Seeds.
T. terristris.—Herb and seeds. *Dec. Fam.* "The small spiny fruits of
this plant are said to have constituted the chief food of the people
during the Madras Famine. *Econ. Prod. of India, Part VI.* See also
Lisboa, U. P. B., p. 196.
Trichosanthes cucumerina.—Fruit. *Shortt, Ind. For., III., 238.*
Triticum sativum.—(The chaff in famine.) *Lisboa, U. P. B., p. 208.*
Typha elephantina.—Pollen *Dec. Fam.*
T. latifolia. Seeds. *Dec. Fam.*
Typhonium bulbiferum.—
T. divaricatu m.—   } Bulb and leaves. *Lisboa, U. P. B., p. 207.*
Urginea indica.—Leaves. *Dec. Fam.*
Vangueria edulis.—Green fruit. *Dec. Fam.*
Vitis quadrangularis.—Leaves. *Dec. Fam. Shortt, Ind. For, III., 236.*
Zea Mays.—Grain. *Lisboa, U. P. B, p. 208.* (The cobs in famine.)
Zizyphus nummularia, *W. & A.*—Fruit.
Z. Jujuba.—Dry fruit powdered. *Dec. Fam.*
Z. rugosa.—Fruit. *Dec. Fam.*

Fan Palms, see Borassus flabelliformis, *Linn.; Vol. I., 495.*

FARSETIA, *Desv.; Gen. Pl., I., 72.*

A genus of under-shrubs or herbs, comprising about 20 species; natives of
South Europe. West Asia, and North Africa. There are three Indian species
which have much the same habitat, possess the same economic properties,
and are known to the natives by the same vernacular names; they may therefore be considered collectively.

33 Farsetia ægyptica, *Turr.; Fl. Br. Ind., I., 140;* CRUCIFERÆ.

34 F. Hamiltonii, *Royle; Fl. Br. Ind., I., 140.*

35 F. Jacquemontii, *H. f. & T.; Fl. Br. Ind., I., 140.*
Vern.—*Mulei, fárid búti, láthin, fárid múli,* PB.
References.—*Stewart, Ph. Pl., 13; Murray, Pl. and Drugs, Sind, 49;
Baden Powell, Pb. Pr., 328; Spons, Encyclop., 1079; Gazetteer, N. W. P.,
IV., lxvii. i.; Punjab, Montgomery Dist., 20; Settlement Report of the
Montgomery Dist., 20.*
Habitat.—F. ægyptica is found in the Salt Range of the Panjáb;
F. Hamiltonii in the Upper Gangetic plain and the Panjáb, also from

Agra westwards; and **F. Jacquemontii**, in sandy places in the Panjáb and Sind.

Medicine. —According to writers on the plants of the Panjáb, all three species have a pleasant pungent taste, are pounded and taken as a cooling medicine, and are considered specific for rheumatism.

Food. —The. Settlement Report of the Montgomery District says of **F. Hamiltonii** : " The SEEDS are said to be poisonous, but were habitually used by Bábá Faríd, Shakarganj, when he was hungry."

The plant is described by **Mr. A. O. Hume** as a favourite food of the large bustard.

MEDICINE.
36

FOOD.
Seeds.
37

FEATHERS AND BIRDS USED FOR ORNAMENTAL PURPOSES.

38

Dr. Forbes Watson, in his list of Indian Products, drawn up in connection with a proposed Industrial Survey of India, enumerates some 68 birds, the plumage of which are used for ornamental purposes. · It is, perhaps, unnecessary to republish that list, but it may be said to include many of the honey-suckers, herons, bitterns, king-fishers, storks, jays, rollers, egrets, water-hens, bee-eaters, orioles, shrikes, bulbuls, snake-birds, grebes, and the hoopoes. These birds are systematically killed either for certain special feathers obtained from them, or on account of their entire skins. The following may be specially mentioned :—

1st -**Ceryle rudis.** The pied king-fisher.
2nd—**Coracias indica.** The roller, vulgarly known as the Blue Jay.
3rd—**Herodias alba.** The large Egret.
4th—**Houbara macqueeni.** The Houbara Bustard.
5th—**Leptoptilos argala.** The Adjutant or Gigantic Stork. The feathers of this bird are known in trade as Marabout.
6th—**L. javanica.** The Small Adjutant.
7th—**Pavo cristatus.** The Peafowl.
8th—**Plotus melanogaster.** The Snake-bird.
9th—**Upupa nigripennis.** The Hoopoe.

39
40
41
42
43
44
45
46
47

In works treating of feathers, the subject is generally referred to Common Feathers used in Upholstery; Down; Ornamental Feathers; and Quills.

In India the feathers of domesticated birds are universally destroyed by the indolent, though expeditious system, of removing them after immersion of the bird in hot water. Were an effort made to remedy this defect, India might afford a large supply annually of upholstery feathers. The same remark is practically applicable to the collection of down. Of ornamental feathers there are generally said to be two classes—(*a*), those like ostrich, in which the barbules are long and loose, giving beauty of form; and (*b*) those that manifest beauty and brilliancy of colour. Within the past few years India may be said to have entered on a new industry— that of Ostrich Farming. In another volume (under **Ostrich**) will be found some account of this industry, but it is believed the Trade Returns of Feathers, at the present date, refer mainly if not entirely to the second class of ornamental feathers. Prior to the year 1879-80 the exports from India of ornamental feathers were valued at about 1½ lakhs of rupees. Since that year, however, they seem to have steadily increased. In 1880-81 they were valued at R2,69,447; in 1882-83 at R3,04,253; in 1884-85 at R6,33,017; and last year (1887-88) at R5,70,495. The imports are unimportant, the highest record having been in 1886-87, when the imports of foreign feathers were valued at R1,068.

Little or nothing can be learned regarding the total number of birds thus annually destroyed to meet this large export trade. A missionary

Y

once mentioned to the writer that were he to adopt the system pursued by traders in capturing and destroying the blue roller, he could from the proceeds easily render his charge self-supporting. Dr. Balfour gives in his *Cyclopædia of India* an interesting account of the industry in ornamental feathers. A passage from that work may be here given: "Commercolly, in Bengal, is celebrated for its egret's feathers for head-dresses, tippets, boas, and muffs, and some of them are exceedingly beautiful, and not inferior in quality to those imported into Great Britain from Africa. The down of the young adjutant bird is also made into ladies' boas and victorines. The under-tail coverts are collected and sold in considerable quantity. Many are procured at Trichoor in Malabar. In the Panjáb the narrow black wing feathers of the onkar are used to make the *kalgi* or plumes for the *khod* or helmet. These plumes have a very elegant appearance; they stand about 6 or 8 inches above the helmet. The feathers of the bustard are similarly used. In Madras, dealers in birds' feathers carry on their trade on an extensive scale. One dealer had nearly 100 sets of hunters, each composed of four or five shikaris and one cook; most of these people are Korawa (basket-makers) who live in and about Madras. Each set has its headman, who is responsible for the others. These sets are sent out once a year, each receiving from R20 to R100, together with a certain number of nets, a knife, &c. They traverse all India, collecting the feathers of king-fishers, and return after six or eight months to Madras, each set bringing from 1,000 to 6,000 feathers, which are taken by the dealers at R14 per 100, and shipped to Burma, Penang, Singapore, and Malacca, bringing 10 to 13 dollars the 100." "The blue feathers of the jay, the king-fisher, and other blue-feathered birds are largely used in China for ornamentation, pasted on silver gilt."

Feather Grass, see Stipa.

48

FEL—BILE.

Vern.—*Safra,* HIND.; *Pitta,* SANS.; *Safral,* ARAB.; *Zahrahe,* PERS.

MEDICINE.
Bile.
49

Medicine.—The BILE of the buffalo, wild boar, goat, peacock, and the rohituka fish are used in medicine, as laxatives, and also in place of water in which to soak powders intended to be made into pills (*U. C. Dutt*).

Gall.
50

GALL is an absorbent and purgative; it is used along with antimony as a stimulant for the eye. In 1 drachm doses, mixed with 1 drachm of wax, when taken internally, it is said to cause abortion. Bile made into an ointment is used in inflammatory swellings (*Dr. Emerson*).

SPECIAL OPINIONS.—§ "Bile of fish or of the goat is given in night-blindness" (*J. N. Dey, Jeypore*). "Black pepper soaked in the bile of pigs for 40 days is given to cure madness" (*V. Ummegudien Mettapollian, Madras*). "Pigment calculi from the gall bladder of the cow, *gorochana,* are much valued by the natives as a medicine, and fetch a very high price" (*W. Dymock, Bombay*).

Felis, see Tiger.

FELSPAR.

51

Felspar.—The felspar group of minerals is the most important of all the rock-forming materials. Granitic rocks may be said to consist of quartz, mica, and felspar. The disintegration of granite frequently results in the quartz and mica being washed away, with the decomposed materials of felspar left in a more or less state of purity. This constitutes the finest of all known pottery clays. Impure clay may be said to be pure clay adulterated with organic and metallic substances.

F. 51

| *Ferns.* | (*G. Watt.*) | FERNS. |

Several works refer to felspar as an economic mineral, such as the Manual of the Coimbatore District, pages 23 and 453; the Manual of Trichinopoly District, page 67; and *Mason's Burma and its People*, pages 583, 734, &c. Since, however, felspar is employed almost entirely in the art of pottery, the reader is referred for further particulars to the article Clay in this work (*Vol. II, pp. 360 to 368*).

Fennel, see **Fæniculum vulgare,** *Gærtn.;* UMBELLIFERÆ.

Fennel, Flower, see Nigella sativa, *Linn.;* RANUNCULACEÆ.

Fennel, Giant, see Ferula below.

Fenugreek, see Trigonella Fænum-grœcum, *Linn.;* LEGUMINOSÆ.

FERNS.

52

Ferns.—**Beddome,** in his *Ferns of British India,* describes over 700 species and varieties. This may be accepted as an enumeration of only the better-known Indian forms. Out of that large assemblage of highly ornamental plants, however, only some 10 or 12 are of interest economically. A very large number are grown as rockery and foliage plants, but none are cultivated for food or medicine. One, **Asplenium ensiforme,** *Wall.,* yields a bright-red dye, which stains the mounting paper. The most important food-product afforded by this great family—the young underground stems and young fronds of the Common Braken Fern (**Pteris aquilina**)—are not, in the writer's opinion, eaten by the hill tribes of India. He has pointedly asked the Himálayan, as also the hill tribes of Manipur and the Nilghiris, but has invariably got the same reply, *viz.,* that no part of that very plentiful plant is eaten. **Stewart,** however, would lead one to suppose that he had found the people eating it, and **Cleghorn** states that when cooked it is juicy but rather insipid. The latter writer may be referring to his personal experience and not to the verdict of the people. At most hill stations, however, young fronds are regularly offered for sale, and in Simla these appear, for the most part, to be those of **Asplenium (Anisogonium) esculentum** (*Vol. I., No. 1583 A.*). This is doubtless the plant which **Madden** speaks of as "**Nephrodium eriocarpum.**" **Botrychium virginianum,** *Swartz.,* also forms an article of food among the Himalayan tribes. (*See Vol. I., 517.*) In New Zealand and other islands of the South Sea, where Tree Ferns abound, the centre of the stem of an **Alsophila** and of a **Cyathea** consists of a mucilaginous pith, which is used as food. In Sikkim one or two of the tree ferns are similarly eaten, especially **Alsophila latebrosa.**

Several ferns are employed medicinally, but in India the merits of the Male Fern (**Lastrea Filix-mas**) do not appear to have been discovered, although it is one of the most plentiful species on the hills from 4,000 to 10,000 feet above the sea. The various species of 'Adiantum are, however, extensively employed medicinally, the one most generally to be seen in the drug-shops being **A. venustum** (*see Vol. I., pp. 110 to 114*). The Rev. A. Campbell mentions the fact that the Santals employ **Cheilanthes tenuifolia** (*see Vol. II., p. 265*). An officinal root, the *bisfáij,* is by **Stewart** referred (probably incorrectly) to **Polypodium vulgare.** He wrote: "I have no clue as to which of our Himálayan ferns this is generally derived from, or whence it is brought, but Kábul is given by one authority, and **Honigberger** says 'the hills.' It is used as an alterative." **Polypodium vulgare** does not occur in India, though met with in Europe and Turkey in Asia. Dr. **Dymock** refers the *basfáij* to that species, however, but does not mention the region from which it is obtained. He says the rhizomes are aperient and deobstruent, and are considered to act as an expellant of peccant humours: they are also used as an alterative in a variety of

disorders and are frequently combined with Cassia pulp and honey.
He identifies the *basfáij* with the Polupodion (περὶ πολυποδίου) of the
Greeks, and the *Azrás-el-kalb* of the Arabs. Among the other medicinal
ferns may be mentioned **Actiniopteris dichotoma** (*Vol. I.*, *No. 448 A.*),
which *is* used as an anthelmintic and styptic. **Dymock** mentions a species
of **Asplenium** (known at Goa as *Káli pándan*) which is employed as an
alterative in cases of prolonged malarious fever. **Asplenium fimbriatum**
is said to be given by the natives of British Garhwal as a remedy for
snake-bite.

FERONIA, *Correa; Gen. Pl., I., 305.*

53

Feronia elephantum, *Correa; Fl. Br. Ind., I., 516;* Rutaceæ.
THE ELEPHANT- OR WOOD-APPLE, *Eng.;* BALONG, *Port.;* POMMIER
D' L'ELEPHANT, *Fr.*

Syn.—CRATÆVA VALLANGA, *Kœnig.*

Vern.—*Kaith, bilin, kait (kowit), kat-bél, kavitha,* HIND.; *Kath-bel, kait-
kát-bél,* BENG.; *Kainta, koch-bel,* SANTAL; *Koeta,* URIYA; *Kyth* (SHAJE-
HANPUR), N. W. P.; *Kait, bilin,* PB.; *Keiri* (AJMIR-MERWARA), RAJ.;
Katori, kavatha, SIND; *Kabit,* BERAR; *Kavit, kowit,* BOMB.; *Kawat,
kavith, kavatha, kovit,* MAR.; *Kotha, kavit,* GUZ.; *Vellam,* MADURA;
Vilám, vallanga, velá, kavit, kairt, TAM.; *Thana, kavit,* KONKAN;
Velagá, or néla velaga, elaka, yellanga, kapidh, TEL.; *Bilwar, byala da
hannu, byala, bélada, bél,* KAN.; *Vilám,* MALAY.; *Hman, mahan,*
BURM.; *Divul or diwul (meladi-kurundu,* TAM. in CEYLON), SING.;
Kapittha, kapipriya (dear to monkeys), *bilin (dadhiphala*—the fruit),
SANS.; *Kabit,* ARAB.; *Kabit,* PERS.

References.—*Roxb., Fl. Ind., Ed. C.BC.,* 374; *Brandis, For. Fl.,* 56,
Kurz, For. Fl. Burm., I., 198; *Gamble, Man. Timb.,* 62; *Dals. & Gibs,
Bomb. Fl.,* 30; *Stewart, Pb. Pl.,* 29; *Sir W. Elliot, Fl. Andh.,* 83, 133,
145, 159, 190; *Rev. A. Campbell's Report on Econ. Prod. Chutia Nagpur,
No.* 8211; *Mason, Burma and Its People,* 452; *Stocks, Report on Sind; Sir
W. Jones, V.,* 119, *No.* 42; *Pharm. Ind.,* 48; *Ainslie, Mat. Ind., I.,* 161;
II., 82; *O'Shaughnessy, Beng. Dispens.,* 14; *Moodeen Sheriff, Supp.
Pharm. Ind.,* 140; *Pereira, Mat. Med. II., p.* 550; *U. C. Dutt, Mat.
Med. Hind.,* 131, 303; *Dymock, Mat. Med. W. Ind., 2nd Ed.,* 142; *Phar-
macographia Indica,* 281; *Flück. & Hanb., Pharmacog.,* 131, 239; *S.
Arjun, Bomb. Drugs,* 22; *Murray, Pl and Drugs, Sind,* 79; *Moodeen
Sheriff's new work on Materia Medica, South India (Proof Copy), pp.
79—81; Baden Powell, Pb. Pr.,* 334; *Atkinson, Him. Dist.,* 736; *Econ·
Prod., V.,* 44, 52; *Drury, U. Pl.,* 212; *Lisboa, U. Pl. Bomb.,* 34, 148, 250,
291; *Birdwood, Bomb. Pr.,* 13, 142, 259, 324; *Cooke, Gums and Gum-
resins,* 17; *Atkinson, Gums and Gum-resins,* 5, 7, 16; *Liotard, Dyes,* 33;
Watson, Report on Gums, 4, 18, 20, 34, 65, 68; *Spons' Encyclop.,* 793,
1414, 1621, 1668, 1692-3; *Balfour, Cyclop.,* 1086; *Smith, Dic.,* 163;
Treasury of Bot., 490; *Kew Off. Guide to the Mus. of Ec. Bot.,* 25; *Kew
Off. Guide to Bot. Gardens and Arboretum,* 68; *Journ., As. Soc., II.,
ii.,* 1867, 79; *Home Dept. Cor. in connection with the Pharm. of India,*
238; *Indian Forester, III.,* 200; *V.,* 13; *XI.,* 388; *XIII.,* 119; *Gazetteers
of Bengal (Orissa), II.,* 180; *Of N.-W. P., I.,* 79; *IV., p.* LXIX;
X., 307; *Of Mysore and Coorg, I.,* 49; *Of Bombay, IV.,* 24, 285, 360;
VI., 13; *VII.,* 39, 40, 42; *XIII.,* 25; *XV.,* 69; *XVII,* 25; *XVIII ,* 47; *Of
Burma, I.,* 133; *Of C. P.,* 136; *Settlement Reports of C. P., Mandla,* 89;
Chanda, VI.; Chindwara, 110; *Upper Godavery,* 38; *Madras Manuals,
Cuddapah Dist.,* 263; *Trichinopoly Dist.,* 78; *Coimbatore Dist.,* 41; *Special
Reports furnished for this work by the Conservator of Forests, Southern
Circle, Bombay; Northern Circle, Bombay; Berar; Coorg; N. W. P.;
Ajmir; and Northern Circle, Madras.*

Habitat.—A medium-sized tree, found in the sub-Himálayan forests,
from the Rávi eastward; throughout the greater part of the plains of India,
being more plentiful in the moister tracts of Bombay, Madras, Bengal,
and Burma than in Northern India. To a considerable extent cultivated
as a road-side tree near villages. **Stewart** says he has not seen it wild

| The Wood-Apple. | (*G. Watt.*) | FERONIA elephantum. |

in the Panjáb, and though also scarce in the North-Western Provinces, the fruits obtained in Bundelkhand are spoken of as exceptionally fine. It flowers in February to May, and the fruits ripen about October, and often remain a considerable time on the trees.

Gum.—Dr. Moodeen Sheriff (in his forthcoming work, proofs of which have been obligingly furnished to the writer) gives perhaps the best account of this gum. He describes it as occurring "in small, roundish, oblong or tapering tears, or in broken pieces, varying in size from a pea to that of a soap-nut; generally colourless and transparent, sometimes opaque, with numerous minute cracks on the surface; odourless, bland, and mucilaginous in taste. This gum," he continues, "is very frequently confounded with the Indian Gum Arabic, for it not only bears a great resemblance to it, but there is also a great similarity between the pronunciation of the Tamil names of both, the former being called '*Vilám-pishin*' and the latter '*Vélam-pishin*' (Gum-*pishin*). Feronia gum being rather scarce and comparatively very dear, the native druggists take advantage of the above facts, and generally pick out the whiter and more transparent pieces from the Indian gum arabic and sell these for the former. The only ready and practical difference between these gums is that the gum of F. elephantum is invariably much whiter and more transparent than that of Acacia arabica." The *Pharmacopœia of India*, confirmed by several writers on Economic Products, describes the gum as occurring "in the form of irregular, semi-transparent, reddish-brown tears. Treated with water, it affords a brownish, tasteless mucilage, not less adhesive than that of gum arabic, for which it may be used as a substitute." Dr. Dymock says: "The gum is in tears or irregular masses, yellow or brownish; dissolved in water it forms an almost tasteless mucilage, much more viscid than that of gum arabic made in the same proportions."

The chemistry of the gum does not appear to have been worked out. Flückiger and Hanbury (*Pharmacog.*, *239—240*), however, give some interesting facts regarding it which have been reproduced by Dr. Dymock in his *Materia Medica of Western India* and in the *Pharmacographia Indica*. Flückiger and Hanbury say that dissolved, in two parts of water, Feronia gum "affords an almost tasteless mucilage, of much greater viscosity than that of gum arabic made in the same proportions. The solution reddens litmus paper and is precipitated, like gum arabic, by alcohol, oxalate of ammonium, alkaline silicates, perchloride of iron, but not by borax. Moreover, the solution of Feronia gum is precipitated by neutral acetate of lead or caustic baryta, but not by potash. If the solution is completely precipitated by neutral acetate of lead, the residual liquid will be found to contain a small quantity of a different gum, identical, apparently, with gum arabic, inasmuch as it is not thrown down by acetate of lead." A large proportion of Feronia gum, they continue, is therefore by no means identical with gum arabic. It deviates polarized light O·4° to the right, instead of 5° to left as with gum arabic. "Gum arabic may be combined with oxide of lead; the compound (arabate of lead) contains 30·6 per cent. of oxide of lead, whereas the plumbic compound of Feronia gum, dried at 110° C., yielded only 14·76 per cent. of Pb O." "Feronia gum, repeatedly treated with fuming nitric acid, produces abundant crystals of mucic acid." And concluding their brief notice of this substance they add: "We found our sample of the gum to yield 17 per cent. of water, when dried at 110° C. It left 3·55 per cent. of ash."

Dye.—At the beginning of the century Dr. Ainslie wrote of the GUM "that a celebrated painter mentioned" to Roxburgh "that it answers better for mixing with colours than gum arabic."

GUM.
54

DYE.
Gum.
55

F. 55

FERONIA **elephantum.**	**The Wood-Apple.**

DYE.

Dr. Warden, in a note to the writer on this gum, repeats the above statement, and *Spons' Encyclopœdia (page 1693)* puts the matter even stronger. "For preparing water colours, it has a reputation beyond all other gums. It is much cheaper than gum arabic, while apparently equal to it for all purposes." This statement of the price of the gum would at least appear to be incorrect, and the reputation of the gum as used with paints would seem to rest alone on Dr. Ainslie's original statement.

Balfour gives two sentences which probably allude to one and the same substance. These are : " When an incision is made in the trunk, a transparent oily fluid exudes, which is used by painters for mixing their colours." " It yields a large quantity of a clear white gum, much resembling gum arabic in its sensible properties." So, again, Cooke, in his Report on Gums, &c., of India, while referring to this reputed property, writes that " Dr. Ainslie says that the wood-apple gum is used by dyers and painters, particularly the miniature and chintz painters; it is also employed in making ink and certain varnishes, and by the brick-layers in preparing a fine kind of whitewash." No modern writer has, however, confirmed the frequently-repeated statement of its use to painters. Dr. McCann, for example, in his *Dyes and Tans of Bengal ;* Mr. Liotard, in his *Memorandum on Dyes and Dyeing ;* and Mr. Wardle, in his recent *Report on the Dyes of India,* make no mention of **Feronia** gum. So also Sir E. C. Buck (in his work on the *Dyes of the North-Western Provinces),* while dealing fully with the art of calico-printing and distinguishing the properties of the gums used, does not allude to **Feronia** gum.

OIL.
56

Essential Oil.
57

Oil.—One or two writers mention an OIL, but in such general terms that very little can be compiled of a definite nature on this subject. In the Settlement Report of the Chanda District, for example, it is stated that "oil extracted from the fruit is a remedy against itch." Cooke, in his *Oils and Oil-seeds of India,* says that the seeds are reputed to afford an oil. The authors of the *Pharmacographia Indica* write : " the leaves yield to distillation a small quantity of ESSENTIAL OIL similar to that obtained from *bael* leaves."

MEDICINE.
Ripe Fruit.
58

Medicine.—The RIPE FRUIT, made into a sort of *chátní,* with oil, spices, and salt, is esteemed by the natives. The fruit itself is an aromatic anti-scorbutic, and in the form of a sherbet is sometimes given to children, alone or in combination with *bél* fruit, as a stomachic stimulant. It is supposed to increase the appetite and to possess alexipharmic properties.

Pulp.
59

The PULP is reputed to be especially useful in cases of affections of the gums and throat. It is also often applied externally as a remedy in snake-bite or employed to remove the pain caused by venomous insects. But for this purpose the powdered RIND may be employed if the pulp be not pro-curable. The Hindus regard the UNRIPE FRUIT as a useful astringent in diarrhœa and dysentery, and Muhammadan authors, for example the writer of the *Makhzan-el-Adwiya,* affirm that the fruit is cold and dry in the second degree, refreshing, astringent, cardiacal, and tonic, a useful remedy in salivation and sore-throat, strengthening the gums and acting as an astringent. Elephant-apple is often used to adulterate *bél* fruit, but the two fruits should be easily enough distinguished.

Rind.
60
Unripe Fruit
61

Leaves.
62

The LEAVES are aromatic and carminative, and have the odour of anise (*Ainslie*). The author of the *Makhzan-el-Adwiya* describes them as very astringent and as possessing the taste and odour of Tarragon. Ainslie remarks that the native practitioners of South India (in his day) prescribed the leaves "in the indigestions and slight bowel affections of children."

Bark.
63

The BARK is said to be sometimes prescribed for biliousness.

Gum.
64

The GUM has already been alluded to. Ainslie was the first writer to affirm that in medicinal properties the gum of this tree came nearest of

| The Wood-Apple. | (*G. Watt.*) | FERONIA elephantum. |

all Indian gums to the true gum arabic. "The Tamool practitioners prescribe a solution of gum arabic," he says, "to relieve tenesmus in bowel affections, and as we do in other cases requiring demulcents," and he states that for this purpose **Feronia** gum "is commonly used for medicinal purposes by all the practitioners of Lower India."

A fatty OIL has been incidentally referred to, and although its exact source and nature have not been determined, it may here be stated that, according to some writers, this oil is not only useful in itch and other skin diseases, but in leprosy. A medicated oil is, however, also employed for these purposes which would be more correctly described as sweet oil impregnated with the pulp or powdered rind. It is probable that this preparation may be the so-called **Feronia** oil of medical writers, unless indeed the essential oil distilled from the leaves be the substance alluded to. Considerable ambiguity, it must be admitted, exists in the literature of **Feronia** oil.

SPECIAL OPINIONS.—§ "Unripe fruit astringent. Gum—Gum Arabic" (*Thomas Ward, Apothecary, Madanapalle, Cuddapah*). "Very common in the Mysore jungles. The unripe fruit is much used for dysentery and diarrhœa" (*Surgeon-Major John North, Bangalore*). "The ripe fruit is by some said to promote digestion, by others is regarded as deleterious, bringing on rheumatism and chest complaints" (*Assistant Surgeon Shib Chunder Bhuttacharji, Chanda, Central Provinces*).

Food.—This tree produces a round, hard-shelled FRUIT, of the size of a large apple, which has a strong odour when ripe, and a very acrid taste not unlike that of the Bengal quince. The natives sometimes eat the raw fruit with sugar. A jelly, much resembling black-currant, is prepared from the pulp of the fruit, which, however, has a very astringent taste. **Surgeon-Major Robb** informs the writer that the fruit is used as a condiment. Under the paragraph "Medicine" above it has been stated that a *chátní* is also made of it. In the *Medical Topography of Dacca* it is said that the name Elephant-apple proceeds from the fact that the elephant is very fond of the fruit. "It is," **Dr. Taylor** adds, "prepared by the natives as an article of diet, by mixing the pulp with salt, oil, and pepper." **Dr. Buchanan-Hamilton**, in his account of Dinajpur, says, the fruit is eaten by the natives "but is very poor." On the other hand, many writers speak of the fruit in much higher terms. The Conservator of Forests, Northern Division, Madras, in a recent communication, says:— "This tree is common and of good size in the Northern Circars. It is planted throughout the Circars and Carnatic. The fruits are eaten, and may usually be seen on sale in the bazars." In the Trichinopoly Manual it is said, "the fruit is eaten by all classes." In the Settlement Report of Chanda, it is affirmed that "the fruit is much eaten and the leaves and the bark are used in cases of bilious illness."

Structure of the Wood.—Yellowish-white, hard. Annual rings distinctly marked by a white line. Weight about 50℔ per cubic foot. It is used for house-building, naves of wheels, oil-crushers, and agricultural implements. Somewhat contradictory opinions are given regarding this timber. **Dr. Buchanan-Hamilton** (*Statistics of Dinajpur, p. 153*) says that "the wood is not applied to any use." The Conservator of Forests, Southern Circle, Bombay, has recently reported that "the wood, which is hard, strong, and lasting, is used for various purposes." In the *Trichinopoly District Manual* it is stated that "the wood is white, hard, durable, and fine-grained;" and in the *Mysore and Coorg Gazetteer* it is added, to a similar description, that the wood is "suited for ornamental carving."

Domestic Uses.—The hard dry shells of small FRUITS are used as snuff-boxes.

MEDICINE.

OIL.
65

FOOD.
Fruit.
66

TIMBER.
67

DOMESTIC.
Fruit.
68

Ferrum, see Iron.

69

FERULA, *Linn.; Gen. Pl., I., 917.*

A genus of umbelliferous herbs, comprising some sixty species, a few of which, though growing on perennial root-stocks, attain annually a height of from 8 to 10 feet. Interest in the species of FERULA is mainly centred on the sub-arborescent forms—the Giant Fennels,—which may be said to be characteristic of the dry semi-desert tracts of Central Asia. From these are obtained the various forms of Asafœtida, Galbanum, Sambul, &c. So much confusion even still exists, however, in the literature of these famed drugs, that the writer has thought it the preferable course to give a concise review of the history of Asafœtida and rest satisfied with brief notices under the individual species of FERULA. But even in so far he will touch only on the species that can be regarded as connected with the Trade and Commerce of India.

HISTORY.
70

History of Asafœtida.—When **Dr. Falconer,** in 1838, discovered **Narthex Asafœtida** in the valley of Astor, North Kashmír, it was at first supposed that the problem of the source of the drug asafœtida had been solved. The roots procured by him were planted in the Saharanpur Botanic Garden. Seeds were subsequently sent to the Royal Botanic Gardens at Edinburgh. In 1842 these germinated, and in 1859 several of the plants flowered, yielding seeds which were distributed to the various botanical gardens throughout the world. From this source the so-called asafœtida plant in cultivation was derived. It must be observed, however, that while this species yields an asafœtida-like substance, it has by no means been demonstrated that any portion of the asafœtida of European commerce is derived from it. **Sir J. D. Hooker** figured the plant in the *Botanical Magazine,* No. 5168. He *then* wrote that it "yields excellent asafœtida in the form of copious milky juice." But he added, "It would be impossible to discuss here the vexed question of the history of the origin of all the asafœtidas, nor would the discussion be very profitable." Long anterior to **Dr. Falconer's** discovery, the German traveller **Kœmpfer,** in the year 1687, saw asafœtida being extracted from a species of **Ferula** in Lauristan in Persia. He brought to Europe samples of the resin and a fragmentary specimen of the plant from which that resin had been obtained. These specimens were described by **Linnæus** under the name of **Ferula Asafœtida.** But **Kœmpfer's** collections are in the Sloane Herbarium at the British Museum, and were carefully examined by **Dr. Falconer,** with the result that he entertained a strong suspicion that **Ferula Asafœtida,** *Linn.,* was not the plant he had discovered in Northern Kashmír. He accordingly named his plant **Narthex Asafœtida.** Hooker (*Bot. Mag., l. c.*) wrote that "it is certain that **Kœmpfer** had two plants (species or varieties) in view, from different countries, that his descriptions and drawings and specimens (in the British Museum) do not tally, and that though **Dr. Falconer** considers his plant one of **Kœmpfer's,** other botanists do not." The discovery, in the Steppes east of the Caspian, of the plant Bunge named **Scorodosma fœtidum** is also referred to by Hooker. Borszczow, who devoted some attention to the genus **Ferula,** also examined **Kœmpfer's** specimens, and came to the conclusion that they should rather be referred to **Scorodosma.** Royle, on the other hand, held the opinion that **Kœmpfer's** plant should be assigned to the genus **Narthex.** More recently **Boissier** referred an asafœtida-yielding species, discovered by him in Persia, to **Ferula Asafœtida,** *Linn.,* and *that* modern writers regard as **Scorodosma fœtidum,** *Bunge,* a synonym for **Ferula fœtida,** *Regel,* but view it as most probably not **Ferula Asafœtida,** *Linn.* **Dr. Dymock,** however, writes to the author that he is disposed to think that **Ferula Asafœtida,** *Linn.,* may prove the same as **Ferula fœtida,** *Regel.*

The learned authors of the *Pharmacographia* are careful to say that it has not been proved that either of the plants reputed to yield the

Asafœtida of European commerce is actually the source of that drug. The species they allude to are **Ferula Narthex**, *Boiss.* (the **Narthex Asafœtida**, *Falconer*), and **Ferula Scorodosma**, *Bentl. & Trim.* (the **Scorodosma fœtidum**, *Bunge*, and **Ferula Asafœtida**, *Linn.*, in *Boiss. Fl. Or.*). **Dr. Dymock** has the honour of having been one of the first writers who pointedly drew attention to the fact that the Asafœtida most highly prized in India is distinct from the Asafœtida of European commerce. This was noticed some time previously, however, by **Guibourt**, *Hist. des Drogues, III., 220 (1850)*, and named by **Vigier Asafœtida nausseux**—*Gommes-résines des Ombelliféres, Paris, 1869.* **Dr. Dymock** restricted the vernacular names (which, prior to his study of the subject, were viewed as synonymous), assigning to the Indian most highly-prized drug the name of *Hing*, and that of *Hingra* to the European Asafœtida. In a letter to the writer he says, however, that "the name *Hing* may be applied to any choice asafœtida. *Hingra* means common asafœtida, just as *Rai* in Guzerathi means Mustard and *Raira* Rape. With the public generally all kinds of asafœtida are *Hing*." **Fluckiger & Hanbury**, in their *Pharmacographia*, speak of *Hingra* as if it were an inferior quality of the European asafœtida instead of the Indian name for that drug. There are, however, many qualities of both *Hing* and *Hingra*, and adulteration with foreign materials is carried to a great extent. But it would seem also that there are, apart from adulteration, different qualities, the result, perhaps, of more careful preparation, or due to being derived from different parts of the plant, or to being collected at different seasons, or from different species of **Ferula**. **Dr. Dymock** was fortunate in procuring, from a merchant at Yezd, specimens of the plant which affords the Khorassan asafœtida—the drug which, on arrival in India, is designated *Hing*. These specimens he forwarded to the late **Mr. D. Hanbury**, and that gentleman submitted them to **M. Boissier**, who identified them as **Ferula alliacea**, an opinion which Hanbury entirely concurred in. Thus, so far, a definite conclusion seemed to have been arrived at. The Indian Asafœtida or *Hing* was established as obtained from a distinct species from the article *Hingra* or European asafœtida. The "**Ferula, sp.,** *Hingra*" of the first edition of his work, **Dr. Dymock** in his second edition identified as obtained from "**Ferula Narthex**, *Boiss.*, and **Scorodosma fœtidum**, *Bunge.*" In his account of this product he there says :—"Commercial Asafœtida is collected by the Kákar Pathans in Western Afghánistan; in May the mature roots begin to send up a flowering stem, which is cut off and the juice collected in the manner described by **Kœmpfer**, who witnessed its collection in the province of Láristan in Persia." **Dr. Dymock** obtained this information, together with a specimen of the plant, from **Dr. Peters**; but in a correspondence on this subject, he authorises the writer to say that he is now convinced **Dr. Peters'** plant is **Ferula fœtida**, *Regel.*

Turning to the more recent botanical publications regarding Afghánistan—**Dr. Aitchison's** various official reports—it is somewhat surprising that that author makes no mention of having seen **Ferula Narthex**. He deals, however, with **Ferula fœtida**, *Regel*, and under that species he places the following synonyms :—F. **Scorodosma**, *Bent. & Trim.*, **Scorodosma fœtidum**, *Bunge*, and **Ferula Asafœtida**, *Boiss.* He affirms that the resin obtained from that species is "the drug of commerce called Asafœtida—*Anguza, Hing.*" Before the Pharmaceutical Society of Great Britain, **Dr. Aitchison** also read a paper dealing with the economic products of Afghánistan, and was highly complimented for the valuable services he had rendered in clearing up many obscure points regarding Asafœtida, Galbanum, &c., &c. The opinion seemed to have been formed that the whole difficulty regarding Asafœtida had been removed.

FERULA.	Two forms of Asafœtida.

HISTORY.

In the correspondence with **Dr. Dymock** (to which reference has been made above), there occurs the following passages, which may fitly be quoted in concluding this brief review : "I think," he writes, "we may regard it as settled that the asafœtida of commerce in Europe is all derived from F. fœtida, *Regel,* growing in Persia and Afghánistan." **Dymock** retains two species, however, as yielding—the one the Indian, the other the European—asafœtida and (following **Holmes**) gives the synonymy of these species as follows :—

" 1. **Ferula alliacea,** *Boiss.*
　　Syn.—F. ASSAFŒTIDA, *Boiss. et Bunge, non Linn.*
" This produces the *Hing* of Bombay markets—the kind of asafœtida preferred. as a condiment in India.

" 2. **F. fœtida,** *Regel.*
　　Syn.—F. SCORODOSMA, *Bentl. & Trim.* (wrongly lettered in their plate No. 127 as FERULA FŒTIDA, *Benth. & Hook. f.*), also SCORODOSMA FŒTIDUM, *Bunge,* and F. ASAFŒTIDA, *Boiss.* (? *Linn.*)
" The selected gum from the bud is called Kandaharí *Hing,* and fetches a high price. The thick opaque gum afterwards obtained from the root is the asafœtida of European commerce."

Presumably, therefore, the opaque gum is the *Hingra,* but according to the above notes, the same species furnishes a superior form of *Hing* also. It may accordingly be suggested that perhaps after all certain species of **Ferula** yield either *Hing* or *Hingra,* or both these drugs—the superior and inferior qualities of Asafœtida. Future research may reveal the fact that, as with **Cannabis sativa** in affording various resinous substances, so with certain species of **Ferula,** different systems of extraction and manipulation, or diversified conditions of climate and soil, produce both *Hing* and *Hingra.* It is difficult to believe that only two species contribute to the supply of these products, while perhaps half a dozen are alluded to by travellers as affording a milky sap which, on drying, possesses at least the physical properties of Asafœtida. It may, however, be safe to affirm that the bulk of the Persian drug imported into India by sea is the *Hing,* derived from **Ferula alliacea,** but that a considerable proportion of the *Hingra* comes also from Persia and Turkistan. The whole of the asafœtida that enters India by the frontier land routes from Afghánistan is now satisfactorily proved to be derived from **F. fœtida.** This conclusion would seem to be borne out by the trade returns of India, where a far larger quantity of *Hingra* (European Asafœtida) is shown to be exported to Europe and other countries than would appear to be imported from Afghánistan by road, rail, and river.

TRADE.
71

TRADE IN ASAFŒTIDA.

In the statement of the Trade and Navigation of British India, Asafœtida was apparently first separately returned (apart from other minor drugs) in the year 1876-77. Since, however, almost the entire traffic takes place with Bombay, the Asafœtida statistics of that Presidency, for earlier years, may be accepted as representing the whole of India. In the report for 1868-69 two forms of asafœtida are separately recorded in the Presidency Statistics ; these were :—

(*a*) *Hing*—

Imports from the Persian Gulf . . . 1,538 cwt., valued at R		85,118
" " Madras . . . 7 " "		412
" " Sind (Karáchi) . . 695 " "		18,455
These give a total of *Hing* imported into Bombay of . . . . . 2,240 " "		1,03,985

Trade in Asafœtida.	(*G. Watt.*)	FERULA.

(*b*) *Hingra*—
Imports from the Persian Gulf . . . 1,893 cwt., valued at R 18,935
  ,,   ,, Sonmeanee and Meckran . 20 ,,   ,,   114

These give a total of *Hingra* imported of . 1,913 ,,   ,, 19,049

The *Pharmacographia* quotes the similar returns for 1872-73, *viz.*, 3,367 cwt. of *Hing* and 4,780 cwt. of *Hingra*, but the authors of that work would appear to have regarded the former as the asafœtida of European commerce and the latter a crude article, since they write "the value of the latter is scarcely a fifth that of the genuine kind." Later on they deal with *Hing*, remarking that "among the natives of Bombay, a peculiar form of asafœtida is in use that commands a much higher price than those just described." This mistake is. here pointedly alluded to, as it is current in the literature of asafœtida. As stated above, there are doubtless many qualities of both *Hing* and *Hingra*, but the asafœtida of European commerce is *Hingra*, not *Hing*. In 1876-77 the total imports by sea into India (of *Hing* and *Hingra* collectively) were 4,472 cwt., valued at R2,16,638, and from that year to the present date all but a few cwts. of the imports by sea have come from Persia. Madras and Bengal occasionally receive small parcels from Ceylon or Aden, but with these exceptions the entire traffic takes place between the Persian Gulf and Bombay. Asafœtida is not separately returned in the statement of coastwise traffic (*e.g.*, between province and province), but it would appear that a much larger share in this trade is yearly being taken by the railways. For example, an important item of the coastwise traffic in asafœtida used formerly to consist in the supplies drawn by Bombay from Karáchí. A very considerable slice of the Indus river trade. has doubtless been taken by the Kandahar State Railway· (tapping the Kandahar source), and by the North-Western Railway at Peshawar, draining the Kábul market. The following may be given as the IMPORTS of Asafœtida into India *by land* during the past five years :—

1884-85	.	.	. .	. 1,218 cwt., valued at R1,04,023
1885-86	.	.	. .	. 1,775 ,, ,, 95,652
1886-87	.	.	. .	. 1,090 ,, ,, 53,310
1887-88	.	.	. .	. 1,030 ,, ,, 47,192
1888-89	.	.	. .	. 907 ,, ,, 37,615

Of these land imports the major portion comes from Kábul, and is presumably therefore derived from F. *fœtida*,—the *Hingra*.

The IMPORTS *by sea* during the corresponding periods were :—

1884-85	.	.	. .	. 10,340 cwt., valued at R3,50,076
1885-86	.	.	. .	. 7,228 ,, ,, 2,69,883
1886-87	.	.	. .	. 5,704 ,, ,, 2,53,303
1887-88	.	.	. .	. 4,521 ,, ,, 1,70,973
1888-89	.	.	. .	. 9,504 ,, ,, 4,31,502

The figures for the last of these years relate to Bombay: as a rule Sind is the only other province that receives asafœtida by sea (except small quantities imported by Bengal and Madras from Ceylon or Aden), and the imports into Sind were last year 50 cwt., valued at R797. During the same periods the foreign EXPORTS (drawn from the above imports) were :—

1884-85	.	. .	.	. 2,638 cwt. valued at R57,471
1885-86	.	. .	.	. 2,530 ,, ,, 49,026
1886-87	.	. .	.	. 1,865 ,, ,, 42,543
1887-88	.	. .	.	. 1,553 ,, ,, 27,451

The figures for the year 1888-89 have not as yet been published. It will thus be seen that, deducting these exports from the total imports (in

| FERULA. | Trade in Asafœtida. |

round figures), about two-thirds of the imported drug remain in India, so that India is itself perhaps the largest asafœtida-consuming country in the world. The highest exports on record were in 1883-84, *viz.*, 4,065 cwt., valued·at R86,457, and the following year showed the highest imports, *viz.*, 10,340 cwt., valued at R3,50,076.

In the statement of the Trade and Navigation of British India, however, a trade is shown in exporting asafœtida, which is returned as "Indian produce and manufacture." The writer is utterly at a loss to understand what this can mean. He is not aware that any asafœtida is produced in India, and therefore (as with camphor) it seems probable that the drug undergoes some process of "manufacture," more probably a systematic adulteration than a purification. There are two features of this so-called Indian asafœtida that may be here mentioned. It goes entirely to the United States of America, Australia, and Mauritius; none of it to Europe or China. It is exported from Calcutta or Madras; none of it from Bombay—the port that supplies Europe and China. The trade in the so-called Indian asafœtida fluctuates very considerably, but it seems to have been steadily declining for some years back. In 1879-80, however, it amounted to 1,130 cwt., valued at R23,698, and of this the United States took 943 cwt. In 1884-85 it amounted to 1,343 cwt., but the average of the past ten years does not much exceed 300 cwt., and in 1887-88 the trade had decreased to 4 cwt., 3 of which went to Australia.

PRICES, DESCRIPTION, &c.—The declared value of products in trade statistics are not often of much practical importance, since dealers may be presumed to give a valuation of their articles which best suits their own interests. Viewing the figures given above, remarkable fluctuations in the declared values will be observed which are, to some extent, doubtless due to the reason given above. The article varies much, however, according to supply and purity. **Dr. Dymock** says of *Hingra* "the imports into Bombay are about 2,500 cwts. annually from Persia and Afghanistan. Value R10 to R20 per Surat maund of 37½lbs." There would seem to be some mistake as to this estimate of the extent of the Bombay imports of *Hingra*. Last year (1888-89) the imports by sea were 5,042 cwt. and from Kábul 907 cwt. An average of 5,000 cwt. of *Hingra* would thus appear a safer estimate. **Dr. Dymock** next deals with Khandahari *Hing*, which he concludes is derived from the same plant as *Hingra*. He says it comes into "the Bombay market in small quantities; it is sewn up in goat skins, forming small oblong bales, with the hair outside. When it first arrives it is in moist flaky pieces and tears, from which a quantity of reddish-yellow oil separates on pressure; the gum-resin is also of a dull reddish-yellow colour, soft and somewhat elastic; with an odour recalling that of garlic and oil of caraways. By keeping, it gradually hardens and becomes brittle and of a rich red-brown colour; the odour also becomes more purely alliaceous, and approaches to that of the commercial kind." "This kind of *Hing* is entirely consumed in Bombay by the manufacturers of adulterated asafœtida, its strong odour and flavour making it especially valuable for this purpose. The average value is R25 per Surat maund of 37½lb." The ordinary form of *Hingra* (good quality) "occurs in tears or flat pieces, upon the under-surface of which particles of sand often adhere; the external surface is yellowish, but the fresh fracture is of a pearly white, which, by exposure to the air, becomes bright pink and finally dirty yellow. Inferior samples consist of agglutinated tears, with a certain proportion of moist brown clammy gum-resin filling up the interspaces between them. Sometimes the asafœtida which comes from Persia is a homogeneous, soft, white, mass like clotted cream; these parcels upon exposure to the air develop

Asafœtida—Hing.	(*G. Watt.*)	**FERULA alliacea.**

an unusually bright pink colour. The drug has a powerful but not purely alliaceous odour, and a bitter acrid taste " (*Dymock*).

Of *Hing*, Dr. Dymock also furnishes an admirable description. It is known in the Bombay market, he says, " as *Abu-shaheri-Hing ;* it arrives in skins which contain about 100℔; latterly some boxes- have been received. The quality varies greatly ; inferior parcels contain an undue proportion of the root ; in Bombay it is often still further adulterated by mixing it with gum arabic in different proportions, according to the priced article required. To do this the package is broken up and moistened, the gum is then added, and the whole trodden together by men with naked feet upon a mat. When sufficiently mixed, it is sewn up in skins to imitate the original packages. Recently, adulteration with sliced potato has been observed. *Hing* of good quality is worth about R80 per cwt. in Bombay." In an earlier passage Dr. Dymock gives additional facts regarding this form of asafœtida. He writes : " The collected mass, consisting of alternate layers of root and gum-resin, when packed in a skin (in quantities of about 100℔) forms the *Hing* of Indian commerce : it is imported into Bombay in large quantities (about 2,500 cwts. annually), and is valued at the Custom House for assessment at R55 per cwt., commercial asafœtida, *Hingra*, being only valued at R20." It may here be added that the imports of *Hing* for many years past have never been below 3,500 cwt., and last year they were 4,462 cwt. In a report on the Land Trade of Sind, it is stated that Afghánistan asafœtida is valued at R50 per maund, "while that imported from Beluchistan is only R14 per maund, the latter having been of a very inferior or coarse description." Dr. Aitchison came across a root of asafœtida in Northern Beluchistan after much difficult searching which he believed to belong to another species, *i.e.*, not F. fœtida. He found many leaves in traversing the plains, where he believes during summer the plant must have grown in abundance. There are only one or two other isolated references to a Beluchistan asafœtida, but nothing of a definite nature can be learned regarding it. The imports by the Kandahar State Railway are valued very much higher than those that appear in the other commercial returns. But in concluding this statement of the Indian trade in asafœtida, the reader's attention may be directed to the fact shown in the statement of the imports from Karáchí to Bombay (quoted in the opening paragraph above), *viz.*, that *Hing* and not *Hingra*, as might have been expected, appears in the early official returns.

(*J. Murray.*)

Ferula alliacea, *Boiss.*

76

Syn.—F. ASSAFŒTIDA, *Boiss. et Bunge (non Linn.), Fl. Or., II.,* 995.

Vern.—*Hing,* HIND. ; *Anjudán,* KASHMIR ; *Hing,* BOMB. ; *Hing,* GUZ. ; *Kyam, perungayam,* TAM. ; *Hingu,* SANS. ; *Hiltut,* ARAB. ; *Angusa, anguseh,* PERS.

As explained above, the name *Hing* literally means pure or superior *Hingra*. It is thus probable that all the vernacular names for this and the next species are vulgarly applied to the resinous substance obtained from any of these **Ferulas.**

References.—*Pharm. Ind., 102 ; Ainslie, Mat. Ind., I., 20 ; O'Shaughnessy, Beng. Dispens., 362 ; Moodeen Sheriff, Supp. Pharm. Ind., 61 ; Dymock, Mat. Med. W. Ind., 2nd Ed., 381 ; Flück. & Hanb., Pharmacog., 319 ; S. Arjun, Bomb. Drugs, 66 ; Waring, Bazar Med., 21 ; Birdwood, Bomb. Pr., 40 ; Cooke, Gums and Gum-resins, 52, 55 ; Spons' Encyclop., 1634 ; Kew Off. Guide to the Mus. of Ec. Bot., 76.*

Habitat.—A herb of much the same appearance as F. fœtida, but smaller, growing only to a height of from 2 to 4 feet, the diameter of the crown of the root seldom attaining more than 2 inches. Found in Eastern

| FERULA alliacea. | Asafœtida—Hing. |

Persia in the neighbourhood of Djendack and Yezd, and in Khorassan near Seharud, Nischapur, Meshed, Dehrachtindjan, and Kerman (Buhse). Called *Angusheh* in Khorassan and *Zendebuj* in Kirman (*Boiss. Fl. Or., 995*). It grows on stony arid soil, and to an altitude of 7,000 feet.

CULTIVATION.
77

Cultivation.—Grows wild ; is not cultivated.

GUM-RESIN.
Collection.
78

GUM-RESIN.

COLLECTION.—The following description is given by **Dr. Dymock** (*vide Mat. Med. of Western India, p. 382*), òn the authority of a merchant of Yezd, who had personally seen the process going on :—

"The hill-men collect the gum-resin, taking an advance from the merchants. The time for collecting it is in the spring." "The collectors protect each plant by building a small cairn of stones round it ; they also remove the soil from the upper portion of the root, making a kind of circular basin. When the stem begins to grow it is cut down, and the upper part of the root being wounded, a small quantity of very choice gum is collected, which seldom finds its way into the market. Afterwards, a slice of the root, about ¼ inch thick, is removed every two or three days with the exudation adhering to it, until the root is exhausted. The collected mass, consisting of alternate layers of root and gum-resin, when packed in a skin (in quantities of about 100℔), forms the *Hing* of Indian commerce."

Characters.
79

CHARACTERS.—The gum-resin, as found in the market, consists of a blackish-brown, originally translucent, brittle mass of extremely fœtid alliaceous odour, unadulterated by earth, or gypsum, but always containing slices of the root. **Dr. Dymock** mentions that in Bombay it is often adulterated by the addition of gum arabic, and that the cheaper sorts contain an undue proportion of the root. This is produced by the exhausted root being cut up and mixed with the gum-resin and water. Recently, adulteration with sliced potato has been observed.

The term "*hira-hing*" is said to be applied to a liquid of treacly consistence, often found in the 'centre of the bales, which is squeezed out and sold at a high price. (*Spons' Encyclop.*)

Chemistry.
80

CHEMICAL COMPOSITION.—The essential oil is very abundant, and differs from that of *Hingra* in having a reddish hue, being of higher specific gravity, and having a much stronger rotatory power.

An alcoholic tincture is not precipitated by acetate of lead, nor is the sulphuric acid solution fluorescent. In all these respects there is consequently a well-marked difference between *Hing* and *Hingra* (*Flückiger and Hanbury*).

MEDICINE.
Gum-resin.
81

Medicine.—This drug is very much used in India, and has, from the earliest times, been held in great esteem by eastern doctors. It is reputed a carminative and antispasmodic, and therefore as useful in colic, cholera, &c., and when taken daily it is said to ward off attacks of malarial fever. Hindú medical writers direct it to be fried before being used. The Muhammadans place asafœtida amongst their aphrodisiacs and hypnotics,

Fruit.
82

and consider the FRUIT to be stimulant (*Dymock*). Waring, in the *Pharmacopœia of India*, writes, "it produces excellent effects in the advanced stages of pneumonia and bronchitis in children." Information collected from medical men in various parts of the country shew that the drug is considered useful as a carminative in colic and flatulent dyspepsia, as an anthelmintic in cases of round worm, and as an emetic. It is also described by two writers as a useful local anæsthetic in hemicrania and dental caries.

SPECIAL OPINIONS—§ "*Hing* is said to be used internally in guineaworm and colic. Dose 5 to 15 grains, made into a paste with water ; it is used as an external application to frontal headache" (*Joseph Parker, M.D.,*

F. 82

MEDICINE.

Deputy Sanitary Commissioner, Poona). "It is also an aphrodisiac; and is very useful in rendering *dál* digestible—an important article of native dietary" (*Surgeon-Major A. S. G. Jayakar, Muskat*). "Useful in dyspepsia with indigestion" (*Surgeon J. C. H. Peacocke, I.M.D., Nasik*). "Given as an emetic in poisoning by opium and other substances. Also used to expel round worms. Very useful in flatulent colic" (*Assistant Surgeon Shib Chundra Bhuttacharji, Chanda, Central Provinces*). "An emulsion (grs. 5 to 1 drachm) dropped into the nostril is useful in cases of hemicrania. In caries of the teeth a mixture of opium and *hing* may be put into the hollow tooth" (*Surgeon James McCloghey, Poona*). "The utility of asafœtida in the early stages of cholera appears to me to be undoubted. It should be given in combination with camphor and black pepper, opium being added if the disease is not fully developed" (*Surgeon S. H. Browne, M.D., Hoshangabad, Central Provinces*). "The native midwife uses this to encourage the lochial discharge after childbirth. The gum-resin is first fried, a small quantity is then mixed with garlic and palmyra jaggery, a bolus is thus made, and given to the patient every morning" (*Surgeon W. F. Thomas, Madras Army, Mangalore*). "I have found it very useful in reducing the irritant properties of purgatives when they have to be continued long as in spleen diseases" (*Surgeon K. D. Ghose, M.D., M.R.C.S., Khoolna*).

Food.—The GUM-RESIN is employed by the natives of all parts of India as a condiment, and is especially prized by the vegetarian Hindu classes. It is mixed in various ways with rice, *dál*, &c. There is no mention of the stem or leaves of this species being used as food or fodder.

FOOD. Gum-Resin. **83**

Trade.—See article Trade under the account of the genus.

Ferula fœtida, *Regel.*

84

Syn.—FERULA SCORODOSMA, *Bent. & Trim., Med. Pl, No. 127;* SCORODOSMA FŒTIDUM, *Bunge;* FERULA ASAFŒTIDA, *Boiss., Fl. Or. II., 99 (non Linn.).*

Vern.—*Hingra* (also *Hing*), HIND.; *Angusa-kema, kurne-kema, khorakema* (the plant), *Hing* (the Resin) (according to Aitchison in Afghan Delim. Com. Report), AFG.; *Vaghayani,* SIND; *Hingra,* BOMB.; *Hingu* SANS.

References.—*Aitchison's Afghan Del. Com. Rept., p. 68; Irvine, Med. Top. Ajmir (F. Narthex), 136; Fleming, Med. Pl. & Drugs (F. Narthex), in As. Res.; Vol. XI., 185; Pharm. Ind., 102; O'Shaughnessy, Beng. Dispens., 37; Dymock, Mat. Med. W. Ind., 385; Flück. & Hanb., Pharmacog., 314; S. Arjun, Bomb. Drugs, 67; Jour. and Trans. Pharmac. Soc., 3rd Ser., XVII., 465; Birdwood, Bomb. Pr., 41; Cooke, Gums and Gum-resins, 50; Dr. F. Watson's Report on Gums (pb. by P. W. D.), p. 26; Review, in the Chemist and Druggist, of Dr. Aitchison's paper on Plants and Plant-products of Afghanistan, delivered before the Pharm. Soc. of Great Britain; also the same reprinted in the Indian Forester, XIII., 90—95.*

NOTE.—Many of the references above are to passages describing **Ferula Narthex** or **Narthex Asafœtida**, which are presumed to be in reality accounts of **F. fœtida**, *Regel.*

Habitat.—A herb, with a circular mass of foliage, which may grow to the extent of 6 feet in diameter, springing annually from the perennial rootstock; the flowering plant shoots up a stem, peculiarly massive and pillar-like, to the height of 4 to 5 feet. It has been described by Lehman as growing over the whole of Southern Turkistan as far north as the river Syrdarja; by Bunge it was found in the sandy deserts and arid hills of Eastern Persia, in Khorassan, and the neighbouring parts of Afghánistan near Herat; and by **Dr. Aitchison** (with the Afghan Boundary Commission of 1884-85) in the same region. It has also been collected further north in Central Asia between the Caspian and Sea of Aral by Borszczow.

FERULA foetida.	Asafœtida—Hingra.

CULTIVATION.
85

Cultivation.—It is described by Aitchison and others as growing freely of itself, without any cultivation, in the sandy deserts of the countries given above. Dr. Aitchison in his paper on "Some Plants of Afghanistan, and their Medicinal Products" writes,—"The country in which these UMBELLIFERÆ flourish consists of the great shingle and conglomerate plains lying between the hills and the beds of the rivers, which are broken up by numerous ravines and traversed by what are usually dry water-courses, which once in every two or three years, on the occurrence of heavy falls of snow on the hills above, or local showers of rain, suddenly become roaring torrents. The altitude of these plains above the sea-level ranges from 2,000 to 4,000 feet. Theseplains during winter are perfectly treeless and bare, the only signs of a past vegetation being the gnarled remains scarcely over a foot in height, of a few shrubs." "In early spring great cabbage-like heads are to be seen distributed at intervals amongst the asafœtida plants. Their peculiar forms represent the primary stages of the flower-heads, enclosed and completely covered up by the large sheathing stipules of its leaves." From these the tall flowering stalk arises and the circular mass of foliage springs out, after which the plant assumes its fully grown appearance. Only about one plant in a hundred is said by Aitchison to bear a flowering stem. The only localities in India offering the natural conditions required for the growth of **F. foetida** are perhaps parts of the sandy deserts of Rajputana, Sind, and the Panjáb. The remark, therefore, in *Spons' Cyclopœdia*, drawing the attention of planters in India to the simplicity of its cultivation, seems rather out of place.

GUM-RESIN.
86

Collection.
87

Gum-resin.—Forms the drug of commerce known in Europe as asafœtida—in India as *Hingra*. The process of collection has been variously described by Kœmpfer, Bellew, and others. Dr. Aitchison's account, being the most recent, is here given at length :—

"The method of collecting the drug, as far as I could learn, was as follows. A few men, employed for the purpose by some capitalist at Herat, are sent to these asafœtida-bearing plains during June. These take with them provisions consisting of flour and several donkey-loads of water-melons, the latter in lieu of water, which is not only scarce there, but usually saline. The men begin their work by laying bare the root-stock to a depth of a couple of inches of those plants only which have not as yet reached their flower-bearing stage. They then cut off a slice from the top of the root-stock, from which at once a quantity of milky juice exudes, which my informant told me was not collected then. They next proceeded to cover over the root by means of a domed structure of from 6 to 8 inches in height called a *khora*, formed of twigs and covered with clay, leaving an opening towards the north, thus protecting the exposed root from the rays of the sun. The drug collectors return in about five or six weeks' time, and it was at this stage that the process of collecting came under my personal observation. A thick gummy, not milky, reddish substance now appeared in more or less irregular lumps upon the exposed surface of the root, which looked to me exactly like the ordinary asafœtida of commerce as employed in medicine. This was scraped off with a piece of iron hoop, or removed along with a slice of the root, and at once placed in a leather bag,—the tanned skin of a kid or goat. My guide informed me that occasionally the plant was operated upon in this manner more than once in the season. The asafœtida was then conveyed to Herat, where it usually underwent the process of adulteration with a red clay *táwah*, and where it was sold to certain export traders called *Kákri-log*, who convey it to India. On August 17th, when I crossed the great asafœtida plains, where this drug is chiefly collected, except for the small domes over each root there was not a leaf or a stem or anything left to point to the fact

F. 87

| Asafœtida—Hingra. | (*J. Murray*) | **FERULA fœtida.** |

GUM-RESIN.

that any such plant had ever existed there, the heat and winds of July and August having removed every trace" (*The Pharmaceutical Journal and Transactions, December 11th, 1886*).

Bellew, in *his* account, says that after cutting the plant through, above the root, three or four incisions are made in the stump. The operation of incision is repeated every three or four days, so long as the sap continues to exude. Bellew also describes the quantity of asafœtida obtained from each root as varying from a few ounces to two pounds, according to the thickness of the roots, which vary from the size of a carrot to that of a man's leg. The resin is called by the natives near Herat *angusa*. A particular sort is mentioned by Bellew as being "obtained solely from the node or leaf-bud in the centre of the root-head of the newly sprouting plant." This kind is never adulterated, and sells for a much higher price than the ordinary adulterated form. This is probably the fine quality of the drug known as *Khandahari-hing*.

The common form or *Hingra* is much adulterated by the *táwah* above mentioned, by wheat or barley flour, and by powdered gypsum. It is also mixed with slices of the root. The asafœtida obtained from this species of plant, with the exception of the Khandahari Hing, is not used in India. It is nearly all exported to Europe, where it forms the drug of commerce.

GENERAL CHARACTERS.—The purest kind (*Khandahari-hing*) consists chiefly of slightly or not agglutinated tears. *Hingra*, or the coarser form exported to Europe, varies much in appearance in different samples, owing chiefly to adulteration. The pure tears display, when fractured, a conchoidal surface, which changes from milky white to purplish pink in the course of some hours All samples of the drug have a powerful and persistent alliaceous odour and a bitter acrid alliaceous taste.

Characters.
88

CHEMICAL COMPOSITION.—Asafœtida consists of resin, gum, and essential oil in varying proportions; but the first generally amounts to more than half. The resin is partly soluble in ether or chloroform. The essential oil constitutes about 5 to 9 per cent. of the drug, and may be separated by distillation. It is light yellow, and has a pungent odour of asafœtida; when exposed to the air it evolves sulphurated hydrogen.

The gum occurs in small quantity and is unimportant. An alcoholic tincture of the drug is precipitated by acetate of lead. A solution in sulphuric acid is fluorescent.

Chemistry.
89

Medicine.—Asafœtida is used in Europe as an antispasmodic and stimulant, but is in much greater demand on the Continent than in Great Britain. In India, unlike the allied *Hing* obtained from F. alliacea, it is neither used as a condiment nor as a drug.

MEDICINE.
90

Food and Fodder.—According to Drs. Bellew and Aitchison the plant is used as a food by the natives. Bellew says—"The fresh LEAVES of the plant, which have the same peculiar stench as the secretion, when cooked, are commonly used as an article of diet by those near whose abode it grows, and the white inner part of the STEM of the full grown plant, which reaches the stature of a man, is considered a delicacy when roasted and flavoured with salt and butter. Aitchison writes : "He" (a native) "will take out his knife, remove the head, cut the stem from its base, strip off the few sheathing stipules that are still adherent to the stem, and in his hand you see what looks like a very large cucumber ; from this he will remove the dark-green cuticle, and then slice away at the deliciously cool, soft, crisp, copiously milky stem, and eat slice after slice." Burns, in his "Travels in Bokhara," states that the YOUNG PLANT is eaten with relish by the people, and that sheep crop it greedily.

FOOD. Leaves.
91

Stem.
92

Young Plant.
93
FODDER.
94

Trade.—See the account given under the generic heading.

z

| FERULA galbaniflua. | Galbanum. |

(G. Watt.)

95

Ferula galbaniflua, *Boiss et Buhse.*

The drug known from historic times as GALBANUM is now believed to be derived from one or two species of **Ferula**, chiefly **F. galbaniflua,** *Boiss et Buhse*; **F. rubricaulis,** *Boiss.,* according to BORSZCZOW, is also a source of the drug.

Vern.—*Bireja, ganda-birosa* (the last name is also given to the turpentine of Pinus **longifolia**), HIND.; ⏐*Badra-kéma, bi-ri-jeh* (the gum, *jao-shir*), AFG.; *Barsad, kuineh,* ARAB.; *Jawashir, khassuch, gaoshir, birees,* PERS.

According to some Muhammadan writers this is the *Khalbani* of the Greeks (περι χαλβανηs of Dioscorides).

References.—*Aitchison, Pharm. Jour. and Trans., 3rd Ser., XVII., p. 466 (London 1887), also Delim. Comm. Report (Trans. Linn. Soc. III. (2nd Series), p 68); Dymock, Mat. Med. W. Ind., 2nd Ed., 390; Flück. & Hanb., Pharmacog., 320; Bent. & Trim., Med. Pl., 128; Kew. Off. Guide to the Mus. of Ec. Bot., 75.*

Habitat.—A native of Persia, from which the gum is imported into Bombay and re-exported to Egypt and Turkey. Dr. Aitchison says this is one of the most characteristic plants of certain tracts of the Badghis, specially common around Gulran. No other plants are to be seen for miles, the young leaves on the top of the perennial stems appearing like cushions of moss.

GUM-RESIN.
96

Gum-Resin.—The *Jao-shir* resin, as met;with in India, is not dry agglutinated tears, but a yellow or greenish semi-fluid resin, generally mixed with the stems, flowers, and fruits of the plant. It has an odour between that of Levant Galbanum and Sagapenum It is not used in India. Dr. Aitchison remarks: "The stem, on injury from its earliest stage of growth, yields an orange-yellow gummy fluid, which very slowly consolidates, usually forming on the stem, like the grease on a guttering candle, and possessing, in common with the whole plant, when crushed, a strong odour resembling that of celery. The gum is commonly found adhering to the lower portions of the stem, and is so tenacious that when subsequently examined pieces of the plant are frequently found attached to it. No artificial means are employed to my knowledge in the collection of this drug. It is stated to be an article of export through Persia, *viz.,* the Gulf of Arabia and India."

MEDICINE Gum-resin.
97

Medicine.—The *Jawashir* or (*Gaoshir*) was not identified by the Arabs and Persians with the Galbanum of the Greeks. The *Ganda-birosa* of the Indian bazárs is the turpentine of Pinus **longifolia** (which see). Muhammadan writers (*e.g.,* the *Makhsan)* describe the Persian *Gaoshir* as a fœtid gum-resin, and say it is used medicinally as an attenuant, detergent, antispasmodic, and expectorant; prescribed in paralytic affections, hysteria, and chronic bronchitis (*Dymock*). Aitchison writes that "in Persia and Afghánistan it is said to be administered to parturient women, and the entire shrub is hung round the house to keep off evil spirits whilst parturition is actually taking place."

The ordinary Galbanum of European commerce is the Levant resin—for the chemistry of which see the *Pharmacographia.*

SPECIAL OPINIONS.—§ "Oil distilled from the gum is used in gonorrhœa; it is an excellent substitute for Copaiba" (*Surgeon Anund Chunder Mukerji, Noakhally*). *Ganda Biroja,* I have been told, is useful as a topical agent to promote the absorption of inflammatory products; it may be employed thus with advantage in bubo and inflammatory enlargements generally" (*Surgeon J. Ffrench Mullen, M.D., I. M. S., Saidpore*).

TRADE.
98

Trade.—According to Dymock *Jawashir* is imported into Bombay from Persia, where it is said to be collected between Shiráz and Kirmán.

F. 98

		FERULA
Sumbul.	*(G. Watt.)*	**Sumbul.**

The imports are irregular : sometimes large quantities arrive. Most of it is re-exported to Egypt and Turkey. Value R8 per maund of 37½ ℔.

Ferula Jaeschkiana, *Vatke ; Fl. Br. Ind., II., 708.*

Gum-resin.—The *Flora of British India* remarks on this species. "Regel and Schmalh think that this plant probably produces the Asafœtida of Commerce; this may be so, as it is an abundant species in Kashmír, and very abundantly supplied with oil; but it is not the Asafœtida of Linnæus."

It has become customary of late for writers on Materia Medica to abandon all idea of **Falconer's** Kashmír plant-yielding asafœtida. This view has been followed above, but at the same time it must be admitted that the reports of trade between Kashmír and India regularly show a considerable amount of the drug as obtained from that State. This fact may be merely in consequence of its being conveyed from more northern and western regions to India *viá* Kashmír. On the other hand, so many writers speak of the drug as produced in Kashmír, that it may be as well to add that perhaps after all a certain amount of alliaceous resin may be derived from **F. Narthex** or **F. Jaeschkiana** and be employed as a substitute or adulterant for the true drug.

Dr. Stewart, in his *Punjab Plants,* mentions that he found **Ferula Asafœtida,** *Linn.,* in Khágán, Jhelam basin, at about 6,000 feet, and **Cleghorn** states that specimens of that plant were brought to him on the Upper Chenab at over 8,000 feet The plant these authors allude to is doubtless **F. Jaeschkiana.** At all events, it was found by **Aitchison** while with the Kuram Valley Force. He describes it as covering the ground, in the forests between Dukalla and Karatigah, and as common on all the hills to the north of Hariáb district, at 10,000 to 11,000 feet.

Medicine.—Yields a GUM-RESIN which, **Aitchison** says, is applied to wounds and bruises by the inhabitants of Kuram Valley.

F. Narthex, *Boiss. ; Fl. Br. Ind., II., 707.*

Syn.—**Narthex Asafœtida,** *Falconer ; Bent. & Trim., Med. Pl., t. 126 ; (the description of production and properties of drug, there given, however, most probably chiefly refer to* **F. fœtida,** *Regel) ; Bot. Mag., t. 5168 ; Balfour, Trans. R. Soc. Edinb., XX., 366, tt. 21, 22.*

Habitat.—Found by **Dr. Falconer** in Astor, Baltisthan, but apparently never since re-collected.

Medicine.—It is significant that this species has never been found in Afghánistan, a fact which may be assumed as proving the authors incorrect who ascribe to it the Áfghán asafœtida. **Aitchison** witnessed the collection of that drug in Afghánistan and brought samples to Europe, but the plant from which it was obtained was **F. fœtida,** *Regel.* Modern writers have accordingly accepted that discovery as establishing the true source of the Afghánistan asafœtida. Acting on this opinion, the writer, in the above account of the drug, has transferred to **F. fœtida** the economic facts hitherto recorded under **F. Narthex.**

F. (§ Euryangium) suaveolens, *Aitch. et Hansl., Afghan Delim. Comm. Report.*

Reference.—*Aitchison in Pharm., Soc. Journ. 3rd Ser. XVII., 407.*
Habitat.—Khorasan, on the hills to the south of Bezd.

F. (§ Euryangium) Sumbul, *Hook. f., in Bot. Mag., t. 6196 (1875).*

References.—*Flück and Hanb., 312 ; Bent. and Trim., Med. Pl., 129.*
Habitat.—Found on the mountains to the south-east of Samarkand.
Medicine.—This and the preceding species are the chief plants which afford the musk-scented medicinal root—Sumbul—exported from Persia by the Persian Gulf into Bombay and thence distributed over India.

99
GUM-RESIN.
100

MEDICINE.
Gum-Resin.
101

102

MEDICINE.
103

104

105

MEDICINE.
106

FESTUCA rubra.	The Fescue Grasses.

(*F. F. Duthie.*)

FESTUCA, Linn. ; Gen. Pl., III., 1189.

A large genus, and widely distributed in temperate and alpine regions. Some of the species, such as Meadow Fescue and Sheep's Fescue, are reckoned amongst the most valuable of European pasture grasses. The generic name is said to be derived from the Celtic word *fest,* meaning pasture or food.

107

Festuca elatior, *Linn.*

References.—*Mueller, Select Pl., 173 ; Sutton, Permanent and Temporary Pastures, pp. 36 and 44; Stebler and Schrotter, The Best Forage Plants (Eng. Ed.), 35.*

Habitat.—A tufted perennial species, with stems upwards of 3 feet, met with occasionally on the North-West Himálaya. **Professor Hackel,** in his monograph of the Genus, divides **F. elatior** into two sub-species, *viz.,* **pratensis** (the true Meadow Fescue), and **arundinacea,** which is a taller and coarser plant.

FODDER.
108

Fodder.—Meadow Fescue has a great reputation, both in Europe and America, as being one of the most valuable grasses for pasture as well as hay. It thrives best in soils rich in humus and where the climate is damp. Cattle are said to prefer it even to Fox-tail **(Alopecurus pratensis).**

109

F. gigantea, *Vill.*

Syn.—BRORNUS GIGANTEUS, *Linn.*

Habitat.—This species is found at moderate elevations on the North-West Himálaya.

Mueller (*Select Plants, page 174*) describes it as a good perennial forest grass.

110

F. ovina, *Hackel.* (This includes **F. ovina,** *Linn.*)

References.—*Treasury of Bot., I., 490 ; Sutton, Permanent and Temporary Pastures, 45, 47; Stebler and Schrotter, The Best Forage Plants, 88 ; Müeller, Select Pl., 174.*

Habitat.—This species is easily distinguished by its compact growth and close-tufted bristle-like foliage. It occurs abundantly on the Himálaya, up to 15,000 feet, and in Kashmír. It is extremely variable and has been divided by **Professor Hackel** into five sub-species and several varieties, of which the following are represented in India.

Sub-species eu-ovina, *Hack.,* var. **vulgaris.** This is **Linnæus's F. ovina,** and the true Sheep's Fescue. According to **Sutton** it is the smallest grass cultivated for agricultural purposes. Owing to its hard wiry foliage it is useless for hay, but being nutritious, it affords excellent pasturage for sheep. Another variety of this sub-species is **durinscula (F. durinscula,** *Koch***)** or Hard Fescue, so called on account of the hard nature of the florets when ripe. It has stouter stems, larger spikelets, and thicker leaves than those of the preceding variety, and is altogether a more robust plant. It is also a most valuable constituent of sheep pastures in localities where the soil is too poor for the growth of better grasses,

Sub-species **subcata,** var. **Valeriaca,** (*Hack.*), is distinguished from the above varieties by having glaucous leaves and stems, and the leaves when dry become furrowed.

111

F. rubra, *Linn.*

CREEPING OR RED SHEEP'S FESCUE.

References.—*Sutton, Permanent and Temporary Pastures, 48 ; Stebler and Schrotter, The Best Forage Plants, 107.*

Habitat.—A perennial grass, distinguished from the other species of fescue by its creeping habit. It occurs on the Himálaya at moderate elevations.

FODDER.
112

Fodder.—This is said to be one of the few grasses which improve as they get older, the leaves and stems being actually more nutritious, as well

F. 112

as of superior bulk, at the time of ripening seed than earlier in the season. (*Sutton, l. c.*). It thrives on various kinds of soil; and on loose sandy ground and railway embankments it spreads rapidly by means of its underground stems, and serves to bind the soil. Royle says that, owing to the greater produce it affords, it is more valued than Sheep's fescue.

(*G. Watt.*)

FIBRAUREA, *Lour. ; Gen. Pl. I., 960.*

113

Fibraurea tinctoria, *Lour. ;* MENISPERMACEÆ ; *Fl. Br. Ind. I., 98.*

Syn.—FIBRAUREA TINCTORIA, FASCICULATA, AND CHLOROLEUCA, *Miers. ;* COCCULUS FIBRAUREA, *DC. ;* MENISPERMUM TINCTORIUM, *Spreng.*
Vern.—*Tien-sien-tan* and *hoang-ten*, CHINESE ; *Cay-vang-dang*, COCHIN CHINA.
References.—*Lour., vol. II., 627 ; Agri.-Horti. Soc· Ind. Jour., XI., 142.*
Habitat.—An extensive climber, found in the forests of Penang, Malacca, Cochin China, and Borneo.
Dye.—According to many writers the STEMS of this plant afford a permanent yellow dye, which is said to be used along with Indigo to form one of the green dyes of China. It is interesting that the new species indicated below is in Manipur used for a similar purpose.

DYE.
Stems.
114

F. Trotterii, *Watt, MS.*

115

Vern.—*Napoo*, MANIPUR.
Habitat.—An extensive climber, common in the forests of Manipur. The writer (in his Calcutta Exhibition Catalogue) took the liberty of provisionally naming this curious plant in honour of its discoverer, the late Major Trotter, Political Agent, Manipur. Not having seen flowering specimens he was, however, unable to give a detailed description of the plant, but as only one species (**F. tinctoria,** *Lour.*) has been hitherto published, there seems no doubt this will prove distinct.
Dye.—Major Trotter narrated the process of dyeing from this plant as follows :—" Five chittacks of dry ROOT of the *napoo* plant to be washed clear and beaten into long shreds ; then soaked in 2½ quarts of water for 15 or 20 minutes, when it will be found that the water has become of a yellow colour ; this water should be put aside, as it will be required later on. Take out the pounded roots and re-steep in the same quantity of fresh water and let stand for 24 hours. Then wash the cloth to be dyed clean, thoroughly soak it in the first solution and take out and repeat the process in the second water, leaving the cloth to soak in it for about half an hour ; then wring out and steep in half a pint of *heiboong* (**Garcinia pedunculata**) water, pressing and flopping it about in the vessel, so that every part of it may become thoroughly saturated with this water, then wring out and dry in the shade."

DYE.
Root.
116

FIBRES.

117

A detailed list of the fibres and fibrous plants of India will be found in the appendix. (See the explanation made under FOODS and also DOMESTIC and SACRED PRODUCTS.) It may be here stated that fibres are classified into :—

I.—Vegetable Fibres.

A.—Bark fibres suitable for the higher textile purposes, *e.g.*, Rhea (See *Vol. I., 461—484; also Selections from Records of the Government of India, Vol. I., 283—312*). Callotropis (See *Vol. II., 33—49*). **Marsdenia** (See *Selection from Records of the Government of India, Vol. I, 320—322*). Flax, Hemp (**Cannabis sativa,** *Vol. II., 103—126*), &c.

F. 117

FICUS annulata.	**The Banyan Tree.**
FIBRES.	B.—Bark fibres suitable for the lower textile purposes, *e.g.*, Jute (*See* **Corchorus**, *Vol II.*, *534—562*). Sun-hemp (*See* **Crotalaria juncea**, *Vol. II.*, *595—614*), and Coir (*Vol. II.*, *415—459*). Manilla-hemp, **Bauhinia** (*See Vol. I.*, *424–425* ; *also Selections from Records of the Government of India, Vol. I.*, *183—186*), **Hibiscus**, &c.

C.—Bark fibres suitable for Cordage and Ropes (*See Vol. II.*, *p. 566*).
D.—Paper materials.
E.—Flosses, *e.g.*, Cotton, Silk Cotton, Kapok, &c., &c. (*See Selections from the Records of the Government of India, I.*, *323—339*).

II.—Animal Fibres.

F.—Wool (*See Selections from the Records of the Government of India, Vol. I.*, *23—52*).
G.—Silk.
H.—Hair, Pashm, &c.

III.—Mineral Fibres.

I.—Asbestos, &c. (*See Vol. I.*, *338*).
Certain information will be found under each of these sectional headings, in their respective places in this work ; but to discover the descriptions of all the fibrous material of India the enumeration given in the appendix must be consulted, which will afford the key to the numerous articles on fibres scattered throughout the Dictionary.

(*Murray & Watt.*)

FICUS, *Linn. ; Gen. Pl., III.*, *367.*

A genus of trees, shrubs, or climbers, sometimes epiphytic, comprising about 600 species, mostly tropical, of which, according to **Hooker's** *Flora of British India*, 112 are Indian.
The chief interest, economically, in the species of FICUS, arises from the fact of their having a milky sap which contains Caoutchouc,—**F. elastica** being one of the sources of the India-rubber of Commerce.

118	**Ficus altissima**, *Bl. Bijd. ; Fl. Br. Ind., V.*, *504 ; Wight, Ic., t. 656 ; King, Ficus, 30, t. 30, 30A, 31, 82,* *82*[a1] *; URTICACEÆ.*

    Syn.—F. LACCIFERA, *Roxb. ; Wight, Ic. ; Brandis ; Kurz ; Bedd. ;* UROSTIGMA ALTISSIMUM, and U. LACCIFERUM, *Miq.*
    Vern.—*Bur*, ASSAM. ; *Kathal bat*, SYLHET ; *Yokdúng*, LEPCHA ; *Prab, phegran*, GARO ; *Nyaung* (F. *laccifera* according to **Kurz**), BURM.
    References.—*Roxb., Fl. Ind., Ed. C.B.C., 641 ; Brandis, For. Fl., 418 ; Kurz, For. Fl. Burm., II., 441, 442 ; Beddome, For. Man., 223 ; Gamble, Man. Timb., 332.*
    Habitat.—A large spreading tree, with few aerial roots. Found in the Tropical Himálaya from Nepal to Bhutan ; in the plains and lower hills of the Deccan and Ceylon ; and from Assam to Burma and the Andaman Islands. According to Gamble this tree is epiphytic.

CAOUTCHOUC **119**	**Caoutchouc.**—In the *British Burma Gasetteer*, its Caoutchouc is said to be as good as that of **F. elastica**. Brandis remarks of it that it merits further examination. Gamble says it yields caoutchouc more sparingly than **F. elastica**, and of inferior quality. In Sylhet lac is collected from the branches of the tree.
TIMBER. **120**	**Structure of the Wood.**—White, coarse, and soft, perishable (*Kurz*, under F. *laccifera*). [*V.*, *502.*
121	**F. annulata**, *Blume ; King, Ficus, p. 25, Pl. 23, 81 t. ; Fl. Br. Ind,*

    Syn.—FICUS FLAVESCENS and VALIDA, *Bl. ;* UROSTIGMA ANNULATUM and FLAVESCENS, *Miq.*
    Reference.—*Kurz, For. Fl. Burm., II.*, *443.*

		FICUS
The Banyan Tree.	(*Murray & Watt.*)	**bengalensis.**

Habitat.—A large stem-clasping tree, semi-scandent. Found on the plains and lower hills of Burma.

Caoutchouc.—Kurz writes that it yields a rather good quality of Caoutchouc.

CAOUTCHOUC
122

Structure of the Wood.—Yellowish, turning pale brown, rather heavy, soft, and perishable.

TIMBER.
123

Ficus asperima, *Roxb.; Fl. Br. Ind., V., 522; Wight, Ic., t. 633.*

124

    Syn.—F. HISPIDISSIMA, *Wight, MS.;* F. POLITORIA, *Morn., Cat. Ceyl. Pl.*

    Vern.—*Kál-ambar,* GUJ.; *Karakarbúda,* TEL.; *Kharoti, khoréti, karwat,* BOMB.; *Khargas,* KAN.; *Kharwat,* MAR.

    References.—*Roxb., Fl. Ind., Ed. C.B.C., 644; Dals. and Gibs., Bomb. Fl., 243; Bedd. For. Man., 224; Bomb. Gas., III. (Gujrat), 202; XV., Pt. I. (Kánara), 69; King, Ficus, pp. 80 & 81; Pl., 100; Dymock, Mat. Med. W. Ind., 2nd Ed., 746; Thwaites. Enum. Ceyl. Pl., p. 266.*

Habitat.—A tree or shrub with scabrid shoots. Found in Central India and the Deccan and distributed to Ceylon. It ascends the hills to a height of 3,000 feet.

MEDICINE.
Juice.
125
Bark.
126

Medicine.—Dymock says the JUICE and the BARK are in Bombay "well known remedies for glandular enlargements of the abdomen, such as liver and spleen."

Domestic Uses.—LEAVES very rough, and used in place of sand paper, both in Gujrat and Ceylon. In Kánara they are employed to polish horns. The YOUNG BRANCHES are said to be jointed and hollow.

DOMESTIC.
Leaves.
127
Young Branches.
128
129

F. bengalensis, *Linn.; Fl. Br. Ind., V., 499; Wight, Ic., t. 1989; King, Ficus, pp. 18 & 19; Pl., 13, 81°.*

    THE BANYAN TREE, *Eng.*; ARBOR DE RAIS (a tree of roots), *Port.*

    Syn.—F. INDICA, *Linn. in part (Amæn.);* UROSTIGMA BENGALENSE, *Gasp.*

The word Banyan, according to Yule and Burnell, appears to have been first bestowed popularly on a famous tree of this species growing near Gombroon, under which the Banyans or Hindu traders, settled at that port, had built a pagoda. Tavernier speaks of it as the Banyan's tree, and describes the village with its pagoda and bathing tanks at which the Hindu traders dwelt. Many other early writers describe this as especially the favourite tree of the Banyans or Hindu traders.

    Vern.—*Bor, bar, ber, bargat,* HIND.; *Bar, but,* BENG.; *Bai,* KOL.; *Boru,* URIYA; *Bare,* SANTAL; *Ranket,* GARO; *Bot,* ASSAM; *Borhar,* NEPAL; *Kangji,* LEPCHA; *Bor,* MAL (S.P.); *Barelli,* GOND; *Wóra, kurku,* N.-W. P.; *Bera, bor, bohir, bohar, bargad* (milky-juice *shir,* the fibres of aerial roots are *rish bargad*); PB.; *Baagat, bar,* PUSHTU; *Phagwari,* HAZARA; *Wur, bur,* SIND; *Wad, vad, war, barghat,* BOMB.; *War, vada,* MAR.; *Ala,* TAM.; *Mari, peddi mari (marri),* TEL.; *Ahlada, alada, ala, alava,* KAN.; *Peralu, peralin,* MALAY; *Pyi-nyoung (panyaung* or *pyinyaung),* BURM.; *Maha-nuga (a'l,* TAM. in Ceylon), SING.; *Vata,* SANS.

    References,—*Roxb, Fl., Ind., Ed. C.B.C., 639; Brandis, For. Fl., 412; Kurs, For. Fl. Burm., II., 440; Beddome, For. Man., 222; Gamble, Man. Timb., 333; Dals. & Gibs., Bomb. Fl., 240; Stewart, Pb. Pl., 213; Sir W. Jones, V., p. 160; Cleghorn, 147, 197; Rheede, Hort. Mal., I., t. 28; Trimen, Cat. Cey. Pl., p. 84; Elliot, Fl. Andh., 113; Mason, Burma and Its People, 450, 776; Voyage of John van Linschoten, II., 53 to 58; Pharm. Ind., 217; Ainslie, Mat. Ind., II.; 10; O'Shaughnessy, Beng. Dispens., 577; Moodeen Sheriff, Supp. Pharm. Ind., 142; U. C. Dutt, Mat. Med. Hind., 323; Dymock, Mat. Med. W. Ind., 2nd Ed., 745; Murray, Pl. and Drugs, Sind, 31; Med. Top., Oudh, 4, 6; Report on the Fibres of India, by Cross, Bevan, King, and Watt, p. 53; Baden Powell, Pb. Pr., 377; Atkinson, Him. Dist., 737; Drury, U. Pl., 212; Lisboa, U. Pl. Bomb., 129, 204, 235, 261, 278, 279, 283, 290, 291; Liotard, Paper-making Mat., 34, &c.; Watson, Report on Gums, 65; Indian Forester,*

I., 274; III., 205, 236; V., 15, 212; VI., 218, 240; IX., 247; X., 33, 325; XII., App. 21; XIII., 121, 551; Balfour, Cyclop., I, 1100; Smith, Dic., 36; Kew Reports, 1879, 34; Kew Off. Guide to the Mus of Ec. Bot., 122; Kew Off. Guide to Bot. Gardens and Arboretum, 41; Journ. Agri. Hort., Vol. IV., 128; V., 29; VI., 71; VIII., 102; Journ. Agri. Hort. Soc., 1885, Vol. VII., New Series, 263, 276; For. Ad. Rep., Ch. Nagpore, 1885, 6, 33; Gazetteers, Orissa, II., 179; App. VI.; Bomb., II., 39; V., 23, 285, 360; VI., 13; VII., 36, 38, 43; XI., 26; XII., 25; XIII., 27; XV., 69; XVI., 16; XVII., 25; Panjáb—Karnal, 16; Muzaffargarh, 23; Hoshiarpur, 10; Hazara, 13; Ludhiana, 10; Jhang, 17; N.-W. P.,— Bundelkhand, I., 84; Agra, IV., lxxvii.; Mysore and Coorg, Vol. I., 49, 70; II., 8; III., 24; Manuals, Cuddapah Dist., p. 263; Buchanan's Statistical Account of Dinajpur, 164; Settlement Report, South Arcot, 34; Kohat, 29; Guzrat, 134; Peshawar, 113; Kangra, 22; Shajehanpur, IX.; Seonee, 10; Baitool, 127; Chindwara, 110; Nimar, 307.

Habitat.—A large tree, wild in the Sub-Himálayan tracts and lower slopes of the Deccan; planted throughout India. **Mr. J. Cameron** writes that this tree is so common in Mysore that it may be said to be character-istic of the arboreal vegetation in many parts of that province.

It attains a height of 70 to 100 feet, and sends down roots from its branches, thus indefinitely expanding its horizontal growth. The branches from which these roots descend may often be seen to increase in thickness as they spread away from the central axis, and here and there this occurs to such an extent as to form auxiliary stems. The tree originates usually from the germination of seeds dropped by birds on other trees. Very often, owing to the natural receptacles formed by the axils of the leaves of palms (particu-larly the Palmyra and Date), this Fig may be observed embracing, until it strangles, a crown of palm leaves which are seen to grow from the centre of the Banyan. The death of the supporting palm, leaves a decaying central mass, which in time (or with the maturity of the Banyan) results in the death of the original axis, but the daughter axes continue their forest-like expansion until an area is embraced sufficient to afford shade for many thousand people. **Colonel Sykes** described a very large Banyan which grew on an island in the River Nerbudda. This was known as the Kabir-bar, and was probably the large tree described by **Nearchus**. In the *Poona Gazetteer* (*Vol. XVIII., Pt. I., p. 54*) a Banyan is spoken of in the Andhra valley, so large as to afford shade to 20,000 people. **Forbes** describes its circumference as of 2,000 feet, and its overhanging branches, beyond the daughter stems, as stretching over a much larger area. It had about 320 large trunks and over 3,000 smaller, and was capable of sheltering 7 000 men. High floods have, however, since carried away portions of the island, and with these sections of this great tree. Better known examples are the famous Banyan in the Royal Botanic Gardens, Cal-cutta, and the Satara one in Bombay. **Dr. King** describes the Calcutta banyan as about 100 years old, and as possessing 232 aerial roots. The main or parent trunk of this remarkable tree, he says, has a girth of 42 feet; the circumference of its leafy crown being 857 feet. It is, however, still growing vigorously, and as **Dr. King** remarks, "there is no reason why it should not go on increasing indefinitely." It is known to have taken its birth about the year 1782, on a sacred date-palm. **Mr. Warner** describes the Satara banyan—a still larger example than the Cal-cutta one. In 1882 its circumference was 1,587 feet, its length from north to south 595 feet, and from east to west 442 feet.

The Banyan is a favourite road-side tree and is accordingly largely planted for shade. In the Panjáb the young trees are said to require pro-tection from frost. Both this tree and **F. religiosa** effect serious destruc-tion to buildings, especially in Bengal. Bird-droppings, containing the seeds from the fruits, germinate on the walls of temples and other buildings,

| The Banyan Tree. | (*Murray & Watt.*) | FICUS bengalensis. |

and owing to the superstition of the people, these can only be removed provided injury be not done to the plants (*Buchanan*). **Valentia** (1809) speaks of this tree as the greatest enemy to buildings.

In Ratnagiri the Banyan trees were subjected to a tax owing to the number of the oil-bearing seeds of **Calophyllum inophyllum**, dropped by the flying-foxes who lived in the Banyan trees – the owners of these trees not being allowed to participate free of duty, while the owners of **Calophyllum** trees were taxed. (*Bombay Gazetteer, Vol. X, 39*).

Caoutchouc.—It yields an inferior rubber, and the milk is by the natives made into bird-lime. Lac is often collected from the tree. Dr. Buchanan describes the preparation of this bird-lime. "The milky juice," he says, "coagulates into a kind of elastic gum. It is collected by making incisions in the branches, is strained, and mixed with $\frac{1}{4}$ of its weight of mustard seed oil. It is then fit for use." **CAOUTCHOUC 130**

Fibre.—A coarse rope is prepared from the BARK and from the AERIAL ROOTS. Paper is also reported to have been formerly largely prepared in Assam from the bark, and to a small extent it is still so prepared at Lakhimpore and in Bellary in Madras. This fibre was used by the Sikhs as a slow-match. The length of the ultimate fibres has, by Cross, Bevan, and King been ascertained to be 1—3 m.m. The fibres obtained from the genus FICUS contain from 40 to 60 per cent. of cellulose, and under hydrolysis lose from 20 to 40 per cent. of their weight. Chemically they are therefore worthless fibres. (*See* **F. infectoria** *and* **F. religiosa**). **FIBRE. Bark. 131 Aerial roots. 132**

SPECIAL OPINIONS AS TO FIBRE — § "The inner bark is an article of common use for cordage, &c., in the rural districts" (*J. Cameron, Superintendent, Botanical Garden, Bangalore*). "Used for tying bundles of wood, &c." (*Dr. Dymock, Bombay*).

Medicine.—The MILKY JUICE is externally applied for pains and bruises, and as an anodyne application in rheumatism and lumbago. It is considered as a valuable application to the soles of the feet when cracked or inflamed, and is also applied to the teeth and gums as a remedy for tooth-ache. An infusion of the BARK is supposed to be a powerful tonic and is considered to have specific properties in the treatment of diabetes. The SEEDS are deemed cooling and tonic The LEAVES are applied, heated as a poultice, to abscesses, and after they have turned yellow are given with roasted rice in decoction as a diaphoretic. The ROOT fibres are given in gonorrhœa in the Panjáb, being considered by *Bedaks* to resemble **Sarsaparilla**. **MEDICINE. Juice. 133 Bark. 134 Seeds. 135 Leaves. 136 Root. 137**

SPECIAL OPINIONS.—§ "An infusion of the small branches is useful in hœmoptysis" (*Civil Surgeon J. Anderson, M B., Bijnor*). "The tender ends of the hanging roots are given for obstinate vomiting" (*Surgeon-Major Robb, Civil Surgeon, Ahmedabad*). "The concentrated juice is much used by natives in combination with fruit as an aphrodisiac, also in spermatorrhœa and gonorrhœa" (*Narain Misser, Kothe Bazar Dispensary, Hazaribagh,*). "The young buds are said to be astringent and useful in diarrhœa" (*Civil Medical Officer U. C. Dutt, Serampore*). "Really useful in cracked heels" (*Assistant Surgeon Shib Chunder Bhattacharji, Chanda, Central Provinces*). "A small quantity of the milky juice is taken early in the morning in dysentery. The milky juice is a good astringent" (*Surgeon W. F. Thomas, Madras Army, Mangalore.*)

Food and Fodder.—The small red FIGS are often eaten by the poorer people, especially during times of scarcity. Though much eaten by birds, they are said to be poisonous to horses. (*Bomb. Gaz., XVIII., Pt. I., p. 54 ; Vol. XVII., 27*). The LEAVES and YOUNG TWIGS are greedily eaten by elephants, and cattle are also said to eat them. **Linschoten** alludes to the fact that, in his time, the leaves of this tree were given to elephants (*viz.*, in 1596). **FOOD. Fruit. 138 FODDER. Leaves. 139 Young Twigs. 140**

FICUS Benjamina.	The Banyan Tree.

TIMBER.
141

Structure of the Wood.—Grey, moderately hard; no heartwood. Weight about 37℔. It is of little value, but is durable under water, and therefore used for well-curbs. It is sometimes employed for boxes and door panels. The wood of the drops is stronger and is used for tent-poles, cart-yokes, and banghy-poles (*Gamble*). **Kurz** and **Brandis** describe the wood as whitish, open-grained, and soft.

DOMESTIC.
Leaves.
142
Juice.
143
SACRED.
144

Domestic Uses.—The LEAVES are much in demand as plates. The milky JUICE is in Lahore employed to aid in the oxidation of copper.

Sacred Uses. — According to Hindu mythology, Bráhma was transformed into a *Vada* tree. **Dr. Buchanan** says that the Banyan is viewed as the male to the Peepul. It is regarded as a sin to destroy either of these trees, but more especially the male. It is meritorious to plant a young male close to the female, and this is done with a ceremony somewhat similar to that of marriage. It is customary, he adds, to place a piece of silver money under the roots of the young Banyan tree. So superstitious are the Hindus against cutting down the Banyan tree, that a Mr. T. Marsden, of the Madras Engineers, is said to have been poisoned by the Brahmans of Triplasore, in 1771, because he had cut down a Banyan tree during the construction of the fort. **Lisboa** writes that the dry twigs are used as *Samidhas* for producing sacred fire. The leaves are employed as one of the *Panch pallavs* or platters, and also for pouring libations. In the *Vratrág* females are ordered to worship this tree on *Jesht shudh* 15th (May), to water it, to wind a thread round it, and to worship it with *gandh* flowers, &c. (? the Indian Marigold—see *Vol. II., p. 24 and p. 272;* also **Tagetes erecta.** On the Himálaya the introduced but now completely naturalised **Dahlia** is similarly used). They are further ordered to make *Pradakshanas* (*i.e*, to go round it a certain number of times, to praise it, and to pray to it for the survival of their husbands and for the fulfilment of their wishes). They are told that by worshipping this tree they attain one of the heavens—*Shivloke.* They are encouraged to this worship by the tradition that *Savitri*, the wife of *Satyawan*, got back her deceased husband through the adoration of this tree. They are recommended to perform the thread ceremony of this tree and its marriage with the *Durva* plant— **Cynodon Dactylon.**

The umbrella poles often used at ceremonies are made of the wood of the aerial roots, and the young thin roots are by the Santals and other aboriginal tribes of Chutia Nagpur wound around the neck as a charm to ensure conception.

145

Ficus Benjamina, *Linn.; Fl. Br. Ind., V., 508; Wight, Ic., t. 658.*

Syn.—FICUS COMOSA, *Roxb.*; *Beddome*; *Wight, Ic.*; F. PENDULA, *Link.*; F. PAPYRIFERA, *Griff.*; *Icon. Pl. As., t. 554;* UROSTIGMA NUDUM, *Miq.*; U. BENJAMINA, *Miq.*; FICUS NUDA, *Miq.*; F. BENJAMINA, *Linn.*; *var.* COMOSA, *Kurz.*

Vern.—*Sunonijar,* SANTAL; *Juripakri,* ASSAM; *Kabra,* NEPAL; *Kunhip,* LEPCHA; *Pimpri,* BOMB.; *Jili,* CHUTIA NAGPUR; *Pútra-jauvi,* TEL.; *Jili,* MALAY; *Nyaung-thabieh,* BURM.

References.—*Roxb., Fl. Ind., Ed. C.B.C., 644; Brandis, For. Fl., 417; Kurz, For. Fl. Burm., II., 446; Beddome, Fl. Sylv., II., 223; Gamble, Man. Timb., 338; Dals. & Gibs., Bomb. Fl., 242; King, Ficus, 43; Elliot, Flor. Andhr., 161; Drury, U. Pl., 214; Gamble, Trees, Shrubs, &c., Darjeeling, 74.*

Habitat.—A moderate-sized, evergreen, often epiphytic tree, cultivated in the Malay Peninsula, wild (*var.* comosa only) along the base of the Eastern Himálaya, to Assam, Chittagong, Burma, the Andaman Islands, and the Deccan.

MEDICINE.
Leaves.
146

Medicine.—According to **Drury** a decoction of the LEAVES mixed with oil is believed in Malabar to be a good application to ulcers.

F. 146

		FICUS
The Fig.	(*Murray & Watt.*)	Carica.

Structure of the Wood.—Grey, beautifully mottled, moderately hard. Weight, 34℔ per cubic foot.

Lac.—Gamble writes that lac is produced on this species in Assam.

Ficus Carica, *Linn.; Brandis, For. Fl., 418; Aitchison, Afgh.*
THE FIG. [*Delim. Rept., Pl., 46.*

Vern.—*Anjír,* HIND.; *Anjír,* BENG.; *Kimri, fagu, fagúri, fagári,* PB.; *Anjíra,* BOMB.; *Anjír,* GUZ.; *Anjura,* or *anjúri,* KAN.; *Tie-thie,* BURM.; *Anjíra,* SANS.; *Ten,* ARAB.; *Anjír,* PERS.

References.—*Roxb., Fl. Ind., Ed. C.B.C., 635; Gamble, Man. Timb., 333; Stewart, Pb. Pl., 211; DC., Origin Cult. Pl., 295; King, Ficus, 147; Elliot, Flora Andhr., 15; Stocks' Account of Sind; Aitch., Afgh. Del. Com., 109; Lace, Quetta, Pl.; Ainslie, Mat. Ind., I., 131; U. C. Dutt, Mat Med. Hind., 291; Dymock, Mat. Med. W. Ind., 2nd ed., 745; Flück. & Hanb., Pharmacog., 542; S. Arjun, Bomb. Drugs, 127; Irvine, Mat. Med., Patna, 117; Atkinson, Him. Dist., 736; Lisboa, U. Pl., Bomb., 130, 172; Birdwood, Bomb. Pr., 176; Atkinson, Ec. Prod., N. W. P., Pt. V., 44, 83; Mason, Burma and its People, 459, 776; Ayin-i-Akbári (Gladwin's Trans.), I., 83; Smith, Dic., 172; Kew Off. Guide to the Mus. of Ec. Bot., 122; Kew Off. Guide to Bot. Gardens and Arboretum, 145; Settlement Reports: Peshawar, 13; Kohat, 39; Hazara, 94; Kangra, 22; Gujrat, 135; Delhi, 27; Port Blair, 33; Gasetteers: N.-W. P., Bundel-khand, I., 84; Agra, IV., lxxvii; Bombay: Kathiawar, VIII., 184; Poona, XVIII., 41; Mysore and Coorg, I., 53, 70; Bannu, 23; Dera Ismail Khan, 19; Peshawar, 28; Special Reports: from Govt. of Burma; from Collector, Bellary; from Collector, Cuddapah; Madras Board of Revenue, Jan. I., 1889, No. 266, p. 5.; and Director of Land Records and Agriculture, Bombay.*

Habitat.—Cultivated in many parts of India, more especially in the North-Western Provinces, the Panjáb, the Western Himalaya, Sind, and Beluchistan. Reports have been received of its cultivation in Bombay, Madras, Burma, and the Andaman Islands. In some of the references, however, room for doubt seems to exist as to their really referring to this species. **Dr. Aitchison** thinks **F. Carica** is probably a native of Afghánistan and Persia. It is indigenous, he says, in the Badghis country and Eastern Persia. According to **DeCandolle,** "the prehistoric area of the fig tree covered the middle and southern part of the Mediterranean basin from Syria to the Canaries." He further mentions the fact that "leaves and even fruits of the wild **Ficus Carica,** with teeth of **Elephas primigenius,** and leaves of plants of which some no longer exist, and others, like **Laurus canariensis,** which have survived in the Canaries," were found by **Planchon** in the quaternary tufa of Montpellier, and by de **Saporta** in those of Aygalades near Marseilles, and in the quaternary strata of La Celle near Paris.

Cultivation.—In the Bombay Experimental Farm reports repeated mention is made of the cultivation of this fig, but the following special report by the Director of Land Records and Agriculture, Bombay, gives the results of the experience gained at Poona :—

"In 1887-88 the area under figs amounted to 271 acres. With the exception of a few acres in Surat, Ahmadnagar, and Belgaum, almost the whole area was confined to Poona. There are two varieties—dark-purple and greenish. The tree grows from 6 to 7 feet high.

"The fig tree does not require very rich soil. Alluvial or loamy soil of yellow or reddish brown colour, with a rocky or *murum* bed 3 or 4 feet below the surface, is best suited to its growth. The rocky or *murum* bed prevents the roots from penetrating deep into the soil, and favours the side growth of rootlets, which is very desirable. Fig trees also thrive in clayey soil, but the land must not be water-logged. Rich black soil is

FICUS Carica.	The Fig.

CULTIVA-TION.

unsuited to fig trees. In it the plant grows tall and runs to leaf, and the fruit is much inferior both in size and taste.

"The crop requires a mixed manure, about 10 to 12 cart-loads for the first year. The ingredients of the mixture are town sweepings, sheep-droppings, cowdung, and ashes. The use of each of these ingredients separately is considered prejudicial. Sheep-droppings make the skin of the fruit render, so that it comes off at a touch. Cowdung causes a disease which injures the tree. The use of ashes by themselves is considered injurious to the plant. Dry fish forms a very good manure but is not easily procurable. Poudrette has not yet been tried.

"The plants are raised from cuttings ⅓ to 1 inch thick and a foot and a half long, planted in rows 10 to 12 inches apart in a richly manured and watered plot. The cuttings should be put into the ground in June after the monsoon has set in and should be watered every eighth, sixth, or fifth day as necessary. In about two months they begin to throw roots and shoots and make a few leaves. If they are properly taken care of the plants after a year become fit for transplantation, otherwise they take from 10 to 12 months more. The best season for transplanting is July-August. To allow of free growth and to prevent the tangling of branches and injury from shade, the plants should be at least 12 to 14 feet apart. About 200 plants go to an acre.

"At the end of every August, when there is a break in the rains, the soil at the roots of the plants should be turned up and loosened, the outstretching roots cut and the remaining roots exposed to the sun for four or five days. The roots should then be covered with a little earth and one or two basketfuls of manure, and the plants watered. The whole operation should not extend over a fortnight. A little manure is sometimes applied but none should be given after October. From the beginning of March to the end of May the soil should be slightly turned and cleared every fortnight. In this way the soil should be dressed about 20 to 25 times a year. If the plant turns to wood and leaf and does not bear, it should be pruned, slightly manured, and watered every eighth day.

"The fig tree requires careful watering. In the fruiting season, the failure of a single weekly watering reduces the outturn. The quantity of water should also be gradually increased, and the period between two waterings should begin with four days and end with eight days, having an intermediate period of six days. The watering should commence with September and end with the fruiting season. During the first two years light crops may be raised between the lines of fig trees. In the first year, onions, garlic, and other vegetables may be cropped, and in the second year, radish and fenugreek. But from the third year, when the plant begins to bear, no crop should be raised. The plant begins to bear in the second or third year after transplantation. But the full crop can be gathered only from the fourth year. The tree continues to bear from 12 to 15 years; and 20 years is the utmost limit, after which the tree generally dries up. Vigorous growth of the plant in September-October is a sure indication of a good crop. The tree fruits twice a year. The first season commences in June-July, but the crop is not allowed to ripen, as it, besides being sour in taste, injures the second crop, which is by far the most valuable. The first crop is gathered green and is sold as an inferior vegetable. The second season commences in January and lasts till the commencement of the monsoon. The first takes about two months to mature. If a tree has fruited too thickly to allow all the fruits to attain good size, the crop is thinned. But this thinning must be done by experienced hands. A full-grown tree, which is 6 to 9

F. 150

			FICUS
The Fig.	*(Murray & Watt.)*		**Cunia.**

<table>
<tr><td></td><td>CULTIVA-
TION.</td></tr>
</table>

feet high, yields according to season from 2 to 20℔ of fruit. Excessive heat or cold and cloudy weather cause great injury to the tree and fruit. Two blights, locally called *dhui* and *mooa*, often cause considerable injury to fig gardens from October to December.

"The first year's expenses for an acre of fig garden, as shown below, amount to R75 to R90 :—

		R	to	R
Cost of 200 plants	. . . . . .	12	„	14
Planting charges	. . . . „ .	5	„	6
Manure 10 to 12 cart-loads	. . . .	12	„	15
Watering and other charges	. . . .	46	„	55
		75	to	90

"If no hired labour were employed these charges would be reduced to R50. In the second year about R50 and in the third year about R80 to 100 are required. In the third year the produce is worth from R50 to R100, from the fourth to the tenth year the income of an acre of fig garden varies from R300 to R400 against an expenditure of about R100. In the Poona bazar fresh figs sell from 4 to 12℔ per rupee.

"Figs are eaten fresh. They are preserved in sugar, but are never dried. Large quantities of Poona figs are exported to Bombay. They are believed to increase blood and to have a cooling effect on the system."

Medicine.—The dried FRUIT is demulcent, emollient, nutritive, and laxative. It is, however, only rarely employed medicinally. Persons suffering from habitual constipation find it useful as an article of diet. The fruit is also used in the form of a poultice to effect suppuration. The PULP of the fig, mixed with vinegar and sugar, is very useful in bronchitic affections, principally in children (*Dr. Emerson*). Flückiger and Hanbury say the dry fig contains about 60 to 70 per cent. of grape sugar and the unripe fruit starch. Ainslie remarks that the *Vytians* prescribe figs in consumptive cases. The Arabians place them among their *Mobehyat* or aphrodisiacs, and *Muzijat* or suppurantia. Smyrna figs are deemed the best.

SPECIAL OPINIONS §.—"The JUICE of the LEAVES is of use when applied locally in the early stages of leucoderma" (*Narain Misser., Kathe Bazar Dispensary, Hoshangabad, Central Provinces*). "Largely imported from the Persian Gulf ports" (*W. Dymock, Bombay*).

Food.—From Afghanistan, FIGS of a better quality than those grown in India are imported into the Panjáb in considerable quantities annually. The fruit is, however, not uncommonly offered for sale but it is eaten chiefly by the Natives. The fresh figs of India are inferior to those of Western Asia.

	MEDICINE. Fruit. **151**
	Pulp. **152**
	Juice. **153** Leaves. **154**
	FOOD. Fruit. **155**

Ficus chittagonga, *Miq.*, see F. glomerata, *Roxb.*

✠ F. cordifolia, *Roxb.*, see F. Rumphii, *Kurz.*

F. Cunia, *Ham.; Fl. Br. Ind., V., 523; Wight, Ic., t. 669.*

156

Syn.—FICUS CONGLOMERATA, *Roxb.*

Vern.—*Khewnau, khurhur, kassa, ghwi, khenan, ghui,* HIND.; *Dumbur, jajya-domur,* BENG.; *Riu, ain,* KOL; *Porok podha,* CHUTIA-NAGPUR; *Horpodo,* SANTAL; *Kanhya,* NEPAL; *Sangji,* LEPCHA; *Kanai, palkai taikrau,* MICHI; *Poroh, perina, teregam,* MAL (S.P.); *Kunia,* KUMAON; *Kathjular, trumbal, karndol, kuri,* PB.; *Porodumer,* KHARWAR; *Ye-kha-ong, ye-kha-ong,* BURM.; *Jonua, sodoi,* MAGH.

References.—*Roxb., Fl. Ind., Ed. C.B.C., 646; Brandis, For. Fl., 421; Kurz, For. Fl. Burm., II, 461; Beddome, Fl. Sylv., 224; Gamble, Man. Timb., 339; Stewart, Pb Pl., 212; Rev. A. Campbell's Report on the Econ. Prod., Chutia Nagpur; Atkinson, Him. Dist., 318; Ec. Prod.,*

FICUS foveolata.	**The Caoutchouc of Indian Commerce.**

N.-W. P., Pt. V., 84; Report on the Shan States by Mr. Aplin; For. Ad. Report, Chutia Nagpur, 1885, 6, 33; Ind. For., III., 205; VI., 218; VIII., 82; X., 222, 325; XI., 4; Bom. Gaz., III., 202.

GUM.
157
FIBRE.
Bark.
158
MEDICINE.
Fruit.
159
Bark.
160
Roots.
161
FOOD.
Fruit.
162
TIMBER.
163
DOMESTIC.
Leaves.
164
165

Habitat.—A moderate-sized tree of the sub-Himálayan tract, from the Chenab eastward to Bengal and Burma; ascending to 4,000 feet in altitude.

Gum.—Lac is produced on the tree.

Fibre.—The BARK is used to tie the rafters of native houses. **Mr.** Campbell says it affords a good strong fibre useful for ropes.

Medicine.—The FRUIT is given in aphthous complaints. A bath made from the fruit and BARK is a cure for leprosy (*Rheede*). The juice from the ROOTS is given in bladder complaints, and boiled in milk in visceral obstructions (*Rev. A. Campbell*).

Food.—The FRUIT is eaten and is said to be good, though somewhat insipid. According to **Stewart**, however, it is not eatable.

Structure of the Wood.—Rough, moderately hard, greyish-brown. Weight 31℔ per cubic foot. It is not used economically.

Domestic Uses.—The LEAVES are rough, and are consequently employed in place of sand-paper.

54; Wight, Ic., 663.

Ficus elastica, *Roxb.; Fl. Br. Ind., V., 508; King, Ficus, p. 45, Pl.*

Syn.—UROSTIGMA ELASTICUM, *Miq.;* VISIANIA ELASTICA, *Gasp.*

Vern.—*Bor, attah bar,* BENG.; *Kagiri, kasmir,* KHASIA; *Bar, attah bar,* ASSAM; *Rauket,* GARO; *Lesu,* NEPAL; *Yok,* LEPCHA; *Nyaung bawdi,* BURM.

References.—*Roxb., Fl. Ind., Ed. C.B.C., 640; Brandis, For. Fl., 417; Kurz, For. Fl. Burm., II., 444; Gamble, Man. Timb., 336; Stewart, Pb. Pl., 212; Mason, Burma and its People, 523, 776; Lisboa, U. Pl. Bomb., 130; Christy, Com. Pl. and Drugs, VI., 53; VII., 25; Liotard, Dyes, 33; Watson, Report on Gums, 34; Kew Off. Guide to the Mus. of Ec. Bot., 122; Kew Off. Guide to Bot. Gardens and Arboretum, 69; Bomb. Gaz., 404; Burm. Gaz., 124; Trans. Agri. Hort. Soc., Vol. IV., 221; Indian Forester, I., 86, 124, 126, 127, 129, 132, 133, 134, 136, 138, 139-141, 188; III., 46; IV., 40, 41; V., 190; VI., 49, 50; VII., 101, 241-243; VIII., 203; IX., 225; X., 403; XI., 256, 354, 485, 487; XII., 563; XIII., 550; XIV., 297; Special Reports: Conserv., Forests, South Circle, Madras; Conserv., South Circle, Bombay; Conserv. of Sind; Conserv. of Bengal (Chittagong); Official Correspondence and Reports: Assam Forest Reports from 1873-74 to 1887-88.*

Habitat.—A large evergreen tree, usually epiphytic, throwing down numerous aerial roots from the branches. It occurs in damp forests from the base of the Sikkim Himálaya eastward to Assam and Arracan. There are large Government plantations in Assam, and it is also being cultivated in other provinces. **Kurz** remarks that it is frequent in Upper Burma, "where whole forests of the species are said to exist in the valley of Hookhoom."

For the cultivation of this and other Caoutchouc-yielding plants, see the account under INDIA-RUBBER.

GUM.
166
TIMBER.
167
168

Gum.—The tree yields the Caoutchouc of Indian commerce.

Structure of the Wood.—White or light brown. Weight 43℔ per cubic foot. It is not used.

F. foveolata, *Wall.; Fl. Br. Ind., V., 528; King, Ficus, p. 133-135; Pll. 166, 167, and 168; Griff., Icon. Pl. As., t. 561.*

Syn.—FICUS PUBIGERA, *Wall.; Brandis, Kurz;* F. EREATA, *Miq. (non Thunb.);* F. THUNBERGII, *Maxim.;* F. IMPRESSA, *Benth.;* F. LUDUCCA, *Roxb.;* F. LUDENS, *Wall.;* F. WRIGHTII, *Benth.*

Vern.—*Dudíka,* NEPAL; *Taksot,* LEPCHA; *Bat phagár, nágár jamán, thaur, phogri, dúdagrú, mambre, dúgurú, shiruli, mathágar, karmbal,*

F. 168

---

garelú (these names are given by **Stewart** for **F. reticulata**, *Miq.*, which **Brandis** regards as **F. foveolata**, *Wall.*), Pb.; *Grelu*, Simla; *Makreru*, Kunawar.

References.—*Brandis, For. Fl., 423, 424; Kurz, For. Fl. Burm., II., 450; Gamble, Man. Timb., 339; Stewart, Pb. Pl., 214.*

Habitat.—An evergreen scandent shrub found in the Himálaya, from Chumba to Bhutan; altitude 2,000 to 7,000 feet; also the Khasia Hills and Burma.

Fodder.—Stewart says of his **F. reticulata** that it is browsed by goats.
Structure of the Wood.—Light brown, soft, very porous. Weight 38℔ per cubic foot.

	FODDER.
	169
	TIMBER.
	170
	171

[*2; Wight., Ic., t. 650, 651, 652.*

Ficus gibbosa, *Blume; Fl. Br. Ind., V., 495; King, Ficus, 4, Pl.*

Syn.—F. unigibba, *Miq.*; F. rigida, paradoxa, and cuneata, *Blume*; F. altimeraloo, *Roxb., MSS.*; F. excelsa, *Vahl.? in Roxb., Fl. Ind.; Kurz, For. Fl. Burm.*

The *Flora of British India* describes four varieties of this plant as follows:—

α. **F. gibbosa,** *Blume.* Malay Peninsula.
β. **F. cuspidifera,** *Miq.* Throughout India.
   Syn.—F. excelsa, *Wall.*; F. reticulosa, *Miq.*
γ. **F. parasitica,** *Koen.* Central India, Behar, &c.
   Syn.—F. ampelos, *Koen*; F. sclerophylla, *Roxb.*; Urostigma volubile, *Dals.*; U. ampelos, *Dals. & Gibs.*
δ. **F. tuberculata,** *Roxb.* Western Ghâts.
   Syn.—F. angulata, *Miq.*

	172
	173
	174
	175

Vern.—*Datír*, Bomb.; *Umbar*, Guz.; *Kouda-júvee, tellabarinka*, Tel.; *Attiméralú*, Malay; *Udumber*, Sans.

References.—*Roxb., Fl. Ind. Ed., C.B.C. (under four specific names), 641, 643, 644; Brandis, For. Fl., 420; Kurz, For. Fl. Burm., II., 451; Beddome, Fl. Sylv., 224; Dals. & Gibs., Bomb. Fl., 242, 315; Dymock, Mat. Med. W. Ind., 2nd Ed., 746; Drury, U. Pl., 216; Balfour, Cyclop., 1100.*

Habitat.—This protean species the *Flora* describes as a tree met with at the bases of the hill ranges throughout India, from Kumaon eastward to Burma and southward to the Malay Peninsula, Andaman Islands, and Ceylon. Distributed to the Malay Islands, Hong-Kong, &c.

Medicine.—The decoction of the root acts as a powerful aperient. The root-bark is stomachic and gently aperient (*Dymock*).
Domestic Uses.—Leaves used to polish ivory (var. **parasitica**, *Roxb.*).

	MEDICINE.
	Root.
	176
	Root-bark.
	177
	DOMESTIC
	Leaves.
	178
	179

F. glomerata, *Roxb., Fl. Br. Ind., V., 535; Roxb., Corom. Pl., II., No. 123; Wight, Ic., 667; King, Ficus, pp. 173-174, Pll., 218, 219; Brandis, For. Fl. Pl., 49.*

Syn.—F. chittagonga, *Miq.*; F. racemosa, *Wall (non Roxb.)*; F. mollis, *Miq. (non Vahl.)*; F. goolereea, *Roxb.*; Covellia glomerata, *Miq.*

Vern.—*Gúlar, paroa, lelka, umar, umrái, tue, dimeri*, Hind; *Jagya dumar* (Gamble), *Yajnadumbar* (U. C. Dutt), Beng.; *Lowa, lóa*, Kol; *Lowa, loa*, Santal; *Dumer*, Chutia Nagpur; *Dimeri*, Uriya; *Dumri*, Nepal; *Tchongtay*, Lepcha; *Dumer*, Mal (S.P.); *Thoja*, Gond; *Alawa*, Kurku; *Dumer*, Kharwar; *Gúlar, panwa, lelka*, N. W. P.; *Kathgúlar, krumbal, rumbal, batbar, palák, kakammal, dadhuri*, Pb.; *Ormul*, Pushtu; *Umbar gular*, C. P.; *Umbar*, Bomb.; *Umbara, atti, rumadi*, Mar.; *Umbar*, Guz.; *Atti*, Tam.; *Moydi, atti, bodda, paidi, mari, medi*, Tel.; *Kulla-kith, atti* (the gum is called *Chandarasa*), Kan.; *Ye-tha-pan*, (*yae-tha-phan*, Mason), Burm.; *Atteeka*; Sing.; *Udumbara*, Sans.

FICUS
glomerata.   The Chándarasa Gum.

References.—*Roxb., Fl. Ind, Ed. B.C.C., 646, 639 ; Brandis, For. Fl., 422 ; Kurz, For. Fl. Burm., II., 458 ; Beddome, Fl. Sylv., 224 ; Gamble, Man. Timb. 339 ; Thwaites, En. Ceylon Pl., 267 ; Dals. & Gibs., Bomb. Fl., 243 ; Stewart, Pb. Pl., 212 ; Rev. A. Campbell, Rep. Econ. Prod., Chutia Nagpur, No. 7531 ; Elliot, Fl. Andh., 18, 28, 114, 141 ; Mason, Burma and its People, 460, 776 ; Sir William Jones, V., 159, No. 72 ; Ainslie, Mat. Ind., II., 30 ; U. C. Dutt, Mat. Med. Hind. 235, 321, 324 ; Dymock, Mat. Med. W. Ind., 2nd Ed., 744 ; Baden Powell, Pb. Pr., 377 ; Atkinson, Him. Dist, 317, 737. ; N. W. P. Econ. Prod., Pt. V., 84 ; Lisboa, U. Pl. Bomb., 131, 204, 278, 282, 290 ; McCann, Dyes and Tans, Beng., 136, 144 ; Watson, Report on Gum, 61 ; Special Report, Baroda Durbar, No. 109 ; Balfour, Cyclop, I., 1100 ; Journ. Agri.-Hort., 1885, VII., (New Series), 276 ; Indian Forester: I., 23, 273 ; III., 205, 236 ; IV., 321 ; V., 471 ; VII., 232 ; VIII., 35, 411 ; IX., 222, 325 ; XII., App., 21, 28 ; XIII., 121 ; XIV., 144, 371 ; Settlement Reports : N. W. P., Shahjehanpur, p. ix. ; C. P.: Chindwara, 110 ; Seonee, 10 ; Baitool, 127 ; Chanda, App. VI. ; Bhundara, 19 ; Hoshungabad, 179 ; Nimar, 307 ; Raepur, 76, 77 ; Punjab: Simla, App. II., p. xliv ; Kohat, 29—30 ; Peshawar, 26 ; Manuals and Gazetteers : Trichinopoly, 78 ; Coimbatore, 247 ; Orissa, II., 179, App. VI. ; Bombay, III., 199 ; V., 283 ; VII., 38, 40, 43 ; XI., 24 ; XII., 28 ; XIII., 27 ; XV, 69 ; XVI., 16 ; XVII 26 ; Mysore and Coorg, I., 70, 434 ; N. W. P., III., 33, 248 ; IV., lxxvii ; For. Admn. Rep., Ch. Nagpur, 1885, 633.*

Habitat.—A large tree of the Salt Range and Rajputana along the sub-Himálayan tracts to Bengal, Central and South India, Assam and Burma.

GUM.
180

Gum.—In Chanda it is said a gum (*sic.*) is obtained from this tree (*Settle. Report*). The *Mysore and Coorg Gazetteer*, referring probably to the same substances, says a gum known as *Chandarasa* is prepared from the milky juice. In both these passages the word Caoutchouc should probably be substituted for *gum*.

The lac insect is reported to occasionally frequent the tree. Brandis remarks that it abounds in a milky juice from which bird-lime is prepared.

DYE.
181

Dye.—This tree is said to afford a dye (*C. P. Gaz., 419*). McCann says that the bark, under the name of *goolur*, is mentioned as one of the in gredients used in Lohárdagá in preparing a good black dye.

MEDICINE.
Leaves.
182
Bark.
183
Fruit.
184
Root.
185
Galls.
186
Milky Juice.
187

Medicine.—The LEAVES, BARK, and FRUIT are employed in native medicine. The bark is given as an astringent and a wash for wounds. It is also employed to remove the poison from wounds made by a tiger or cat. The ROOT is useful in dysentery, and a fluid obtained from it by incision is administered as a powerful tonic. Ainslie speaks of this fluid as *attie vayrtannie*—a powerful tonic when drunk for several days together. The leaves reduced to powder and mixed with honey are given in bilious affections. The small blister-like GALLS common on the leaves, soaked in milk and mixed with honey are given to prevent pitting in small-pox (*Atkinson*). The figs are considered astringent, stomachic and carminative, and are given in menorrhagia and hæmoptysis. The MILKY JUICE is administered in piles and diarrhœa, and in combination with sesamum oil in cancer. The fresh juice of the ripe fruit is used as an adjunct to a metallic prepara-tion which is given in diabetes and other urinary diseases. In the Trichi-nopoly Manual it is said " a juice is extracted from the trunk which is used by the natives in cases of diabetes." In the Baroda Durbar report of the drugs shown at the Colonial and Indian Exhibition, "the SAP" is said to be used " locally applied to mumps and other inflammatory glandular en-largements." **Dr.** Dymock also alludes to this application, and adds that it is employed in gonorrhœa in doses of four *tolas*. The Settlement Report of the Chanda district adds that it is used as an application to wounds.

Sap.
188

The bark is given to cattle when suffering from rinderpest. It is ground with onions, cummin, and cocoa-nut spathes and mixed with vinegar. (*Coimbatore Dist. Man.*).

F. 188

	FICUS
The Gular or Umbar Fig. *(Murray & Watt.)*	**heterophylla.**

SPECIAL OPINIONS §.—"Used in cases of spongy gums to harden them" (*Surgeon-Major Ratton, M.D., Salem*). "An infusion of the bark is much employed by the Tamil-speaking people for menorrhagia" (*Surgeon W. F. Thomas, Madras Army, Mangalore*). "The sap of the root is used in diabetes" (*Native Surgeon T. Ruthnam Moodelliar, Chingleput, Madras Presidency*). "The tree grows very commonly in Mysore, and the bark is frequently given as an astringent" (*Surgeon-Major John North, Bangalore*).

Food.—The FRUIT (which ripens from April to July) is very inferior, but is occasionally, says Stewart, eaten raw and in curries by the poor. Campbell remarks that the Santals cook the unripe figs in their curries. Gamble, however, writes that the ripe fruit is eaten, and is good either raw or stewed. Atkinson adds that the fruit affords a valuable food resource in seasons of scarcity, and Dr. Dymock that it was eaten in the famine of 1877. Brandis confirms this observation, enlarging that the unripe fruit is pounded, mixed with flour and made into cakes.

Fodder.—The FRUIT is greedily eaten by cattle. The LEAVES are collected as cattle and elephant fodder.

Structure of the Wood.—Grey, soft, mottled on a longitudinal section. Weight 25 to 30℔ (*Gamble*). Pale brown, coarsely fibrous, light and perishable (*Kurz*). It is not durable, though it lasts well under water, and is consequently used for well-frames (*Stewart*). In Trichinopoly it is said to be used for building purposes, but it is described as brittle and coarse-grained. It is spoken of as one of the timbers of the Puri district, Orissa. In Kolaba (Bombay) the wood is reported to be used for rice-mortars. In Khandesh the wood is used for shoring wells, and in Kanara it is described as often employed for doors and well-frames. In Ahmadnagar it is said to be employed for planks and shutters.

Sacred.—In the *Baroda Gazetteer* it is stated that there is a common belief that near every *umbar* tree there runs a hidden stream. The tree is regarded as sacred.

[636, 659, 661; King, Ficus, pp. 75-77, Pl., 94.

Ficus heterophylla, *Linn. f.; Fl. Br. Ind., V., 518; Wight, Ic., t.*

Syn.—FICUS TRUNCATA, DENTICULATA, RUFESCENS, *Vahl.*; F. TRUNCATA, REPENS, RUFESCENS, *Ham.*; F. AQUATICA, *Kœnig.*; F. SCABRELLA and HETEROPHYLLA, *Roxb.*; F. REPENS, *Willd., Roxb.*; F. RUBIFOLIA, *Griff.*

Vern.—*Gaori-shiora, balábahulá, balálátá, ghoti-suara, bhui-dúmúr,* BENG.; *Ballam dúmúr,* CHITTAGONG; *Pakhur,* C. P.; *Buróni,* TEL.; *Valli-teragam,* MALAY.; *Wal-ehetú,* SING.; *Tráyamáná,* SANS.

References.—*Roxb., Fl. Ind., Ed. C.B.C., 637, 638; Brandis, For. Fl., 424; Kurz, For. Fl. Burm., II., 455, 456; Dals. & Gibs., Bomb. Fl., 243; Elliot, Flora Andh., 32; Trimen, Cat. Fl. Pl., Ceylon, 84; U. C. Dutt, Mat. Med. Hind., 321; Settle. Rept., Seone, 10; Gazetteer, Mysore & Coorg, I., 70; Gazetteer, N. W. P. (Bundelkhand), I., 84.*

Habitat.—A creeping pubescent shrub common along the banks of larger rivers throughout the hotter parts of India and Burma from Chittagong and Ava down to Upper Tenasserim. Distributed southward to Perak and Ceylon.

The *Flora of British India* refers the polymorphous forms of this species to two varieties:—

F. scabrella, *Roxb.*: characteristic of Chittagong.—*Roxburgh.*

F. repens, *Willd.*

Medicine.—The JUICE of the ROOT of this shrub is internally administered in colic pains, and the juice of the LEAVES mixed with milk in dysentery. The BARK of the root, which is very bitter, pulverised and

MEDICINE.	
FOOD.	
Fruit.	
189	
FODDER.	
Fruit.	
190	
Leaves.	
191	
TIMBER.	
192	
SACRED.	
193	
194	
195	
196	
MEDICINE.	
Juice.	
197	
Root.	
198	
Leaves.	
199	
Root-bark.	
200	

2 A

A Useful Emetic.

mixed with coriander seed, is considered a good remedy in coughs and asthma, and similar affections of the chest (*Rheede*).

FOOD.
Fruit.
201

Food.—The FRUIT of **scabrella** is eaten by the natives of Chittagong in curries (*Roxb.*).

202

[*154, 155; Wight, Ic., t. 638, 641.*

Ficus hispida, *Linn., f.; Fl. Br. Ind., V., 522; King, Ficus, Plates*

Syn.—FICUS OPPOSITIFOLIA, *Willd.; Roxb., Corom. Pl., t. 124;* F. PRO-MINEUS, *Wall.;* F. DŒMONUM, *Kœnig;* F. MOLLIS, *Willd.;* COVELLIA DŒMONUM, *Miq.; Dalz & Gibs.*

Vern.—*Kagsha, gobla, totmila, kat-gularia, konea-dumbar,* HIND.; *Dumar, kako-dumar, kak-dumar,* BENG.; *Bhudoi,* CHUTIA NAGPUR; *Kotang, sosokera,* KOL; *Sita pordóh,* SANTAL; *Khoskadumar,* ASSAM; *Shakab,* GARO; *Koreh,* KURKU; *Kharwa,* NEPAL; *Kharwa,* PAHARI; *Taksot,* LEPCHA; *Poksha,* MICHI; *Maiu-lok,* MAGH; *Bhudoi,* MAL (S.P.); *Katu-mer, bomair,* GOND; *Kagsha, kagoha, dhúra, gobla, tomila,* KUMAON; *Dadúri, degar, rúmbal,* PB.; *Katumbri,* C. P.; *Rambal, dumbar, mira, dhedu,* BOMB.; *Kharawat,* MAR.; *Dhe daumaro, jangli angir,* GUZ.; *Dhedumeia,* PANCH MEHALS; *Pe-attiss* (**Moodeen Sheriff**), TAM.; *Boda-mamadi, bomma-médi, brahma-médi, bummarri, bamari, korasana,* TEL.; *Adavi-atti,* KAN.; *Pe-yatti paraka,* MALAY.; *Kadut, kadot,* BURM.; *Kota-dimbula,* SING.; *Kikadumbar, ummiatto-dumbara,* SANS.; *Tine-barri,* ARAB.; *Anjir-dashte,* PERS.

References.—*Roxb., Fl. Ind., Ed. C.B.C., 647; Brandis, For. Fl., 423; Kurz, For. Fl. Burm., II., 460; Beddome, For. Man, 224; Gamble, Man. Timb., 340; Trees, Shrubs, &c., Darjeeling, 76; Dalz. & Gibs., Bomb. Fl., 243, 244; Elliot, Flora Andh., pp. 28, 30, 31, 77, 98; Trimen, Syst. Cat. Pl. Ceylon, 84; Pharm. Ind., 217; Moodeen Sheriff, Supp. Pharm. Ind., 143; U. C. Dutt, Mat. Med. Hind., 301; Dymock, Mat. Med. W. Ind., 2nd Ed., 745; Atkinson, Him. Dist., 737; Drury, U. Pl., 216; Lisboa, U. Pl. Bomb., 131; Balfour, Cyclop, I., 1101; Home Dept. Cor. regarding Pharm. Ind., p. 240; Indian Forester, X., 325; XIV., 391.*

Habitat.—A moderate-sized tree or shrub, common throughout the outer Himálaya from the Chenab eastward, ascending to 3,500 feet; Bengal, Central and South India, Burma, and the Andaman Islands. Distributed to Malacca, Ceylon, China, and Australia.

FIBRE.
Bark.
203

Fibre.—Dr. Dymock informs the writer that in Bombay (especially near the coast) a fibre is prepared from the BARK which is used for tying bundles.

MEDICINE.
Fruit.
204
Seeds.
205
Bark.
206
Milk.
207

Medicine.—The FRUIT, SEEDS, and BARK are possessed of valuable emetic properties followed by more or less purging. This property was first brought to notice by Dr. **Moodeen Sheriff**. The acrid MILK obtainable from this species is used medicinally in Kangra. The bark, in doses of from 15 to 30 grains, three or four times daily, is stated to act effectually as an antiperiodic, and in half these quantities as a good tonic (*Pharm. Ind.*). In Bombay and the Concan the powdered fruit heated with water to form a poultice is applied to buboes. It is also given to milch cattle to dry up their milk (*Dr. Dymock*).

SPECIAL OPINIONS.—§ " According to Sanskrit writers the figs of this plant promote the secretion of milk. They are also supposed to preserve the fœtus in the womb" (*U. C. Dutt, Civil Medical Officer, Serampore*). "I have been using the fruit, seeds, and bark of **Ficus hispida** occasionally in my practice ever since I first found them in 1867 to possess the emetic property. They are good emetics, and act efficiently if assisted with warm water and tickling of the throat. The seeds of the ripe fruit should be dried and preserved from moisture in stoppered bottles, reduced to a powder when required, and administered in one-drachm doses. The bark is a stronger emetic, but its action is sometimes attended with more or less purging. Its dose is from forty grains to a drachm. The dose of

The Citron-leaved Ficus. (*Murray & Watt.*)	**FICUS infectoria.**

the ripe and fresh fruit is from four to six " (*Honorary Surgeon Moodeen Sheriff, Khan Bahadur, Triplicane, Madras*).

Food and Fodder.—The FRUIT, which is small and covered with short white hairs, is, according to Gamble, edible. The LEAVES are lopped for cattle fodder and are good for elephants.

FOOD.
Fruit.
208
FODDER.
Leaves.
200

Structure of the Wood.—Soft, dirty-grey, no heartwood, no annual rings. Weight 25 to 35 ℔. Put to no economic use.

TIMBER.
210

Domestic Uses.—According to Balfour this is one of the most destructive of figs to buildings.

DOMESTIC.
211

[*39, 40, Pll. 45, 83b.*

Ficus indica, *Linn., Sp. Pl.; Fl. Br. Ind., V., 506; King, Ficus, pp.*
Syn.—FICUS SUNDAICA & RUBESCENS, *Bl.;* UROSTIGMA RUBESCENS, SUNDAICUM, PSEUDO-RUBRUM, *Miq.;* F. LONGIFOLIA, *Ham.;* F. INDICA, *Linn.; Kurz, For. Fl., Barm., II., 442.*

212

Habitat.—A large spreading tree of Burma and the Andaman Islands. It seems probable that some of the economic information recorded under **F. bengalensis** may probably refer to this species. Until recently, in popular works, **F. indica** has been treated as a synonym for **F. bengalensis.**

[*t. 665; King, Ficus, 60, t. 75 to 79.*

213

F. infectoria, *Roxb.* (*non Willd.*), *Fl. Br. Ind., V., 515; Wight, Ic.,*
Syn.—F. TJELA, *Wall.;* F. VENOSA, *Wall.;* F. LACOR, *Ham.;* F. LUCESCENS, *Blume;* UROSTIGMA INFECTORIA, *Miq.*

The *Flora* remarks that "several geographical forms occur, of which three are Indian " : —

F. infectoria, proper.
F. Lambertiana, *Miq.*
Syn.—UROSTIGMA LAMBERTIANUM, *Dalz. & Gibs.*
A tree of Western and Central India.

214
215

F. Wightiana, *Wall.; Bedd., For. Man., 222.*
A tree of the south edge of the Gangetic plain and Western Ghâts.

216

Vern.—*Pilkhan, kahimal, ramanjir, pákhar, pákri, keol, kaim, khabar, pakur,* HIND.; *Pákar, pakur,* BENG.; *Buswesa,* KOL; *Prab,* GARO.; *Safed kabra,* NEPAL; *Kangji,* LEPCHA; *Pepere,* KURKU; *Serilli,* GOND.; *Pákhar,* MELGAT; *Pakur,* N.-W. P.; *War, palkhi, batbar, jangli pipli, palákh, pakhar, pilkin, trimbal,* PB.; *Killah,* KONKAN; *Pipli, bassari, pakri, kaim,* BOMB.; *Pepar, gándhaumbara, dhedumbara,* MAR.; *Pepri,* GUZ.; *Jooi, kall-alun, pepre, kurku,* TAM.; *Jewi, yuri, bassari,* TEL.; *Kari, basri, bassari,* KAN.; *Tsjakela,* MALAY.; *Nyaungchin, nyoungchin,* BURM.; *Kalaha, kiripella,* SING.; *Plaksha, parkati,* SANS.

References.—*Roxb., Fl. Ind., Ed. C.B.C., 643; Brandis, For. Fl., 414; Kurz, For. Fl. Burm., II., 446; Beddome, For. Man., 222; Gamble, Man. Timb., 334; Dalz. & Gibs., Bomb. Fl., 241; Stewart, Pb. Pl., 214; Sir William Jones, V., 159; U. C. Dutt, Mat. Med. Hind., 235, 312, 313; Atkinson, Him. Dist., 317; Lisboa, U. Pl. Bomb., 129, 235; Liotard, Dyes, 33; Watson, Report on Gums, 61; Kew Reports, 1879, 34; For. Ad. Report, Ch. Nagpur, 1885, 33; Journ. Agri-Hort. Soc., XIV., (Stewart on Hazara) p. 29; VII., 1885, New Series, 263, 276; Indian Forester, Vol. I., 274; VI., 218; VIII., 82; X., 33, 325; XIII., 121; Gazetteers, N.-W. P. (Bundelkhand), I., 84; (Agra), IV., lxxvii; Hoshiarpur, II.; Jalandhar, 5; Ludhiana, 10; Karnal, 16; Settle. Repts., Shahjehanpur, IX.*

Habitat.—A large tree (*Gamble*); a deciduous low tree (*Fl. Br. Ind*); found in the Suliman and Salt Ranges, the outer Himálaya; the plains and hills of India, Bengal, Assam, Burma, Central India, and specially the Western Coast Forest. Commonly planted; rarely met with wild.

Fibre.—The BARK yields a fibre which is said to be good for ropes (*Gamble*).

FIBRE.
Bark.
217

2 A 2

F. 217

FICUS **oppositifolia.**	A Burmese Caoutchouc-yielding Plant.

MEDICINE.
Bark.
218

FOOD.
Young shoots.
219
FODDER.
Leaves.
220
TIMBER.
221
DOMESTIC.
222

223

TIMBER.
224

225

FODDER.
Leaves.
226
TIMBER.
227

228

GUM.
229

Medicine.—The BARK of this, along with the barks of other four species of Ficus and of Melia Azadarachta, pass by the name of *Panchaval kaīa* (or the five barks); they are used in combination. A decoction is much employed as a gargle in salivation, as a wash for ulcers, and as an injection in leucorrhœa.

Food and Fodder.—The YOUNG SHOOTS are said to be eaten in curries by the natives. The LEAVES make good elephant and cattle fodder. (*Brandis*).

Structure of the Wood.—Grey, moderately hard. Weight about 35℔; not durable. It is used in Assam and Cachar to make charcoal, but according to Roxburgh it is useless even for firewood.

Domestic Uses.—A good avenue tree and planted for ornamental purposes.

Ficus laccifera, *Roxb.* see **F. altissima.**

[*14, 15, 81.*

F. mysorensis, *Hevne, Fl. Br. Ind., V., 500; King, Ficus, 19, t.*
Syn.—F. INDICA, *Linn.*, in part; F. COTONIÆFOLIA, *Vahl.*; F. CITRIFOLIA, *Willd.*; F. GONIA, *Ham.*; UROSTIGMA MYSORENSE, *Miq.*; U. DASY-CARPUM, *Miq.*; F. SUBREPANDA, *Wall.*; F. TOMENTOSA, *Hort. Madr.*; *Rheede, Hort. Mal., III., t. 57.*

References.—*Beddome, For. Man., 222; Kurz., For. Fl. Burm., II., 440; Dals. & Gibs., Bomb. Fl., 242; Gamble, Cat. Trees, Shrubs, &c., Darjeeling, 73; Trimen, Cat. Ceyl. Pl., 84; Lisboa, U. Pl. Bombay, p. 129; Bomb. Gaz., Kanara, XVI., Pt. I., 443.*

Vern.—*Goni,* KAN.; *Sunkong-kúng,* LEPCHA; *Búnuga,* SING.

Habitat.—A large umbrageous tree met with in the forests at the base of the Himálaya from Sikkim eastward; Khasia Hills, Burma, the Deccan Peninsula, and Ceylon.

Structure of the Wood.—Enumerated among the timber trees of .Bombay.

F. nemoralis, *Wall.; Fl. Br. Ind., V., 534.*
Syn.—F. GEMELLA and F. BINATA, *Wall.*; F. DENSA, F. TRILEPIS, and F. FIELDINGII, *Miq.*

References.—*Brandis, For. Fl., 424; Gamble, Man. Timb., 338.*

Habitat.—A moderate-sized tree of the outer Himálaya from the Hazára to Bhutan, ascending to 7,000 feet; Khasia Hills, Assam

Fodder.—The LEAVES are lopped for cattle fodder (*Gamble*).

Structure of the Wood.—White, moderately hard, close-grained. Weight 38℔ per cubic foot.

F. nitida, *Thunb.* See **F. retusa,** *Linn.*

[*King, Fic., 42, t. 49, 83*.

F. obtusifolia, *Roxb.; Fl. Br. Ind., V., 507; Wight, Ic., t. 662;*
Syn.—F. LONGIFOLIA, *Ham.*; UROSTIGMA OBTUSIFOLIUM, *Miq.*

Vern.—*Krapchi,* MICHI; *Date,* MAGH; *Nyaunggyat,* SHAN; *Nyoung-kyap,* BURM.

References.—*Roxb., Fl. Ind., Ed. C.B.C., 641; Kurz, For. Fl. Burm., II., 443.*

Habitat.—A small-leaved, large epiphytic tree, of the tropical forests at the base of the Eastern Himálaya, from Sikkim to Manipur, Assam, Chittagong, Burma, and Perak.

Gum.—Yields a rather good quality of caoutchouc (*Gamble, Man. Timbers*). Gives an India-rubber of inferior quality (*Gamble, List of Trees and Shrubs, &c., of Darjeeling*).

F. oppositifolia, *Willd.* See **F. hispida,** *Linn. fil.*

F. 229

| The Peepul Tree. | (*Murray & Watt.*) | **FICUS religiosa.** |

Ficus parasitica, *Kœn.* See **F. gibbosa,** *Blume.*

[*Fic.,* *146, t. 185.*

F. palmata, *Forsk. ; Fl. Br. Ind.,* V., *530; Wight, Ic., t. 649; King,*   230
  Syn.—F.. CARICOIDES, *Roxb.*; F. VIRGATA, *Roxb., Wright, Brandis;*
  F. PSEUDO-SYCAMORUS, *Dcne.*
  Vern.—*Gúlar, khabára, anjiri, beru, bedu,* HIND. ; *Phagwara, kák, kok,*
  phedú, insar, phag, kirmi, phagorú, fágú, phog, khabáre, phegra,
  thapur, jamir, dhúru, dhúdi, daholia, PB. ; *Phagwara* (HAZARA),
  PUSHTU ; *Angír, insar,* AFG. ; *Kembri* (MARWARA), RAJ. ; *Dhoura,*
  C. P. ; *Pepri,* GUZ. ; *Fagwara, thapur* (PLAINS OF UPPER INDIA).
  References.—*Roxb., Fl. Ind., Ed. C.B.C., 636 ; Brandis, For. Fl., 419 ;*
  Gamble, Man. Timb., 338 ; Stewart, Pb. Pl., 212 ; Boiss., Fl. Orient.,
  IV., 1155 ; Baden Powell, Pb. Pr., 377 ; Atkinson, Him. Dist., 317 ;
  Econ. Prod. of N.-W. P., V., 84 ; Balfour, Cyclop., I., 1102 ; Gazetteers :
  Simla, 9 ; Hoshiarpur, 11 ; Amritsar, 4 ; Agra, IV., lxxvii. ; Indian
  Forester, Vol. VI., 218 ; VIII., 82 ; XII., App. XXI. ; Settlt. Report,
  Hasara, 12 ; Stewart, Journal of Tour in Hasara (Journ. Agri. Hort.
  Soc., Vol. XIV., 7.
  Habitat.—This may be called the Indian representative of **Ficus Carica.**
It is a bush or moderate-sized tree, and is found in the Suliman and Salt
Ranges, and in the outer Himálaya of the Panjáb, eastward to Nepal
and Oudh, ascending to 6,000 feet. It also occurs on Mount Abu.
  Medicine.—The FRUITS contain chiefly sugar and mucilage, and   MEDICINE.
accordingly act as a demulcent and laxative. They are principally used   Fruits.
as diet in cases of constipation and in diseases of the lungs and bladder.   **231**
They are also used like the fruits of **Carica** as poultices (*Baden Powell*).
  Food and Fodder.—The FRUIT is eaten by the natives in the Panjáb   FOOD.
hills. Stewart says that at 5,000 feet he has found it excellent, though   Fruit.
generally poor fruit. It is largely eaten by the natives, and is even export-   **232**
ed to the plains (*Atkinson*). It ripens from June to October. The LEAVES   FODDER.
are given to cattle as fodder.   Leaves.
  233
  Structure of the Wood.—White, close- and even- grained, moderately   TIMBER.
hard. Weight 39℔ per cubic foot. According to the Revenue Settle-   **234**
ment Report of Belaspore, this is one of the timbers most commonly used
in that district for building.

F. pomifera, *Wall. ; Fl. Br. Ind.,* V., *535 ; King, Fic., 171, Pl., 215.*   235
  Syn.—F. HAMILTONIA, *Wall.* ; F. OLIGODON, *Miq.* ; F. REGIA, *Miq., Kurs.*
  Vern.—This seems to be the *Neverra* of Nepal.
  It seems probable that the bulk of the economic information published
by popular writers under **F. regia,** *Miq.,* should be relegated to this
species, but according to **King** some of the botanical writers who deal
with **F. regia** refer to **F. pomifera,** others to **F. Roxburghii.**

[*King, Fic., 55, t. 67A, 84" ; Bedd., Fl. Syl., t. 314.*

F. religiosa, *Linn. ; Fl. Br. Ind.,* V., *513 ; Wight, Ic., t. 1967 ;*   236
  The PEEPUL TREE.
  Syn.—F. AFFINIOR, *Griff.* ; UROSTIGMA RELIGIOSUM, *Gaspar, Dals. &*
  Gibs. ; U. AFFINE, *Miq.*
  Vern.—*Pipal,* HIND. ; *Ashathwa, aswat, asúd, asvattha,* BENG. ; *Hesar,*
  pipar, KOL. ; *Hesak,* SANTAL ; *Jári,* URIYA ; *Bor-bur,* CACHAR ; *Pipli,*
  NEPAL ; *Ali,* GOND ; *Pipri,* KURKU ; *Pipal, bhor,* PB. ; *Pippal,* PUSHTU ;
  Pipur, SIND ; *Jári, pimpal, piplo* (SURAT), BOMB. ; *Pimpala,* MAR. ;
  Pipul, GUZ. ; *Arasa, aswartham,* TAM. ; *Rái, raiga, ragi, rávi, or kulla*
  rávi, TEL. ; *Rangi, basri, arali, arle, haspath, rági, asvalta,* KAN. ;
  Nyaungbaudi, nyoungbaude, nyoungbaudi, nyaungbawdi, BURM. ; *Bo,*
  (*Arasa,* TAM.), SING. ; *Aswaththamu, asvattha,* SANS.
  References.—*Roxb., Fl. Ind., Ed. C.B.C, 642 ; Brandis, For. Fl., 415 ;*
  Kurs, For. Fl. Burm., II., 448 ; Gamble, Man. Timb., 334 ; Dals. &

F. 236

The Peepul Tree.

*Gibs., Bomb. Fl., 241 ; Stewart, Pb. Pl., 213; Campbell, Report Econ.
Prod., Chutia Nagpur, No. 7548 ; Cleghorn, 199 ; Mason, Burma and
its People, 424, 776 ; Trimen, Cat. Pl. Ceylon, 83 ; Sir W. Jones,
V., 159 ; Flora Andh., Elliot, 17, 162, 163 ; Ainslie, Mat. Ind., II.,
25 ; O'Shaughnessy, Beng. Dispens., 577 ; U. C. Dutt, Mat. Med.
Hind., 292 ; Dymock, Mat. Med. W. Ind., 2nded., 743 ; S. Arjun,
Bomb. Drugs, 198 ; Murray, Pl. and Drugs, Sind, 31 ; Med. Top.,
Oude, 4 ; Baden Powell, Pb. Pr., 377 ; Atkinson, Him. Dist., 317,
737 ; Drury, U. Pl., 217 ; Lisboa, U. Pl. Bomb., 130, 204, 279, 283,
290, 291 ; McCann, Dyes and Tans, Beng., 50, 136, 144, 159, 165 ;
Liotard, Dyes, 33 ; Liotard, Paper-making Mat., 31 ; Report on Indian
Dyes, by Wardle, 24 ; Watson, Report by, 34, 43, 44, 61, 65 ; Balfour,
Cyclop., I., 1101 ; Kew, Off. Guide to the Mus. of Ec. Bot., 122 ; Kew Off.
Guide to Bot. Gardens and Arboretum, 29, 42 ; Journ. Agri. Hort.
Soc., 1885, VII. (New Series), 263—276 ; Indian Forester, I., 273 ; III.,
205, 236 ; V., 212 ; VI., 218, 240 ; VII., 277 ; X., 63, 325 ; XII., App.
XXI., XXVIII.; XIII., 58, 69, 121 ; XIV., 391 ; Bomb. Gaz., II., 39,
355 ; III., 199 ; IV., 24 ; V., 28, 285 ; VI., 13, 183 ; VII., 37, 39, 40,
43 ; X., 39 ; XII., 26 ; XIII., 26 ; XV., Pt. I., 69 ; XVIII., Pt. I., 51 ;
XX., 13 ; XXIII , 64; Panjáb Gazetteers: Sialkot, 11 ; Ludhiana, 10 ;
Julundar, 5 ; Meerut, 33 ; Delhi, 18 ; Hoshiarpur, 10 ; Karnal, 16 ;
Rawalpindi, 15 ; Jhang, 17 ; Montgomery, 18 ; N.-W.P. Gazetteers :
Agra, IV., p. lxxvii. ; Mozuffarghur, 22 ; Oudh Gaz., Vol. II., 345 ;
Mysore and Coorg, Vol. I., 47, 70; III., 25; Manual, Trichinopoly Dist.,
78 ; Man. Chindwara Dist., 110.*

Habitat.—A large glabrous usually epiphytic tree, found wild in the
sub-Himálayan forests in Bengal and Central India. Extensively cultivated in most provinces of India, though less frequently so in Burma.

**GUM.
237**

Gum.—The bark yields a tenacious milky juice which hardens into
a substance resembling Caoutchouc.

"Its stem gives out a resinous gum which is used as sealing-wax,
and is also employed by artificers to fill up the cavities of hollow ornaments" (*Gaz. Bomb., VII., 37*). This same curious fact is alluded to
in the Ahmedabad Gazetteer (IV., 24). It is there stated that " The *piplo*
(**Ficus religiosa**) and the *bordi* (**Zizyphus Jujuba**) yield a wax much used
by goldsmiths for staining ivory red. It may here be pointed out, however, that these trees are the chief source of lac, and that the so-called
gum mentioned above may be only the waxy excretion caused by the lac
insect, and not a gum at all. The Rev. A. Campbell remarks that the
milky sap is known among the Santals as *lòré*. Lac is abundantly produced on this tree; indeed, according to many writers, this is its chief
use. A bird-lime is prepared from the milky juice, which is in the Deccan
called *shelim*.

SPECIAL OPINIONS.—§ " Juice used as a bird-lime. One-fourth seer
pipal juice, 2 chittacks linseed oil (castor oil will not do) ; simmer over
fire for five minutes, let cool " (*W. Forsyth, Civil Medical Officer,
Dinajpore ; U. C. Mukerji, M.B., C.M., Civil Medical Officer, Dinajpore*).

**DYE and TAN.
Bark.
238
Leaves.
239**

Dye and Tan.—The BARK is said to be sometimes used in tanning.
Drury mentions that the LEAVES are employed by the Arabs for this purpose. Wardle, however, says it contains little or no tannin, but yields to
boiling water a reddish pale-brown colouring substance which by the employment of various processes gives to tasar, mulberry silk, and woollen
fabrics, faint reddish fawn colours. The amount of colouring matter in
the bark is small, but it might prove a convenient dye where faint shades
are required or for modifying the colours produced by other dye-stuffs.
McCann wrote that the bark of this tree is also mentioned as being used
along with other barks when preparing a permanent black in Bengal.
Liotard says the roots, on being boiled in water, produce with alum on
cotton cloth a pale pink colour.

F. 239

The Peepul Tree.	*(Murray & Watt.)*	**FICUS religiosa.**

Fibre.—A fibre is extracted from the BARK. In Burma this was former-ly made into the paper used in the construction of the peculiar green umbrellas of that province; but the manufacture is rapidly dying out, and the umbrellas in use by Burmans are now mainly imported from China.

According to **Cross, Bevan** and **King**, the chemical composition of this fibre is—Moisture, 10·0; Ash, 7·9; Hydrolysis by (*a*) process (*i.e.*, boiling in alkali for five minutes), loss 22·6; by (*b*) process boiling for one hour), loss 46·8; Cellulose, 41·2. Chemically, therefore, the fibre may be pronounced worthless. The percentage composition of cellulose is very low, and the loss by weight due to alkali purification is ruin-ously high.

FIBRE.
Bark.
240

Medicine.—The BARK is astringent and is used in gonorrhœa. It has also maturative properties. An infusion is given internally in scabies. The ROOT-BARK is one of the five barks used by the Sanskrit physicians. The FRUIT is laxative and helps digestion. Dried and powdered, if taken in water for 14 days, it is said to remove asthma and make women fruitful (*Bartolomeo*). The SEEDS are said to be cooling and alterative. The LEAVES and YOUNG SHOOTS are used as a purgative, and have the reputa-tion of being useful in skin diseases (*Ainslie; Wight*). A paste of the pow-dered bark is employed as an absorbent in inflammatory swellings (*Dr. Emerson*).

MEDICINE.
Bark.
241
Root-bark.
242
Fruit.
243
Seeds.
244
Leaves.
245
Shoots.
246

SPECIAL OPINIONS.—§ "Water in which the freshly-burnt bark has been steeped is said to cure cases of obstinate hiccup" (*Civil Surgeon J. H. Thornton, B A., M.B., Monghyr*). "Ashes of the growing shoots when well sifted are sprinkled on chronic unhealthy ulcers to bring them into a healthy condition" (*Surgeon-Major Bankabihari Gupta, M.B., Pooree*). "In cracked foot the JUICE is employed, which is very sticky" (*Assistant Surgeon T. N. Ghose, Meerut*). "The powder of the dried bark is used in fistula in ano. I have seen a hakim use it with benefit, in the following way: he introduced a metallic tube, something like a blow-pipe, into the fistula, and putting a small quantity of the powder into it, blew the same into the fistula" (*Assistant Surgeon Nobin Chun-der Dutt, Durbhanga*).

Juice.
247

Food and Fodder.—The small, smooth, elliptical LEAVES and BRANCHES are good elephant and buffalo fodder. According to **Campbell**, the leaves are extensively lopped as cattle fodder. The young leaf-buds are eaten in Central India in famine times (*Gamble*). According to some writers, the small FIGS of this tree are eaten, but possibly during famine times only. Mr. **Campbell** says they ripen in the cold weather and are regularly eaten by the Santals. The *gori* silk-worms are fed in Assam on the leaves of this tree.

FOOD AND FODDER.
Leaves.
248
Branches.
249
Fruit.
250

SPECIAL OPINIONS.—§ "The leaves are used as a vegetable by the Gonds" (*Narain Misser, Kathe Bazar Dispensary, Hoshangabad, Cen-tral Provinces*).

Structure of the Wood.—Greyish white, moderately hard. Weight 30 to 45℔ (*Gamble*). "Uniformly yellowish white, very light, coarsely fibrous, perishable, takes an inferior polish" (*Kurz*).

In the *Indian Forester* the following is given as the analysis of the ash:—Soluble potassium and sodium compounds, 0·15; Phosphate of iron, calcium, &c., 2·25; Calcium carbonate, 1·96; Magnesium carbonate, 1·07; Silica with sand and other impurities 0·05; total ash, 5·48 (*Vol. X., 63*). It is used for fuel, for packing-cases, and in Cachar for charcoal.

TIMBER.
251

Domestic and Sacred Uses.—Largely planted as an avenue and road-side tree, especially near temples. It is held sacred by the Hindus, being viewed as the female to the Banyan. **Lisboa**, however, says that accord-ing to the *Valkhilya* the marriage of the *peepul* with the *tulas* (*Ocymum*

DOMESTIC AND SACRED.
252

FICUS retusa.	The Peepul Tree.

SACRED.

sanctum) is ordered. He further remarks that it is the transformation of the gods, *Guru,* and is termed *Ashwath.* It is specially worshipped on every Saturday of the month *Shrávan,* and on every *Somvati, i.e.,* on every Monday on which a new moon falls. The Hindu who plants a peepul tree does so expecting that just as he thereby affords shade to his fellow-creatures in this world, so after death he will not be scorched by excessive heat in his journey to the-kingdom of Yama (*Oudh Gaz., III., 345*). There are five sacred trees among the Hindus, *viz., peepul, gulár, bargad, pákar,* and *mango,* but of these the first is by far the most reverenced. A good Hindu who on a journey sees a peepul tree will take off his shoes and walk five times round the tree from right to left (*pardachna*). While doing so he repeats the verse which may be translated "The roots are Brahma, the bark Vishnu, the branches the Mahadeos. In the bark lives the Ganges, the leaves are the minor deities. Hail to thee, king of trees" (*Elliott, Chronicles of Ornao*).

The peepul is "believed to be inhabited by the sacred triad, Brahma, Vishnu, and Shiv. It is used at the thread investiture and at the laying of the foundation of a building. Vows are made to it and it is worshipped; male offspring is entreated for under its shade, pious women moving round its trunk 108 times. So sacred is it that none will destroy it, even when it grows on the crevices of walls and buildings, pulling down the strongest masonry. Of its wood the spoons are made with which to pour clarified butter on the sacred fire" (*Bomb. Gaz., V., 37*).

[*Fic., 50, t. 61, 62, 84*.

253

Ficus retusa, *Linn.; Fl. Br. Ind., V., 511; Wight, Ic., t. 642; King,*
 Syn.—F. DILATATA, *Miq.;* F. NITIDA, *Thumb.; Wight, Ic.;* F. RUBRA,
  Roth.; F. LITTORALIS, *Blume;* F. MICROCARPA, *Linn.;* F. BENJAMINA,
  Willd.; Roxb., Fl. Ind.; UROSTIGMA RETUSUM, NITIDUM, MICRO-
  CARPUM, and OVOIDEUM, *Miq.*
 Vern.—*Kamrup, sir,* BENG.; *Butisa,* KOL.; *Sunumjon,* SANTAL; *Jili,*
  CHUTIA NAGPUR; *Jamu,* NEPAL; *Sitnyok,* LEPCHA; *Jili,* MAL (S.P.);
  Nandruk, MAR.; *Yerrajuvi, nandiréka,* TEL.; *Pilála, pinval,* KAN.;
  Nyaungok, nyoungthabyeh, BURM.
 References.—*Roxb., Fl. Ind., Ed. C.B.C., 643; Brandis, For. Fl., 417;
  Kurz, For. Fl. Burm., II., 444; Beddome, For. Man., 223; Gamble,
  Man. Timb., 336; List, Trees and Shrubs, &c., of Darjeeling, 75; Dáls.
  & Gibs., Bomb. Fl., 241, 242; Trimen, Cat. Ceyl. Pl., 84; Elliot, Fl.
  Andh., 27, 68; Dymock, Mat. Med. W. Ind., 2nd Ed., 745; Lisboa, U.
  Pl. Bomb., 130; Balfour, Cyclop., I., 1101; For. Ad. Report, Ch. Nagpore,
  1885, 33; Bomb. Gas.: XIII., 26; XV., Pt. I., 69; XVI., 16; Indian
  Forester: III., 205; VIII., 332; IX., 516.*

Habitat.—A large evergreen tree, having a few aerial roots : met with at the base of the Eastern Himálaya from Kumaon to Bengal, Assam, South India the Deccan Peninsula, Burma, and the Andaman Islands. Distributed to the Malay Islands, China, and New Caledonia.

The *Flora* describes two varieties of this species :—
 α F. retusa, *Linn.*—The *Nandruk* of the Deccan Peninsula.
 β F. nitida, *Thunb.*—The tree of the trans-Gangetic regions.

254
255
MEDICINE.
Root-bark.
256
Root
257
Leaves.
258

Medicine.—The bark of the ROOT, the root itself, and the LEAVES boiled in oil form good applications for wounds and bruises (*Rheede*). In rheumatic headaches the leaves and bark pounded are applied as a poultice. In flatulent colic the leaf-juice is used, mixed with that of *tulsi,* and *ghí* (equal parts), applied externally and accompanied by fomentation with a hot brick (*Dymock; Rheede*). The juice of the bark in doses of one tola in milk has a reputation in liver disease.

TIMBER.
259

Structure of the Wood.—Light reddish-grey, close-grained, moderately hard, beautifully mottled. Weight 40℔ per cubic foot. It is used

| **The Peepul Tree.** | *(Murray & Watt.)* | **FICUS Rumphii.** |

for fuel, but as it is very prettily grained it might be found valuable for tables, door panels, and other purposes. A valuable avenue tree, as it affords dense shade.

[*King, Fic., 168, t. 211.*

Ficus Roxburghii, *Wall.; Fl. Br. Ind., V., 534; Wight, Ic., t. 673;*

Syn.—F. MACROPHYLLA, *Roxb.*; F. SELEROPTERA, *Griff.*; F. REGIA, *Miq.*; CERELLIA MACROPHYLLA, *Miq.*

Vern.—*Trimmal, timal, timla,* HIND.; *Demúr, doomoor,* BENG.; *Sapai,* MAGH.; *Kotang,* KOL; *Kasrekan,* NEPAL; *Kundoung,* LEPCHA; *Urbúl, urmúl, barbaru, túsi, trimbal, trímal, tírnal, dadúri, tremal, tirmí, tiamb, timbal, burh,* PB.; *Ber* (fruit = *hurmal*) (HAZARA), PUSHTU; *Sin-tha-hpan,* BURM.

References.—*Roxb., Fl. Ind., Ed. C.B.C., 645; Brandis, For. Fl. 422; Kurz, For. Fl. Burm., II., 460; Gamble, Man. Timb., 340; Stewart, Pb. Pl., 214; Atkinson, Him. Dist., 317; Tropical Agricult., 1889, 566; For. Ad. Rep. Ch. Nagpore, 1885, 33; Gazetteers: Simla, 11; Hazara, 13; Hoshiarpur, 11.*

Habitat.—A moderate-sized tree of the outer Himálaya from the Indus eastward to Bhutan, ascending to 6,000 feet; Sylhet, Khasia hills, Chittagong, and Burma.

Fibre.—In the Sutlej valley a coarse rope is made from the bark.

Food and Fodder.—The FRUIT is eaten in curries. It is described as handsome, of a russet-red colour, and of the shape and size of a Dutch turnip. They are carried in enormous bunches on the stem, especially near its base, and in smaller bunches on the main branches. A specimen which fruited in the Botanic Gardens, Calcutta, produced about 1 cwt. of figs. These are said to be unpalatable, insipid, and sloppy (*Gardener's Chronicle*).. Stewart, however, remarks that the fruit is sweet and of a pleasant flavour. According to the Kangra and Simla Gazetteers, it is regularly brought to market. The LEAVES are used as fodder.

Structure of the Wood.—Reddish grey, moderately hard. Weight 34℔.

[*Ficus, p. 54, t. 67B, 84ᵗ; Brandis, t. 48.*

F. Rumphii, *Bl.; Fl. Br. Ind., V., 512; Wight, Ic., 640; King,*

Syn.—FICUS CORDIFOLIA, *Roxb.* (*non Bl.*); UROSTIGMA RUMPHII, *Miq.*; U. CORDIFOLIUM, *Miq.*; FICUS, *Sp., Griffith, Icon. Pl. As., t. 549, Itin. Notes, III., n. 145.*

Vern.—*Kabar, gajna, pipul, gajiún, pipal, gagjaira, pakar, khabar,* HIND.; *Gaiaswát,* BENG.; *Suman-pipar,* KOL.; *Sunamjor,* SANTAL; *Pakri,* ASSAM; *Sat-bur,* CACHAR; *Pakar,* NEPAL; *Prab,* GARO; *Kabai pipal,* KUMAON; *Pulákh, rúmbal, badha, palák, pilkhan,* PB.; *Parás pipal,* RAJ.; *Pair, páyar, asht* (*ashta*), MAR.; *Kabai pipal, ganjar, suman, pipar,* LOHARDUGGA; *Nyaung byu,* BURM.

References.—*Roxb., Fl. Ind., Ed. C.B.C., 642; Brandis, For. Fl., 416; Kurz, For. Fl. Burm., II., 448; Gamble, Man. Timb., 335; Stewart, Pb. Pl., 212; Mason, Burma and its People, 424, 776; Rev. A. Campbell, Rep. Econ. Prod., Chutia Nagpore, No. 8497; Dymock, Mat. Med. W. Ind., 2nd Ed., 744; Atkinson, Him. Dist., 317; Lisboa, U. Pl. Bomb., 130, 279, 284, 291; Indian Forester: I., 86; IX., 562; X, 325; XII., App. XXI.; Smith, Dic., 1099; For. Adm. Rep., Chutia Nagpore, 1885, 33; Gazetteers: Thana, XIII., 26; Kanara, XV., Pt. 1., 443; Ahmadnagar, XVII., 26.*

Habitat.—A large deciduous tree of the outer Himálaya, closely resembling **F. religiosa**, occurs on the dry lower slopes of the mountains of the Panjáb; and in Northern, Western, and Central India, Assam, Burma, and the Malay Peninsula, ascending to 5,000 feet. It is generally epiphytic and accordingly very destructive to timber trees. It is said in the *Bombay Gazetteer* (Ahmadnagar) to frequent teak-wood forests and the regions of heavy rain. In Thana it is remarked that it is an unshapely tree,

260

FIBRE.
261
FOOD.
Fruit.
262

Leaves.
263
TIMBER.
264

265

FILICIUM decipiens.	**The Peepul Tree.**

thus being less suited for avenue and road-side planting than **F. retusa,** which is spoken of as the best of the road-side trees. In Oudh it seems to be specially associated with the *Sál* (**Shorea robusta**). The fruits ripen in May to June.

GUM.
266

Gum.—Roxburgh remarks that the milky juice flows abundantly from fresh wounds, and is very tenacious.

RESIN.
267

Resin.—The lac insect is reared extensively on **F. Rumphii** in Assam. This tree is specially cultivated for that purpose, and is remarkable on account of the insect not destroying it, though crops are taken annually.

FIBRE.
Bark.
268

Fibre.—According to the Rev. A. Campbell the BARK yields a cordage fibre of good quality.

MEDICINE.
Fruit.
269
Juice.
270

Medicine.—The Santals use the FRUIT as a drug. Dymock writes of this species: "The JUICE is used in the Concan to kill worms and is given internally with turmeric, pepper and *ghí,* in pills, the size of a pea, for the relief of asthma; it causes vomiting. The juice is also burned in a closed vessel, with the flowers of *umdar,* and 4 *gunjás* weight of the ashes mixed with honey, is given for the same purpose."

FOOD.
Fruit.
271
Leaves.
272
FODDER.
Branches.
273

Food and Fodder.—The FRUIT is eaten by the natives. The LEAVES and BRANCHES are used for cattle fodder.

TIMBER.
274

Structure of the Wood.—Very soft, spongy. Weight 27℔ per cubic foot. The wood is used in Cachar to make charcoal, and is also employed in tea manufacture and as fuel.

DOMESTIC
AND SACRED.
275
276

Domestic and Sacred.—The leaves are used in *panch-pallavs.*

[*Fic.,* 59, *t.* 74, 84ˣ·².

Ficus Tsiela, *Roxb.; Fl. Br. Ind.,* V., 515; *Wight, Ic., t.* 668; *King,*
　　　Syn.—F. AMPLISSIMA, *Smith;* F. INDICA, VAR., *Linn.;* F. BENJAMINA,
　　　Wall.; UROSTIGMA PSEUDO-TJELA, and PSEUDO-BENJAMINA and TJIELA,
　　　Miq.
　　　Vern.—*Jari,* HIND.; *Pimpri,* BOMB.; *Juvvi, ichchi,* TAM.; *Juvvi*
　　　(? *jovi*), TEL.
　　　References.—*Roxb., Fl. Ind., Ed. C.B.C.,* 642; *Beddome, For. Man.,* 314;
　　　Thwaites, En. Ceylon Pl., 265; *Dals. & Gibs., Bomb. Fl.,* 241; *Cleghorn,*
　　　196, 199; *Elliot, Fl. Andh.,* 75; *Lisboa, U. Pl. Bomb.,* 130; *Indian*
　　　Forester: III., 205; *XII., App.* 21; *Mans.: Coimbatore Dist.,* 39;
　　　Cuddapah, 263; *Bombay Gazetteer, Vol. XVII.,* 26.
　　　Habitat.—A large spreading tree without aerial roots, met with in the Deccan Peninsula from the Concan southward. Roxburgh regards it as next to **F. religiosa,** the largest species of Indian fig. It is a handsome tree, with smooth bark, wholly glabrous, and is met with in cultivation along roads throughout India.

FIBRE.
Bark.
277

Fibre.—The BARK gives a good fibre.

TIMBER.
278

Structure of the Wood.—No author seems to have specially described this, but it is used as firewood.

F. virgata, *Roxb.;* see **F. palmata,** *Forsk.*

Filberts, see **Corylus Colurna,** *Vol. II., p. 575, No. 1988.*

FILICIUM, *Thw.; Gen. Pl., I.,* 325.

279

Filicium decipiens, *Thwaites; Fl. Br. Ind., I.,* 539; BURSERACEÆ.
　　　Vern.—*Katu puveras,* TAM.; *Pehimbia,* SING.
　　　Habitat.—A tree with elegant fern-like leaves, found in the Western Ghâts up to 4,500 feet, also in Ceylon.

TIMBER.
280

Structure of the Wood.—Heartwood red, moderately hard. Pores small, in groups or short radial lines. Medullary rays fine, numerous, at unequal distances. Weight 68℔ per cubic foot.
　　　The wood is strong and valuable for building (*Gamble, Man. Timb,* 68).

F. 280

Filix-mas, see **Nephrodium Filix-mas,** *Richard.;* FELICES.

FIMBRISTYLIS, *Vahl.; Gen. Pl , III.,* 1048 ; CYPERACEÆ.

281

The species of sedges referred to this genus do not appear to be of much economic value. **F. Kysoor** in *Dals. & Gibs.,* Bomb. Fl., p. 288 (**Scirpus Kysoor,** *Roxb., Fl. Ind. Ed. C.B.C.,* 77) is said to be eaten in times of famine (*Lisboa, U. Pl. Bomb.,* 208). It is the *Kysur* or *Kesúri* of Bengal. This should not be confused with *kesúria*—**Eclipta alba.**

Fimbristylis junciformis, *Kunth,* is the *Bindi muthi* of the Santals, the roots of which, according to the **Rev. A. Campbell,** are given in dysentery.

282

F. monostachya, *Hassk.,* is known to the Santals as *Nanha bindi mutha.*

283

Fir, see **Abies** and **Pinus,** CONIFERÆ.

(*J. Murray.*)

FISH, *Day, Fishes, in Fauna of British India.*

284

For the purposes of a description, such as the following, the Fish of India may be divided into two great classes—THE MARINE and the FRESH-WATER—both of which are not only very large, but owing to their forming an extremely important source of the animal food of the Natives of this country, are well worthy of careful attention. The question of the best means of protecting and stimulating the large fishing industries of India has always attracted much attention, and the natural history of the subject has been the object of careful and laborious research on the part of many learned zoologists. Of all the provinces of India a Fisheries Act exists in Burma alone, but the question of framing an Act to embrace all the provinces is at present under the consideration of Government.

References.—*Day, Fishes of India ; Fresh-water Fishes of India and Burma ; Rep. on Sea Fish and Fisheries of India and Burma ; Rep. on Fish and Fisheries of the Fresh-waters of India ; Rep. on Fisheries of Assam; Indian Fish and Fishing, in the Internat. Fisheries Exhb. Lit., Vol. II., Pt. II.,* 441 ; *Condensed Rep., Vol. VIII.,* 345 ; *Catal. of Ind. Sec. Fish. Exhibit.; Beavan, Fresh-water Fishes of India ; Thomas, Rod in India; Rep. on Pisciculture in South Cánara ; Tennent, Nat. Hist. of Ceylon,* 323 ; *Rep. on the Fisheries of the Hensada Dist., Burma ; Seaton, Rep. on Fisheries in British Burma ; Rev. and Agric. Dept., Proceedings on Fisheries, Bill* 1 *to* 13, *June* 1888 ; 1 *to* 10, *Jany.* 1889 ; *Robinson, Fishes of Fancy, in Fish Exhb. Lit., Vol. III., Pt. I ; Walpole, Official Rep. on the Internat. Fish. Exhb. in Lit. of same, Vol. XIII.,* 15 ; *Simmonds, Commercial Products of the Sea; Balfour, Cyclop., I.,* 1107 ; *Forbes Watson, Ind. Survey,* 346—366, 392, 400, 404 ; *Bidie, Cat. Raw Prod. of Southern India, Paris Exhb.,* 96 ; *Ainslie, Mat. Ind., I.,* 227, 395 ; *Irvine, Mat. Med. of Patna,* 69, 100 ; *Gazetteers of Bengal, Central Provinces, Madras, Bombay, North-West Provinces, Panjáb, in many passages.*

Distribution in India.—The whole of the seaboard of India and Burma, computed at about 4,611 English miles, is washed by waters more abundantly stocked with fish than are even those which yield the great fish harvests of the British Isles. Fish abound also in the rivers, tanks, irrigation canals, ditches, and marshes of this country,—in fact, wherever water exists, from the sea-level to almost the highest elevations.

DISTRIBU-
TION.

Food.—The value of such well-stocked fisheries naturally depends to a great extent on the degree to which the production is utilized as food

FOOD.
285

F. 285

FISH.	Fishing Classes aud Fisheries.

by the people of the country. In considering this question it is, therefore, necessary to observe first of all what proportion of the people of India and Burma can consume fish as food without infringing religious prejudices. In the Panjáb and North-Western Provinces comparatively few of the inhabitants are thus prohibited; the large Muhammadan population eat fish, except those without scales and fins (such as the eel), while the Hindus, with the exception of certain Brahmans, Thakurs, Baniyas and Bhagats, consume fish of all kinds. Similarly, in Hyderabad, Mysore, and Coorg, more than half the population are permitted by their religion to consume fish; in Oudh the majority can do so; and in Sind nearly all except the Brahmans. Varying statements are made regarding Bombay in the District Gazetteers, from three-fourths in Khandesh, to 25 per cent in Bijapur, but the former figure probably represents more nearly the actual average; only Brahmans, high-caste Súdras, Márwár Vanís, Lengáyats, Jains, and a few others being prevented by their castes from eating fish. In Madras about a similar percentage; in Bengal proper from 90 to 95 per cent.; in Assam and Chittagong almost the entire population are permitted to eat fish; while in Burma the use of fish diet is universal, notwithstanding that the Burmans as Buddhists profess the greatest horror at taking the lives of the lower animals They console their consciences, however, with the idea that the sin lies entirely with the fishermen, and in Burman temples are depicted vivid representations of the terrible tortures the latter will have to endure in a future existence.

Notwithstanding the enormous market for fish, and the teeming waters in and around India, the supply appears to be everywhere insufficient to meet the demand, while the fishing classes are wretchedly poor. Dr. Day, in commenting on this fact, writes: "Investigating how the local markets were supplied with fish up to 1873, the replies from native officials gave the following results. In the Panjáb one in ten markets was sufficiently supplied, in the North-West Provinces one in three, in Oudh one in four. In Bombay the amount was stated to be insufficient in all, and similar reports came from Hyderabad, Mysore, and Coorg. In Madras near the sea the quantity was sufficient, but inland it was only so in one out of ten." In a further passage he writes: "The most casual observer cannot fail to perceive how numerous are the varieties, and vast the number of the finny tribes in the seas of India, but from some cause,—whether due to legislative enactments and local obstructions, or native apathy and impecuniosity,—the harvest has, up to within the last few years, been comparatively untouched, an enormous amount of food still remains uncaptured, while famines are devastating the contiguous shores."

Fishing Classes and Fisheries.—The MARINE FISHING CLASSES of India present many features of great interest, showing, as they do, survivals of manners and customs dating from very remote times. According to ancient Hindu legislation they belonged to the Súdra or servile caste. In most places they still maintain that they were, of old, divided into two distinct classes: (1) those who captured fish in the deep sea, (2) those who pursued their avocation from the shore, fishing in back-waters and creeks. Nowadays, however, owing to the depressed condition of the fishing industry, the deep sea fishermen (except where salt is cheap or a good local market exists) have taken to the less expensive occupation of plying their work inshore, and earn part of their living by work of other sorts. In Sind the fishermen are Muhammadans and are termed Mohanís. They are probably partly immigrants from Arabia, and partly Hindus converted to Islam. In Bombay they are chiefly Machhís, Márátha Bhoís, Káche Bhoís, Menjage Bhoís, Bagdi Bhoís, and Kolís, but many other classes occasionally fish. In the Madras Presidency they have customs of a patriarchal nature, which are, however, more strictly observed on the Coromandel than on the Western coast. The present organization in those parts is probably the remains of a very ancient system, as it is difficult on any other supposition to account for the immense hereditary power held by certain individuals. Not only have they hereditary and elective headmen of villages, but also hereditary priestly chiefs, who are the final referees in all family and caste disputes. Regarding these fishing tribes Dr. Day writes : " The condition of the sea-fishermen in Sind about ten years ago when investigations were made, showed that they were fairly well off ; miserably poor in Bombay except in the vicinity of large towns ; in a prosperous condition from South Cánara down the western coast of the Madras Presidency, but once round Cape Comorin they again appeared as a poverty-stricken race of people, and continued so up the Coromandel coast, except when residing near large centres of population."

The FISHERMEN OF FRESH WATERS are, as a rule, members of fish-eating castes, who engage in fishing as an occasional and subsidiary occupation, only a very few of the original fishing castes still restricting their means of livelihood to their hereditary industry. Under native rule in India this was not so ; fishing having then been in the hands of distinct castes, but as British rule has given up taxes on the industry, and of recent years fishing rents as well, it is now no one's interest to prevent undue depletion of the fisheries, and as a consequence fishing is no longer generally remunerative.

Classification of Fisheries—SALT-WATER.—Many and various methods of fishing are employed along the coasts of India and Burma, of which it is impossible, within the scope of the present article, to give a complete account. The chief characteristics of the systems may, however, be briefly adverted to, the information being chiefly compiled from Dr. Day's elaborate account in his Fisheries' Exhibition Report. *1st, Tidal Fisheries* —May consist of simple tidal ponds, into which fish are carried by the flood of the tide, and are left impounded by the ebb. They are then removed by scoop, lave, cast or other nets, or screens may be constructed of stonework, bamboo, rattan, or reed, to allow of the escape of the water while retaining the fish. Another common contrivance for tidal fisheries is the labyrinth, composed of wicker-work placed at right angles to the shore, generally at the head of an estuary. *2nd, Stake Nets*—Are probably an evolution of later date, but now constitute one of the chief means of obtaining a supply of fish on certain parts of the Indian coast. The stakes, which are generally made of the stems of certain palms, and may have a height of as much as 100 feet, are driven into the sand or mud at a distance of about 25 feet apart. To these long bag nets are affixed, into which the fish are carried by the currents running along the shore. *3rd, Moveable Nets*—Are of many forms—purse-nets used in shallows, cast-nets, drag-nets, and special nets for particular purposes varying in size, shape, and diameter of mesh according to the fish they are intended to capture. *4th, Wicker Traps*—Are very extensively employed in all parts of the East. They may be cone- or bell- shaped with both ends open, in which case they are employed in shallows, the fisherman placing the larger end over the fish and extracting them from the smaller ; or they may be built like a rat trap, baited and simply placed in tideways. *5th, Miscellaneous Methods.*—Diving, spearing, shooting with arrows, and fishing with hooks and lines with natural or artificial bait, are all employed in various parts of the country. *6th, Deep-sea Netting*—Is, as already stated, carried on to a very limited extent only, not only because of the insufficiency of a remunerative market, but also because the necessary appliances, boat, net,

FISH.	Classification of Fisheries.
FISHERIES.	&c., are expensive, and the fisher class is a miserably poor one. For instance, Dr. Day informs us that in Sind, a boat costs about £100, and a net suitable for deep-sea fishing involves an outlay' of from £40 to £50. The purchase of such an expensive plant, therefore, necessitates the borrowing of the money, on which the fisherman has to pay an exorbitant interest, leaving but a poor margin of profit as the reward of his labour.
FRESH-WATER, 289	FRESH-WATER.—With the establishment of British rule the fishing on rivers, which at one time was restricted either by the imposition of taxes or by leasing out to contractors the monopoly of fishing, has become, in most parts of the country, free and unrestricted. The natural result has been that every fish consumer is at liberty to capture his own fish, and the old fresh-water industry has necessarily declined. But an evil outcome of this has been that every endeavour is now made to catch as many fish, of all sorts and sizes, as quickly and cheaply as possible, and for this purpose all kinds of appliances are used. Rivers are dragged with nets having infinitesimally small meshes, or with coarse cloths; or a similar
Contrivances. 290	apparatus is even placed across a stream from bank to bank, and another dragged down stream to it, thus clearing every living thing out of the tract netted. At the same time the agricultural classes catch fish for themselves by means of wicker traps, baskets, and nets. Neither breeding fish nor fry are respected, everything caught is killed and eaten or destroyed, and no close season anywhere exists; hence as a natural result the supply of fish is everywhere diminishing. This is especially so in the case of the finer migratory hill fishes, such as the *mahasír.* Owing to the immense number of wicker-work and net weirs now to be found in most mountain streams at every few miles, the water is literally strained, with the inevitable consequence that the fish are rapidly decreasing in the lower reaches. In some places, more especially in the Doon hill tracts, streams are also frequently diverted in part of their course by damming them up, the large fish are extracted from the pools in the old bed of the river, and the fry are left to die as the water dries up. Not only are these and many other of the poaching practices so strongly condemned in England carried
Poisons. 291	on day after day, but poisoning the water is also frequently resorted to as a means of ready and wholesale destruction. The principal plants employed for this purpose are:—Strychnos Nux-vomica, Lasiosiphon speciosus, Balanitis Roxburghii, Tephrosea suberosa, Euphorbia Tirucalli, Hydocarpus Wightiana, H. venenata. Of recent years, also, a still more powerful agent of destruction has been found in dynamite, to the use of which natives employed in mines, and on tea, coffee, and cinchona estates have become habituated. They find no difficulty in possessing themselves of their employers' cartridges on off-days, and employ them freely, with the result that the place dynamited is denuded of all fish life, full grown, fry and ova. Besides these methods of directly killing fish, there are many other artificial agencies which indirectly, but to a very great
Explosives. 292	degree, affect fisheries in many districts. Perhaps the most important of these is the large irrigation works now existing in many parts of the country, formed by diverting a large amount of the water of a river down a canal. Where these canals are not constructed for navigation as well as irrigation, falls frequently exist, down which the fish can pass, but cannot return. The canal is thus converted into a vast fish trap, wherein all the fish are destroyed when run dry to examine it for necessary repairs. In the same way the small tributary irrigation canals act as traps from the main channel, all the fish entering them being invariably killed. The yearly inundations attendant on the rains, and the annual drying up of many tanks, must also be fertile sources of mortality. Dr. Day, in summing up the consideration of this subject in his admirable report,

Curing of Fish.	(*J. Murray.*)	FISH.

writes: "Thus it has come to pass that among the animal productions of India, fresh-water fish meet with the least sympathy, and the greatest persecution, many forms having to struggle for bare existence, in rivers which periodically diminish to small streams, or even become a mere succession of pools, or in tanks from which the water totally disappears. They have their enemies in the egg stage, in their youth, and during their maturity; but among these man is their greatest foe, as any one who desires a fish diet captures these creatures, whenever and wherever he gets the chance, irrespective of season, age, and size. In certain districts they simply appear to exist solely because man and vermin have been unable to destroy them."

Many suggestions have at different times been made to remedy this wholesale and indiscriminate destruction, by such means as preventing poisoning, regulating the size of net mesh, guarding the mouths of irrigation canals against the entrance of fish, levying taxes on the use of fishing implements, &c. As above stated these are at present under the consideration of Government, with a view to the introduction of a Fisheries Bill.

Rent of Fisheries.—The available amount of information regarding the proportion of fisheries either rented out by Government or owners is very meagre; but from a few statistics derived from the Gazetteers of different districts, it appears that the amount thus annually realised at the time of report must have been a large one. Thus in Bengal alone 27 districts are mentioned as yielding a revenue to Government or proprietors, the total of which was £6,417. In only a few were the value of the fisheries and the rent paid both given, but a calculation based on these shows the percentage rent to have been ·17 on the value of the property farmed. The revenue derived from Sind Fresh-water Fisheries in 1882-83 was R92,541, and from Burma in 1883, 12 to 13 lakhs of rupees—a not uninteresting evidence in favour of a Fisheries Act for the other provinces of India.

Rent.
293

Salt and Dried Fish.—It is apparent that in a tropical country such as India, the prosperity of sea fisheries must to a very great extent depend on the facilities afforded for curing fish thoroughly, and at a sufficiently small cost to meet the demand. In olden times this was possible, as salt was allowed duty-free in British territory for salting fish, but this privilege was withdrawn, because the excise officers found that it facilitated smuggling. As a consequence the fishermen and fish curers have done their best to escape from the tax, and in many localities employ salt earth, which imparts a bitter and unpleasant flavour to the fish and is liable to engender disease, while in other districts the fish are simply cleaned, dipped in the sea, and dried in the sun. Fish thus prepared are very inferior, often half putrid, and are only used as food by the poorest classes, while fish prepared by taxed salt are only bought by the rich, and for exportation. It is to be hoped, however, that means may be found to remedy this state of matters; indeed, during the last few years, the system of bonded enclosures, within which fish may be cured, with free salt, has been tried at Madras and with a fair amount of success. In Burma a putrescent preparation of fish is largely eaten called *nga-pí*. It is prepared as follows: "A quantity of semi-putrid fish is put into a jar with some salt and suffered to rot, until it is crowded with maggots; it is then baked, worms and all, over the fire and potted for after-use. The Burmans can no more live without *nga-pí* than others without fish. A better and cleaner sort of *nga-pí* is prepared at, and procured from, Penang by the Anglo-Burmese, which, though far superior, is still excessively unbearable" (*Fenwick*).

SALT FISH.
294
DRIED FISH.
295

Nga-pí.
296

Trade in Cured Fish.—A large import and export trade exists, the former doubtless due to the difficulties in the way of the Indian curer. Thus in

TRADE.
297

FISH.	Industrial Products from Fish.

TRADE IN CURED FISH.

the five years ending 1887-88 the total average imports were 12,088,846℔, valued at R10,82,836, In comparison with the five years ending 1882-83, this shews a considerable increase, the average for that period having been 8,921,583℔, value R7,85,557. Not only is there an increase in imports but a larger proportion of the fish thus obtained is consumed in the country, the re-exports shewing a decrease from an average of 444,447℔ in the five years ending 1882-83 to 176,361℔ in the later period. The countries which form the chief sources of supply are Mekran and Sonmiani, the Straits Settlements, Arabia, Persia, Ceylon, and Turkey in Asia. The exports appear to have remained very steady during the past ten years, though fluctuating considerably year by year. Thus in the latter half of that period the total average quantity was 4,096,074℔, value R3,55,756, while in the former it was 3,393,634℔, value R1,82,857. The port from which much the largest proportion was exported last year was Madras, which shipped 4,560,858 out of a total of 4,870,944℔, while Ceylon formed the principal market, importing 4,384,034℔ of the whole. It would be interesting to know to what extent the enlightened efforts to supply cheap salt had influenced the formation of the large Madras export trade as compared to any other province.

**FISH OIL.
298**

Fish Oil.—The manufacture of oil from fish is carried on all along the Western coast of India and also in other parts. It is obtained chiefly from the livers of sharks, skates, saw-fishes, cat-fishes, oil sardines, and other kinds, also from the heads, intestines, and even the whole body of some species. The process of manufacture as carried out in India is very crude; the livers are not washed, but fresh or putrid, clean or foul, they are put into a pot and heated up to boiling point, when the oil separates, floats on the top, and is skimmed off. It undergoes no straining and is consequently impure, and frequently rancid. At Rangoon a large amount is manufactured, the average quantity being said to exceed 77 tons a month. The ordinary oil thus obtained is employed for the purposes of cooking, lighting, and in tanning leather, while that extracted from the livers of species of Carcharias or shark is said to be an efficient substitute in medicine for cod-liver oil. Fish oil is a commercial article of considerable importance, large quantities being exported to Europe. In the official trade returns, however, no separate statistics exist, so that definite information as to its extent cannot be furnished.

**FISH ROES.
299**

Fish Roes—Obtained from several species, are largely employed as an article of food in many parts of India, and are sold in nearly every bazár of South and East Asia.

**FISH SKIN.
300**

Fish Skin.—The rough skins of species of Sharks, Skates, and Rays are employed for polishing in several parts of the country. Shagreen or shark's skin is chiefly used to cover scabbards.

**FISH MAWS.
301**

Fish Maws—Along with sharks' fins form an important article of foreign trade. See SHARKS' FINS, FISH MAWS, &c., in another volume.

**FISH SCALES.
302**

Fish Scales.—The scales of the Mahasir (**Barbus tor**) are employed in the manufacture of playing-cards. The scales are cut in a circular form about 1½ inch in diameter, and painted as required The principal seat of their manufacture is at Shahabad in Bengal.

**MEDICINE.
303**

Medicine.—Generally speaking, fish diet is considered by Hindu writers to be less heating than animal flesh, less likely to excite an inordinate flow of bile, more easily digested, and to be particularly indicated in cases of diabetes. Certain forms of dried fish are also held to be powerfully aphrodisiac, and in Patna, Dr. Irvine informs us, a concretion from the head of a fish called "*Sung-sir-mahí*" is supposed to have the same property. The oil of the liver of the **Gadus morrhua**, or common Cod, has well-known properties as a nutritive tonic and alterative, and, as already mentioned, it

F. 303

appears that the oil derived from the liver of species of Carcharias possesses similar valuable properties. The bile of certain species has fanciful properties ascribed to it by Natives in many localities, such as that of causing abortion, of being a specific in night blindness, &c.

Agricultural Uses.—Fish rendered useless as food through putrefaction, and the offal resulting from fish-curing, form valuable manure.

Sacred Uses.—Hindú religion and mythology contain many references to the fish, and certain species are employed in religious ceremonies.

The following LIST OF FISHES, for the names and properties of which the writer is chiefly indebted to *Day's Fishes of India*, comprises those of chief economic value as sources of food, oil, isinglass, or shagreen. When common to all the species of a genus, the economic properties will be found described in the remarks under the first. Subsequent to the receipt of first proof the writer obtained, however, the *Faunia of British India—* FISHES—to which he has consequently been able to give references only.

Ætobatis narinari ; *Day, Fish. Ind., 743 ; Fau. Br. Ind., I., 59.*

Vern.—*Currúway tiriki,* TAM.; *Il-tenki,* TEL.; *Teherrundi,* MALAY.; *Il-tenki,* VIZAG.; *Pari-lung,* MALAYS ; *Ra-ta-charm-dah,* ANDAMANS.

Habitat.—The Red Sea, seas and estuaries of India, to the Malay Archipelago and beyond.

Eaten raw and salted ; the livers are also employed to produce oil, and the fins are exported to China with those of other rays, skates, and sharks.

Ailia coila ; *Day, Fish. Ind., 488 ; Fau. Br. Ind., I., 134.*

Vern.—*Puttuli, buns putta, bounce-puttri,* URIYA ; *Man-gli-ah-ni,* SIND; *Vella kalada,* TEL.; *Kajoli,* RANGPUR ; *Basanguti,* GORAKPUR ; *Bátausi,* BHAGULPUR.

Habitat.—From the Kistna, and Orissa, throughout the Indus, Jumna, and Ganges, after they leave the hills to their termination ; also the rivers of Assam.

This fish is excellent eating.

Ambassis baculis ; *Day, Fish. Ind., 51 ; Fau. Br. Ind., I., 485.*

Vern.—*Kung-gi,* PB.; *Nga-kóun-mah, nga-sin-sat,* BURM.

Habitat.—Fresh waters of Bengal, Orissa, and as far north as the Panjáb, also in Burma.

All the species of this genus, though dry and insipid, are eaten either fresh or sun-dried, by the poorer classes of Natives. They are valuable as a diet for these people, since their structure allows of their being cured without the use of expensive salt.

A. commersoni ; *Day, Fish. Ind., 52 ; Fau. Br. Ind., I., 488.*

Vern.—*Selintan,* MADRAS.

Habitat.—Seas of India, ascending rivers and estuaries.

A. gymnocephalus ; *Day, Fish. Ind., 54 ; Fau. Br. Ind., I., 489.*

Vern.—*Chandi,* URIYA.

Habitat.—Seas of India.

A. nama ; *Day, Fish. Ind., 50 ; Fau. Br. Ind., I., 484.*

Vern.—*Cart-kana, goa-cháppi,* URIYA ; *Son dah,* ASSAM ; *Buck-ra, pom-pi-ah,* N.-W. P. ; *Muckni, ched-du-ah,* PB.; *Pud-du, put-to-lak,* SIND.; *Ak-ku-rati,* TEL.

Habitat.—Throughout the fresh waters of India, Assam, and Burma.

A. ranga ; *Day, Fish. Ind., 51 ; Fau. Br. Ind., I., 485.*

Vern.—*Chandi,* BENG.; *Chandi, lál-chandi,* URIYA ; *Chandi,* N.-W. P.; *Pi-dak,* SIND.; *Gandrichri,* MAR.; *Nga-tenyet,* BURM.

Habitat.—Throughout India and Burma.

Amblypharyngodon atkinsonii ; *Day, Fish. Ind., 555 ; Fau. Br. Ind., I., 290.*

Vern.—*Nga-pan-ma,* BURM.

2 B

FISH.	Indian Fishes

Habitat.—Rivers throughout Burma. The species of this genus though bony, where abundant, enter largely into the diet of the Natives.

314   **Amblypharyngdon melettina ;** *Day, Fish. Ind.*, *555 ; Fau. Br. Ind., I.*, *292.*
       Vern.—*Kali-korafi*, HIND.; *Ulari*, TAM.; *Wumbú*, MALAY.; *Paraga*, KAN.
       Habitat.—The fresh waters of the Malabar coast, and Southern India from the Nilghiris to Madras, also Ceylon (Bombay, according to **Cuv.** and **Val**).

315   **A. mola ;** *Day, Fish. Ind.*, *555 : Fau. Br. Ind., I.*, *291.*
       Vern.—*Kavdi*, BENG.; *Morara, patia kerundi*, URIYA; *Moah*, ASSAM; *Mukni*, PB.; *Talla-maya*, TEL.; *Nga-beh-byú, nga-zen-zap*, BURM.
       Habitat.—Ponds and fresh-water rivers from Sind, throughout India (except the Malabar coast), Assam, and Burma.

316   **Amphipnous cuchia ;** *Day, Fish. Ind.*, *656 ; Fau. Br. Ind., I.*, *69.*
       EEL, *Eng.*
       Vern.—*Cuchia*, BENG., URIYA; *Dondu-paum*, MADRAS; *Nga-shin*, BURM.
       Habitat.—The fresh and brackish waters of the Panjáb, extending to Bengal, Orissa, Assam, and Burma.
       Natives reject this as food, and imagine that its bite is fatal to cattle.

317   **Anabas scandens ;** *Day, Fish. Ind.*, *370 : Fau. Br. Ind., II.*, *367.*
       CLIMBING FISH, *Eng.*
       Vern.—*Coi*, BENG.; *Coi, cown*, URIYA|; *Coi*, ASSAM ; *Sennal, pauni-eyri*, TAM.; *Undi-colli*, MALAY; *Kavaya*, or *kawhy-ya*, SING.; *Nga-pri*, MUGH; *Nga-byays-ma*, BURM.; *Harúan*, MALAYS.
       Habitat.—Estuaries and fresh waters of India, Ceylon, and Burma.
       This fish is most remarkable for its powers of living in the air, and can travel a long distance on land. The boatmen of the Ganges carry them in moist earthen pots, killing and cooking them as required. They are highly esteemed as a nourishing food.

318   **Apocryptes bato ;** *Day, Fish. Ind.*, *302 : Fau. Br. Ind., II*, *278.*
       Vern.—*Rutta*, URIYA.
       Habitat.—Rivers of Orissa and Lower Bengal, within tidal reach.

319   **A. lanceolatus ;** *Day, Fish. Ind.*, *301 ; Fau. Br. Ind., II.*, *277.*
       Vern.—*Changua*, BENG.; *Pitallu*, URIYA; *Nullah-ramah*, TEL.
       Habitat.—Seas of India.

320   **Arius burmanicus ;** *Day, Fish. Ind.*, *458 : Fau. Br. Ind., I.*, *173.*
       Vern.—*Nga-young*, BURM.
       Habitat.—Tidal rivers of Burma. The several species are employed as food, though of an inferior quality. On the Western coast they are largely salted, and a considerable amount of isinglass is prepared by drying their air-vessels.

321   **A. gagora ;** *Day, Fish. Ind.*, *465 ; Fau. Br. Ind., I.*, *185.*
       Vern.—*Gagora*, BENG.; *Nga-youn, nga-yeh*, BURM.
       Habitat.—Seas, estuaries, and tidal rivers of Orissa and Bengal, to Siam.

322   **A. jatius ;** *Day, Fish. Ind.*, *466 Fau. Br. Ind. I.*, *186.*
       Vern.—*Jat-gagora*, BENG.; *Nga-youn, nga-yeh*, BURM.
       Habitat.—Estuaries and rivers of Bengal and Burma, ascending far above tidal reach.

323   **A. macronotacanthus ;** *Day, Fish. Ind.*, *465 ; Fau. Br. Ind. I.*, *184.*
       Vern.—*Ikau-saludu*, MALAYS.
       Habitat.—Rivers of India.

	of Economic Value. (*J. Murray.*)	FISH.

Arius sagor; *Day, Fish. Ind., 461 ; Fau. Br. Ind., I., 178.*
  Vern.—*Sagor*, BENG.
  Habitat.—From Bombay, through the seas and estuaries of India; very common at Batavia, where it is largely consumed.

324

A. thalassinus; *Day, Fish. Ind., 463; Fau. Br. Ind., I., 181.*
  Vern.—*Cuntea*, URIYA; *Deddi-jella*, VIZAGAPATAM.
  Habitat.—From the Red Sea, through those of Africa and India, entering tidal rivers.

325

Aspidoparia morar; *Day, Fish. Ind., 585 ; Fau. Br. Ind., I., 338.*
  Vern.—*Chippuah, chelluah*, HIND.; *Morari, morar*, BENG.; *Bayi*, URIYA; *Chula, mou-ah, boreala*, ASSAM; *Pa-o-char, chilwa*, PB.; *Ka-rir-re*, SIND; *Amli*, DEC.; *Ulsa*, TEL.; *Nga-hpyen-bú, yen-boung-sa*, BURM.
  Habitat.—Sind, the Panjáb, continent of India, except the Western coast and localities south of the Kistna river.
  Eaten by the Natives of mány districts.

326

Atherina forskalii; *Day, Fish. Ind., 345 ; Fau. Br. Ind., II., 338.*
  WHITEBAIT of Europeans in Malabar.
  Vern.—*Ko-re-dah*, ANDAMANS.
  Habitat.—Seas of India.
  " It only reaches to a few inches in length, and is most commonly captured during the cold season. It is one of several genera, certain species of which are indiscriminately termed 'whitebait' by Europeans, and are dressed for the breakfast table " (*Day*).

327

Badis buchanani; *Day, Fish. Ind., 128 ;Fau. Br. Ind., II., 80.*
  Vern.—*Kahli-poi, búndei, kahli-bundahni*, URIYA ; *Nabat, ran-doh-ni*, ASSAM ; *Kundala, ka-sundara*, TEL.; *Kala-pú-ti-ah, chiri*, PB.; *Pin-lay-nga-ba-mah, nga-mi-loung*, BURM.
  Habitat.—Fresh waters of India and Burma.

328

Bagarius yarrellii; *Day, Fish. Ind., 495; Fau. Br. Ind., I., 194.*
  FRESH-WATER SHARK, *Eng.*
  Vern.—*Búnch, gúnch*, HIND.; *Baag-aari*, BENG.; *Sah-lun, cart-cuntea*, URIYA; *Goreah*, ASSAM; *Rahti-jellah*, TEL.; *Guwch, khird, múlandah*, MAR.
  Habitat.—Large rivers of India and Java, descending to the estuaries.
  This fish attains 6 feet or more in length, and though it takes a live bait, is difficult to kill, as it is sluggish, goes to the bottom and generally escapes by destroying the tackle. Like other SILURIDÆ it is more eaten by the poorer than the richer classes, partly because the members of the family are forbidden to Muhammadans, and partly because they are very foul feeders.

329

Barbus ambassis; *Day Fish. Ind., 576 ; Fau. Br. Ind., I., 324.*
  Vern.—*Bunkuai*, URIYA; *Kalay*, TEL.
  Habitat.—The rivers of Bengal, Orissa, Madras, and Assam. A small species, attaining only about 3 inches in length. The larger species of this genus are generally termed *Mahasír*, though this name is more correctly applied to **Barbus tor** only. The species enumerated in this list are all employed as food.

330

B. amphibius; *Day, Fish. Ind., 574; Fau. Br. Ind., I., 322.*
  Vern.—*Uli perli*, MALAY.
  Habitat.—A fish, generally attaining the length of 6 inches, of the rivers of Central India, Deccan, Bombay, the Western coast of India, Madras, and up the coast as high as Orissa.

331

B. apogon; *Day, Fish. Ind., 575; Fau. Br. Ind., I., 324.*
  Vern.—*Nga-ta-zee, nga-lay-toun*, BURM.

332

FISH.	Indian Fishes
	Habitat.—The rivers of Tenasserim and throughout Burma (certainly as high as Mandalay) to the Malay Archipelago.
333	**Barbus carnaticus**; *Day, Fish. Ind., 563 ; Fau. Br. Ind., I., 304.* Vern.—*Giddi-kaoli*, HIND.; *Poari candi, saal candi, shelli*, TAM.; *Gidpakke*, KAN. Habitat.—Rivers along the bases of the Nilghiris, Wynaad, and South Cánara Hills. This is a large species, attaining the weight of at least 25℔.
334	**B. chagunio**; *Day, Fish. Ind., 559 ; Fau. Br. Ind., I., 299.* Vern.—*Chaguni, jerruah*, BENG.; *Chaguni*, BEHAR; *Púti-keintah*, ASSAM. Habitat.—The rivers of Bengal, Orissa, Behar, North-Western Provinces, Panjáb, and Assam. A medium-sized fish, attaining the length of at least 18 inches.
335	**B. chola**; *Day, Fish. Ind., 571 ; Fau. Br. Ind., I., 317.* THE BITTER CARP. Vern.—*Katcha karawa*, HIND.; *Karrundi, chola*, BENG.; *Pittha-kerrundi*, URIYA; *Korún*, TAM.; *Chuddu paddaka*, TEL.; *Nga-khonma, nga-lowah*, BURM. Habitat.—The rivers of Bengal, Orissa, the Gangetic Provinces, the Panjáb, the Central Provinces, ⁚Madras, Malabar, and Wynaad, also Akyab and Burma to Mergui.
Food. 336	As food, this fish is bitter; in some localities in Burma oil is obtained from it, during the breeding season.
337	**B. chrysopoma**; *Day, Fish. Ind., 561 ; Fau. Br. Ind., I., 301.* Vern.—*Mundutti*, MALABAR. Habitat.—Fresh waters along the coast of India, from Kutch to Bengal.
338	**B. conchonius**; *Day, Fish. Ind., 576 ; Fau. Br. Ind., I., 325.* Vern.—*Kunchon pungti*, BENG. Habitat.—The rivers of Assam, Lower Bengal, Orissa, Behar, the North-Western Provinces, the Panjáb, and the Deccan.
339	**B. cosuatis**; *Day, Fish. Ind., 581 ; Fau. Br. Ind., I., 332.* Vern.—*Koswati*, BENG.; *Pangut*, MAR. Habitat.—The rivers of Bengal, North-Western Provinces, Deccan, Bombay, and down the Western coast as low as Cottayam in Travancore.
340	**B. filamentosus**; *Day, Fish. Ind., 582 ; Fau. Br. Ind., I., 333.* THE RED-TAILED CARP. Vern.—*Sawaal-candi, chevalle*, TAM.; *Curroah*, MALAY. Habitat.—Western coast and Southern India. A very curious change occurs in this fish immediately after death, the whole body becoming scarlet.
341	**B. gelius**; *Day, Fish. Ind., 577 ; Fau. Br. Ind., I., 327.* Vern.—*Gili, pungti*, BENG.; *Cutturpoh*, URIYA. Habitat.—The rivers of Ganjam, Orissa, Bengal, and Assam.
342	**B. guganio**; *Day, Fish. Ind., 579 ; Fau. Br. Ind., I., 328.* Vern.—*Gugani*, BENG.; *Nga-hkon-mahgyi, nga-chong*, BURM. Habitat.—The Gangetic Provinces and Assam.
343	**B. hexastichus**; *Day, Fish. Ind., 565 ; Fau. Br. Ind., I., 308.* Vern.—*Parrah-perli*, MALAY.; *Lobura*, ASSAM. Habitat.—A large fish, attaining 3 feet in length, of the rivers in and around the Himálaya, Kashmír, Sikkim, and Assam.
344	**B. kolus**; *Day, Fish. Ind., 573 ; Fau. Br. Ind., I., 319.* Vern.—*Nilusu*, TEL. Habitat.—The Central Provinces, and the Deccan, throughout the Kistna, Tambúdra, and Godaveri rivers.

of Economic Value. (*J. Murray.*)	FISH.

Barbus micropogon; *Day, Fish. Ind., 563 ; Fau. Br. Ind., I., 304.* | 345
  Vern.—*Coati candi*, TAM.
  Habitat.—The rivers around the base of the Nilghiris, Wynaad, and
South Cánara range of hills; also of Mysore.

B. neilli; *Day, Day, Fish. Ind., 569; Fau. Br. Ind., I., 314.* | 346
  Vern.—*Khudri*, MAR.
  Habitat.—Kurnúl on the Tambúdra river.

B. phutunio; *Day, Fish. Ind., 578 ; Fau. Br. Ind., I., 327.* | 347
  Vern.—*Phutini pungti*, BENG.; *Kudji-kerundi*, URIYA.
  Habitat.—The rivers of Ganjam, Orissa, and throughout Bengal and
Burma.

B. punctatus; *Day, Fish. Ind., 577 ; Fau. Br. Ind., I., 326.* | 348
  Vern.—*Putter perli*, MALAY.
  Habitat.—The rivers of Malabar and the Coromandel coast.

B. puntio; *Day, Fish. Ind., 582 ; Fau. Br. Ind., I., 334.* | 349
  Vern.—*Pungti*, BENG.
  Habitat.—Ponds and ditches of Bengal and Lower Burma.

B. sarana; *Day, Fish. Ind., 560 ; Fau. Br. Ind., I., 300.* | 350
  Vern.—*Durhie, giddi-kaoli, potah*, HIND.; *Sarana-pungti, sarana*, BENG.;
    Sarana, URIYA; *Sen-ni*, ASSAM; *Jundúri*, PB.; *Pap-pri, kuh-nah-ni*,
    SIND.; *Pungella, kunnaku*, TAM.; *Kannaku*, TEL.; *Panjiri*, MADRAS;
    Gid-pakke, KAN.; *Nga-khon-mah-gyi, nga-chong*, BURM.
  Habitat.—Rivers and tanks throughout India, Assam, and Burma.

B. sophore; *Day, Fish. Ind., 566 ; Fau. Br. Ind., I., 309.* | 351
  Vern.—*Pungti*, BENG.; *Chadu-perigi*, TEL.; *Sophore*, SANS.
  Habitat.—The rivers and ponds of Assam and the Khásia Hills.

B. stigma; *Day, Fish. Ind., 579 ; Fau. Br. Ind., I., 329.* | 352
  Vern.—*Katcha-karawa, pottiah*, HIND.; *Patia-kerundi*, URIYA; *Chadu-
perigi*, TEL.; *Katch-ku-rawa*, KAN.; *Nga-khún-ma*, BURM.
  Habitat.—The rivers of Sind, throughout India and Burma, as high as
Mandalay. Though employed as food this fish is bitter.

B. stoliczkanus; *Day, Fish. Ind., 577 ; Fau. Br. Ind., I., 326.* | 353
  Vern.—*Nga-thine-glay*, BURM.
  Habitat.—Eastern Burma.

B. terio; *Day, Fish. Ind., 580 ; Fau. Br. Ind., I., 330.* | 354
  Vern.—*Tripungti*, BENG.; *Kakachia-kerundi*, URIYA.
  Habitat.—The fresh waters of Bengal and Orissa, to the Panjáb.

B. tetrarupagus; *Day, Fish. Ind., 572 ; Fau. Br. Ind., I., 318.* | 355
  Vern.—*Til-pungti*, BENG.; *Borajali*, ASSAM; *Pet-toh-i*, SIND.
  Habitat.—The rivers, tanks, and ponds of Bengal, Orissa, the North-
Western Provinces, the Panjáb, Sind, the Deccan, and Assam.

B. ticto; *Day, Fish. Ind., 576 ; Fau. Br. Ind., I., 325.* | 356
  Vern.—*Kaoli, kotri*, HIND.; *Kudji-kerundi*, URIYA; *Kah-nipotiak*, ASSAM.
  Habitat.—Rivers and tanks throughout India and Ceylon.

B. tor; *Day, Fish. Ind., 564; Fau. Br. Ind., I., 307.* | 357
  Vern.—*Naharm*, HIND.; *Mahasir, mahasaula, jora*, BENG.; *Burapatra,
bura-hetea, mahsir, lobura*, ASSAM; *Kukhiah*, PB.; *Joon-gah, petiah,
kurreah*, SIND.; *Pú-min-candi*, TAM.
  Habitat.—This fish, the celebrated "Mahasir" of sportsmen in India,
is found generally throughout India, but grows to the largest size and is
most abundant in mountain or rocky streams.

FISH.	Indian Fishes
358	**Barbus vittatus ;** *Day, Fish. Ind., 582 : Fau. Br. Ind., I., 333.* Vern.—*Kúli,* HIND. ; *Putti,* URIYA. Habitat.—The rivers of Kutch, Mysore, Madras, Wynaad, Malabar, and Ceylon.
359	**Barilius barila ;** *Day, Fish. Ind., 591 : Fau. Br. Ind., I., 384.* Vern.—*Perci,* HIND. ; *Gilland, chaedri, barili,* BENG. Habitat.—Rivers of the North-Western Provinces, Central Provinces, Bengal, Orissa, and Lower Assam. The several species of this genus, like most other carps, are largely employed as food by the Natives.
360	**B. barna ;** *Day, Fish. Ind., 592 ; Fau. Br. Ind., I., 350.* Vern.—*Barna, bali-bhola, bareli,* BENG. ; *Bahri,* URIYA ; *Balisundri, oz-o-la,* ASSAM. Habitat.—Assam, the Ganges and its branches ; rivers of Bengal and Orissa.
361	**B. bendelisis ;** *Day, Fish. Ind., 590 : Fau. Br. Ind., I., 347.* Vern.—*Khoksa,* BENG. ; *Bahgra-bahri,* URIYA ; *Pak-tah, kunnul, dah-rah, burreah, puck-wah-ri,* PB. ; *Aguskitti,* TAM. ; *Johra,* MAR. Habitat.—Rivers of Assam, the Himálaya, through the continent of India, as far as the Western Ghâts.
362	**B. bola ;** *Day, Fish. Ind., 594 ; Fau. Br. Ind., I., 352.* THE TROUT of Europeans in India. Vern.—*Buggarah,* HIND. ; *Bola, goha,* BENG. ; *Buggush,* URIYA ; *Korang,* ASSAM. Habitat.—Rivers of North-Western Provinces, Orissa, Bengal, Assam, and Burma. This is a very game fish, generally called " Trout " by the English in India, takes the fly well, and is one of those termed " *Raja mas,*" or " chief of the fishes," in the Assam rivers.
363	**B. gatensis ;** *Day, Fish. Ind., 592 ; Fau. Br. Ind., I., 349.* RIVER CARP, OR NILGERRY TROUT of Europeans in India. Vern.—*Choari, árt-candi,* TAM. Habitat.—Rivers of the Western Ghâts, Malabar, and the Nilghiri hills, up to about 5,000 feet above the level of the sea.
364	**B. guttatus ;** *Day, Fish. Ind., 593 : Fau. Br. Ind., I., 351.* Vern.—*Nga-la-wah,* BURM. Habitat.—River Irrawadi, from Prome to Mandalay.
365	**Belone annulata ;** *Day, Fish. Ind., 510 ; Fau. Br. Ind., I., 419.* Vern.—*Pahmum kolah,* TAM. ; *Wahlah-kuddera,* VIZAGAPATAM ; *Toda,* MALAYS. Habitat.—Seas and estuaries of India. The several species of **Belone,** or " Gar-fish," though generally of indifferent quality, are employed as food by the Natives.
366	**B. cancila ;** *Day, Fish. Ind., 511 ; Fau. Br. Ind., I., 420.* Vern.—*Kangkila,* BENG. ; *Gungituri,* URIYA ; *Coco-min,* TAM. ; *Coahlan, morrahlú,* MALAY ; *Nga-ohpoung-yoh, nga-phou-yo,* BURM. Habitat.—Fresh waters of India, Ceylon, and Burma.
367	**B. strongylura ;** *Day, Fish. Ind., 512 ; Fau. Br. Ind., I., 421.* THE LONG-NOSED FISH. Vern.—*Cungúr,* SIND ; *Ushi-collarchi, coco-min,* TAM. ; *Wodlah-muku,* TEL. ; *Coplah,* MALAY ; *Kuddera,* VIZAGAPATAM ; *Toda,* MALAYS ; *Thúk-o-dú-nú-dah,* ANDAMANS. Habitat.—Seas and coasts of India.
368	**Callichrous bimaculatus ;** *Day, Fish. Ind., 476 ; Fau. Br. Ind., I., 131.* THE BUTTER FISH.

F. 368

of Economic Value. (*J. Murray.*)	FISH.

Vern.—*Kani-pabda, chechra,* BENG.; *Gúng-wah-ri, puf-ta,* HIND.; *Pob-tah,* URIYA; *Pah-boh,* ASSAM; *Pufta, gúngwah, pallu,* PB.; *Dimmon,* SIND; *Chelahwahlah, chotah-wahlah,* TAM.; *Dúka-dúmú,* TEL.; *Gúgli, gúgul, purwa,* MAR.; *Godla,* KAN.

Habitat.—Fresh waters throughout India, Ceylon, and Assam. Although rarely exceeding a foot in length, the species of **Callichrous** are excellent as food, and are considerably used by Europeans.

Callichrous macrophthalmus; *Day, Fish. Ind., 478; Fau. Br. Ind., I., 132.*
   Vern.—*Nga-nú-than, nga-myin-bouk,* BURM.
   Habitat.—Fresh waters of Madras, Assam, and Burma. | 369 |

C. malabaricus; *Day, Fish. Ind., 478; Fau. Br. Ind., I., 133.*
   Vern.—*Chota-wahlah,* TAM.; *Mungi-wahlah,* MALAY.
   Habitat.—Malabar coast of India. | 370 |

Caranx affinis; *Day, Fish. Ind., 219; Fau. Br. Ind., I., 158.*
   THE HORSE MACKEREL.
   Vern.—*Warriparah,* TAM.; *Battaparra,* MALAY.
   Habitat.—Seas of India. | 371 |

C. oblongus; *Day, Fish. Ind., 222; Fau. Br. Ind., I., 163.*
   Vern.—*Ro-thul-dah,* ANDAMANS.
   Habitat.—Seas of India. | 372 |

C. rottleri; *Day, Fish. Ind., 213; Fau. Br. Ind., I., 150.*
   Vern.—*Komara-parah,* TAM.; *Sora-parah,* TEL.; *Woragú,* VIZAG.
   Habitat.—Seas of India. | 373 |

Carassius auratus; *Day, Fish. Ind., 552; Fau. Br. Ind., I., 283.*
   THE GOLDFISH, or GOLDEN CARP.
   Vern.—*Nukta,* MAR.
   Habitat.—River Inderani, above Púna (*Watson*). Not indigenous to India, or only possibly so in Upper Burma (*Day*). | 374 |

Carcharias acutidens; *Day, Fish. Ind., 713; Fau. Br. Ind., I., 11.*
   Habitat.—Coasts of Sind and the Indian Ocean. All the species of this genus are valued for the oil obtained from their livers, their gelatinous fins, their skin, which is employed as shagreen, and by the poor for their flesh, which is extensively eaten both fresh and salted. | 375 |

C. acutus; *Day, Fish. Ind., 712; Fau. Br. Ind., I., 10.*
   Vern.—*Parrúmai sorrah,* TAM.; *Som sorrah,* TEL.; *Parl sorrah,* MALAY.
   Habitat.—Seas of India. | 376 |

C. ellioti; *Fish. Ind., 716; Fau. Br. Ind., I., 15.*
   Vern.—*Paducan, adugu-pal-sorrah,* TAM.; *Pal-sorrah,* VIZAG.
   Habitat.—The Seas of India, not uncommon at Karachi. | 377 |

C. gangeticus; *Day, Fish. Ind., 715; Fau. Br. Ind., I., 13.*
   Habitat.—Seas of India, ascending rivers to above tidal influence. This is one of the most ferocious of Indian sharks and frequently attacks bathers in the Hooghly at Calcutta. | 378 |

C. limbatus; *Day, Fish. Ind., 716; Fau. Br. Ind., I., 17.*
   Habitat.—Very common along the sea borders of India, extending through the Indian Ocean. It attains at least 6 feet in length. | 379 |

C. macloti; *Day, Fish. Ind., 713; Fau. Br. Ind., I., 12.*
   Vern.—*Pala sorrah,* TEL.
   Habitat.—A small shark of the Indian seas. | 380 |

C. melanopterus; *Day, Fish. Ind., 715; Fau. Br. Ind., I., 14.*
   Vern.—*Caval sorrah, nella vekal sorrah, raman sorrah, mukhan sorrah, boka sorrah, ran sorrah,* TAM. | 381 |

F. 381

FISH.	Indian Fishes

Habitat.—Seas of India. A very large shark, the liver of one of which is said by Day to have weighed 270℔. It is, perhaps, of all the species the most prized as an oil-yielding fish.

382 Carcharias menisorrah ; *Day, Fish. Ind., 716 ; Fau. Br. Ind., I., 16.*
Vern.—*Karamúti sorrah, ciga sorrah,* TEL.
Habitat.—The seas of India. A large shark attaining 12 feet or more in length.

383 C. (Odontaspis) tricuspidatus ; *Day, Fish. Ind., 713 ; Fau. Br. Ind., I., 27.*
Habitat.—A large shark abounding in the seas of Sind, and attaining a length of at least 20 feet.

384 Catla buchanani ; *Day, Fish. Ind., 553 ; Fau. Br. Ind., I., 287.*
Vern.—*Catla,* HIND.; BENG., PB.; *Barkur,* URIYA; *Boassa,* N.-W. P.; *Tambra,* BOMB.; *Botchi,* TEL.; *Tay-li,* SIND; *Nga-thaing,* BURM.
Habitat.—Rivers and tanks of Sind, through India to the Kistna, and eastwards through Bengal and Burma to Siam.
This fish is largely employed for stocking tanks, and is much esteemed as an article of food when not over 2 feet in length ; larger ones are coarse.

385 Chætodon vagabundus ; *Day, Fish. Ind., 105 ; Fau. Br. Ind., II., 4.*
Vern.—*Pah-nú-dah,* ANDAMANS.
Habitat.—The seas of India.

386 Chanos salmoneus ; *Day, Fish. Ind., 651 ; Fau. Br. Ind., I., 403.*
The MILK FISH or WHITE MULLET.
Vern.—*Tulu-candal,* TAM.; *Palah, bontah,* TEL.; *Hu-min,* KAN.; *Pu-min,* TULU.
Habitat.—The seas of India and tanks of fresh and brackish water in South Cánara. It was introduced into the latter artificial habitat by Hyder Ali, and still thrives.

387 Chatoessus chacunda ; *Day, Fish. Ind., 632 ; Fau. Br. Ind., I., 386.*
Vern.—*Chacunda,* BENG.; *Muddirú,* TEL.; *Kore-paig-dah,* ANDAMAN.
Habitat.—The seas and estuaries of India and Burma. The several species of this genus, along with other members of the CLUPEIDÆ or herrings, are captured in great quantity, and largely consumed by the native population.

388 C. manminna ; *Day, Fish. Ind., 633 ; Fau. Br. Ind., I., 386.*
Vern.—*Mackundi,* URIYA.
Habitat.—Fresh waters of Sind, and the districts watered by the Indus and its branches ; also the main streams of the Ganges, Jumna, Brahmaputra, and Mahanuddi ; through the tanks and estuaries of India and Assam, except the Deccan, South and Western India, and Ceylon.

389 C. modestus ; *Day, Fish. Ind., 633 ; Fau. Br. Ind., I., 386.*
Vern.—*Nga-la-pay,* BURM.
Habitat.—Along the Bassein River, as high as the In-gay-gyí lake, also the Salwein at Moulmein.

390 C. nasus ; *Day, Fish. Ind., 634 ; Fau. Br. Ind., I., 387.*
Vern.—*Kome,* URIYA; *Muddu-candai,* TAM.; *Kome,* TEL.; *Núnah,* MALAY.; *Pedda-kome,* VIZAG.
Habitat.—Seas of India. This fish is good eating, but bony.

391 Chela argentea ; *Day, Fish. Ind., 601 ; Fau. Br. Ind., I., 364.*
WHITE CARP.
Vern.—*Chaya-vellachi, vellachi-candi,* TAM.
Habitat.—Bowany river (at the base of the Nilghiris), Cauvery river, and the rivers of Mysore. This and the other species enumerated below are eaten by the Natives.

F. 391

of Economic Value. (*J. Murray.*)	FISH.

Chela bacaila; *Day, Fish. Ind., 603; Fau. Br. Ind., I., 367.*
    Vern.—*Chelliah,* HIND.; *Bacaila,* BENG.; *Jellahri,* URIYA; *Badishaya,* TEL.
    Habitat.—The rivers and tanks of India, except those of Malabar, Madras, Mysore, and parts of the Deccan. | 392 |

C. clupeoides; *Day, Fish. Ind., 602; Fau. Br. Ind., I., 366.*
    Vern.—*Tikani,* DEC.; *Balúki,* MAR.; *Netteli, vellache-kende,* TAM.
    Habitat.—The rivers of Cutch, Jubbulpur, the Deccan, Madras, Mysore and Burma.
    This species is specially good eating. | 393 |

C. gora; *Day, Fish. Ind., 600; Fau. Br. Ind., I., 362.*
    Vern.—*Chel-hul,* HIND.; *Ghora chela,* BENG.; *Hum-catchari,* URIYA; *Bounchi, kundul,* PB.
    Habitat.—Rivers of Sind, the Panjáb, the North-Western Provinces, Bengal, Orissa, and Assam. | 394 |

C. jorah; *Day, Fish. Ind., 599; Fau. Br. Ind., I., 361.*
    Vern.—*Jorah,* MAR.
    Habitat.—Beema river near Pairgaon in the Deccan. | 395 |

C. phulo; *Day, Fish. Ind., 602; Fau. Br. Ind., I., 365.*
    Vern.—*Dunnahrí,* HIND.; *Phul-chela,* BENG.; *Sel-konah,* ASSAM; *Túk, bung-ka-chael,* PB.; *Muk-ka,* SIND.
    Habitat.—The rivers and ponds of Bengal, Orissa, Central India, and the Deccan, as far southwards as the Tambádra and Kistna. | 396 |

C. sardinella; *Day, Fish. Ind., 600; Fau. Br. Ind., I., 363.*
    Vern.—*Nga-kún-nyat,* BURM.
    Habitat.—Irrawadi river at Rangoon, also the Salwein at Moulmein. | 397 |

C. sladoni; *Day, Fish. Ind., 600; Fau. Br. Ind., I., 363.*
    Vern.—*Nya-yin-boun-sa,* BURM.
    Habitat.—Irrawadi river, as far north as Mandalay. | 398 |

C. untrahi; *Day, Fish. Ind., 601; Fau. Br. Ind., I., 364.*
    Vern.—*Untrahi,* URIYA.
    Habitat.—Mahanaddi river in Orissa, also the Cauvery and Colerún in Southern India. | 399 |

C. alkootee; *Day, Fish. Ind., 599; Fau. Br. Ind., I., 362.*
    Vern.—*Alkúti,* MAR.
    Habitat.—Rivers of the Deccan. (Doubtful species.) | 400 |

Chiloscyllium indicum; *Day, Fish. Ind., 726; Fau. Br. Ind., I., 34.*
    Vern.—*Corangan sorrah,* TAM.; *Etti,* MALAY; *Boki-sorrah, ra-sorrah,* VIZAG.; *Yu-tokay,* MALAYS; *Pús-hi,* BELUCH.
    Habitat.—The seas of India. | 401 |

Chirocentrus dorab; *Day, Fish. Ind., 652; Fau. Br. Ind., I., 368.*
    Vern.—*Kunda, kundah,* URIYA; *Kiru-wahlah, mulú-alley,* TAM.; *Wah-lah,* TEL.; *Párang-párang,* MALAYS.
    Habitat.—The seas of India. | 402 |

Chorinemus lysan; *Day, Fish. Ind., 231; Fau. Br. Ind., II., 175.*
    Vern.—*Parah,* HIND.; *Toal-parah,* TAM.; *Aken-parah,* VIZAG.; *Tallang-raya,* MALAYS.
    Habitat.—The seas of India. Though considerably employed as food, the members of this genus are dry and rather tasteless. | 403 |

C. moadetta; *Day, Fish. Ind., 230; Fau. Br. Ind., II., 174.*
    Vern.—*Tol-parah,* VIZAG.
    Habitat.—Red Sea and seas of India. | 404 |

FISH.	Indian Fishes

405 **Chrysophrys berda**; *Day, Fish. Ind., 140 ; Fau. Br. Ind., II., 44.*
BLACK ROCK-FISH of Europeans in Malabar.
    Vern.—*Kala madwan,* HIND. ; *Dun-de-a, jarras,* SIND. ; *Currie, currapu-mattawa,* TAM. ; *Kalamara,* TEL. ; *Ari,* MALAY ; *Mú-rú-ki-dah,* ANDA-MAN.
    Habitat.—The seas of India to the Malay Archipelago and beyond. This fish is excellent eating, greatly excelling the other species, and is common in Malabar until July.

406 **C. sarba** ; *Day, Fish. Ind., 142 ; Fau. Br. Ind., II., 47.*
    Vern.—*Suffada-maddawa,* HIND. ; *Vellamattawa,* TAM. ; *Chitchilli,* TEL. ; *Tin-til,* BELUCH.
    Habitat.—The seas of India, especially abundant on the Madras coast. As food it is inferior to the **berda**.

407 **Cirrhina cirrhosa** ; *Day, Fish. Ind., 547 ; Fau. Br. Ind., I., 277.*
    Vern.—*Ven-kándi,* TAM. ; *Aruzu,* TEL.
    Habitat.—Godavery, Kistna, and Cauvery rivers, and generally in Southern India. A very active fish, fair eating, but bony.

408 **C. fulungee** ; *Day, Fish. Ind., 549 ; Fau. Br. Ind., I., 280.*
    Vern.—*Fulungi,* MAR.
    Habitat.—Rivers of Poona and the Deccan.

409 **C. latia** ; *Day, Fish. Ind., 548 ; Fau. Br. Ind., I., 279.*
    Vern.—*Kala-batta,* BENG. ; *Behrah, tellarri,* PB. ; *Curru,* SIND. ; *Wattu-nah,* MAR.
    Habitat.—The rivers of Bengal, Orissa, the North-West Provinces, the Panjáb, Sind, the Deccan, and along the Himálaya.

410 **C. mrigala** ; *Day, Fish. Ind., 547 ; Fau. Br. Ind., I., 278.*
    Vern.—*Mrigala, naim,* HIND. ; *Rewah,* BENG. ; *Mrigale, mirrgah,* URIYA ; *Mor-ah-ki,* SIND ; *Nga-kyin, nga-gyein,* BURM. ; *Mirgal, mrigala,* SANS.
    Habitat.—The rivers and tanks of Bengal, the North-West Provinces, the Panjáb, Sind, Kutch, the Deccan, and Burma. An excellent fish for stocking tanks.

411 **C. reba** ; *Day, Fish. Ind., 549 ; Fau. Br. Ind., I., 279.*
    Vern.—*Rewah,* HIND. ; *Batta,* BENG. ; *Chetchua-porah,* URIYA ; *Sunni,* PB. and SIND ; *Pil-aringan,* TAM. ; *Ilemose, chittahri, pullarazu,* TEL. ; *Lassim,* ASSAM ; *Boggut, kólis,* MAR.
    Habitat.—Rivers throughout India.

412 **Clarias magur** ; *Day, Fish. Ind., 485 ; Fau. Br. Ind., I., 115.*
    Vern.—*Magúr, mah-gur,* BENG. ; *Mangri,* PATNA and MONGHIR ; *Magu-rah,* URIYA ; *Kug-ga,* PB. ; *Yerri-vale,* TAM. ; *Marpú,* VIZAG. ; *Nga-khú,* BURM. and MUGH.
    Habitat.—Fresh and brackish waters of the plains of India, Burma, Ceylon, and the Malay Archipelago. As food this fish is deemed highly nourishing, and is extensively salted in Burma.

413 **Clupea fimbriata** ; *Day, Fish. Ind., 637 ; Fau. Br. Ind., I., 273.*
SARDINE, of Europeans in India.
    Vern.—*Charri-addi,* HIND. ; *Kich-uk-lonar,* SIND ; *Púnduringa,* TAM. ; *Cuttay-charlay,* MALAY.
    Habitat.—Red Sea and the seas of India. Employed extensively as food and also in the preparation of fish-oil. All the members of this genus are much captured for food by the Natives, and some are considered delicious by Europeans.

414 **C. ilisha**; *Day, Fish. Ind., 640 ; Fau. Br. Ind. I., 276.*
THE SABLE or SHAD FISH ; HILSA.

F. 414

of Economic Value. (*J. Murray.*)	FISH.

Vern.—*Hilsa, ilisha,* BENG.; *Rúri* of the Ganges; *Dumra* of the Indus; *Pulla,* SIND; *Ulum,* TAM.; *Pulasa, pulasu,* or *palasah,* TEL.; *Olam-min,* MADRAS; *Nga-tha-louk,* BURM.; *Ikan-truboh,* MALAYS.

Habitat.—Persian Gulf and coasts of India and Burma, passing up the large rivers to breed.

These fish are excellent as food until they have deposited their ova, when they become thin and positively unwholesome. Their flavour has been compared to a combination of that of the salmon and herring; but, though highly esteemed for the table, they are rather rich and difficult of digestion.

Clupea longiceps; *Day, Fish. Ind.,* 637; *Fau. Br. Ind., I.,* 373.   **415**
THE MALABAR OIL SARDINE.

Vern.—*Mutthi, charlay, karlay,* MAL.; *Mutthi,* KAN.; *Lonar,* SIND; *Li-gur,* BELUCH.

Habitat.—Sind and the Western coast of India, more rarely found on the Eastern, Ceylon, and Andaman coasts. Large quantities of oil are made from this species in Malabar.

C. variegata; *Day, Fish. Ind.,* 639; *Fau. Br. Ind., I.,* 375.   **416**
Vern.—*Nga-la-bi,* BURM.
Habitat.—The Irrawaddi and its branches.

Coilia ramcarati; *Day, Fish. Ind.,* 631; *Fau. Br. Ind., I.,* 396.   **417**
Vern.—*Urialli,* URIYA.
Habitat.—The rivers and estuaries of Bengal.

Corica soborna; *Day, Fish. Ind.,* 642; *Fau. Br. Ind., I.,* 378.   **418**
Vern.—*Cut-wál ursi, god-hai;* URIYA.
Habitat.—The rivers of Bengal and Orissa.

Cybium commersonii; *Day, Fish. Ind.,* 255; *Fau. Br. Ind., II.,* 211.   **419**
THE SEER or SEIR FISH.

Vern.—*Konam, mah-wu-laachi, ah-ku-lah,* TAM.; *Chambam,* MALAY; *Ikantanggiri,* MALAYS.

Habitat.—Seas of India. The species of this genus, when of the proper size, are considered amongst the most delicate of all marine fishes. If under a foot in length they are dry, from 1½ to 2½ feet they are most excellent, while above this they become coarse.

C. guttatum; *Day, Fish. Ind.,* 255; *Fau. Br. Ind., II.,* 210.   **420**
THE SEER or SEIR FISH.

Vern.—*Wingeram,* VIZAG.; *Arrakiah,* MALABAR.
Habitat.—The seas of India. Good eating, especially if cooked when quite fresh; salts well.

C. lineolatum; *Day, Fish. Ind.,* 256; *Fau. Br. Ind., II.,* 212.   **421**
TBE SEER or SEIR FISH.

Vern.—*Barim-kúti,* MALAY.; *Tanggiri,* MALAYS.
Habitat.—Seas of India.

Cynoglossus lingua; *Day, Fish. Ind.,* 433; *Fau. Br. Ind., II.,* 445.   **422**
SOLE, of Europeans in India.

Vern.—*Kot-aralu,* TAM.; *Ikan-ledah,* MALAYS.
Habitat.—Seas and estuaries of India. Highly esteemed for the table. It is mentioned by Ainslie as light, nutritious, delicate, and one of the fish that may be safely given to invalids.

Danio dangila; *Day, Fish. Ind.,* 596; *Fau. Br. Ind., I.,* 356.   **423**
Vern.—*Dhani,* BENG.
Habitat.—The rivers of Bengal, Behar, and the Himálaya, at Darjeeling, also of the hills above Akyab. The prettily-marked fish con-

F. 423

FISH.	Indian Fishes

stituting this genus, which are nearly allied to the Tench, are considerably used as food.

424　　**Danio devario** ; *Day, Fish. Ind.*, 595 ; *Fau. Br. Ind., I.*, 354.
Vern.—*Debari*, BENG. ; *Bonkuaso*, URIYA ; *Da-bah, dukri-e*, N.-W. P.; *Khan-ge, mál-le, pur-ran-dah*, PB. ; *Chay-la-ri*, SIND.
Habitat.—The ponds and rivers of Bengal, the North-West Provinces, the Panjáb, Sind, Orissa, the Deccan, and Assam.

425　　**D. malabaricus** ; *Day, Fish. Ind.*, 595 ; *Fua. Br. Ind., I.*, 355.
Vern.—*Poarah-cunjú-candi*, TAM.
Habitat.—The Western coast of India and Ceylon.

426　　**D. neilgherriensis** ; *Day, Fish. Ind.*, 597 ; *Fau. Br. Ind., I.*, 357.
Vern.—*Cowlie*, TAM.
Habitat.—Rivers on the Nilghiri Hills.

427　　**D. rerio** ; *Day, Fish. Ind.*, 597 ; *Fau. Br. Ind., I.*, 358.
Vern.—*Poncha-geraldi*, ÚRIYA.
Habitat.—Rivers of Bengal and of the country extending down the Coromandel coast to Masulipatam.

428　　**Diagramma crassispinum** ; *Day, Fish. Ind.*, 78 ; *Fau. Br. Ind., I.*, 514.
BLACK ROCK-FISH of Europeans in Malabar.
Vern.—*Tawúlú pinnel*, TEL.
Habitat.—The seas of India. It attains 2 feet or more in length and is good eating.

429　　**Discognathus lamta** ; *Day, Fish. Ind.*, 527 ; *Fau. Br. Ind., I.*, 246.
HILL TROUT of Europeans.
Vern.—*Korafi-kaoli*, HIND. ; *Choak-si*, BENG. ; *Putter-chettah*, N. W. P.; *Dhoguru, kúrka*, PB. ; *Kul-korava*, TAM. ; *Pandi-pakke*, KAN.
Habitat.—Rivers and mountain streams throughout India and Ceylon. This fish is good eating, but putrefies very rapidly after death.

430　　**Drepane punctata** ; *Day, Fish. Ind.*, 116 ; *Fau. Br. Ind., II.*, 21.
Vern.—*Pulli, torriti*, TAM. ; *Thetti*, TEL. ; *Pündthi*, MALAY ; *Latte-terla*, VIZAG. ; *Punnur*, SIND ; *Nga-shengna*, BURM. ; *Rúpi-chanda*, CHITTAG. ; *Shengna-roét*, ARRAK. ; *Shuk*, BELUCH. ; *Gun-na-to-dash*, AND.
Habitat.—Seas of India. It is in most places esteemed as food.

431　　**Dussumieria acuta** ; *Day, Fish. Ind.*, 647 ; *Fau. Br. Ind., I.*, 399.
SARDINE of Europeans in Malabar.
Vern.—*Púnduouringa*, TAM. ; *Kúrie*, MALAY. ; *Tamban-bulat*, MALAYS ; *O-pul-dah*, AND.
Habitat.—From Sind, through the seas of India. Cantor says this species, like the true Sardine, may be preserved *à huile*. It is very common in Malabar, and is excellent eating.

432　　**Echeneis naucrates** ; *Day, Fish. Ind.*, 257 ; *Fau. Br. Ind., II.*, 214.
Vern.—*Putthú-muday*, MALAY. ; *Ubbay*, TAM. ; *Ala-mottah*, VIZAG. ; *Guddimi*, MALAYS.
Habitat.—Seas of India. The Malays consider these fish to be a valuable manure for fruit trees.

433　　**Elacate nigra** ; *Day, Fish. Ind.*, 256 ; *Fau. Br. Ind., II.*, 213.
Vern.—*Cuddul-verarl*, TAM. ; *Pedda-mottah*, VIZAG.
Habitat.—Seas of India, to Japan.

434
435　　**Eleotris butis** ; *Day, Fish. Ind.*, 315 ; *Fau. Br. Ind., II.*, 296.
Vern.—*Kullahray*, MALAY.
Habitat.—Seas and estuaries of India.

436　　**E. fusca** ; *Day, Fish. Ind.*, 313 ; *Fau. Br. Ind., II.*, 293.
Vern.—*Bundi, balah-kera*, ÚRIYA ; *Cul-cúndallum*, TAM. ; *Púllan*, MALAY.
Habitat.—Brackish and fresh waters of the whole coast of India.

F. 436

of Economic Value. (*J. Murray.*)	FISH.

Eleotris ophiocephalus tumifrons ; *Day, Fish. Ind.,* *312 ; Fau. Br.* [*Ind.,* | **437**
II., 293.
   Vern.—*A-rig-dah, mu-tùk-dah,* AND.
   Habitat.—The coasts of the Andamans.

Elops saurus ; *Day, Fish. Ind., 649 ; Fau. Br. Ind., I., 401.* | **438**
   Vern.—*Ullahti,* TAM. ; *Jallugu, jinnagow,* TEL.
   Habitat.—Seas of India.

Engraulis hamiltonii ; *Day, Fish. Ind., 625 ; Fau. Br. Ind., I., 389.* | **439**
   Vern.—*Púrawah,* VIZAG.
   Habitat.—Found throughout the seas of India. The species of this
genus are largely consumed by the Natives.

E. indicus ; *Day, Fish. Ind., 629 ; Fau. Br. Ind., I., 394.* | **440**
   WHITEBAIT, of Europeans in India.
   Vern.—*Nettelli, teran gúni,* TAM. ; *Nattú,* TEL. ; *Bunga-ayer, badah,*
      MALAYS ; *Jú-rú-cart-dah,* AND.
   Habitat.—Seas and tidal rivers of India. It is extensively employed as
food, cooked in the same way as whitebait.

E. malabaricus ; *Day, Fish. Ind., 625 ; Fau. Br. Ind., I., 389.* | **441**
   Vern.—*Púr-rolan,* TAM. ; *Monangú,* MALAY ; *O-pul-doh,* AND.
   Habitat.—Coasts of Sind, and through the seas of India.

E. purava ; *Day, Fish. Ind , 628 ;. Fau. Br. Ind., I., 393.* | **442**
   Vern.—*Phasa,* BENG. ; *Pussai, tampara,* ÚRIYA ; *Pedda-púrawah,* VIZAG.
   Habitat.—Seas and estuaries of both sides of India.

E. telara ; *Day, Fish. Ind., 627 ; Fau. Br. Ind., I., 392.* | **443**
   Vern.—*Phasa, phasah, fessah, pencha,* BENG. ; *Tampara,* URIYA ; *Telara,*
      DINAJPUR ; *Nga-hta-yawet,* BURM.
   Habitat.—Rivers of Orissa, Bengal, Cachar, and Burma.

Ephippus orbis ; *Day, Fish. Ind., 115 ; Fau. Br. Ind., II., 20.* | **444**
   Vern.—*Nalla torriti,* TAM. ; *Kol-liḍ-dah, kow-lid-dah,* AND.
   Habitat.—Seas of India.

Equula daura ; *Day, Fish. Ind., 240 ; Fau. Br. Ind., II., 188.* | **445**
   Vern.—*Dacer-karah,* VIZAG. ; *Rama karé,* TAM.
   Habitat. — Ceylon and the Coromandel coast. The small fish constituting this genus are eaten fresh or sun-dried after being soaked in sea-water.
Their thin and bony structure renders them easily cured without the application of strong brine or salt, but they are very apt to putrify in moist
weather, and if consumed during the monsoon months tend to set up
visceral irritation, resulting in diarrhœa or dysentery.

E. insidiatrix ; *Day, Fish. Ind., 242 ; Fau. Br. Ind., II., 191.* | **446**
   Vern.—*Paarl cúrchi,* MALAY.
   Habitat.—Seas of India. Like the former species it is dried on the
Malabar coast.

E. ruconius ; *Day, Fish. Ind., 242 ; Fau. Br. Ind., II., 192.* | **447**
   Vern.—*Ruconi-chanda,* BENG. ; *Tunka-chandi,* ÚRIYA.
   Habitat.—Seas and tidal rivers of India.

Etroplus maculatus ; *Day, Fish. Ind., 415 ; Fau. Br. Ind., II., 429.* | **448**
   Vern.—*Cundahla,* ÚRIYA ; *Shellel,* TAM. ; *búrakas, chella kassu,* TAM ;
      Pullattay, MALAY ; *Rallia,* SING.
   Habitat.—Fresh waters, along the coast of Madras, and from South
Canara along Malabar ; also found in Ceylon. It extends from the sea at
least 60 or 80 miles inland.

F. 448

FISH.	Indian Fishes
449	**Etroplus suratensis;** *Day, Fish. Ind., 415; Fau. Br. Ind., II., 430.* Vern.—*Pitul-kas*, HIND.; *Cundahla*, ÚRIYA; *Karsaar, pillinchan*, TAM.; *Senel-kas, cashi-mara*, TEL.; *Corallia*, SING. Habitat.—Fresh and brackish waters, along the coasts of Ceylon and India, as far as Orissa.
450	**Eutropiichthys vacha;** *Day, Fish. Ind., 490; Fau. Br. Ind., I., 128.* Vern.—*Ni-much*, HIND.; *Váchá*, BENG.; *Butchua, nandi-butchua*, ÚRIYA; *Chel-lí*, SIND; *Nga-myen-kouban, katha-boung*, BURM. Habitat.—From the Panjáb, through the large rivers of Sind, Bengal, and Orissa; and variety **E. burmanicus**, in Burma. This species attains upwards of a foot in length, and is good eating.
451	**Gagata cenia;** *Day, Fish. Ind., 492; Fau. Br. Ind., I., 208.* Vern.—*Jungla*, BENG.; *Puttuh-chettah*, ÚRIYA; *Cenia*, SIND; *Nga-nan-joung*, BURM. Habitat.—Rivers of Bengal and Orissa, the Jumna, Ganges, and Indus, also those of Burma.
452	**Gerres filamentosus;** *Day, Fish. Ind., 98; Fau. Br. Ind., I., 537.* Vern.—*Udan*, TAM.; *Jaggari*, TEL.; *Wúdaahwah, wúdan*, VIZAG.; *Po-ra-chal-dah*, AND.; *Nga-wet-sat*, ARRAK. Habitat.—Seas of India. This is the best eating of all the species of GERRES, though some of the others are also used as food to a small extent. They are mostly eaten by the indigent classes, being little esteemed whilst fresh, on account of their numerous bones and deficiency in flavour. As they salt and dry well, however, large numbers are thus prepared in many parts of the country for future use or export.
453	**Glyphidodon sordidus;** *Day, Fish. Ind., 385; Fau. Br. Ind., II., 386.* Vern.—*Calamoiapota*, TEL.; *Chák-mud-dah*, AND. Habitat.—Seas of India. Used for food.
454	**Glyptosternum lonah;** *Day, Fish. Ind., 496; Fau. Br. Ind., I., 196.* Vern.—*Lonah*, MAR. Habitat.—The rivers of the Deccan. Eaten, like other SILURIDÆ, by the poorer classes.
455	**Gobius giuris;** *Day, Fish. Ind., 294; Fau. Br. Ind., II., 266.* Vern.—*Gúlú*, HIND.; *Gulah, bali gulah*, ÚRIYA; *Ulúway*, TAM.; *Issaki-dúndú, tsikideondoa*, TEL.; *Kurpah*, MAR.; *Wartí-pú-lah. púan, kurdán*, MALAY; *Ab-bro-ny*, KAN.; *Gú-lú-wah, boul-la*, PB.; *Gúlú*, SIND; *Pú-dah*, AND.; *Nga-tha-boh*, BURM. Habitat.—Fresh waters throughout the plains of India, Ceylon, and Burma. The small variety (? species), *kokius*, never exceeds a span, and appears to be entirely confined to the sea and estuaries along the coast of India and the Andamans.
456	**G. striatus;** *Day, Fish. Ind., 292; Fau. Br. Ind., II., 262.* Vern.—*Mahturi, naolli* (=young), ÚRIYA; *Cúndallum, ulúway*, TEL.; *Cún-dallum*, TAM. Habitat.—Fresh and back-waters of Madras and Kanara.
457	**Haplochilus panchax;** *Day, Fish. Ind., 523; Fau. Br. Ind., I., 417.* Vern.—*Pang-chak*, BENG.; *Kana-kuri, bar-ro-gaddi*, ÚRIYA; *Cho-to-dah*, AND.; *Nga-saki*, MUGH. Habitat. – From Orissa, through the Lower Province of Bengal, Burma, and Siam to the Malay Archipelago, also the Andamans.
458	**Harpodon nehereus;** *Day, Fish. Ind., 505; Fau. Br. Ind., I., 412.* THE BOMBAY DUCK. Vern.—*Nehare, bumalo, bummaloh*, BENG.; *Cucah sawahri, coco mottah*, TEL.; *Bummelo*, MALAY; *Wangara-was*, MADRAS; *Wana-motta*, VIZAG.; *Luli*, MALAYS.

of Economic Value.	(*J. Murray.*)	FISH.

Habitat.—Seas and estuaries of India, most common at Bombay, but decreasing in numbers down the Malabar coast. This fish is highly esteemed as food, whether fresh or salted; in the latter form it is extensively employed as a relish with curries, and is known as "Bombay duck."

Hemirhamphus buffonis ; *Day, Fish. Ind., 516 ; Fau. Br. Ind., I., 427.*   459
    Vern.—*Ku-dú-rock-o-dah,* ANDAMANS.
    Habitat.—The seas and tidal rivers of Bombay, Bengal, and the Andamans. The roes of the fishes of this genus are collected largely on the Malabar coast of India, where they are esteemed a great delicacy.

H. cantori ; *Day, Fish. Ind., 514 ; Fau. Br. Ind., I., 423.*   460
    THE GUARD FISH of the Straits Settlements.
    Vern.—*Toda-pendek,* MALAY.
    Habitat.—Bombay, Malabar, Madras, and the seas of India.

H. ectuntio ; *Day, Fish. Ind., 517 ; Fau. Br. Ind., I., 427.*   461
    Vern.—*Gungituri,* URIYA ; *Nga-phoung-yo,* BURM.
    Habitat.—The river Hooghly, and the tidal streams of Akyab, Burma, and Siam.

H. reynaldi ; *Day, Fish. Ind., 515 ; Fau. Br. Ind., I., 425.*   462
    Vern.—*Morrul,* MALAY.
    Habitat.—The seas of India (*Day*). Malabar and the tanks around Calcutta (*Watson*).

Labeo angra ; *Day, Fish. Ind., 541 ; Fau. Br. Ind., I., 267.*   463
    Vern.—*Kharsa, mochna,* HIND. ; *Paungsi, morala,* BENG. ; *Lassim,* ASSAM ; *Nga-lu,* BURM.
    Habitat.—The rivers of Bengal, Orissa, Assam, and Burma. The several species of this genus, enumerated below, are employed as food by the Natives. Some, such as the ROHÚ, are also highly esteemed by Europeans.

L. ariza ; *Day, Fish. Ind., 544 ; Fau. Br. Ind., I., 272.*   464
    Vern.—*Ariza,* BENG. ; *Coal,* TAM. ; *Nga-lú,* BURM.
    Habitat.—The Wynaad and Bowany rivers at the foot of the Nilghiri hills, also the Cauvery river.

L. boga ; *Day, Fish. Ind., 543 ; Fau. Br. Ind., I., 269.*   465
    Vern.—*Bangum-batta, boga,* BENG. ; *Kala-battali,* URIYA ; *Ariza,* TEL. ; *Kinda-min, coal-arinza-candi,* TAM. ; *Kyouk-nya-lu,* BURM.
    Habitat.—The rivers and tanks of the Gangetic Provinces, Madras, and Burma.

L. calbasu ; *Day, Fish. Ind., 536 ; Fau. Br. Ind., I., 259..*   466
    Vern.—*Kala-beinse,* HIND. ; *Kalbasu, kundna, cuggera,* BENG. ; *Nulla-gandu-menu,* TEL. ; *Kala-beinse,* URIYA ; *Di,* PB. ; *Di-hi,* SIND ; *Dai,* CUTCH ; *Kurri-minu,* KAN. ; *Mahli,* ASSAM ; *Nga-nek-pya, nga-nú-than, nga-ong-tong,* BURM.
    Habitat.—The fresh waters of the Panjáb, Sind, Cutch, the Deccan, Southern India, and Malabar, and from the Kistna through Orissa, Bengal, and Burma.

L. diplostomus ; *Day, Fish. Ind., 540 ; Fau. Br. Ind., I., 265.*   467
    Vern.—*Mohaylí, gaywah,* HIND. ; *Kul-ka-batta,* BENG. ; *Gid, giddah,* PB. ; *Nepura,* ASSAM.
    Habitat.—Along the Sind hills and Himálaya, also a native of the Brahmaputra in Assam.

L. dussumieri ; *Day, Fish. Ind., 538 ; Fau. Br. Ind., I., 262.*   468
    Vern.—*Túli,* MALAY.
    Habitat.—Rivers of South Malabar, Ceylon, and perhaps Bombay.

FISH.	Indian Fishes

469   Labeo fimbriatus ; *Day, Fish. Ind.*, 536 : *Fau. Br. Ind., I.*, 258.
    Vern.—*Bahrum*, URIYA ; *Vencandi, shaal*, TAM. ; *Ruchu, gandu-menu*,
       TEL. ; *Bobri*, MAR.
    Habitat.—The rivers of the Panjáb, Sind, and the Deccan, also of South-
ern India, at least as far as Orissa. It is a fairly large fish, attaining a
length of 1½ feet, and though bony, is good eating.

470   L. gonius ; *Day, Fish. Ind.*, 537 ; *Fau. Br. Ind., I.*, 261.
       Vern.—*Cursa, collúse*, HIND. ; *Kurchi, kursi, goni*, BENG. ; *Cursua*,
          URIYA ; *Courie, bahtur*, ASSAM ; *Mosúl*, TEL. ; *Cir-re-oh*, SIND ; *Nga-
          pay, nga-dane, nga-hú*, BURM.
    Habitat.—The Indus in Sind, through the North-Western Provinces,
Bengal, and Orissa, to Ganjam, as low as the Kistna ; also Assam and
Burma. It is a large fish attaining the length of 5 feet, and is much used
for stocking tanks.

471   L. kontius ; *Day, Fish. Ind.*, 539 ; *Fau. Br. Ind., I.*, 264.
       Vern.—*Carramanni, carú-múli-candi*, TAM.
    Habitat.—The rivers along the base of the Nilghiris, and the Cau-
very and Coleroon in all their branches down to the coast.

472   L. nandina ; *Day, Fish. Ind.*, 535 : *Fau. Br. Ind., I.*, 258.
       Vern.—*Nandin*, BENG. ; *Nga-ohn-don, nga-ne-pyah, nga-yın-pounsa*,
          BURM.
    Habitat.—The fresh waters of Bengal, Assam, and Burma.

473   L. pangusia ; *Day, Fish. Ind.*, 541 : *Fau. Br. Ind., I.*, 266.
       Vern.—*Loanni, pengusiya*, BENG.
    Habitat.—Rivers and tanks of the Himaláya'; found also generally
throughout Sind, the Deccan and the North-West Provinces, Bengal,
Cachar, and Assam.

474   L. rohita ; *Day, Fish. Ind.*, 538 ; *Fau. Br. Ind., I.*, 262.
       THE ROHO, or ROHÚ.
       Vern.—*Rui, rowi, rohita, rui-mutchli*, BENG. ; *Ruhu*, URIYA ; *Rui*,
          ASSAM ; *Nga-myit-chin, nga-myit-tsan-ni*, BURM.
    Habitat.—Fresh-waters of Sind, and from the Panjáb through India
and Assam to Burma. A large fish of 3 feet or more in length, esteemed
excellent as food, and propagated with care in ponds in Bengal. Yields
oil, for which it is principally employed in the North-West Provinces.
U. C. Dutt remarks that the bile of this species is employed in medicine
by the Hindús.

·475   Lactarius delicatulus ; *Day, Fish. Ind.*, 245 : *Fau. Br. Ind., II.*, 196.
       Vern.—*Sudumu*, TELUGU ; *Purruwah*, MALAY. ; *Chundawah*, VIZAG.
    Habitat.—Seas of India. It is insipid, but is eaten, either fresh or salted,
by the Natives.

476   Lates calcarifer ; *Day, Fish. Ind.*, 7 : *Fau. Br. Ind., I.*, 440.
       COCK-UP, *Calcutta* ; NAIR FISH, *Malabar*.
       Vern.—*Begti, bhekti*, BENG. ; *Durruah, bekkut*, URIYA ; *Dangara*, SIND ;
          Painni-min, koduwa, karona, TAM. ; *Pandu kopah, pandu-menu*, TEL. ;
          Nuddi-min, nair-min, MALAY. ; *Padúmenú*, VIZAG. ; *Kuduva*, MADRAS ;
          Nga-tha-dyk, ARRAC. ; *Koral baor*, CHITTAGONG ; *Todah*, AND. ;
          Kakadit, BURM ; *Ikan siyakup*, MALAYS.
    Habitat.—Seas, back-waters, and mouths of tidal rivers. This fish is
excellent eating, when obtained from the vicinity lof large rivers. It salts
well, and from it some of the best "Tamarind fish" is prepared.

477   Lepidocephalichthys guntea ; *Day, Fish. Ind.*, 609 ; *Fau. Br. Ind., I.*, 220.
       Vern.—*Gúnteah, gúteah, bilgagora*, BENG. ; *Kondaturi, gupkari, jubbi-
          cowri*, URIYA ; *Nga-tha-ley-doh*, BURM.

F. 477

of Economic Value. (*J. Murray*.)	FISH.

Habitat.—The rivers and tanks of India; except those along the Malabar coast, Mysore, and south of the Kistna. Eaten by Natives.

Lethrinus rostratus (miniatus) ; *Day, Fish. Ind., 134 ; Fau. Br. Ind., II., 37.* **478**
    Vern.—*Po-tang-dah*, AND.
Habitat.—Seas of India.

Lobotes surinamensis ; *Day, Fish. Ind., 84 ; Fau. Br. Ind., I., 519.* **479**
    Vern.—*Chota bekkut*, URYAH ; *Musalli*, TAM. ; *Parrandi*, MALAY. ; *Ikan-batu*, MALAYS.
Habitat.—East coast of Africa, and seas of India. It is excellent as food.

Lutjanus argentimaculatus ; *Day, Fish. Ind., 37 ; Fau. Br. Ind., I., 472.* **480**
    THE RED ROCK-COD of the Straits Settlements.
    Vern.—*Rangú*, TEL. ; *Singara, senan karawa*, MADRAS ; *To-go-re-dah*, ANDAMANS.
Habitat.—Throughout the seas of India. This fish attains upwards of 2 feet in length and is good eating. The other species of the genus are good as food, though some are insipid ; and are extensively salted and dried in many localities.

L. decussatus ; *Day, Fish. Ind., 47 ; Fau. Br. Ind., I., 481.* **481**
    Vern.—*Jú-win-dah*, ANDAMANS.
Habitat.—Seas of India ; especially abundant on the coasts of the Andamans, where it is readily captured by bait.

L. erythropterus (annularies) ; *Day, Fish. Ind., 32 ; Fau. Br. Ind., I., 466.* **482**
    Vern.—*Sústa*, URIYA ; *Chirtah*, VIZAG. ; *An-na-kah-ro-dah*, ANDAMANS.
Habitat.—Seas of India. It is captured all the year round at Madras, but is most abundant during the cold months.

L. fulviflamma ; *Day, Fish. Ind., 41 ; Fau. Br. Ind., I., 475.* **483**
    Vern.—*Shemhara, currumay*, TAM. ; *Vella-chembolay*, MALAY. ; *Antika-dúndiawah*, VIZAG.
Habitat.—Seas of India, especially abundant off the coasts of Madras.

L. jahngarah ; *Day, Fish. Ind., 40 ; Fau. Br. Ind., I., 474.* **484**
    Vern.—*Purruwa*, URIYA ; *Sillaú*, VIZAG.
Habitat.—Seas of India. It attains two feet or more in length, is esteemed as food, and is extensively cured by drying on the coast of Orissa.

L. johnii ; *Day, Fish. Ind., 42 ; Fau. Br. Ind., I., 476.* **485**
    Vern.—*Chembolay*, MALAY ; *Dúndiawah*, VIZAG. ; *Nga-pá-ni*, BURM.
Habitat.—Seas of India.

Macrones aor ; *Day, Fish. Ind., 444 ; Fau Br. Ind., I., 479.* **486**
    Vern.—*Aor*, BENG. ; *Alli, or addi, arriah-alli, gugah-alli*, URIYA ; *Singala, sang-go-ah*, PB. ; *Cambú-kelleti*, TAM. ; *Mukul-jellah, muti-jella*, TEL. ; *Singhari*, SIND ; *Singhala*, MAR. ; *Nga-joung*, BURM.
Habitat.—Rivers throughout Sind and India to Burma. The species of Macrones here enumerated are employed as food by the poorer classes, but are of inferior quality, being rather insipid.

M. cavasius ; *Day, Fish. Ind., 447 ; Fau. Br. Ind., I., 155.* **487**
    Vern.—*Kavasi tengara*, BENG. ; *Guntea, cuntea*, URIYA ; *Vella kelleti, cutta*, TAM. ; *Muti-jella, nahra-jella*, TEL. ; *Singti, surah*, MAR. ; *Nga-sin-sine*, BURM.
Habitat.—Rivers from Sind, throughout India, Assam, and Burma.

M. corsula ; *Day, Fish. Ind., 446 ; Fau. Br. Ind., I., 153.* **488**
    Vern.—*Punjah-gaggah*, URIYA ; *Nga-ike*, BURM.
Habitat.—Rivers, from Orissa through Bengal and Assam.

2 C

FISH.	Indian Fishes

489   Macrones leucophasis ; *Day, Fish. Ind.,* 449 *; Fau. Br. Ind., I., 158.*
Vern.—*Nga-pet-lek, nga-nouk-thawa,* BURM.
Habitat.—Rivers of Burma.

490   M. malabaricus ; *Day, Fish. Ind.,* 450 *; Fau. Br. Ind., I., 160.*
Vern.—*Cutti mín,* TAM.
Habitat.—Malabar, coast of India, and the Wynaad, extending inland to the gháts in South Cánara.

491   M. punctatus ; *Day, Fish. Ind.,* 445 *; Fau. Br. Ind., I., 153.*
Vern.—*Sholang kelleté, psetta-kelleté,* TAM.
Habitat.—The Bowany river at the base of the Nilghiris.

492   M. tengara ; *Day, Fish. Ind.,* 447 *; Fau. Br. Ind., I., 156.*
Vern.—*Kuttahrah,* HIND. ; *Tengara, tengrah,* BENG. ; *Bikuntia,* URIYA ; *Ting-ga-rah,* ASSAM ; *Karaal, ting-ga-rah,* PB. ; *Saku-jella,* TEL. ; *Nga-sin-sine,* BURM.
Habitat.—Northern India, the Panjáb, and Assam.

493   Mastacembelus armatus ; *Day, Fish. Ind.,* 340 *; Fau. Br. Ind., II., 334.*
THE SPINED EEL, or THORNY-BACKED EEL.
Vern.—*Barua,* HIND. ; *Bahm, bummi, gouti,* BENG., URIYA ; *Bahm, kahm, gro-age,* PB. and SIND ; *Kul-aral, sha-ta-rah,* TAM. ; *Mudi-bom-mi-day,* TEL. ; *Nga-maway-doh-nga,* BURM.
Habitat.—From Sind, throughout the fresh and brackish waters of the plains and hills of India, Ceylon, and Burma. It attains 2 feet or more in length, and is good eating, especially when curried, or fried.

494   M. pancalus ; *Day, Fish. Ind.,* 340 *; Fau. Br. Ind., II., 333.*
THE SMALL SPINED EEL.
Vern.—*Ju-gar,* HIND. ; *Turi, bahru,* URIYA ; *Tu-rah,* ASSAM ; *Par-pa-raal,* TEL. ; *Chen-da-la, gürchi, gro-age,* PB.
Habitat.—Deltas of large rivers of India, and localities near the sea. Good eating, whether fresh or salted.

495   Megalops cyprinoides ; *Day, Fish. Ind.,* 650 *; Fau. Br. Ind., I., 402.*
Vern.—*Punnikaú, naharn,* URIYA ; *Moran cundai,* TAM. ; *Cunnay,* MALAY. ; *Kundinga,* VIZAG. ; *O-pul-dah,* AND. ; *Nga-tan-youet,* BURM.
Habitat.—Fresh waters and estuaries of India and Ceylon. It is occasionally captured in rivers, but much more frequently in tanks.

496   Mugil corsula ; *Day, Fish. Ind.,* 354 *; Fau. Br. Ind., II., 349.*
THE MULLET.
Vern.—*Undala,* HIND. ; *Corsula, in-ge-li,* BENG. ; *Kakunda,* URIYA ; *Hurd-wah-re,* PB. ; *Nga-sen,* BURM.
Habitat.—Rivers and estuaries of Bengal and Burma, extending far above tidal influence in the fresh water. It attains 1½ foot in length and is considered excellent eating. Ainslie remarks regarding this genus : "they are the most excellent fish in India, but are perhaps a little too fat and rich for those who are delicate. They are used both in the fresh and salted state and are much prized by the natives. The spawn salted and dried forms a kind of *cavier,* called by the Italians *boborágo*" (*Mat. Ind., I.,* 227). The same objection to its use, however, exists as with the OPHIOCEPHALIDÆ, certain classes refusing to eat the mullet, owing to the resemblance of its head to that of a serpent.

497   M. cunnesius ; *Day, Fish. Ind.,* 349 *; Fau. Br. Ind., II., 342.*
THE MULLET.
Vern.—*Mahlah,* MALAY ; *Cunnesi,* VIZAG. ; *Sada-parauda,* MADRAS.
Habitat.—Seas of India.

498   M. hamiltonii ; *Day, Fish. Ind.,* 354 *; Fau. Br. Ind., II., 349.*
THE MULLET.
Habitat.—Rivers of Burma.

of Economic Value. (*J. Murray.*)	FISH.

Mugil ocur; *Day, Fish. Ind., 353; Fau. Br. Ind., II., 384.*
    MULLET.
    Vern.—*Kola-kende, mahlah,* MALAY.
    Habitat.—Seas of India and China. The season for capturing these fish along the western coast commences about the middle of November, when they swarm close inshore in order to enter estuaries and the mouths of large rivers to deposit their ova, and extends to about February. The roes are collected and dried in the sun, with or without the use of salt.

499

M. parsia; *Day, Fish. Ind., 350; Fau. Br. Ind., II., 344.*]
    MULLET.
    Vern.—*Tarui,* BENG.; *Pasi-kende, paranda,* MADRAS.
    Habitat.—Seas and estuaries of India. It attains at least 1½ feet in length and is commonly captured for food in the Hooghly at Calcutta.

500

M. planiceps (tade); *Day, Fish. Ind., 350; Fau. Br. Ind., II., 344.*
    MULLET.
    Vern.—*Bangon,* BENG.; *Jumpul,* MALAYS.
    Habitat.—Seas, estuaries, and tidal rivers of India. Common in the Hooghly.

501

M. poicilus; *Day, Fish. Ind., 351; Fau. Br. Ind., II., 345.*
    MULLET.
    Vern.—*Cunnumbú,* MALAY.
    Habitat.—Rivers of Bombay and the Western coast of India, especially common during the colder months.

502

M. seheli; *Day, Fish. Ind., 355; Fau. Br. Ind., II., 350.*
    MULLET.
    Vern.—*Magi,* URIYA.
    Habitat.—Seas of India.

503

M. waigiensis; *Day, Fish. Ind., 359; Fau. Br. Ind., II., 356.*
    FRESH-WATER MULLET, *Eng.*
    Vern.—*Do-dah,* ANDAMANS.
    Habitat.—Throughout the seas of India, ascending rivers to the limit of tidal influence during the monsoon. It attains a foot or more in length, and is good eating.

504

Muræna sathete; *Day, Fish. Ind., 668; Fau. Br. Ind., I., 77.*
    Vern.—*Sathete,* BENG.
    Habitat.—Bay of Bengal and Penang, especially affecting estuaries.

505

M. tile; *Day, Fish. Ind., 668; Fau. Br. Ind., I., 76.*
    THE EEL.
    Vern.—*Tile,* BENG.; *Vellangú,* TEL.; *Ahir,* MAR.; *Chemlú-pamú,* MADRAS; *Palug-dah,* ANDAMANS.
    Habitat.—Seas and estuaries of Bengal, ascending tidal rivers, and common in the Hooghly at Calcutta.

506

Murænesox telabon; *Day, Fish. Ind., 661; Fau. Br. Ind., I., 90.*
    THE BAMBOO FISH.
    Vern.—*Kotah, kulivi-pambú,* TAM.; *Culim-poun,* TEL.; *Tala-bon,* VIZAG.; *Boschi,* ANDAMANS.
    Habitat.—Seas of India, attaining 10 feet or more in length.

507

Nandus marmoratus; *Day, Fish. Ind., 129; Fau. Br. Ind., II., 82.*
    Vern.—*Vádhul,* HIND.; *Latha, gudtha,* BENG.; *Bodosi, gossiporah,* URIYA; *Gad-gud-di, bad-vád-hi,* ASSAM; *Mussoassah,* PB.; *Septi, isoppitay,* TEL.; *Mútahri,* MALAY.
    Habitat.—Fresh and brackish waters of India and Burma, common in ditches and inundated fields.

508

FISH.	Indian Fishes
509	**Nemacheilus zonatus**; *Day; Fish, Ind., 618; Fau. Br. Ind., I., 233.* Vern.—*Mugah*, BENG. Habitat.—Throughout the Jumna and Ganges and their affluents, Bírbhúm, Assam, and Orissa.
510	**Notopterus chitala**; *Day, Fish, Ind., 654; Fau. Br. Ind., I., 407.* Vern.—*Chitala, chitol,* BENG.; *Chitul,* URIYA; *Si-tul,* ASSAM; *Gundun,* SIND. Habitat.—A large fish attaining 4 feet or more in length, found in the fresh waters of Sind, Lower Bengal, Orissa, Assam, Burma, and Siam. Hamilton-Buchanan writes : "The belly is uncommonly rich and well flavoured, but the back contains numerous small bones, and a strong prejudice exists against using this fish as food, owing to its being supposed to live on human carcasses.
511	**N. kapirat**; *Day, Fish., Ind., 653; Fau. Br. Ind., I., 406.* Vern.—*Moh,* HIND; *Pholoe,* BENG.; *Pulli,* URIYA; *Ambutan-wahlah, chota wahlah,* TAM.; *Kau-dú-li,* ASSAM; *Moh, but, purri,* PB.; *Nollak-tattah,* MYSORE; *Nga-hpeh, nga-phe,* BURM. Habitat.—Fresh and brackish waters of India. It grows to 2 feet or more in length and is salted in Burma.
512	**Ophichthys boro**; *Day, Fish, Ind., 664; Fau. Br. Ind., I., 94.* Vern.—*Boro, harancha, hijala,* BENG. Habitat.—Seas and estuaries of India. The natives in some parts of Bengal imagine that this fish proceeds from the ear of a porpoise.
513	**Ophiocephalus barca**; *Day, Fish, Ind., 365; Fau. Br. Ind., II., 361.* THE WALKING FISH. Vern.—*Barca,* BENG. ; *Bora-chang,* BUTAN. Habitat.—Large rivers of the Bengal Presidency. All the fish of this genus have hollow cavities in their heads, an amphibious system of respiration, are able to exist for a lengthened period out of water, and can travel some distance over the ground, especially where it is damp. They are all useful as food, and the possibility of carrying them in moist vessels for a long distance renders them extremely valuable. Some classes of natives, however, object to them on account of the resemblance of their heads to those of serpents.
514	**O. gachua**; *Day, Fish. Ind., 367; Fau. Br. Ind., II., 364.* THE WALKING FISH. Vern.—*Dheri dhok,* HIND.; *Chenga, chayung,* URIYA; *Chen-gah,* ASSAM; *Doarruh,* PB.; *Para korava, munrú,* TAM.; *Karavu,* MALAY.; *Mah korava,* KAN.; *Korah-mottah,* VIZAG.; *Chad-dah,* AND. Habitat.—Fresh waters, throughout India, Ceylon, Burma, and the Andamans. Described by Thomas as an excellent live bait.
515	**O. marulius**; *Day, Fish. Ind., 363; Fau. Br. Ind., .II, 360.* THE WALKING FISH, or MURREL. Vern.—*Pu murl,* HIND.; *Sál,* URIYA; *Ha-al,* ASSAM ; *Kubrah sál, daulah,* PB.; *Pu verarl,* TAM. ; *Pula chapa,* TEL.; *Choari verarl, curavu,* MALAY. ; *Húvina murl,* KAN.; *Murrul,* MAR.; *Sowarah,* VIZAG.; *Nga-yan-dyne,* BURM. Habitat.—Fresh waters (principally rivers), from Ceylon and India to China. This fish is described by Thomas as affording excellent sport either with live bait or fly. It is one of the best of the OPHIOCEPHALIDÆ as a food fish, and is excellent for stocking tanks.
516	**O. punctatus**; *Day, Fish. Ind., 367; Fau. Br. Ind., II., 364.* THE BLACK CABOOSE. Vern.—*Phúl dhok,* HIND. ; *Gorissa, gurrie, cartua-gorai,* URIYA and ASSAM ; *Dullinga,* PB.; *Dhoali,* SIND. ; *Korava, para-kora wa,* TAM. ; *Muttah,* TEL. ; *Beli-korava,* KAN.; *Nga-vin,* MUGH.

F. 516

of Economic Value.	(*J. Murray.*)	FISH.

Habitat.—Commonly found in fresh waters, of the plains, preferring stagnant ponds to streams.

Ophiocephalus striatus; *Day, Fish. Ind., 366; Fau. Br. Ind., II., 363.* 517
  THE WALKING FISH, or MURREL.
  Vern.—*Morrul, murl, dheri murl,* HIND.; *Sol, chena,* BENG.; *Sola,* URIYA; *Verarlu, currupu verarl,* TAM.; *Sowarah, kora, muttageddasa,* TEL.; *Verarl, wrahl,* MALAY.; *Muttah,* VIZAGAPATAM; *Kúchina murl,* KAN.; *Lúlla,* SING.; *Nga-ain-di,* MUGH; *Nga-yaw,* BURM.; *Ikan-haruan,* MALAYS.
  Habitat.—Fresh waters, throughout the plains of India. Like O. marulius, it affords excellent sport, is good as food, though bony, and is a very good stock for tanks. The Telaings are said to employ this fish in one of their religious ceremonies.

Opisthopterus tartoor; *Day, Fish. Ind., 646; Fau. Br. Ind., I., 384.* 518
  Vern.—*Tartoore,* VIZAGAPATAM.
  Habitat.—From Sind, through the seas of India.

Oreinus plagiostomus; *Day, Fish. Ind., 530; Fau. Br. Ind., I., 250.* 519
  THE KASHMIR TROUT.
  Habitat.—Rivers of Afghánistan, Kashmír, and Bútan. All the species of Orcinus are used as food.

O. richardsonii; *Day, Fish. Ind., 530; Fau. Br. Ind., I., 250.* 520
  THE KEMAON TROUT.
  Vern.—*Asla,* NEPAL.
  Habitat.—The rivers of Nepál, Bútan, and the Sub-Himálayan range.

O. sinuatus; *Day, Fish. Ind., 529; Fau. Br. Ind., I., 248.* 521
  TROUT of Europeans.
  Vern.—*Gúl-gúlli, saul,* PB.; *Jis,* KASH.
  Habitat.—Afghánistan and Himálayan rivers, not extending to the plains far from the base of the hills. It attains 2 feet in length, and is pretty good eating, but bony; it is too rich for some people, but does not deleteriously affect those accustomed to it.

Osphromenus nobilis; *Day, Fish. Ind., 372; Fau. Br. Ind., II., 370.* 522
  Habitat.—Rivers of North-eastern Bengal and Assam, extending into those of the hills. Like the next species it is excellent eating, and good for stocking tanks, but as it is a very promiscuous feeder, care must be taken to prevent its obtaining access to foul substances.

O. olfax; *Day, Fish. Ind., 372; Fau. Br. Ind., II., 369.* 523
  THE GOURAMY.
  Habitat.—A native of China and the Malay Archipelago, but introduced into tanks near Calcutta, Madras, and the Nilgiris. It attains 20℔ or more in weight and is excellent eating when kept in clean water.

Osteogeniosus militaris; *Day, Fish. Ind., 469; Fau. Br. Ind., I., 190.* 524
  Vern.—*Poné kelíti,* TAM.; *Poné-ketti,* MALAY.
  Habitat.—Seas, estuaries, and tidal rivers of India. It is eaten by the poorer classes, and is one of the species which furnish "fish maws" from which isinglass is manufactured.

Otolithus maculatus; *Day, Fish. Ind., 196; Fau. Br. Ind., II., 127.* 525
  Vern.—*Birralli,* URIYA.
  Habitat.—Seas of India. Both species of this genus are eaten, and their air-vessels collected for isinglass.

O. ruber; *Day Fish Ind., 196; Fau. Br. Ind., II., 128.* 526
  PÊCHEPIERRE, *French* at Pondicherry.

FISH.	Indian Fishes
	Vern.—*Jarang-gigi*, MALAYS. Habitat.—Seas of India. A large fish, attaining 2⅓ feet or more in length, and fairly good for the table.
527	**Pangasius buchanani**; *Day, Fish. Ind., 470; Fau. Br. Ind., I., 142.* Vern.—*Cula-kelletti*, TAM.; *Banka-jella*, TEL.; *Jellum*, URIYA. Habitat.—The large rivers and estuaries of India, Assam, and Burma. It attains upwards of 4 feet in length and is eaten, though a foul feeder.
528	**Pellona motius**; *Day, Fish. Ind., 643; Fau. Br. Ind., I., 381.* Vern. —*Ursi, alise*, URIYA. Habitat.—Rivers of Assam, Bengal, and Orissa, descending as low as the coast. Used as food.
529	**P. sladeni**; *Day, Fish. Ind., 645; Fau. Br. Ind., I., 383.* Vern.—*Nga-sen-bya*, BURM. Habitat.—River Irrawaddi, as high as Mandalay. It is eaten by the Burmans.
530	**Perilampus atpar**; *Day, Fish. Ind., 598; Fau. Br. Ind., I., 359.* Vern.—*Kachhi, atpar*, BENG.; *Bonkuaso*, URIYA; *Mor-ri-ah*, PB.; *Bidah*, SIND; *Arku-konissi*, TEL.; *Nga-man-dan, ya-paw-nga, nga-phyin-gyan*, BURM. Habitat.—Rivers of Sind, throughout India and Burma. The carps of this genus are eaten by natives.
531	**P. laubuca**; *Day, Fish. Ind., 598; Fau, Br. Ind., I., 360.* Vern.—*Dannahrah*, HIND.; *Layubuka, dankena*, BENG.; *Bankoe*, URIYA; *Moh-do-ni-konah, her-bag-gi*, ASSAM; *Cún-che-li-e*, N.-W. P.; *Nga-meloung*, BURM. Habitat.—The rivers of Bengal, Orissa, Central India, Ganjam, Assam, and Burma.
532	**Plagusia bilineata**; *Day, Fish. Ind., 431; Fau. Br. Ind., II., 452.* Vern.—*Aralu*, TAM.; *Ikan-ledah*, MALAYS; *Jerri-potú*, VIZAGAPATAM. Habitat.—Seas of India. Used as food.
533	**Platax teira**; *Day, Fish. Ind., 235; Fau. Br. Ind., II., 182.* Vern.—*Cha-la-dah, gú-na-dah*, ANDAMANS. Habitat.—Seas of India. Russell and Cantor both remark that the flavour of this fish is excellent.
534	**Platycephalus insidiator**; *Day, Fish. Ind., 276; Fau. Br. Ind., II., 238.* CROCODILE FISH of Europeans in Malabar. Vern.—*Ulpathy*, TAM.; *Irrwa*, TEL.; *Nga-paying-ki*, MUGH.; *A-rawud-dah, chau-ur-dah*, AND. Habitat.—Seas of India. Eaten by the lower classes of natives, but much dreaded on account of the severe irritative wounds caused by its spines.
535	**Plotosus arab**; *Day, Fish. Ind., 483; Fau. Br. Ind., I., 113.* Vern.—*Ingeli*, VIZAG.; *Múrghi*, MALAY; *Similáng-karong*, MALAYS. Habitat. —Seas of India. Wounds from the pectoral spines of this fish are much dreaded, as they occasion phlegmonous inflammation, or even tetanus.
536	**P. canius**; *Day, Fish. Ind., 482; Fau. Br. Ind., I., 113.* Vern.— *Kani-magur*, BENG.; *Irung-kell-etti*, TAM.; *Li-mi-dah, bondah*, ANDAMANS. Habitat.—The estuaries of India, Burma, and the Malay Archipelago. A large fish 3 feet or more in length, the flesh of which is supposed by the Malays of Batavia to have emmenagogue properties.
537	**Polyacanthus cupanus**; *Day, Fish. Ind., 371; Fau. Br. Ind., II., 368.* Vern.—*Punnah*, TAM.; *Heb-bu-ti*, TEL.; *Ta-but-ti*, KAN.; *Caringanah, wúnnutti*, MALAY.

F. 537

Habitat.—Fresh waters of Malabar and the Coromandel coasts, often found in ditches, paddy-fields, and other shallow waters. Although of small size, it is employed as food by the lower classes of Natives. Jerdon remarks that wounds from the spines of this fish cause severe burning pain, which lasts for two or three hours.

Polynemus indicus; *Day, Fish. Ind., 179; Fau. Br. Ind., II., 105.* | 538
ROWBALL of Europeans at Vizagapatam.
Vern.—*Selé, sulea, suliah, selliah,* BENG.; *Dara,* BOM.; *Tahlun-kala,* TAM.; *Bhát,* MAR.; *Póle-kala,* MADRAS; *Maga-boshi,* VIZAG.; *Yeta,* MALAY.; *Lukwah,* ARRAKAN; *Kwey-yeng,* TAVOY; *Ikan-kurow,* MALAYS; *Katha,* or *ka-ku-yan,* BURM.
Habitat.—The seas of India. All the fish of this genus are excellent as food, and also form one of the principal sources of "fish maws."

P. paradiseus; *Day, Fish. Ind., 176; Fau. Br. Ind., II., 102.* | 539
MANGO FISH of Europeans in Calcutta.
Vern.—*Tupsi, tupsi muchi,* BENG.; *Toposwi,* HIND.; *Nga-púngna,* BURM·
Habitat.—The Indian seas, Bay of Bengal, at least as low as Coconada, also along the coasts of Burma to the Malay Archipelago. It enters rivers for spawning purposes, during the south-west monsoon and the cold months. Though a small fish attaining only 9 inches in length, it is much prized as an article of food.

P. tetradactylus; *Day, Fish. Ind., 180; Fau. Br. Ind., II., 106.* | 540
THE ROWBALL.
Vern.—*Teriya-bhanggan,* BENG.; *Polun-kala,* ΓAM.; *Yerra-kala,* MADRAS; *Maga-jelli,* VIZAG.; *To-bro-dah,* ANDAMANS; *Py-tha-corah,* MALAY.; *Ikan-sálangan, sinanghi* or *salanghi,* MALAYS.
Habitat.—The seas of India. This is a very large fish, 6 feet or more in length; indeed, Buchanan records a specimen which formed a load for six men. It is excellent eating, and is salted on the Madras coast.

Pristipoma guoraka; *Day, Fish. Ind., 75; Fau. Br. Ind., .I, 512.* | 541
Vern.—*Guoraka,* VIZAG.
Habitat.—The seas of India; said also to have been captured in fresh water. All the species of this genus are fair as food, but are not much esteemed; the air vessels also are in some places collected for isinglass.

P. hasta; *Day, Fish, Ind., 73; Fau. Br. Ind., I, 510.* | 542
Vern.—*Caroua, corake,* TAM.; *U-rug-nud-dah, kúr-kú-to-dah,* ANDA-MANS.
Habitat.—The seas of India.

P. maculatum; *Day, Fish. Ind., 74; Fau. Br. Ind., I, 510.* | 543
Vern.—*Currutche,* TAM.; *Erruttum corah,* MALAY.; *Caripe,* TEL.; *Ur-ung-dah,* ANDAMANS.
Habitat.—Seas of India.

Pristis cuspidatus; *Day, Fish. Ind., 728; Fau. Br. Ind., I., 37.* | 544
THE SAW-FISH.
Vern.—*Yahla,* VIZAG; *Ikan-garagaji,* MALAYS; *Vela min,* TAM.
Habitat.—The seas of India, ascending rivers. A huge fish, attaining 20 feet in length, and of great economic value. The flesh is highly esteemed, the fins are prepared for exportation to China, oil is extracted from the livers, and the skins are useful for sword scabbards, or for smoothing down wood.

Psettus argenteus; *Day, Fish. Ind., 235; Fau. Br. Ind., II., 180.* | 545
Vern.—*Nga-pus-súnd,* MUGH.; *U-chra-dah,* ANDAMANS.
Habitat.—Seas of India. Used as food.

FISH.	Indian Fishes

546

Pseudeutropius atherinoides ; *Day, Fish. Ind., 473 ; Fau. Br. Ind., I., 141.*
  Vern.—*Put-tah-re,* HIND. ; *Battuli, bopotassi, jemmi carri,* URIYA ; *Boh-du-ah, pátási, doyá,* ASSAM ; *Put-tul, chel-li,* PB. ; *Ah-hí,* SIND ; *Akku-jella,* TEL. ; *Nga-than-chyeik,* BURM.
  Habitat.—Throughout the rivers of India and Assam. All the species of this genus are excellent as food, but in some localities are to be avoided, as they consume offal.

547

P. garua ; *Day, Fish. Ind., 474 ; Fau. Br. Ind., I., 141.*
  Vern.—*Buchua,* HIND. ; *Puttosi, garua, pultosi,* BENG. ; *Punia buchua,* URIYA ; *Dhon-ga-nu,* SIND.
  Habitat.—Found generally throughout the larger rivers of India, Assam, and Burma.

548

P. goongwaree ; *Day, Fish. Ind., 471 ; Fau. Br. Ind., I., 137.*
  Vern.—*Gúgli, gúngwari,* MAR. ; *Nga-myen-oke-hpa,* BURM.
  Habitat.—The rivers of Bengal, the Deccan, and Burma.

549

P. murius ; *Day, Fish. Ind., 472 ; Fau. Br. Ind., I., 139.*
  Vern.—*Butchua,* HIND. ; *Muri-vacha, motusi,* BENG. ; *Muri-vacha,* URIYA ; *Ke-raad,* PB. ; *Chhotká váchoyá,* KUSI.
  Habitat.—The rivers of Sind, Bengal, Orissa, and Assam.

550

P. taakree ; *Day, Fish. Ind., 471 ; Fau. Br. Ind., I., 138.*
  Vern.—*Tákrí,* MAR. ; *Salava-jella,* TEL. ; *Nga-zin-sap, nga-myin,* BURM.
  Habitat.—The fresh waters of Púna, the Deccan, and the rivers Kistna and Jumna. This fish attains upwards of a foot in length, and is one of the best of the genus as a food.

551

Pseudorhombus arsius ; *Day, Fish, Ind., 423 ; Fau. Br. Ind., II., 441.*
  Vern.—*Ikan-siblah,* MALAYS ; *Ky-tha-thong-dah,* ANDAMANS.
  Habitat.—Through the seas and estuaries of India. Used as food.

552

Psilorhynchus balitora ; *Day, Fish. Ind., 527 ; Fau. Br. Ind., I., 244.*
  Vern.—*Balitora,* BENG.
  Habitat.—Hill streams and rapids in North-east Bengal and Assam. Employed as food by Natives.

553

Pseudoscarus rivulatus ; *Day, Fish. Ind., 413 ; Fau. Br. Ind., II., 426.*
  Vern.—*Ar-dah,* ANDAMANS.
  Habitat.—Seas of India. Eaten by Natives of some parts of the coast.

554

Pterois volitans ; *Day, Fish. Ind., 154 ; Fau. Br. Ind., II., 62.*
  Vern.—*Parrúah,* MALAY ; *Kodipungi,* VIZAG. ; *Chib-ta-ta-dah,* AND.
  Habitat.—Throughout the seas of India. Employed as food in some parts of the country.

555

Pteroplatea micrura ; *Day, Fish. Ind., 741 ; Fau. Br. Ind., I., 56.*
  Vern.—*Perúm tiriki,* TAM. ; *Tappu cúti,* TEL. ; *Tenki-kunsul,* VIZAG. ; *Lek kyouk temengnee,* BURM.
  Habitat.—The seas of India. Used as food.

556

Raconda russelliana ; *Day, Fish. Ind., 646 ; Fau. Br. Ind., I., 384.*
  Vern.—*Potassah-fessah, phasah,* BENG.
  Habitat.—The Bay of Bengal ; the young are common in the Sunderbans. Largely consumed by the native population.

557

Rasbora buchanani ; *Day, Fish. Ind., 584 ; Fau. Br. Ind., I., 337.*
  Vern.—*Rasbora,* BENG.
  Habitat.—The rivers of India, Assam, and Burma. Most common in the valley of the Ganges and along the Coromandel coast. Used as food by the Natives.

558

R. daniconius ; *Day, Fish. Ind., 584 ; Fau. Br. Ind., I., 336.*
  Vern.—*Mile-lo-ah,* HIND. ; *Danikoni, angjani,* BENG. ; *Jilo, dundikerri,* URIYA ; *Doh-ni-ko-nah,* ASSAM ; *Chin-do-lah, raan-kaal-le, charl,* PB. ;

F. 558

of Economic Value. (*J. Murray.*)	FISH.

Ovaricandi, purruvu-kende, TAM.; *Kokanutchi,* MALAY; *Jonir,* KUTCH; *Neddean, jubbo,* KAN.; *Nga-doung-zi, nga-nauch-youn,* BURM.

Habitat.—The rivers of India and Ceylon. Much more common than R. buchanani.

Rhynchobatus ancylostomus; *Day, Fish. Ind., 730 ; Fau. Br. Ind., I., 41.* **559**
THE MUD-SKATE.
   Vern.—*Manu-ulavi,* TAM.; *Manu ulava, naladindi,* TEL.
   Habitat.—Throughout the seas of India. The species of this genus are valued, like other skates, for their skins, fins, and livers.

R. djeddensis; *Day, Fish. Ind., 730 ; Fau. Br. Ind., I., 40.* **560**
   Vern.—*Ulavi, tipi ulavi,* TEL.; *Walawah-tenki,* VIZAG.; *Ranja,* MAR.
   Habitat.—Seas of India. A large fish attaining 6 feet or more in length, the flesh of which is considered nourishing, whether eaten salted or fresh, and the oil from its liver is much esteemed.

Rhynchobdella aculeata; *Day, Fish. Ind., 338 ; Fau. Br. Ind., II., 331.* **561**
THE SAND or SPINED EEL.
   Vern.--*Bara, thuri, gutti,* URIYA; *Tou-rah,* ASSAM; *Aral, cul, monah-aral,* TAM.; *Bommiday, bomri,* TEL.; *Theluja,* SING.; *Nga-mawaydoh-nya,* BURM.
   Habitat.—Brackish waters within tidal influence, also throughout the deltas of the large rivers of India, Burma, and Sind ; but apparently absent from the northern portions of the Panjáb and Malabar coasts. It is excellent as food, though objected to by certain classes owing to its resemblance to a snake. Buchanan remarks, " They have less of a disgusting appearance than the **Muræna,** and are more sought after by Natives, the highest of whom in Bengal make no scruple in eating them ; and by Europeans they are esteemed the best of the eel kind." It salts well, but the flesh is reputed to be slightly heating.

Rita buchanani; *Day, Fish. Ind., 454; Fau. Br. Ind., I., 165.* **562**
   Vern.—*Rita,* BENG.; *Muss-ayahri, cunta-gagah,* URIYA; *Gudla-jella,* TEL.; *Nga-htway,* BURM.
   Habitat.—The Rivers Indus, Jumna, Ganges, and Irrawaddi.
This fish, though a very foul feeder, is esteemed as food by the Natives. All the species of this genus are employed for food by the lower classes, and are valuable from their capability of retaining life long subsequent to their removal from water, owing to which they can be carried fresh for long distances.

R. hastata; *Day, Fish. Ind., 456; Fau. Br. Ind., I., 168.* **563**
   Vern.—*Kuterni,* MAR.
   Habitat.—The rivers of the Deccan and Púna, and the Tambudra and Kistna.

R. pavimentata; *Day, Fish. Ind., 455 ; Fau. Br. Ind., I., 167.* **564**
   Vern.—*Pilah-gokundu,* HIND.; *Banki yeddu,* TEL.; *Gograh, khirurh, putturchattah,* MAR.
   Habitat.— Rivers of Puna and the Deccan, and affluents of the Kistna.

Rohtee belangeri; *Day, Fish. Ind., 587 ; Fau. Br. Ind., I., 342.* **565**
   Vern.—*Kilay,* TEL.; *Nga-hpeh-oung, nga-net-pya,* BURM.
   Habitat.—The Godavery river, and throughout Burma. Employed as food by the Natives.

R. cotio; *Day, Fish. Ind., 587 ; Fau. Br. Ind., I., 340.* **566**
   Vern.—*Gúrdah, chen-da-lah, muckni,* HIND.; *Roti, gunta,* BENG.; *Gunda, gollund,* URIYA; *Puttu, duh-rie,* SIND; *Phenk,* MAR.; *Nga-hpan-ma,* BURM.
   Habitat.--Found in rivers, ponds, and ditches from Sind throughout India (except the Malabar coast and south of the Kistna) and Burma.

FISH.	Indian Fishes
567	**Rohtee ogilbii**; *Day, Fish. Ind., 588 ; Fau. Br. Ind., I., 342.* 　　Vern.—*Kunninga,* TEL.; *Rohti,* MAR. Habitat.—The Kistna and Godavery, and the rivers of the Deccan.
568	**Saccobranchus fossilis**; *Day, Fish. Ind., 486 ; Fau. Br. Ind., I., 125.* 　THE SCORPION FISH. 　　　Vern.—*Bitchu-ka-mutchi, singi,* HIND.; *Singgi, singhi,* BENG.; *Singi,* 　　　*URIYA ; Singi, shin-i,* ASSAM; *Lo-har,* SIND; *Lahùrd* (young), *nullie* 　　　(adult), PB.; *Thay-li, tharli,* TAM.; *Marpu,* TEL.; *Kahri-min,* MALAY.; 　　　*Nga-gyi, nga-kyi,* BURM. and MUGH. 　Habitat.—The fresh waters of India, Ceylon, Burma, and Cochin-China. attaining 1 foot or more in length. It is considered exceedingly wholesome and invigorating by Natives, though in some places deemed impure by the Brahmins. In Burma it is salted.
569	**Saurida tumbil**; *Day, Fish. Ind., 504 ; 　Fau. Br. Ind., I., 410.* 　　Vern.—*Ulùway, cul-nahmacunda,* TAM.; *Arranna,* MALAY.; *Badimottah,* 　　VIZAG. 　Habitat.—Seas of India. Though rather dry and insipid it is consi- derably used as food.
570	**Sciæna bleekeri**; *Day, Fish. Ind., 185; Fau. Br. Ind., II., 112.* 　　Vern.—*Soh-li,* BENG. 　Habitat.—Bombay. This species is extensively salted at Gwadur.
571	**S. coitor**; *Day, Fish. Ind., 187 ; Fau. Br. Ind., II., 115.* 　　Vern.—*Coitor,* BENG.; *Botahl, putteriki,* URIYA; *Vella-ketcheli,* TAM.; 　　*Nga-ta-dun, nga-pok-thin,* BURM. 　Habitat.—Throughout the larger rivers of India and Burma, descend- ing to the sea at certain seasons.
572	**S. cuja**; *Day ; Fish. Ind., 187 ; Fau. Br. Ind., II., 115.* 　　Vern.—*Cuja,* BENG. 　Habitat.—The estuaries of the Ganges.
573	**S. diacanthus**; *Day, Fish. Ind., 189; Fau. Br. Ind., II., 118.* 　　Vern.—*Chaptis,* BENG.; *Katcheli, nalla-katcheli,* VIZAG.; *Ikan sam- 　　bareh,* MALAYS. 　Habitat.—The seas of India, ascending tidal rivers and estuaries. It is found in the Hooghly as high as Calcutta.
574	**S. maculata**; *Day, Fish. Ind., 190 ; Fau. Br. Ind., II., 119.* 　　Vern.—*Cùrùwa, vari katcheli,* TAM.; *Cutlah,* MALAY.; *Sari-kullah,* 　　VIZAG.; *Taantah,* BEL. 　Habitat.—The seas of India. It is not considered such a good food fish as the other species.
575	**Scomber microlepidotus**; *Day, Fish. Ind., 250 ; Fau. Br. Ind., II., 203.* 　THE MACKEREL. 　　Vern.—*Karah,* BENG.; *Karna-kita,* or *karnang-kullutan,* TAM.; *Kana- 　　gurta,* TEL.; *Cunny-ila,* MAD.; *Ila,* MALAY.; *Kanagurta,* VIZAG.; *Nga- 　　congri,* MUGH.; *Lùk-wa-dah,* ANDAMANS. 　Habitat.—Indian seas. A small fish rarely exceeding 10 inches in length, very common throughout the cold season in Malabar. It is extensively salted and dried, but although good eating is seldom brought to the tables of Europeans, as it rapidly taints, and if eaten in that condi- tion gives rise to visceral irritation.
576	**Semiplotus mc'clellandi**; *Day, Fish. Ind., 550 ; Fau. Br. Ind., I., 281.* 　　Vern.—*Sundari, sentori, lah-bo-e, rajah-mas* (="King's fish"), ASSAM. 　Habitat.—The rivers of Assam, especially the upper portions of that district, but found as low as Goalpara; also in Burma. It is asserted that this fish obtained the vernacular name of "king fish," owing to the

F. 576

of **Economic Value.**	(*J. Murray.*)	**FISH.**

fact that in olden times, when captured, it had always to be taken to the Rajas for their own consumption. Day, however, remarks that, as it is very common, this explanation is improbable, and it is more likely that it was so named from a tax being levied on its capture. Very varying accounts of the value as food of the **Semiplotus** exist. McClelland states that it is the most delicious in Assam, while Day records from personal experience that it is rich and liable to set up intestinal irritation.

Serranus diacanthus; *Day, Fish. Ind., 17; Fau. Br. Ind., I., 449.*
    Vern.—*Damba,* SIND.; *Chándcha,* BELUCH.; *Killi-min,* MALAY.
    Habitat.—Seas of India. All the species of this genus of the PERCIDÆ are good as food, though coarse when very large. A small amount of isinglass also is obtained from their air vessels.

    577

S. lanceolatus; *Day, Fish. Ind., 18; Fau. Br. Ind., I., 450.*
    Vern.—*Gussir,* SIND.; *Commári, wutla-callawah* (=Perch with a sore-head), TAM.; *Kurrupu,* MALAY.; *Ikan-krapu,* MALAYS; *Suggalahtú-bontú,* VIZAG.; *Bole,* CHITTAGONG.; *Nga-towktú-shweydú,* ARRAK.
    Habitat.—Seas of India and east coast of Africa.

    578

S. malabaricus (pautherinus); *Day, Fish. Ind., 19; Fau. Br. Ind., I., 451.*
    Vern.—*Punni-calawah,* TAM.; *Bontú, madinawah bontú,* TEL.; *Búl,* CHITTAGONG; *Nga-towktú,* ARRAK; *Kyouk-theyga-kakadit,* BURM.; *Ráb-nadah, o-ro-tam-dah,* ANDAMANS.
    Habitat.—Seas of India and China.

    579

Sillago sihama; *Day, Fish. Ind., 265; Fau. Br. Ind., II., 224.*
    WHITING of Europeans in Madras.
    Vern.—*Gudji-curama,* URIYA; *Kulingah, kilinjan, kigingan,* TAM.; *Soring, tella-soring, arriti-ki,* TEL.; *Cudirah,* MALAY.; *Nga-rui,* MUGH; *Thol-adah,* ANDAMANS.
    Habitat.—Seas of India, ascending tidal rivers. Native women who have young babies are advised to eat it, as it is said to be even more nourishing than shark's flesh, and to have special milk-forming properties.

    580

Silundia gangetica; *Day, Fish. Ind., 488; Fau. Br. Ind., I., 145.*
    Vern.—*Jil-lung, silond,* BENG. and URIYA; *Silond,* PB.; *Wallaho-kellette, púnatti,* TAM.; *Wangon, wanjon,* TEL.; *Parri, sillum,* MAR.
    Habitat.—Estuaries of India and Burma, ascending high up the larger rivers to nearly their sources. It is a large and extremely voracious fish attaining a length of 6 feet or more, and is hence called a "shark" by the natives. It is eaten by the poorer classes, and its air vessels are collected for isinglass. In the Gazetteer of the North-Western Provinces it is stated that it is also employed in the manufacture of fish oil for burning.

    581

Sphyræna jello; *Day, Fish. Ind., 342; Fau. Br. Ind., II., 335.*
    Vern.—*Chilahú,* MALAY.; *Jellow,* VIZAG.; *Thal-lib-dah,* ANDAMANS.
    Habitat.—Seas of India. A large fish attaining 5 feet or more in length, used as food, although not much esteemed.

    582

Stromateus cinereus; *Day, Fish. Ind., 247; Fau. Br. Ind., II., 198.*
    SILVER POMFRET (immature), GREY POMFRET (mature).
    Vern.—*Vella voval,* TAM.; *Sudi-sandawa, telli-sandawa,* VIZAG.
    Habitat.—The seas of India, attaining one foot or more in length. The adult or "grey pomfret" is considered superior to the immature or "silver pomfret" for the table, and is excellent eating. It is also salted along the coasts of India and Burma.

    583

S. niger; *Day, Fish. Ind., 247; Fau. Br. Ind., II., 199.*
    THE BLACK POMFRET.
    Vern.—*Baal,* URIYA; *Karúpú-voval,* TAM.; *Nalasandawah,* TEL.; *Karapu-voval,* MADRAS; *Nala-sandawah,* VIZAG.; *Kar-arwúli,* MALAY.; *Ko-lig-dah,* ANDAMANS; *Bawar, bawal-tumbak,* MALAYS.

    584

F. 584

FISH.	Indian Fishes

Habitat.—The seas of India, growing to two feet in length. It is excellent eating and is extensively salted, though in certain parts the natives dislike it, because a species of parasite, like a woodlouse, is often found in its mouth.

585   **Stromateus sinensis**; *Day, Fish. Ind., 246 ; Fau. Br. Ind., II., 197.*

THE WHITE POMFRET.

Vern.—*Mogang voval*, TAM.; *Vella arwúli*, MALAY.; *Atúkoia*, VIZAG.; *Mowe*, MADRAS; *Bawal-chirmin*, MALAYS.

Habitat.—Seas of India, common in Malabar during the south-west monsoon. The young abound round the coasts, and ascend estuaries. It is the finest of the genus for eating, and should be cooked when quite fresh. Like the other species it is extensively salted wherever it is captured on the coasts of India and Burma.

586   **Synaptura orientalis**; *Day, Fish. Ind., 429 ; Fau. Br. Ind., II., 449.*

Vern.—*Sappati*, MALAY.

Habitat.—Sind, Western coast of India, Andamans, and the China seas. Used as food.

587   **Teuthis concatenata**; *Day, Fish. Ind., 167 ; Fau. Br. Ind., II., 90.*

Vern.—*Thar-oar-dah*, ANDAMANS.

Habitat.—The Andaman and Malayan seas. All the species of this genus, enumerated below, are eaten by Natives.

588   **T. java**; *Day, Fish. Ind., 165 ; Fau. Br. Ind., II., 88.*

Vern.—*Ottah*, TAM.; *Worahwah*, TEL.; *Thar-oar-dah*, ANDAMANS.

Habitat.—The seas of India.

589   **T. vermiculata**; *Day, Fish. Ind., 166 ; Fau. Br. Ind., II., 88.*

Vern.—*Kut-e-rah*, MALAY.; *Chow-lud-dah*, ANDAMANS.

Habitat.—The seas of India.

590   **T. virgata**; *Day, Fish. Ind., 166 ; Fau. Br. Ind., II., 89.*

Vern.—*Tah-mir-dah*, ANDAMANS.

Habitat.—The Andaman and Malayan seas.

591   **Toxotes jaculator**; *Day, Fish. Ind., 117; Fau. Br. Ind., II., 23.*

Vern.—*Cha-ra-wud-dah*; ANDAMANS; *Ikan-sumpit*, MALAYS.

Habitat.—Seas of India. Used as food.

592   **Trachynotus ovatus**; *Day, Fish. Ind., 234 ; Fau. Br. Ind., II., 179.*

Vern.—*Kútili*, TAM.; *Múkali-parah*, VIZAG.

Habitat.—Seas of India. This fish salts well, but when fresh is dry and insipid.

593   **Trichiurus haumela**; *Day, Fish. Ind., 201 ; Fau. Br. Ind., II., 134.*

Vern.—*Puttiah*, URIYA; *Sona-ka-wahlah*, TAM.; *Sawala*, TEL.; *Wale*, MADRAS; *Pa-pa-dah*, ANDAMANS; *Ikan-puchuk*, MALAYS.

Habitat.—Seas and estuaries of India. All three species of this genus are employed for food, but are held in various estimation in different places. In Baluchistan and where salt is cheap, no one will touch them, but along the coasts of India, where the salt-tax has ruined the fish-curer's trade, they are more esteemed, mostly because, being thin and ribbon-shaped, they can be dried without salting. Russell observed that in his time they were esteemed by European soldiers, and Jerdon states that they afford very delicate eating when fresh, though never brought to the table of Europeans (*Day*).

594   **T. muticus**; *Day, Fish. Ind., 200 ; Fau. Br. Ind., II., 134.*

Vern.—?

Habitat.—Seas of India; very common in Orissa.

F. 594

of Economic Value.	(*J. Murray.*)	FISH.

Trichiurus savala ; *Day, Fish. Ind., 201 ; Fau. Br. Ind., II., 135.*   595
  Vern.—*Droga-puttiah*, URIYA ; *Sa-vale*, MADRAS.
  Habitat.—The seas and estuaries of India.

Trichogaster fasciatus ; *Day, Fish. Ind., 374 ; Fau. Br. Ind., II., 372.*   596
  Vern.—*Kolisha*, BENG. ; *Kussuah, coilia*, URIYA ; *Koh-li-hona*, ASSAM ;
  Kun-gi, PB. ; *Pich-ru*, SIND ; *Ponundi*, TEL. ; *Nga-pin-thick-kouk,*
  nga-phyin-thaleb, BURM.
  Habitat.—Fresh waters of the Panjáb, North-Western Provinces,
Sind, Cachar, Assam, the Coromandel coast as far south as the river
Kistna, and the estuaries of the Ganges and Burma. It is extensively
dried in various parts of the country, and in Burma is made into *nga-pí.*

Trygon sephen ; *Day, Fish. Ind., 740 ; Fau. Br. Ind., I., 50.*
  Vern.—*Adavalan tiriki*, TAM. ; *Volugiri tenki*, TEL. ; *Wolga-tenki,*   597
  VIZAG.
  Habitat.—Through the seas of India, growing to a large size. Wounds
inflicted by the spine of its tail are considered dangerous. All the species
are valuable on account of their skins, from which shagreen may be prepared, or which may be employed for sand-paper ; their fins which are exported to China ; and their livers, from which oil is extracted.

T. uarnak ; *Day, Fish. Ind., 737 ; Fau. Br. Ind., I., 53.*
  RAY, *Eng.*   598
  Vern.—*Sankush*, URIYA ; *Sona-kah-tiriki*, TAM. ; *Puli-tenke*, TEL.
  Habitat.—Seas and estuaries of India, attaining a large size,—5 feet
or more across the disk. As in the former species, the caudal spines are
capable of inflicting severe wounds. In addition to possessing the properties detailed under **T. sephen**, this species is of value as food, and is
dried in several places along the coasts.

Umbrina russellii ; *Day, Fish. Ind., 183 ; Fau. Br. Ind., II., 110.*   599
  Vern.—*Qualar katcheli*, MAD. ; *Ikan-gulama*, MALAYS.
  Habitat.—Seas of India. The best food fish of the genus, though
like other SCIÆNIDÆ, its flesh is rather tasteless when young, and coarse
when large. The sounds or air vessels are a valuable source of isinglass.

Upeneoides vittatus ; *Day, Fish. Ind., 120 ; Fau. Br. Ind., II., 25.*   600
  MULLET, *Eng.*
  Vern.—*Chirul*, MALAY. ; *Bandi-gúlivinda*, VIZAG. ; *Chah-ti-ing-ud-dah,*
  AND.
  Habitat.—Red Sea and the seas of India. Like many other species
of the family MULLIDÆ, the flesh of this fish is most excellent eating.

Wallago attu ; *Day, Fish. Ind., 479 ; Fau. Br. Ind., I., 126.*   601
  Vern.—*Boyari*, BENG. ; *Boalli, ballia, moinsia-ballia*, URIYA ; *Mul-la,*
  pi-i-ki, jer-i-ki, SIND ; *Purram, worshúrah*, MAR. ; *Wahlah, tele*, TAM. ;
  Wallagú, valaga, TEL. ; *Wahlah*, MALAY. ; *Nga-batt*, BURM.
  Habitat.—Throughout the fresh waters of India, Ceylon, and Burma.
It attains at least 6 feet in length, and though a voracious and not very
cleanly feeder, is good eating.

Zygæna malleus ; *Day, Fish. Ind., 719 ; Fau. Br. Ind., I., 22.*   602
  HAMMER-HEADED SHARK, *Eng.*
  Vern.—*Koma-sorra*, TEL. ; *Nga-man thanwoot*, BURM.
  Habitat.—Tropical and temperate seas of India. The adult fish
is a large and extremely dangerous one, but the young are captured
along the shores in large numbers on account of their flesh, which is supposed by the poorer classes to be very nourishing, of the oil which is obtained from their livers, of their gelatinous fins, and of the skin, which is
used for the manufacture of shagreen.

F. 602

FLACOURTIA, *Comm. Gen., Pl. I., 128.*

A genus of trees or shrubs, often spinous, containing about twelve species, natives of the Old World, of which some are cultivated in tropical countries. There are eight Indian species, of which five are of economic interest.

603

Flacourtia Cataphracta, *Roxb. ; Fl. Br. Ind., I., 193.*

MANY-SPINED FLACOURTIA, *Eng. ;* PRUNNIER D'INDE, *Fr.*

Vern.—*Talispatri, pániámalak, pani-aonvola,* HIND.; *Pæniálá,* BENG.; *Jan-gama, támbath, jaggam,* BOMB.; *Tambat,* MAR.; *Tálispálra,* GUZ.; *Tálisapatri,* TAM.; *Tálisapatri,* TEL.; *Naydwéd,* BURM.; *Práchinamalaka, talisha,* SANS.; *Zarnab,* ARAB.; *Talis-patar,* PERS.

References.—*Roxb., Fl. Ind., Ed. C.B.C.,739; Kurz, For. Fl. Burm., 74; Gamble, Man. Timb., 17; Pharm. Ind., 27; Ainslie, Mat. Ind., II., 407; O'Shaughnessy, Beng. Dispens., 9; Dymock, Mat. Med. W. Ind., 2nd Ed., 74; Pharmacographia Indica, I., 152; Irvine, Mat. Med., Patna, 87; Lisboa, U. Pl. Bom., 7, 146, 277; Birdwood, Bom. Pr., 8; Balfour, Cyclop., I., 1126; Journ. As. Soc., 1867, 80, II., 2; Home Dept. Cor., 239; Journ. Agri-Hort. Soc., XII., 345.*

Habitat.—A small tree of Assam, Bengal, Burma, Bombay, and the Western Gháts. Commonly cultivated in India.

OIL.
Seeds.
604

Oil.—The SEEDS yield an oil, of which little is known, but further information regarding it might lead to the opening up of a trade in an article which even the poorest cultivator might supply from the wild plant.

MEDICINE.
Leaves.
605
Shoots.
606
Bark.
607
Fruit.
608

Medicine.—The LEAVES and YOUNG SHOOTS taste like rhubarb, and are supposed to possess astringent and stomachic properties. They are prescribed in diarrhœa, weakness, and consumption. An infusion of the BARK is also given for hoarseness. The FRUIT is said by **Dymock** to be recommended as useful in bilious conditions.

Compare with **Abies Webbiana.**

SPECIAL OPINIONS.—§ " The leaves are said to have diaphoretic properties " (*Deputy Sanitary Commr. Joseph Parker, M.D., Poona*). "Used as a powder in chronic bronchitis" (*Surgeon-Major J. J. L. Ratton, M.D., Salem*). " Sold in Mysore bazars and used in combination with other drugs for cough, &c." (*Surgeon-Major John North, Bangalore*). " Under the name *Talispatri* are sold in the bazar the leaves of a pine (**Abies Webbiana**)" (*Asst. Surgeon Sakharam Arjun Ravut, L.M., Gorgaum, Bombay*) *Talispatri* is probably this plant and not **Abies.**—*Ed.*

FOOD.
Fruit.
609

Food.—Taylor in his "Topography of Dacca" writes: "The FRUIT of this tree, which is of a purple colour, and of the size and appearance of a plum, is sold in the city during the rains." **Dr. Watson** reports that the fruit is eaten in Allahabad. It is also generally used as an article of food in Assam.

TIMBER.
610

Structure of the Wood.—Heavy, brown, close-grained, rather hard and brittle, and takes a fine polish (*Kurz, For. Fl. Burma*).

611

F. inermis, *Roxb. ; Fl. Br. Ind., I., 193.*

Vern. —*Tomi-tomi,* MAL. (S.P.); *Tambat, jaggam,* BOMB ; *Ubbolu,* KAN.

References.—*Roxb., Fl. Ind., Ed. C.B.C., 739; Kurz, For. Fl. Burm., 74; Gamble, Man. Timb., 17; Lisboa, U. Pl. Bomb., 7, 146.*

Habitat.—A middling-sized tree, probably introduced from the Moluccas. At present found in Sylhet, South India, and Martaban. It blossoms during the dry season, and ripens its fruit towards the close of the rains.

FOOD.
Fruit.
612

Food.—The FRUIT, says **Roxburgh,** is too sour to be eaten raw, but makes very good tarts. In the Moluccas, however, it is eaten.

613

F. montana, *Grah., Fl. Br. Ind., I., 192.*

Vern.—*Attak-ke-jar, attak,* BOMB.; *Champer,* MAR.; *Hannu sampige,* KAN.

FOOD.
Fruit.
614

Habitat.—A very thorny tree found in Kanara and the Concan.

Food.—" The FRUIT—used as a fruit " (*Birdwood, Bom. Products*).

F. 614

Flacourtia Ramontchi, *L'Herit., Fl. Br. Ind., I., 193.*

615

Syn.—F. SAPIDA, *Roxb.*

Vern.—*Bilangra, bhanber, kanjú, handi, kattár, katti, kundayi, bunj, bowchi,* HIND.; *Bincha, katái, tambat,* BENG.; *Katail,* PALAMOW; *Serali, merlec, sarlarkha,* KOL.; *Merlee,* SANTAL; *Bonicha, baili, baincho,* URIYA; *Arma-suri, katien,* GOND; *kúkai, kakoa, kangú, handei, kukoa,* PB.; *Bhutankas, bávaché,* SIND; *Kánk, kánki, biláti,* C. P.; *Swadu, kantaka, tambat, kaikun, pahar, bhekal, kakad,* BOMB.; *Kundayee, bunj, bowchee,* DEC.; *Pahar, bhekal, kakei, kaker, aturni,* MAR.; *Kaikun,* MHAIRWARA; *Gurgoti,* KURKU.; *Kanregu, pedda-kanru, kaka, nakka-naregu,* TEL.; *Na-yuwai,* BURM.; *Ugurassa,* SING.

References.—*Roxb., Fl. Ind., Ed., C.B.C., 739 ; Brandis, For. Fl., 18 ; Kurz, For. Fl. Burm., 75 ; Gamble, Man. Timb., 17 ; Stewart, Pb. Pl., 18; Rev. A. Campbell, Rep. on Ec. Prod., Chutia Nagpur, No. 8441 ; Lisboa, U. Pl. Bomb., 6, 146, 277 ; Birdwood, Bomb. Pr., 7 ; For. Adm. Report, Chutia Nagpur, 1885, 28 ; Raj. Gas., 27.*

Habitat.—A small thorny deciduous tree met with in dry hills throughout India and the Prome District of Burma.

Medicine.— Native inoculators in the Panjáb use the THORNS for breaking the pustule of small-pox on the 9th or the 10th day. After child-birth among natives in the Deccan the SEEDS are ground to a powder with turmeric and rubbed all over the body to prevent rheumatic pains from exposure to damp winds. The GUM is given along with other ingredients for cholera. The BARK is applied to the body along with that of **Albizzia** at intervals of a day or so during intermittent fever in Chutia Nagpur.

SPECIAL OPINIONS.—§" According to Sanskrit writers the FRUITS are sweet, appetising, and digestive. They are given in jaundice and enlarged spleen " (*U. C. Dutt, Civil Medical Officer, Serampore*).

Food and Fodder.—The FRUIT and the LEAVES are eaten. The former is of the size of the plum, has a sharp but sweetish taste, and is used either raw or cooked. The leaves are employed as cattle fodder.

Structure of the Wood.—Red, hard, close and even-grained, splits, but does not warp, and is durable. Weight about 53℔. Is used for turning and agricultural implements.

MEDICINE.
Thorns.
616
Seeds.
617
Gum.
618
Bark.
619
Fruits.
620
FOOD.
Fruit.
621
Leaves.
622
TIMBER.
623

F. sepiaria, *Roxb., Fl. Br. Ind., I., 194.*

624

Vern.—*Kondai,* HIND.; *Sherawane, sargal, dajkar, jidkar, khatái, kingro'* PB.; *Bainch,* C.P.; *Atrúna, támbat,* BOMB.; *Kanru, kána régu,* TEL.

References.—*Roxb., Fl. Ind., Ed. C.B.C., 739 ; Kurz, For. Fl. Burm., I., 75 ; Gamble, Man. Timb., 17 ; Stewart, Pb. Pl., 18 ; Lisboa, U. Pl. Bomb., 146, 277 ; Kew Off. Guide to Bot. Gardens and Arboretum, 68.*

Habitat.—A small, stiff, spiny shrub, found in dry jungles throughout Bengal, the Western Peninsula, and Ceylon. It also occurs about Delhi, in the Salt Range, and on the skirts of the Sulimans. Is extensively employed for making hedges.

Medicine.—An infusion of the LEAVES and ROOTS is supposed to be an antidote to snake-bite. The BARK triturated in sesamum oil is used as a liniment in rheumatism (*Wight ; Ainslie ; Rheede*).

Food.—The FRUIT is said to be eaten by the natives of the Panjáb tracts where it is found, but it is small, hard, and insipid ; it is, however, sometimes described as " pleasant, refreshing, and sub-acid." The LEAVES are thrashed out for cattle fodder.

MEDICINE.
Leaves.
625
Roots.
626
Bark.
627
FOOD.
Fruit.
628
Leaves.
629
630

FLAME TREES.

Different trees, having brilliant flowers, which in most cases appear before the leaves when seen at a distance, they have the appearance of

F. 630

being on fire—hence the popular name Flame Trees. The principal trees of this nature are :—

Amherstia nobilis.
Bombax malabaricum.—Silk Cotton Tree.
Butea frondosa and superba.—Tésú Flowers.
Cæsalpinia pulcherrima.—Barbadoes Pride or Gold Mohur Tree (a corruption of the Hind. name *Gulmor* or Peacock Flower).
Cochlospermum Gossypium.—White Silk Cotton Tree.
Lagerstrœmia Flos-Reginæ.
Poinciana regia.—The Mascarene.
Pterospermum acerifolium.
Rhododendron arboreum, &c., &c.

Flax, Common, *see* **Linum usitatissimum,** *Linn. ;* LINEÆ.

631   **Flax (New Zealand).** The fibre of **Phormium tenax.**

632   **Flea-bane.**
A powder made of the dried flowers or seeds of several species of plants for the destruction of, or rather driving away of, fleas.

In Persia the flowers of three species of **Pyrethrum** are employed. In India the flea-bane, commonly used, is the *Purple Flea-bane* or seed of **Veronia anthelmintica** (*Willd*). See **Pyrethrum** and **Veronia.**

Fleece of Sheep, *see* **Skins.**

(*G. Watt*).

FLEMINGIA, *Roxb. ; Gen. Pl., I, 544.*

633   **Flemingia congesta,** *Roxb., Fl. Br. Ind., II., 228 ; Wight, Ic., t. 390 ;*
[LEGUMINOSÆ.

Vern.—*Bara-salpan, bhalia, supta, cusunt,* HIND. ; *Bara-salpan, bhalia,* BENG. ; *Buru ekasira nari, bir but,* SANTAL ; *Batwasi,* NEPAL ; *Mipit-múk,* LEPCHA ; *Dangshukop,* MICHI ; *Dowdowlá,* BOMB. and MAR. ; *Tha kya nai,* BURM.

References.—*Roxb., Fl. Ind., Ed. C.B.C., 572 ; Gamble, List of Trees, Shrubs, &c., of Darjeeling, 28 ; Dals. & Gibs., Bomb. Fl., 75 ; Rev. A. Campbell's Report on Econ. Prod., Chutia Nagpur, No. 8465 ; Atkinson, Econ. Prod. N.-W. P., Pt V., 94 ; Kew Reports, 1881, 50 ; Kew Off. Guide to the Mus. of Ec. Bot., 45 ; Report, Bot. Gardens, Nilgiri, 1883-84, 10.*

Habitat.—An erect, woody shrub, common in the thickets and forests of the warmer parts of India.

The *Flora of British India* reduces to this species the following forms described by Roxburgh as distinct (see *Ed. C. B. C., pp. 571-72*) :—

F. procumbens, F. prostrata, F. nana, F. congesta, and F. semialata, forming four varieties :—

634   *Var. 1*—semialata (*sp. Roxb. ;* syn. F. stricta, *Wall. ;* F. prostrata, *Roxb.*)—Central Himálaya, ascending to 5,000 feet in altitude.

635   *Var. 2*—latifolia (*sp. Benth.*)—Khasia Hills, altitude 2,000 to 3,000 feet.
636   *Var. 3*—Wightiana (*sp. Grah.*)—Nilghiris, Bhutan, Ava.
637   *Var. 4.*—nana (syn. F. procumbens *Roxb. ;* F. capitata, *Ham.*)—Central and Eastern Himálaya and the Concan.

HISTORY.
638   **Modern Commercial History of *Waras* Dye.**—In a correspondence forwarded by the Secretary of State for India to the Revenue and Agricultural Department, Sir J. D. Hooker communicated certain facts regarding the *waras* drug and dye of Africa which led to the suggestion that that substance was obtained from a **Flemingia** and probably one of the forms of the common Indian species F. congesta. Roxburgh nearly a century before had drawn attention to the garnet-coloured hairs on the pods of

F. 638

HISTORY.

that plant, but was apparently ignorant of the fact that these yielded a valuable dye. In the Kew Report for 1881 further information was published regarding *waras*, and it was there suggested that it was in reality obtained from the African species **F. rhodocarpa**. The Director of Kew, however, suggested to Mr. **M. A. Lawson**, Botanist to the Madras Government, that he should ascertain if the pods of the Indian species yielded the dye. This resulted in Mr. **Lawson** procuring a sample of the powder which was sent to Kew, and ultimately tested by Mr. **Wardle**, of Leck. About the time these experiments were being performed, Major **F. M. Hunter**, of Aden, forwarded to Kew a report which threw still further light on the subject. The specimens furnished by Major **Hunter** led to Mr. **W. T. Thiselton Dyer's** writing: "There can be now no sort of doubt that the 'waras' plant is really that described by Mr. **J. G. Baker**, F.R.S., in the 'Flora of Tropical Africa,' as **Flemingia rhodocarpa**. But my colleague, Professor **Oliver**, F.R.S., whose kindness is only equalled by his sagacity, has made the curious discovery that a **Flemingia** apparently confined to South India, **F. Grahamiana**, *W. & A.*, is not specifically distinguishable from **F. rhodocarpa**; the pods are in fact clothed with the same peculiar epidermal glands so characteristic of that species. The '*waras*' plant is therefore really to be found in India after all. In creating a new species for the '*waras*' plant, Mr. **J. G. Baker** pardonably neglected the comparison of the material he was working upon with specimens of the species occurring in so remote and botanically widely severed an area as the southern part of the Indian peninsula" (*Jour. Pharm. Soc., May 31st, 1884*). Shortly before the date of appearance of the above passage, Mr. **Lawson**, in his Annual Report for 1883-84, while dealing with his efforts to procure a sample of *waras* from an Indian **Flemingia** wrote: "From specimens which I sent to Kew, *waras* turns out to be the produce of **Flemingia Grahamiana** and **F. congesta**. With respect to the distinctive characters of these two species, I pointed out that, after studying the plants in their living condition, I did not think them sufficiently constant to allow of the two species being kept separate, and in this opinion both Mr. **Thiselton Dyer** and Professor **Oliver** now concur." If this position be confirmed by future research, then apparently both **F. rhodocarpa** and **F. Grahamiana** would have to be referred, along with **F. congesta**, to forms of one species. It is on the probability of such a rearrangement and as a matter of economy of space that the writer has thrown the present account of the African *waras* into one place and under one species instead of attempting to discuss it under several.

Dye.—Mr. **Lawson** wrote of his experiments with the Indian powder procured by him from **F. Grahamiana** and **F. congesta**: "The *waras* yields a beautiful dye when applied to animal substances such as silk or wool, but it is inferior as a dye when used for the purpose of colouring vegetable products such as cotton or linen. Mr. **Thiselton Dyer** has kindly obtained for me a London expert's opinion upon the value of *waras*, and I regret to say that it is not such as is likely to lead one to believe that it will ever become an object of commercial interest. I may mention that when I was in Madras last winter, I saw at the Agri.-Horticultural Gardens flower-show a specimen of *waras* in a native dyer's collection which was being exhibited, and from which it would appear that *waras* is not unknown as a dye in India." It would be interesting to know if the sample alluded to by Mr. **Lawson** was critically examined so as to remove any doubt as to its being in reality *waras* and not *kamála* (see **Mallotus philippinensis**). One other notice occurs, however, regarding an Indian knowledge of the dye property of the **Flemingias**. The Rev. **A. Campbell**, in his Report on the Economic Products of Chutia Nagpur,

DYE.
639

| FLEMINGIA congesta. | The Waras Dye. |

DYE.

writes of **F. congesta** : " The pods are said to yield a dye." It would thus appear that the Santals are familiar with the dye, and as Mr. Campbell does not call this *waras*, there is no room for doubting but that he alludes to a fact, the interest of which, beyond the limits of his own province, Mr. Campbell was in all probability not aware of.

It may serve a useful purpose to reproduce here **Major Hunter's** description of the collection and purification of the dye as pursued in Africa, at Harrar :—

" In the neighbourhood of the city ' *wars* ' is not now raised from seed sown artificially, and it is left to nature to propagate the shrub in the surrounding terraced gardens. The plant springs up, among jowari, coffee, &c., in bushes scattered about at intervals of several yards, more or less. When sown, as among the Gallas, it is planted before the rains in March. If the soil be fairly good a bush bears in about a year. After the berries [pods] have been plucked, the shrub is cut down to within six inches of the ground. It springs up again after rain and bears a second time in about six months, and this process is repeated every second year until the tree dies. Rain destroys the berry [pod] for commercial purposes ; it is therefore only gathered in the dry season ending about the middle of March. The bush grows to a maximum height of six feet, and it branches close to the ground. The growth is open and the foliage sparse. Each owner has a few acres of land.

" In the middle of February 1884, the following processes were observed :—

" The leaves [? fruiting shoots] of some plants were plucked and allowed to dry in the sun for three or four days. (The picking is not done carefully and a considerable quantity of the surrounding twigs, &c., is mixed with the berries [pods].) The collected mass was placed on a skin heaped up to about six or eight inches high and was tapped gently with a short stick about half an inch thick. After some time the pods were denuded of their outer covering of red powder which fell through the mass on to the skin. The upper portion of the heap was then cleared away and the residual reddish green powder was placed in a flat woven grass dish with a sloping rim of about an inch high. This receptacle was agitated gently and occasionally tapped with the fingers, the result being the subsidence of the red powder and the rising to the surface of the chaffy refuse, which latter was carefully worked aside to the edge of the dish and then removed by hand. This winnowing was continued until little remained but red powder. (No great pains are even taken to eliminate *all* foreign matter.) A *rotl* was sold in 1884 for about 13 piastres=1 rupee 10 as. nearly.

" ' *War* ' is sent to Arabia, chiefly to Yemen and Hadhramaut, where it is used as a dye, a cosmetic, and a specific against cold. In order to use it, a small portion of the powder is placed in one palm and moistened with water ; the hands are then rubbed smartly together, producing a lather of a bright gamboge colour, which is applied as required " (*W. T. Thiselton Dyer, Pharm. Jour., May 31, 1884*).

Mr. Wardle regards *waras* as a distinctly inferior dye to *kamála* (**Mallotus philippinensis**). The latter has been exported from India to Europe for many years past as an adulterant or substitute for the former. Mr. Wardle writes of *waras :* " This substance contains only a small amount of colouring matter compared with the vegetable yellow dyes of commerce, and no colour can be obtained from it which will bear comparison in depth and richness with those produced from *kamála* or *kapila*, for which, as stated in the Kew Report for 1880, it is used as a substitute, and which is certainly a very much more valuable dye-stuff.

F. 639

The Waras Dye.	(*G. Watt.*)	FLEMINGIA vestita.

DYE.

"As far as my observations have gone, *waras* is inferior to *kamála* in permanence, as regards the action of light." "The colour produced with *waras* is easily turned brown by alkaline solutions, whilst *kamála* is only slightly reddened. Both dyes, however, resist the action of acids very well." "I corroborate the statement made by **Professor Lawson** that *waras* is suitable for a dye for silk rather than for wool, and that it is quite useless as a dye for cotton. I have tried it on cotton with most of my mordants, as well as without mordants, and the result is a pale-yellow shade."

In Bombay the word *waras* (as a pure coincidence probably) is given to a Bignoniaceous plant—**Heterophragma Roxburghii**,—but a far more likely error would be to mistake *kamála* for *waras*. That substance is alluded to by some of the early Arabic writers, its Sanskrit name being corrupted into *kinbíl*. The author of **Kámús** who wrote A. H. 768 notices both *kinbíl* and *waras*, but treats them as distinct substances. The latter, he says, is only found in Arabia, and it does not possess the anthelmintic properties of the former. So again the *Makhsan* distinguishes the two plants, the one being the pulp, as it is called, from the fruits of a tree, while the other is obtained from the pod of a pea-like *másh* (**Phaseolus**) (*Dymock*). It would thus appear clear that from whatever cause has proceeded the confusion which till recently existed in modern literature, the early writers fully understood the properties and sources of the two plants—*kamála* and *waras*.

(For further information consult the account of **Mallotus philippinensis**.)

Medicine.—The POWDER from the PODS constitutes the African drug *waras* or *wars*. This does not appear to be employed in India, though much of the obscurity into which the anthelmintic drug *kamála* has been thrown, is doubtless due to *waras* having been substituted. The ROOTS of **Flemingia congesta**, the Rev. Mr. **Campbell** informs us, are used by the Santals as an external applicant to ulcers and swellings mainly of the neck.

MEDICINE.
Powder.
640

Roots.
641

Food.—According to **Atkinson** (*Econ. Prod. N.-W.P., V., 94*) the PODS are eaten. Mr. **Campbell** says that the Santals also eat them.

FOOD.
Pods.
642
643

Flemingia Grahamiana, *W. & A.; Fl. Br. Ind., II., 228.*

This Nilghiri plant, according to Mr. **Baker's** account of it in the *Flora of British India*, differs from **F. congesta** mainly in the leaflets being longer, more obtuse, and borne on shorter petioles, and in the rigid subpersistent bracts. Mr. **Lawson** in the passage quoted above regards this species, however, as doubtfully distinct. It is probable that whether it be regarded as a species or only as a variety, this plant yields the *waras* powder more freely than other known Indian forms.

Several species of **Flemingia** are occasionally mentioned by authors on Indian Economic Botany, but none of them (except **F. vestita**) seem of sufficient merit to deserve separate notice in this work. It is somewhat remarkable that practically none of these Leguminous plants are recorded as being eaten by cattle, sheep, or goats.

F. Strobilifera, *R. & M.*

Is repeatedly mentioned for its medicinal properties. It is the *sim busak* of the Santals, the roots of which, the Rev. A. Campbell informs us, are sometimes given in epilepsy; it is the *Bolu* of the Darjeeling hill tribes and the *Phá-tán-phyu* of Pegu. In the Central Provinces buffaloes are said to eat this species.

644
MEDICINE.
645
FODDER.
646

F. vestita, *Benth.; Fl. Br. Ind., II., 230.*

A small creeping plant with dark brick-red flowers which appear in July to August. This is said to be cultivated in "some parts of North-West India for the sake of its edible tuberous roots, which are nearly

647
FOOD.
648

elliptical and about an inch long " (*Lindley and Moore's Treasury*). The writer has never seen it cultivated, nor can he discover any Indian author who alludes to this fact, but around Simla the plant is very plentiful, and along with **Vigna vexillata**—the *gúlái* or *ban* (wild) *mung*, of the N.-W. Himálaya—the roots are regularly collected and eaten, especially by herd boys attending on cattle. They have a sweet, agreeable, nutty flavour, and if systematically cultivated might come to afford a useful new vegetable somewhat of the character of the Jerusalem Artichoke. The Himálayan form has few flowers, much less crowded than in the variety described as **nilgheriensis**; *Wight, Ic., t. 987.*

(*J. Murray.*)

649

FLESH, Animal.

In India the flesh of animals is not only used as a food, but from very early times has been much employed medicinally by native practitioners, both internally as *ghritas*, and externally as *taila páka*.

FOOD.
650

Food.—Sanskrit writers describe the different properties of the flesh of various animals in great detail. By them the flesh of the goat, domestic fowl, peacock and partridge, is said to be easily digested and suited for the sick and convalescent; the meat of the deer, sambar, hare, quail, and partridge is recommended for habitual use; while beef and pork are viewed as hard to digest and unsuited for daily use.

MEDICINE.
651

Medicine.—Medicinally the goose, fowl, jackal, goat, snail, and mungoose are principally employed, their flesh being prescribed for many forms of disease, but chiefly those of the nervous system.

The *ghrita* and the *taila páka* into which they are compounded contain in addition a great variety of vegetable drugs (*U. C. Dutt, Hind. Mat. Med., p. 286*).

652

FLINT.

Vern.—*Chakmak*, HIND.; *Chakimuki*, TAM., TEL.

Flint is a massive compact form of almost pure silica and is generally of a dark-brownish colour. It breaks with a conchoidal surface, and forms sharp cutting edges. True flints are of rare occurrence in India, but in the manufacture of implements in prehistoric times, horses' bones, agates, &c., were substituted, and some of these form efficient gun-flints, or flints for flint and steel.

Flints are said to be found at Coorchycolum in the Trichinopoly district of Madras (*Manual of the Trichinopoly District, p. 67*); in the Dharwar district of Bombay (*Madras Jour. of Lit. and Sci., Vol. XI., p. 46*); in the Bannu district of the Panjáb (*Baden Powell's Pb. Prod., p. 46 ;* and in Afghánistan immediately across the Kurram (*Records, G. S. I., XII., p. 111*).

Owing to the extensive use of the chalcidonic quartzes in place of the true flint it is difficult to decide whether the mineral reported to be found in the above situations is real flint or not.

USES.
653

Uses of.—Flint when calcined and ground, is used in the manufacture of pottery, and in the natural condition for gun-flints.

Flour, see **Triticum sativum,** *Lamk.,* and **Oryza sativa,** *Linn.;* GRAMINEÆ.

Flower Fence, see **Cæsalpinia pulcherrima,** *Swarts.,* Vol. II., 10.

Flower oil, see **Sesamum indicum,** *D.C.;* PEDALINEÆ.

654

FLUOR-SPAR.

Derbyshire Spar.—This mineral consists of calcic fluoride. Found in India only in very small quantities, probably owing to the small number of metal mines at present worked.

F. 654

It has been recorded as found at Chicholi in the Raipur district of the Central Provinces (*Rec., G. S. I., Vol. I., 37*), in the Rewah State (*Mem., G. S. I., VII., 122*), and at Spiti in the Panjáb (*Mem., G. S. I., V., 166*).

Uses of.—Are few, principally employed in the preparation of hydrofluoric acid, for the etching of glass; and for making a flux sometimes used in the reduction of the ores of copper and other metals.

Fluoride of calcium, see Fluor-spar.

FLYING-FOX.

655

Flying fox is the name given by Europeans in India to several species of BATS constituting the genera **Pteropus** and **Cynopteris.** Those commonly found in this country are **Pteropus medius,** *Tem.,* the large flying-fox, and **Cynopteris marginatus,** the small flying-fox; but **C. affinis** and **P. minimus** are also natives of India.

The habits of the whole family are very similar, and as they are indifferently termed Flying-fox, and the vernacular names for all seem the same, they may be described collectively.

 Vern.—*Gadal, chamgidar,* HIND.; *Cham-guddri, chidgu,* BENG.; *Kanka-pati,* KAN.; *Gabbday, jiburai,* TEL.
 Reference.—*Jerdon's Mam. of Ind., 18.*

Habitat.—Common bats found throughout India, Burma, and Ceylon. They roost in large colonies in trees during the day, often numbering two or three hundred on a single tree—generally the *pipal* (**Ficus religiosa**); at night they roam over the district, doing incalculable harm to fruit trees.

Food.—The natives of Bengal catch this animal in the following manner :—A string is tied to the very topmost branch of a tree, likely to be visited during the night, while a man sits below holding the string. A bat coming in contact with the string closes its wings around it, in order to save itself from falling. The man then jerks the string sharply and the bat glides down into his hands.

FOOD.
656

The trees usually selected for this operation are the favourite avenue tree **Polyalthia longifolia** (the nuts of which form a favourite food of the flying-fox) and **Terminalia Catappa.**

The flesh of these bats is eaten by the lower-class Bengalis, also by the natives of Madras.

Medicine.—The FLESH is recommended by native practitioners in cases of diabetes, and when muscular energy is deficient. The FAT boiled down is a favourite remedy in rheumatism of the joints.

MEDICINE.
Flesh

657
Fat.
658

FŒNICULUM, *Adans. ; Gen. Pl., I., 902.*

A genus of glabrous herbs, belonging to the Natural Order Umbelliferæ, having 3 or 4 species, which are widely distributed, from the Canaries to Western Asia. **F. vulgare** is extensively cultivated.

Fœniculum vulgare, *Gærtn. ; Fl. Br. Ind., II., 695 ; Bentley & Trim., Med. Pl., No. 123 ; Wight, Ic., t. 570 ;* UMBELLIFERÆ
 THE FENNEL.

659

 Syn.—FŒNICULUM PANMORIUM, DC.; *Wight, Ic.;* F. OFFICINALE, *Allion ;* ANETHUM FŒNICULUM, *Linn.;* A. PANMORIUM, *Roxb.;* F. CAPILLACEUM, *Gilib. ; Bentley & Trim.;* F. DULCE, *C. Bauh.;* OZODIA FŒNICULACEA, *W. & A., Prodr.*
 Vern.—*Saunf, bari saunf, sonp, sont,* HIND.; *Mauri, pan-muhori,* BENG.; *Bari-shopha, panmohuri,* BOMB. ; *Variari, wariaree, variyali,* GUZ.; *Badishep,* MAR.; *Badisopu,* KAN.; *Aspa, badyan,* TURKI ; *Sohikire,* TAM.; *Pedda-jila-kurra,* TEL.; *Madhuriká,* SANS.
 References.—*Roxb., Fl. Ind., Ed. C.B.C., 272 ; Stewart, Pb. Pl., 107 ; Ainslie, Mat. Ind., I., 129; O'Shaughnessy, Beng. Dispens., 36 ; U. C.*

FŒNICULUM. vulgare.	Fennel.

Dutt, Mat. Med. Hind., 173; Dymock, Mat. Med. W. Ind., 2nd Ed., 372; S. Arjun, Bomb. Drugs, 64; Murray, Pl. and Drugs, Sind, 197; Irvine, Mat. Med. Patna, 88; Atkinson, Him. Dist., 705, 737; Lisboa, U. Pl. Bomb., 161; Birdwood, Bomb. Pr., 41, 665; Home Dept. Cor., 231.

Habitat.—This perennial attains a height of 5 to 6 feet, and is commonly cultivated throughout India at all altitudes up to 6,000 feet; but is sometimes also found wild. Several cultivated races seem to grow in India, but these do not appear to have been botanically recognised. The seed is smaller and straighter than that of the European fennel, but is otherwise similar.

CULTIVA-TION.
660

Cultivation.—This plant seems in India generally to be grown only in small patches in homestead lands, as a cold-weather crop. The method of cultivation is that of an ordinary market-garden crop. In Bombay, however, it appears to be cultivated to a larger extent. The following account has been received from the Director of Land Records and Agriculture, dated September, 1889:—" In 1887-88 Fennel occupied 1,454 acres, of which 834 acres were in Khándesh. It is grown in some districts of Gujarat and the Deccan. In the former district it is grown in *gorat,* light soil moderately manured—10 cart-loads to the acre. The land is ploughed, harrowed, and rolled three times between June and October. About 9℔ of seed per acre are scattered by hand into beds, which are irrigated once a fortnight until January. The crop is cut in rather a green state, and allowed to lie in the ground for five days. The acre yield varies from 280 to 1,120℔—720℔ being a good average crop. In the gardens in the Deccan it is sown at any time. It is also sown on the edges of dry crops in July and August. The probable total yearly outturn is estimated at 13,000 maunds, and the price realised varies from R6 to R8 per Indian maund."

CHEMISTRY.
661

Chemistry.—Fennel fruit yields about 3 per cent. of volatile oil, anethol or anise camphor, and a variable proportion of a liquid isomeric with turpentine. Anethol (the constitutent, important medicinally and as a flavouring agent) may be obtained either as a liquid or crystal, as it takes the latter form at a moderately low temperature (*Pharmacographia, 275*).

OIL.
662

Oil.—The fruit contains a volatile OIL, pale yellow, with a pleasant aromatic odour. It is used in Europe in the manufacture of cordials, and enters into the composition of fennel water which is employed medicinally, but chiefly as a vehicle for other drugs. This water is distilled largely in India and sold under the name of *Arak bádián.*

MEDICINE.
Fruits.
663
Root.
664
Leaves.
665

Medicine.—The FRUITS are used medicinally as a stimulant, aromatic, and carminative, and are prescribed in colic, diarrhœa, and dysentery. The ROOT is regarded as purgative, and the LEAVES as diuretic.

Besides these properties it is believed, in some parts of the country, that the fruits have a specific value. Thus in Madras they are said to be used as a medicine in venereal diseases.

SPECIAL OPINIONS.—§" Stimulant, aromatic, and carminative in colic" (*Assistant Surgeon Nehal Sing, Saharunpore*). "The infusion of the seeds is used as a cooling drink in fever, &c." (*Civil Surgeon J. H. Thornton, B.A., M.B., Monghyr*). "The seeds fried and powdered are used in dysentery with sugar" (*Assistant Surgeon T. N. Ghose, Meerut*). "Cold infusion of seeds very useful in colic and indigestion of children, and an excellent vehicle for other medicines. Used also to relieve thirst in fever" (*Assistant Surgeon Shib Chunder Bhattacharji, Chanda, Central Provinces*).

FOOD.
Pot-herb.
666
Leaves.
667

Food.—The plant is frequently cultivated as a pot-herb in the plains. Its LEAVES are strongly aromatic and are used in fish sauces. **Roxburgh** wrote: "This plant is cultivated in various parts of Bengal during the cold season for the seed, which the natives eat with their betel, and also use in their curries." Seed time, the close of the rains, or about the end of October. Harvest time, March.

Trade.—The principal amount of fennel fruit sent to Bombay is from Jubbulpore, Kupperwanj, and Khándesh. The value of the fruit in Bombay is R3 to R4 per Surat maund of 37⅛℔. The export trade has been increasing during the past ten years. Thus, in 1881-82 the total exports were 2,201 cwt., in 1887-88 they were 4,353 cwt., valued at R31,260. Almost the whole quantity was exported from Bombay in the latter year, *viz.*, 4,337 cwt., Madras sending 15 cwt. and Sind 1 cwt. Great Britain received only 221 cwt. of this amount, France 975 cwt., Belgium and Austria each 200 cwt.; the rest went to Eastern ports.

The root is said by Irvine in his *Mat. Med. of Patna* to be worth R1-8 per ℔.

FOOD, Human.

In the account of any one product, it has been the system in this work to follow uniformly an established skeleton. Thus if it affords (1) a GUM, that forms the subject of the first important paragraph and is followed in their order by (2) its DYES, TANS, or MORDANTS; (3) its FIBRE; (4) its OIL; (5) its MEDICINE; (6) its FOOD or edible material; (7) its TIMBER; and, last of all, (8) the DOMESTIC AND SACRED uses to which the product is put.

It has already been explained (under DOMESTIC AND SACRED) that it is intended to give in an Appendix a detailed classification of the substances which in a museum might be grouped according to these eight headings. The reader is referred to FOOD AND FODDER in the Appendix, but it may here be explained that food for men and animals may be grouped into—

　　　I. Animal food materials.
　　　II. Mineral 　„　　　„
　　　III. Vegetable „　　　„

Each of these is capable of a separate classification. Thus under Vegetable food materials the chief sections might be given as (*a*) Cereals, (*b*) Pulses, (*c*) Vegetables and Tubers, (*d*) Fruits and Nuts, (*e*) Spices, (*f*) Starches and Sugars, and (*g*) Oils. The reader will find a partial elaboration of (*a*) to (*g*) in their respective alphabetical positions in this work.

(J. F. Duthie.)

FOOD AND FODDER FOR CATTLE.

The following enumeration in four sections may be given as the chief trees, shrubs, herbs, and grasses known to afford food or fodder for cattle in the various parts of India.

For geographical distribution, vernacular names, and other information, reference should be made to the several articles relating to these plants in their alphabetical positions. A review of the Indian Fodder question, together with lists of fodders suited for different animals, will be found in the Appendix.

I.—FODDER PLANTS OF THE PLAINS.

Acacia arabica, *Willd.* VERN. *Batul* or *kikar.* The tender shoots, leaves, and green pods are much liked by cattle, and the tree is greatly valued in regions affected by drought.

A. Catechu, *Willd.* Cattle eat the lower and small branches (*R. Thompson*).

A. ferruginea, *DC.* Cattle eat the lower and small branches (*R. Thompson*).

A. Intsia, *Willd..* var. **cæsia.** Cattle eat the leaves (*R. Thompson*).

A. Jaquemonti, *Benth.* A shrub thriving on rocky and sandy soils. The branches are cut and the leaves thrashed out and given as fodder.

| FOOD & | Food and Fodder. |

Acacia lenticularis, *Ham.* Cattle eat the leaves and small branches.

A. leucophlœa, *Willd.* Leaves and pods. The latter, however, are considered by some to be poisonous and should be used with caution.

A. modesta, *Wall.* The leaves and fallen blossoms are collected as cattle fodder.

A. Suma, *Kurz.* VERN. *Safed khair.* Leaves and young branches (*R. Thompson*).

Achyranthes aspera, *Linn.* According to Mr. T. N. Mukharji, the young plants are given to cattle in Bengal in times of scarcity.

Adhatoda Vasica, *Nees.* The Conservator of Forests, Northern Circle, Bombay, states that the leaves supply fodder for cattle. This plant is abundant in Northern India, but appears to be there used only as a medicine for cattle for the cure of colic.

Adina cordifolia, *Hook. f.* Leaves (*R. Thompson*).

Ægle Marmelos, *Correa.* Bael tree. Brandis mentions that the twigs and leaves are lopped for cattle fodder.

Ærua javanica, *Juss.* Plant (*R. Thompson*).

Ailanthus excelsa, *Roxb.* Leaves (*R. Thompson*).

Albizzia amara, *Boivin.* Leaves (*R Thompson*).

A. Lebbek, *Benth.* VERN. *Siris.* In Mysore the leaves of this tree are considered to be good fodder for cattle.

A. odoratissima, *Benth.* The branches are lopped for cattle fodder.

A. procera, *Benth.* Leaves (*R. Thompson*).

A. stipulata, *Boiv.* The branches are lopped for cattle fodder.

Allium Cepa, *Linn.* Boiled onions are given with other food to milch cows and buffaloes in the Nasik District.

Alysicarpus rugosus, *DC.* This and other species are eaten by cows and buffaloes in Bundelkhand.

Amarantus spinosus, *Linn.* This common wayside weed is often given to milch cows in Bengal. **Mr. T. N. Mukharji** says that chopped up and mixed with the boiled ends of rice-stems, "the preparation is considered highly lactiferous." Other species of **Amarantus,** many of which are cultivated as pot-herbs, might be substituted with advantage.

Amorphophallus campanulatus, *Bl.* According to Mr. T. N. Mukharji this plant when dead and dry is greedily eaten by cattle in Bengal, and householders occasionally collect it for their cows.

Anogeissus acuminata, *Wall.* }
A. latifolia, *Wall.* } Leaves (*R. Thompson*).

A. pendula, *Edgew.* Bhai Sadhu Singh, Forest Officer to the Jeypur State, says that buffaloes and cattle eat the dry leaves of this tree.

Anthocephalus Cadamba, *Benth. & Hk. f.* The leaves are sometimes used as cattle fodder.

Antidesma diandrum, *Roth.* "Cattle eat the leaves" (*Rev. A. Campbell, Chutia Nagpur*).

Arachis hypogæa, *Linn.* The ground-nut is cultivated in many parts of India, especially in the Bombay and Madras Presidencies. The stems and leaves, fresh or dry, are greedily eaten, and the oilcake is an excellent food for fattening cattle and increasing the quantity of their milk.

Argyreia speciosa, *Sweet.* Leaves (*Rev. A. Campbell*).

Artocarpus integrifolia, *Linn.* The leaves of the jack-fruit tree are considered fattening for cattle; and according to Mr. T. N. Mukharji the rind of the ripe fruit is "greedily eaten by cattle as the greatest of luxuries."

A. Lakoocha, *Roxb.* Extensively lopped for cattle fodder (*R. Thompson*).

Atriplex nummularia, *Lindl.* Baron Von Mueller in his *Select Plants,*

I.
**FODDER
PLANTS OF
THE PLAINS.**

p. 52, describes this as "one of the tallest, and most fattening and whole-some of the Australian pastoral salt-bushes. Sheep and cattle pastured on salt-bush country are said not only to remain free of fluke, but to re-cover from this and other allied ailments." Experiments are still being undertaken to test the suitability of this species for planting on the *reh*-infected tracts in Northern India, the successful establishment of which in such localities would prove an undoubted gain.

Atylosia mollis, *Benth.* Mentioned by the Rev. **A. Campbell** as yielding fodder for cattle in Chutia Nagpur.

Balanites Roxburghii, *Planch.* The young twigs and leaves are said to be browsed by cattle.

Barringtonia acutangula, *Gærtn.* Brandis says that the bark of this tree mixed with chaff and pulse is given as cattle fodder.

Basella alba, *Linn.* According to **Mr. T. N. Mukharji** the plant is given raw to cattle in Bengal.

Bassia latifolia, *Roxb.* The leaves, flowers, and fruit of the mahua tree are eaten by cattle. The flowers are said to be very fattening.

Bauhinia purpurea, *Linn.* The leaves are lopped for cattle fodder (*Brandis*).

B. racemosa, *Lamk.* The leaves of this tree are said to be eaten by buf-faloes in parts of Northern India.

B. retusa, *Roxb.*
B. Vahlii, *W. & A.* }The branches of these plants are often lopped for cattle fodder.
B. variegata, *Linn.*

Bischoffia javanica, *Bl.* Buffaloes eat the leaves (*R. Thompson*).

Bœrhaavia repanda, *Willd.,* and **B. diffusa,** are both occasionally eaten by cattle, and in Bengal the latter is supposed to increase the quantity of milk. Another species, **B. verticillata,** *Povi.,* is used in Rajputana as fodder.

Bombax malabaricum, *DC.* (Semal or Cotton-tree). The twigs and leaves are lopped for fodder in the Hoshiarpur district and elsewhere.

Borassus flabelliformis, *Linn.* (Palmyra Palm). "The shell enclosing the fruit, and the yellow pulpy mass around the stones, are eaten by cattle in Bengal. This food is considered fattening. The green calyx of the unripe fruit is also given to cattle" (*T. N. Mukharji*).

Boswellia serrata, *Roxb.* Buffaloes eat the leaves (*R. Thompson*).

Brassica campestris, *Linn.,* var. **glauca.** VERN.—*Sarson.* Largely grown in Northern India for the oil contained in the seed. The refuse, after extracting the oil, is given to cattle. In many parts of the Panjáb it is grown mainly as a fodder crop, and cattle and camels are allowed to graze on it early in the season. In the Montgomery district it is grown either for fodder or for its seed. When used as fodder it is treated much in the same way as turnips. It is cut in January in order that it may yield a second crop. The pods, after the removal of the seeds, are given to cattle in Bengal. The extensive cultivation of *sarson* for oil production in Upper India renders its use for fodder of great value. The early fruiting variety called *Toria* is often plucked as green food for cattle in the Karnal district, and probably elsewhere.

B. campestris, *Linn.,* var. **Rapa.** (Turnip.) Turnips constitute a most important crop in many of the Panjáb districts, where cattle are largely fed on the tops and roots. In the Jhang district the turnip and jowar crops afford strengthening food to the heavily-worked well-oxen during the wheat sowings and the first waterings. If the turnips fail, or are late owing to the failure of first sowings, the working power of the bul-locks is weakened, and the wheat suffers from insufficient waterings. Sowings commence in September and go on till November. The crop

FOOD &	Food and Fodder.

ripens in three months. A first-class crop is that which yields a good fodder crop of leaves first and a heavy root-crop afterwards (see *Gaz. of Jhang District, p. 111*). In the Gujranwala district turnips are largely grown, often amongst the wheat in the highly-cultivated lands bordering the Chenab, grass being very scarce. A dry season is favourable to a good crop of turnips, and an extension of their cultivation would alleviate one of the worst dangers of a drought, the failure of fodder for cattle (*Gaz. of Gujranwála, p. 54*). In the district of Dera Ismail Khan turnips are grown principally as cattle fodder, and in the Kachi tracts as a head rather than as a root crop. They are extensively cultivated in the Montgomery district, and from the middle of November the crop is used as fodder. In Muzaffargarh they are mostly used as fodder, and ripen just in time to relieve the failing stocks of other kinds of fodder. In the Multan district cattle are fed on turnips from 15th November to the 1st February.

Brassica campestris, *Linn.*, var. **Toria**. Often used as green fodder in the Karnál district (Panjáb).

B. juncea, *H.f. & T.* Sometimes given as green fodder, when other kinds of food are scarce.

Briedelia montana, *Willd.* The leaves are lopped for cattle fodder (*Brandis*).

B. retusa, *Spreng.* The leaves are valued as fodder and the tree is frequently lopped (*Brandis*). Cattle fed on these leaves are said to be cured of worms.

Broussonetia papyrifera, *Vent.* (Paper Mulberry) This valuable fibre plant can be easily cultivated in almost any kind of soil, and the foliage will probably be found to be a useful fodder for cattle.

Buchanania angustifolia, *Roxb.* Buffaloes eat the leaves (*R. Thompson*).

B. latifolia, *Roxb.*—The leaves of this tree are said to be given as fodder in the Savantvadi district, Bombay; and according to **Mr. R. Thompson** they are similarly employed in the Central Provinces.

Bursera serrata, *Colebr.* Buffaloes eat the leaves (*R. Thompson*).

Butea frondosa, *Roxb.* The *Dhák* tree. Buffaloes are very fond of the leaves, and their milk is said to be improved thereby. They are said to be more wholesome if given when not quite fresh. Camels and goats will not touch this tree.

B. superba, *Roxb.* A large climbing shrub, the leaves of which, according to the **Rev. A. Campbell**, are eaten by cattle in Chutia Nagpur, and by buffaloes as stated by **Mr. R. Thompson**.

Buettneria herbacea, *Roxb.* Plant (*R. Thompson*).

Cajanus indicus, *Spreng.* VERN. *Arhar.* Largely cultivated in most parts of India. The leaves and pod-shells are considered excellent feeding for cattle. The husks and broken grain soaked in water are sometimes given to cattle to keep them quiet when being milked.

Calendula officinalis, *Linn.* A weed of cultivation in the Western Panjáb and Sind. It is supposed to increase the flow of milk in cows.

Calotropis gigantea, *R. Br.* In Chutia Nagpur cattle eat the leaves (*Rev. A. Campbell*).

C. procera, *R. Br.* Cattle will eat the dried leaves.

Careya arborea, *Roxb.* The fruit is said to be eaten by cattle in the Kánara district of Bombay.

Carthamus tinctorius, *Linn.* The chaff of this plant is said to be sold as fodder for cattle in the Bulandsharh district. The oil-cake is rather bitter, and is apt to taint the milk.

Cassia Fistula, *Linn.* VERN. *Amaltás.* The twigs and leaves are lopped for cattle fodder in Oudh and Kumaon (*Brandis*).

| Food and Fodder. | (*J F. Duthie.*) | FODDER. |

Cedrela Toona, *Roxb.* (*Toon* tree.) In some parts of the hills the young shoots and leaves are lopped as cattle fodder (*Brandis*). The seeds also are sometimes given to cattle as a fattening food.

Celastrus paniculata, *Willd.* Cattle eat the leaves in Chutia Nagpur. (*Rev. A. Campbell*) ; also eaten by buffaloes (*R. Thompson*).

Ceratonia Siliqua, *Linn.* (The Carob tree.) Cattle are fond of the sweet pods and will also browse on the foliage if allowed to do so. Baron Von Mueller states that "in some of the Mediterranean countries horses and stable cattle are almost exclusively fed upon the pods. The fattening properties of these pods, which contain about 66 per cent. of sugar and gum, are twice those of oil-cake. To horses and cattle 6℔ a day are given of the crushed pods, raw or boiled, with or without chaff."

Cicer arietinum, *Linn.* (Chick Pea, or Bengal gram.) Largely grown as a rabi crop in Northern and Central India. The grain, known generally as *chana*, is staple food in Northern India for cattle and horses. In some districts of the Panjáb cattle and horses are allowed to graze on the young plants. If after this the crop gets rain, the plants grow up all the stronger for having been grazed over; they tiller better. The custom of allowing cattle to graze on the green crops is very prevalent in some of the Panjáb districts, especially that of Jhang, where the agricultural population depend so much on their cattle for their sustenance. (See *Gaz. of Jhang, p. 109.*) In the Montgomery district the dry stalks and leaves of gram are considered injurious to milch cattle. In the Karnál district the *bhusa* or straw is considered admirable fodder, and is also very well thought of in the Hoshiarpur district. In Bengal it is said to be not liked by cattle on account of its bitter taste. In the Ahmednagar district (Bombay) the *bhusa* is carefully preserved as cattle food. When the grain is thrashed, or trodden out by cattle, the pod shells are separated by winnowing and used as manure or burnt, as they are considered, owing to their sharpness, liable to injure the mouths of cattle.

Cistanche tubulosa, *Wight.* A curious and rather handsome herb, parasitical on the roots of **Ærua javanica** and **Calligonum polygonoides**, and found on sandy ground in parts of the Panjáb and in Sind. **Stewart**, under its synonym **Phelipæa Calotropidis,** *Walp.*, says that the upper portion of the plant is given as fodder to oxen.

Citrullus Colocynthis, *Schrad.* The fruit is said to be relished by buffaloes.

C. vulgaris, *Schrad.*—(Water-melon). In the Dera Ismail Khán district cattle are sometimes fed on the raw fruit, and the seeds are carefully preserved as cattle food during the winter. The seeds are also given to cattle in the Malláni district of Rájputána.

Clerodendron phlomoides, *Linn. f.* ⎫ Buffaloes eat the leaves (*R. Thomp-*
C. serratum, *Spreng.* ⎬ *son*).

Cocos nucifera, *Linn.* (Cocoa-nut Palm.) In the Thána district of Konkan the refuse, after the oil has been pressed out, is sometimes given to cattle.

Cocculus villosus, *DC.* Cows and buffaloes eat it (*Roxburgh* under **Menispermum hirsutum**).

Cochlospermum Gossypium, *DC.* Buffaloes eat the leaves and flowers (*R. Thompson*).

Colebrookia oppositifolia, *Smith.* Buffaloes are said to be fond of the leaves of this shrub.

Colocasia antiquorum, *Schott.* In Bengal, according to Mr. T. N. Mukharji, yams are cut into small pieces and boiled either alone or mixed with rice ends or with portions of **Amarantus spinosus**, and given to cattle.

FOOD &	Food and Fodder.

Combretum ovalifolium, *Roxb.* Buffaloes eat the leaves and young shoots (*R. Thompson*).

Commelina bengalensis, *Linn.* This plant is said by **Bhai Sadhu Singh** to be given as fodder to cattle in the Jeypur State.

Convolvulus arvensis, *Linn.* Gathered by village children as fodder for cattle (see *Vol. II., p. 519*).

C. pluricaulis, *Chois.* Mentioned by **Stewart** as being eaten by cattle, and considered cooling.

Corchorus olitorius, *Linn.* (Jute.) Leaves eaten by cattle after the plant is cut for fibre.

Cordia Macleodii, *H. f. & T.* Buffaloes eat the leaves (*R. Thompson*).

C. Myxa, *Linn* The leaves are given to cattle.

C. Rothii, *R. & S.* Buffaloes eat the leaves (*R. Thompson*).

Cratæva religiosa, *Forst.* Buffaloes eat the leaves and fruit (*R. Thompson*).

Crinum, *sp.* The flowers of this (apparently undescribed) species are, according to the Rev. A. Campbell, eaten by cattle.

Crotalaria juncea, *Linn.* (Sunn Hemp.) Cultivated for its fibre, and also, according to Roxburgh, in parts of the Northern Circars as a fodder plant for milch cows. The stems are used as fodder in the Kistna district (Madras), and also in Godaveri, where they are stored in bundles, and covered over with palmyra leaves to protect them from rain. The seeds are also collected and given to cattle in some parts of India.

C. linifolia, *Linn. f.* An annual, common throughout India. **Roxburgh** says that cattle eat it.

C. medicaginea, *Lamk.* The plant is eaten by cattle in Bundelkhand and Rájputána.

Croton oblongifolius, *Roxb.* Cattle eat the leaves in Chutia Nagpur (*Rev. A. Campbell*).

Cyamopsis psoralioides, *DC.* Vern. *Guár.* Cultivated during the rains in various parts of India. The pods are used as human food, and the seeds are given to cattle and horses; in the former case it is grown as a garden crop, and in the latter as a field crop, being often sown with *bájra.* It is largely grown for cattle in the Meerut Division; also in some of the Panjáb districts in light soils. It is sometimes given green to bullocks.

Cyanotis axillaris, *R. & S.* "Cattle are very fond of this plant" (*Roxburgh*).

C. tuberosa, *R. & S.* Cattle eat the plant (*Rev. A. Campbell*).

Cyperus longus, *Linn.* Mr. T. N. Mukharji says that in Bengal this plant is weeded out from fields and given to cattle.

C. rotundus, *Linn.* Vern. *Mothá.* Cattle eat this plant. Other species of Cyperaceæ, known under the general name of *dila,* are eaten by cattle, and especially by buffaloes.

Dalbergia lanceolaria, *Linn.* Buffaloes eat the leaves (*R. Thompson*).

D. latifolia, *Roxb.* In Oudh, according to Brandis, the tree is pollarded for cattle fodder. In the Bombay Presidency also it is said to be used for fodder.

D. Sissoo, *Roxb.* (The Shisham tree.) Cattle are fond of the young shoots and leaves, and will browse freely on them if allowed to do so.

D. volubilis, *Roxb.* Cattle eat the leaves in Chutia Nagpur (*Rev. A. Campbell*), and in the Central Provinces (*R. Thompson*).

Daucus Carota, *Linn.* (The Carrot.) A most valuable crop for tracts affected by periodical droughts. Cattle eat both the tops and the roots, and in Kolhápur they are frequently given to milch cows. In Cutch carrots are largely grown both for fodder and for pickling.

Food and Fodder.	(*J. F. Duthie.*)	FODDER.

Derris scandens, *Benth.* Cattle eat the leaves and pods (*R. Thompson*).

Desmodium Cephalotes, *Wall.* Cattle eat the leaves of this shrub in Chutia Nagpur (*Rev. A. Campbell*).

D. diffusum, *DC.* "Cattle eat this species greedily, and as it grows quickly and with luxuriance it might be cultivated with advantage" (*Roxburgh* under **Hedysarum quadrangulatum**).

D. parvifolium, *DC.* A trailing herbacerous perennial, common in the plains; it is eaten by cattle and other animals.

D. pulchellum, *Benth.* Cattle eat the leaves and shoots (*R. Thompson*).

D. triflorum, *DC.* Similar in habit to the preceding, and equally abundant. Roxburgh (under **Hedysarum triflorum**) remarks that it is "very common on pasture ground and helps to form the most beautiful turf we have in India." He also says that cattle are very fond of it. Baron Von Mueller recommends this species "for places too hot for ordinary clover, and as representing a large genus of plants, many of which may prove of value for pasture." Forty-nine species are described in the *Flora of British India.*

Dichrostachys cinerea, *W. & A.* Buffaloes eat the leaves and pods (*R. Thompson*).

Digera arvensis, *Forsk.* Mainly used as a fodder for cattle in South Baluchistán (*Dr. R. P. Banerji*).

Dillenia aurea, *Smith.* Buffaloes eat the young leaves and fruit (*R. Thompson*).

D. pentagyna, *Roxb.* Young leaves and fruit (*R. Thompson*).

Dioscorea bulbifera, *Linn.* Leaves (*Rev. A. Campbell*, Chutia Nagpur).

D. oppositifolia, *Linn.* Plant (*Rev. A. Campbell*, Chutia Nagpur).

Diospyros Embryopteris, *Pers.* } Buffaloes eat the young leaves (*R.*
D. melanoxylon, *Roxb.* } *Thompson*).

D. montana, *Roxb.* Leaves (*R. Thompson*).

Dolichandrone falcata, *Seem.* Buffaloes eat the young leaves in the Central Provinces (*R. Thompson*).

Dolichos biflorus, *Linn.* (Horse gram of Madras.) Chiefly grown in South India for its grain, which is largely used for feeding horses. The stems and leaves, green or dry, are considered to be good fodder for cattle. In parts of the Panjáb and in Káthiawar, it is grown only for fodder, and is given to cattle, green or dry.

D. Lablab, *Linn.* (Cow gram of Mysore.) The leaves and stalks are considered a valuable fodder for milch cows, and the pulse is given to cattle in the Madras Presidency.

Dregea volubilis, *Benth.* Cattle eat the leaves in Chutia Nagpur (*Rev. A. Campbell*).

Ehretia acuminata, *Br.* Buffaloes eat the leaves (*R. Thompson*).

E. lævis, *Roxb.* Leaves (*Brandis*).

Elæodendron glaucum, *Pers.* Leaves (*R. Thompson*).

Equisetum debile, *Roxb.* Sometimes given to cattle as fodder (*Stewart*).

Eriodendron anfractuosum, *DC.* (Kapok tree) Oil-cake.

Erioglossum edule, *Bl.* Buffaloes eat the leaves in Oudh (*R. Thompson*).

Eriolæna Hookeriana, *W. & A.* Buffaloes eat the leaves in the Central Provinces (*R. Thompson*).

Eruca sativa, *Lamk.* Largely used in the Panjáb as a green fodder for cattle, and often specially cultivated for this purpose. The oil-cake is also given to cattle.

Erycibe paniculata, *Roxb.* Buffaloes eat the leaves (*R. Thompson*).

Erythrina indica, *Lamk.* } Buffaloes eat the leaves (*R. Thompson*).
E. suberosa, *Roxb.* }

FOOD &	Food and Fodder.

Eugenia Jambolana, *Lamk.* ⎫
E. operculata, *Roxb.*　　　⎬ Buffaloes eat the leaves and fruit (*R.*
E. Heyneana, *Wall.*　　　 ⎭　*Thompson*).

Euphorbia helioscopia, *Linn.* Cattle eat this plant in Beluchistan (*Dr. R. P. Banerji*).

Farsetia Jacquemontii, *Hk. f. & T.* Eaten by cattle in the Panjáb.

Feronia elephantum, *Correa.* Buffaloes eat the leaves and fruit (*R. Thompson*).

Ficus Cunia, *Buch.* Buffaloes eat the leaves.

F. glomerata, *Roxb.* Leaves and fruit.

F. hispida, *Linn.* Much lopped for cattle fodder (*Brandis*).

F. infectoria, *Roxb.* Leaves (*Brandis*).

F. palmata, *Forsk.* (=F. virgata, *Roxb.*) Leaves.

F. Roxburghii, *Wall.* Leaves (*Brandis*).

F. Rumphii, *Bl.* (=F. cordifolia, *Roxb.*) Leaves (*Brandis*). Buffaloes eat the leaves (*R. Thompson*).

F. saemocarpa, *Miq.* Leaves (*Madden*).

Flacourtia Ramontchi, *L'Hérit.* Leaves (*Brandis*).

F. sepiaria, *Roxb.* Leaves.

Flemingia strobilifera, *R. & Br.* Buffaloes eat the plant in the Central. Provinces (*R. Thompson*).

Flueggia Leucopyrus, *Willd.* Buffaloes eat the leaves in the Central Provinces (*R. Thompson*).

Gardenia latifolia, *Ait.* Leaves eaten by cattle in Chutia Nagpur (*Rev. A. Campbell*), and by buffaloes in the Central Provinces (*R. Thompson*).

Garuga pinnata, *Roxb.* Leaves (*R. Thompson*).

Gmelina arborea, *Roxb.* Cattle are fond of the fruit (*Gas., Kolaba Dist., Bombay, p. 24*).

Gossypium herbaceum, *Linn.* (Cotton.) The seed is a valuable food for milch cattle. The oil-cake is also largely given. In some districts cattle are allowed to graze on the leaves and shoots after the cotton-picking is over.

Grewia lævigata, *Vahl.* Twigs and leaves in North-Western Provinces (*Brandis*).

G. tiliæfolia, *Vahl.* Leaves (*Brandis*).

G. vestita, *Wall.* Leaves.

Guazuma tomentosa, *Kunth.* Leaves valued for fodder in the Bombay Presidency.

Guizotia abyssinica, *Cass.* The oil-cake is much prized for milch cattle.

Hamiltonia suaveolens, *Roxb.* Buffaloes eat the leaves (*R. Thompson*).

Hardwickia binata, *Roxb.* VERN. *Anjan.* "Cattle are exceedingly fond of the leaves. In the Cauvery forests, Northern Mysore, and Berar, the trees were formerly, and are still to a great extent, pollarded for cattle fodder" (*Brandis*).

Helicteres Isora, *Linn.* Buffaloes eat the leaves (*R. Thompson*).

Heterophragma Roxburghii, *DC.* Leaves much eaten by cattle (*Gas., Thana Dist., p. 27*).

Hibiscus cannabinus, *Linn.* In the Poona district the seed is sometimes given to cattle.

Hippocratea arborea, *Roxb.* Buffaloes eat the leaves (*R. Thompson*).

Hiptage Madablota, *Gærtn.* Leaves (*R. Thompson*).

Holarrhena antidysenterica, *Wall.* Leaves (*R. Thompson*).

Holoptelea integrifolia, *Planch.* (=Ulmus integrifolia, *Roxb.*) "The leaves are lopped for cattle fodder, and the tree is often used to stock fodder for winter supply" (*Brandis*).

F. 671

Food and Fodder.	(*J. F. Duthie.*)	FODDER.

Holostemma Rheedii, *Wall.* Cattle eat the plant in Chutia Nagpur (*Rev. A. Campbell*).

Hymenodictyon excelsum, *Wall.* Leaves.

Indigofera cordifolia, *Heyne.* Buffaloes are fond of this plant.

I. enneaphylla, *Linn.* Helps to form the best pasture lands in Bengal, where it is always found in plenty (*Roxburgh*).

I. glandulosa, *Willd.* Cattle are fond of the plant (*Roxburgh*).

I. linifolia, *Retz.* Plant. The seeds of this and other species of wild indigo are highly nitrogenous.

I. paucifolia, *Del.* Plant.

I. pulchella, *Roxb.* Cattle eat the leaves in Chutia Nagpur (*Rev. A. Campbell*).

Ipomæa aquatica, *Forsk.* VERN. *Kalmi.* "This plant is given to cattle in Bengal dried and smoked like *nár* grass, and is considered lactiferous" (*T. N. Mukharji*).

I. Batatas, *Lamk.* (Sweet potato.) The stems are considered excellent fodder for cattle.

Ixora parviflora, *Vahl.* Buffaloes eat the leaves in the Central Provinces (*R. Thompson*).

Kydia calycina, *Roxb.* Buffaloes eat the leaves in the Central Provinces (*R. Thompson*).

Kyllingia monocephala, *Linn.* The plants are given to cattle in Bengal (*T. N. Mukharji*).

Lagenaria vulgaris, *Seringe.* (Bottle-gourd.) In Bengal the fruit chopped up with rice-ends (*khud*) is often given to milch cows. In the Kolhapur district of Bombay the fruit when grown in abundance is chopped up and given to buffaloes.

Lagerstrœmia parviflora, *Roxb.* Buffaloes eat the leaves (*R. Thompson*).

Lathyrus Aphaca, *Linn.* A cold-season weed of cultivated ground in Northern India. It is often pulled up and given as fodder to cattle.

L. imphalensis, *Watt, MS.* Used as fodder in Manipur.

L. sativus, *Linn.* Grown in the Panjáb chiefly as green fodder for cattle. In the Montgomery district, however, the dry stalks and leaves are considered good cattle-fodder. In some parts of Bengal, according to Mr. T. N. Mukharji, it is sown broadcast among transplanted rice after the rains, when the land is still wet. The plants grow up luxuriantly after the rice has been reaped, and then the cattle are allowed to graze upon them. It is also sown in this way on river-banks or silts deposited by the annual inundations, and the crop is either grazed or allowed to ripen its seed.

Lens esculenta, *Mœnch.* (Lentil.) The dry stalks and leaves are sometimes given to cattle, though considered by some to be a heating form of food.

Lepidium Draba, *Linn.* A common wayside weed at Quetta. Judging from the extent to which it is used as green fodder for cattle and other animals in that neighbourhood, it deserves attention.

Linum usitatissimum, *Linn.* Linseed cake is given to cattle in Bengal, but to a limited extent, as most of the seed is exported. Cattle are fed, however, on the empty capsules (*T. N. Mukharji*).

Litsea sebifera, *Pers.* Buffaloes eat the young leaves (*R. Thompson*).

Mallotus philippinensis, *Müll.* VERN. *Kamela.* Buffaloes eat the leaves (*R. Thompson*).

Mangifera indica, *Linn.* (Mango.) In Bengal the rinds and stones are sometimes given to cattle. The latter when ripe are swallowed entire, and after becoming soft in the stomach they are brought up as a cud; the kernels are then pressed out and eaten, and the refuse rejected

**I.
FODDER
PLANTS OF
THE PLAINS.**

F. 671

FOOD &	Food and Fodder.

(*T. N. Mukharji*). In dry seasons buffaloes eat the leaves (*R. Thompson*).

Marsdenia tenacissima, *W. & A.* Buffaloes eat the leaves (*R. Thompson*).

Medicago denticulata, *Willd.* A cold-season weed, largely used as green fodder in the Panjáb, and considered good for milch cows.

M. lupulina, *L* (Hop-trefoil.) A cold-season weed of Northern India often collected for fodder, and worthy of cultivation in the Panjáb. Sutton, in his "Permanent and Temporary Pastures," p. 71, says that the herbage is more nutritious than that of Red clover, and helps to make a good bottom to a pasture, and that it is supposed to impart colour and good flavour to butter.

M. sativa, *Linn.* Lucerne is now well known all over India as a very valuable green fodder crop, especially for horses. It should be given, however, only as a supplement to the ordinary food, as animals will always suffer if allowed to eat as much of it as they will. Mixed with the chopped straw of oats, barley, or wheat, it forms a very wholesome feed. For further information see the article on **Lucerne** under **Medicago.**

Melia Azadirachta, *Linn.* (The Neem tree.) ⎫ The leaves are said to be

M. Azedarach, *Linn.* (Persian Lilac.) ⎬ given as fodder to cattle in the Ahmednagar district of Bombay.

Melilotus parviflora, *Desf.* Very common in Northern India as a cold-season weed of cultivation, and largely used in the Panjáb as green fodder for cattle. It is said to be cultivated in some districts for this purpose. An allied species with white flowers (**M. alba,** *Lamk.*) has been known to give colic to cattle; but all plants, especially of the clover kind, if eaten in excess in the green state, are liable to cause this complaint.

Miliusa velutina, *Hk. f. & T.* VERN. *Dom-sál.* Buffaloes eat the leaves (*R. Thompson*).

Millettia auriculata, *Baker.* This climber is extensively lopped to afford fodder to buffaloes (*R. Thompson*).

Mimusops hexandra, *Roxb.* VERN. *Khirni.* Buffaloes eat the leaves in the Central Provinces (*R. Thompson*).

Morinda tinctoria, *Roxb.* VERN. *Ál.* In the Rewa Kántha district of Bombay, the leaves are given to cattle when grass and forage are scarce.

Morus indica, *Linn.* The leaves are said to be a good fodder for cattle.

Musa paradisiaca, *Linn.* (The Plantain.) Chopped into small pieces it is largely used as fodder in many parts of India, and according to Mr. T. N. Mukharji it forms the staple food of cattle in parts of the Hughli district. It cannot, however, be very nutritious, and is apt to cause diarrhœa. Mr. Mukharji also says that the white portion of the root is chopped fine and given to cattle, and is a more substantial food than the stems. Cattle are very fond, too, of the skin of the fruit, and the flowers when available.

Nyctanthes Arbor-tristis, *Linn.* VERN. *Harsinghar.* Buffaloes eat the young leaves (*R. Thompson*).

Ochna squarrosa, *Roxb.* Buffaloes eat the leaves in the Central Provinces (*R. Thompson*).

Ocimum canum, *Sims.* Cattle eat the leaves in Chutia Nagpur (*Rev. A. Campbell*).

Odina Wodier, *Roxb.* This tree is often lopped and pollarded, the leaves and branches being a favourite fodder of cattle (*Brandis*).

Olea cuspidata, *Wall.* The leaves are said to be good fodder for cows and milch buffaloes (*Gaz., Rawal Pindi, p. 80*).

Opuntia Dillenii, *Haworth.* (Prickly Pear.) This is the kind which grows so plentifully in Southern India, and were it not for the spines it might be used with advantage as a profitable adjunct to the ordinary food of cattle, especially in times of scarcity. **Dr. Shortt** (*Indian Forester, Vol. III. p. 233*) refers to a spineless form on which **Mr. H. S. Thomas**, then Collector of Tanjore, fed his cattle. It has been found possible to preserve the stems and leaves in silos, and the product mixed with grass was pronounced by **Mr. Hooper** to be wholesome food for cattle. (See *Bulletin of Useful Information, Royal Gardens, Kew, 1888, p. 173.*)

Ougeinia dalbergioides, *Benth.* The branches are often lopped for cattle fodder.

Oxalis corniculata, *Linn.* Cattle eat this plant in Chutia Nagpur (*Rev. A. Campbell*).

Pæderia fœtida, *Linn.* [*Jour. Agri. Hort. Soc. Ind., VII., 224.* In the publication referred to, it is stated that this climber is greedily eaten by elephants.—*Ed.*]

Papaver somniferum, *Linn.* Poppy-seed cake is given to cattle in Bengal, but the supply is insignificant (*T. N. Mukharji*).

Pavetta indica, *Linn.,* var. **tomentosa.** Buffaloes eat the leaves in the Central Provinces (*R. Thompson*).

Phaseolus aconitifolius, *Jacq.* VERN. *Moth.* The grain is often given to cattle, and is said to be very fattening. It is believed, however, to reduce the flow of milk if given to milch cattle. The stems and leaves, green or dry, are highly valued as fodder.

P. calcaratus, *Roxb.* Is mentioned as yielding fodder for cattle.

P. Mungo, *Linn.* VERN. *Mung.* The grain is considered fattening for horned cattle, and is sometimes given boiled and mixed with *ghí.* Roxburgh says that cattle do not like the straw, and in Mysore it is looked upon as useless. In the Panjáb, however, it is thought highly of, though valued less than that of *moth* and *urd.*

P. Mungo, *Linn.,* var. **radiatus.** VERN. *Urd.* The grain of this is also given as a fattening food to cattle. Roxburgh says that cattle eat the straw, and that it is considered nourishing. In Mysore it is thought to be harmful to cattle, and is therefore used as manure, or for feeding camels. The husks are much valued in the Madras Presidency.

P. trilobus, *Ait.* The grain, which is sometimes called Red gram, is used for feeding cattle, and in Coimbatore is sown chiefly for that purpose.

Phœnix dactylifera, *Linn.* In Káthiawar cattle feed on the local dates called *khalela,* and the refuse of the distilleries is eagerly eaten by them.

Phyllanthus Emblica, *Linn.* VERN. *Ámla.* Buffaloes eat the leaves and fruits (*R. Thompson*).

P. urinaria, *Linn.* Roxburgh says that cattle eat this herb.

Piptadenia oudhensis, *Brand.* The tree is pollarded for cattle fodder (*Brandis*).

Pisum sativum, *Linn.* (Common Pea.) In many parts of the Panjáb this, and probably also the Field Pea (**P. arvense,** *Linn.*), is grown only as a fodder crop for cattle. It is considered excellent fodder, whether green or dry. The straw is also used as cattle fodder in Berar and in the Bombay Presidency. In Bengal, according to Mr. T. N. Mukharji, the seeds are given to cart-bullocks, but only in towns.

Pithecolobium dulce, *Benth.* (Manilla Tamarind.) Introduced from Mexico. Cattle eat the pods.

P. saman, *Benth.* (Rain tree of South America.) Thrives in localities free of frost. The sweet pods are relished by cattle.

Poinciana elata, *Linn.* Planted near villages in Western India, and the foliage is given as fodder to cattle.

Polyalthia cerasoides, *Benth. & Hk.f.* **P. suberosa,** *Benth. & Hk.f.* } Buffaloes eat the leaves in the Central Provinces (*R. Thompson*).

FOOD &	Food and Fodder.

Polygonum barbatum, *Linn.* ⎫ Roxburgh says that cattle are fond of
P. chinense, *Linn.* ⎭ these plants.

P. tomentosum, *Willd.* Cattle eat it greedily (*Roxburgh*).

Pongamia glabra, *Vent.* (Indian Beech.) Cattle are said to be fond of the leaves of this tree. It is almost evergreen, and is much used for planting along road-sides. Grass grows well under its shade.

Populus euphratica, *Oliv.* ⎧ The leaves afford fodder for cattle, and the
P. nigra, *Linn.* ⎨ tree is lopped occasionally for that pur-
⎩ pose (*Brandis*).

Portulaca oleracea, *Linn.* Cattle eat this herb in Chutia Nagpur (*Rev. A. Campbell*).

Premna integrifolia, *Linn.* ⎫ The leaves are a good fodder for cattle
P. latifolia, *Roxb.* ⎭ (*Brandis*).

Prosopis juliflora, *Benth.* (Mesquite Bean.) Introduced from Texas. The sweet pods are much liked by cattle. It thrives well in Upper India even on poor soils.

P. spicigera, *Linn.* VERN. *Jand.* The pods are eaten by cattle. Though not so nutritious as the fresh pods of the *babul*, they can be kept good longer.

Psoralea corylifolia, *Linn.* The plant is eaten by cattle in Bundelkhand.

Pterocarpus Marsupium, *Roxb.* (Bastard Teak.) The leaves are a favourite food of cattle.

Pueraria tuberosa, *DC.* The leaves are considered to be good fodder for cattle.

Putranjiva Roxburghii, *Wall.* The leaves are lopped for cattle fodder (*Brandis*).

Randia dumetorum, *Lamk.* The leaves are lopped and used as cattle fodder (*Brandis*).

R. uliginosa, *DC.* The leaves are browsed by cattle (*Brandis*).

Raphanus sativus, *Linn.* (Radish.) The oil-cake, although much liked, is given to cattle only in certain parts of Northern Bengal (*T. N. Mukharji*).

Rhizophora mucronata, *Lamk.* The leaves of the Mangrove tree are largely used in Káthiawar to feed cattle, and the berries are said to increase their milk-giving powers. In the Kistna district of Madras cattle eat the dried leaves.

Ricinus communis, *Linn.* (Castor.) The oil-cake is given to cattle in Sind, according to Stocks. In Bengal it is used as manure (*T. N. Mukharji*).

Saccopelatum tomentosum, *Hk. f. & T.* The leaves are used as cattle fodder.

Salix acmophylla, *Boiss.* About Quetta the tree is much lopped for cattle fodder (*Brandis*).

S. tetrasperma, *Roxb.* The tree is often lopped for cattle fodder (*Brandis*).

Salvadora oleoides, *Dcne.* VERN. *Jál.* The fruit is said to be eaten by cattle of the highlands of the Rohtak district.

Sapindus Mukorossi, *Gærtn.* The leaves are given as fodder to cattle (*Brandis*).

Schleichera trijuga, *Willd.* In Oudh this tree is lopped, and the twigs and leaves are used as cattle fodder. Mr. Smythies says that the fruit also is eaten by cattle.

Scirpus barbatus, *Roxb.* The plant is used as fodder for cattle in the Jeypur State (*Bhai Sádhu Singh*).

S. maritimus, *Linn.* Fair forage for cattle.

Sesamum indicum, *DC.* (Gingelly or Til.) The oil-cake is a fattening food for milch cattle, and by those who can afford it is often given to

| Food and Fodder. (*J. F. Duthie.*) | FODDER. |

hard-working oxen. The empty capsules are also given to cattle. In the Baroda State bruised sesamum is given mixed with bruised gram.

Sesbania ægyptiaca, *Pers.* VERN. *Jait.* Cattle are very fond of the foliage.

S. grandiflora, *Pers.* Cattle eat the leaves and tender parts (*Roxburgh*).

Shorea robusta, *Gærtn.* VERN. *Sál.* Cows and buffaloes are fond of the young leaves; the Sál trees of the Government forests in Garhwál used to be extensively lopped for feeding buffaloes, but this practice is now forbidden.

Smithia sensitiva, *Ait.* Makes excellent hay (*Roxburgh*).

Sonchus oleraceus, *Linn.* Cattle are fond of this plant.

Soymida febrifuga, *A. Juss.* Buffaloes eat the young leaves in the Central Provinces (*R. Thompson*).

Spondias mangifera, *Pers.* (Hog Plum.) Cattle eat the leaves (*R. Thompson*), and according to Mr. Smythies, the fruit.

Stereospermum chelonoides, *DC.*
S. suaveolens, *DC.*
S. xylocarpum, *Wight.* } Buffaloes eat the young leaves (*R. Thompson*).

Stephegyne parvifolia, *Korth.* VERN. *Kaddam.* Cattle eat the leaves.

Sterculia colorata, *Roxb.* Twigs and leaves lopped for cattle fodder (*Brandis*).

S. villosa, *Roxb.* The leaves are given to cattle in the Savantvádi district of Bombay.

Streblus asper, *Lour.* The leaves are lopped extensively for cattle fodder (*Brandis*).

Strobilanthes callosus, *Nees.* VERN. *Kárvi* (Bombay). This shrub flowers profusely about every eight or nine years, and then becomes covered with a sticky exudation (*mel*). Herds of cattle gather from all sides to feed on it (*Gaz., Thána district, p. 43*). This plant is abundant on Mount Abu, where it flowered abundantly in 1887.

Symphytum peregrinum, *Ledeb.* (Prickly Comfrey.) Yields excellent fodder for milch cattle, but requires too expensive treatment for general use. A hill climate, such as that of the Nilghiris, appears to suit it best.

Tecoma undulata, *G. Don.* The leaves are greedily browsed by cattle. Recommendable for tracts subject to droughts.

Tectona grandis, *Linn. f.* (Teak.) In the Baroda State cattle are said to be often fed on its twigs and leaves.

Tephrosia purpurea, *Pers.* Cattle feed on this plant.

Terminalia Arjuna, *Bedd.* Cattle eat the leaves in Chutia Nagpur (*Rev. A. Campbell*), and the young leaves are eaten by buffaloes (*R. Thompson*).

T. belerica, *Roxb.* VERN. *Bahera.* In the Kángra district the leaves are considered to be the best fodder for milch cows.

T. Chebula, *Retz.* VERN. *Harar.* Cattle are said to eat the leaves of this tree.

T. tomentosa, *W. & A.* The leaves are lopped for cattle fodder (*Brandis*).

Thespesia Lampas, *Dalz. & Gibs.* Buffaloes eat the leaves in the Central Provinces (*R. Thompson*).

Tiliacora racemosa, *Colebr.* Buffaloes eat the leaves in Oudh (*R. Thompson*).

Tinospora cordifolia, *Miers.* VERN. *Golancha.* This twining plant, which is common on trees in Bengal villages, is greedily eaten by cattle. People gather it occasionally, and give it to their animals cut into small pieces. It is said to increase the flow of milk in milch cows, but it gives a smell to the milk (*T. N. Mukharji*).

2 E 2

F. 671

FOOD &	**Food and Fodder.**

Trewia nudiflora, *Linn.* Buffaloes eat the leaves in Oudh (*R. Thompson*).

Trianthema pentandra, *Linn.* Eaten by cattle.

Trigonella Fœnum-græcum, *Linn.* (Fenugreek.) Is grown extensively in the Panjáb, where it is used chiefly as a green fodder for cattle. It yields only one cutting.

Turpinia pomifera, *DC.* The leaves are used as fodder.

Vangueria spinosa, *Roxb.* The leaves are said to be a useful fodder in the Tháná district of Bombay.

Ventilago calyculata, *Tulasne.* Buffaloes eat the leaves in the Central Provinces (*R. Thompson*).

Vicia Faba, *L.* VERN. *Bákla.* The seeds are sometimes given to cattle.

V. hirsuta, *Koch.* Cultivated locally for cattle fodder.

Vicoa auriculata, *Cass.* Buffaloes are said to be fond of this plant.

Vigna Catiang, *Endl.* VERN. *Lobiya.* The leaves and stems are sometimes used as cattle fodder. In Mysore the straw is said to be useful only as manure.

V. pilosa, *Baker.* The straw of this plant is said to be used as a cattle fodder.

V. vexillata, *Benth.* Cattle eat the plant in Chutia Nagpur (*Rev. A. Campbell*).

Vitex leucoseylon, *Linn. f.* Buffaloes eat the leaves in the Central Provinces (*R. Thompson*).

Wendlandia exserta, *DC.* Cattle eat the leaves (*R. Thompson*).

Woodfordia floribunda, *Salisb.* Cattle eat the leaves in Chutia Nagpur (*Rev. A. Campbell*).

Wrightia tinctoria, *R. M.* Leaves eaten by buffaloes and other cattle in the Jeypore State (*Bhai Sádhu Singh*), and by buffaloes in the Central Provinces (*R. Thompson*).

W. tomentosa, *R. & S.* Cattle eat the leaves in Chutia Nagpur (*Rev. A. Campbell*). Leaves eaten by buffaloes in Central Provinces (*R. Thompson*).

Xylia dolabriformis, *Benth.* Buffaloes eat the leaves in the Central Provinces (*R. Thompson*).

Zizyphus Jujuba, *Lamk.* VERN. *Ber.* The leaves are much valued as cattle fodder.

Z. nummularia, *W. & A.* VERN. *Jhárberi.* Cattle are largely fed on the leaves of this bush in many parts of India, and it is often a most useful stand-by when other sources of fodder fail.

Z. rugosa, *Lamk.* Cattle eat the leaves in Chutia Nagpur (*Rev. A. Campbell*).

Z. xylopyra, *Willd.* VERN. *Katber.* The young shoots, leaves, and fruit serve as fodder for cattle (*Brandis*).

II. INDIAN FODDER GRASS—EXCLUDING HIMÁLAYAN SPECIES.

Æluropus littoralis, *Parl.*, var. **repens.** Sandy and saline tracts in the Western Panjáb, resembling *dúb* (**Cynodon Dactylon**), which it replaces.

Alopecurus pratensis, *Linn.* (Meadow Fox-tail.) A common European grass occurring also on the Himálaya and descending to the Panjáb plains. Abundant at Quetta, where it is largely used for feeding horses. Might be cultivated with advantage as a winter grass in many parts of the Panjáb.

Andropogon annulatus, *Forsk.* An abundant and excellent fodder grass. A variety with the outer glumes 3-toothed, **A. Bladhii,** *Retz.*, is also plentiful.

Food and Fodder. (*J. F. Duthie.*)	**FODDER.**

Andropogon caricosus, *Linn.* Plentiful in Bundelkhand and the Central Provinces, and largely used as fodder.

A. foveolatus, *Del.* Abundant on sandy and stony ground, and generally considered to be a good fodder grass.

A. Isthæmum, *Linn.* A good fodder grass resembling **A. annulatus,** but less abundant.

A. laniger, *Desf.* Common in North-Western India. Cattle eat this grass readily when it is young and tender, but horses are liable to suffer from colic after feeding on it. It is strongly aromatic, and the scent is often communicated to the milk of cows.

A. micranthus, *Kunth.,* var. villosulus (*Hack. Mongr., 490*). On Mount Abu, where it is called *Ballak,* and is much valued for fodder. It occurs also on the Himálaya.

A. muricatus, *Retz.* (**A. squarrosus,** *Linn. f.* in *Hack. Mongr., 542*) This is the *khas-khas* grass the roots of which are employed in making tatties. It thrives best on damp low-lying land, where, when young, it affords abundance of fodder for buffaloes and in seasons of excessive drought it is cut and given to cattle.

A. pachyarthrus, *Hack.* Common in Central India on black soil, also on saline and sandy tracts. A good grass for cattle, but not for horses.

A. pertusus, *Willd.* VERN. *Palwa.* An excellent grass for grazing and stacking, and very abundant.

A. Schœnanthus, *Linn.* A sweet-scented species abundant in Northern and Central India. Although largely used as fodder it is not considered very wholesome. In Rajputana it usually forms the roofing portion of the stacked hay. The essential oil, *rusa ka tel,* is supposed to exercise a preservative action when this grass is stacked with others.

A. serratus, *Thunb.,* var. nitidus. (*Hack. Mongr.*) = **Sorghum muticum,** *Nees.*) Hilly parts of India; occasionally used for fodder.

Anthistiria anathera, *Nees.* (**Themeda anathera,** *Hack.*) Abundant on the Himálaya and descending to the Panjáb plains. According to Captain Wingate, it is much liked by the horses of the British cavalry and artillery at Rawal-Pindi.

A. ciliata, *Linn. f.* (**Themeda ciliata,** *Hack.*) Common in hilly parts of India and on the Himálaya. Though rather a coarse grass, it is much used for fodder in Central India.

Apluda aristata, *Linn.* (**A. varia** *Hack.,* var. aristata.) Abundant in India amongst bushes, and in forest land often forms a large portion of the undergrowth. Considered to be good fodder when young.

Aristida depressa, *Retz.* Abundant on sandy and stony ground, where it affords good grazing when young.

A. hystrix, *Linn. f.* Met with in similar situations, and probably ot equal value.

Arthraxon lanceolatus, *Hochst.* (**Andropogon lanceolatus,** *Roxb.*). Common on rocky ground, and said to be a good fodder grass in Rájputana.

Avena sativa, *Linn.* (Oats.) First-rate fodder, both green and as hay, especially for horses.

Bambusa arundinacea, *Retz.* A favourite fodder of elephants. The leaves are given to horses as a medicine.

Bromus uniloides, *H. B. & K.* (Prairie-grass of Australia.) Much valued both in Australia and America as a nutritious fodder grass, whether green or dry. Has been tried in India, but as a crop was found inferior to oats.

Cenchrus catharticus, *Del.* VERN. *Bhurt.* A characteristic desert grass, and much valued for grazing purposes on account of the early appearance of its foliage.

INDIAN
FODDER
GRASS.

FOOD &	Food and Fodder.

Cenchrus montanus, *Nees.* VERN. *Anjan.* Flourishes on sandy soils. Very good for grazing and makes excellent hay.

C. pennisetiformis, *Hochst. & Steud.* A tall succulent grass, growing in bushy places and often assuming a climbing habit.

Chloris barbata, *Swartz.* Considered good for cattle up to the time of flowering.

C. digitata, *Steud.* Amongst bushes and under the shade of trees. Is used as fodder in Rájputana.

C. tenella, *Roxb.* Said to be a good fodder grass in Rájputana.

C. tetrastachys, *Hack. MS.* Apparently confined to the saline usar tracts of the North-Western Provinces, where over considerable areas it constitutes the only vegetation.

Chrysopogon serrulatus, *Trin.* **(Andropogon Trinii,** *Steud.* in *Hack. Monogr.*) Common in hilly parts of India, a very good fodder grass, and much liked by horses.

Coix lachryma, *Linn.* Largely eaten by cattle in Oudh, and said to be very fattening.

Cynodon Dactylon, *Pers.* VERN. *Dúb.* Universally recognised to be the most nutritious and useful fodder grass in this country, whether green or dry, especially for horses.

Dendrocalamus strictus, *Nees.* Affords abundant fodder for elephants.

Dinebra arabica, *Beauv.* Plentiful in Central India on cultivated ground.

Diplachne fusca, *Beauv.* Common on low-lying ground, especially where the soil is saline. Buffaloes appear to be very fond of it.

Eleusine ægyptiaca, *Pers.* VERN. *Makra.* A common grass, especially on cultivated ground. Said to be very good for cattle, but not for horses.

E. Coracana, *Gærtn.* VERN. *Mandua* or *ragi.* Cultivated as a grain crop in most parts of India, and largely so in Mysore, where it affords abundance of fodder, both green and as straw. Ragi straw is there considered to be the best fodder for cattle, which are said to work and thrive on it alone, without requiring grass. Horses also are sometimes fed on it when grass is scarce. It is said to improve by keeping.

E. flagellifera, *Nees.* VERN. *Chhimbar.* A nutritious perennial species resembling *dúb,* common on sandy ground. In Bikanir it is said to be the best grass for cattle and sheep.

E. indica, *Gærtn.* Rather a coarse grass, though liked both by horses and cattle.

E. scindica, *Duthie.* Like a slender form of *makra.* Said to be a good fodder grass. Found on sandy ground.

E. verticillata, *Roxb.* Considered to be a good fodder grass in the Panjáb and Rájputana.

Elionurus hirsutus, *Munro.* **(Rottbœllia hirsuta,** *Vahl.* in *Hack. Monogr.*) A characteristic desert grass affording excellent grazing when young. Said to be liked by elephants.

Eragrostis abyssinica. Introduced from Abyssinia, where it is grown as a cereal under the name of *Teff.* Affords exexcellent green fodder.

E. bifaria, *W. & A.* Common on sandy and rocky ground. Eaten by cattle in Rájputana.

E. Brownei, *Nees.* Wet ground. Valued as fodder in Australia.

E. ciliaris, *Link.* Sandy ground, good for grazing.

E. cynosuroides, *R. & S.* VERN. *Dáb.* A coarse deeply-rooting grass frequent on low-lying waste lands. It is much liked by buffaloes.

E. elegantula, *Nees.* Frequent on wet ground. Eaten by cattle.

E. megastachya, *Link.* Used as fodder.

E nutans, *Nees.* Plentiful on damp clay soils. Cattle readily eat it when other grasses fail.

F. 672

| Food and Fodder. | (*J. F. Duthie*). | **FODDER.** |

Eragrostis pilosa, *Beauv.* Relished by buffaloes.

E. plumosa, *Link.* A fairly good fodder grass, varying according to the soil. A dwarf variety with denser flowering spikes is abundant on sandy and saline tracts.

E. tenella, *Beauv.* Common on cultivated ground. It is eaten by cattle both when fresh and as hay, and the seeds which it bears in profusion are said to render it all the more nutritious.

Eriochloa annulata, *Kunth.* Grows in wet places. In Australia it is said to be much relished by stock.

Euchlæna luxurians, *Ascher.* (*Teosinte.*) A native of Guatemala. A quick-growing nutritious annual, but too expensive to cultivate for fodder on a large scale.

Hemarthria compressa, *Kunth.* (**Rottbœllia compressa,** *Linn. f.* in *Hack. Monogr.*) Cattle are fond of this grass. Is said to be highly esteemed in Australia for moist pastures.

Heteropogon contortus, *R. & S.* (Spear-grass.) (**Andropogon contortus,** *Linn.* in *Hack Monogr.*) Common all over India, and up to 7,000 feet on the Himálaya. Largely used as fodder when young, and after the spears have fallen. In Rájputana and Bundelkhund it is regularly stacked after the rains are over. In Australia it is considered to be a splendid grass for a cattle run.

Hordeum vulgare, *Linn.* (Barley.) The grain is often given to horses, and also to cattle when gram is scarce. The *bhus*, or broken-up straw, is considered to be a good fodder, but inferior to that of wheat.

Hygrorhiza aristata, *Nees.* A jhil grass, and usually found floating on the surface of the water. Roxburgh says that cattle are fond of it.

Imperata arundinacea, *Cyrill.* When young, it is relished by cattle, especially after being fired. This grass forms the greater portion of the pasturage in Bengal.

Isachne australis, *R. Br.* Horses and cattle are said to be fond of this grass. It is found usually on wet ground.

Ischæmum angustifolium, *Hack.* (**Pollinia eriopoda,** *Hance.*) Vern. *Bhábar.* Eaten by cattle when young.

I. ciliare, *Retz.* Common in Central India, and occasionally used as a fodder grass.

I. laxum, *R. Br.* Vern. *Sairan* (Rájputana); *Sira* (C. Prov.). Abundant in Rájputana and the Central Provinces, where it is much valued for fodder.

I. pilosum, *Hack.* A common black-soil grass and considered to be good for fodder.

I. rugosum, *Salisb.* Found on wet ground and in paddy-fields, and in its young state is hardly distinguishable from rice. Cattle and horses eat it when young.

Iseilema laxum, *Hack.* Common in Northern and Central India, usually on low-lying land. It is also a characteristic black-soil grass and in Bundelkhund, where it is called *musel*, it is greatly prized for fodder. Buffaloes are very fond of it.

I. Wightii, *Anders.* Associated with the preceding, and apparently not recognized by the natives as distinct.

Kœleria phleoides, *Pers.* A common Mediterranean grass, extending through Afghánistan to the Panjáb. Dr. Aitchison recommends its cultivation in Northern India as a winter fodder grass.

Leptochloa chinensis, *Nees.* Used more or less for fodder.

Lolium perenne, *Linn.* (Perennial Rye-grass.) A well-known and very important fodder grass in Europe, would probably thrive in the Panjáb as a cold-weather crop. It is found wild on the Himálaya.

II.
INDIAN
FODDER
GRASS.

F. 632

FOOD &	Food and Fodder.

Manisuris granularis, *Swartz.* According to Coldstream it is prized and stacked at Hissar, but is not much relished by cattle, though at Ajmere it is considered to be a good fodder grass.

Melanocenchris Royleana, *Nees.* Common on sandy ground, and said to afford good grazing when young. It is, however, too small to be of much account.

Ophiurus corymbosus, *Gærtn.* A common black-soil grass, eaten by cattle when young, or when other grasses fail.

O. perforatus, *Linn.* Found on low-lying pastures. Cattle eat it when it is young and green.

Oplismenus Burmanni, *Retz.* Found usually in shady places. Cattle eat it when young, and it is said to make good hay.

Oryza sativa, *Linn.* (Rice.) Rice-straw is the chief fodder in the Madras Presidency, and is stacked in every district. It is usually kept for a few months to season, and will remain good for three years. It is also very largely used as fodder in Bengal and parts of the Bombay Presidency. In Northern India it is less valued. The young shoots after the rice has been harvested afford good pasturage for sheep in the Ratnagiri district. The husks mixed with oilcake are sometimes given to buffaloes. In Burma and Manipur unhusked rice is frequently given to horses.

Panicum antidotale, *Retz.* VERN. *Ghamur.* A tall coarse-looking grass found in clumps, and often associated with other herbage which, like itself, seeks shelter under prickly bushes. **Wingate** says that more than three fourths of the grass growing in the Changa Manga plantation consists of this species, and that the natives feed their cattle on the green fodder. In the Sirsa Settlement Report it is stated that cattle are apt to be poisoned if they eat it green. At Hissar, however, according to Coldstream, it is grazed only when young, as it afterwards acquires a bitter and saltish taste.

P. colonum, *Linn.* VERN. *Sawánk.* A common weed on cultivated land. Is greedily eaten by all kinds of cattle, both before and after it has flowered, the abundant crop of grain yielded by it adding materially to its nutritive value. Aitchison says that it is sometimes cultivated at Jhelum.

P. Crus-Galii, *Linn.* VERN. *Sánwak.* A coarser plant than the preceding, and usually found near water. Is said to be cultivated in the Lahore district. Cattle, especially buffaloes, are fond of it. In America, where it is known as Barn-yard grass, it is said to be much liked by horses, both when green and dry.

P. distachyum, *Linn.* Common in Northern India. In Australia this species is grown for hay, and is said to be an immense yielder.

P. erucæforme, *Sibth. & Sm.* Common on black and sandy soils in Bundelkhund and Central India, especially on cultivated ground. Yields an abundance of grain.

P. flavidum, *Retz.* Plentiful in the plains, and much liked by cattle and horses. It yields an abundance of grain which contains twice as much oil or fat as that of any other species examined by **Professor A. H. Church.**

P. fluitans, *Retz.* A water-grass. An abundant grain-yielder.

P. frumentaceum, *Roxb.* VERN. *Sánwan.* Grown as a rainy-season crop chiefly for its grain, but occasionally for fodder. The straw is a good fodder and is much used in parts of Mysore and in the Madras Presidency, though ranked below that of *ragi* and rice.

P. helopus, *Trin.* VERN. *Kurí.* Considered to be a very good fodder grass for horses and cattle. It is a common weed of cultivated ground in the plains, and is found also on the Himálaya at moderate elevations.

F. 672

Food and Fodder. (*J. F. Duthie*).	**FODDER.**

Panicum humile, *Nees.* Common in Central India, where it is used as fodder.

P. jumentorum, *Pers.* (Guinea Grass.) A very valuable fodder plant, easily cultivated in the plains, and capable of yielding seven or eight cuttings during the year under irrigation. A single cutting will yield as much as 180 maunds of green fodder. All kinds of stock will thrive on it.

P. miliaceum, *Linn.* Vern. *Chena.* Yields excellent green fodder, and is largely grown for this purpose in some districts of the Panjáb. In parts of Mysore the straw is considered better fodder than that of rice.

P. miliare, *Lamk.* Vern. *Kutki.* A small kind of millet grown largely by the poorer classes in Central and South India on inferior soils. Cattle are fond of the straw.

P. Petiverii, *Trin.* A fairly good fodder grass, but said to be unsuited for hay.

P. prostratum, *Lamk.* A good fodder grass and a heavy seed-yielder.

P. psilopodium, *Trin.* Vern. *Mijhri.* Resembles *Kutki* (**P. miliare**), and is cultivated and utilised in the same manner.

P. repens, *Linn.* A perennial glaucous species occurring in swampy ground. Both Roxburgh and Royle state that cattle are fond of this grass.

P. sanguinale, *Linn.* Vern. *Takri.* Common all over the plains, and up to moderate elevations on the Himálaya. It is largely used as fodder. In America it is known as Crab-grass, and is much valued for pasture as well as hay. A variety with hairy glumes (**P. ciliare,** *Retz.*) is also common, particularly on dry sandy soils, and is largely used for fodder.

Paspalum Kora, *Willd.* Common on wet ground, and eaten by buffaloes.

P. scrobiculatum, *Linn.* Vern. *Koda.* A rainy-season crop yielding a coarse kind of grain used mostly by the poorer classes of people. Cattle should be prevented straying into the fields when this crop is ripening, as the grain, until it has been washed several times, is most unwholesome. The straw is sometimes given to cattle.

Pennisetum cenchroides, *Rich.* Vern. *Dháman* or *anjan.* A most excellent fodder grass, thriving best where the soil is sandy. In the Múltan district it is considered to be the best grass to give to milch cows. Would probably repay cultivation.

P. typhoideum, *Rich.* Vern. *Bájra.* The chopped stalks are considered a good fodder in many parts of India, though inferior to *juár.* In some districts the stalks are left standing after the heads have been removed, and are eaten by cattle. In Káthiawár *bájra* grain is thought better for horses than gram.

Phragmites communis, *Trin.* According to Aitchison this grass is largely collected in Afghánistan for fodder.

Poa annua, *Linn.* A common weed of irrigated ground in West Panjáb and Beluchistan, abundant also on the Himálaya. The foliage is very nutritious, though scanty.

Pollinia argentea, *Trin.* A characteristic black-soil grass. Affords excellent fodder for cattle when young.

Saccharum ciliare, *Anders.* Vern. *Múnj.* The young shoots are eaten by cattle in the Panjáb and are regarded as good fodder for milch cows.

S. officinarum, *Linn.* (Sugarcane). The green tops and the stalks when juicy are sometimes given to cattle.

S. spontaneum, *Linn.* Vern. *Káns.* A tall coarse grass, abundant by the sides of rivers and on low-lying ground. It is much relished by buffaloes, and when young is given to elephants. In the Rohtak district of the Panjáb it is considered good fodder for horses.

II.
INDIAN
FODDER
GRASS.

FOOD &	Food and Fodder.

Setaria glauca, *Beauv.* VERN. *Bandra.* Very common, especially in damp ground. A moderately good fodder, but unsuited for hay.

S. italica, *Beauv.* VERN. *Kangni.* Cultivated for its grain. In parts of Mysore the straw is reckoned as next in quality to that of *ragi.* In the Montgomery district the *bhusa* is considered a strengthening food. It is known in the United States as "Hungarian grass" and is much valued as forage; also in Australia.

S. verticillata, *Beauv.* A coarse grass common in shady places. Cattle eat it when young.

Sorghum halepense, *Pers.* (**Andropogon Sorghum,** *Brot.,* var. **halepensis** in *Hach. Monogr.*). VERN. *Baru.* Said to be good for grazing and for hay, but not considered wholesome until after the rains are over. Opinions, however, are at variance on this point. In Australia it is much valued for pasturage and hay, also in the United States, where it is called "Johnson grass."

S. saccharatum, *Pers.* (**Andropogon Sorghum,** *Brot.,* var. **saccharatus** in *Hack. Monogr.*) Two varieties were introduced into India about 30 years ago, one called *Sorgho,* from China, and the other from Africa, called *Imphi. Sorgho* has taller stems and looser panicles of flowers. It is cultivated in tropical countries for its grain, and in temperate regions for fodder and sugar. The Chinese grow it chiefly for making alcohol. As a fodder plant it is greatly valued. It was first tried in India in 1858, and the result of the experiment showed that though it could not be compared with the ordinary sugarcane of the country as a sugar-yielder, it would prove of great value as a forage plant. Subsequent trials. undertaken chiefly in South India, have confirmed this opinion. The Chief Commissioner of Mysore in his report for 1871 observes:— "With respect to the value of *Sorgho* as an article of fodder, there appears to be no doubt that it will grow fairly in this province as a dry crop, *i.e.,* on land not irrigated during the rainy season, and that if cut for fodder before seeding, it is well suited for cattle, especially milch cows, their milk being enriched to an extraordinary degree by its use in small quantities." **Mr. Phillips'** experiments with *Sorgho* at Allahabad in the years 1872, 1873, 1874 gave some wonderful results in the way of yield and profit. The United States Agricultural Department has declared that the value of *Sorgho* for feeding stock cannot be surpassed by any other crop, as a greater amount of nutritious fodder can be obtained from it in a shorter time, within a given space, and more cheaply. The African *Imphi* is a smaller plant, and though on this account less profitable as a crop it appears to be equally nutritious.

S. vulgare, *Pers.* (**Andropogon Sorghum,** *Brot.,* in *Hack. Monogr.*) VERN. *Juár.* Yields excellent fodder, green or dry, which is largely used in various parts of India. It is often specially grown as a fodder crop under the name of *chari*; in which case it is sown earlier and more thickly than when cultivated for the grain. The stalks of certain juicy varieties afford valuable feeding for milch cattle. The chopped-up straw (*karbi*) is much used as cattle food in Northern India. In the Madras Presidency the straw is less valued than that of *ragi,* but is considered superior to that of rice.

Sporobolus diander, *Beauv.* Said to be eaten by horses and cattle.

S. indicus, *R. Br.* A good pasture grass for horses, also given as fodder when young.

S. pallidus, *Nees.* VERN. *Palengi.* A gregarious species common in moist sandy ground, and affording a considerable amount of forage. A variety called *kálusra* constitutes the greater part of the grass vege-

F. 672

Food and Fodder.	(*J. F. Duthie.*)	FODDER.

tation of the *usar* tracts in the North-Western Provinces, and is always a sure indication of the presence of *reh* salts.

Tetrapogon tetrastachys, *Hack.* (*MS.*). A characteristic usar grass accompanying **Sporobolus pallidus,** var., and often constituting the entire vegetation.

T. villosus, *Desf.* A common Panjáb and Rájputana species, said to be a good fodder grass at Ajmere.

Tragus racemosa, *Hall.* Occurs on sandy ground. According to Coldstream it is much grazed at Hissar, and is very nutritious, but is too small to stack.

Triticum sativum, *Lamk.* (Wheat.) In Northern India green wheat is largely used as fodder. In the Jhang district sheep and goats are allowed to graze on the wheat crops once in order to strengthen the stalks and prevent their being laid by wind. The straw is often given as fodder, but in Mysore it is said to cause distemper. The chaff or *bhúsa* is a well-known form of food. It is sometimes mixed with gram chaff to render it more wholesome.

Zea Mays, *Linn.* (Maize or Indian Corn.) Often given as green fodder, or dried and mixed with other green fodder.

III. HIMALAYAN FODDER PLANTS—EXCLUDING GRASSES.

The following trees, shrubs, and herbs have been recorded as affording food or fodder for cattle, sheep, and goats on the Himálayan Ranges. The fodder-yielding trees of the tropical and temperate zones of India are often severely lopped for the supply of winter fodder to village cattle, especially those of tracts within the region of snowfall. The vegetation of the Alpine tracts form irregular belts above the limits of the upper forests, and chiefly consists of grass herbage, which becomes available for cattle and sheep, during the summer months. The majority of the grasses found on these elevated pastures belong to European genera, and many of the speciest are even botanically identical with those which constitute the finest pasture lands of Great Britain and the Continent of Europe.

Abelia triflora, *R. Br.;* Caprifoliaceæ. Temperate region. Browsed by goats.

Abies Webbiana, *Lindl.;* Coniferæ. Temperate region. On the Panjáb Himálaya the twigs and leaves are cut and stored for use in winter.

Acer pictum, *Thunb.,* and **A. villosum,** *Wall;* Sapindaceæ. Temperate region. The branches are lopped for fodder.

Achillea millefolium, *Linn.;* Compositæ. Temperate and Alpine regions. A perennial herb affording excellent fodder for sheep.

Æsculus indica, *Colebr.;* Sapindaceæ. Temperate region. The foliage is largely used as fodder for cattle, and is sometimes stored for winter use. Cattle and goats feed on the nuts, and these latter are ground and given to horses and mules.

Allardia glabra, *Dcne.;* Compositæ. Alpine region. A perennial herb, browsed by sheep and goats.

Alnus nitida, *Endl.;* Cupuliferæ. Temperate region. The leaves are used as fodder.

Aralia cachemirica, *Dcne.;* Araliaceæ. Temperate region. The leaves of this shrub are eaten by goats.

Artemisia parviflora, *Roxb.;* Compositæ. Temperate region. A perennial herb, browsed by sheep and goats.

A. sacrorum, *Ledeb.* Temperate and Alpine region. Eaten by sheep.

Astragalus multiceps, *Wall.;* Leguminosæ. Temperate and Alpine regions. A shrub occasionally eaten by cattle.

FOOD &	Food and Fodder.

Bauhinia variegata, *Linn.;* LEGUMINOSÆ. Tropical region. The leaves are eaten by cattle.

Betula utilis, *Don.* SYN.—B. **Bhojpattra,** *Wall;* CUPULIFERÆ. Temperate and Alpine regions. Lopped for cattle fodder.

Buxus sempervirens, *Linn.;* EUPHORBIACEÆ. Temperate region. Eaten sparingly by goats, poisonous to other animals.

Caragana pygmæa, *DC.;* LEGUMINOSÆ. Temperate and Alpine regions. A prickly shrub, browsed by goats.

Cedrela serrata, *Royle;* MELIACEÆ. Tropical and temperate regions. The shoots, leaves, and seeds are given to cattle.

Cedrus Libani, *Barrel,* var. **Deodara;** CONIFERÆ. Temperate region. The shoots and young plants of the deodar are browsed by goats.

Celtis australis, *Linn.;* URTICACEÆ. Temperate region. Planted for shade and fodder, and the winter supply of hay is often to be seen stored among its branches.

Cicer soongaricum, *Steph.;* LEGUMINOSÆ. Temperate and Alpine regions. An annual, said to be very fattening for cattle.

Colebrookia oppositifolia, *Smith;* LABIATÆ. Tropical region. Buffaloes eat the leaves of this shrub.

Coriaria nepalensis, *Wall.;* CORIAREÆ. Temperate region. Sheep browse on this shrub.

Cornus capitata, *Wall.;* CORNACEÆ. Temperate region. Eaten by cattle, goats and sheep.

C. macrophylla, *Wall.* Temperate region. The leaves are eaten by goats.

Cotoneaster acuminata, *Lindl.;* ROSACEÆ. Temperate region. Cattle, goats, and sheep eat the leaves.

Cratægus crenulata, *Roxb.;* ROSACEÆ. Temperate region. Sheep and goats eat the leaves of this shrub.

Debregeasia hypoleuca, *Wedd.;* URTICACEÆ. Tropical region. Sheep browse on this shrub.

Desmodium tiliæfolium, *G. Don;* LEGUMINOSÆ. Temperate region. Cattle feed on this shrub.

D. triflorum, *DC.* Tropical region. According to **Roxburgh** cattle are very fond of this herb. **Mueller,** in his *Select Plants, 7th Ed., p. 132,* alludes to this species as "recommendable for places too hot for ordinary clover, and as representing a large genus of plants, many of which may prove of value for pasture." Doubtless several other Himálayan species will be found capable of affording nutritious fodder.

Dolichos biflorus, *Linn.* (*Kulthi* or *Kulath*); LEGUMINOSÆ. Tropical region. A cultivated rainy season crop. The straw is given to cattle.

D. Lablab, *Linn.* Tropical region. Cultivated. The stalks and leaves are excellent fodder for cattle.

Dracocephalum heterophyllum, *Benth.;* LABIATÆ. Alpine region. This herb is browsed by sheep and goats.

Elæagnus latifolia, *Linn.;* ELÆAGNACEÆ. Tropical and temperate regions. The leaves are used as fodder in Jaunsár.

E. umbellata, *Thunb.* Temperate region. The leaves are used as fodder.

Engelhardtia Colebrookiana, *Lindl.;* JUGLANDEÆ. Temperate region. Cattle and goats eat the leaves.

Ephedra vulgaris, *Rich.;* GNETACEÆ. Temperate and Alpine regions. This shrub is browsed by goats.

Eruca sativa, *Lamk.;* CRUCIFERÆ. Cultivated in tropical and temperate regions. Often given as green fodder.

Euonymus fimbriatus, *Wall.;* CELASTRINEÆ. Temperate region. Young shoots and leaves lopped for goats.

F. 673

Food and Fodder. (*J. F Duthie.*)	FODDER.

**III.
HIMALAYAN
FODDER
PLANTS.**

Euonymus Hamiltonianus, *Wall.* Temperate region. Young shoots and leaves lopped for cattle.

Ficus foveolata, *Wall.*; URTICACEÆ. Tropical and temperate regions. Browsed by goats.

F. hispida, *Linn.* Tropical region. Lopped for cattle fodder.

F. nemoralis, *Wall.* Tropical and temperate regions. Used as cattle fodder.

F. palmata, *Forsk.* Tropical region. Used as cattle fodder.

F. religiosa. *Linn.* Tropical region. A favourite fodder of elephants.

F. Roxburghii, *Wall.* Tropical region. The leaves are valued as fodder for cattle and elephants.

F. Rumphii, *Blume.* SYN. **F. cordifolia,** *Roxb.* Tropical region. The leaves are eaten by cattle, goats, and elephants.

F. saemocarpa, *Miq.* Tropical region. The leaves of this shrub are used to feed cattle (*Madden*).

Fraxinus xanthoxyloides, *Wall.*; OLEACEÆ. Temperate region. Much lopped for sheep and goats.

Glycine Soja, *Sieb. & Zucc.*; LEGUMINOSÆ. Cultivated in the tropical region, under the name of *bhat*. The stems and leaves afford excellent fodder for all kinds of stock. [The cultivated plant may be **G. hispida,** *Maxim. Ed.*]

Grewia lævigata, *Vahl.*; TILIACEÆ. Tropical region. Lopped for cattle.

G. oppositifolia, *Roxb.* Tropical and temperate regions The leaves and twigs are stored as winter fodder for sheep and goats.

G. tiliæfolia, *Vahl.*, and **G. vestita,** *Wall.* Tropical region. Both these trees are lopped for fodder.

Hedera Helix, *Linn.*; ARALIACEÆ. Tropical and temperate regions. Goats are fond of ivy leaves.

Heracleum, sp.; UMBELLIFERÆ. Temperate region. Collected in Bissahir and Chamba as winter fodder for goats.

Hiptage Madablota, *Gærtn*; MALPIGHIACEÆ. Tropical region. This climbing shrub is said to afford very good fodder.

Holmskioldia sanguinea, *Retz.*; VERBENACEÆ. Tropical region. Eaten by sheep and goats.

Holoptelea integrifolia, *Planch.*; URTICACEÆ. SYN.—**Ulmus integrifolia,** *Roxb.* Tropical region. Yields fodder for cattle.

Hymenodictyon excelsum, *Wall.*; RUBIACEÆ. Tropical region. The leaves are given to cattle as fodder.

Ilex dipyrena, *Wall.*; ILICINEÆ. Temperate region. The leaves are sometimes given to sheep.

Indigofera pulchella, *Roxb.*; LEGUMINOSÆ. Tropical and temperate regions. Eaten by cattle and goats.

Iris, sp.; IRIDACEÆ. Alpine region. The leaves are used as fodder in Ladak.

Juglans regia, *Linn.*; JUGLANDEÆ. Temperate region. The twigs and leaves of the walnut mixed with hay are often stored in the boughs of trees for winter use.

Limnanthemum nymphæoides, *Link.*; GENTIANACEÆ. This aquatic herb is largely used as fodder in Kashmír, and is said to increase the milk of cows feeding on it.

Lonicera hypoleuca, *Dcne.*; CAPRIFOLIACEÆ. Temperate region. Goats are said to fatten on the leaves of this shrub.

L. quinquelocularis, *Hardw.* Temperate region. The leaves of this shrub are used as cattle fodder.

Lotus corniculatus, *Linn.* (Bird's-foot Trefoil); LEGUMINOSÆ. Temperate region. Valued for grazing and for hay in Europe and Australia.

F. 673

FOOD &	Food and Fodder.

**III.
HIMALAYAN
FODDER
PLANTS.**

Marlea begoniæfolia, *Roxb.;* CORNACEÆ. Tropical and temperate regions. The leaves are collected for sheep fodder.

Medicago falcata, *Linn.;* LEGUMINOSÆ. Wild and cultivated on the Western Himálaya.

M. sativa, *Linn.* Lucerne is cultivated to a small extent at most of the Himálayan stations as green fodder for horses.

Morus serrata, *Roxb.;* URTICACEÆ. Temperate region. The branches are lopped for cattle fodder.

Myricaria elegans, *Royle,* and **M. germanica,** *Desr.;* TAMARISCINEÆ. Temperate and Alpine regions. Sheep are said to browse on these shrubs.

Olea cuspidata, *Wall.;* OLEACEÆ. Tropical and temperate regions. The leaves are bitter and are considered to be one of the best kinds of fodder for goats and sheep. Also said to be good for cows and milch buffaloes, both increasing the quantity and improving the quality of their milk.

O. glandulifera, *Wall.* Tropical and temperate regions. The leaves are eaten by cattle, sheep, and goats.

Otostegia limbata, *Benth.;* LABIATÆ. Tropical region. Goats are said to browse on this bush on the Panjáb Himálaya.

Ougeina dalbergioides, *Benth.;* LEGUMINOSÆ. Tropical region. The branches are lopped as fodder for cattle and sometimes for elephants.

Oxalis corniculata, *Linn.;* GERANIACEÆ. A common weed in the tropical and temperate regions. Cattle, sheep, and goats eat the plant.

Oxybaphus himalaicus, *Edgew.;* NYCTAGINEÆ. Dry temperate region. This herb is collected for winter fodder.

Oxytropis microphylla, *DC.;* LEGUMINOSÆ. Alpine region. Sheep and yaks are said to browse on this perennial herb.

Phaseolus aconitifolius, *Jacq.;* LEGUMINOSÆ. (VERN. *Moth.*) This, as well as *mung* (**P. Mungo**), *urd* (**P. radiatus**), and **P. trilobus,** are cultivated to some extent by the villagers in the warmer regions of the Himálaya, and, as in other parts of India, the leaves, stems, and chaff are available as cattle food.

Physochlaina præalta, *Hook. f.;* SOLANACEÆ. Dry Alpine region. Used as cattle fodder in Lahoul.

Picea Morinda, *Link.* SYN.—**Abies Smithiana,** *Forbes;* CONIFERÆ. Himálayan Spruce. Temperate region. Affords fodder for sheep and goats.

Picrasma quassioides, *Benn.;* SIMARUBEÆ. Tropical and temperate regions. The leaves are eaten by sheep and goats.

Pistacia integerrima, *Stewart;* ANACARDIACEÆ. Tropical and temperate regions. The twigs and leaves are a favourite food of buffaloes and camels.

Pisum sativum, *Linn.;* LEGUMINOSÆ. The common pea is cultivated on the Western Himálaya up to 13,000 feet; at the higher elevations it does not ripen its seed, and is then used as fodder.

Polygonum aviculare, *Linn.;* POLYGONACEÆ. Temperate region. Sheep and goats are said to fatten when fed on this plant.

P. chinense, *Linn.* Tropical and temperate regions. Cattle are fond of this species. Many other kinds of **Polygonum** are found at various elevations on the Himálaya, and are used more or less as fodder.

Populus balsamifera, *Linn.;* SALICINEÆ. Inner ranges of Western Himálaya. The branches are often lopped for cattle fodder.

P. ciliata, *Wall.* Temperate region. Affords fodder for goats.

P. nigra, *Linn.* (Lombardy Poplar.) Is cultivated in the temperate regions of the Western Himálaya, and the branches are often lopped for cattle fodder.

| Food and Fodder. | (*J. F. Duthie.*) | FODDER. |

Potamogeton crispus, *Linn.;* NAIDACEÆ. This aquatic plant is said to be used as fodder in Ladák. **P. gramineus, P. lucens,** and **P. natans,** are similarly used in other parts.

Potentilla fruticosa, *Linn.;* ROSACEÆ. Temperate and Alpine regions. This shrub is browsed by sheep.

P. Salessovii, *Steph.* Dry Alpine region. Is browsed by sheep.

Prunus Padus, *Linn.;* ROSACEÆ (Bird cherry). Temperate region. Yields excellent fodder for cattle.

Pueraria tuberosa, *DC.;* LEGUMINOSÆ. Tropical region. The leaves are considered to be very good fodder for horses. The tubers, chopped up, are also sometimes given.

Pyrus Pashia, *Ham.;* ROSACEÆ. Tropical and temperate regions. Cattle and goats eat the leaves.

Quercus dilatata, *Lindl.;* CUPULIFERÆ. Temperate region. The leaves are prized for feeding sheep and goats.

Q. Ilex, *Linn.* Temperate region. The leaves are stored for winter fodder.

Q. incana, *Roxb.* Temperate region. The leaves are given to cattle and sheep.

Q. lanuginosa, *Don.* Temperate region. The leaves are used as fodder.

Q. semicarpifolia, *Smith.* Temperate region. The leaves are stored as winter fodder for cattle.

Randia dumetorum, *Lamk.;* RUBIACEÆ. Tropical region. The leaves are used as fodder for cattle, sheep, and goats.

R. uliginosa, *DC.* Tropical region. The leaves are browsed by cattle.

Rhus parviflora, *Roxb.;* ANACARDIACEÆ. Tropical region. Cattle and goats eat the leaves.

Salix acmophylla, *Boiss.;* SALICINEÆ. Tropical region. The tree is often lopped for cattle fodder.

S. daphnoides, *Vill.* Temperate and dry Alpine regions. Yields fodder for cattle.

S. elegans, *Wall.* Temperate region. Cattle are fond of the leaves.

S. tetrasperma, *Roxb.* Tropical and temperate regions. This tree is often lopped for cattle fodder.

Sapindus Mukorossi, *Gærtn.;* SAPINDACEÆ. Tropical region. The leaves are given to cattle.

Saurauja napaulensis, *DC.;* TERNSTRÆMIACEÆ. Tropical and temperate regions. The leaves are lopped for cattle fodder.

Smithia sensitiva, *Ait.;* LEGUMINOSÆ. Tropical region. A small annual, said to make excellent hay.

Sonchus oleraceus, *Linn.;* COMPOSITÆ. Tropical and temperate regions. Cattle are fond of this plant.

Streblus asper, *Lour.;* URTICACEÆ. Tropical region. Lopped extensively for fodder.

Syringa Emodi, *Wall.;* OLEACEÆ. Temperate and Alpine regions. The leaves are eaten by goats.

Tanacetum senecionis, *Gay;* COMPOSITÆ. Alpine and Western Himálaya. Browsed by goats.

Taxus baccata, *Linn.;* CONIFERÆ. Temperate region. In Europe goats, sheep, and rabbits eat the leaves of the Yew freely. Brandis says that the leaves are considered poisonous, but not everywhere, nor under all circumstances.

Terminalia Chebula, *Retz.,* and **T. tomentosa,** *Bedd.;* COMBRETACEÆ. Tropical region. Afford fodder for cattle.

Trifolium fragiferum, *Linn.;* LEGUMINOSÆ. Temperate region. Used in Kashmír as fodder for cattle.

FOOD &	Food and Fodder.

**III.
HIMÁLAYAN
FODDER
PLANTS.**

Trifolium pratense, *Linn.* Temperate region. Well known in Europe as Red or Broad Clover. It grows wild on the Himálaya and is occasionally collected for fodder.

T. repens, *Linn.* Dutch or White Clover. Temperate and Alpine regions. An essential constituent of every good pasture in Europe. It is plentiful on the Himálaya as a wild plant.

Tulipa stellata, *Hook.;* LILIACEÆ. Tropical and temperate regions. The bulbs are eaten by cattle.

Ulmus Wallichiana, *Planch.;* URTICACEÆ. Temperate region. Lopped extensively for cattle fodder.

Vicia hirsuta, *Koch.;* LEGUMINOSÆ. Tropical and temperate regions. Occasionally cultivated as a fodder plant under the name of *masur chana* up to 5,000 feet in Kumáun. Cattle and goats eat it.

Vigna Catiang, *Endl.;* LEGUMINOSÆ. A variety called *Lobiya-riánsh* is cultivated in the tropical region and affords fodder for cattle.

V. vexillata, *Benth.* Temperate region. Cattle and goats eat this plant.

Wendlandia exserta, *DC.;* RUBIACEÆ. Tropical region. Cattle eat the leaves.

Woodfordia floribunda, *Salisb.;* LYTHRACEÆ. Tropical region. Cattle and goats eat the leaves.

Wrightia tomentosa, *R. & S.;* APOCYNACEÆ. Tropical region. The leaves are eaten by cattle.

Xanthium strumarium, *Linn.;* COMPOSITÆ. Tropical region. A common weed of cultivated ground. Probably introduced from America, where it is said that cattle eat the young plants.

Zizyphus oxyphylla, *Edgew.;* RHAMNEÆ. Tropical and temperate regions. Goats are fond of the leaves.

Z. xylopyra, *Willd.* Tropical region. The young shoots, leaves, and fruit are eaten by cattle and goats.

**IV.
HIMÁLAYAN
GRASSES.
674**

IV. HIMÁLAYAN GRASSES.

The gradual changes which determine the character of the Flora at different altitudes on the Himálayan Ranges is well exemplified in the case of grasses. As we ascend from the plains, the sub-tropical forms are gradually lost sight of, other species and genera taking their place. On reaching an elevation of about 7,000 or 8,000 feet, the majority of the species are found to be characteristic of a temperate climate, many European genera, such as **Avena, Brachypodium, Bromus, Dactylis,** and **Festuca,** being represented. At still higher elevations, and up to the limit of melting snow, we meet with many species identically the same as occur on the mountains of Europe and America and along the shores of countries within the Arctic region.

Although very little is known concerning the nutritive value of Himálayan fodder grasses individually, it is, nevertheless, certain that excellent pasturage is obtainable at every elevation during certain seasons of the year. The wide open stretches of grass land (maidáns) extending from the upper limits of the forests towards the snow line constitute the finest feeding grounds for cattle and sheep during the summer months. Many of the grasses which flourish in these elevated meadows are known to be highly prized constituents of the best European pastures, and with them are found many allied species which analysis would no doubt prove to be equally valuable.

**A.
SUB-TROPI-
CAL.
674**

A.—The following is a list of the more important plains or sub-tropical fodder yielding species which are found at various elevations approaching the temperate region :—

Andropogon annulatus, *Forsk.*

Food and Fodder.	(*J. F. Duthie.*)	FODDER.

Andropogon intermedius, *R.Br.* Var. **punctata.**
A. Ischæmum, *Linn.*
A. Schœnanthus, *Linn.*
A. serratus, *Thunb.* (Syn.—A. tropicus, *Spreng.*)
A. Trinii, *Steud.* (Syn.—Chrysopogon serrulatus, *Trin.*)   Largely used as fodder.
Anthistiria ciliata, *Linn. f.*
Apluda aristata, *Linn.* Abundant and largely used as fodder.
Arthraxon ciliaris, *Beauv.*
A. echinatus, *Hochst.*
A. lanceolatus, *Hochst.*
A. microphyllus, *Hochst.*
Arundinella nepalensis, *Trin.* Largely represented in the bundles of grass supplied for horses and cows at Simla.
A. Wallichii, *Nees.*
Arundo madagascariensis, *Kunth.*
A. mauritanica, *Desf.*
Chionachne barbata, *R. Br.*
Chloris digitata, *Steud.*
Coix lachryma, *Linn.*
Cynodon Dactylon, *Pers.* (*Dúb.*)
Eleusine ægyptiaca, *Pers.* (*Makra*).
E. Coracana, *Gærtn.* (*Mandua*). Cultivated.
E. indica, *Gærtn.*
Eragrostis Brownei, *Nees.*
E. elegantula, *Nees.*
E. megastachya, *Link.*
E. pilosa, *Beauv.*
E. plumosa, *Link.*
E. poæoides, *Beauv.*
E. tenella, *Beauv.*
E. uniloides, *Nees.*
Heteropogon contortus, *R. & S.*   (Spear-grass.)
Imperata arundinaceæ, *Cyrill.*
Isachne australis, *R. Br.*
Ischæmum rugosum, *Gærtn.*
Manisuris granularis, *Swartz.*
Ophiurus perforatus, *Trin.*
Oplismenus Burmanni, *Retz.* Grows well under the shade of trees.
Oryza sativa, *Linn.* (Rice.) Cultivated.
Panicum ciliare, *Retz.*
P. colonum, *Linn.*   (*Sawánk.*)
P. Crus-Galli, *Linn.*
P. flavidum, *Retz.*
P. frumentaceum, *Roxb.* Cultivated.
P. helopus, *Trin.* (*Kuri.*)
P. miliaceum, *Linn.* (*Chena.*) Cultivated up to 11,000 feet. It yields very nutritious fodder in the green state.
P. Petiverii, *Trin.*
P. psilopodium, *Trin.*
P. sanguinale. *Linn.* (*Takria.*)
Paspalum scrobiculatum, *Linn.* (*Kodon.*)   Cultivated.
Pennisetum typhoideum, *Rich.* (*Bájra.*) Cultivated.
Pogonatherum saccharoideum, *Beauv.*
Pollinia argentea, *Trin.*
Rottbœllia exaltata, *Linn. f.*

IV.
HIMALAYAN
GRASSES.
A.
SUB-TROPI-
CAL.

2 F

F. 634

FOOD &	Food and Fodder.

IV. HIMALAYAN GRASSES.

Saccharum spontaneum, *Linn.* (*Káns.*)
Setaria glauca, *Beauv.*
S. intermedia, *R. & S.*
S. italica, *Beauv.* (*Kangni.*) Cultivated.
S. verticillata, *Beauv.*
Sorghum halepense, *Pers.* (*Baru.*)
Sporobolus diander, *Beauv.*
S. indicus, *R. Br.*
Zea Mays, *Linn.* (Indian Corn.) Cultivated.

B. TEMPERATE.

B. The names of the species included in the list which follows, are, more strictly speaking, those of Himálayan grasses, excepting a few, growing within the temperate region, which occur also on the more elevated portions of Central and Southern India. Our knowledge of the grass vegetation of the Himálaya is by no means complete, and several species have yet to be determined botanically.

Agropyrum caninum, *R. & S.* Alpine region.
A. longiaristatum, *Boiss.* Alpine region.
A. semicostatum, *Nees.* Temperate and Alpine regions.
Agrostis alba, *Linn.* (Fiorin, or Creeping Bent grass) Temperate region. A variety of this (stolonifera) is a well-known fodder grass in Europe and is useful for mixing with other grasses. See *Sutton's Permanent and Temporary Pastures, p. 25;* and *Stebler, and Schröter, Best Forage Plants, p. 65 (Eng. Ed.).*
A. ciliata, *Trin.* Alpine region.
A. Hookeriana, *Munro.* Temperate and Alpine regions.
A. pilosula, *Trin.* Temperate region.
A. Roylei, *Trin.* Temperate and Alpine regions
Alopecurus pratensis, *Linn.* (Meadow Fox-tail Grass.) Temperate and Alpine regions. One of the best of English pasture grasses. See *Sutton's Permanent and Temporary Pastures, p. 26; Stebler and Schröter, Best Forage Plants, p. 65 (Eng. Ed.).*
Andropogon distans, *Nees.* Temperate region.
A. Gryllus, *Linn.* SYN.—Chrysopogon Gryllus, *Trin.* Sub-tropical and temperate regions.
A. micranthus, *Kunth.*, var. villosulus. Sub-tropical and temperate regions; also on Parasnáth and Mount Abu. [Abu.
A. montanus, *Roxb.* Sub-tropical and temperate regions; also on Mount
A. Nardus, *Linn.*, var. exsertus. Sub-tropical and temperate regions.
A. tristis, *Nees.* Temperate region.
Anthistiria anathera, *Nees.* Sub-tropical and temperate regions. It is much thought of by the hillmen as a good fodder grass.
Anthoxanthum odoratum, *Linn.* Temperate region. Probably introduced. A perennial grass, thriving in all kinds of soil.
Arthraxon submuticus, *Nees.* Sub-tropical region.
Arundinaria falcata, *Nees.*
A. Falconeri, *Benth. & Hk. f.* } Temperate region.
A. spathiflora, *Trin.*
Arundinella setosa, *Trin.* Sub-tropical and temperate regions.
Avena pratensis. *Linn.* (Meadow Oat Grass.) Alpine region. Recommended in Europe for dry soils.
A. pubescens, *Linn.* (Downy Oat Grass.) Temperate region. Grown in Europe for fodder.
A. sativa, *Linn.* (Oats.) Cultivated up to the Alpine region.
A. virescens, *Nees.* Alpine region.
Brachypodium pinnatum, *Beauv.* Temperate region.

F. 674

Food and Fodder.	(*J. F. Duthie.*)	FODDER.

Brachypodium sylvaticum, *R. & S.* Temperate region.

Briza media, *Linn.* (Quaking grass.) Temperate and Alpine regions. A familiar ingredient in English pastures, especially on a dry soil.

Bromus arvensis, *Linn.*

B. asper, *Murray.* Temperate region. Recommended in Europe for wooded localities.

B. confertus, *Bieb.*

B. confinis, *Nees.* Temperate and Alpine regions.

B. crinitus, *Boiss.* Alpine region.

B. Danthoniæ, *Trin.* Temperate and Alpine regions.

B. inermis, *Leyss.* Temperate region.

B. japonicus, *Thunb.* Temperate and Alpine regions.

B. membranaceus, *Jacqm.* Temperate region.

B. mollis, *Linn.* Temperate and Alpine regions.

B. patulus, *Mert. & Koch.* Alpine region.

B. squarrosus, *Linn.* Temperate region.

B. tectorum, *Linn.* Temperate region.

Calamagrostis nepalensis, *Nees.* Temperate region.

C. scabrescens, *Griseb.*, var. elatior, and var. humilis. Alpine region.

Dactylis glomerata, *Linn.* (Cock's-foot Grass.) Temperate region. Highly valued in Europe as a fodder grass for cattle. See *Suttons' Permanent and Temporary Pastures, p. 34: Stebler and Schröter, Best Forage Plants, p. 30 (Eng. Ed.).*

Danthonia kashmiriana, *Jaub. & Spach.* Alpine region. Considered by the hill-men to be a good fodder grass. Some of the Australian species of **Danthonia** are much valued.

Deschampsia cæspitosa, *Beauv.* Alpine region.

Elymus dasystachyus, *Trin.* Alpine region.

E. nutans, *Griseb.* Temperate region.

E. sibiricus, *Linn.* Alpine region.

Festuca dura, *Vill.* Kashmír.

F. elatior, *Linn.* Temperate region. (Tall|Fescue.) Much used in Europe for fodder and considered very nutritious. See *Suttons' Permanent and Temporary Pastures, p. 40.*

F. filiformis, *Jacqm.* Alpine region.

F. gigantea, *Vill.* Temperate region.

F. ovina, *Hack.* (Sheep's Fescue.) Alpine region. Well known in Europe as affording excellent grazing for sheep, but unsuitable for hay. There are several varieties, of which the following are Himálayan :—**F. ovina,** *Linn.*, the true Sheep's Fescue; F. duriscula, *Linn.*, or Hard Fescue; **F. valesiaca,** *Schleich,* and F. supina, *Hack.*, all occurring within the Alpine region. See *Suttons' Permanent and Temporary Pastures, p. 45 : Stebler and Schröter, Best Forage Plants, p. 88.*

F. rubra, *Linn.* (Red or Creeping Fescue.) Temperate region. Differs from F. ovina by its stoloniferous habit and the reddish brown foliage. It is cultivated in Europe, and is found to stand drought well.

F. scaberrima, *Nees.* Temperate region.

F. spadicea, *Linn.* Alpine region.

Garnotia adscendens, *Munro, MS.* Temperate region.

Glyceria aquatica, *Presl.;* var. caspica. Temperate region.

G. fluitans, *R. Br.* (Manna grass.) Temperate region.

Graphephorum nutans, *Munro.* Alpine region. Evidently a good fodder grass.

Hierochloe laxa, *R. Br.* Alpine region. It emits during the process of drying a perfume like that of the English hay-scented grass Anthoxanthum odoratum. H. borealis of Western Europe and H. redolens,

FOOD &	Food and Fodder.

inhabiting the mountains of Australia and New Zealand, have the same properties.

Hordeum murinum, *Linn.* Temperate region, descending to the plains in North-Western Panjáb.

H. pratense, *Linn.*
H. sylvaticum, *Huds.* } Alpine region.

H. vulgare, *Linn.* (Barley.) Cultivated up to the Alpine region. There are many varieties, including **H. ægiceras,** a beardless kind found in Tibet, and Siberian barley (**H. cæleste**). A third variety known in North Kumaun as *oi jáu* is cultivated for the manufacture of a strong spirit.

Isachne albens, *Trin.* Temperate region.

Ischæmum Hugelii, *Hack.* Temperate region.

I. notatum, *Hack.* *Monogr., p. 246.* Temperate region of East Kumáon.

Kæleria cristata, *Pers.* Temperate region. Regarded in Europe as a fairly nutritious grass.

Lolium perenne (Perennial Rye-grass). Alpine region. Largely cultivated in Europe, and a valuable constituent of the best pasture land. There are very many varieties. See *Suttons' Permanent and Temporary Pastures, p. 49 ; Stebler and Schröter, Best Forage Plants, p. 20* (*Eng. Ed.*).

L. temulentum, *Linn.* (Darnel). Temperate region ; also occurring as a weed of cultivation in the plains of North-Western Panjáb. The grain is very liable to become ergotized.

Melica ciliata, *Linn.* Temperate and Alpine regions. Mueller says "a perennial fodder grass particularly desirable for sheep." The following species are also recorded as occurring in the Alpine region:—**M. Jacquemontii,** *Dcne.,* **M. micrantha,** *Nees,* **M. persica,** *Kunth.,* **M. secunda,** *Regel,* and **M. vestita,** *Boiss.*

Milium effusum, *Linn.* (Millet Grass.) Temperate region. It is said to be relished by cattle in Europe, and the grain can be used like millet.

Muehlenbergia Hugelii, *Trin.*
M. geniculata, *Nees.*
M. sylvatica, *Trin.* } Temperate region.
M. viridissima, *Nees.*

Oplismenus acuminatus, *Nees.* Temperate region.

O. compositus, *R. & S.* Sub-tropical region.

O. undulatifolius, *R. & S.* Temperate region.

Oryzopsis paradoxa, *Nutt.* Temperate region. Besides the above are four or five other species, not satisfactorily determined, some of which are found within the Alpine region.

Panicum excurrens, *Trin.* Sub-tropical and temperate regions. Foliage like that of P. plicatum.

P. neurodes, *Schult.* Sub-tropical region.

P. vestitum, *Nees.* Sub-tropical and temperate regions.

Paspalum jubatum, *Griseb.* Temperate region.

P. minutiflorum, *Steud.* Sub-tropical region.

Pennisetum flaccidum, *Griseb.* Temperate and Alpine regions. Often a weed of cultivation at high elevations.

P. lanatum, *Klotsch.* Dry temperate region.

P. triflorum, *Nees.* Sub-tropical and temperate regions ; abundant.

Phleum alpinum, *Linn.* (Alpine Catstail.) Alpine region.

P. arenarium, *Linn.*
P. asperum, *Vill.* } Temperate region.

P. pratense, *Linn.* (Timothy, or Meadow Catstail.) Extensively cultivated in Europe and much valued for pastures on a heavy soil. **Royle** records it from the Chor Mountain. See *Suttons' Permanent and Temporary Pastures, p. 58; Stebler and Schröter, Best Forage Plants, p. 52 (Eng. Ed.)*.

Phragmites communis, *Trin.* On the inner Panjáb Himálaya up to 14,000 feet; also in the plains of the North-Western Panjáb, and Afghánistan, where, Dr. Aitchison states, it is largely collected for fodder.

Poa alpina, *Linn.* (Alpine Meadow grass.) Alpine region.

P. annua, *Linn.* Sub-tropical and temperate regions, reaching the plains in the North-Western Panjáb. Common in Europe, where it is considered good for early pasturage.

P. arctica, *Br.*
P. attenuata, *Trin.*
P. bulbosa, *Linn.*
P. cenisia, *All.* } Alpine region.
P. compressa, *Linn.*
P. laxa, *Hænke.*
P. nemoralis, *Linn.*

P. pratensis, *Linn.* (Smooth-stalked Meadow Grass.) Alpine region. This species is much valued in Europe for early hay. It is the Blue Kentucky grass of the United States. See *Suttons' Permanent and Temporary Pastures, p. 60; Stebler and Schröter Best Forage Plants, p. 72 (Eng. Ed.).*

P. soongarica, *Boiss.*

P. trivialis, *Linn.* (Rough-stalked Meadow Grass.) Has been found in Western Tibet. This grass is valued in Europe for rich moist pastures. See *Suttons' Permanent and Temporary Pastures, p. 62; Stebler and Schröter, Best Forage Plants, p. 77 (Eng. Ed.)* There are many other Himálayan species which have not yet been botanically determined.

Pollinia ciliata, *Trin.* Temperate region.

P. hirtifolia, *Hack. Monogr. p. 165.* Temperate region.

P. japonica, Syn.—*Miscanthus sinensis, Anders.,* in *Hack., Monogr. p. 105.* Temperate region.

P. Lehmanni, *Nees* } Temperate region.
P. mollis, *Hack.*

P. nepalensis. Syn.—*Miscanthus nepalensis, Hack. Monogr. p. 104.*

P. nuda, *Trin.*

P. phæothrix, *Hack., Monogr., p. 168.*
P. velutina, *Hack.* Syn.—*Erianthus velutinus, Munro, MS.* } Temperate region.

Polypogon fugax, *Nees.* Sub-tropical and temperate regions, in wet ground.

Rottbællia speciosa, *Hack.* Syn.—*Ischæmum speciosum, Nees;* **Vossia speciosa.** Temperate region.

Setaria viridis, *Beauv.* Temperate and Alpine regions, usually occurring as a weed of cultivation.

Sporobolus ciliatus, *Presl.* Sub-tropical and temperate regions.

Stipa (Orthoraphium) Roylei, *Nees.*) Temperate and Alpine regions.

S. sibirica, *Lamk.* Temperate region. A poisonous grass, abundant in Kashmír and Hazára, extending east to Kumáon.

S. (Lasiagrostis) splendens, *Kunth.* Alpine region.

Tripogon bromoides, *R. & S.* } Sub-tropical and temperate regions.
T. filiformis, *Nees.*

Trisetum aureum, *Nees.* } Alpine region.
T. subspicatum, *Beauv.*

Triticum sativum, *Lamk.* Wheat is cultivated at various elevations, and in Tibet has been observed at 16,000 feet above the sea.

FRAGARIA vesca.	Strawberries.

Forbidden Fruit, see **Citrus decumana,** *Linn.;* Vol. II., 348.

Forest Trees, see **Timbers.**

(J. Murray.)

FORSKOHLEA, *Linn.; Gen. Pl., III., 393.*

675

Forskohlea tenacissima, *Linn.; Fl. Br. Ind., V., 593;* Urticaceæ.

Habitat.—Said to be a native of India, occurring at Simla (*Stocks*) and in the Panjáb (*Jacquemont, Fleming*), extending to Afghánistan and Beluchistan.

FIBRE.
Bark.
676

Fibre.—The bark yields a strong fibre: hence the origin of the specific name, but no definite information is obtainable regarding its economic use.

Fourcroya, *Schult.,* see **Furcrœa,** *Vent.*

Foxglove Purple, Digitalis purpurea, *Linn.;* Scrophularineæ.

677

A European plant, naturalised in gardens in the temperate regions of India.

FRAGARIA, *Linn.; Gen. Pl., I., 620.*

A genus of perennial herbs, belonging to the Natural Order Rosaceæ, of which the swollen fleshy receptacle forms the Strawberry. Distributed through the temperate regions of the Northern Hemisphere, South America, the Sandwich Islands, and Bourbon.

[Rosaceæ.

678

Fragaria indica, *Andr.; Fl. Br. Ind., II., 343; Wight, Ic., t. 989;* The Indian Strawberry.

Syn.—F. malayana, *Roxb.;* F. nilgirica, *Zenker;* F. arguta, *Lindl.;* F. roxburghii, *W. & A.;* Duchesnea fragariodes, *Sm.;* D. chrysantha, *Miq.;* D. fragiformis, *Don.;* Potentilla denticulora and Wallichiana, *Ser.;* P. Durandii, *Torr. & Gr.;* P. fragariæfolia, *Klotsch;* P. trifida, *Lehm.*

Vern.—*Paljor, kansars, ingrach, yangtarsh, búnún musrini, bana-phal, tawai,* Pb.

References.—*Roxb., Fl. Ind., Ed. C.B.C., 409; Stewart, Pb. Pl., 80; Atkinson, Ec. Prod., N.-W. P., Pt. V., 68, 69; Gazetteer, N.-W. P., X., 309; Balfour, Cyclop., 1149.*

Habitat.—This plant (a small yellow-flowered Fragaria) grows on the Himálaya from east to west, at altitudes of 5,000 to 8,000 feet; also on the Khásia Hills and Nilghiris.

FOOD.
Fruit.
679
680

Food.—The indigenous strawberry yields abundantly a very insipid fruit, which, however, can be much improved by cultivation.

F. nilgerrensis, *Schld.; Fl. Br. Ind., II., 344; Wight, Ic., t. 988.*

Syn.—F. elatior, *W. & A.*

Habitat.—A species, which may turn out to be only a variety of F. vesca, found on the Khásia and Nilghiri mountains.

It is a robust form and bears a large strawberry, globose in form but inclined to be conoidal in the Nilghiris and flattened in the Khásia hills; is of a pale pinkish-white colour.

FRUIT.
681

Fruit.—There is no account of its cultivation, but it might when crossed with F. vesca yield a fine variety of strawberry.

682

F. vesca, *Linn.; Fl. Br. Ind., II., 344.*

The Strawberry.

Vern.—*Kansars, ingrach, bunun, tawai, tash, fraga, bana-phal,* Pb.

References.—*Stewart, Pb. Pl., 80; DC., Origin, Cult. Pl., 203; Firminger, Manual of Gard. for Ind., Part II., 252; Atkinson, Him. Dist., 309, 713; Lisboa, U. Pl. Bomb., 155; Birdwood, Bomb. Pr., 150; Balfour, Cyclop.,*

F. 682

| Strawberries. | (*J. Murray.*) | FRAGARIA vesca. |

III., 744; Smith, Dic., 394; Treasury of Bot., I., 504; Gasetteer of the Simla Dist., 12; Trans. Agri.-Hort. Soc., I., 21 (Proc.), 241; IV., 106; V. (Proc.), 5; VI., 247, 235; Jour. Agri.-Hort. Soc. (Old Series), IV. (App.), 83; VII., 285; VIII., 214; (New Series), III., 114.

Habitat.—Found wild in the temperate Himálaya from Murree and Kashmír, altitude 5,000 to 10,000 feet, to Sikkim, altitude 6,000 to 13,000 feet (*Hooker*); 6,000 to 10,000 feet in Manipur; also found in the Ruby Mines and Bhamo districts of Burma. The plant was quite neglected by the natives of India till its cultivation was commenced in the gardens of Europeans. It is significant that in the *Ain-i-Akbari*, a work which treats in the utmost detail with the fruits cultivated, during the reign of Akbar, in India, Kashmír, and Afghánistan, no mention is made of the strawberry.

Dr. Stewart says that the fruit of the Himálayan plant is excellent when gathered dry, and improves by cultivation. It is one of the most wholesome of fruits.

CULTIVATION.

CULTIVATION.

HISTORY OF.—Since the first introduction of the cultivation of the strawberry into India, the plant has spread in the most remarkable way in the plains, from Behar in the south, to Peshawar in the north. At first the experiment of its cultivation was tried only in the hills, where the temperature and natural conditions resembled those enjoyed by the fine fruit-producing plant in Europe; but it has since been grown with marked success in the Panjáb, the North-Western Provinces, and Behar. It withstands remarkably well the great heat of the hot weather, and produces fruit abundantly and of very good quality from February to May; the season of ripening varying in different parts. The *Madras Manual of Administration* (II., 27, 85, 124) reports **F. vesca** as thriving fairly well on parts of the Western Ghâts, and in the Shevaroys. In Lower Bengal, and the plains of Madras and Bombay, on the other hand, the plant does not thrive; it is seemingly unable to withstand the *moist* heat of those provinces.

The earliest obtainable record of successful cultivation in the plains is one in the *Trans. Agri.-Hort. Soc.* (I., 21) by **Dr. Tytler**, in which he refers to the plant as growing to perfection on the banks of the Jumna near Allahabad. It is not, however, definitely mentioned whether the plants alluded to were English stock or the indigenous **F. vesca**, but subsequent records show that both have been tried, and that the strawberry of the Indian market now probably contains a strain of both.

METHOD OF.—The strawberry thrives best in a light soil with old stable and vegetable manure at first, but as soon as it begins to flower it ought to have goat's or sheep's dung applied round the roots.

The following is the method laid down by **Firminger** in his *Manual of Gardening for India:*—

" The time for planting out young strawberry plants is about the beginning of October. I have put them out a month earlier than this, but without advancing the growth of the plants in the slightest degree. The finest fruit in England is obtained from plants of two years old, but in this country it seems all but universally agreed that young plants only of the current year's growth can be employed with success.

" Having chosen a piece of ground fully exposed to the sun, dig rows of holes in it eight inches in diameter, and six inches deep, the holes a foot apart, and the rows also a foot asunder. Fill the holes with a mixture of equal parts of old cow-manure, leaf mould, and common soil, and in each put down a strawberry plant. Water the plants at the time, and as often afterwards as they seem to require. When they become well

History.
683

Method of.
684

FRAXINUS excelsior.	**Strawberries ; the Common Ash.**

CULTIVA-
TION.

established they will perhaps begin to send out runners. Then it would be well to remove, though some persons are of opinion that the doing so causes a larger development of leaves than is favourable to the productiveness of the plant. By February they will have become good large plants, and may be expected then to be in full blossom."

The strawberry may be propagated either by seed or by rooted runners, but varieties can only be obtained from sports in seedlings or by hybridization.

Regarding its cultivation in Bombay, the Director of Land Records and Agriculture has furnished the following report, dated September 1889 : —" Though it is much met with in gardens above the Ghâts it can only be successfully grown on the two hill stations of Mahábleshwar and Panchgani, where the fruit develops to a good size. The climate of the plains does not seem to agree with the plant. In Gondal and Kathiawár the plant was twice or thrice tried without success. Towards the end of 1887 about 2,000 strawberry plants were sent from Saharanpur to Mahábleshwar and were distributed amongst cultivators. The plants have taken kindly to the soil, and the plantations are in a flourishing condition. The cultivation of the strawberry has not, however, gone as yet beyond the experimental stage."

The history of the ready adaptability of **F. vesca** to the intense dry heat of the plains of Behar, the Central Provinces, and Upper India, and of the greatly increasing production of the fruit, encourages the hope that the cultivation of the strawberry, in the vicinity of hill stations, and of towns in the plains of which the climatic conditions are favourable, may become a large branch of market gardening. The outturn on even a very small area is very great in comparison to the outlay of money required; but the crop is one that absolutely demands a great deal of attention. It is said that in the Bombay Dekkan, where the plant is peculiarly difficult to grow, a bed of a few square yards will bring in from £15 to £20 the season.

It also appears probable, when one considers the history of the cultivated strawberry in Europe, that a judicious system of crossing the indigenous **F. vesca** with European stock, or with the fine large **F. nilgerrensis,** might produce varieties of fruit in no way inferior to those obtained in Europe.

The success that has already attended the efforts of private and market gardeners in many parts of the country, perhaps especially in the large strawberry gardens at Siri near Simla, ought to encourage similar endeavours on the part of Natives near other large centres of demand.

Francœuria crispa, *Cass.;* see **Pulicaria crispa,** *Benth.;* COMPOSITÆ.

Frankincense, see **Boswellia,** Vol. I., 511.

FRAXINUS, *Linn.; Gen. Pl., II., 676.*

A genus of trees consisting of 30 species found in the north temperate regions of both hemispheres, of which 4 are natives of India.

685   **Fraxinus excelsior,** *Linn.; Fl. Br. Ind., III., 606;* OLEACEÆ.

THE COMMON ASH.

Syn.—F. HETEROPHYLLA, *Vahl.;* F. MOORCROFTIANA, *Wall.;* ORNUS MOORCROFTIANA, *G. Don.*

Vern.—*Súm, kúm,* PB.

References.—*Brandis, For. Fl., 303; Gamble, Man. Timb., 256; Pharm. Ind., 136; Ainslie, Mat. Ind., I., 209; O'Shaughnessy, Beng. Dispens., 435; Flück. & Hanb., Pharmacog., 409.*

| The Common Ash. | (*J. Murray.*) | **FRAXINUS excelsior.** |

Habitat.—A large tree of the temperate West Himálaya and Western Tibet from 4,000 to 9,000 feet; distributed from the Caucasus westward to Britain (*Fl. Br. Ind.*)

According to **Brandis** " Basin of the Jhelam, Chenab and Ravi rivers, between 4,000 and 6,000 feet."

Cultivation.—**Brown,** in his *Forester (page 193)*, gives the following description of the propagation and cultivation of the Ash in England :—

CULTIVA-TION. 686

" It is propagated by seeds, and varieties are extended by grafting and budding on plants of the same species. The seeds are enclosed in what are termed 'samaras,' or keys, which are generally ripe for gathering about the end of October. When gathered for the purpose of sowing, the seeds should be mixed with a quantity of dry sand or light dry earth, in which they should be kept for eighteen months, in order to rot off the outer coat; and in order the more effectually to ensure this, the whole mass of seeds and sand should be turned over every three months. The mass should not be much over one foot in depth, as, if more, it will be liable to heat, and in consequence the vitality of the seed would be injured. In the second March, after they are gathered, the seeds should be sown in rows rather thinly, and upon any moderately well pulverised soil. They are sure to come up thickly and injure one another if not sown thin—say one seed to every three square inches, and the covering of earth should not exceed ¾ inch. In the following spring the plants will be ready for being transplanted into the nursery rows, which may be 15 inches from one another, and 4 inches plant from plant in the rows.

" When the plants have stood two years in the nursery rows they may be removed into the forest ground; but if wanted of a larger size they may be left a year longer.

" The ash is in all respects a hardy tree and accommodates itself to most soils and situations not too high-lying and exposed, but to grow it to large dimensions of timber, and to have that of good quality, the tree must be planted in a rather low-lying situation, and on a strong loamy soil, but not a retentive one, nor on one wet in the sub-soil. There is no situation so well fitted for the profitable growth of the ash as the sides of ravines having a good strong loamy soil, where there is a constant supply of water for the roots from the ground above."

Brandis says that the tree requires much light, and that, like the teak, it grows best in a mixed forest.

Medicine.—A small quantity of saccharine matter exudes on incision from its bark. This only constitutes, however, a very small part of the Manna of European commerce, and does not appear to be used in India at all. The bark is bitter and astringent, and was at one time, though very undeservedly, called *European Cinchona.* The leaves are purgative.

MEDICINE. Manna. 687 Bark. 688 Leaves. 689 TIMBER. 690

Structure of the Wood.—Whitish with a distinct brown, often mottled, heartwood, thus differing from that of **F. floribunda.** According to **Brandis,** its weight varies between wide limits, slowly-grown wood being sometimes lighter than wood which has grown more rapidly. **Treagold** gives the weight as from 43·1 to 50·7℔ per cubic foot, but **Brandis** says he has seen English ash weighing as much as 55℔.

It is of very great value on account of its toughness and elasticity, which renders it highly useful for such purposes as the making of wheels, oars, handles of tools, and furniture. The young wood is valuable for the manufacture of hop-poles, hoops, baskets, &c. From the literature obtainable on the subject, it seems that the timber of the Indian-grown tree has not been thoroughly examined, therefore it is not as yet known whether it possesses all the good qualities of the European ash. It is to be hoped, however, that this question may soon be cleared up, as there

The Flowering Ash.

FOOD.
601
DOMESTIC.
692
693

would seem no very great reason why the Ash should not become an important cultivated timber in this country.

Food. — The fruit in England is preserved in vinegar as a pickle.

Domestic, &c. — The ash coppices well (*Brandis*).

Fraxinus floribunda, *Wall.; Fl. Br. Ind., III., 605.*

Syn. — FRAXINUS UROPHYLLA, *Wall.*; ORNUS FLORIBUNDA, *Dietr.*; O. UROPHYLLA, *G. Don.*

Vern. — *Kangu, tuhasi,* NEPAL; *Angan, angú, dakkúri,* N.-W. P.; *Angú, súm, sunnu, shun, húm, hamer, túnnú,* PB.; *Banarish,* AFG.

References. — *Brandis, For. Fl.,302; Gamble, Man. Timb., 256; Stewart, Pb. Pl., 138; Ainslie, Mat. Ind., I., 209; O'Shaughnessy, Beng. Dispens., 434; Atkinson, Him. Dist., 737; Gazetteers:—Rawalpindi Dist., 15; N.-W. P., X., 313; Gurdaspur Dist., 55; Hazara Dist., 14; Indian Forester, VI., 146; IX., 290; X., 317; XIII., 67.*

Habitat. — A large deciduous tree of the Himálaya, from the Indus to Sikkim, between 5,000 and 8,500 feet.

MEDICINE.
Manna.
694

Medicine. — A concrete, saccharine exudation (manna) is obtained from the stem, by incision, and is employed as a substitute for the officinal manna.

The sugar contained in this exudation, called mannite, differs from cane and grape sugar in not being readily fermentable; though under certain conditions it does ferment, yielding a quantity of alcohol varying from 13 to 33 per cent. (*Dr. Warden*). Like the officinal manna, this is used for its sweetening and slightly laxative properties.

TIMBER.
695

Structure of the Wood. — White, with a light-red tinge, no heartwood, soft to moderately hard. Weight 48lb per cubic foot. It is very similar in structure to the wood of the European ash, from which, however, it differs in having no heartwood.

It is very valuable, possessing most of the qualities of European ash, and is used for oars, jampan poles, ploughs, platters, spinning-wheels, and other purposes.

The Conservator of Forests, Panjáb, writes: "In 1879 samples were supplied to the Timber Ordnance Agent, Fattehgarh, for sponge staves."

696

F. ornus, *Linn.; DC., Prodr., VIII., 274.*

THE FLOWERING ASH.

Syn — ORNUS EUROPÆA, *Pers.*

This, though not an Indian species, may be briefly considered, as it is the principal source of the drug known officinally in Europe as "Manna." F. rotundifolia and F. excelsior are, however, to a smaller extent also manna-yielding ashes.

Vern. — *Shir-khist,* HIND.; *Shir-khist,* DEC.; *Méná,*TAM., TEL.; *Manna,* MALAY; *Mann, shir-khisht,* ARAB.; *Shir khihst,* PERS.

References. — *Pharm. Ind., 136; Ainslie, Mat. Ind., I., 208; O'Shaughnessy, Beng. Dispens., 434; Flück. and Hanb., Pharmacog., 409; Irvine, Mat. Med., Patna, 101; Birdwood, Bomb. Pr., 52; Smith, Dic., 26; Kew Off. Guide to the Mus. of Ec. Bot., 94.*

Habitat. — A small tree of the mountains of South Europe and Asia Minor, extending in the Mediterranean region westwards to Corsica and Eastern Spain.

MEDICINE.
Manna.
697

Medicine. — The name MANNA is applied to the saccharine exudation obtained, by incision, from this tree as well as to other substances. Originally the name was applied to the miraculous food provided for the Israelites during their journey from Egypt, but since then it has come to be used for most saccharine exudations. The officinal manna of European medicine is the production of the three species of ash above mentioned, principally of F. ornus, and is frequently known from that circumstance as Calabrian manna. It appears that the manna of Indian medi-

| The Flowering Ash : Manna. | (*J. Murray.*) | FRAXINUS ornus. |

cine is derived from a wholly different source. The true *shir-khist* of the bazárs of North-Western India is imported from Afghánistan, Turkestan, and Persia, and is probably the exudation of **Cotoneaster nummularia**, and to a lesser extent of **Araphaxis spinosa**. Flückiger and Hanbury have examined fragments of this *shir-khist* and pronounce it to be indisputably derived from **Cotoneaster**. They write : " It is in irregular roundish tears, from about ¼ up to ¾ inch in greatest length, of an opaque, dull, white colour, slightly clammy, and easily kneaded in the fingers. With water it forms a soapy solution with an abundant residue of starch granules." According to Ludwig, *Shir-khist* was found to consist of an exudation analogous to tragacanth, but containing at the same time two kinds of gum, and an amorphous lævogyre sugar besides starch and cellulose.

MEDICINE.
Imported manna.
698

There is, however, a certain amount of manna obtained in India from indigenous plants, other than **Fraxinus**, but to what extent this is actually used medicinally, has not been determined, nor indeed can it be said that we know definitely all the plants from which Indian Manna is derived. [See **Alhagi** (*Vol. I.*, *165*); **Calotropis** (*Vol. II.*, *37*, *47*); and **Tamarix**.] A sample of manna has recently been received by the Reporter on Economic Products from the Central Provinces, the source of which is being at present investigated. Dr. **Dymock** to whom a specimen has samples have been sent has obligingly drawn the writer's attention to an interesting passage in the *Makhzan-el-Adwiya*, the author of which speaking of *Shir-khisht* writes, "and they say that in the towns of the Subeh of Behar, Patna, and Bhágulpur, a substance like *shir-khisht* is obtained from a plant, called, in Hindi, *Katra*; and they prepare it in this manner : the tree is cut down and fire applied to the root, which causes a flow of boiling juice which concentrates into lumps like white sugar sweetmeats, and this sugar has all the properties of the *shir khisht Harlálu*. Hakim Mir Muhammad Abdul Hamid writes, ' I have myself used it as *Shir-khisht*.' "

Indigenous manna.
699

The manna alluded to in the above passage cannot possibly be the substance obtained from the Central Provinces, which is evidently a natural exudation, which, falling in a shower, incrustates leaves, twigs, stones, &c., with a deposit often ½ an inch in thickness.* It may be added that the writer has presently under examination another Indian manna. As a probable consequence of an exceptionally dry autumn the pines of the Western Himálaya, more especially **Pinus excelsa**, have been exuding manna from the tips of the twigs, which, cementing the needles into clotted masses, and melting through the heat of the sun, has encrusted with a varnish-like covering the leaves, twigs, and stones around the trees. This was apparently last mentioned by Major Madden, and according to native opinion, although **Pinus excelsa** sheds manna every now and then to a limited extent, a large exudation, participated in by Pinus longifolia and **Cedrus Libani**, † only occurs once in twenty or thirty years. It is not reputed to be used medicinally, but is collected and eaten, or employed in adulterating honey.

For the chemical composition of European officinal manna, which not being an Indian economic product need not here be further discussed, the reader is referred to the *Pharmacographia* of Flückiger and Hanbury. Therapeutically the Indian manna and the officinal article seem very similar. They are both employed as sweetening agents and as slight laxatives.

* **Dymock** reports that it does not appear to agree with any known manna. It, however, contains glucose and a crystalline sugar-like mannite.

† **Rhododendron arboreum** has since been observed to be exuding manna as a result of **Aphides**.

FROGS.	The Ash; Frogs.

Dr. Ainslie writes : "The Hindus know and care little about manna ; the Muhammadans of India prescribe it as a laxative to children and delicate women, in doses from ℥ 2 to ℥ ½; and the Arabians give it a place amongst their *Mushilat-sufra* (cholagogues)."

700 **Fraxinus xanthoxyloides,** *Wall. ; Fl. Br. Ind., III.,* 606.

Syn.—F. Moorcroftiana, *Brand. ;* Ornus xanthoxyloides, *G. Don.*

Vern.—*Auga, gaha,* N.-W. P.; *Hanus, nuch, shilli, chuj, thum, shangal, kanóch, hanóch,* Pb.; *Shang, hagai,* Pushtu.

References.—*Brandis, For. Fl., 304; Gamble, Man. Timb., 256 ; Stewart, Pb. Pl., 139; Atkinson's Flora of the Kuram Valley, 79 ; Baden Powell, Pb. Pr., 581 ; Balfour, Cyclop., I., 1151 ; Indian Forester, V., 185, 478 ; Rawalpindi Gazetteer, 15 ; Simla Gazetteer, 11.*

Habitat.—A small tree, or more often a shrub, met with in Afghánistán, the Trans-Indus, and from the Jhelum to Kumaon in the North-West Provinces (*Gamble*). Aitchison in his *Kuram Valley Flora* mentions it as being found on the ascent to Péwar Kotal, and occasionally all over the Hariab district to Drékalla and Kárátigah. Brandis gives its distribution as the North-Western Himálaya from Kashmír to Kumaon, between 3,000 and 9,000 feet, and Lace mentions the shrub as growing near Quetta.

TIMBER.
701
Structure of the Wood.—A good elastic wood of small size, suitable for staves, jampan poles, walking-sticks, and employed for making ploughs in Kághán (*Baden Powell*). Used for agricultural implements (*Lace, Quetta*).

FODDER.
702
Fodder.—Dr. Stewart says its leaves are used as fodder, and Mr. Lace writes that in Southern Afghanistan the tree is never allowed to attain f ll size, owing to its young branches being continually lopped and the leaves given to sheep and goats, which are very fond of them.

French Bean, see Dolichos Lablab, pp. 184, 185, also Phaseolus.

703 **French Honeysuckle,** see Hedysarum coronarium, *Linn. ;* Legumi-
[nosæ.

FROGS.

Vern.—*Renak,* Hind ; *Bheng,* Beng.

Amphibians of the sub-class Batrachia and order Anura, of which they constitute the family Ranidæ. They occur very commonly in all parts of India, and are especially noticeable during the rains, when their deafening croaking resounds on all sides. Several species are peculiar to definite localities, and many are characterised by the peculiar sounds they produce. Amongst these one may be noticed, an inhabitant of the Khasia Hills, which has a croak so exactly similar to the tinkling of a hammer on an anvil, that even some of the most accurate observers appear to have been deceived by it (*Him. Journ., II., 295*). But perhaps the most amusing record of frogs in Indian literature occurs in the *Ain-i-Akbari,* the writer of which remarks : " Frogs also may be trained to catch sparrows. This looks very funny." Adams, in his *Wanterings of a Naturalist in India,* mentions that at Poona having shot a sun-bird which fell on the margin of a pool, he saw it seized and devoured by a large green frog. This lends a certain support to the somewhat extraordinary statement made by Abul Fazl. Mr. Edgar Thurston, the Superintendent of the Central Museum, Madras, in a recent exhaustive monograph on the Batrachia, Salientia, and Apoda of Southern India, has described six genera as natives of that region and Ceylon ; *viz.,* 1, Rana ; 2, Rhacophorus ; 3, Ixalus ; 4, Nyctibatrachus ; 5, Nannebratrachus ; and 6, Nannophrys ; of which the first comprises 19 ; the second 14 ; the third 19; the fourth 2 ; the fifth 1 ; and the sixth 2 species. Scientific information regarding the occurrence and distribution of the species of

The Chief Fruits of India. (*G. Watt.*)	**FRUITS.**

this family in other parts of India appears to be meagre, nor is there any record of the exact species, or number of species used as food.

Food.—Certain species are eaten by some of the lowest caste natives in India, and by many of the Burmese. In the bazars of the latter country, boiled frogs are exposed for sale amongst other articles of food (*Mason*).

FOOD.
704

(*G. Watt.*)

FRUITS.

705

The fruits of the East, it is believed, are much overrated in Europe. Many of the best of Indian fruits have been introduced from Europe, China, the West Indies, and America. The most characteristic modern fruits of India are the mango, guava, lichi, pine-apple, and plantain. The mangosteen is common in the Straits, and is regarded as the most delicately-flavoured fruit of the East.

It is remarkable that while the wild forms of many of the fruits of Europe are abundant, as indigenous plants on the Himálaya, a very few only were cultivated before the arrival of Europeans; and the gooseberry, the currant, and the bramble, which have been carried to such perfection in Europe, are still uncultivated in India. The peach succeeds in the plains of India, but the effect of climate upon it is marked. In Bengal excellent peaches are to be had, attaining much of their European flavour and ripening into a soft pinkish separable pulp. They reach the market just before the mangos, or at the beginning of the hot season. In the Panjáb this soft condition is rarely attained, and the pulp adheres firmly to the stone, which breaks readily on the peach being cut open. On the Western Himálaya peaches do not succeed well, the rains apparently prevent the ripening of the fruit, while on the Nilgiris, at the same altitude, peaches are wonderfully good. The apricot shows a somewhat similar behaviour. In Afghánistán, Kashmír, and Chamba, excellent apricots are obtained, and indeed the tree, if not indigenous to Afghánistán, is quite naturalised, at an altitude between 6,000 and 9,000 feet. It is grown in the Panjáb, although not in the plains of India generally; but in the Panjáb, and along the Himálayan chain, the fruit is very inferior to the Kashmír and Afghán apricot. Even at Simla, only a few miles east of Chamba, the apricots are very inferior, and this degeneration increases on passing further east and south-east. In the moister mountain regions of Sikkim, Assam, and the Nilgiris, the apricot cannot even be cultivated.

The grapes of Kashmír and Afghánistán are famous, but, owing to the period of plucking and the method of packing, they have lost their natural flavour before they reach the plains of India. A very considerable trade is, however, done by the Kabuli merchants in small circular boxes of grapes. His Highness the Maharaja of Kashmír has successfully introduced the wine grape into Kashmír from which wine and brandy of good quality are obtained.

The foreign trade in fruits is comparatively small, the cocoa-nut being the chief article of commerce, but in the present work that is viewed as a NUT not a FRUIT. The following enumeration may be given of the chief fruits of India, those bearing a * being introduced (*i.e.*, non-indigenous). For further information regarding the individual fruit-yielding plants, the reader is referred to the articles regarding each in its respective alphabetical position in this work.

*Achras Sapota, *Linn.*; THE SAPODILLA PLUM or SAPOTA. SAPOTACEÆ.

*Adansonia digitata, *Linn.*;

    THE BAOBAB TREE, SOUR GOURD, MONKEY-BREAD. MALVACEÆ.

Ægle Marmelos, *Correa.*; THE BEL or BAEL FRUIT. RUTACEÆ.

FRUIT-
YIELDING
PLANTS.
706

FRUITS.	The Chief Fruits of India.
FRUIT-YIELDING PLANTS.	***Ananas sativa,** *Linn.;* THE PINE-APPLE. BROMELIACEÆ.

There are many forms of this fruit, and these improve in quality on passing eastward. They are fairly good in Bengal, but are excellent in Burma and the Malaya, where the plant seems to have become completely naturalised. Abul Fazl (in the *Ain-i-Akbari, p. 68*) alludes to the pine-apple calling it *Kat' hal-i-Safári* or the Jack fruit of travellers. And in the *Túzuk-i-Jahán-gírí* it is stated that the pine-apples at the time of Akbar's son came from the harbour towns of the Portuguese.

*Anona reticulata, *Linn.;* BULLOCK'S HEART. ANONACEÆ.

*A. squamosa, *Linn.* THE CUSTARD APPLE or SWEET SOP.

*Artocarpus incisa, *Linn.;* THE BREAD-FRUIT TREE. URTICACEÆ.

A. integrifolia, *Linn.* THE JACK-FRUIT:
An important fruit with the natives of the plains of India: rarely eaten by Europeans.

A. Lakoocha, *Roxb.* THE LAKUCHA.

*Averhoa Carambola, *Linn.;* THE KARMAL. GERANIACEÆ.

*A. Bilimbi, *Linn.* THE BILIMBI.

Bassia butyracea, *Roxb.,* SAPOTACEÆ.

B. latifolia, *Roxb.* THE BUTTER or MÁHWA TREE.
The ripe corolla tubes constitute an important article of food with the people of the central table-land of India.

Borassus flabelliformis, *Linn.;* THE PALMYRA PALM. PALMÆ.
A common palm in Bengal and other parts of the plains. It produces its fruits in the cold season, in the interior of which exists a cold, insipid, gelatinous pellucid pulp eaten by the natives but only rarely by Europeans.

Capparis spinosa, *Linn.;* THE CAPER BERRY. CAPPARIDEÆ.

Carica Papaya, *L.;* THE PAPAW or PAPAYA TREE. PASSIFLOREÆ.
It is significant that it is not mentioned in the *Ain-i-Akbari*, a fact that fixes its introduction into India as after the reign of Akbar.

Carissa Carandas, *Linn.;* THE CARENJA FRUIT. APOCYNACEÆ.
The unripe fruit is pickled, the ripe fruit made into tarts.

Celtis australis, *Linn.;* URTICACEÆ.
Supposed by some to be the Lotus fruit of [the ancients. *Conf.* with Diospyros Lotus, Vol. III., pp. 136—156.

Cephalandra indica, *Nand.;* CUCURBITACEÆ.

Citrullus Colocynthis, *Schrad.;* ENGLISH COLOCYNTH. CUCURBITACEÆ.

C. vulgaris, *Schrad.* THE WATER-MELON.
Var. fistulosus. THE TANDUS.

* Citrus Aurantium, *Linn.;* THE ORANGE. RUTACEÆ.

* C. decumana, *Willd.* THE SHADDOCK, or POMELO, or FORBIDDEN FRUIT.

C. Medica, *Linn.* THE CITRON, LEMON, LIME.
Var. 1.—Medica proper. The Citron.
Var. 2.—Limonum. The Lemon.
Var. 3.—acida. The Sour Lime of India.
Var. 4.—Limetta. The Sweet Lime.
Var. 5.—Lumia. The Sweet Lemon.

| The Chief Fruits of India. | (*G. Watt.*) | FRUITS. |

Cordia Myxa, obliqua, and **Rothii** yield edible fruits often pickled. In Sind **C. Rothii** is viewed as a regular fruit-tree.

Cornus capitata, *Wall.*, is generally classed as one of the Himálayan wild fruits ; eaten and made into preserves.

Cucumis Melo, *Linn. ;* THE MELON. CUCURBITACEÆ.
There are many forms of this fruit met with in India, some being used as dessert fruits, others as vegetables. **Dr. Aitchison** found the melon wild in Afghánistan.

Cucurbita moschata, *Duchesne ;* THE MUSK MELON. CUCURBITACEÆ.
Eaten mostly as a vegetable.

Cydonia vulgaris, *Tour. ;* THE QUINCE. ROSACEÆ.

Dillenia indica, *Linn. ;* THE CHALTA. DILLENIACEÆ.

Diospyros Kaki, *Linn. f. ;* EBENACEÆ.
THE CHINESE FIG and PLUM ; THE KEG FIG of JAPAN.

D. Lotus, *Linn.* THE AMTOK or DATE PLUM.
These and other species of **Diospyros** yield edible fruits, for which they are often cultivated.

Durio Zibethinus, *DC. ;* DURIAN, or CIVET-CAT FRUIT-TREE. MALVACEÆ.

Elæagnus ; ELÆAGNEÆ.
One or two species of this genus are cultivated by the hill tribes, especially in Baluchistan. They yield an edible fruit often known as the Wild Olive.

* **Eriobotrya japonica,** *Lindl. ;* LOQUAT or JAPAN MEDLAR. ROSACEÆ.

Eugenia Jambolana, *Lam. ;* THE JAM. MYRTACEÆ.

E. Jambos, *Linn.* THE ROSE-APPLE.

Flacourtia Cataphracta, *Roxb. ;* BIXINEÆ.
Yields a fruit eaten by the natives. It tastes like an inferior plum.

* **Ficus Carica,** *Linn. ;* THE COMMON FIG. URTICACEÆ.

Fragaria vesca, *Linn. ;* THE STRAWBERRY. ROSACEÆ.

Garcinia Cowa, *Roxb. ;* THE COWA FRUIT. GUTTIFERÆ.
This is a native of Eastern Bengal and yields an acid fruit which makes a remarkably fine preserve. It ripens in the beginning of June.

* **G. Mangostana,** *Linn.* THE MANGOSTEEN.
This is by most writers held to be the most deliciously flavoured fruit of the East. It is a native of the Malay Peninsula, and while it may be grown in Bengal and Madras, it fails to produce good fruit anywhere beyond the limits of Burma.

Grewia asiatica, *L. ;* THE PHALSA. TILIACEÆ.
A common wild tree which yields an edible fruit, often cultivated near villages on this account.

Hibiscus Sabdariffa, *Linn. ;* THE ROZELLE or INDIAN SORREL. MALVACEÆ.
There are two kinds, differing in the colour of the succulent calyx—red and white—which forms the edible part.

* **Lycopersicum esculentum,** *Miller ;* THE LOVE-APPLE or TOMATO. SoLANACEÆ.

FRUITS.	The Chief Fruits of India.

FRUIT-YIELDING PLANTS.	**Mangifera fœtida,** *Lour.;* ANACARDIACEÆ.
	M. indica, *Linn.* THE MANGO TREE.
	    The number of cultivated and distinct forms of this fruit are probably as great as that of the European apple.
	M. sylvatica, *Roxb.* THE WILD MANGO.
	Mimusops hexandra, *Roxb.;* THE KHIRNI. SAPOTACEÆ.
	    Cultivated in Western India, especially at Goa, as a fruit. It is said to be agreeable and subacid.
	Morus indica, *Linn.;* THE MULBERRY. URTICACEÆ.
	    A favourite fruit in many parts of India, but especially so with the hill tribes.
	Musa paradisiaca, *Linn.;* THE PLANTAIN. SCITAMINEÆ.
	M. sapientum, *Linn.* BANANA.
	    The number of Plantains and Bananas is very great. The reader is referred to the account of them given under Musa in another volume. The *chumpa* plantains of Bengal and Burma are perhaps the finest in flavour.
	Myrica sapida, *Wall.;* THE KAPHUL. MYRICACEÆ.
	    A fruit of the Lower Himálaya and the Khasia Hills; ripening about May. Though largely eaten by the hill tribes the tree does not appear to be cultivated.
	*****Nephelium Litchi,** *Camb.;* THE LITCHI. SAPINDACEÆ.
	    This tree is supposed to have been recently introduced into India from China. There are various forms, differing in thickness and flavour of pulp. The fruit comes into season in April and May. It succeeds best in the hot damp areas, such as in Bengal.
	N. Longana, *Camb.* THE LONGAN FRUIT.
	    This fruit, which ripens about the end of June, is in Calcutta about the size and form of a marble, borne in great branches like grapes. The fleshy aril is, as in the Litchi, the edible portion.
	***** Olea europæa,** *Linn.;* THE OLIVE. OLEACEÆ.
	***** Opuntia Dillenii,** *Haw.;* THE PRICKLY PEAR. CACTEÆ.
	***** Passiflora ;** PASSIFLOREÆ.
	    Several species of Passion-flower yield edible fruits—the GRANADILLA fruit—especially **P. quadrangularis, P. laurifolia,** and **P. edulis.** Though several species flower profusely on the Himálaya, none appear to be eaten in India.
	***** Phœnix dactylifera,** *Linn.;* THE DATE PALM. PALMÆ.
	P. sylvestris, *Roxb.* THE WILD DATE.
	Phyllanthus Emblica, *Linn.;* THE EMBLIC MYROBALAN. EUPHORBIACEÆ.
	    Yields a useful fruit in the cold season which is pickled and made into jelly.
	P. distichus, *Muell.;* THE OTAHEITE GOOSEBERRY.
	    Yields a fruit which, when cooked with sugar, greatly resembles green gooseberries. It is a native of India, though only rarely met with in cultivation.
	***** Physalis peruviana,** *Linn.;* THE CAPE GOOSEBERRY or TIPÁRÍ. SOLANACEÆ.
	    Extensively cultivated in the plains of India and eaten in dessert or made into jam and chutney. Become quite acclimatised in some parts of the country.

| The Chief Fruits of India. | (*G. Watt.*) | FRUITS. |

* **Prunus armeniaca,** *Linn.;* THE APRICOT, MISHMUSH, or ' MOON OF THE FAITHFUL. ROSACEÆ.

* **P. Avium,** *Linn.;* THE SWEET or BIRD CHERRY.

* **P. Cerasus,** *Linn.;* THE SOUR CHERRY.

The *Flora of British India* states that both species of cherry occur on the North-West Himálaya in a state of cultivation at altitudes up to 8,000 feet. Of **P. Avium** it is added that it is almost naturalised. The writer has never seen it except in gardens, and the Himalayan wild cherry is **P. Puddum,** *Roxb.*

* **P. communis,** *Huds.* THE PLUM.

    Var. domestica. ALUCHA.

    Var. Insititia. THE BOKHARA PLUM.

The plum, although most successfully grown in the gardens of Upper India, as Delhi, Saharanpur, &c., is much less successful on the plains than the peach. On the Himálaya it also succeeds admirably and becomes of such flavour as to admit of its being classed as a dessert fruit. The plums of the plains make admirable preserves.

* **P. persica,** *Benth. & Hook.;* THE PEACH.

The peach has a greater claim to be regarded as indigenous on the Himálaya than any other member of this series of fruit trees (except perhaps the cherry). It occurs near every village in the North-West Himálaya, the fruit often never even eaten by the people, though in many cases of good quality. In the neighbourhood of towns where Europeans reside, it is cared for and the fruit brought to market, but even in such cases the natives do not themselves seem to appreciate it. Throughout the plains it is also frequent, and even in the neighbourbood of Calcutta produces admirable peaches. It is in fact the only **Prunus** that appears to be able to withstand tropical influences. It yields in fact more freely, and the fruit is of much finer flavour in the plains than on the Himálaya. The North-West Himálayan peach (where the tree is probably indigenous) is small, green, and seems never to ripen, the fruits remaining on the trees from May to November. In the plains, on the other hand, it does not last more than three weeks or a month, the fruits coming into season in the middle of May.

The **Nectarine** is a glabrous form of the peach. A flattened peach is also common, but, what is perhaps more significant, the green semi-wiid fruit of the Himálayas is a *clingstone* fruit, while that of the greater part of the plains and Nilghiri hills is *freestone.*

P. Puddum, *Roxb.*

Commonly known as the WILD or HIMALAYAN CHERRY.

A plentiful small tree in the Temperate Himálaya (3,000 to 7,000 feet), becoming covered with its elegant pink flowers in October and ripening its yellow, orange or pink fruits in March. These are not or only rarely eaten by the Natives, but are sold to the Europeans to be used in the preparation of cherry-brandy.

Pyrus baccata, *Linn.;* THE SIBERIAN CRAB. ROSACEÆ.

P. communis, *Linn.* THE COMMON PEAR.

The hard round pear of the North-West Himálaya is quite distinct from the modernly introduced pyriform fruit, and it is probably an indigenous production. In Kullu and other parts of the Himálaya large, yellow, soft, luscious pears are grown which compare favourably with any of the pears produced in Europe.

2 G

FRUITS.	The Chief Fruits of India.

FRUIT-YIELDING PLANTS.

Pyrus Malus, *Linn.;* THE APPLE.

On the North-West Himálaya there are many forms of this fruit, some admittedly of modern introduction, and others, by **Brandis,** &c., spoken of as "apparently wild." The Afghan apple is a peculiar oblong fruit with pink marblings and wooly flavour. This is met with in many parts of the Western Himálaya, often becoming less than an inch in length, while preserving all its other characters. A flattened dark-green apple, which when ripe colours faintly on one side, is also frequent on the Himálaya occurring in the gardens of the poorest peasants, and forming a neglected shrub of enclosures. It is probable that these forms represent the so-called wild fruit, but the writer would be much more disposed to accept the round pear as indigenous, than to admit any of the apples as such. A small yellow pippin is common in Delhi, Saharanpur, and other Panjáb plains stations. It comes into season about April and May. At Kullu and also near Simla, large orchards have recently been established where apples, almost equal to the best produced in Europe, may now be purchased. The credit of having developed this new industry is mainly due to Sir **E. C. Buck.**

P. Pashia, *Ham.*

This indigenous plant (cultivated in Kullu and elsewhere on the Himálaya) yields a fruit which is edible on falling from the tree in an over-ripe state. (See Fungoid Pests, p. 457.)

* **Psidium Guyava,** *Raddi;* THE GUAVA TREE. MYRTACEÆ.

* **Punica Granatum,** *Linn.;* THE POMEGRANATE; GRENADES, *Fr.;* GRANATS, *Ger.* LYTHRACEÆ.

Rhododendron arboreum, *Sm.;* ERICACEÆ.

The flowers of the tree **Rhododendron** are regularly collected and made into a pleasant subacid jelly. They appear in February to May.

Rhodomyrtus tomentosa, *Wight;* THE NILGHIRI HILL GOOSEBERRY. MYRTACEÆ.

This elegant shrub yields a berry which is largely collected and in South India is made into a jelly resembling apple jelly.

Ribes; SAXIFRAGACEÆ.

The Gooseberry and Currant, though wild plants on the Himálaya, do not appear to be cultivated.

Rubus; ROSACEÆ.

Various species of Bramble and Raspberry are collected from the wild source; none are cultivated like **R. Idæus**—the Raspberry—of Europe. **R. ellipticus** is the yellow raspberry, the fruits of which are collected and sold at bazárs on the Himálaya; it comes into season in May to June.

Sambucus nigra, *L.;* THE ELDER BERRY. CAPRIFOLIACEÆ.

Though two or three species of Elder occur on the Himálaya, they do not appear to have been grown for their berries, nor does the true Elder-berry appear to have been introduced.

Spondias dulcis, *Willd.;* THE OTAHEITE APPLE. ANACARDIACEÆ.

S. mangifera, *Pers.* THE HOG PLUM.

Tamarindus indica, *Linn.;* THE TAMARIND. LEGUMINOSÆ.

Triphasia trifoliata.

* **Vitis vinifera,** *Linn.;* THE GRAPE. AMPELIDEÆ.

The early records of Kashmír (such as the *Ain-i-Akbari*) shew that grape cultivation was once upon a time more extensive than at the present

The Bladder Wrack.	(*J. Murray.*)	**FUCUS** vesiculosus.

day. The fruit is described, two centuries ago, as having been carried from the northern hilly tracts of India in basket-loads and sold in the plains at R3 to 4 a basket. At the present day the better class of grapes obtained in the plains of India are those imported by Kabul merchants, preserved in cotton wool in small circular boxes. At hill stations, as at Simla, grapes of a very superior quality are grown from recently-imported European stock. At one time a large trade was done in Bashahr in growing grapes for the Simla market and raisins into Tibet. A disease, however, appeared in the form of a destructive insect and the cultivation has in consequence been almost completely abandoned. A small grape, which also occurs wild, is collected and sold in the bazárs. It yields a peculiarly-flavoured fruit, very refreshing, but which bears little resemblance to the European grape. It appears to be the produce of **Vitis parvifolia**, but it is probable the cultivated states of this small grape may have a strain of hybridization, possibly with **V. vinifera**. Throughout the plains of India, in favourable situations, grape cultivation occurs as a garden curiosity, but the fruits obtained are small, green, and unpalatable, though in some parts of Upper India, *e.g.*, in Peshawar, the results are much more satisfactory.

Zizyphus Jujuba, *Lam.;* THE BAER OR JUJUBE; THE CHINESE DATE. RHAMNEÆ.

Z. vulgaris, *Lamk.*
The long or round plum, the *Kúl-phul*, is largely cultivated by the natives of the plains of India.
For further information see NUTS.

(*J. Murray.*)
FUCUS.

The typical genus of the family FUCACEÆ belonging to the Natural Order ALGÆ. It is characterized by having plane, compressed, or linear fronds, generally of a brownish colour, which in some species grow to a great length. The only two species which have been described as Indian are **F. nodosus** and **F. vesiculosus.**

Fucus amylaceus, *O'Sh.* 707
The name under which **O'Shaughnessy** described and brought to notice the plant yielding the "CEYLON MOSS," **Gracilaria lichenoides,** *Grev.* (*which see*).

F. nodosus, *Linn.* 708
THE KNOBBED SEA WRACK.
Habitat.—A very common sea-weed in the northern temperate seas, said by **Murray** (*Plants and Drugs of Sind*) to be found commonly along the sea-shore.
Similar in properties to the following species:—

F. vesiculosus, *Linn.; Bent. & Trim., t. 304.* 709
THE BLADDER WRACK.
Syn.—F. SPIRALIS, *Linn.*; F. DIVARICATUS, *Linn.*; F. DISTICHUS, *Lightf.*; F. BALTICUS, *Ag.*; F. PLATYCARPUS, *Thuret.*
Habitat.—Very common on the shores of the United Kingdom, also along the North Atlantic Ocean, from Norway and Greenland to the West Indies, and on the North Pacific coast of America. It is said by **Murray** in his *Plants and Drugs of Sind* to be found on the Manora Rocks.
Medicine.—The entire alga is used in the manufacture of a medicine. Since the introduction of Iodine, however, it has gone greatly out of use, and is not now to be found in the British Pharmacopœia, nor in those of

MEDICINE. 710

FUEL.	Fuel and Firewood.

India and the United States. To the natives of India the plant as a medicinal substance is unknown. Its therapeutic properties are very similar to those of iodine, being deobstruent, and considered of specific value in scrofulous affections, rheumatism, and glandular swellings, particularly goitre.

Iodine.
711
Bromine.
712
Kelp.
713
FODDER.
714
MANURE.
715

716

In 1862, **Dr. Duchesne Dupare** described it as having a marked effect in diminishing obesity, and it is said to be an ingredient in the extensively-advertised nostrum,—"Anti-Fat." In Europe this plant for a long time formed a considerable source of soda alkalis, but its importance for this purpose has diminished in recent years. Its principal value is now in the manufacture of IODINE and BROMINE, as it with **F. nodosus** forms the greater part of the sea-weed burned to form KELP.

Fodder and Manure.—It is said by Greville to form an article of FODDER and SHEEP FOOD in some of the islands of Scotland. It is also a valuable MANURE.

It is possible that both species of **Fucus** may be found in greater quantity than is generally known along the northern shores of the Indian Ocean, in which case it is well to remember their important economic properties.

FUEL & FIREWOOD.

With very few exceptions all the timber trees of India might be used as firewood. Certain timbers, however, emit an objectionable odour, and on that account are rarely used; others are too valuable. The heat-giving property is a point of great importance in fuel-supply, and it seems probable that a thorough investigation of the heat evolved from given weights of timber would greatly narrow the list of plants which should be enumerated as suitable for steam purposes, whether railway or machinery.

FUEL AND FIREWOOD. TIMBERS, &c., USED FOR—

Abies Smithiana (=Picea Morinda).
Acacia arabica.
A. Catechu (firewood for steamers).
A. leucophlœa.
A. melanoxylon.
A. planifrons.
Adhatoda Vasica (brick-burning)
Adina sessilifolia.
Ægiceras corniculata.
Alangium Lamarckii.
Albizzia amara.
Amoora cucullata.
Anogeissus latifolia.
Avicennia officinalis.
Balanites Roxburghii.
Berberis aristata.
B. vulgaris.
Betula cylindrostachys.
Boswellia serrata.
B. thurifera.
Briedelia stipularis.
Bruguiera gymnorhiza.
Calligonum polygonoides.
Capparis aphylla.
Carissa diffusa.
Cassia siamea (Ceylon locomotive fuel).

Castanopsis tribuloides.
Casuarina equisetifolia.
Ceratonia Siliqua.
Cerbera Odollam.
Ceriops Candolleana.
Cordia Myxa.
C. Rothii.
Coriaria nepalensis.
Cornus capitata.
Croton caudatus.
Crypteronia paniculata.
Cynometra ramiflora.
Dalbergia Sissoo (Railway fuel).
Dillenia indica.
Ekebergia indica.
Elæagnus hortensis.
Ephedra vulgaris.
Eucalyptus Globulus.
Eurya japonica.
E. symplocina.
Excæcaria Agallocha.
E. indica.
Ficus religiosa.
F. retusa.
Fraxinus xanthoxyloides
Garuga pinnata.
Helicteres Isora.

Fuel and Firewood; Fuller's Earth.   (*J. Murray.*)   FULLER'S EARTH.

Heritiera littoralis.	Premna mucronata.	**TIMBERS USED FOR FUEL AND FIREWOOD.**
Hibiscus tiliaceus.	Prinsepia utilis.	
Hippophæ rhamnoides.	Prosopis spicigera.	
Hydnocarpus alpina.	Prunus armenica.	
Juniperus communis.	Pygeum zeylanicum.	
J. excelsa.	Quercus acuminata.	
J. recurva.	Q. Ilex.	
Kandelia Rheedii.	Q. incana.	
Lebedieropsis orbicularis.	Q. lanuginosa.	
Lonicera quinquelocularis.	Q. semecarpifolia.	
Lumnitzera racemosa.	Randia dumetorum.	
Lycium europæum.	Rhamnus virgatus.	
Mæsa montana.	Rhazya stricta.	
Mallotus philippinensis.	Rhododendron arboreum.	
Meliosma Wallichii.	Rhus mysorensis.	
Mimosa dulcis.	Salix (species).	
Myricaria elegans.	Salvadora oleoides.	
M. germanica.	S. persica.	
Myrsine semiserrata.	Securinega leucopyrus.	
Nyctanthes Arbor-tristis.	Sesbania ægyptiaca.	
Olea ferruginea.	S. grandiflora.	
Phyllanthus Emblica.	Sonneratia acida.	
Pieris ovalifolia.	Streblus asper.	
Pinus longifolia (bark as fuel).	Symplocus lucida.	
Pithecolobium dulce.	Tamarix dioica.	
Pongamia glabra.	Taxus baccata (burnt as incense).	
Populus balsamifera.	Terminalia tomentosa.	
P. euphratica.	Teucrium macrostachyum.	
Premna integrifolia.	Xylosma longifolium.	
P. latifolia.	Zizyphus rugosa.	

FULLER'S EARTH; *Ball.; In. Man. Geol. of India, Vol. III., 570.*

The following brief note on this subject has been obligingly furnished by Mr. H. B. Medlicott for this work :—

Fuller's earth.   717

TERRE À FOULON, *Fr.;* WALKERERDE, *Ger.;* CRETA DA SODARE I PANNI, *Ital.*

As regards the distribution of Fuller's Earth in India, information is very incomplete; but it is known to be carried for long distances from certain localities where it occurs. In the Bhagalpore division of Bengal, in the neighbourhood of Colgong, a *sabun-mitti* or soap-earth is obtained. In Rajputana a fuller's earth used to be obtained in fissures of quartz and schistose rocks, with carbonate of lime, near Ajmír. At the village of Meth, near Kolath in the Bikanír State, fuller's earth is excavated. In some parts of Western Sind a pale-greenish clay is found which is used for washing cloth, &c.; it is also eaten by pregnant women. In the Panjáb, in the Dera Ghazi Khan and Multan districts, a clay resembling fuller's earth is imported from the interior of the Suleman Range: the so-called *Multani-mitti* imported into Multan is of three qualities :—

(1) White mitti, called " *khajrú*," or edible, from Bikanír and Jessalmír;   718

(2) Yellow mitti, or " *bhakri*," for dyeing cloths, from the same   719
     localities;

(3) Light green or " *sabuz mitti*," for cleaning the hair, from Vadur   720
     in the Dera Ghazi Khan district.

FUMARIA parviflora.	The Fumitory.

At Nilawan, in the Salt Range, a lavender-coloured clay or decomposed rock, which is found with volcanic rock at the above locality, is used as fuller's earth by the natives. The reader is referred for further information to the article CLAY (Vol. II., pp. 360—368, but especially paragraph No. 1319 on *Edible and Medicinal Earths*).

Fulwa Butter, see **Bassia butyracea,** *Roxb.;* Vol. I., 405.

FUMARIA, *Linn.; Gen. Pl., I., 56, 965.*

A genus which belongs to the Natural Order FUMARIACEÆ, having about eight species; usually weeds of cultivation in the temperate regions of the Old World. Only one of these is indigenous to India, namely, **F. parviflora,** but **F. officinalis,** *Linn.,* may be also briefly considered, as it yields the true Fumitory and is employed in Native medicine.

721

Fumaria officinalis, *Linn.;* FUMARIACEÆ.

Vern.—*Pit-pápará,* HIND.; *Shátrá,* DEK.; *Turu,* TAM.; *Chata-rashi,* TEL.; *Baglatul mulk, shateraj,* ARAB.; *Sháhtara,* PERS.

References.—*Pharmacographia Indica, I., 114; Ainslie, Mat. Ind., I., 138; O'Shaughnessy, Beng. Dispens., 184; Moodeen Sheriff, Supp. Pharm. Ind., 273; Dymock, Mat. Med. W. Ind., 52.*

Habitat.—A weed of cultivation in Persia. Two varieties of Fumitory are described in the *Makhzan-el-Adwiya,* one with violet coloured flowers, and the other and larger kind with white flowers.

F. officinalis was mentioned by **Dr. Stewart** in 1859 as occurring as a field weed near Abbottabad, but it is probable that the plant he collected was really **F. parviflora,** since **F. officinalis** has not been found by other botanists in India.

MEDICINE.
Fumitory.
722

Medicine.—The entire plant except the root is used medicinally, constituting FUMITORY which has long been known, and was highly esteemed by the Greeks and Romans. It is, however, not now employed by European practitioners, and is not to be found in the Pharmacopœia of England, America, or India, although still much used in this country by native practitioners. The fumitory sold in Bombay is this species (*Dymock*), and is imported from Persia, while in Upper India the indigenous plant is substituted.

The vernacular terms are used indiscriminately, and as the medicinal properties are similar, the uses of both species may be detailed in the account of the Indian plant.

723

F. parviflora, *Lamk.; Fl. Br. Ind., I., 128.*

Vern.—*Pitpapara (Pitpápra),* HIND.; *Ban-sulpha,* BENG.; *Sháhtara, pit-papra, pápra,* PUSHTU; *Shatra,* SIND; *Pitpápra,* BOMB.; *Pitpapda,* GUZ.; *Pitpapara, shátrá,* DEC.; *Turá,* TAM.; *Cháta-ráshi,* TEL.; *Bukslat-ul-mulik, baglatul-mulk,* ARAB.; *Shatra, sháhtarah,* PERS.

References.—*Roxb., Fl. Ind., Ed. C.B.C., 531; Stewart, Pb. Pl., 11; Pharmacographia Indica, I., 115; O'Shaughnessy, Beng. Dispens., 184; Moodeen Sheriff, Supp. Pharm. Ind., 273; Dymock, Mat. Med. W. Ind., 52; S. Arjun, Bomb. Drugs, 9; Murray, Pl. and Drugs, Sind, 77; Irvine, Mat. Med. Patna, 90; Moodeen Sheriff, Mat. Med. Madras, 22; Atkinson, Him. Dist., 737; Birdwood, Bomb. Pr., 7; Aitchison, Afgh. Del. Com. Rep., 128; Balfour, Cyclop., I., 1155; Bomb. Gaz., VI., 14; Raj. Gaz., 30.*

Habitat.—Found in rice-fields during the cold season; in the Indo-gangetic plain, Lower Himálaya (up to 8,000 feet), and Nilghiri hills. It is described by **Dr. Aitchison** as generally distributed over the whole of Afghánistán.

MEDICINE.
724

Medicine.—Fumitory has long been regarded as laxative, diuretic, alterative, tonic, diaphoretic, and febrifuge. It has consequently been

F. 724

| Fungi and Fungoid Pests. | (*J. Murray.*) | FUNGI, &c. |

much used by native practitioners in India, and is still highly esteemed by the Muhammadans. It is, however, very little used by European practitioners, and its value has probably been overestimated by the natives. **Dr. Thornton**, however, is of opinion that the drug is useful in leprous affections, and in the recently published *Pharmacographia Indica* fumitory is described as beneficial in dyspepsia due to torpidity of the intestines, and as a valuable remedy in scrofulous skin affections.

SPECIAL OPINIONS.—§ " The leaves and stems given in the form of infusion in doses of 1 to 2 ounces are much used as a febrifuge and alterative" (*Lal Mahomed, Hospital Assistant, Hoshangabad, Central Provinces*).

FUNGI AND FUNGOID PESTS.

725

The FUNGI of India are very numerous, and comprise many species of economic interest. Several are used as food, others as medicine, while certain microscopic forms are of importance, since they produce the rusts, moulds, smuts, and other pests which infest many of our crops, fruits, and timber trees. The writer is much indebted to **Dr. Barclay** for having kindly revised the following brief article.

Vern.—(For large, mushroom-like fungi), *Kúmbh samarogh, herar* (Bazar names), HIND., BENG.; *Ot*, SANTAL; *Kat phula*, ASSAM; *Mopsha*, CHAMBA; *Manskhol*, KASHMIR; *Shirian, bat-bakri, buin-phal, kunba, kánakach, kangach, kanha bichu, girchhatra, máns kel, moksha, khúmba, khámbúr, chattri*, PB.; *Samarogh*, AFG.; *Kuti bubhá, khumba*, SIND; *Alombe, kalambe*, BOMB.; *Kagdana chhatra*, GUZ.; *Chattrak*, SANS.; *Kullalie-dio* (Fairies), *chatr-i-mar, samárugh*, PERS.

References.—*Stewart, Pb. Pl., 267; Barclay's Descriptive List of the Uredineæ of the Western Himálaya, also in Sc. Memoirs by Med. Officers of the Army of India, Parts II., III., IV., V.; Dymock, Mat. Med., W. Ind., 865; Flüch. & Hanb., Pharmacog., 740; S. Arjun, Bomb. Drugs, 84; Balfour, Agricultural Pests of India, 59; Baden Powell, Pb. Pr., 257, 384; Balfour, Cyclop., I., 1156; Smith, Dic., 183; Treasury of Bot., I., 512; Jour. Agri.-Hort. Soc., Vol. V., Pl. 1, pp. 51-53; Indian Forester, XIII., 290, 389.*

Medicine.—For an account of the medicinal uses of the different forms of **Agaricus** and **Polyporus**, the reader is referred to the article on the former in Vol. I., at page 129. **Balfour** mentions a fungus found growing on the roots of a bamboo in Burma, which is regarded by the natives as a valuable anthelmintic.

MEDICINE.
726

The spores of a fungus, probably of **Lycoperdon gemmatum**, are sold in the bazárs of the Panjáb, and are considered to act like **Agaricus** and **Polyporus** by expelling cold and bilious humours. A medicinal truffle, **Melanogaster durissiums**, *Cooke*, is found in abundance near Simla, and is much used by the natives (see **Truffle**). **Schrotium stipatum**, *Curr.*, which occurs in the nests of white-ants, is also supposed to possess medicinal virtues (*Balfour*).

Food.—For a description of the principal edible forms in India, namely, **Agaricus campestris, Morchelia esculenta, Helvella crispa**, and **Hydnum coralloides**, see the article Mushroom under the heading **Agaricus campestris**; also the description of the Indian Truffle under the heading **Truffle**.

FOOD.
727

Besides these, the more important species, there are no doubt many other forms widely used as food by certain classes of natives in India, but it is to be regretted that, owing to the meagreness of the literature on the subject, a complete list cannot be given. The Muhammadans will only eat **Morchella agaricus**, as they consider the others impure food. Most Hindus eat any mushroom which has a pleasant taste and odour. **Mr. Gibbon**, in the *Journal of the Agri.-Horti. Society, Ind.* (*N.S.*), Vol. V., pp. 51—53, describes a species of **Lapiota** as being found in the nests of white-ants and eaten with relish by the natives. **Stewart** also mentions

FUNGI, &c.	Fungi and Fungoid Pests.

FOOD.

another species as being freely eaten in the Panjáb, which is known as *shirían* in the Jhelam, and *bat-bakrí* in the Kair valley. He describes it as "a thin, flat, ragged-looking Fungus, yellow above and with white gills below, which is got on dead trees in various parts of the Panjáb Himálaya at 8,000 to 8,500 feet. The natives slice and cook it either fresh or dry, and eat it as a relish with bread. I have tried this species in stews, &c., but found it leathery and flavourless."

The same author also mentions an "underground mushroom" of doubtful species, found in cultivated ground near Multan and known as *boinphal* in the vernacular. This, he says, is also eaten by the natives.

Balfour in his "*Agricultural Pests of India*," p. 61, describes an underground fungus, **Mylitta**, as occurring in the Nilgiri hills. and considers it probably closely allied to the so-called native-bread of Tasmania; but gives no record of its being eaten by the natives.

FERMENTS.
728

Ferments.—Some of the microscopic forms seem to be useful as substitutes for yeast (see **Cerevisiæ Fermentum**, Vol. II., 257).

FUNGOID PESTS.
729

Fungoid Pests, the characters of which can generally be made out by the use of the microscope only, are small fungi, which attack and injure the plants or animals on which they are parasitic. Among the more hurtful in India are species of **Æcidium**, Capundium, Chætomium, Clarterisporum, Diplodia, Dothidea, Eurotium, Glenospora, Hemileia, Hendersonia, Hydnum, Isaria, Leutinus, Pellicularia, Pestalozzia, Puccinia, Russula, Septoria, Uromyces, and Ustilago (*Balfour's Agricultural Pests of India*). Chionyphe Carteri, *Berkeley* (Mycetoma sp. of *H. Vandyke Carter*) is the fungus whose ravages cause the deeply-seated disease known as the **MADURA FOOT.**

730

Polyporus anthelminticus, *Berkeley*, grows at the root of old bamboos, and is employed as an anthelmintic in Burma.

Ergot.
731

Ergot—Is the sclerotoid condition of **Claviceps purpurea** (see Vol. II., 359).

Mildew.
732
Mould.
733
Rust.
734
Smut.
735

Fungi attacking plants produce an appearance on the leaves, stems, &c., known as MILDEW, MOULD, RUST or SMUT. These small parasites present many features of great interest both to the botanist and agriculturist; but owing to the difficulty of determining their life-histories, little is as yet known regarding them. The following forms, however, are those which are at present recorded as attacking the more important crops and trees of India :—

Peridermium Thomsoni, *Berkeley*, is a fungus found on the **Picea Morinda** of the Himálaya. The leaves under the growth of the parasite become reduced one half in length, curved, and sprinkled, sometimes in double rows, with Æcidia. The growth in time proves fatal to its host. Dr. Barclay has recently described three species of URIDINEÆ which attack the same tree in the North-Western Himálaya—two species of **Æcidium** and one of **Chrysomyxa**. One of the Æcidia causes general pseudo-hypertrophic distortion of the needles of its host, while the other attacks only the youngest shoots. The first of these may be the same as that described above, but the data given in the description of **Peridermium Thomsoni** are not sufficient to allow of a decision being arrived at. Dr. Barclay, while regretting that he has not had the time nor opportunity to fully work out the life-history of his first species, writes, "A continued study of it is much to be desired, if only from an economic point of view, for the affection must prove very destructive to these valuable timber trees. Apart from the diversion of nutriment it must occasion, the habit it has of attacking new shoots, and so completely involving them as to destroy them must be most injurious to these trees." A similar æcidial parasite has also been found on **Cedrus Libani**, *var.* **Deodara**, by the same investigator. **Pinus longifolia** and **P. excelsa**, particularly the former,

FUNGOID PESTS.

are largely attacked in certain parts of the Himálaya by an æcidial parasite found on the needles only.

Acacia eburnea, *Willd.*, is attacked largely in the Poona district by a species of Æcidium which Dr. Barclay has named A. esculentum to indicate its edibility, a rare property in this group of fungi, the only other one known to be eaten being A. Urticæ, *Schum.*, var. himalayense, *Barclay*. Mr. Wroughton, Forest Officer of the Poona Division, informed Dr. Barclay that the fungus is universally eaten in that region, after being cooked, as a relish.

736

Gymnosporangium.—Dr. Barclay has kindly furnished the following information: "The only URIDINE occurring on fruit trees that I have come across, is a species of **Gymnosporangium**, on **Pyrus Pashia**. This, I believe, is a new species, and I am describing it in a forthcoming paper as **G. Cunn'nghamianum** (*Scientific Memoirs by Medical Officers of the Army of India, Part V.*). It has some resemblance to G. clavariæformæ, *Jacq.*, and I provisionally named it so in my list of Simla Uridineæ."

737

Puccinia graminis, *Pers.*, is assumed to be the form of "CORN MILDEW" which occurs commonly on the cereals of the Himálaya, where three species of **Barberry** occur, on two of which the æcidium-bearing parasite has been found by Dr. Barclay. The same **Puccinia** is generally believed to be the cause of rust and mildew in other parts of India also; but as no species of **Barberry** occurs in the plains, it is probable that the parasite in such regions has a different life-history. It has been suggested that the WHEAT RUST of the plains is due to a species of Æcidium reared on a Euphorbia (see article on **Ergot**, Vol. II., 359).

738

Wheat-rust.
739

Melampsora.—Flax crops are often attacked in some localities, especially the Central Provinces, with rust, which has been supposed to be the same species as that attacking cereal crops; but Dr. Barclay informs the writer that this parasite is a species of **Melampsora** and probably M. Lini, *Pers.* It is probable that the Rust on Mustard, which is also largely prevalent, is a species of the same genus; but its identity has not been established.

740

Chrysomyxa.—A species of this genus (C. himalense, *Barclay*) is extensively prevalent in the Simla region on **Rhododendron arboreum**, *Linn.*; giving rise to conspicuous witches-brooms. Another species (C. Piceæ, *Barclay*) occurs on **Picea Morinda**.

741

Ravenelia.—Two species of this fungus, R. sessilis, *Berk.*, and R. stricta, *Berk. & Br.*, are noted by Dr. D. D. Cunningham to be very common in the neighbourhood of Calcutta, the former on **Albizzia Lebbek**, and the latter on **Pongamia glabra.**

742
743

Hemileia vastatrix, *Berk. & Br.*, as is well known, has been immensely destructive to the coffee plantations of Ceylon and Southern India.

744

Perinospora.—The POTATO crops of Assam have been largely attacked by P. infestans. Dr. D. D. Cunningham has noted the occurrence of P. arborescens as a destructive parasite on the POPPY. It is quite possible that the cause of the destruction of the VINE industry of Basahr was due to P. viticola, but unfortunately there is no sufficient evidence to show what was really the cause of that vine disease. It may very possibly have been due to **Oidium (Erysiphe) Tuckeri.**

745

Dr. D. D. Cunningham reports the existence of a root blight in the Darjeeling district TEA gardens. The blight was undoubtedly due to a fungus, but the specimens at his disposal did not enable him to determine its nature.

746

Tilletia caries, or BUNT is a fungus which attacks WHEAT and occupies the whole farinaceous portion of the grain. SORGHUM and the SMALL MILLETS are liable to attacks from allied parasites.

Bunt.
747

Ustilago or SMUT has been described by Dr. Cooke as attacking BARLEY and many GRASSES in the Panjáb, also the male flowers of the

Smut.
748

FURS.	Fur-bearing Animals.

FUNGOID PESTS.

MAIZE. In 1870, a form of **Ustilago** made its appearance on RICE, and is said to have affected a considerable portion of the crop in the neighbourhood of Diamond Harbour in Bengal. The mycelium of this fungus grows into the tissues of its host, forming a whitish, gummy, interlaced thread-like net, in which the spores form. These become at length a more or less coherent mass, dirty-green on the exterior of the infected grain, but of a bright orange-red colour inside. **Dr. Barclay,** in a note kindly furnished on this subject, writes, " The SMUT on wheat, barley, and oats in Europe is**Ustilago segetum,** *Bull.***;** and **Dr. Brefeld** informs me that the Indian species is identical with it.   That on Maize is **U. Maydis,** *DC.*"

In concluding these brief notices of Fungoid Pests the hope may be expressed that the present active researches of **Dr. Barclay** in the Simla District, and of others, into the interesting life-history of these fungi, may clear up many points which are at present very obscure, and so perhaps open a way to fresh exertions in devising methods for the prevention of the destruction effected by these pests.

For further information regarding Fungoid Pests see **Coffee,** Indigo, Rice, Wheat, &c.

749

FURCRŒA, *Vent.; Gen. Pl., III., 739.*

An American genus of Amaryllidaceous plants containing some 10 or 15 species. These are closely allied to the Agaves and indeed are commercially viewed as identical, the fibres derived from the two genera being collectively designated American Aloe fibres. **Furcrœa gigantea,** the best known fibre-yielding species of this genus, was formerly known as **Agave fœtida,** and by some writers **Agave vivipara** is spoken of as **Furcrœa Cantala.**

There is very little that need be said here regarding these plants. A few of them are cultivated in India, and these have been experimentally tested for their fibres. In this country, however, their cultivation as sources of fibre has up to this time been very unimportant and insignificant, compared with the degree to which they are utilized in Mauritius. The fibre of F. gigantea is, in fact, commercially designated *Mauritius Hemp*. The reader is referred to the article **Agave** in Vol. I., pp. 133—144.

Furniture. See **Cabinet Work,** Vol. II., I.

750

FURS.

The following list of the principal fur-bearing animals of India, compiled principally from **Forbes Watson's** report on a proposed Industrial Survey of India, may be given, leaving the reader, for further information regarding trade, description, and qualities of fur, &c., to refer to the articles on the animals grouped under their popular or commercial names, (Deer, &c.) and to that on "SKINS." The writer is indebted to **Major Ward** for having kindly revised and supplemented this enumeration.

751 **Aliurus fulgens,** *F. Cuv.* The Red Cat Bear.
    Vern.—*Wah,* TIBET.

752 **Arctictis binturong,** *Raffles.* The Black Bear-Cat.
    Vern.—*Myouk, kya,* BURM.

753 **Arctomys bobac,** *Schuler.* The Marmot.
    Vern.—*Kandia-piu,* TIBET.

754 **A. hemachalanus,** *Hodgson.* The Red Marmot.
    Vern.—*Drin,* KASH.

755 **Canis aureus,** *Linn.* The Jackal.
    Vern.—*Gidar, kola,* HIND.

756 **C. lupus,** *Elliot.* The Tibet Wolf ; or Black Wolf.
    Vern.—*Chanco, hakpo.chanko,* TIBET.

F. 756

Fur-bearing Animals. (*J. Murray.*)	FURS.
Canis pallipes, *Sykes.* The Indian Wolf. Vern.—*Bhera, laudgah,* HIND.	757
Capra hircus, *Linn.* The Domestic Goat. Vern.—*Jumnapari, bakra,* HIND.	758
Cuon rutilans, *Temm.* The Wild Dog. Vern.—*Jangli-kuta, sona-kuta, ram-kuta, ban-kuta,* HIND.; *Kosla,* MAR.; *Resa-kutta,* TEL.	759
Felis bengalensis, *Blyth.* The Leopard Cat. Vern.—*Chita-billi,* HIND.	760
F. chaus, *Guld.* The Common Jungle Cat. Vern.—*Jangli-billi,* HIND.	761
F. caracal, *Schrebor.* The Caracal. Vern.—*Siagosh,* HIND.	762
F. jubata, *Schrebor.* The Cheetah or Hunting-Leopard. Vern.—*Chita,* HIND.	763
F. leo, *Linn.* The Lion. Vern.—*Singha, sher, babbar-sher,* HIND.	764
F. lynx. The Lynx. (includes **F. isabellina,** The Tibet Lynx.) Vern.—*Es,* TIBET.	765
F. nebulosa, *Griffith, vel* diardi, *Hodgson.* The Clouded Leopard. Vern.—*Zik,* BHOT.	766
F. pardus, *Linn.* The Pard. Vern.—*Tendua, chita,* HIND.	767
F. tigris, *Linn.* The Tiger. Vern.—*Bagh, sher, sela-vagh, nahar,* HIND.	768
F. torquata, *F. Cuv.* The Spotted or Desert Cat.	769
F. uncia, *Schreber.* The Ounce or Snow-Leopard. Vern.—*I'ker,* TIBET; *Burrel-hay,* SIMLA.	770
F. viverrina, *Bennet.* The Tiger Cat or Fishing Cat. Vern.—*Mach-bagrul,* HIND.	771
Galeopithecus volans, *Linn.* The Flying Lemur. Vern.—*Kabong,* MERGUI.	772
Halicon dugong, *Erxl.* The Dugong. Vern.—*Talla-maha,* CEYLON.	773
Herpestes pallidus. The Common Mungoose. Vern.—*Mangús, newul, newra, nyul,* HIND.	774
H. jerdoni, *vel* monticolus. The Long-tailed Mungoose. Vern.—*Konda-yeutawa,* TEL.	775
Lagomys roylii, *Ogilby.* The Himálayan Mouse Hare. Vern.—*Abra,* NEPAL.	776
Lepus nigricollis, *Cuv.* The Black-naped Hare. Vern.—*Khargosh,* HIND.	777
L. pallipes, *Hodgson.* The Tibet Hare. Vern.—*Rek, rigong,* TIBET.	778
L. ruficaudatus, *Geoffr.* The Indian Hare. Vern.—*Khargosh,* HIND.	779
Loris gracilis, *Shaw.* The Slender Lemur; Sloth. Vern.—*Dewantsi-pilli,* TEL.	780

FURS.	Fur-bearing Animals.
781	**Lutra leptonyx,** *Horsf.* The Clawless Otter. Vern.—*Chusam,* BHOT.
782	**L. nair,** *F. Cuv.* The Common Indian Otter. Vern.—*Pani-kúta,* HIND.
783 784	**Macacus silenus,** *Anderson.* The Black Lion-tailed Monkey. **Martes flavigula,** *Bodd.* The Indian Marten. Vern.—*Tuturala,* N.-W. HIM.; *Mal-sampra,* NEPAL.
785	**M. toufaeus,** *Hodgson.* The Tibet or Beech Marten. Vern.—No name. Major Ward writes: **M. toufaeus,** *Hodgson,* is found in Ladak, Baltistán, Tibet, &c. I have seen skins brought to Simla, and have killed it in many places in Baltistan. It is a highly-priced fur. I think **M. erminea** has been confused with **M. toufaeus** in its winter coat.
786	**M. kathiah,** *Hodgson.* The Yellow-bellied Weasel. Vern.—*Kathia-nyal,* NEPAL.
787	**M. strigidorsa,** *Hodgson.* The Striped Weasel. Vern.—No name.
788	**M. subhemachalana** *Hodgson.* The Himálayan Weasel. Vern.—*Krau or grau,* KASH.
789	**Nycticebus tardigradus,** *Geoffr.* The Slow-paced Lemur; Sloth. Vern.—*Sharmindi billi,* HIND.
790	**Ovis aries,** *Linn.* The Domestic Sheep. Vern.—*Hunich, kago, silingia, peluk,* NEPAL.
791	**Paradoxurus bondar,** *Gray.* The Tree Cat. Vern. *Chinghar,* HIND; *Bondar, baum,* BENG.
792	**P. musanga,** *Raffles.* The Common Tree Cat. Vern.—*Mennie, lakati,* HIND.
793	**Pœphagus grunniens,** *Linn.* The Yak. Vern.—*Yak, ban-chur,* HIND.
794	**Pteromys alboniger.** The Black and White Flying Squirrel. Vern.—*Piam piyu,* BHOT.
795	**P. caniceps.** The Grey-headed Flying Squirrel. Vern.—*Biyom-chimbo,* LEPCHA.
796	**P. inornatus,** *Geoffr.* The White-bellied Flying Squirrel. Vern.—*Rusi-gugar,* KASH.
797	**P. magnificus,** *Hodgson.* The Red-bellied Flying Squirrel. Vern.—*Puraj-blakut,* NEPAL.
798	**P. petaurista,** *Pallas.* The Brown Flying Squirrel. Vern.—*Pakya,* MAHR.
799	**P. spadiceus.** The Red Flying Squirrel. Vern.—*Kywet-shov-byan,* ARAKAN.
800	**Rhizomys badius,** *Hodgson.* The Bamboo Rat. Vern.—*Yewcron,* NEPAL.
801	**Scuirus giganteus.** The Black Hill Squirrel. Vern.—*Sheu,* TENASSERIM.
802	**S. indicus.** The Bombay Squirrel. Vern.—*Shekra,* MAHR.

F. 802

Fur-bearing Animals. (*J. Murray.*)	FUSTIC.
Scuirus lokriah, *Hodgson.* The Red-bellied Grey Squirrel. Vern.—*Lokriah,* NEPAL.	803
S. maclellandi, *Horsf.* The Himálayan Squirrel. Vern.—*Kalli-gangdin,* LEPCHA.	804
S. macrourus, *Forster.* The Grizzled Hill Squirrel. Vern.—*Rookerah,* CINGH.	805
S. maximus, *Schreber.* The Red Squirrel. Vern.—*Karrat,* HIND.	806
S. palmarum, *Gmelin.* The Common Indian Ground Squirrel. Vern.—*Gilheri,* HIND.	807
Semnopithecus johni, *Anderson,* The Nilghiri Langúr. Vern.—*Turuni, kodan, pershk,* TODA.; *Korangu,* BUDUGA & KURUMBA; *Karing-korangu,* MALAY.	808
S. schistaceus, *Hodgson.* The Himálayan Langúr. Vern.—*Langúr,* HIND.	809
Talpa micrura, *Hodgson.* The Mole. Vern.—*Biyu-kantyem,* BHOT.	810
Vulpes bengalensis. The Indian Fox. Vern.—*Lumri, lokri,* HIND.	811
V. ferrilatus, *Hodgson.* The Tibetan Grey Fox. Vern.—*Iger,* TIBET.	812
V. flavesceus, *Gray.* The Persian Fox. Vern.—*Wamer,* NEPAL.	813
V. fuliginosus, *Hodgson.* Tibet Fox. Vern.—*Theske.*	814
V. griffithii. The Afghanistan Fox. **V. leucopus,** *Blyth.* The Desert Fox. **V. montanus,** *Pearson.* The Hill Fox. Vern.—*Wamoo,* NEPAL.	815 816 817
V. pusillus, *Blyth.* The Panjáb Fox.	818
Ursus isabellinus, *Horsf.* The Brown Bear. Vern.—*Barf-ka-rich, bhalu,* HIND.	819
U. labiatus, *Blainv.* The Black Bear, or Sloth Bear. Vern. *Bhalu, rich,* HIND.	820
U. malayanus, *Raffles.* The Malayan Sun Bear. Vern.—*Bruang,* MALAYAN.	821
U. torquatus, *vel* tibetanus. The Himálayan Black Bear. Vern.—*Bhalu,* HIND.	822
Urva cancrivora, *Hodgson.* The Crab-eating Mungoose. Vern.—*Urva,* NEPAL.	823

Fustic, see Maclura tinctoria and Rhus Cotinus.

F. 823

GALLS.	The Cod; The Cheese Rennet; Galls.

GADUS.

1

Gadus morrhua, *Linn. ;* Pisces.

THE COMMON COD.

The fish from which the officinal Cod Liver oil is obtained, is not a native of the Indian seas; it abounds on the coasts of Norway, France, Britain, Ireland, and is specially common in the seas along the coast of Newfoundland. The oil extracted from the liver is imported into this country for medicinal purposes. It is a valuable alterative and nutritive tonic, especially beneficial in scrofulous and tuberculous affections, rickets, and other diseases due to impaired nutrition.

SUBSTI-
TUTES.
2

Substitutes.—Several Indian fish yield oil, which is, however, owing to carelessness in methods of manufacture, generally rancid and unfit for medicinal use. Dr. Bidie states that the best of these oils, and one that might be substituted for the officinal OLEUM MORRHUÆ, is that obtained from the livers of certain species of **Carcharias**, which abound off the Western Coast. See **Carcharias,** Vol. II., 155; also **Fish,** Vol. II., 368-397.

Galangal ; see **Alpinia Galanga,** *Willd. ;* Vol. I., p. 192.

3

Galbanum. A gum-resin, probably obtained from two species of **Ferula,** *viz.,* F. **galbaniflua** and F. **rubicaulis.** See Vol. III., 338.

Galena or Sulphide of Lead, see **Lead,** Vol. IV.

GALIUM, *Linn. ; Gen. Pl., II., 149.*

A genus of small, weak herbs of the Natural Order RUBIACEÆ, comprising about 150 species, mostly temperate. Of these 20 are natives of India, and occur chiefly on the Temperate Himálaya.

4

Galium verum, *Linn. ; Fl. Br. Ind , III., 208 ;* RUBIACEÆ.

THE CHEESE RENNET.

References.—*Boiss., Fl. Orient., III., 62; Balfour, Cyclop., I., 1163; Smith, Dic., 107 ; Treasury of Bot., I., 517.*

Habitat.—A perennial herb with erect or rambling stems from 1 to 3 feet high, found in the Western Himálaya at altitudes of 5,000 to 10,000 feet.

DYE.
Roots.
5

Dye.—Smith mentions that the ROOTS are extensively collected in Europe for the dye which they yield, which is said by him to be equal to madder. Several other species of the genus yield a purple dye, but no mention appears to be made by Indian writers of their utilisation in this country.

DOMESTIC.
Plant.
6

Domestic Uses.—The PLANT was formerly extensively employed in Europe as a reagent for curdling milk, from which property it has derived its popular name, but in India the best known vegetable rennet is **Withania coagulans.**

7

Galium, sp.

An undetermined species of GALIUM, mentioned by **Aitchison** as very common in the shade of rocks on the low hills near Badghis, in Afghánistan, which was observed by him to dye the hands a yellow-green on collecting it. (*Botany of Afgh. Del. Com., 73.*)

GALLS.

8

Galls.

By the term gall is commonly understood a deformity or excrescence, due to a parenchymatous hypertrophy of the structure of a plant caused

G. 8

by insects. The exciting cause of these local growths appears, in most cases, to be a minute quantity of some liquid irritant, introduced within the tissues, by the female insect, through the puncture made by her ovipositor. Subsequent irritation, however, must be kept up by the presence of the ovum, or later, of the larva, and this, without doubt, plays an important part in the formation of many galls.

Galls vary greatly in character with the plant on which they occur, and with the insect by which they are produced ; but all possess many qualities in common, qualities which render them of great economic value. It is unnecessary, in an article such as the present, to enter into the subject of the different insects which give rise to these hypertrophies, but it may be mentioned that the **Hymenoptera, Diptera, Hemiptera, Homoptera,** and **Coleoptera** all comprise several gall-forming genera. A list of the chief Indian gall-yielding trees is appended. Not only are the galls of these trees largely employed, but the parasitical excrescences of **Quercus infectoria,** *Oliver,* are also largely imported into India, from Basra and the Persian Gulf ports. They are used as adjuncts in several processes of dyeing shades of brown, grey, and lavender, for tanning leather, and for medicinal purposes.

For information regarding the vernacular names and special economic properties of the several Indian galls, the reader is referred to the articles on the plants they infest, in their respective alphabetical positions in this work.

References.—*Roxb., Fl. Ind., Ed. C.B.C., 381 ; Brandis, For. Fl., 23, 120, 123, 170, 171, 184, 224, 226, 302, 316, 381, 481 ; Kurz, For. Fl. Burm., I., 207 ; Stewart, Pb. Pl., 47, 54, 74, 91, 92 ; Pharm. Ind., 29, 59, 89, 209 ; Ainslie, Mat. Ind., I., 144, 602 ; O'Shaughnessy, Beng. Dispens., 607 ; Moodeen Sheriff, Supp. Pharm. Ind., 145, 239 ; U. C. Dutt, Mat. Med. Hind., 298, 319 ; Dymock, Mat. Med. W. Ind., 2nd Ed., 76, 78, 191, 194, 319, 729 ; Flück. & Hanb., Pharmacog., 167, 595 ; Bent. & Trim., Med. Pl., 249 ; Murray, Pl. and Drugs, Sind, 46, 47, 189 ; K. L. Dey, Indigenous Drugs of India, 99 ; Irvine, Mat. Med. of Patna, 68 ; Baden Powell, Pb. Pr., 471, 472 ; Drury, U. Pl., 413, 419 ; Lisboa, U. Pl. Bomb., 241, 259 ; Birdwood, Bomb. Pr., 9, 19, 83, 309, 313 ; McCann, Dyes and Tans, Beng., 162 ; Buck, Dyes and Tans, N.-W. P., 23, 36 ; Liotard, Dyes, 11, 13, 14, 17 ; Spons' Encyclop., 1983 ; Balfour, Cyclop., I., 1164 ; Treasury of Bot., I., 518 ; District Manual, Trichinopoly, 16 ; Special Reports from For. Dept. in Panjáb, N. W. P., Ajmere-Merwara, Sind, N. Circle, Bombay, S. Circle, Madras, and from J. H. Lace, Esq., Quetta.*

LIST OF THE CHIEF GALL-BEARING PLANTS OF INDIA.

Acacia leucophlœa, *Willd.*
Areca Catechu, *Linn.*
Cinnamomum zeylanicum, *Breyn.*
Fraxinus floribunda, *Wall.*
Garuga pinnata, *Roxb.*
Litsœa polyantha, *Juss.*
Pistacia integerrima, *Stewart.*
P. mutica, *var.* cabulica, *Stocks.*
Pongamia glabra, *Vent.*

Prosopis spicigera, *Linn.*
P. Stephaniana, *Kunth.*
Quercus Ilex, *Linn.*
Salvadora oleoides, *Dcne.*
Tamarix articulata, *Vahl.*
T. dioica, *Roxb.*
T. gallica, *Linn.*
Terminalia Chebula, *Retz.*
T. tomentosa, *Bedd.*

9

GAMBIER.

Gambier.

This resinous extract is prepared *from* **Uncaria Gambier,** *Roxb.,* much in the same manner in which Cutch or Catechu is made. The plant is a native of Malacca, Penang, and Singapore ; distributed to Java and Sumatra. The extract made |up in small (one inch) cubes, is of a pale

10

greyish yellow colour, and has a bitter taste. It is largely imported into India to be eaten in *pán*, but the yellow semi-crystalline form of Cutch prepared in Kumáon is to a large extent used for the same purpose, and is even made up in cubes to resemble Gambier. Gambier is an officinal drug in the British Pharmacopœia, and is known in medicine as pale catechu. In the United States Dispensatory Catechu (**Acacia Catechu**) is officinal, while Gambier is rejected. In the Indian Pharmacopœia both drugs are officinal.

A certain re-export trade in Gambier takes place from India, but the official designation (in Trade Returns) of "Cutch and Gambier" should be understood to refer almost exclusively to the dark or Pegu form of Cutch, and to the pale or Kumáon form of so-called Gambier. See **Acacia Catechu**, Vol. I., pp. 29 to 40; also **Uncaria**.

Gamboge, see the various species of **Garcinia**.

Game Birds, see **Ducks**, &c., also **Peacock**, **Pheasant**, **Pigeon**, and **Snipe**.

Gao-zaban, see **Echium, sp.,** p. 200; also **Onosma bracteatum,** *Wall.,* BORAGINEÆ.

GARCINIA, *Linn.; Gen. Pl., I., 174.*

A genus of trees, usually yielding yellow juice, which belongs to the Natural Order GUTTIFERÆ, and comprises in all some 50 species, which are distributed over Tropical Asia, Africa, and Polynesia. Of these, about 30 are natives of India, and several possess features of considerable economic interest.

[FERÆ.

11 **Garcinia anomala,** *Planch. & Trian; Fl. Br. Ind., I., 266;* GUTTI-
Syn.—GARCINIA AFFINIS, *Wall.* (*in part*).
References.—*Kurz, For. Fl. Burm., I., 89; Kurz, Prelim. For. Rep. on Pegu, App. A., xii.; Indian Forester, IV., 241; XI., 392.*
Habitat.—A small erect tree found in the beds of torrents, in the Jaintia Hills and Khásia Mountains, between altitudes of 3,000 and 5,000 feet; also not uncommon in the damp and dry hill forests of Martaban east of Tounghoo at elevations of from 4,000 to 6,000 feet.

GUM.
12   Gum-Resin.—The tree yields "an inferior gamboge" (*Kurz*).
TIMBER. Structure of the Wood.—Sapwood white, soft (*Kurz*).
13
14 **G. Cambogia,** *Desrouss; Fl. Br. Ind., I., 261.*
Syn.—GARCINIA ZEYLANICA, *Roxb.;* G. AFFINIS, *Wight & Arn.* (*not of Wall.*); G. ELLIPTICA, *Wall.*
Var. 1.—CONICARPA, *Wight. Ic., 121* (excl. o).
Var. 2.—PAPILLA, *Wight. Ic., t. 960, 961* (sp.).
Vern.—*Vilaiti-amli,* BOMB.; *Híla,* BURGHERS (NILGHIRIS); *Aradal, manthulli,* KAN.; *Goraka,* SING.
References.—*Wight & Arn., Prod., I., 561; Roxb., Fl. Ind., Ed. C.B.C., 442; Corom. Pl., III., t. 298; Beddome, Fl. Sylv., t. 85; Gamble, Man. Timb., 24; Thwaites, En. Ceylon Pl., 48; U. S. Dispens., 15th Ed., 1183; Mason, Burma and its People, 480, 515; Drury, U. Pl., 220; Lisboa, U. Pl. Bomb., 10, 147, 241; Cooke, Gums and Gum-resins, 41; P. W. D., Report on Gums, 2, 7-9, 34; Balfour, Cyclop., I., 1175; Treasury of Bot., I., 206; Indian Forester, II., 20, 58, XI., 379, 392; Madras Manual of Administration, II., 65, 135; Gazetteers:—Bombay, XV., 427; Mysore and Coorg, I., 68; Special Reports from the Conservators of Forests of Southern Circle, Madras and Bombay.*
Habitat.—A small evergreen tree of the mountains of the Western Peninsula from Concan to Travancore; also met with in Ceylon.

GUM.
15   Gum-Resin.—Thwaites states that this species yields (in Ceylon) a yellow, insoluble, very adhesive gum, which is valueless as a pigment on account of its insolubility in water. It is, however, easily soluble in

G. 15

spirits of turpentine, and is likely to prove useful as a varnish. A considerable amount of confusion exists in the descriptions of various writers regarding this gum-resin. Thus, in the P. W. Dept. Report, above cited, **Mr. Broughton** writes that the substance appears very similar to true Gamboge and of very fine quality. It appears probable, however, that the gum-resin he examined, which was collected by **Beddome**, was really the exudation of **Garcinia Morella**, *Desrouss.* Recent reports received from the Conservators of Forests in Madras and Bombay confirm **Thwaites'** statement as to the uselessness of the gum-resin as a pigment.

OIL.
16

Oil.—Mr. **Cherry** mentions that this species affords an oil which is used in medicine (*Gamble*).

FOOD.
Rind.
17

Food.—The acid RIND of the fruit is employed as food, and when dried is eaten as a condiment in curries.

TIMBER.
18

Structure of the Wood.—"Grey, cross-grained, shining, hard; weight 54℔ per cubic foot" (*Gamble*). "An excellent, straight-grained, lemon-coloured, slightly elastic wood, which is easily worked, and would answer for common furniture" (*Beddome*).

Garcinia cornea, *Linn.; Fl. Br. Ind., I., 260 ; Wight, Ic., t. 105.*

19

Syn.—GARCINIA AFFINIS, *Wall. Cat., 4852, 4853, and 4854 in part, not of Wight & Arn.*; DISEOSTIGMA FABRILE, *Miquel.*

References.—*Roxb., Fl. Ind., Ed. C.B.C., 444; Kurz, For. Fl. Burm., I., 88; Kurz in As. Soc. Journ. Beng., XXXIX., 64; Prelim. For. Rep. on Pegu, App. A., xii.; Balfour, Cyclop., I., 1175; Ind. Forester, XI., 392.*

Habitat.—An evergreen tree, from 40 to 60 feet in height; met with in Eastern Bengal and Burma.

GUM.
20

Gum-Resin.—Yields an inferior kind of Gamboge.

TIMBER.
21

Structure of the Wood.—Brown, heavy, of a coarse unequal fibre, hard, rather close-grained (*Kurz*).

G. Cowa, *Roxb. ; Fl. Br. Ind., I., 262 ; Wight, Ic., tt. 104 & 113.*

22

Syn.—GARCINIA KYDIA, *Roxb.*; G. ROXBURGHII, *Wight*; G. UMBELLIFERA, *Roxb.*; G. WALLICHII, *Chois.*; G. LOBULOSA, *Wall.*; OXYCARPUS GANGETICA, *Ham.*

Vern.—*Cowa,* HIND.; *Taungthálé, toung-da-lai, ma-dow,* BURM.

References.— *Wight and Arn., Prodr., I., 101; Roxb., Fl. Ind., Ed. C.B.C., 442; Kurz, For. Fl. Burm., I., 90; Prelim. For. Rep. on Pegu, App. A., xii.; Gamble, Man. Timb., 24; Mason, Burma and its People, 480, 482, 751; Cooke, Gums and Gum-resins, 42; Liotard, Dyes, 91; Balfour, Cyclop., I., 1175; Burm. Gas., I., 132; Indian Forester, XI., 392.*

Habitat.—A tall evergreen tree of Eastern Bengal, Assam, Chittagong, Burma, and the Andaman Islands.

GUM.
23

Gum-Resin.—This species produces a kind of gamboge but of a paler colour than that of **G. Morella**, and, according to Mason, insoluble in water. In the *Burma Gazetteer* it is described as forming, with spirits of turpentine, a very beautiful and permanent yellow varnish for metallic surfaces.

DYE.
Bark.
24

Dye.—Liotard mentions that the BARK is employed in the Pegu District to produce a light yellow colour, principally in the colouring of cloth for the garments of Buddhist monks. It is cut up into small pieces, boiled in water and strained, the acid liquid of applewort bark being used as a mordant.

FOOD.
Fruit.
25

Food.—Roxburgh describes the FRUIT as edible "though not the most palatable."

TIMBER.
26

Structure of the Wood.—Greyish-white, moderately hard. Weight 37 to 47℔ per cubic foot (*Gamble*). White, turning yellow, rather heavy, coarsely fibrous, loose-grained, very perishable (*Kurz*).

GARCINIA indica.	The Gamboge Trees.

27

Garcinia echinocarpa, *Thw.; Fl. Br. Ind., I., 264.*

Vern. —*Madol,* SING.

References.—*Beddome, Fl. Sylv., Anal. Gen., xxi.; Thwaites, En. Ceylon Pl., 49; Indian Forester, X., 33.*

Habitat.—A tall tree of the Central and Southern Provinces of Ceylon.

OIL. Seeds.
28

Oil.—" A thick oil, extracted from the SEEDS, is used by the Singhalese for burning in their lamps, but it gives a very indifferent light" (*Thwaites*).

29

G. eugeniæfolia, *Wall.; Fl. Br. Ind., I., 268.*

Habitat.—A small tree of the Eastern Peninsula, found in Singapore, by Wallich, and in Malacca, by Griffith.

GUM.
30

Gum-Resin.—Helfer mentions that the stem exudes a green varnish, and Griffith that the juice of the fruit is milky. No further information, in confirmation of these interesting statements, is, however, available.

31

G. heterandra, *Wall.; Fl. Br. Ind., I., 265.*

Syn.—HEBRADENDRON WALLICHII, *Chois.* Kurz considers this Burmese species to be identical with the Sylhet specimen, G. ELLIPTICA, *Wall.,* and he retains the latter name for both. The *Flora of British India,* however, reduces the Sylhet plant to G. Morella, *Desrouss.,* a synonymy that has been here followed. It appears probable, that the information, given by writers on the resources of Burma, regarding the plant they call G. elliptica, *Wall.,* really refers to the species at present under consideration, and will, consequently, be detailed in this article.

Vern.—*Thanat-tau, tha-nat-dau,* BURM.

References.—*Kurz, For. Fl. Burm., I., 92; in As. Soc. Jour. Beng., XLIII., pt. II., 87; Prelim. Forest Report on Pegu, App. A., xiii.; Gamble, Man. Timb., 22; Mason, Burma and its People, 480-82, 751; Jour. Agri.-Hort. Soc. Ind., X. (old series), pro. cxxi.; Balfour, Cyclop., I., 1175; Indian Forester, XI., 393.*

Habitat.—An evergreen tree of the forests of Pegu and Tenasserim, ascending to 4,000 feet.

GUM.
32

Gum-Resin.—Mason, and later, Kurz, have both described this tree as yielding a superior kind of Gamboge, so similar to the Gamboge of commerce that the former writer considered it identical. He wrote, " In its appearance to the eye and in its properties as a pigment, I have failed to discover the slightest difference between the exudation of this tree and the Gamboge of commerce." It readily forms an emulsion with water.

An interesting account is given, in the Agri-Horticultural Society's Journals, Vol. X. (old series), of an analysis of a gamboge obtained from a tree in Burma, called *Tanatan* (probably a misprint for *Tanatau,* the vernacular name of this species). **Mr. D. Hanbury,** the analyst, writes, " I find this gum-resin to be, in its chemical characters, precisely like the ordinary Siamese gamboge; it is, however, much mixed with impurities, and is, in fact, but rudely prepared. If carefully collected and cast in bamboos (like the Siam drug), I cannot but think that it would equal the finest gamboge we get."

DYE.
33

Dye.—Mason states that the Burmese priests occasionally employ the gamboge obtained from this species to dye their robes, and the Karens to colour their thread; and that it serves equally well as a pigment.

MEDICINE. Gum-resin.
34

Medicine.—The GUM-RESIN is occasionally, though not extensively, employed as a medicine by Burman native practitioners (*Mason*).

TIMBER.
35

Structure of the Wood.—White, soft.

36

G. indica, *Chois.; Fl. Br. Ind., I., 261; Wight, Ill., I., 125.*

COCUM or KOKAM BUTTER, MANGOSTEEN OIL, BRINDONIA TALLOW, *Eng;* BEURRE DE COCUM, HUILE DE MADOOL, *Fr.;* BRINDÁO, *Port.*

G. 36

| Kokam Butter. | (*J. Murray.*) | **GARCINIA indica.** |

Syn.—G. PURPUREA, *Roxb.*; G. CELEBICA, *Desrouss*; BRINDONIA INDICA, *Dupetit-Th.*

Vern.—*Kokam, kokam-ká-tél* (the oil), HIND.; *Rátambi, kokamb,* DEC.; *Kokam, amsúl* (the fruit), *kokam chatel* (the oil), *ratambu-sála* (the bark),* BOMB.; *Bhirand, chirand, kokam, katambi, amsúl, rátambi* (fruit), *bhirandel* (oil), MAR.; *Kokan,* GUZ.; *Múrgal mara,* TAM.; *Rátambi,* KONKAN; *Múrgala, múrgal, margina-huli mara, dhúpadi-enné* (the oil), KAN.; *Brindáo,* GOA.

References.—*Roxb., Fl. Ind., Ed. C.B.C.,* 443; *Beddome, Fl. Sylv. Gen., xxi., For. Man., xx.; Gamble, Man. Timb.,* 22; *Dals. & Gibs., Bomb. Fl.,* 31; *Grah., Cat. Bomb. Pl.,* 25; *Pharm. Ind.,* 31; *Moodeen Sheriff, Supp. Pharm. Ind.,* 146; *Mat. Med. Madras,* 42; *Dymock, Mat. Med. W. Ind.,* 2nd Ed., 78; *Pharmacog. Indica, I.,* 163; *Flück. & Hanb., Pharmacog.,* 86; *Bent. & Trim., Med. Pl.,* 32; *S. Arjun, Bomb. Drugs,* 199, 23; *Murray, Pl. and Drugs, Sind,* 68; *Fleming, Med. Pl. & Drugs in As. Res., Vol. XI.,* 188; *Lisboa, U. Pl. Bomb.,* 10, 146, 213, 241; *Birdwood, Bomb. Pr.,* 14, 218, 278; *Cooke, Oils and Oilseeds,* 13; *Voyage of John Huyghen van Linschoten,* 1596, *II.,* 34; *Agri.-Hort. Soc. India, Trans., VII.,* 75; *Journ. (old series) IV.,* 204; *As. Soc. Beng. Journ., II.,* 592; *Spons' Encycl., II.,* 1395; *Balfour, Cyclob., I.,* 1176; *Kew Reports,* 1881, 13; *Kew Off. Guide to the Mus. of Ec. Bot.,* 16; *Kew Rep.,* 1882, 13; *Indian Forester, XI.,* 328; *Gazetteers:—Bombay, XIII., pt. I.,* 25; *XV., pt. I.,* 70; *XVIII.,* 57; *Home Dept. Cor., regarding Pharm. Ind.,* 307; *Madras Board of Rev. Procgs., June 1st,* 1889, *No.* 2; *Special Reports from F. C. Ozanne, Esq., Bombay,* 1886 *and* 1889; *Conservators of Forests, Northern and Southern Circles; Bombay; Conservator of Forests, S. Circle, Madras.*

Habitat.—A slender tree with drooping branches, found on the Ghâts of the Konkan and Kanara, most commonly in the Southern Konkan, and considerably cultivated in gardens of that district. It bears a conspicuous spherical purple fruit, the size of a small orange, which ripens about April.

Dye.—The juice of the FRUIT has long been employed as a mordant by dyers in South-Western India. Thus Linschoten, in his *Voyage to the East Indies,* in 1596, noticed the fact, mentioning that "the dyers do use this fruit." Lisboa states that it is chiefly employed as a mordant with iron.

DYE. Fruit. 37

Oil.—A valuable oil, "Kokam butter," is obtained from the SEEDS of the fruit to the extent of about 30 per cent. The process of preparation is described in an interesting communication by the Director of Land Records and Agriculture, Poona, as follows:—" The oil, or butter as it is called, is, as a rule, extracted in the cool season, by one of three methods: *1st, Boiling process.*—The seed is cracked and the shell removed. The white kernel is then pounded in a large specially-made stone mortar by a cone-shaped pestle. The pulp is put into an earthen or iron pan with some water and boiled. After some time it is poured out into another vessel and allowed to cool. The oil which rises to the surface on cooling becomes gradually solid, and is roughly moulded by hand into egg-shaped balls, or concavo-convex cakes; *2nd, Churning process.*—The kernel is pounded as described above, and the pulp with some water is kept in a large vessel and allowed to settle for the night. During the night the oil rises to the surface and forms a white layer which is removed in the morning. The mixture is then churned, and the oil which, like butter, rises to the surface in a solid form, is removed by the hand. This process gives the best results, and is most favourably performed in the cold season; *3rd, Pressing process.*—In this process the kernels are pressed in an ordinary oil mill, like other oil-seeds, and the oil is extracted."

OIL. Seeds. 38

DESCRIPTION AND CHEMICAL COMPOSITION.—Kokam butter, as found in the bazárs of India, consists of egg-shaped or concavo-convex cakes of a dirty white or yellowish colour, friable, crystalline, and with a greasy feel like spermaceti. When fresh it has a faint, not unpleasant, smell, and a

CHEMISTRY. 39

GARCINIA **indica.**	Kokam Butter.

CHEMISTRY.

bland oily taste. It melts in the mouth like butter, and leaves a sensation of cold on the tongue. When long kept it is apt to become rancid, and acquires a browner colour, while an efflorescence of shining tufted crystals appears on the surface of the mass. As ordinarily met with, it contains a considerable amount of impurity, chiefly particles of the seed. As above stated, the purest quality is that obtained by the second process (churning). By filtration under the influence of heat it may be obtained perfectly pure, in which condition it is quite transparent, and of a very light yellowish colour, but at lower temperatures it becomes white and crystalline. The butter of commerce melts at about 40°C. **Flückiger** and **Hanbury** give the following account of its chemical composition : " Purified kokám butter boiled with caustic soda yields a fine hard soap, which, when decomposed with sulphuric acid, affords a crystalline cake of fatty acids weighing as much as the original fat. The acids were again combined with soda, and the soap having been decomposed, they were dissolved in alcohol of about 94 per cent. By slow cooling and evaporation crystals were first formed, which, when perfectly dried, melted at 69·5°C. : they are consequently Stearic acid. A less considerable amount of crystals, which separated subsequently, had a fusing point of 55°, and may be referred to Myristic acid. A portion of the crude fat was heated with oxide of lead and water and the plumbic compound dried and exhausted with ether, which after evaporation left a very small amount of liquid oil, which we refer to oleic acid." It contains no volatile fatty acid.

HISTORY.
40

HISTORY.—Kokam butter has doubtless been employed by the Natives, of at least South-Western India, since remote times, but it does not appear to have attracted the notice of Europeans till about the year 1830. In 1833 a writer in the Journal of the Asiatic Society of Bengal, described its employment medicinally by the Natives and advocated its trial by Europeans. It was adopted as officinal during the compilation of the *Indian Pharmacopœia* in 1868, and is now generally recognised as a solid oil of considerable value.

MEDICINE.
Fruit.
41

Medicine.—The FRUIT has been long employed in South-Western India as a semi-medicinal article of diet. The authors of the *Pharmacographia Indica* state that its virtues were first recognised by the English at the end of the eighteenth century, when it was employed as an anti-scorbutic in the Bombay Army. It is acid, slightly astringent, and is considered by native physicians to be superior to tamarind for the preparation of acidulous drinks. **Dymock** states that the apothecaries of Goa prepare a very fine red syrup from the juice of the fruit, which they administer in " bilious affections." The OIL or Kokam butter, already described, is considered demulcent, nutrient, and emollient. **Moodeen Sheriff,** in his forthcoming work on the *Materia Medica* of Madras, writes, " I have used it internally in my practice, and have found that its best medicinal properties are its usefulness in phthisis pulmonalis and some scrofulous diseases, and in dysentery and mucous diarrhœa. In the former, its action is something like that of cod-liver oil, of which it is a pretty good, and very cheap and pleasant, substitute; and in the latter, it is of great service in relieving tormina and tenesmus when employed as an adjuvant to other medicines." He recommends doses of from ½ to 1 ounce as a nutritive tonic in place of cod-liver oil, and 1 to 2 drachms as an emollient adjuvant to other drugs in dysentery and mucous diarrhœa. It is employed externally by the natives as a remedy for excoriations, chaps, fissures of the lips, &c., by partly melting it and rubbing the affected part. It was introduced into the Pharmacopœia, however, chiefly with the purpose of bringing it into use for the preparation of ointments, suppositories, and other similar preparations. **Dymock** considers it an excellent substitute for spermaceti, and recommends its employ-

Oil.
42

ment with equal parts of lard in the preparation of nitrate of mercury ointment. The BARK is said to be astringent, and **Dymock** mentions that the "YOUNG LEAVES, after having been tied up in a plantain leaf and stewed in hot ashes, are rubbed in cold milk and given as a remedy for dysentery."

SPECIAL OPINIONS.—§ "Kokam" is a useful application in the fissures of the skin of the feet so common among natives in the cold weather" (*Surgeon-Major H. W. E. Catham, M.D., Ahmednagar*). "The fruit is made into a sherbet, and as such is useful in fever as a cooling drink. It is also anti-scorbutic" (*Surgeon-Major A. S. G. Jayakar, Muskat*). "Half an ounce of kokam butter, melted and mixed with a little boiled rice, is used in dysentery. The dose is repeated once daily" (*Surgeon James McCloghry, Poona*).

Food.—The purple FRUIT has an agreeable flavour, and has long been esteemed as an article of diet. It is mentioned by **Garcia DeOrta** (1563) as known to the Portuguese of Goa by the name of *Brindones*. A little later (1596), **Paludanus** wrote in connection with **Linschoten's** note regarding the fruits of India: "There is also in East India a fruite called *Brindoijus*, which outwardly is a little red, and inwardly bloud red, verye sowre of taste. There are some also that are outwardly blackish, which proceedeth of the ripenesse, and not so sowre as the first, but yet as red within. Many Indians like well of this fruit, but because of its sowreness, it is not so well accepted of. The barkes of these trees are kept and brought over sea (hither, and are good) to make vinegar withall, as some Portingales have done." The last statement is interesting, as, if correct, this utilisation of the bark appears to have fallen entirely into disuse. In Vol. X. of the *Bombay Gazetteer* it is stated that, "In the Collector's garden in Ratnagiri, some trees, said to have been grafted from plants brought from the Straits, yield delicious fruit, just like the imported mangosteen." By the natives, however, the fruit is chiefly employed in the form of a preparation called *kokam*, which is prepared as follows: "When the fruit begins to ripen it is gathered and kept in shade for three or four days to ripen completely; after this it becomes soft and pulpy, the outer skin is removed and dried in the sun. The seeds and pulpy substance are then put in a bamboo basket, which is kept in a boat-shaped wooden trough. The juice is allowed to trickle down the basket for some time into the trough, and when it ceases to trickle, the seed and pulp are stirred and pressed by the hand, and the whole juice is drained off into the trough. The pieces of the outer skin as they are dried are dipped into the juice, and again dried in the sun. In this way they receive three or four coatings of juice. The pieces of rind are now ready for use and are stored in bamboo baskets. Sometimes a little salt is added to the juice. In Goa the pulp is sometimes made into large globular masses. There are very few separate establishments to prepare *kokam*, the preparation being generally left to the women of a family. They keep as much as is wanted for the household, and sell the rest to the village grocer, who in his turn disposes of it to the exporter." "The seeds, after being thoroughly dried, are stored for the four rainy months, to be used in the preparation of *kokam* butter, and to guard against the attacks of weevils and other insects, soft ashes are sprinkled over them as they are being dried in the sun" (*Report from Director of Land Records and Agriculture, Poona*.

In the same interesting communication it is stated that the *kokam* or dried rind is largely used in the Southern Konkan as an ingredient of curries, taking the place of tamarind, while in other parts of Bombay it is principally employed as a semi-medicinal diet.

CHEMICAL COMPOSITION OF KOKAM.—Dr. **Lyon** of Bombay has analysed the prepared rind and found it to contain neither tartaric nor citric

MEDICINE.
Bark.
43
Young leaves.
44

FOOD.
Fruit.
45

COMPOSITION.
46

GARCINIA Mangostana.   Kokam; Mangosteen.

COMPOSITION OF KOKAM.

acid, but 13·53 per cent. of malic acid. The hot water extract formed 42·9 per cent., and the ash 7·88 per cent., of which 5·92 per cent. was soluble in water. The alkalinity of the ash calculated as potash was 79 per cent.

FOOD. Oil. 47 Young leaves. 48 Seed. 49

The concrete OIL is occasionally employed in native cookery, and is said to be largely used in Goa for the purpose of adulterating *Ghí*, a statement which is, however, contradicted by Dymock. Rumphius mentions that the YOUNG LEAVES were employed in Amboyana in cooking fish. The Collector of South Kanara, in a communication to the Government on the subject, published in 1889, states that "the SEED of the ripe fruit is swallowed raw by the natives as a delicacy."

TRADE. 50

Trade.—The average annual value of a full crop from a well grown tree is said in the Southern Konkan to amount to R7; and in the same locality the *Kokam* sells at 35℔ per rupee, and the oil at 6 to 8℔ per rupee (*Dir. Land Rec. and Agric., Poona*). Dymock states that the dried fruit obtained in Bombay comes principally from Goa, Hingoli, and Malwan, and is sold for R40 per kandy of 28 Bombay maunds of 28℔ each, while the *Kokam* butter, which is principally obtained from Goa, fetches R5 to 7 per Surat maund of 37½℔. A small quantity of the latter is annually exported from Bombay, but the quantity cannot be accurately ascertained, since for statistical purposes it is not registered separately from other sorts of vegetable oils.

INDUSTRIAL USES.

Candles. 51

Industrial and Agricultural Uses.—*Kokam* butter yields stearic acid in larger quantities, more easily, and in a purer state than do most other fats, and therefore appears to be particularly suitable as a substance for candle-making. The learned authors of the *Pharmacographia*, commenting on this fact, write, "But that it is possible to obtain it in quantities sufficiently large for important industrial uses, appears to us very improbable." In connection with this remark it is worthy of notice, that the Director of Land Records and Agriculture, Poona, states that in Ratnagiri alone the number of trees is estimated at 13,000, and that they abound in other parts of the Southern Konkan. It, therefore, appears that the supply need not be so limited as Flückiger and Hanbury supposed, and that the preparation of *Kokam* butter may be an industry capable of considerable and profitable development.

MANURE. Oil-cake. 52 53

MANURE.—The OIL-CAKE obtained as a by-product in the preparation of the concrete oil is considered excellent manure.

Garcinia lanceæfolia, *Roxb.; Fl. Br. Ind., I., 263 ; Wight, Ic., t. 163.*

Syn.—GARCINIA PURPUREA, *Wall., Cat., 4862, and Chois.* (not of *Roxb.*)
Vern.—*Kirindur,* SYLHET.
References.—*Roxb., Fl. Ind., Ed. C.B.C., 442 ; Kurz, For. Fl. Burm., I., 91 ; Gamble, Man. Timb., 22 ; Balfour, Cyclop., I., 1176 ; Agri-Hort. Soc. India, Trans., VII., 75 ; Journal (Old Series), IV., 204.*
Habitat.—A small tree, with dark, rough bark, inhabiting the forests of Assam and Sylhet. It flowers in February, and the fruit ripens in July.

FOOD. Fruit. 54 55

Food.—Roxburgh states that it is cultivated by the natives of Sylhet for its FRUIT, of which they are fond.

G. Mangostana, *Linn.; Fl. Br. Ind., I., 260.*

THE MANGOSTEEN.

Vern. —*Mangústán,* HIND.; *Mangustán,* BENG.; *Mangostín, mangustan,* BOMB.; *Mangastín,* MAR.; *Manggusta,* MALAY.; *Mengkop, mimbu, mengut, mangkob, youngsalai,* BURM.; *Manggis,* MALAYS.
References.—*Roxb., Fl. Ind., Ed. C.B.C., 441 ; Kurz, For. Fl. Burm., I., 87 ; Gamble, Man. Timb., 22 ; DC., Origin Cult. Pl., 188 ; Mason, Burma and its People, 447, 750 ; Pharm. Ind., 31 ; O'Shaughnessy, Beng. Dispens., 236 ; Moodeen Sheriff, Supp. Pharm. Ind., 145 ; Dymock, Mat. Med. W. Ind., 2nd Ed., 82 ; Pharmacog. Ind., I., 167 ; U. S. Dis-*

| The Mangosteen. | (*J. Murray.*) | GARCINIA Mangostana. |

pens., 15th Ed., 281 ; S. Arjun, Bomb. Drugs, 23 ; Year Book Pharm., 1873, 285 ; Lisboa, U. Pl. Bomb., 146 ; Birdwood, Bomb. Pr., 14, 142 ; Cooke, Gums and Gum-resins, 41 ; Smith, Dic., 263 ; Kew Reports, 1871, 91 ; Kew Off. Guide to the Mus. of Ec. Bot., 16 ; Kew Off. Guide to Bot. Gardens and Arboretum, 71 ; Agri-Hort. Soc. Ind., Trans. :—II. (App.), 299 ; V., Pro., 88 ; VI., 127, Pro. 112 ; VII., 75, 108 ; Journ., VII., 72 ; Special Reports from Conservators of Forests, Burma and of S. Circle, Madras ; Burma Gazetteer, II., 230 ; Settlement Report, Port Blair, 1870-71, 33, 42.

Habitat.—An evergreen tree, native of the Straits ; cultivated in British Burma on account of its fruit. Of recent years, it has also been successfully cultivated at a few places in the Madras Presidency. The attempts made by Roxburgh, in Bengal, and by several individuals in Bombay, to introduce this fruit tree into these presidencies have been unsuccessful. The former observes, "The plant has uniformly become sickly when removed to the north or west of the Bay of Bengal, and rarely rises beyond the height of two or three feet before it perishes." De Candolle, remarking on the poor results which have followed attempts to familiarize the mangosteen to other countries than those in which it naturally occurs, writes, "Among cultivated plants it is one of the most local, both in its origin, habitation, and in cultivation. It belongs, it is true, to one of those families in which the mean area of the species is most restricted."

Cultivation in India.—The mangosteen is extensively cultivated in Southern Tenasserim, and, as already remarked, has of late years been successfully introduced into Madras. A congenial amount of heat and moisture throughout the year seems to be necessary for its successful cultivation, a condition which on the main peninsula appears to be met with in the Madras presidency only. Recent reports from the Madras Government contain the information that its cultivation in the hot valleys to the east of the Nilghiris has proved successful, while attempts made in the open plains have resulted in failure. The Conservator of Forests, Southern Circle, further reports (May 1889), that one tree in the Government Gardens at Burliar on the Nilghiris produced a hundred dozen fruits ; also that a considerable number of young plants have recently been distributed from Ootacamund, but that they are still too young to bear fruit.

CULTIVA-TION. 56

Gum-resin.—Kurz mentions that this species exudes gamboge of inferior quality. A specimen sent to the London Exhibition in 1862 from Malacca, somewhat resembling gamboge externally, was in small semi-opaque, smooth, rounded tears, but would not easily form an emulsion and could not be used as a pigment (*Cooke*). O'Shaughnessy states that he obtained small quantities of fine gamboge from the rind of the fruit.

GUM-RESIN. 57

Dye & Tan.—The RIND is employed, in combination with the fruit of Terminalia Catappa, *Linn.*, for dyeing black, and is also said to yield a valuable tan.

DYE. Rind. 58 TAN. 59

Medicine.—The dried RIND or ENTIRE FRUIT is largely employed by natives as a remedy for diarrhœa, dysentery, and affections of the genitourinary tracts. According to Rumphius the BARK and YOUNG LEAVES are employed by the Macassars for the same purposes, and also as a wash for aphthæ of the mouth. The *Pharmacopœia of India* includes "the thick fleshy pericarp" amongst its non-officinal drugs, the Editor remarking that he has found it of manifest advantage, when administered with aromatics, in cases of advanced dysentery and chronic diarrhœa. A strong decoction has also been recommended as an external astringent application (*Waltz*). This fruit, prepared like kokam, is said to have come into use of late years in European medicine as a substitute for Bael.

MEDICINE. Fruit. 60 Bark. 61 Young leaves. 62

Chemical Composition.—An analysis made by Schmidt in 1855 proved that the rind contains *tannin, resin,* and a crystallizable principle *man-*

CHEMISTRY. 63

gostine. As the physiological actions of the two latter constituents have not as yet been separately studied, it is impossible to say whether the effect caused by the drug is due simply to the tannin it contains, or whether the *resin* and *mangostine* may not possess peculiar therapeutic properties. The unanimity of opinion as to the efficacy of Mangosteen rind, evidenced in the following special opinions, would seem to indicate that it is a remedy of decided value, and that it probably does possess some property in addition to the simple astringency of tannin.

SPECIAL OPINIONS.—§ "The powder of the dried rind has been administered in intermittent fever with varying success" (*Honorary Surgeon P. Kinsley, Chicacole, Ganjam, Madras*). "The rind contains a good deal of tannic acid. In fine powder it is largely and effectively used in Burma for diarrhœa and dysentery, but I found it very efficacious in diarrhœa only. A wine of mangostin (ʒI to ʒI) is the best method of administration; dose for an adult, ½ dr. to ʒI" (*Devendro Nath Roy, Campbell Medical School, Calcutta*). "The rind is used with benefit in cases of chronic diarrhœa in children" (*Bolly Chund Sen, Campbell Medical School, Calcutta*). "A decoction of the rind is a good astringent in chronic dysentery and diarrhœa" (*Surgeon D. Picachy, Purneah*). "This fruit is brought here in large quantities from the Straits Settlements in July and August. Natives suffering from gonorrhœa and gleet use it largely, as it lessens urethral irritation, and the discharge is in many instances completely arrested. It is, therefore, classed by them as a cooling and refrigerant fruit. A small quantity of the rind, steeped over night in cold water and taken in the early morning as a draught, is a valuable remedy for long standing diarrhœa, both in adults and children" (*Honorary Surgeon A. E. Morris, Tranquebar*).

FOOD.
Fruit.
64

Food.—The FRUIT is highly esteemed both by Europeans and Natives, and is indeed considered by many to be the most palatable of fruits. It is about the size of a small apple, with a thick, succulent, astringent rind, of a reddish-brown colour externally, but bright crimson on section. Within this are placed the 4 to 12 large seeds each surrounded with its juicy white aril, sweet and acidulous, with a delicate flavour like the odour of the primrose.

TRADE.
65

Trade.—A large quantity of the fruit, both fresh and dried, is annually imported from the Straits and may be purchased on the streets of Calcutta in small baskets, though it is customary to find the fruits of **Achras Sapota** passed off on the ignorant as Mangosteens The fruit comes into season in May and June.

66

Garcinia Morella, *Desrouss; Fl. Br. Ind., I., 264; Wight, Ic., tt.*
THE GAMBOGE TREE.　　　　　　　　　　　　　　*[102, 120.*

Syn.—GARCINIA LOBULOSA, *Wall.*; G. PICTORIA, *Roxb.*; G. ELLIPTICA; *Wall (in part)*; G. ACUMINATA, *Planch. & Trian.*; G. GUTTA, *Wight*; G. CAMBOGIOIDES, *Royle*; HEBRADENDRON CAMBOGIOIDES, *Graham.*

Vern.--The tree=*Tamál*, the drug=*ghótághaubá, gótá ganbá, tamál*, HIND.; the tree=*Tamál*, the drug=*tamál*, BENG.; the drug=*Ausarahe-revan*, DEC., C. P.; the tree=*Tamál*, the drug=*revachinnisírá, tamál*, MAR.; the drug=*Makki, iréval-chinip-pál,* the oil=*makki*, TAM.; the drug=*Révalchini-pál*, TEL.; the tree=*Arsinagurgi mara, aradal, punar puli*; the drug=*Tamál*, KAN.; the tree=*Darámba*, MALAY.; the tree= *Tha-men-gút*, the drug=*sanato-sí, tanato así*; the oil=*parawa, ballowa*, BURM.; the drug=*Gotakú, gotakú-melliyam, kanagoraka*, SING.; the drug=*rubbi-revánd, ausháre révand, farfirán*, ARAB. and PERS. The literal meaning of the above Arabic, Persian, Hindustáni, Dekhani, Telugu, and Mahratti synonyms for the gum-resin is explained by **Moodeen Sheriff** to be the "juice or extract of Rhubarb," but they have become, according to the usages of the languages, the correct names of Gamboge.

| The True Gamboge Tree. | (*J. Murray.*) | GARCINIA Morella. |

References.—*Roxb., Fl. Ind., Ed. C.B.C., 444; Beddome, Fl. Sylv., tt. 86 & 87; Gamble, Man. Timb., 24; Thwaites, En. Ceylon Pl., 49; Hooker, in Journ. Linn. Soc. Lond., XIV., 485; Hanbury, Trans. Linn. Soc., XXIV., 487, t. 50; Mason, Burma and its People, 397, 482, 483, 534 & 751; Ainslie, Mat. Ind., I., 147, 602; O'Shaughnessy, Beng. Dispens., 235; Moodeen Sheriff, Supp. Pharm. Ind., 83, 145; Mat. Med. of Madras, 40, 41; Dymock, Mat. Med. W. Ind., 2nd Ed., 83; Pharmacog. Indica, Pt. I., 168; Flück. & Hanb., Pharmacog., 83; U. S. Dispens., 15th Ed., 327; Bent. & Trim., Med. Pl., 33; S. Arjun, Bomb. Drugs, 23; K. L. Dey, Indig. Drugs of India, 56; Med. Top. Ajmir, 148; Irvine, Mat. Med. Patna, 29; Drury, U. Pl., 221; Birdwood, Bomb. Pr., 14; Cooke, Gums and Gum-resins, 43, 46; Cooke, Oils and Oilseeds, 13; Watson, Report on Gums, 14, 34, 67; Watts' Dic. of Chemistry (Ed. 1882), II., 770; Milburn's Oriental Commerce, Ed. 1825, 483; Spons' Encyclop., 1551, 1651; Balfour, Cyclop.. I., 1176; Smith, Dic., 189; Kew Off. Guide to the Mus. of Ec. Bot., 16; Indian Forester, II., 20; VI., 125; XI., 327, 392; Agri.-Hort..Soc. of India, Trans., V., 41, 75, 79, pro. 40; VI., 127; VII., 76; Journ., II., Sel., 377; VIII., Sel., 140; X. pro., 121; Gazetteers:—Bombay, XV., 56, 70; Mysore and Coorg, I., 46, 68; III., 16.; Special Report from Conservator of Forests, S. Circle, Bombay, 1888.*

Habitat.—A small evergreen tree, found in the forests of Eastern Bengal, the Khásia Mountains, the Western Peninsula (Malabar and Kanara), the Eastern Peninsula (Malacca and Singapore), also in Ceylon and Siam. **Garcinia pictoria**, *Roxb.*, is considered by **Beddome** to be distinct from this species, but in the *Flora of British India* it has been reduced as a synonym.

Gum-Resin.—This species produces the true Gamboge of medicine and of the arts. The chief trade supply is obtained from Siam in the form of cylindrical pieces or sticks. Until very recently the exact source of the Gamboge of Commerce was obscure, the gum-resin of Siam being referred to **Garcinia cochin-sinensis**, and that of Ceylon to **Hebradendron cambogioides**, while that of Southern India was supposed to be the produce of G. **pictoria.** These have now, however, been reduced to one species, namely, G. **Morella**, *Desrouss.*, so that the gum-resins of the Malay Peninsula, India, Ceylon, and Burma may be considered one and the same.

HISTORY.—According to the learned authors of the *Pharmacographia*, Gamboge was known to the Chinese as early as the end of the thirteenth century, but was employed by them almost entirely as a pigment. **Pereira** states that the first notice of the occurrence of the gum-resin in Europe is in the writings of **Clusius** (1605), who received it in Amsterdam from a Dutch traveller. Its medicinal virtues were quickly recognised, as is evidenced by the fact that records exist of its use by **Reuden**, a physician of Bamberg, as early as 1611. In 1615 a considerable quantity was offered for sale in London by the East India Company, the entries in the Court Minute Book describing it as "Cambogium, a drug unknown here," "a gentle purge" (*Flückiger and Hanbury*). Notwithstanding the fact that Gamboge has for many years formed an important article of commerce, there appears to be no doubt that it has never been collected in India as an article of trade, even in the districts where the tree abounds. Thus in the Report on the Destruction of Tropical Forests by a committee appointed to investigate that subject in 1851, the following paragraph appears: "The Coorg or Wynád gamboge tree has an extensive range; we have seen it along all the higher parts of the Malabar Gháts for fully 120 miles from north to south, and in some parts it is very abundant; yet the produce for the most part is made little use of, and the tree is considered of so small value, that we have seen the supports and scaffolding of bridges, &c., entirely composed of the stems of **Garcinia pictoria** (*Agri-Hort. Soc. of India, Journal VIII. (Old Series), Sel., 140.*

The gum-resin of Burma, however, has long been used as a yellow dye for the silk robes of the Buddhist priests, and **Dr. Dymock** states that

GUM-RESIN.
67

HISTORY.
68

GARCINIA Morella.	The True Gamboge Tree.

HISTORY.

"the juice of the tree under the Sanskrit name of *Tamála* has long been employed as a pigment for making sectarial marks on the forehead by the Hindús of Kanara and Mysore." Towards the middle of the present century, specimens of Gamboge procured from Indian trees were carefully analysed and critically compared with pure Siam gamboge by chemists both in this country and in England, with the result that the two were declared to be practically identical. Notwithstanding this no attempt appears to have been made to collect the exudation free from impurities, and in such a state that it could compete with success with the pure "pipe gamboge" from Siam.

COLLECTION.

69

COLLECTION.—The gamboge of commerce, which is imported into Europe from Singapore, Bangkok, and Saigon, and is the produce of Siam, Cambodia, and the Southern parts of Cochin-China, is collected in the following way: "At the commencement of the rainy season a spiral incision is made in the bark round half the circumference of a full-sized tree, and the juice, which then slowly exudes for several months, is received into a joint of bamboo, which is placed at the lower end of the incision for that purpose. When the juice has hardened the shell of bamboo is removed, and the gamboge is thus obtained in the form of a roll or cylinder" (constituting the *Roll* or *Pipe Gamboge* of commerce). "According to **Spencer St. John**, a tree will yield on an average, in a season, sufficient gamboge to fill three joints of bamboo 20 inches in length, by about 1½ inches in diameter. The trees should be incised in alternate years" (*Bentley & Trimen*). "*Cake*" or "*Lump Gamboge*" is obtained either from a similar incision, or by breaking the leaves and twigs, the yellow juice which exudes being collected either on the leaves of the tree or in cocoanut shells. A slightly different account is given by **Flückiger** and **Hanbury** quoting **Dr. Jamie** of Singapore. "The best time for collecting is from February to March or April. The trees, the larger the better, are wounded by a *parang* or chopping knife, in various parts of the trunk and large branches, when prepared bamboos are inserted between the wood and the bark of the trees. The bamboo cylinders being tied or inserted are examined daily till filled, which generally takes from fifteen to thirty days. Then the bamboos are taken to a fire, over which they are gradually rotated till the water in the gum-resin is evaporated, and it gets sufficiently hard to allow of the bamboo being torn off." These methods appear to have been untried in India, answers from forest officers to questions regarding the amount collected and methods of preparation shewing that as a rule minute incisions only are made from which small tears of the gum-resin are obtained. In Ceylon it is usually collected by cutting a thin slice of the size of the palm of the hand off the bark, here and there. On the flat space thus exposed, the gum collects and is scraped off when sufficiently dry. As a consequence only cake gamboge, or the gum-resin in small particles, is obtained, both of which forms are always much less pure than the Siam pipe gamboge. The District Forest Officer of North Malabar reports that, by making small incisions in the bark of a tree 16 inches in diamater, ⅜℔ of first class pigment was obtained, but the method appears much more laborious, less productive, and more liable to result in the admixture of impurities than that of collecting in bamboos. A consideration of the literature on the subject indicates the advisability of giving the Siam method at least a fair trial.

Another method is reported from Madras, which consists in partially stripping the bark, pounding and boiling it, straining the resulting liquor, and inspissating it over a slow fire. This necessarily laborious and expensive process is said to yield an inferior article though in large quantities. But since gamboge to be of commercial value must be pure, and as the pure

| The True Gamboge Tree. | (*J. Murray.*) | GARCINIA Morella. |

article can be obtained by the bamboo method much more readily and cheaply, the experiment above described might naturally have been expected to prove unprofitable.

DESCRIPTION AND CHEMICAL COMPOSITION.—The "pipe" gamboge of commerce is found in the form of cylinders 1 to 2½ inches in diameter and 4 to 8 inches in length, with striations lengthwise, caused by impressions from the inside of the bamboo used in collecting. These cylinders may be distinct and covered externally with a yellowish brown dust, or may be agglutinated into masses of various sizes. The best samples are of a rich brownish orange colour externally, dense and homogeneous, brittle, with a conchoidal fracture of an opaque reddish yellow colour, odourless and taste-less at first, then acrid. Mixed with water, or wetted by the finger, they form at once a yellow emulsion. The powder of pure gamboge is fine yellow. The more impure forms of "pipe" gamboge, and "lump" or "cake" gamboge, contain starch, fragments of leaves, twigs, &c., and are harder and more earthy in fracture than the pure gum-resin. The specimens of Indian gamboge which have been examined, have been as a rule in tears, or in irregular fragments collected on leaves, and have varied much in character. The authors of the *Pharmacographia Indica* state that in a specimen they recently examined, obtained from South Kanara, the finer pieces had the colour and consistence of Siam gamboge, but contained many impurities, while fully half the sample was of a dirty yellow brown colour, and had a spongy structure, caused by admixture with a substance which appeared to be chlorophyll. There is no doubt, however, that the gum-resins of Siam and India are identical, and that the adoption of the method of collection practised in the former country would result in an equally valuable product.

CHEMISTRY. 70

Chemically, gamboge consists of a mixture of resin with 15 to 20 per cent. of gum. The resin dissolves easily in alcohol, forming a clear liquid of a fine yellowish-red hue and acid reaction. Buchner assigns to it the formula $C_{60}H_{85}O_{12}$. Flückiger and Hanbury state that the gum (which they obtained to the extent of 15·8 per cent. by completely exhausting the gum-resin with alcohol and ether), was found to be readily soluble in water, not acid in reaction, and therefore not identical with gum arabic. As already stated, impurities are of common occurrence—rice-flour, sand, or the pulverised bark of the tree being amongst the most common. These mechanical impurities are readily recognised in the residue left after exhausting the gum-resin, while the starchy adulterants are easily detected by adding a solution of iodine to the decoction, a green colour being produced.

Dye and Tan.—The GUM-RESIN is employed by the Burmans for dyeing silks of a yellow colour, and by the Karens for their thread. The RIND OF THE FRUIT may be employed as a tan. As already stated, gamboge is employed by the Hindús in parts of India as a pigment in making caste marks on the forehead. In Europe it is largely used as a pigment, especially for water colour drawing.

DYE. Gum-Resin. 71
TAN. Rind. 72

Oil.—A semi-solid fat of a yellow colour is procurable in moderate quantities from the SEEDS, by similar processes to those followed in the preparation of *Kokam* butter. Cooke states that two and a half measures of seed should yield one seer and a half of butter, and that in the Nagar District of Mysore it is sold at the rate of 1 to 4 annas per seer of R24 weight, or at £36 a ton. It is employed as a lamp oil by the better classes of natives, and as a substitute for *ghí* by the poor. No reliable analysis of this fat is obtainable, but should it, like that obtained from the allied species. *G. indica*, *Chois.*, contain a large proportion of stearic acid, it might prove of value to the candle-maker.

OIL. Seeds. 73

GARCINIA pedunculata.	The Garcinias or Gamboge Trees.

MEDICINE.
Gamboge.
74

Medicine.—GAMBOGE is largely employed as a hydragogue and drastic cathartic and anthelmintic. It is particularly valuable in cases of ana-sarca and other dropsical affections, and in obstinate constipation. In over-doses it is a violent gastro-intestinal irritant poison, and ought to be ad-ministered with caution, especially to children. When prescribed alone, it is liable to cause severe griping, and is therefore almost always given in combination with other purgatives and carminatives. **Moodeen Sheriff** states that Mysore gamboge must be given in half larger doses than the officinal drug, doubtless because it contains a proportion of inert impuri-ties. It is also employed by the natives as an external application to re-lieve pain and swelling, and **Dr. Gray** reports that broken pieces of the BRANCHES rubbed up with water are used as a household remedy for boils.

Branches.
75

SPECIAL OPINIONS.—§ "Siam gamboge is one of the best purgatives in India, and a much stronger drug than jalap. Like the latter, it acts very satisfactorily in combination with other purgatives or laxatives; but not so well when used alone. During the last twenty years, I have used this medi-cine in Triplicane Dispensary with cream of tartar, whenever jalap was out, and never felt the want of the latter. The cheapness and abundance of Siam gamboge in this country is another advantage which it possesses over jalap (*Honorary Surgeon Moodeen Sheriff, Khan Bahadur, Tripli-cane, Madras*). "Mixed with other medicines and applied over sprains and contusions, it relieves pain and swelling" (*Surgeon-Major A. S. G. Jayakar, Muskat*). "The stem rubbed with water is a household remedy amongst natives, as a local application to rising pimples and boils, and often cuts them short" (*Civil Surgeon R. Gray, Lahore*).

FOOD
Oil.
76
TIMBER.
77
DOMESTIC.
Oil.
78
TRADE.
79
80

Food.—The OIL obtained from the seeds is employed by the poor as a substitute for *ghí*.

Structure of the Wood.—Yellow, hard, mottled. Weight about 56℔ per cubic foot. Might be useful for cabinet-making.

Domestic Uses.—The OIL is largely employed, with that of **G. indica,** *Chois.,* for illuminating purposes.

Trade.—In the Indian markets the ordinary pipe gamboge is alone met with, value R1-4 per ℔ (*Pharmacographia Indica*).

Garcinia paniculata, *Roxb.; Fl. Br. Ind., I., 266; Wight, Ic., t. 112.*
    Syn.—G. BHUMICOWA, *Roxb.*
    References.—*Roxb., Fl. Ind., Ed. C.B.C., 443; Kurs, For. Fl. Burm., I., 92.*
    Habitat.—A tree about 40 feet high, native of the Khásia Mountains, the Eastern Himálaya at Bhotán, and of Chittagong.

FOOD.
Fruit.
81
82

Food.—"The FRUIT is palatable, its taste more like that of a mangos-teen than anything else I can compare it to" (*Roxburgh*).

G. pedunculata, *Roxb.; Fl. Br. Ind., I., 264; Wight, Ic., tt. 114, 115.*
    Vern.—*Tikúl, tikúr,* BENG.; *Borthekra, kiyi thekera tenga,* ASSAM; *Hei-búng,* MANIPUR.
    References.—*Roxb., Fl. Ind., Ed. C.B.C., 443; Drury, U. Pl., 221; A Note on the Condition of the People of Upper India, Agric. file, No. 6, 1888; Trotter, Report on Ec. Prod., Manipur; Balfour, Cyclop., I., 1176; Agri.-Hort. Soc. India:—Trans., VII., 75; Journ. (Old Series), VI., 27, 39; X., Pro., 40.*
    Habitat.—A tall tree of the forests of North-Eastern Bengal, near Rungpur and Goalpara, and of Sylhet. It flowers from January to March, and its fruit ripens from that time to June.

DYE.
Fruit.
83

Dye.—**Major Trotter,** in his report on the Economic Products of Manipur, stated that the FRUIT of this plant was largely employed by the natives of that country to deepen and render fast saffron dye. He de-scribed the process as follows: " After the cloth has been dyed with saffron

The Garcinias or Gamboge Trees. (*J. Murray.*)	**GARCINIA succifolia.**

wring it out and lay aside for a few minutes; add ¼ of a pint of the *heibúng* water (prepared very simply, *viz.*, by soaking ⅛ seer of the fruit, cut in slices, in a pint of water for 20 to 24 hours) to the dye in the vessel, and mix thoroughly; then steep the *Golap Machoo* (saffron) cloth in it and press and flop it about till it is thoroughly saturated, then take out, wash in clean water, and hang up in the shade to dry." No further information on this subject has been obtained, and it appears probable that the action of the *heibúng* may be less complete than **Major Trotter** believed. It may be of interest, however, to note that **Major Hannay**, in an article on the " Rheeas of Assam," mentions that " **Garcinia** " fruit (probably the fruit of this species) is employed to bleach rheea fibre in that country. (*Jour. Agri-Hort. Soc. Ind., Vol. VII. (Old Series), 225.*)

DYE.

Food.—This tree yields a large, round, smooth, yellow, edible FRUIT, regarding which **Roxburgh** writes:—" The fleshy part of the fruit which covers the seeds and their proper juicy envelope, or aril, is, in large quantity, of a firm texture and of a very sharp, pleasant, acid taste. It is used by the natives in their curries and for acidulating water. If cut into slices and dried it retains its qualities for years, and might be most advantageously employed during long sea voyages as a succedaneum for lemons or limes, to put into various messes where salt meat is employed, &c.

FOOD. Fruit. 84

Structure of the Wood.—The timber is said by **Major Hannay** to be useful when seasoned (*Note on some of the Forest trees of Upper Assam, Jour. Agri-Hort. Soc., Ind., VI. (Old Series), 27*).

TIMBER. 85

Garcinia speciosa, *Wall. ; Fl. Br. Ind., I., 260.*

Vern.—*Palawa, pa-gyay-theing*, BURM.
References.—*Kurz, For. Fl. Burm., I., 88; Gamble, Man. Timb., 23; Mason, Burma and Its People, 751.*
Habitat.—An evergreen tree of Tenasserim, Moulmein, Martaban, and the Andaman Islands.
Gum-Resin.—It is described by **Kurz** as yielding inferior gamboge.
Structure of the Wood.—Uniformly reddish-brown, close-grained, very heavy, weighing from 50 to 70℔ per cubic foot. It is employed for house and bridge posts, and other purposes, and is said to be used by the Andamanese to make bows. **Kurz** describes it as of equally good quality with the " bullet-wood " of the Andamans.

86

GUM-RESIN. 87
TIMBER. 88

G. stipulata, *T. And.; Fl. Br. Ind., I., 267.*

Vern.—*Sana-kadan*, LEPCHA.
References.—*Gamble, Man. Timb., 24; Balfour, Cyclop., I., 1176.*
Habitat.—A tall tree met with in the moist sub-tropical forests of the Eastern Himálaya from Sikkim to Bhotán, ascending to an altitude of 4,000 feet.
Gum-Resin.—The tree and fruit yield a yellow gum, which does rot seem to be used (*Gamble*).
Food.—The FRUIT produced by this species is yellow, and is sometimes eaten by the Lepchas.

89

GUM-RESIN. 90
FOOD. Fruit. 91

G. succifolia, *Kurz, For. Fl. Burm., I., 91.*

The authors of the *Fl. Br. Ind.* (*I., 270*) regard this as a doubtful species, owing to the female flowers and fruits being unknown. It is considered by **Gamble** to be identical with **G. loniceroides**, *T. Anders., Fl. Br. Ind., I., 264.*
References. -*Kurz, Jour. As. Soc., Beng., 1874, pt. 2, 87; 1877, pt. 2, 293; Prelim. For. Rep. on Pegu, App. A., xiii.*
Habitat.—An evergreen tree from 30 to 35 feet in height, frequent in the swamp forests of the alluvial lands adjoining the Sittang and Irrawaddi rivers.

92

G. 92

GARCINIA **Xanthochymus.**	**The Gamboge-yielding Trees.**

GUM-RESIN.
93
TIMBER.
94

95

GUM-RESIN.
96

97

GUM-RESIN.
98

99

GUM-RESIN.
100

DYE.
Bark.
101

Gum-resin.—This species is said by Kurz to yield little and inferior gamboge.

Structure of the Wood.—White, turning yellowish white, rather heavy, coarsely fibrous, very perishable (*Kurz*).

Garcinia travancorica, *Beddome ; Fl. Br. Ind., I., 268.*
  Syn.—GARCINIA, sp., 2, *Beddome, Flor. Sylvat. Gen., xxi.*
  Vern.—*Malampongu,* TINNEVELLY.
  References.—*Beddome, For. Man., xxi., Fl. Sylv., t. 173 ; Gamble, Man. Timb., 23 ; Cooke, Gums and Gum-resins, 48 ; Balfour, Cyclop., I., 1176 ; Indian Forester, III., 21.*
  Habitat.—A highly ornamental tree, confined to the forests of the southern portions of the Travancore and Tinnevelly Gháts, at elevations of from 3,000 to 4,500 feet (*Beddome*).
  Gum-resin.—Beddome states that every portion of the tree yields an abundance of bright yellow gamboge. No information, however, regarding the chemical composition or physical characters of this gum-resin is available, and it is therefore not known to what extent it might be utilised as a pigment, dye, or varnish.

G. Wightii, *T. And. ; Fl. Br. Ind., I. 265.*
  References.—*Gamble, Man. Timb., 22 ; Balfour, Cyclop., I., 1176.*
  Habitat.—A native of the forests of Southern India.
  Gum-resin.—The gamboge of this species is very soluble, and yields a good pigment (*T. Anderson*).

G. Xanthochymus, *Hook. f. ; Fl. Br. Ind., I., 269.*
  Syn.—XANTHOCHYMUS PICTORIUS, *Roxb.* ; X. TINCTORIUS, DC.
  Vern.—*Dampel, tamál,** HIND. ; *Tamál,** BENG. ; *Tepor, tespur, tihur,* ASSAM ; *Manho-la,* GARO ; *Dampel, onth, osth,* BOMB. ; *Jhárámbi,* MAR., *Iwara memadi tamalamu, chitakamraku,* TEL. ; *Matau,* BURM. ; *Tamála,** SANS.
  References.—*Roxb., Fl. Ind., Ed. C.B.C., 445 ; Wight & Arn., Prod., I., 102 ; Kurz, For. Fl. Burm., I., 93 ; Gamble, Man. Timb., 23 ; U. C. Dutt, Mat. Med. Hind., 320 ; Dymock, Mat. Med. W. Ind., 2nd Ed., 81 ; Pharmacog. Indica, I., 166 ; Lisboa, U. Pl. Bomb., 11, 146, 241 ; Cooke, Gums and Gum-resins, 49 ; Liotard, Dyes, 95 ; Darrah, Note on Cotton in Assam, 30 ; Report on Dyes of Assam ; Balfour, Cyclop., I., 1176 ; Indian Forester, XI., 392.*
  Habitat.—A widely distributed species met with in Eastern Bengal, the Eastern Himálaya from Sikkim to the Khásia Mountains, Burma, Southern India, Penang, and the Andaman Islands.
  Gum-resin.—This species yields a large quantity of inferior gamboge, both from the stem and fruit-rind. Roxburgh states that it is of inferior quality, but it is extensively utilised as a dye in Assam. Lisboa describes the gum-resin obtained from the fruit as follows : "From the full-grown, but not ripe, fruit, a quantity of creamy, resinous, yellow, gum-like gamboge is obtained, which makes a tolerably fair water colour, and might be used either by itself or mixed with blue to form green." No definite account exists of the chemical and physical properties of this gum-resin, but it would seem to contain a larger proportion of gum than that derived from the other species.
  Dye.—The BARK is employed by the Phakials of the Lakhimpur district of Assam for dyeing cotton. The process which they employ is described by the Deputy Conservator of Forests of the province, as follows :

---

* U. C. Dutt states that the above Sanskrit, Hindústáni, and Bengali names are applied to this plant, as well as to **Cinnamomum Tamala,** *Nees.*

G. 101

The Garland Gardenia. (*J. Murray.*)	**GARDENIA coronaria.**

"Chips of the bark and the thread, with the leaves of **Symplocos grandiflora** as a mordant, are boiled, and the colour produced is a bright yellow. If the dye thus obtained be mixed with the blue derived from the leaves of **Strobilanthus flaccidifolia**, a green colour is produced." The dyeing property of the bark is doubtless due to the gum-resin which it contains. **DYE**

Medicine.—This species, like **G. indica**, produces a FRUIT which is employed medicinally either fresh, or dried into a kind of *Amsúl* (see **G. indica**). Dymock states that a sherbet made by mixing about 1 oz. of this preparation with a little rock-salt, pepper, ginger, cummin, and sugar, is administered in bilious conditions. **MEDICINE. Fruit. 102**

Food.—The FRUIT is eaten. Lisboa writes, "The fruit, temptingly beautiful, as big as an orange, smooth and bright yellow, is, however, strongly acid, especially in the fleshy rind. The pulp, though less acid, if eaten, puts the teeth out of order for a couple of days, and is, therefore, only used by poorer Natives." **FOOD. Fruit. 103**

Structure of the Wood.—Yellowish white, with a large, darker-coloured heart-wood, turning pale yellowish-brown, rather heavy, fibrous but close-grained, and fairly hard (*Kurz*). **TIMBER. 104**

GARDENIA, *Linn.; Gen. Pl., II., 89.*

A genus of shrubs or trees belonging to the Natural Order RUBIACEÆ and comprising about 60 tropical or sub-tropical species. Of these from 14 to 20 are natives of India.

Gardenia campanulata, *Roxb.; Fl. Br. Ind., III., 118; Wight, Ic.,* [*t. 578;* RUBIACEÆ.

105

Syn.—GARDENIA LONGISPINA, *Wall.,; ?* G. BLUMEANA, *DC.*
Vern.—*Sethanbaya,* BURM.
References.—*Roxb., Fl. Ind., Ed., C.B.C., 238; Kurz, For. Fl., II., 40; Pharm. Ind., 118; O'Shaughnessy, Beng. Dispens., 400.*
Habitat.—A shrub, from 15 to 20 feet in height, met with at the foot of the Sikkim Himálaya, also in Assam, Sylhet, Chittagong, Behar (at the summit of Pareshnáth), and Pegu.

Medicine.—Roxburgh states that the FRUIT is used medicinally by the Natives of Chittagong, who consider it anthelmintic and cathartic. **MEDICINE. Fruit. 106**

Domestic Uses.—The FRUIT is said to be employed in removing stains from silk (*Roxburgh*). **DOMESTIC. Fruit. 107 108**

G coronaria, *Ham.; Fl. Br. Ind., III., 117.*

THE GARLAND GARDENIA.

Syn.—GARDENIA COSTATA, *Roxb.; ?* G. CARINATA, *Griff.*
Vern.—*Yeng-khat, tsaythambyah,* BURM.
References.—*Roxb., Fl. Ind., Ed. C.B.C., 237; Kurz, For. Fl. Burm., II., 43; Gamble, Man. Timb., 229; Mason, Burma and Its People, 414, 785; P. W. D., Report on Gums, 3.*
Habitat.—A tree met with commonly in mixed moist forests all over Burma, from Chittagong, Pegu, and Martaban down to Tenasserim. It bears handsome large flowers, which are white when they expand at daybreak, but change to a deep yellow towards evening.

Oil.—This species is said to yield a wax which, however, does not appear to have been examined and described, nor is there any record of its utilisation by the Natives. **OIL. 109**

Structure of the Wood.—Pale brown, or white, of an unequal fibre, rather brittle and very close-grained. Weight 51℔ per cubic foot. It is employed for making combs, and for turning, but has the disadvantage of being very liable to crack. **TIMBER. 110**

G. 110

GARDENIA gummifera.	The Cape Jasmine; Dikamali Resin.

III

MEDICINE.
Bark.
112
Pulp.
113
Fruit.
114
Root.
115
116

Gardenia florida, *Linn.; DC., Prodr., IV., 379.*

THE CAPE JASMINE.

A handsome shrub, which, though a native of China, is now extensively cultivated for ornamental purposes in India. In Hindústáni, it is known as *Gúndha-raj*, and in Burmese as *Thong-sin-pan.*

Medicine.—The Japanese are reported to employ its BARK (*routinachi*) and the PULP of its FRUIT as a yellow dye. Dymock states that in the Konkan the ROOT is rubbed into a paste with water, and applied to the top of the head as a remedy for headache during pregnancy; and that it is also given internally in hysteria, alone, or combined with *bhárangi* (Clerodendron Siphonanthus, *Br.*)

G. gummifera, *Linn. f.; Fl. Br. Ind., III., 116.*

Syn.—GARDENIA ARBOREA, *Roxb.;* G. INERMIS, *Dietr.*

Vern.—The gum-resin=*dikmali, dikámli,* HIND.; *Baruri, barúi,* KOL.; *Bruru,* BHUMIJ; *Papra, kamarri, karmarri,* the gum-resin=*dekámáli,* C. P.; the gum-resin=*dikámáli,* BOMB.; *Kamarri, dikámali,* GUZ.; the gum-resin=*Kumbai, diká-málli,* TAM.; *Chittamatta, garaga, chiri-bikki,* the gum-resin=*tella-manga, chinaká-ringuva,* TEL.; *Bikka gida,* the gum-resin=*dikke-malli,* KAN.; the gum-resin=*Kola-lákada,* SING.; the gum-resin=*Kunkham,* ARAB.

References.—*Roxb., Fl. Ind., Ed. C.B.C., 238; W. & A., Prodr., 395; Gamble, Man. Timb., 229; Dals. & Gibs., Bomb. Fl., 120 (Excl. Syn.); Elliot, Fl. Andhr., 41, 44, 58; Pharm. Ind., 118; Ainslie, Mat. Ind., II., 89; Moodeen Sheriff, Supp. Pharm. Ind., 146; Dymock, Mat. Med. W. Ind., 2nd Ed., 411; Murray, Pl. and Drugs, Sind, 195; Year-book of Pharmacy, 1878, 73; Drury, U. Pl., 224; Lisboa, U. Pl. Bomb., 86,162; Birdwood, Bomb. Pr., 44, 269; Cooke, Gums and Gum-resins, 66; P. W. D., Report on Gums, 14, 27, 33, 35; Balfour, Cyclop., I., 1177; Smith, Dic., 154; For. Adm. Report, Ch. Nagpore, 1885, 6, 33; Journal Agri.-Hort. Soc. Ind. (Old Series), X., 10; Indian Forester, III., 203; X., 222; Settlement Reports:—Central Provs., Chanda Dist., App. VI.; Chhindwara Dist., 110; Gazetteers;—Mysore & Coorg, I., 50; Bombay, XV., pt. I., 436.*

Habitat.—A large shrub met with in Central and South India, from the Satpura Range southwards.

GUM-RESIN.
117

Gum-Resin.—The remarkable gum-resin, *dikamali,* or *cumbi-gum*, is obtained from this species and from G. lucida, *Roxb.* The exudation from both species is apparently identical, and in both cases forms transparent tears from the extremities of the young shoots and buds. These shoots and buds are broken off with the drops of gum-resin attached, and exposed for sale either in this form, or after agglutination into cakes or irregular masses.

Characters.
118

CHARACTERS AND CHEMICAL COMPOSITION.—Commercial *Dikámáli* occurs either in the form of the twigs coated with and agglutinated by the gum-resin, or as irregular earthy-looking masses, of a dull olive-green colour which consist of the resin more or less mixed with bark, sticks, and other impurities (*Cooke*). It has a peculiar and offensive odour like that of cat's urine. When carefully collected and free from impurity it is transparent and of a bright yellow colour. The gum-resin has been examined by Flückiger, Dymock, and later by I. Stenhouse and C. E. Groves, and has been found, by the last two investigators, to contain two distinct resins. One of these, an amorphous greenish-yellow substance, is by far the largest constituent; the other occurs only in small proportion, and is obtained in slender, pale, yellow, crystalline needles. To the latter the name of *Gardenin* has been applied. In the investigation referred to, *gardenin* was separated by boiling the *Dikamali* with alcohol, filtering the solution and allowing the filtrate to cool. The needles thus obtained were washed with cold spirit to free them from the green amorphous resin, and then treated

G. 118

with light petroleum to remove a fatty impurity which remained. They were finally purified by alternate crystallisation from hot benzine, in which they are readily soluble, and from alcohol. From the pure *gardenin* thus obtained, a very interesting, brilliant crimson, crystalline substance was derived by treatment with boiling glacial acetic acid, to which the name of *gardenic acid* was provisionally applied. | GUM-RESIN.

Medicine.—Though *Dikámáli* RESIN is produced in great abundance in Western India, it appears to have been little known to ancient Hindu medicine, and is not even mentioned in any of the Sanskrit works on Materia Medica (*Dymock*). It seems, however, to have been known to western civilisation for many centuries, Birdwood referring the κάγ καμον of Dioscorides, and Sprengel, the "concamum" of Pliny, to this drug. In no modern European work, however, does there appear to be any reference to *Dikámáli*, a fact which is the more remarkable when its peculiar and characteristic appearance and odour are considered. Ainslie appears to have been the first to describe its utilisation in India. In his *Materia Indica*, the following passage occurs, "*Cumbipisin* or cumbi-gum is a strong-smelling gum-resin, not unlike myrrh in appearance, and possessing, the Hakims say, nearly similar virtues; it is, however, far more active, and ought, on that account, to be administered in very small doses; as an external application, it is employed, dissolved in spirits, for cleaning foul ulcers, and, where the balsam of Peru cannot be obtained, might be used as a substitute for arresting the progress of sphacelous and phagedenic affections, which that medicine has the power of doing (at least in hot climates) in a very wonderful manner." The drug is considered anti-spasmodic, carminative, and when applied externally, antiseptic and stimulating. It is accordingly employed by the Natives of Southern and Western India, in cases of hysteria, flatulent dyspepsia, and nervous disorders due to dentition in children, also externally as an application to foul and callous ulcers, and extensively to keep away flies from sores. It has also been employed in European practice for the last purpose with marked success, both in hospitals and in veterinary work, and is said to be a successful anthelmintic in cases of round worm. Little is known, however, regarding its exact therapeutic properties as an internal medicine, and it is possible that its virtues may be overestimated by the Natives. | MEDICINE. Resin. 119

SPECIAL OPINIONS.—§ "The powdered gum-resin is said to have diaphoretic and expectorant properties, used internally in guinea-worm, dose from 2 to 16 grains. It is often rubbed on the gums of teething children" (*Deputy Sanitary Commissioner Joseph Parker, M.D., Poona*). "Useful to destroy maggots in old wounds" (*Surgeon-Major and Civil Surgeon G. Y. Hunter, Karachi*). "Used by native farriers. Has a strong aroma, and is used in South India in hospitals to keep away flies from sores" (*Surgeon H. W. Hill, Manbhum*). "A lotion made from *Dikamali* is used to keep maggots from sores. ʒi every morning is given in dyspepsia" (*Surgeon James McCloghry, Poona*). "An infusion is said to be useful in treatment of worms in children" (*Surgeon J. C. H. Peacocke, I.M.D., Nasik*). "The gum of the tree melted in oil is applied to the forehead to check headache" (*V. Ummegudien, Mettapolliam, Madras*).

Food.—The FRUIT is said to be eaten (*Lisboa*). | FOOD. Fruit. 120

Structure of the Wood.—Yellowish-white, hard, close-grained, might serve as a substitute for box-wood. | TIMBER. 121

Agricultural Use.—A solution of the gum-resin has been recommended by Watson as a sheep-wash. | AGRICULT. USE. 122

Trade.—*Dikamali* obtained from Southern India, or imported from Arabia, is sold in Bombay at R3-12 per maund of 37½℔ (*Dymock*). | TRADE. 123

124   **Gardenia latifolia,** *Aiton.; Fl. Br. Ind., III., 116; Wight, Ic., t. 759.*

Vern.—*Pápra, páphar, pepero, ban pindálu,* HIND.; *Papra, papasar, papar,*
KOL.; *Popro,* SANTAL; *Kota-ranga,* URIYA; *Gogar,* BHIL; *Pempri,* MAL.
(S.P.); *Panniabhil, gúngat, bhandra, geggar,* GOND; *Papra, papadar,*
popra, KHARWAR; *Gogar,* C. P.; *Gandru-papura, kariga, phiphar, gho-*
gar, gogarli, BOMB.; *Ghogar, gogarli,* MAR.; *Kumbay,* TAM.; *Pedda*
karinga, pureea, bikki, gaiger, karukiti karinguva, konda manga, TEL.

References.—*Roxb., Fl. Ind., Ed. C.B.C., 237; Brandis, For. Fl., 271;*
Gamble, Man. Timb., 229; Dalz. & Gibs., Bomb. Fl., 120; Elliot, Fl.
Andhr., 27, 77, 83, 92, 96, 104; Rev. A. Campbell, Ec. Prod. Chutia-Nag-
pur, No. 9229; Lisboa, U. Pl. Bomb., 86; For. Ad. Report Chutia-Nag-
pur, 1885, 32; Indian Forester, III., 203; IV., 343, 345; Gazetteers, N.-
W. P., I., 81.

Habitat.—A small deciduous tree, met with in the dry hilly districts of
Western, Central, and South-Western India, also in the North-Western
Himálaya, in Garhwál only, where it ascends to 3,000 feet, and in Behar
and Western Bengal.

FOOD.   **Food.**—"The FRUIT is eaten by the Santals" (*Rev. A. Campbell*).
Fruit.

        Structure of the Wood.—White, with a yellowish tinge, close and fine-
125   grained, weight 52 to 53℔ per cubic foot. It is easy to work and durable,
TIMBER. and has been recommended as a substitute for box-wood, and as likely to
126   be useful for the purposes of the engraver and wood-turner. It is employed
DOMESTIC. by the Natives to make combs.

127   **Domestic, &c.**—The plant is recommended by Roxburgh as worthy of
        attention for ornamental purposes. He writes, "Its large, glossy, green
        leaves, independent of the size, beauty, and fragrance of the flowers, render
        it highly ornamental."

128   **G. lucida,** *Roxb.; Fl. Br. Ind., III., 115; Wight, Ic., t. 575.*

Syn.—G. RESINIFERA, *Roth.*

Vern.—*Dikamali,* HIND.; *Konda manga, kokkita, tetta manga, kúrú,*
C. P.; *Dikamali,* MAR.; *Dikamali,* GUZ.; *Papar,* BIJERAGOGARH; *Kumbi,*
TAM.; *Karinga, karaingi, karung, tella-manga, china karinguva,* TEL.;
[The vernacular names for the gum-resin are the same as those applied to the
exudation of **G. gummifera,** *Linn.* (which see).]

References.—*Roxb., Fl. Ind., Ed. C.B.C., 237; W. & A., Prodr., 395;*
Brandis, For. Fl., 271; Kurz, For. Fl. Burm., II., 42; Beddome, Fl.
Sylv., Anal. Gen., XV., f. 6; Dalz. & Gibs., Bomb. Fl., 120; Elliot, Fl.
Andhr., 39, 177; Pharm. Ind., 188; Ainslie, Mat. Ind., II., 89; Moodeen
Sheriff, Supp. Pharm. Ind., 146; Dymock, Mat. Med. W. Ind., 2nd Ed.,
411; S. Arjun, Bomb. Drugs, 71; Murray, Pl. and Drugs, Sind, 195;
Year-book of Pharmacy, 1878, 73; Drury, U. Pl., 224; Lisboa, U. Pl.
Bomb., 86, 251; Birdwood, Bomb. Pr., 269; Cooke, Gums and Gum-resins,
66; Watson, Report on Gums, 3, 14, 27, 33, 35; Balfour, Cyclop., I., 1177;
Smith, Dic., 154; Kew Off. Guide to the Mus. of Ec. Bot., 79; Indian
Forester, III., 203; VIII., 417; Settlement Reports:—Central Provs.
Upper Godavery Dist., 38; Raepore Dist., 76, 77; Gazetteers:—Bombay,
XV., pt. I., 70, 436; Central Provs., 504.

Habitat.—A small deciduous tree, found in Central and South India
(common from the Konkan southwards), also in Chittagong and Burma.

GUM-RESIN. **Gum-resin.**—This species, along with **G. gummifera,** *Linn.,* yields the
129   *Dikamali* or Cambi resin, for a description of which the reader is referred
        to the article on the latter species.

MEDICINE. **Medicine.**—See description of the properties of *Dikamali* gum-resin in
130   the article on **G. gummifera,** *Linn.*
FOOD.   **Food.**—The FRUIT is said to be an article of food in the Central Prov-
Fruit.   inces.
131

TIMBER. **Structure of the Wood.**—Yellowish-white, close-grained, hard, con-
132   taining no heart-wood, weight 39℔ per cubic foot. It is useful for turn-
        ing, and is employed for making combs by the Natives.

G. 132

Gardenia obtusifolia, *Roxb.; Fl. Br. Ind., III., 116.* — 133
Syn.—G. SUAVIS, *Wall., Cat., 8274;* RUBIACEA, *Wall., Cat., 8294b.*
Vern.—*Veng-khat, yingat, yinkat,* BURM.
References.—*Kurz, For. Fl. Burm., II., 42; Gamble, Man. Timb., 229.*
Habitat.—A small deciduous tree, frequent in the *In* or **Dipterocarpus** dry forests from Prome and Martaban down to Upper Tenasserim.
Resin.—This is said by Kurz and others to yield a fine pellucid yellow resin, which is probably nearly allied in its characters to that derived from G. **gummifera** and G. **lucida.** No information exists, however, regarding its physical and chemical characters, nor is it known to be of any economic value. — RESIN. 134
Structure of the Wood.—White, moderately hard, weight 59℔ per cubic foot. — TIMBER. 135 136

G. turgida, *Roxb.; Fl. Br. Ind., III., 118; Wight, Ic., t. 579.*
Syn.—GARDENIA CUNEATA, *Br.;* G. DONIA, *Ham.*
Var.—MONTANA, *Roxb.* (*Sp.*); leaves orbicular and densely tomentose beneath. G. MONTANA, *DC.*
Vern.—*Thanella, khúrrúr, khuriari, ghúrga, mhaner,* HIND.; *Bamemia, dhobelkirat,* URIYA; *Karhar, duduri,* KOL.; *Phurpata,* KURKU; *Dandu kit, doudouki,* SANTAL; *Kharkar,* MAL. (S.P.); *Panjra, pendra,* GOND; *Thanella,* N.-W. P.; *Karumba,* MERWARA; *Karumba,* RAJ.; *Karhár, khemra,* C. P.; *Khurphendra, pendri, phanda, phetra,* MAR.; *Phetrak,* BHIL; *Manjúnda, telél* (var. montana=*Tella kakkisa*), TEL.; *Bongeri,* KAN.; *Thaminsani,* BURM.
References.—*Roxb., Fl. Ind., Ed. C.B.C., 239; Kurz, For. Fl. Burm., II., 41; Beddome, Fl. Sylv., Anal. Gen., t. 15, f. 6; Gamble, Man. Timb., 228; Rev. A. Campbell, Ec. Prod., Chutia-Nagpur, No. 8495; Duthie, Rep. on a Botanical Tour in Merwara, 15; Atkinson, Him. Dist., 311; For. Ad. Report, Chutia-Nagpur, 1885, 32; Indian Forester:—IV., 322; VIII., 416, 417; IX., 59; X., 325; XII., 419; XIII., 121; XIV., 112; Gazetteer, N.-W. P., IV., lxxiii.*
Habitat.—A small deciduous tree met with in the sub-Himálayan tract from Nepál to the Jumna, ascending to 4,000 feet; also in Rájputana, Burma, and Central and South India.
Gum.—This species is said to yield a hard yellow gum (*E. A. Fraser, Rájputana*). — GUM. 137
Medicine.—The Rev. A. Campbell states that a preparation from the ROOT is employed by the Santals as a remedy for indigestion in children. — MEDICINE. Root. 138
Structure of the Wood.—White with a purplish tinge, no heartwood, close-grained and hard, weight from 54 to 58℔ per cubic foot. It is good and durable, but liable to crack and split in seasoning. — TIMBER. 139
Domestic and Sacred.—The ROOT is regarded as a charm by the Natives of Chutia Nagpúr, who wear it attached to the wrist by a cord. — DOMESTIC. 140 SACRED. Root. 141

Garlic; see **Allium sativum,** *Linn.;* Vol. I., 172.

Garlic Tree; see **Cratæva religiosa,** *Forst.,* var. **Roxburghii,** Vol. II., 585.

Garnets; see **Precious Stones.**

GARNOTIA, *Brongn.; Gen. Pl., III., 1118.*

Garnotia stricta, *Brongn.;* GRAMINEÆ; *Thwaites, En. Pl. Zeyl., 363.* — 142
A grass met with in the more elevated parts of Central Ceylon; said by Thwaites to be much used for thatching.

GARUGA, *Roxb.; Gen. Pl., I., 323.*

Garuga pinnata, *Roxb.; Fl. Br., Ind., I., 528;* BURSERACEÆ. — 143
Syn.—? GARUGA MADAGASCARIENSIS, *DC.*

GARUGA. **pinnata.**	**The Garuga Gum.**

Vern.—*Ghogar, kaikar, túm*, HIND.; *Júm, túm kharpat, nilbhadi*, BENG.; *Mohi*, URIYA; *Nia jowa*, KOL.; *Karúr*, BHUMIJ; *Kekkeda*, KURKU; *Kékur*, KHARWAR; *Gendelipoma*, ASSAM; *Chitopoma*, GARO; *Dabdabbi*, NEPAL; *Maldit*, LEPCHA; *Gia*, MICHI; *Kosramba*, MAL. (S.P.); *Gúpni, kekra, gharri*, GOND; *Karolu, ghogar, kaikar*, OUDH; *Kharpat, gurja, gum=katila*, N.-W. P; *Kilmira, kitmira, kharpat, katula, sarota*, KU-MAON; *Khurpat, katúla, kilmira, sarota*, PB.; *Kúrak, kanghur*, DEC.; *Gurja*, BANDA; *Kankar, kaikra, ghunja, mahárut*, C. P.; *Kekda*, MEL-GHAT; *Kákad, kúrak, kanghur*, BOMB.; *Kúrúk, kuduk*, MAR.; *Karapti*, KATHIAWAR; *Kúsimb*, GUZ.; *Karre vembú, karvambú*, TAM.; *Garugo, kalugudu, garugu, gárgá*, TEL.; *Hala, bálage*, KAN.; *Katu-kalesjam*, MALAY.; *Mroung-shisha*, MAGH.; *Chinyok, chinjop, hsen-youk*, BURM.

References.—*Roxb., Fl. Ind., Ed. C.B.C., 370; W. & A., Prodr., 175; Brandis, For. Fl., 62; Kurz, For. Fl. Burm., I., 207; Beddome, Fl. Sylv., t. 118; Gamble, Man. Timb., 66; Grah., Cat. Bomb. Pl., 43; Stewart, Pb. Pl., 45; Rheede, Hort. Mal., IV., t. 33; Elliot, Fl. Andhr., 58,78; Mason, Burma and Its People, 761; Dymock, Mat. Med. W. Ind., 2nd Ed., 167; Baden Powell, Pb. Pr., 581; Athinson, Him. Dist., 307, 779; Lisboa, U. Pl. Bomb., 38, 149, 241, 278; Birdwood, Bomb. Pr., 147; Cooke, Gums and Gum-resins, 18; Athinson, Gums and Gum-resins, 14; Liotard, Dyes, 33; Athinson, Ec. Prod., N.-W. P. pt. I., 17, part V., 53; Balfour, Cyclop., I., 1182; Indian Forester, I., 83; III., 201; IV., 322; VIII., 414; X., 325; XII., 311; XIII., 120; Gazetteers:—Bombay, VIII., 11; XIII., pt. I., 24; XV., pt. I., 70, 429; N.-W. P., I., 80; IV., lxix.; Burma, I., 137; Aplin, Rep. on Shan States, 1887-88.*

Habitat.—A tree, attaining the height of from 30 to 40 feet, met with in the Sub-Himálayan forest from the Jumna eastwards, where it ascends to 3,000 feet, also in Central and Southern India, Chittagong, and Burma. It flowers from February to March, and the fruit ripens in June and July.

GUM-RESIN.
144

Gum-Resin.—This tree yields a greenish-yellow, translucent exudation in small mamilliform masses, having a mild terebinthinate odour and taste. It has been generally regarded by Indian writers as a true gum, **Watson** and **Cooke**, amongst others, classifying it with Gum acacia, &c. **Dymock**, however, states that it contains small proportions of an oleo-resin, and is in reality a gum-resin. He writes, "Only a small part of it is soluble in rectified spirits, causing a slight turbidity; in water it rapidly disintegrates, forming a tolerably thick mucilage, in which globules of oleo-resin may be seen with the microscope. The insoluble portion is amorphous, flaky, and white; after its removal the mucilage is precipitated milk-white, by rectified spirit." No record exists in economic literature of this exudation being utilised in the arts, but in Bombay it is employed as a medicine. **Mr.** O'Conor mentions **Garuga** in his list of trees on which lac is produced.

DYE & TAN.
Bark.
145
Leaves.
146

Dyes and Tans.—The BARK is used for tanning in many parts of the country, and is said by **Kurz** to be good for that purpose. The same writer mentions that in Burma the LEAVES are frequently invested with large, red, obovate, apiculate galls.

MEDICINE.
Juice.
147
Stem.
148
Fruit.
149
Leaves.
150

Medicine.—Dymock writes: "In Salsette, near Bombay, the JUICE OF THE STEM is dropped into the eye, to cure opacities of the conjunctiva" (? cornea). "The FRUIT, which is greenish yellow, and about the size of a gooseberry, is pickled and eaten as a cooling and stomachic remedy; it is strongly acid. In the Konkan the JUICE OF THE LEAVES, with that of the leaves of **Adhatoda Vasica** and **Vitex trifolia**, mixed with honey, is given in asthma."

FOOD.
Drupe.
151
FODDER.
Shoots.
152
Leaves.
153

Food and Fodder.—The fleshy, smooth, black, acid DRUPE is eaten raw, pickled, or cooked by the natives. As above stated, it is considered a semi-medicinal article of diet. The SHOOTS and LEAVES are collected for fodder, especially for elephants.

TIMBER.
154

Structure of the Wood.—Greyish or yellowish, heartwood dark reddish-brown, rather heavy (about 40℔ per cubic foot), coarsely fibrous, but fairly close grained, takes an indifferent polish, seasons well, but is not durable, and is very liable to the attacks of insects. It is accordingly not much used for construction, but is employed for indoor work, such as beams, rafters, &c., and has been recommended for cabinet work. It is also extensively used as fuel.

TIMBER.

Domestic Uses.—It is stated in the *Thana Gazetteer* that the soft, elastic bark is much employed for flooring cattle-sheds.

DOMESTIC.
155

Geese; see Ducks, p. 196.

Gelatine; see Isinglass, Vol. IV.

Gelidium cartilagineum, *Gaill.*, and

G. corneum, *Lam.;* ALGÆ; see Isinglass, Vol. IV.

GELONIUM, *Roxb.; Gen. Pl., III., 324.*

Gelonium lanceolatum, *Willd.; Fl. Br. Ind., V., 459; Wight, Ic.,* [*t. 1867;* EUPHORBIACEÆ.

156

Syn.—GELONIUM BIFARIUM, *Wight* (not of *Roxb.*).
Vern.—*Kakra*, URIYA; *Káru guggilam, suragada,* TEL.
References.—*Roxb., Fl. Ind., Ed. C.B.C., 738; Beddome, For. Man., 214 (excl. syn.); Gamble, Man. Timb., xxix.; Thwaites, En. Ceylon Pl., 274 (excl. syn.); Balfour, Cyclop., I., 1189.*
Habitat.—A small evergreen tree, found in the Deccan and Ceylon, ascending in the latter locality to 4,000 feet.
Structure of the Wood.—Yellow, smooth, close and even-grained, with a peculiar waxy odour; weight 50℔ per cubic foot. It is well adapted for house-building purposes.

TIMBER.
157

G. multiflorum, *A. Juss.; Fl. Br. Ind., V., 459.*

158

Syn.— GELONIUM FASCICULATUM, *Roxb.;* SUREGADA GLABRA, *Roxb. mss.;* S. BILOCULARIS, *Wall.;* ROTTLERA FASCICULATA and CONGESTA, *Ham.*
Vern.—*Ban naringa,* HIND.; *Sarugáta,* TEL.; *Setahanbaya,* BURM.
References.—*Roxb., Fl. Ind., Ed. C.B.C., 738; Kurz, For. Fl. Burm., II., 409; Elliot, Fl. Andhr., 171; Gazetteer, Mysore and Coorg, I., 65.*
Habitat.—A glabrous tree, from 30 to 40 feet in height. met with from Bengal and the Circars, northwards to the foot of the Sikkim Himálaya; also in Chittagong, Upper and Lower Burma, and Malacca.
Resin.—Roxburgh and Kurz mention that the BUDS of this species exude yellow resin. There is no record, however, of this having been collected or utilised in any way

RESIN.
Buds.
159

Structure of the Wood.—"White, only fit for house-posts and similar purposes" (*Kurz*).

TIMBER.
160

GENIOSPORUM, *Wall.; Gen. Pl., II., 1172.*

Geniosporum prostratum, *Benth.; Fl. Br. Ind., IV., 610;* [LABIATÆ.

161

Syn.—OCIMUM MENTHOIDES, *Burm.;* O. PROSTRATUM, *Linn.;* O. MACRO-STACHYUM, *Poir.;* MENTHA OCIMOIDES, *Lamk.;* THYMUS INDICUS, *Burm.*
Var.—GRACILIS, *Thwaites (sp.)*; G. GRACILE, *Benth.*
Vern.—*Nazel-nagai,* TAM.
References.—*Thwaites, En. Ceylon Pl., 237; Grah., Cat. Bomb. Pl., 148.*
Habitat.—A herb of the Deccan, from the Konkan southwards, and of Ceylon.
Medicine.—In Pondicherry this plant is supposed to have febrifugal properties.

MEDICINE.
162

The Indian Gentian.

GENTIANA, *Linn.; Gen. Pl., II., 815.*

A genus of annual or perennial herbs, comprising about 180 species, chiefly natives of the mountains of the Old World. Of these 37 are met with in India. All the members of the genus are to a greater or less extent characterized by the bitterness of their stems and roots, and many are of considerable medicinal value.

163

Gentiana decumbens, *Linn. f.; Fl. Br. Ind., IV., 117;* GENTIANACEÆ.

Syn.—G. ADSCENDENS, *Pall.*; PNEUMONANTHE ADSCENDENS, *Schmidt;* DASYSTEPHANA ADSCENDENS, *Borkh.*

References.—*Stewart, Pb. Pl., 147.*

Habitat.—Baltistán and Western Tibet, at altitudes of from 11,000 to 15,000 feet, eastwards to Lahoul, common on the Karakorum. Distributed to Dahuria and Siberia.

MEDICINE.
164

Medicine.—A tincture prepared from this plant has been used as a stomachic by the Lahoul missionaries (*Stewart*).

[*68, f. 2, and p. 278.*

165

G. Kurroo, *Royle; Fl. Br. Ind., IV., 117; Royle, Ill. Him. Bot., t.*

Syn.—PNEUMONANTHE KURROO, *D. Don.*

Vern.—*Karú, kútki,* HIND.; *Karú, kútki,* BENG.; *Nilkant, kamalphul, nilakil,* root=*karrú,* PB.; *Pháshánveda, pakánbed,* BOMB.; *Pakhánbhed,* GUZ. According to *Dymock, pakánbed,* though applied by Muhammadan writers to this plant, is the name associated in Bombay with what appears to be the root of an Iris.

References.—*Stewart, Pb. Pl., 147; Clarke in Jour. Linn. Soc., XIV., 440; Pharm. Ind., 149; O'Shaughnessy, Beng. Dispens., 459; Atkinson, Him. Dist., 313, 737; Kew Off. Guide to the Mus. of Ec. Bot., 98; Gazetteer, Panjáb, Simla Dist., 12.*

Habitat.—A small herb, with a handsome blue flower, common in Kashmir and the North-West Himálaya, altitude 5,000 to 11,000 feet.

MEDICINE.
Root.
166

Medicine.—The ROOT is used medicinally as a bitter tonic, and as a substitute for the true Gentian. On the hills it is viewed as a febrifuge, and is largely exported to the plains along with **Picrorhiza Kurrooa,** *Royle,* as the officinal *karrú* or *kútki,* of which Stewart says 36 maunds were, in 1867, brought from Kullu and exposed for sale at Rampúr. Davies' *Trade Report* gives 20 maunds as annually exported from Pesháwar to Kabul, and Atkinson says that five tons are annually exported from the hills to the plains. It appears probable that the root of this species is very similar to that of **G. lutea,** which forms the true Gentian of commerce, the chemical composition and medicinal properties of which will be described below.

SPECIAL OPINIONS.—§ "Used principally as a *masálah* for fattening horses" (*Surgeon-Major C. W. Calthrop, M.D., Morar*). "Acts as an aperient in larger doses" (*Civil Surgeon R. Gray, Lahore*). "Said to diminish the fever of phthisis" (*Surgeon J. C. H. Peacocke, I.M.D., Nasik*). "Used for urinary affections" (*Surgeon-Major S. M. Robb, Civil Surgeon, Ahmedabad*).

167

G. lutea, *Linn.; DC., Prodr., IX., 86.*

COMMON EUROPEAN YELLOW GENTIAN.

Vern.—*Pakhán-béd,** HIND.; *Juntiyánah,* DEC.; *Jintiyáná,* BOMB.; *Jintiyáná,* ARAB.; *Kon-shad,* PERS.

References.—*O'Shaughnessy, Beng. Dispens., 57; Moodeen Sheriff, Supp. Pharm. Ind., 146; Dymock, Mat. Med. W. Ind., 2nd Ed., 543; Flück. & Hanb., Pharmacog., 434; U. S. Dispens., 15th Ed., 707; Bent. & Trim., Med. Pl., 182; S. Arjun, Bomb. Drugs, 90; Year-Book, Pharm., 1874, 627; Irvine, Mat. Med. Patna, 33; Kew Off. Guide to the Mus. of Ec. Bot., 98.*

* See the remarks on this vernacular name under **G. Kurroo.**

The European Gentian.	(*J. Murray.*)	**GENTIANA tenella.**

Habitat.—A handsome perennial herb, native of the alpine and sub-alpine regions of South Europe. The dried root of the plant is imported into India.

Medicine.—The name of the genus is said to be derived from **Gentius,** a King of Illyria, who reigned from 180 to 167 B.C., and by whom, according to **Pliny** and **Dioscorides,** this species was noticed. It has, therefore, been known as a medicine from very remote times, and many of the complex preparations, handed down by the Greeks and Arabians, mention it amongst their ingredients. The Arabian and Persian names show that the knowledge of the plant in this country must have been derived from the Greeks. As above stated, the ROOT is to this day imported to a considerable extent. The drug is an important one in all the Pharmacopœias of Europe and America, and enters into most of the stomachic and tonic prescriptions of modern practice. In India also it is extensively employed both by European and Native physicians, but it appears probable that a more careful and exhaustive examination of indigenous species may lead to the substitution of one or more of them for the imported article. As already stated, **G. Kurroo** appears to be the best known and most widely employed of these native species, and would perhaps, on examination, be found to afford the best substitute.

MEDICINE.
Root.
168

CHEMICAL COMPOSITION.—According to the learned authors of the *Pharmacographia,* the bitter taste of Gentian root is due to a principle, *Gentiopicrin,* $C_{20} H_{30} O_{12}$, which, under the influence of a dilute mineral acid, is resolved into glucose, and an amorphous, yellowish-brown, neutral substance called *Gentiogenin.* Another constituent is *Gentianin,* $C_{14} H_{10} O_5$, a tasteless substance occurring in yellowish prisms. Besides these the root contains pectin to a large extent, and 12 to 15 per cent. of an uncrystallizable sugar, "of which advantage is taken in Bavaria and Switzerland, for the manufacture, by fermentation and distillation, of a potable spirit." The root contains no tannic acid.

CHEMISTRY.
169

ACTION AND USES.—Gentian possesses in a high degree the tonic properties which characterise the simple bitters, of which it is perhaps the most popular and extensively used. It possesses the advantages of being agreeable and slightly aromatic, of being only very slightly astringent owing to the absence of tannin from its composition, and of being a slight laxative and disinfectant. It accordingly excites the appetite, invigorates digestion, slightly increases body-heat by stimulating the circulation, and acts beneficially on the bowels. In very large doses, however, it is apt to cause too great gastro-intestinal irritation, resulting in nausea, and even vomiting and diarrhœa. It is specially indicated in cases of debility, in convalescence after exhausting diseases, and in gouty dyspepsia. It was formerly also held in high repute in India, as a bitter tonic in intermittent fevers. The *United States Dispensatory* contains the information that its powder has been employed as an application to malignant and sloughing sores. The *Pharmacopœia of India* describes four preparations of the root—a Compound Infusion, a Mixture, an Aqueous Extract, and a Compound Tincture.

USES.
170

Trade.—Dymock states that European Gentian root is obtainable in Bombay for about 4 annas per ℔; while Irvine in his *Materia Medica of Patna* writes: "Real Gentian root imported from Turkey, price per ℔ R2-8."

TRADE.
171

Gentiana tenella, *Fries. ; Fl. Br. Ind., IV., 109.*

172

Syn.—GENTIANA PEDUNCULATA, *Royle;* EURYTHALIA PEDUNCULATA, NANA, AND GRACILIS, *Don.*
Vern.—*Tita,* PB.

GERANIUM nepalense.	The Geraniums, or Crane-bills.

References.—*Stewart, Pb. Pl., 148; Atkinson, Him. Dist., 313.*

Habitat.—Common in Kashmír and the Western Himálaya, at altitudes from 10,000 to 14,000 feet. Distributed through Arctic and Alpine Europe, and Northern and Central Asia.

MEDICINE.
Leaves.
173
Stems.
174

Medicine.—Stewart states, on the authority of Atkinson, that in Lahoul a decoction of the LEAVES and STEMS of this and other species is given in fevers.

GEOPHILA, *Don.; Gen. Pl., II., 127.*

[*54;* RUBIACEÆ.

175

Geophila reniformis, *Don.; Fl. Br. Ind., III., 178; Wight, Ic., t.*

Syn.—GEOPHILA DIVERSIFOLIA, *DC.;* PSYCHOTRIA HERBACEA, *Linn.;* CEPHAELIS HERBACEA, *Kurz.*

Vern.—*Kúdi mankúni,* SYLHET; *Karinta kali,* MALAY.

References.—*Roxb., Fl. Ind., Ed. C.B.C., 179; W. & A., Prodr., 436; Kurz, For. Fl. Burm., II., 5; In Jour. As. Soc. Ben., 1877, II., 140; Thwaites, En. Ceylon Pl., 150; Dalz. & Gibs., Bomb. Fl., 111; Rheede, Hort. Mal., X., t. 21.*

Habitat.—A small herb met with in Sylhet, the Khásia Hills, the Western Gháts from the Konkán southwards, Tenasserim, and the Andaman Islands. It is also common in the central parts of Ceylon. Distributed through the Malay Archipelago, Southern China, Polynesia, Tropical Africa, and America.

MEDICINE.
Plant.
176

Medicine.—Kurz writes that this PLANT possesses qualities similar, though inferior, to those of **Cephaelis Ipecacuanha.**

GERANIUM, *Linn.; Gen. Pl., I., 272.*

A genus of herbs or undershrubs, belonging to the Natural Order GERANIACEÆ and comprising about 100 species, of which from 18 to 20 are natives of India. Many species are extensively cultivated as flowering plants. The generic name is derived from the Greek γεραυος (a crane) owing to the supposed resemblance of the fruit to the head of that bird. Certain species appear to have been known from remote times to possess medicinal virtues. Thus Dioscorides mentions a plant called γερανιον as employed for its astringent properties; Pliny alludes to two species.

177

Geranium nepalense, *Sweet.; Fl. Br. Ind., I., 430; Wight, Ill., I.,*

[*153, t. 59;* GERANIACEÆ.

Syn.—GERANIUM RADICANS, *DC.;* G. PALLIDUM and G. PATENS, *Royle;* G. AFFINE, *W. & A.;* G. ARNOTTIANUM, *Stend.*

Vern.—*Bhánda,* HIND.; *Bhánda* (root in bazars=*rowíl, bhand*), PB.

References.—*W. & A. Prod., 133; Stewart, Pb. Pl., 36; Botany of Tour in Hazára in Agri.-Hort. Soc. of Ind. Jour. (Old Series), XIV., 16; Pharmacog. Indica, I., 248; Baden Powell, Pb. Pr., 334; Atkinson, Him. Dist., 307.*

Habitat.—A herbaceous prostrate plant, common throughout the temperate Himálaya at altitudes of from 5,000 to 9,000 feet; found also in the Khásia Hills, the mountains of Southern India, and Ceylon. Distributed to Yunan.

DYE.
Root.
178

Dye.—The ROOT, which greatly resembles that of **Onosma echioides,** affords an abundance of red colouring matter, which is said by Dymock to be employed in colouring medicinal oils. Stewart states that it forms an article of trade, being brought from the hills to the plains of the Panjáb and sold as a dye.

MEDICINE.
Plant.
179

Medicine.—The PLANT possesses the astringent properties of the genus, and is employed, at least in the Panjáb, as an astringent, and in certain renal diseases.

The Geraniums, or Crane-bills. (*J. Murray.*) **GERANIUM Wallichianum.**

Geranium ocellatum, *Camb.; Fl. Br. Ind., I., 433 ; Royle, Ill.,* | **180**
[*149, 150.*

Syn.—GERANIUM BICOLOR and G. CHOORENSE, *Royle.*
Vern.—*Bhánd, bhánda,* HIND. and PB.
References.—*Stewart, Jour. of a Tour in Hasára, in Jour. Agri.-Hort· Soc. of Ind. (Old Series), XIV., 11, 14; Pharmacog. Indica, I., 248 ; Atkinson, Him. Dist., 307, 738.*

Habitat.—A small, straggling herb, met with on the temperate and sub-tropical Himálaya, from the Panjáb to Nepál, and on the summit of the Parisnath in Chutia Nagpúr.

Medicine.—The PLANT possesses astringent and diuretic properties, and is employed medicinally in the Panjáb and North-West Provinces. | MEDICINE. Plant. **181 182**

G. Robertianum, *Linn. ; Fl. Br. Ind., I., 432 ; Royle, Ill., 151, t. 27.*

Syn.—G. LINDLEYANUM, *Royle.*
References.—*Pharmacog. Indica, I., 218; U. S. Dispens., 15th Ed., 1652; Atkinson, Him. Dist., 307.*

Habitat.—A fetid, rather succulent annual or biennial herb, found in the western temperate Himálaya, from Kashmír to Garhwál ; at altitudes of 6,000 to 8,000 feet, distributed to Siberia, Asia Minor, the Caucasus, and Europe.

Medicine.—This herb, though now almost entirely neglected, was formerly much used in European medicine. It has a disagreeable, bitterish, astringent taste, and imparts its virtues to boiling water. It was formerly employed internally in intermittent fever, consumption, nephritic complaints, jaundice, &c., as a gargle in affections of the throat, and externally as a resolvent to swollen breasts and other tumours (*U. S. Dispensatory*). It is somewhat remarkable that while all the species of this genus have been for many years rejected from the European Pharmacopœia, G. maculatum, *Linn.*, is still extensively employed and highly valued in America. It is a domestic remedy in many parts of the United States, and is esteemed as one of the best indigenous astringents contained in their *Dispensatory*, the absence of unpleasant taste, and of other offensive qualities rendering it particularly suitable for administration to children. Diarrhœa, chronic dysentery, cholera infantum, and hæmorrhage are the diseases for which it may be employed with greatest advantage. It appears probable that the nauseous fetid taste and odour of the common European species has led to its rejection, and it may be that one or all of the Indian species, G. nepalense, G. occellatum, and G. Wallichianum, may possess the good properties of the American officinal drug, without having the objectionable qualities of G. Robertianum. | MEDICINE. Plant. **183**

G. Wallichianum, *Sweet.; Fl. Br. Ind., I., 430 ; Wight, Ic., t. 324.* | **184**

Vern.—*Liljahri,* N.-W. P.; Roots=*Mam-i-rán,* AFG.
References.—*Aitchison, Fl. Kuram Valley, 25, 39 ; Pharmacog. Indica, I., 248 ; Atkinson, Him. Dist., 307.*

Habitat.—A herb, with large bluish flowers, native of the temperate Himálaya from Nepál to Marri, at altitudes of 7,000 to 11,000 feet. Aitchison also describes it as met with in the Kuram Valley, " amongst bushes, grass, and boulders, where there is moisture, from 8,000 to 10,000 feet."

Medicine.—This herb evidently possesses the astringent properties of the genus to a marked degree. Aitchison writes : " At Alikhel a native brought me the stems of the plant, which he said was a rare and valuable medicine ; " and in another passage : " The rhizomes of this plant were brought to me (said to be from some hills 30 miles off) as the '*mam-i-ran,*' a good medicine for sore-eyes. This is doubtless a local substitute for the | MEDICINE. **185**

GEUM.	Gerbera; Geum.

true *mam-i-ran, i.e.,* the roots of **Coptis Teeta,** *Wall.*" Duthie states that in the villages of Jumnotri it is employed as a cure for toothache.

GERBERA, *Gronov.; Gen. Pl., II., pt. I., 497.*

186

Gerbera lanuginosa, *Benth.; Fl. Br. Ind., III., 390;* COMPOSITÆ.

Syn.—OREOSERIS LANUGINOSA, *DC.; and Wall, Cat.,* 2929 A. C.; CHAPTALIA GOSSYPINA, *Royle.*

Var.—PULSILLA, OREOSERIS PULSILLA, *DC.;* O. LANUGINOSA, *Wall, Cat.,* 2929 B.

Vern.—*Kapasi, kapasiya,* cloth woven from fibre=*karki, kaffi,* KUMAON; *Gauni,* GARHWAL; *Sung, buchachi,* SIMLA HILLS; *Patpatula, kho, búr, buzlí, kapfí, púrjlú, patola, kapasi, bújlo, tsar, kafí, kúfra, kharebúti,* PB.; *Sokhta,* tomentum=*kaff,* MURREE HILLS.

References.—*Stewart, Pb. Pl.,* 218; *Royle, Ill. Him. Bot.,* 251, *t.* 59, *f.* 2; *Atkinson, Him. Dist.,* 312, 793; *Royle, Fib. Pl.,* 302; *Cross, Bevan, and King, Rep. on Indian Fibres,* 68; *Kew Off. Guide to the Mus. of Ec. Bot.,* 87; *Agri.-Hort. Soc. of India Trans., III.,* 75, *Pro.,* 267; *VIII.,* 272, 276; *Jours. (Old Series), VII., Sel.* 48; *IX.,* 283, *Pro.,* 139; *X., Pro.,* 135; *Gazetteer, Panjáb, Simla District,* 12.

Habitat.—A herbaceous procumbent plant of the Western Himálaya, from Murree to Kumáon, between the altitudes of 4,000 and 8,000 feet. The variety **pulsilla,** which is apparently a starved condition of **G. lanuginosa,** extends to Nepál.

FIBRE.
Leaf.
187

Fibre.—The under-surface of the LEAF is covered with a cotton-like tomentum, which is employed by the natives of the Himálaya as tinder, and for the manufacture of cloth. This tomentum has attracted considerable attention at different times, and has been variously recommended as a cloth-making fibre, as a paper-making material, and as a substitute for cotton in the manufacture of explosive compounds. No practical result, however, appears to have been produced by these suggestions, and the fibre is still employed by the natives only. The tomentum is prepared for use as follows:—About the middle of the rains, when the leaf attains its full size, the plant is gathered, the point of the leaf is severed, and the down stripped off towards the base in an entire layer. It is then without further preparation twisted into a thread, on the common perpendicular "*churka*" of the country. From the thread thus prepared a cloth is woven, from which blankets, sacks, and bags are made by the hill people. This cloth has been described as very highly prized for its strength and durability, and superior to that manufactured from hemp. It is very frequently employed also for making the characteristic bags in which the hillmen carry their *hookahs.* The tomentum can only, however, at best be obtained in very small quantities and is of interest as a curiosity only. It can never prove of commercial value.

GEUM, *Linn.; Gen. Pl., I., 619.*

188

A genus of perennial ROSACEÆ, which derives its name from the Greek γευο, an agreeable taste, on account of the slightly aromatic flavour of the roots of certain of the species. Two are natives of India, **G. elatum,** *Wall.,* and **G. urbanum,** *Linn.* Neither appears, however, to be recognised in this country as of value, a somewhat remarkable fact in the case of the latter, which is the **Aveus,** Radix Caryophyllata, or HERB BENNET, of old European herbalists. The root of this species has a clove-like odour, and, owing to its stringent properties, has been employed in cases of dysentery, diarrhœa, &c. It was also used to flavour ale in olden times, and has been recommended in cases of caries of the teeth, &c., to impart an agreeable odour to the breath. **G. urbanum** (*Linn.; Fl. Br. Ind., II.,* 342) is to be found in India, on the Western temperate Himálaya, from Murree to Kumáon, at an altitude of 6,000 to 11,000 feet.

G. 188

Ghi or Clarified Butter.	(*J. Murray.*)	GHI.

GHÍ.

189

Clarified butter, largely prepared from the milk of the cow and buffalo, and to a smaller extent from that of the sheep and goat, is universally employed for domestic cooking in India, and forms an important article of trade. By far the greatest proportion is prepared from buffalo-milk, not only because that animal yields more highly fatty milk, but because it is cheaper, and more easily reared and fed, than the more delicate cow. As a consequence, cattle-breeding for dairy purposes is mainly confined to buffaloes.

Vern.—*Ghi*, HIND.; *Neyi*, TAM.; *Neyi*, TEL.; *Ghrita, ghruttham*, SANS.

References.—*Ain-i-Akbari, Blochmann's Trans., 63; Voyage of John Huyghen van Linschoten to the East Indies, I., 56, 58, 60, 63, 67; Milburn's Oriental Commerce, Ed. 1825, 288; U. C. Dutt, Mat. Med. Hind., 14, 282; Baden Powell, Pb. Pr., I., 151; Balfour, Cyclop., I., 1198; Settlement Reports:—Central Prows., Chindwara District, 112; Jubbulpur, 87; Panjáb, Jhang District, 63; Gazetteers:—Bombay, III., 74: Central Prows., 516; N.-W. P., IV., 250; Fanjáb, Shahpur, 74; Gujranwala, 60; Dera Ghazi Khan, 91; Amritsar, 48; Bombay Admn. Rep., 1871-72, 394; Revenue and Agricultural Dept. Reports, 1881 to 1886.*

Preparation.—For the following account of the methods of preparing *ghí* in the principal *ghí*-producing districts of India, the writer is indebted to a report drawn up by Mr. **T. N. Mukharji** for the Revenue and Agricultural Department in May 1884.

PREPARA-TION.

190

In Bengal.—The process of manufacture generally followed is thus described : Fresh milk is boiled on a slow fire for five or six hours, being occasionally stirred with an iron spoon to prevent its boiling over; the fuel used is cowdung cake, which gives out a moderate heat. The milk gradually assumes a red-brown colour and a thick crust is formed on the surface, after which it is taken down and allowed to cool. It is then transferred to a separate earthen vessel, and a small quantity of whey introduced, which in about 12 hours causes the milk to coagulate and turn into pure curd. This curd is transferred to a large earthen or metallic vessel, and a quantity of water added, for the purpose of reducing it to a liquid state to facilitate churning. It is then churned by a churning-staff as long as it continues to yield butter, which is every now and then taken out of the vessel, scraped off the staff, and collected in a separate pot containing water, to allow it to remain cool. Sometimes water is added twice or thrice to the curd before it is quite freed from butter. The butter thus obtained is heated until the greater part of the moisture in it evaporates; the oil-like *ghí* then rises on the surface and the half-burnt refuse falls as a sediment. Too much boiling gives the *ghí* an acrid taste, while, on the other hand, imperfect heating renders it liable to putrefaction. People who manufacture *ghí* for sale do not, however, heat it to the full extent, for fear that it might lose in weight; hence the ordinary *ghí* sold in the bazár is generally not of the best sort. Butter loses about 25 per cent. in weight in the process of being made into *ghí*.

Bengal.

191

The vessel (generally earthen) in which milk is boiled is always kept very clean, and is warmed on a fire before being filled with fresh milk, especially in the cold weather. In the cold season whey is introduced into the milk before it is quite cool, since without this addition it does not curdle properly, while in the hot weather the application of acid in the warm state decomposes the milk. One ounce of whey is considered sufficient to coagulate about two gallons of boiled milk. Failing a supply of whey, other acids are used, such as dried mangoe, tamarind, lime-juice, and even a piece of tarnished silver (a rupee), but none of these are so effective as whey. No

GHI.	Ghi or Clarified Butter.

PREPARA-
TION.
Bengal.

measure can be given for the quantity of water to be added to the curd before churning, as it depends upon the consistency of the latter; generally, however, one quart of water is considered sufficient for three quarts of curd. The water ought to be gradually added during the process of churning. In the cold weather hot water is first added until the butter begins to form, after which cold water is dashed in to expedite the process.

It is not absolutely necessary that the fuel should consist of cowdung. Nor is it necessary that the milk should be heated for five or six hours; indeed, the acid whey or curd is in some places put into raw cold milk. By this process, however, a longer time is necessary to curdle the milk. It is stated that the longer the curdled milk is kept unchurned, the larger is the yield of butter, and that the maximum time for which curdled milk can be kept without deterioration is three days. The proper time for churning is the cool morning hours, as after sunrise the butter does not form into good large lumps, and owing to the heat is liable to get thin and to mix with the whey.

Near large towns where there is a great demand for milk and curd, people sometimes take off the crust or cream and sell the milk in a raw or curdled state. The cream is then churned and the butter obtained is melted into *ghí* in the usual way. Generally speaking, however, the manufacture of *ghí* is confined to villages where there are no purchasers for milk, as it is more profitable to sell milk in the raw state than to convert it into *ghí*.

In certain localities, such as the Monghyr and Bhagalpur districts, butter is extracted by churning the raw milk, either fresh or after being boiled. The milk is then sold either raw or curdled, and *ghí* is made by heating the butter.

In the Tippera district milk is first boiled down to the consistency of a thick hard jelly, thus forming a substance known as *khír* in Bengal and *khoya* in Upper India, which is eaten as a delicacy and enters largely into the composition of most of the native sweetmeats. This substance is ground on a stone, placed in an earthen or a metallic vessel, reduced to a liquid state by mixing water with it, and then churned. The butter thus obtained, when melted, is said to yield a superior quality of *ghí*.

Rajputana
192

In Rajputana.—The process adopted differs somewhat from that detailed above, and is thus described by A. Wingate, Esq., C.I.E., Settlement Officer, Meywar :—

"The milk is slowly boiled on a cowdung cake fire, and occasionally stirred with an iron spoon to prevent it boiling over. A little whey is poured in to make the cream rise. The white curds are then skimmed off and kept in earthen or brass pots till a sufficient quantity is collected. These curds are then poured into a large earthen vessel and some *warm* water added. The churn called 'rawai' is at once put in and worked by a woman. From time to time cold water is freely added. The butter is then collected with the hand into a similar earthen pot, and heated till it melts. The melted butter is then laid aside to cool, and is thenceforth known as *ghí*. The best *ghí* is white, like soft lard, and has no smell, and keeps good for almost any length of time.

"Every household makes its own *ghí*, and the 'chach' or watered skim milk is much used for drinking at meals with Indian-corn porridge or baked cakes. The villagers, in making *ghí*, mix all their milk up together, whether obtained from the cow, buffalo, or goat, and the shepherd classes also add the milk of their sheep. Consequently *ghí* sold in the bazárs is frequently very strong in smell and taste, and of reddish-yellow colour.

"The amount of *ghí* from a given quantity of milk depends altogether upon the feeding of the cattle. Most families keep one or two milch kine and buffaloes at home and feed them well. Such cattle, they say, give

| Ghi or Clarified Butter. | (*J. Murray.*) | GHI. |

about two ounces of *ghí* per seer of buffalo's milk, and one ounce or less per seer of cow's milk. Goat's milk gives about four ounces and sheep's milk less than an ounce of *ghí* per seer."

The Agent to the Governor General also states :—

"In Rájputana, *ghí* ordinarily sold in the market is chiefly derived from the milk of the sheep, which, though decidedly lesser in quantity, is thicker in composition and richer in butter than that from the buffalo. The outturn of Rájputana *ghí* chiefly depends on the large flocks of sheep reared in this part of the country by Jats, Gujars, and other agriculturists. A flock of 100 sheep can be maintained at less expense than 10 buffaloes, and yet the outturn of milk and butter is nearly treble. Sheep's milk is said to have medicinal virtues, which give it a superior rank. The butter is whiter than that of the buffalo, and excels it in fragrance and taste."

In Madras.—The process of manufacture has been described as follows by Mr. Robertson :—"In making *ghí*, the first object is to get the butter thoroughly separated from the milk, in as pure a condition as possible. This is secured by placing the can or vessel containing the freshly-drawn milk in an earthenware vessel of boiling water for about 5 minutes. The milk, after being thus exposed to a temperature of about 212 degrees, is poured into another vessel, and butter-milk is added, from two to three drops in hot weather to a tea-spoonful in cold weather per quart of milk. The vessel with the milk is put aside for 24 hours, and the milk is then churned. The yield of butter averages from about $1\frac{1}{2}$ to 2 ounces per quart of milk, but of course varies greatly. The butter is next melted in an open vessel over a slow fire, the heat coagulates the caseine which, with other impurities, sinks to the bottom of the vessel; boiling is continued for from 15 to 20 minutes, when most of the water is evaporated into the air, and the *ghí*, clear and bright, rests on the heavier sediment covering the bottom of the vessel. The *ghí*, when cold, is carefully poured off, leaving the sediment behind, and is fit for immediate use, or for storing for future use. The outturn of *ghí* varies with the quality of the butter and the purity of the *ghí* made—an average outturn of 50 to 60 per cent. of the weight of the butter used, when the butter is made from the milk of the cow. The yield of *ghí* from buffalo butter is higher. *Ghí* is never made when a fair price can be obtained for milk or butter. A *viss* (3℔ 2 ozs.) of *ghí* sells usually for about 1s. $10\frac{1}{2}d$., and to make this not less than 6℔ of butter or 48 quarts of milk of the cow would be needed. In nearly all the large towns of South India, cow's milk will sell readily at $2\frac{1}{2}d$. per quart and butter at 1s. 3d. per pound. Thus, the milk that would be required to make 3℔ 2 ozs. of *ghí* worth 1s. $10\frac{1}{2}d$., would, as fresh milk, sell for about 10s., and if churned, would yield butter worth 7s. 9d."

CHARACTERS, QUALITY, AND YIELD OF GHI.—The ordinary *ghí* of the bazárs is principally derived from buffalo milk, which is not only obtained in greater quantity from one animal, but is richer in butter than that of the cow. One quart of buffalo milk yields about three ounces of *ghí*, while the same quantity of cow's milk only affords about one ounce and a half. Reports from the Panjáb and Bombay, however, appear to indicate that the difference is not always so great, since the quantity obtained from cow's milk is said to be only one fourth less than that derived from buffalo milk. There is no doubt that the food given to the cow is an important element in deciding the amount of butter obtainable, cotton-seed and oil-cake especially making a great difference in the amout of fatty matter in the milk. Careful experiments by Mr. E. J. Kitts, Assistant Commissioner in the Hyderabad Assigned Districts, gave the following results :—"One buffalo in milk gives about $4\frac{3}{4}$ seers ($6\frac{1}{3}$ qts.) of milk per diem, and nearly 9 seers (12 qts.) of milk are required to obtain one seer (32 ozs.) of butter. When

GHI.	**Ghi or Clarified Butter.**

warmed and strained, the butter becomes *ghí* and in the change it loses 25 per cent. of its weight. On the average, therefore, each buffalo in milk gives the equivalent of two fifths of a seer (12⅖ ozs.) of *ghí* per diem."

In Bundelkhand, Rájputana, and other localities, *ghí* is also made from sheep and goat's milk. That of the latter is inferior, owing to the disagreeable odour it possesses; while *ghí* made from the former is said to be better in many ways than that of the buffalo.

In many parts of the country *ghí* obtained from cow's milk is highly esteemed, owing partly to its superior quality, and partly to its greater purity from a religious point of view. It is, however, always dearer than buffalo milk *ghí*, not so easily procurable, and is moreover more liable to deterioration. It is of a yellowish colour, and has a more pleasant odour and agreeable taste than that prepared from buffalo milk.

The following statement of the comparative yield from different kinds of milk drawn up by the Superintendent of the Government Farm, Cawnpore, may be here given :—

Cattle.	Weight of fresh milk.	Weight of boiled milk.	Weight of curdle.	Weight of Matha (curdle & water).	Weight of Nainu (ram ghí).	Weight of ghí.	Percentage of ghí over fresh milk.
	℔ oz.	℔ oz.	℔ oz.	℔ oz.	℔ oz.	℔ oz.	
Buffalo (first testing)	22 8	21 0	20 14	23 7	1 3	0 12⅝	3·47
Buffalo (second testing) . . .	20 0	18 0	17 6	19 2	1 1	0 11	3·43
Cow (first testing) . . .	20 0	17 11	17 0	19 2	0 12	0 8	2·5
Cow (second testing) . . .	20 0	18 1	17 6	19 0	0 13	0 8½	2·34
Cow (third testing) . . .	10 0	9 0	8 10	9 15	0 6	0 4	2·5
Goat (first testing) .	24 4	22 0	21 8	24 5	0 13	0 9½	2·44
Goat (second testing) . . .	20 0	17 13	17 6	19 10	0 14	0 8½	2·65
Sheep . . .	6 0	5 8	5 4	6 15	0 6	0 4½	4·6

ADULTERANTS.—The chief articles employed in adulterating *ghí* are vegetable oils, animal fat, especially mutton-fat, and starches. Of the last the commonest are : rice-flour, flour of *bajra* millet (**Pencillaria spicata**), ripe plantain, and the starch obtained from the boiled tubers of **Ipomœa Batatas** and **Colocasia antiquorum.** Of vegetable oils the oils of cocoa-nut, poppy-seed, sesamum, *mahuá* (**Bassia latifolia**), and *kokam* (**Garcinia indica**) are most frequently employed, and occasionally also raw castor-oil. Besides these other impurities occur, resulting from imperfect heating and careless preparation Several methods of purification are adopted, the commonest being to boil the *ghí*, dash cold water on it while in a state of ebullition, and then to separate the pure oil which on cooling floats on the surface. In Rájputana fresh milk is mixed with the impure *ghí*, in the proportion of one of the latter to four of the former, and the whole process of manufacture is repeated. In certain other localities purification is effected by heating the *ghí* with leaves of lemon.

PACKING.—Formerly all *ghí* was packed for local use in earthen jars, and for transport to a distance in leathern cases called *kuppas*. Of late years, however, old kerosine tins, and new tins of the same shape and size, have come into almost universal use in all cases in which the *ghí* is required for transport by sea or rail. In Madras, Rájputana, and Sind, however,

Ghi or Clarified Butter.	(*J. Murray.*)	GHI.

though the kerosine tin is gradually superseding the older method, skin *kuppas* are still extensively employed, and in Bengal, the only receptacle used for transporting *ghí* to Calcutta by river is the old earthen jar or *matka*. In Madras and Bombay, zinc cases, either shaped in imitation of a *kuppa* or of a kerosine tin, and wooden casks, are also employed, but only to a limited extent.

PRODUCTION AND CONSUMPTION.—The principal *ghí*-producing tracts are the North-West Provinces and Oudh, Bengal, Rájputana, the Central Provinces, and the Panjáb, or, in other words, the most densely populated and highly cultivated parts of the country. Bombay also produces a small quantity, but obtains its chief supply by importation. Regarding consumption, Mr. T. N. Mukharji writes: " Roughly speaking, about a fourth of the total population use *ghi* at an average rate of 8℔ per head per annum. In a population of nearly 300 millions, this rate would give an annual consumption of 267,000 tons, the value of which, at the rate of £45 per ton, would amount to more than £8,000,000. The provincial rates, which are a little in excess of this figure, are as follows:—Madras 33,000 tons, Bombay and Sind 22,000 tons, Bengal 74,000 tons, North-West Provinces 63,000 tons, Panjáb 54,000 tons, Central Provinces 10,000 tons, rest of India 44,000 tons,"—the total amounting to 300,000 tons.

PRODUCTION, &c, 197

Medicine.—*Ghrita* has long been regarded as a substance of medicinal value by Hindú practitioners. U. C. Dutt writes: " That obtained from cow's milk is considered superior to that prepared from the milk of the buffalo, and is preferred for medicinal use. Clarified butter is considered cooling, emollient, and stomachic. It increases the fatty tissues and mental powers, improves the voice, beauty, and complexion, and is useful in eye-diseases, retained secretions, insanity, tympanitis, painful dyspepsia, ulcers, wounds, &c." It is also employed extensively as the basis of a form of medicinal preparation called *ghritapáka*. This is prepared as follows:— " The *ghrita* or clarified butter is first of all heated on a fire, so as to deprive it of any water that may be mixed with it; a little turmeric juice is then added to purify it, as it is said, but the object, I suppose, must be to colour it. *Ghrita* thus purified is placed on a fire in an earthen, copper, or iron pan, and melted with a gentle heat. Then the medicinal paste and fluids to be used are added, and the whole boiled together till the watery parts are all evaporated, and the *ghrita* is free from froth. It is then strained through cloth and preserved for use" (*U. C. Dutt, Mat. Med. of the Hindus*).

MEDICINE. 198

These *Ghritapáka* are prepared in three varieties by different degrees of boiling: the first, *mridupáka*, is a soft paste; the second, *madhyamapáka*, is just soft enough to be made into pills; the third, *kharapáka*, is hard and dry. The underboiled form is said to be useful as snuff, the intermediate is preferred for internal administration, and the overboiled variety is employed for external application. *Purána ghrita*, or *ghí* more than ten years old, is a much-prized external application in Hindú medicine. U. O. Dutt writes: " It has a strong pungent odour and the colour of lac. The longer this old butter is kept, the more efficacious it is said to prove. Clarified butter a hundred years old is often heard of. The richer natives have always a stock of old *ghrita* of this description, which they preserve with care for their own use as well as for distribution to their poorer neighbours." " Old clarified butter is used externally. It is first repeatedly washed with cold water, and then rubbed with it till it is reduced to a soapy, frothy fluid, which is used as a liniment. It is regarded as cooling and emollient, and is much used in nervous diseases such as insanity, epilepsy, neuralgia, paralysis, cephalalgia, and asthma; also in rheumatic affections, stiff joints, burning of the body, hands or feet, affections of the

GHI.	Ghi or Clarified Butter.

MEDICINE.

FOOD.
199

DOMESTIC.
200
SACRED.
201

PRICES.
202

TRADE.
203

Inter-prov-
incial.
204

Trans-
frontier.
205

eyes, &c." It is much valued as an application for reducing the temper-ature in high fever.

Food.—*Ghi* has long been one of the most important articles of diet of all classes who can afford it in India. Linschoten makes frequent reference to its extensive employment in Sindh, Bombay, and other places which he visited along the coast in his travels. The *Ain-i-Akbari* contains it in the list of more important articles of food during the reign of Akbar, and reference is frequently made to *ghrita* in many ancient Sanskrit works. It is used in much the same way as butter in European cookery, being employed in the preparation of vegetables, curries, pulses, meat, rice *palao*, &c. It is also eaten uncooked with bread or boiled rice, and enters into the composition of the sweetmeats and pastry so extensively consumed by the population of all large towns. The poorer classes reserve the use of *ghi* as a luxury for feast days and festivals, and sub-stitute, for ordinary consumption, some of the sweet vegetable oils.

Domestic and Sacred uses.—In parts of India where vegetable oils are expensive, *ghi* is said to be employed by women for dressing the hair, &c. *Ghi* prepared from cow's milk is largely used in many religious and social ceremonies of the Hindús; thus it is burnt as an offering to the fire-god (*Agni*), and with sandal-wood in Bombay to invoke Lakshmi.

Prices and Trade.—Reports submitted at different times to the Revenue and Agricultural Department indicate that, as a rule, the selling price of *ghi* ranges between 5*d*. and 7*d* per ℔. In Bombay, Madras, Calcutta, and other large centres of demand, however, the price ranges as high as 11½ *d*. to 1*s*. for first quality *ghi*; and, as already stated, that prepared from the milk of the cow always fetches a higher price than that made from buffalo milk. Though by far the greatest proportion of *ghi* prepared within the country is consumed in India, a considerable trade exists with trans-frontier countries, and also with foreign ports, princi-pally Mauritius, the Straits Settlements, and other colonies where well-to-do Native emigrants can afford to purchase it. As might be expected from the almost universal consumption of the article, the inter-provincial trade returns shew a large traffic in *ghi*. The following figures indicate the trade by rail and river during the year 1888-89, including the Indus-borne traffic between the Panjáb and Sind, that between Bengal and As-sam, by the Brahmaputra and Megna, and the trade to and from Calcutta by river. The North-Western Provinces and Oudh exported 1,88,521 maunds, Bengal 85,587 maunds, Madras 42,019 maunds, the Panjáb 25,633 maunds, the Central Provinces 20,811 maunds, and Berar, Bombay, Assam, and Sind smaller amounts. Of the large towns excluded in the above figures, Calcutta exported 24,903 maunds, Karachi 10,868 maunds, Bombay 4,498 maunds, and Madras 477 maunds. The largest amounts imported were by Calcutta, 1,43,897 maunds, Bombay town 98,894 maunds, Madras seaports 32,907 maunds, Sind (excluding Karachi) 36,047 maunds, Bengal 31,440 maunds, Bombay 29,380 maunds, Ráj-putana and Central India, 27,840 maunds, and the Panjáb 25,196 maunds.

An extensive import trade is carried on with the frontier states, the amount and value of which for the past three years has remained remark-ably uniform. The figures are:—

	1885-86.	1886-87.	1887-88.
Amount in cwt. . .	63,658	54,073	58,591
Value in R . .	22,56,545	19,39,985	21,20,917

The principal sources of supply are Kashmír and Nepál, the latter of which in 1887-88 supplied 14,995 cwt., the former 34,153 cwt. There is also a small export trade, which, however, is almost entirely confined to Upper Burma, Kashmír, and trans-frontier by the Sind-Pishin Railway, the *ghi*

Ghi or Clarified Butter.	(*J. Murray.*)	**GHI.**

thus exported being consumed almost entirely by Indian troops and followers.

The imports from foreign countries represent a large and constantly increasing trade, but bear a very small proportion to the figures representing the trans-frontier and inter-provincial trades. There appears to be little doubt that if a source of cheap supply could be found, the consumption, and consequently the imports from foreign countries, might become very greatly increased. The average import for the past five years has been 1,980,709℔, value R7,14,122, in comparison with 431,912 ℔, value R1,22,459, of the five years immediately preceding. It may be noted also that the imports of the year 1888-89 increased to 2,731,280℔, in comparison with 1,382,389℔ in 1884-85. The imports are almost entirely into Bombay and Karachi; the sources of supply are Turkey in Asia, the neighbouring pastoral tracts of Southern Baluchistán, and the shores of the Persian Gulf.

As in the case of the trans-frontier trade, the chief foreign markets to which *ghi* is exported are Mauritius, the Straits Settlements, Aden, and other similar colonies, where well-to-do Indian emigrants supply a market. A certain amount is also exported to the neighbouring coasts of Africa and Asia, and a small quantity is despatched to the United Kingdom, possibly for re-export to some of the colonies. The average export for the past five years has been, 1,938,092℔, value R7,10,287, or almost exactly equal to the average import, and shews little change in comparison with that of the five years ending 1883-84, which was 1,659,613℔, value R5,83,423.

The coasting trade is a large and increasing one, but, like the trans-frontier exports, its most remarkable feature is the transport of *ghi* to Indian consumers in non-producing districts. In 1888-89 the total export from Bengal was 1,322,539℔, value R4,87,575; from Bombay 1,181,542℔, value R4,33,303; from Sind 136,465℔, value R49,856; from Madras 2,182,832℔, value R6,88,736; and from Burma 23,068℔, value R9,112. By far the largest importer was Burma, which recorded 3,412,644℔, value R13,05,499, chiefly from Madras and Bengal. The probable reason of this large consumption in a country, the Buddhist inhabitants of which do not employ *ghi* as an article of food to any extent, is probably the large and increasing population of emigrants from Madras, Bengal, and other provinces of the main peninsula.

In 1881 an endeavour was made by the Government of India to give an impetus to Indo-Australian trade, by establishing a return trade, the absence of which greatly augments the price of exported Indian goods by causing high shipping rates. It was considered that the only article besides animals, timber, and metals, which could profitably be thus sent to India was *ghi*, for the production of which the northern portion of South Australia appeared to possess many advantages. **Sir E. O. Buck** accordingly drew out a memorandum drawing the attention of the Australians to the subject, and suggesting the methods by which such a trade might be most profitably and advantageously commenced. As an outcome of this suggestion, buffaloes and *ghi*-makers were asked for, and were supplied by the Government of India in 1883. Experiments were started at Port Darwin, with the result that the buffaloes were found to thrive well and to breed healthy calves, and excellent *ghi* was produced, which obtained a gold medal at the Calcutta Exhibition in 1884. The initial cost was necessarily high in proportion to the smallness of the herd, and, accordingly, the success of the experiment from a commercial point of view is not as yet established. The industry is one, however, that appears to have a hopeful future. The demand is a large and constantly increasing one, the climate of the northern territory of South Australia is admirably suited

| GIRARDINIA heterophylla. | Bamboos ; The Nilghiri Nettle. |

TRADE IN GHI. for buffaloes, and if managed with due attention to the prejudices of the consumers, and by the help of imported Indian labour, there appears to be every likelihood of such an enterprise affording a good return.

GIGANTOCHLOA, *Kurz; For. Fl. Burm.,* II., 555.

A genus of evergreen densely tufted bamboos, which are employed for the same purposes as other members of the Tribe BAMBUSEÆ. For information regarding these the reader is referred to Vol. I., 370. The following are the principal Indian species :—

209 ### Gigantochloa albo-ciliata, *Kurz ; For. Fl. Burm.,* II., 555 ;
[GRAMINEÆ.

Syn.—OXYTENANTHERA ALBO-CILIATA, *Munr.*
Vern.—*Wapyugale,* BURM.
Habitat.—Common in the mixed forests of Pegu and Martaban.

210 ### G. andamanica, *Kurz ; For. Fl. Burm.,* II., 556.
Vern.—*Podák,* AND.
Habitat.—Common in the mixed forests of the Andamans.

211 ### G. auriculata, *Kurz ; For. Fl. Burm.,* II., 557.
Vern.—*Talaguwa,* BURM.
Habitat.—An evergreen arboreous tufted bamboo, found in the low forests of Southern Pegu, but rather rare, cultivated in villages of Arracan and Chittagong. A useful timber with very strong stems.

212 ### G. macrostachya, *Kurz ; For. Fl. Burm.,* II., 557.
Vern.—*Wanet,* BURM.
Habitat.—Not unfrequent in the tropical forests of Martaban and Tenasserim, also cultivated in the villages of the Irrawaddi valley and of Arracan.

Ginger ; see **Zingiber officinale,** *Roscoe ;* SCITAMINEÆ.

Ginger Grass ; see **Andropogon Schœnanthus,** *Linn.,* Vol. I., 249.

Gingelly Oil, a name in India for an oil obtained from **Sesamum indicum,** *DC.;* PEDALINEÆ, which see.

GIRARDINIA, *Gaud. ; Gen. Pl.,* III., 384.

A genus of annual or perennial herbs, belonging to the Natural Order URTICACEÆ.

213 ### Girardinia heterophylla, *Dcne.; Fl. Br. Ind.,* V., 550 ; *Wight, Ic.,*
[*t.* 687 ; URTICACEÆ.

THE NILGHIRI NETTLE.

Syn.—URTICA HETEROPHYLLA, *Vahl.;* U. DIVERSIFOLIA and HORRIDA, *Link.;* U. PALMATA, *Forsk.*
Var. PALMATA, *Gaud.*
Var. ZEYLANICA, *Dcne.* SYN.—URTICA HETEROPHYLLA, *Wight;* U. ZEYLANICA, *Burm.*
Vern.—*Awa, alla, bichua, chichr,* HIND.; *Horú surat,* ASSAM; *Serpa, herpa,* BHUTIA; *Ullo,* NEPAL; *Kasu,* LEPCHA; *Shishuna, awa-bichhu,* KUMAON; *Kali, kubra, jurkunkundalu, kundalu,* GARHWAL; *Ein, keri, kingi, sanoli, ánján, kárla, kal, bhábar,* PB.; *Moti khajati,* MAR.; *Ana schorigenain,* MALAY.; *Betya, bekshá, phetyákyi,* BURM.; *Gass-kaham-bilya,* SING.
References.—*Roxb., Fl. Ind., Ed. C.B.C.,* 655 ; *Brandis, For. Fl.,* 404 ; *Gamble, Man. Timb.,* 323 ; *Dalz. & Gibs., Bomb. Fl.,* 238 ; *Stewart, Pb. Pl.,* 215 ; *Mason, Burma and Its People,* 775 ; *Atkinson, Him. Dist.,* 317, 797 ; *Rheede, Hort. Mal.,* II., *t.* 41 ; *Drury, U. Pl.,* 225 ; *Lisboa, U. Pl.,*

G. 213

Bomb., 234; Royle, Fib. Pl., 367-372; Liotard, Paper-making Mat., 512; Forbes Watson, Rep. on Rheea Fibre, 1875, 39; On the Preparation and Use of Rheea Fibre, 1883, 35; Watt, Sel. from the Rec. of the Govt. of India, 177, 260, 319; Agri.-Hort. Soc. of India, Trans. VIII., 75, 275; Jours. (Old Series), VI., 44; VII., 223; Spons' Encyclop., 999; Balfour, Cyclop., I., 1207; Indian Forester, XII., App., 21; XIV., 269, 273; Special Reports from Forest Department, N.-W. P., Madras, and the Panjáb.

Habitat.—A tall, stout, erect, tufted herb, from 4 to 6 feet high, exceedingly common in the temperate and sub-tropical Himálaya, from Marri eastwards, ascending to an altitude of 5,000 feet. It is also to be met with in Assam, Sylhet, and Burma, and extends from Marwar and Central India to Travancore and Ceylon. The variety **palmata** is a native of the Nilghiri hills and Ceylon, while **zeylanica** is confined to the latter locality and parts of the Deccan.

Fibre.—Considerable confusion exists in the literature of the fibres yielded by this species, apparently owing to a neglect of the fact that the three varieties afford fibres which are perfectly distinct in many of their characters. It is, therefore, necessary in the present article to consider the varieties separately, as far as the fibre of each is concerned.

G. heterophylla proper.—The Himálayan nettle is extensively employed in the localities where it abounds. "Its stems are often employed for making twine and ropes by the dry process, but these are not prized, and perish quickly from wet" (*Stewart*). "Yields a fine strong fibre, much used for cordage and twine, but cannot stand much moisture" (*Atkinson*). Dr. **Forbes Watson**, in his report upon Rheea Fibre, publishes under this species certain facts regarding what appears to be the fibre of *var.* **palmata** and reproduces **Wight's** plate of *var.* **zeylanica** as representing the typical species. This same mistake has been made by other authors, all the economic information regarding **Girardinia** being confused and given under one or other of the above names. Under the heading of "Other Himálayan Nettles," Dr. **Royle** gives certain facts regarding what appears to be the fibre now under consideration. Having arrived at the conclusion that the *horú surat* of Assam was identical with the fibre of the Nilghiri nettle of Southern India, he apparently could not reconcile himself to class with these the *bábar* of the Himálaya, nor the fibre from which the *bangra* cloth of Sikkim was made. Presumably, therefore, he merely classed the Assam fibre with that of the Nilghiri nettle from descriptions he had received, and not from actual observation. Had he actually seen the fibre he must have assigned it a place with the fibres of the North-West Himálaya nettle, with which it is in reality identical. The following description of the method of preparation pursued near Simla, given by **Captain Rainey** and quoted by **Royle**, may be here reproduced, as being the most complete account available: "In August and September, when the plant is in perfection, it can be obtained in any quantity, running from five to six or seven feet in height." "The following is the preparation to which the article is subjected by the natives of the place; but, I doubt not, much of the process might be omitted or simplified:—

"*1st*—Being cut in August or September the weed is exposed for *one* night in the open air.
"*2nd*—The stalk is then stripped of leaves and dried in the sun.
"*3rd*—When well dried it is deposited in an earthen pot which contains water mixed with ashes (the refuse remains of any wood fire) and boiled for four and twenty hours.
"*4th*—The stalk thus boiled is then taken to a stream and well washed.
"*5th*—The hemp is then brought home, and being sprinkled with flour (*atta*) (of the grain called *Koda*) it is again dried in the

FIBRE. 214

FIBRE.

sun, and afterwards spun at any time into cords for nets of every description."

In Garhwál and other localities in the North-West Himálaya, a simpler method appears to be followed. The plant is cut down in the beginning of the cold season, the stalks are washed three or four days in water, and the fibre is stripped off like that of hemp. It is a fine, white, glossy, silky fibre, but is coarser and more brittle than that yielded by *var.* **palmata.** According to **Captain Rainey** it is extensively employed in the preparation of twine for fishing-nets, in consequence of the virtue ascribed to it by the Natives of gaining increased strength by immersion in water, and of resisting decay longer than other fibres. It is also used in Sikkim for the manufacture of a coarse cloth like gunny.

Var. **palmata**—the true Nilghiri nettle—yields a finer and more valuable fibre than **G. heterophylla** proper. **Royle** writes : " The fibre is very long white, soft, and silky, and has been much admired by many of the best judges of fibres. The hill people on the Nilgiri hills prepare the fibre by boiling the twigs. **Dr. Wight** says of it that 'it produces a beautiful fine and soft flax-like fibre, which the Todawars use as a thread material, and if well prepared is fitted to compete with flax for the manufacture of even very fine textile fabrics.'

" At Dundee it was thought a very good fibre, but rather dry. **Mr. Dickson,** who has passed it through his machine and liquid, has rendered it like a beautiful, soft, silky kind of flax, and calls it a wonderful fibre, of which the tow would be useful for mixing with wool as has been done with the China-grass, and the fibre used for the finest purpose." **Dr. Forbes Watson,** speaking of what is apparently this variety, says : " The remarks made with respect to the rough character of the Rheea fibre are still more applicable to those of the Nilghiri nettle. Indeed, so similar to wool is its fibre that when cut short and crumpled up or ' scribbled,' I have known it valued by an experienced broker as wool, and its price stated accordingly. The term ' vegetable wool' which it has already received is, therefore, very suitable." The same writer gives the mean diameter of the ordinary fibre as $\frac{1}{510}$ and the ultimate fibre as $1\frac{1}{300}$ of an inch; and **Cross, Bevan,** and **King** give the following analysis : " Moisture 7·3, ash 1·5, hydrolysis (a) 2·5, (b) 9·7, cellulose 89·6." In Spons' *Encyclopædia,* the Girardinias are spoken of collectively under the name of **G. heterophylla,** but it seems that **G. palmata** alone is meant. The following extract may be found useful : " It succeeds well by cultivation. The bark abounds in fine, white, glossy, strong fibres, which have a rougher surface than those of **Bœhmeria nivea,** and are therefore more easily combined with wool in mixed fabrics." Owing to the high percentage of cellulose and the small loss from hydrolysis, the fibre is chemically one of the best produced in India.

The late **Mr. M'Ivor,** of the Government gardens, Ootacamund, experimented with the Nilghiri plant and submitted a most interesting report to the Madras Government. The following extract from *Drury's Useful Plants* will be found to contain briefly the more important facts from **Mr.** M'Ivor's report :—

**CULTIVA-
TION.
215**

" Cultivation.—The Nilghiri nettle has been described as an annual plant; it has, however, proved, at least in cultivation, to be a perennial, continuing to throw out fresh shoots from the roots and stems with unabated vigour for a period of three or four years. The mode of cultivation, therefore, best suited to the plant, is to treat it as a perennial by sowing the seeds in rows at fifteen inches apart, and cutting down the young shoots for the fibre twice a year,—*viz.*, in July and January. The soil best suited to the growth of this plant is found in ravines which have

The Nilgihri Nettle. (*J. Murray*). **GIRARDINIA heterophylla.**

CULTIVA-
TION.

received for years the deposit of alluvial soils washed down from the neighbouring slopes. In cutting off the first shoots from the seedling crop, about six inches of the stem is left above the ground; this forms 'stools' from which fresh shoots for the succeeding crops are produced After each cutting the earth is dug over between the rows to the depth of about eight inches; and where manure can be applied, it is very advantageous when dug into the soil between the rows with this operation. When the shoots have once begun to grow, no further cultivation can be applied, as it is quite impossible to go in among the plants, owing to their stinging property. The plant is indigenous, growing all over the Nilghiris, at elevations varying from 4,000 to 8,000 feet. This indicates the temperature best suited to the perfect development of the fibre.

" *Produce per acre.*—From the crop of July an average produce of from 450 to 500℔ of clean fibre per acre may be expected. Of this quantity about 120℔ will be a very superior quality; this is obtained from the young and tender shoots, which should be placed by themselves during the operation of cutting. The crop of January will yield on an average 600 or 700℔ per acre; but the fibre of this crop is all of a uniform and somewhat coarse quality, owing to shoots being matured by the setting in of the dry season in December. It might, therefore, be advantageous, where fine quality of fibre only is required, to cut the shoots more frequently, probably three or four times in the year, as only the finest quality of fibres is produced from young and tender shoots.

" PREPARATION OF THE FIBRE.—Our experiments being limited, our treatment of the fibre has been necessarily very rude and imperfect, as in this respect efficient appliances can be obtained only in extensive cultivation.

PREPARA-
TION.
216

" The inner bark of the whole plant abounds in fibre, that of the young shoots being the finest and strongest, while that of the old stems is comparatively short and coarse, though even they produce a fibre of very great strength and of a peculiar silky and wooly-like appearance, and one which no doubt will prove very useful in manufactories.

" For cutting down the crop fine weather is selected; and the shoots when cut are allowed to remain as they fall for two or three days, by which time they are sufficiently dry to have lost their stinging properties; they are, however, pliable enough to allow of the bark being easily peeled off the stems and separated from the leaves. The bark thus taken from the stems is tied up in small bundles and dried in the sun, if the weather is fine; if wet, is dried in an open shed with a free circulation of air. When quite dry, the bark is slightly beaten with a wooden mallet, which causes the outer bark of that in which there is no fibre to break and fall off. The fibrous part of the bark is then wrapped up in small bundles, and boiled for about an hour in water to which a small quantity of wood-ashes has been added in order to facilitate the separation of the woody matter from the fibre. The fibre is then removed out of the boiling water and washed as rapidly as possible in a clear running stream, after which it is submitted to the usual bleaching process employed in the manufacture of fibre from flax or hemp. *Report, April 1862.*" (*Drury's Useful Plants, 225.*)

Var. **zeylanica.**—Little is known regarding the fibre of this variety, although it is used in the Konkan and other parts of Western and South-Western India. It would appear, however, that it is very similar to that produced by the true Nilghiri nettle, described above.

There is no doubt that these fibres are perhaps the strongest, and in many ways most valuable, of any produced in India; a very serious practical difficulty exists, however, against their extensive use in the stinging hairs with which all the varieties are abundantly provided. These cause

| GISEKIA pharnaceoides. | A Valuable Anthelmintic. |

PREPARA-TION.

great annoyance to the persons employed in extracting the fibre, and even after being manufactured into cloth, the irritant property may not be completely destroyed. Indeed, in many cases, it persists to such an extent as to cause great irritation to the person wearing or even touching the cloth.

FOOD.
Leaves.
217

Food.—The LEAVES of G. heterophylla proper are said to be largely used as a vegetable in the hilly tracts of the North-West Provinces.

GIRONNIERA, *Gaud.; Gen. Pl., III., 356.*

218

Gironniera reticulata, *Thw.; Fl. Br. Ind., V., 486;* URTICACEÆ.

Syn.—GIRONNIERA CUSPIDATA, *Kurz;* SPONIA SUBSERRATA, *Kurz;* AP-HANANTHE CUSPIDATA, *Planch;* GALUMPITA CUSPIDATA, *Blume;* CY-CLOSTEMON CUSPIDATUM, *Blume;* HELMINTHOSPERMA GLABRESCENS, *Thwaites, mss.;* CELTIS RETICULATA, *H. f. & T.*

Vern.—*Kho manig;* NILGHIRI HILLS; *Koditáni,* TAM.

References.—*Kurz, For. Fl. Burm., II., 470; Beddome, Fl. Sylv., t. 313; Gamble, Man. Timb., 324; Balfour, Cyclop., I., 1208; Indian Forester, II., 21, 22, III., 23.*

Habitat.—A lofty tree, native of the Sikkim Himálaya, at altitudes of 1,000 to 3,000 feet; also met with in Assam, the Khasia Mountains, Upper Burma, South-Western India, and Ceylon; distributed to Java.

TIMBER.
219

Structure of the Wood. —"Very hard and heavy, a valuable engineering timber" (*Beddome*).

GISEKIA, *Linn.; Gen. Pl., III., 80.*

220

Gisekia pharnaceoides, *Linn.; Fl. Br. Ind., II., 664; Wight, Ic.,*
[*tt. 1167, 1168;* FICOIDEÆ.

Syn.— GISEKIA MOLLUGINOIDES, *Wight;* G. LINEARIFOLIA, *Schum;* PHAR-NACEUM OCCULTUM, *Forsk.*

Vern.—*Manalie kirai, nummnelli kirai,* TAM.; *Isikedunti kúra, isaka dásari kúra,* TEL.; *Aetrilla palla,* SING.

References.—*Kurz, in Journ. Linn. Soc., 1877, pt. II., 111; Elliot, Fl. Andhr., 71; Pharm. Ind., 183; Drury, U. Pl., 227; Lisboa, U. Pl. Bomb., 200; Birdwood, Bomb. Pr., 69; Home Dept. Cor., regarding Pharm. Ind., 240; Gazetteer of the N.-W. P., I., 83; IV., lxxii; Indian Forester, III., 236; Jour. Agri-Hort. Soc. of India (Old Series) IX., 285.*

Habitat.—A glabrous herb found in the Panjáb, Sind, South India, and Ceylon, distributed to Ava, Afghánistan, and Africa.

MEDICINE.
221

Medicine.—The medicinal virtues of this plant were first brought to notice by Captain W. H. Lowther, in the Journal of the Agri-Horticultural Society of India above cited. He claimed for it strong anthelmintic pro-perties, and considered it, when properly administered, a specific for tænia or tape-worm. The treatment is described as follows: " I prefer the administration of the remedy when the plant is forming its seed vessels (all vegetable products being then fullest of their medicinal virtues). An ounce or more of LEAVES, STALKS, and CAPSULES, taken indiscriminately, are ground in a mortar, and sufficient water is added to form a draught. The patient should fast for twelve hours previous to taking the medicine, and three such doses should be given, one every four days. To destroy any latent germs, give for precaution's sake additional doses for two fort-nights following." Captain Lowther's estimate of the drug is very high, and his results with the fresh plant, which he urges must alone be used, since it loses its value on drying, appear to have been good. As yet, how-ever, no medical evidence in favour of the alleged virtues of Gisekia have been adduced, and in the Home Department correspondence, on the ad-visability of bringing out a new edition of the Indian Pharmacopœia, none

Leaves.
222
Stalks.
223
Capsules.
224

of the authorities consulted appear to have recommended the retention of this drug.

Food.—Balfour states that the LEAVES are used by natives in the preparation of *dál*, and Lisboa mentions that in time of famine they are employed as a pot-herb.

MEDICINE.

FOOD.
Leaves.
225

GIVOTIA, *Griff.; Gen. Pl., III., 297.*

Givotia rottleriformis, *Griff.; Fl. Br. Ind., V., 395; Wight, Ic.,* [*t. 1889;* EUPHORBIACEÆ.

226

Syn.—GOVANJA NIVEA, *Wall.*

Vern.—*Vendále, butalli, bulali,* TAM.; *Tella púnki, tella ponuku,* TEL.; *Polki,* MALAY.

References.—*Brandis, For. Fl., 442; Beddome, Fl. Sylv., t. 285; Gamble, Man. Timb., 365; Dalz. & Gibs., Bomb. Fl., 228; Elliot, Fl. Andhr., 178; Lisboa, U. Pl. Bomb., 124; Kew. Off. Guide to the Mus. of Ec. Bot., 118; Indian Forester, III., 204; Bomb. Gaz., XV, 70.*

Habitat.—A small tree of the Dekkan, Mysore, the Eastern Gháts, and Ceylon.

Oil.—The SEEDS yield an oil, which is valuable as a lubricant for fine machinery.

Structure of the Wood.—White, exceedingly light, very soft, but even-grained. Weight 14lb per cubic foot. It is employed for making carved figures, toys, imitation fruit, boxes and other fancy articles; also for catamarans. The Kanára Gazetteer contains the further information that its surface takes paint readily.

OIL.
Seeds.
227
TIMBER.
228

GLASS.

229

Vern.—*Kanch,* HIND.; *Kunnadi,* TAM.; *addannú,* TEL.; *Shíshah,* PERS.; *Kisas,* ARAB.

Glass is a mixture of silicate of potassium or sodium, or of both, with one or more silicates insoluble in water, such as those of the alkaline earths, aluminium, manganese, iron, or lead. The mixture is effected by fusion, which takes place less readily the more silica it contains. Silica for the manufacture of glass is obtained from ground quartz or flint, or from silicious sand, treated with a mineral acid to free it from metallic impurities. The alkali is derived from pearl-ash or wood-ash, carbonate of soda, native or artificially prepared soda, or from other available sources. The necessary insoluble constituent may be obtained from any mineral yielding one of the above-mentioned elements, as desired. India abounds in materials which readily yield these necessary constituents. Perhaps the simplest of these is *reh* which contains soda in the form of carbonate, and a large quantity of silica ready mixed. Notwithstanding the abundance of this, and other glass-forming materials, glass-making in India has not advanced beyond the first and very rudest stage. Too much alkali is employed, and too little heat given, with the not unnatural consequence that the resulting material is a coarse, impure, dirty-coloured mass, full of flaws and air-bubbles, unfitted for any better use than the manufacture of beads, coarse bangles, and other minor and unimportant articles.

One reason of this may probably be found in the fact that glass is very little employed in India for the ordinary purposes for which it is used in other countries. There is very little demand for glass bottles outside the requirements of Europeans; and glass drinking vessels are almost unused by the native population, indeed, by Hindús earthen vessels are preferred on religious grounds. A serious difficulty in the way of the extension of a

HISTORY.
230

GLASS.	Glass and Glass-ware.

HISTORY.

glass making industry in this country is the lack of fuel. Mr. Baden Powell remarks on this subject, "It would probably be cheaper to carry such glass-making materials as are to be found in the Panjáb to the hearths of Staffordshire and bring them back made up into glass than to attempt the manufacture on a large scale here." Evidence, however, exists of glass-making having formerly existed on a much larger scale than it does at present. At the time of the composition of the *Yajur-veda* glass was one of the articles from which the ornaments of females were made. The substance is also mentioned in the *Mahábharata.* In the *Ain-i-Akbari* glass for windows is included in the list given by its author of building materials, and it is said to have cost R1 for 1¼ seer or 4 *dam* for one pane. Abul Fazl in his descriptions of Behar and Agra also mentions glass-making, and writes of Allore, "Here are considerable manufactures of woollen carpets and glass." A glass *guláb* bowl and a *hukka* bowl found in the Muhammadan capital, Bijapur, were shewn by Major Cole, R.E., at one of the Simla Art Exhibitions. These he described as probably of the sixteenth century. They were of thick white glass, cut or moulded in a hexagonal diaper pattern with fluted necks, and of undoubtedly Indian design, though of far superior workmanship to anything produced in this country of late years. Now-a-days, indeed, the glass-making industry is almost entirely confined to a few families in the Lahore, Karnál, Jhelam, and Hoshiarpur districts of the Panjáb; in the Bijnor and Saháranpur districts, N.-W. P.; in Lucknow; in Ahmadnagar, Kaira, and Baroda in Bombay; in Seoni, Central Provinces; in Patna, Bengal; in Jeypore; and in the North Arcot District of Madras. In these localities the glass-makers for the most part, confine their manufactures to rude globes, silvered inside with mercury and tinfoil, small coarse glass toys, rude bottles for attar, and to a greater extent beads and bangles. In Karnál the large, thin, pear-shaped glass retorts or carboys in which the native manufacture of Salammoniac is effected are also prepared.

In some parts of the country, however, the industry appears to have reached a higher development, as will be seen from the following short descriptions taken from the *Journal of Indian Art :* "Very curious coloured glass-ware is made at Patna. The specimens shewn at the Calcutta Exhibition were of considerable excellence. These articles would have an extensive sale if better known, and if proper facilities were afforded to the public for obtaining them." "In Delhi and Lahore glass bangles and lamp chimnies are made; in Karnál, glass globes, pear-shaped glass carboys; and various wares in Hoshiarpur. The art is as yet quite in its infancy. The Hoshiarpur workman is almost the only one of these who works independently with his own materials—independently, that is, of foreign aid—for the few glass-blowers in Lahore collect fragments of white European glass, and melting them down, blow cheap lamp chimnies and bottles. At Karnál the glass globes are made which when silvered inside are broken up into the small mirrors used in *shishadár* ornamental plaster, and run into embroideries known as *shishadár phulkaris.*"

But the following passage is even of greater interest, since perhaps it describes the only branch of the industry worthy of the name of art-manufacture: "Kapadnanj, in the Kaira district, is the only place in the Bombay Presidency where glass is manufactured in its primitive state from a natural earth called *us*, which is a mixture of the Carbonates and Silicates of Soda with several mineral impurities. It is, however, remarkable for its iridescent properties and good colour, resembling the antique Venetian. The shapes are quaint and beautiful. It is said that crude glass of the value of about 3 lakhs of rupees is annually sent to Bombay for foreign export by some Bhoras and Banias, and that it is

Patna ware.
231

Delhi ware.
232
Lahore ware.
233

Kapadnanj.
234

Beads.
235

purified and turned into various shapes in the glass manufactories of Europe. It would be interesting to find out some more definite statistical account of this trade, which, though at present represented by a few pots and bottles, may, if well regulated, develop into an important item of the manufactures of Bombay." With reference to the remark regarding the export of this glass, it may be noticed that the statement is not supported by the Official Trade Returns, which show the value of the exports for the whole of India in 1888-89 to have been only R41,799. "White glass phials and other small articles in various colours, such as cobalt blue, Indian red, marbled and dark green, are made in the School of Art, Jeypore, and by one or two men in the bazár. Glass bracelets or *churis* of different colours are made at Jeypore and in many other places in the State. They are worn by Mussulmans."

About ten years ago endeavours were made by the Department of Agriculture and Commerce to foster and improve the glass-making industry, with the double purpose of utilising the abundant glass-making material available in the "*reh*" lands of Northern India, and of meeting the demand for glass beads from an indigenous source. An engineer was specially deputed to conduct experiments ; beads for patterns, tools, and an account of the methods employed in Venice were obtained, and furnaces constructed on the English pattern were tried.

It was found as the result of these experiments—

(1) That the *reh* was not sufficiently pure to make good colourless window glass.

(2) That the *reh*, when heated in a good furnace, yielded a material very similar to superior bottle glass ; but that the furnace required trained skill both in building and working.

(3) That though good beads could be made, they were much inferior to those obtained from Venice, and that owing to want of skill on the part of the workmen, they could be produced only at a much greater expense. The last result is particularly disappointing, since, as already remarked, beads and bangles are the only form of glass for which a really large demand exists amongst the native population of India.

Trade.—A large and increasing import trade exists in glass. In the year 1888-89, 6,407,266 superficial feet of sheet and plate glass was imported, value R5,61,550; 27,993 cwt. of beads and false pearls, value R19,47,676; 23,848 cwt. of bottles, value R2,32,448, and R38,38,867 of other miscellaneous glass-ware. The total value of glass imports was thus R69,80,541 in comparison with R49,97,005 in the year 1884-85. The sheet and plate glass are obtained chiefly from Belgium and the United Kingdom ; the beads from Italy, Austria, the United Kingdom, and France ; the bottles almost entirely from the United Kingdom, and the miscellaneous glass-ware not included under the above headings from the United Kingdom, China, Belgium, and Austria. As already remarked, the export trade is small, amounting in value to from R29,910 in 1884-85 to R41,810 in 1888-89. Of this R36,956 worth was exported by Bombay, and R11,262 imported by Aden, which appears to be the chief market.

Glazed pottery ; see **Clay,** II., 367.

GLOCHIDION, *Forst. ; Gen. Pl., III., 272.*

A genus of evergreen trees or shrubs belonging to the Natural Order EUPHORBIACEÆ, and comprising about 120 species, chiefly natives of tropical Asia. Of these 55 are Indian ; few are known to be of economic value.

238 | **Glochidion lanceolarium,** *Dalz. ; Fl. Br. Ind., V., 308 ; Wight,*
[*Ic., t. 1905*; EUPHORBIACEÆ.

Syn.—PHYLLANTHUS LANCEOLARIUS, *Muell.*; GLOCHISANDRA ACUMI-
NATA, *Wight*; BRADLEIA LANCEOLARIA, *Roxb.*

Vern.—*Bhoma,* BOMB.

References.—*Roxb., Fl. Ind., 692; Brandis, For. Fl., 453; Kurz, For.
Fl. Burm., II., 343; Beddome, For. Man., 192; Balfour, Cyclop., I.,
1212.*

Habitat.—An evergreen tree from 25 to 30 feet in height, found in the
forests of North-West India from Nepál eastwards to Assam, also in Sylhet
and Chittagong (*Fl. Br. Ind.*). **Beddome** states that it occurs in Malabar,
the Konkán, and South Kanara.

TIMBER, | **Structure of the Wood.**—Hard and durable, employed by the natives
239 | of the Bombay Gháts and Eastern India for house-building.

240 | **G. velutinum,** *Wight; Fl. Br. Ind., V., 322; Wight, Ic., t. 1907-12.*

Syn.—PHYLLANTHUS VELUTINUS, *Muell. Arg.*; P. NEPALENSIS, *Muell.
Arg.*; BRADLEIA OVATA, *Wall.*

Vern.—*Mowa, bakalwa,* N.-W. P.; *Púndna, kalaon, gol kamíla, samá,
bera, amblú, koámil,* PB.; *Kari, koria,* C. P.

References.—*Brandis, For. Fl., 453; Kurz, For. Fl. Burm.,'II., 344;
Beddome, Forester's Man., 195; Stewart, Pb. Pl., 196.*

Habitat.—A small tree or shrub, native of the hot valleys of the Himálaya, Burma, the Khasia Mountains; also the Deccan from the Konkán
to the Nilgiri hills.

TAN, | **Tan.**—The BARK is employed for tanning in the North-Western Himá-
Bark. | laya.
241 | **Structure of the Wood.**—Brownish-white, compact, but soft. Used for
TIMBER, | fuel.
242

GLORIOSA, *Linn.; Gen. Pl., III., 830.*

243 | **Gloriosa superba,** *Linn.; Baker in Linn. Soc. Jour., XVII., 457;*
[*Wight, Ic., t. 2047;* LILIACEÆ.

Syn.—GLORIOSA ANGULATA, *Schum.*; METHONICA SUPERBA, *Lam.*

Vern.—*Kariári, karihári, lánguli, kulhári,* HIND.; *Bishalánguli, ulat-
chandal, bisha,* BENG.; *Siric samano,* SANTAL; *Kúrzhári,* N.-W. P.;
Mulim, kariári, PB.; *Rájahrar,* AJMERE; *Nát-ká-bachhnág,* DEC.;
Karianag, BOMB.; *Nágkaria, indai,* MAR.; *Kalaippaik-kishangu,
kárttikaik-kishangu,* TAM.; *Agni-skikha, kalappa-gadda, Adavi nábhi,
potti dumpa,* TEL.; *Ventóni, mendoni,* MALAY.; *Sima-don, Simmi-dái,
hsee-mee-touk,* BURM.; *Neyangalla,* SING.; *Lángaliká, agnisikhá,
kalikari,* SANS.

References.—*Roxb., Fl. Ind., Ed. C.B.C., 288; Stewart, Pb. Pl., 235; Elliot,
Fl. Andhr., 11, 12; Rev. A. Campbell, Ec. Prod. of Chutia Nagpur, No.
9497; Mason, Burma and Its People, 429, 814; Pharm. Ind., 242; Moodeen
Sheriff, Supp. Pharm. Ind., 147; U. C. Dutt, Mat. Med. Hind., 263,
307; Dymock, Mat. Med. W. Ind., 2nd Ed., 832; S. Arjun, Bomb.
Drugs, 145; Atkinson, Him. Dist., 319, 738; Lisboa, U. Pl. Bomb., 270;
Birdwood, Bomb. Pr., 91; Balfour, Cyclop., I., 1212; Indian Forester,
II., 27; XII., App., 21; Home Dept. Cor. regarding Pharm. Ind., 230, 240,
290; Gazetteers:—Mysore and Coorg, I., 67; II., 7; III., 18; Bombay,
XV., 444; N.-W. P., I., 85; IV., lxxviii.*

Habitat.—A large scandent herb, grasping by the tips of its leaves;
found in the forests of India, Burma, and Ceylon, ascending to 6,000 feet.
It produces a large and very handsome flower, during the rains.

MEDICINE. | **Medicine.**—The ROOT is supposed by Hindu and Muhammadan physi-
Root, | cians to have valuable medicinal properties. Dutt writes, "It consti-
244 | tuted one of the seven minor poisons of Sanskrit writers and had for

Gloriosa.	(*J. Murray.*)	**GLORIOSA superba.**

one of its synonyms '*garbhaghátini,*' or ' the drug that causes abortion,' but I am not aware of its being used as an abortive for criminal purposes. The tuberous root, powdered and reduced to a paste, is applied to the navel, suprapubic region, and vagina, with the object of promoting labour. In retained placenta a paste of the root is applied to the palms and soles, while powdered **Nigella** seeds and long pepper are given internally with wine." Early English writers on Indian botany and materia medica speak of it as a violent poison, but none furnish satisfactory details of a case in which marked ill-effects were produced by its use. It seems highly probable that these ill-effects have been greatly over-estimated, an assumption which is confirmed by experiments recently conducted by **Moodeen Sheriff.** In a special opinion kindly furnished to the editor he writes : " The root is not so poisonous as is generally supposed. I have taken it myself in small quantities, gradually increasing the dose to 15 grains. There were no bad effects, but on the contrary my appetite improved, and I felt distinctly more active and stronger. I have been using it in my practice during the last sixteen or seventeen years, and consider it to be a pretty good tonic and stomachic. Dose from 5 to 12 grains three times daily." In Bombay it is supposed to be an anthelmintic, and is accordingly frequently administered to cattle affected by worms. In Madras it is believed to be specific against the bites of poisonous snakes, and the stings of scorpions, and is also used as an external application in parasitical affections of the skin. **Surgeon-MajorThomson, C.I.E.,** has kindly furnished the following information regarding its utilisation in Madras :—

" There are two varieties of this plant. The root of one plant divides dichotomously, that of the other does not divide at all but appears as a single piece shooting into the ground. The former is supposed by the natives to be the male plant, the latter the female. The male root is gathered during the flowering season, cut up in thin slices and soaked in butter-milk to which a little salt is added. In this composition it is soaked by night and dried by day for four or five days. It is eventually dried well and preserved. By this process its poisonous properties are said to be removed. When so prepared, and administered by giving a piece or two internally in a case of cobra bite, it is said to be an effectual antidote in cobra poisoning. It is called in Tamil '*Katharum cheddy.*' In scorpion and centipede stings and bites relief is obtained from the pain by applying a paste of the root rubbed up with cold water and then warming the part affected over the fire. This paste is applied also for parasitic affections of the skin."

The starch obtained from the root by washing is given internally in gonorrhœa.

Notwithstanding its characteristic appearance, the tuber is occasionally employed by natives as an adulterant of the roots of **Aconitum ferox,** to which, indeed, they believe it to be closely allied in therapeutical properties.

Physical characters and Chemical composition.—The root, flattened or cylindrical, sometimes much pointed at both ends, sometimes consisting of two tubers uniting at right angles. On the upper surface may be seen a circular scar marking the point of origin of the stem, and on the under-surface beneath this another mark to which thin small rootlets are frequently left attached. Covering the tubers is a thin, loose, and wrinkled epidermis of a brownish-gray or pale-brown colour, and on removing this skin, a brown or dark-brown surface is exposed. On cutting the tuber it is found to be dull-white and farinaceous internally. The taste is faintly bitter, the odour slightly acrid. A chemical examination by **Dr. Warden** resulted in the separation of two resins and a bitter principle, *superbine,* which the analyst considered closely allied to, if not identical with, that of **Scilla maritima** (*Dymock*).

MEDICINE.
Root.

Starch.
245

CHEMISTRY.
246

G. 246

GLUTA tavoyana.	Glossocardia ; Glossogyne ; Gluta.

GLOSSOCARDIA, *Cass. ; Gen. Pl., II., 384.*

247

Glossocardia linearifolia, *Cass. ; Fl. Br. Ind., III., 308 ; Wight,*
[*Ic., t. 1110 ;* COMPOSITÆ.

Syn.—GLOSSOCARDIA BOSVALLIA, *DC. ;* VERBESINA BOSVALLIA, *Linn. f. ;*
V. BOSWELLIA, *Roxb. ;* PECTIS MEIFOLIA, *Wall.*
Vern.—*Seri,* HIND ; *Pithapra, phatursuva,* BOMB. ; *Pitta-pápada,* POONA ;
Páthara-suva, MAR. ; *Parapalanam,* TEL. ; *Pithari,* SANS.
References.—*Roxb., Fl. Ind., Ed. C.B.C., 607 ; Dals. & Gibs., Bomb. Fl.,*
129 ; Dymock, Mat. Med. W.Ind., 2nd. Ed., 433 ; Lisboa, U. Pl. Bomb.,
200 ; Gazetteers :—Bomb., XV., 436 ; Mysore and Coorg, I., 56 ; N.-W.P.,
I., 82.
Habitat.—A branched, glabrous annual herb ; native of Rohilkhand,
Banda, Central India, and the Deccan.

MEDICINE.
248

Medicine.—**Dymock** states that this plant is employed medicinally by
the druggists of Poona, and **Dalzell** and **Gibson** mention that it is " much
used in female complaints," the nature of which, however, they do not
specify.

FOOD,
Leaf,
249

Food.—Lisboa includes this in his list of Famine Plants, and writes,
".The LEAF is said to be eaten in ordinary years as a vegetable, and is believed to be perfectly wholesome."

GLOSSOGYNE, *Cass. ; Gen. Pl., II., 288.* [POSITÆ.

250

Glossogyne pinnatifida, *DC. ; Fl. Br. Ind., III., 310 ;* COM-

Syn.—BIDENS RIGIDA, *Hort. Calc. ;* ZINNIA BIDENS, *Retz. ;* BIDENS PIN-
NATIFIDA, *Heyne.*
Vern.—*Barangom bir barangom,* SANTAL.
References.—*Roxb., Fl. Ind., Ed. C.B.C., 604 ; Dals. & Gibs., Bomb. Fl.,*
129 ; Rev. A. Campbell, Ec. Prod., Chutia Nagpur, Nos. 7541, 8424 ; N.-
W. P. Gazetteer, I., 82 ; IV., lxxiii.
Habitat.—A perennial, glabrous herb of the plains of India from Jammu
and Garhwál to Western Bengal and Behar, and southwards to Madras.

MEDICINE,
Root.
251

Medicine.—The Rev. A. **Campbell** states that a preparation from the
ROOT is employed by the Santals as an application to snake-bite and scor-
pion-sting.

GLUTA, *Linn. ; Gen. Pl., I., 421.*

252

Gluta elegans, *Wall. ; Fl. Br. Ind., II., 22 ;* ANACARDIACEÆ.

Vern.—*Thayet-thitsé, khye,* BURM.
References.—*Kurz, For. Fl. Burm., I., 310 ; Prelim. For. Rep. on Pegu,*
App. A., xli. ; Gazetteer, Burma, I., 136.
Habitat.—A small evergreen tree found along the coast of Tenasserim.

DYE.
Wood.
253

Dye.—**Kurz** mentions that the WOOD is used in Burma for dyeing, yielding, with different mordants, various shades of colour from orange to black.
In the *Burma Gazetteer* the colours obtained are described as follows :
"With—1, muriate of tin,—three shades of orange varying with the temperature of the bath and the time of immersion ; 2, acetate of alumina,—
two shades of flame colour ; 3, acetate of iron,—two shades of drab ;
4, acetate of iron, with a weak solution of galls,— a fine black of two
shades."

TIMBER.
254

Structure of the Wood.—" Good for furniture, and when steeped in
ferruginous mud turns jet black, looking like ebony. Used also for building purposes, boxes, &c." (*Kurz*).

255

G. tavoyana, *Wall. ; Fl. Br. Ind., II., 22.*

Vern.—*Thayet-thitsé,* BURM. ; *Ohay, thúmay,* KAREN.
References.—*Kurz, For. Fl. Burm., I., 309 ; Conference on Timbers, Col.*
& Ind. Exhib., July 26th, 1886, p. 2.

G. 255

| Gluta; The Manna Grass; Soy Bean. (*J Murray.*) | GLYCINE. |

Habitat.—A small evergreen tree of the coast of Tenasserim from Tavoy southwards.

Structure of the Wood.—Heart-wood bright-dark-red, close-grained, not so mottled with dark and light streaks as that of **G. travancorica.** When seasoned it floats, and is very durable though brittle. Specimens of the wood of this species and of **G. travancorica** were shown at the Conference on timbers held in connection with the Colonial and Indian Exhibition of 1886, but neither appears to have attracted favourable attention, though their merits were urged by the Indian officials present.

TIMBER,
256

Gluta travancorica, *Beddome; Fl. Br. Ind., II., 22.*

257

Vern.—*Shen kurani, shen curungi,* TINNEVELLY.
References.—*Beddome, Fl. Sylv., t. 60; Gamble, Man. Timb., 109; Indian Forester, III., 22, 23.*
Habitat.—A large evergreen tree abundant in the dense, moist forests of the Tinnevelly and Travancore Gháts.

Structure of the Wood.— Sapwood light-reddish-grey, heartwood, dark-red, very hard and close-grained, beautifully mottled with dark and light streaks. Weight 40℔ (*Beddome*) 46 to 58℔ (*Gamble*) per cubic foot. Gamble remarks, "This wood is little used, but its splendid colour and markings should bring it to notice as a valuable wood for furniture. It seems to season well, and works and polishes admirably."

TIMBER.
258

Gluten of wheat; see Triticum sativum, *Lam.*; GRAMINEÆ.

GLYCERIA, *R. Br.; Gen. Pl., III., 1197.*

Glyceria fluitans, *R. Br.; Duthie, Indigenous Fodder Grasses of the Plains of the N.-W. P., 41;* GRAMINEÆ.

259

MANNA GRASS.

Syn.—FESTUCA FLUITANS, *Linn.*; POA FLUITANS, *Scop.*
References.—*Baron Ferd. von Müeller, Select Extra-tropical Plants, 324; Trans. Agri-Hort. Soc. of India, VIII., 98; Smith, Dic., 265.*
Habitat.—A perennial grass with tender foliage, met with in the Baspa Valley and Pángi. It delights in stagnant water, ditches, pools, ponds, and slow flowing streams, covering their surface.

Food and Fodder.—The FOLIAGE is sweet, tender, and much liked by cattle. The SEEDS are used for food, in many countries, being cooked as a sort of porridge.

FOOD and
FODDER,
Foliage,
260
Seeds.
261

GLYCINE, *Linn.; Gen. Pl., I., 530.*

262

A genus of twining or sub-erect herbs, belonging to the Natural Order LEGUMINOSÆ, and comprising about 12 species; distributed throughout the tropics of the Old World, especially Australia. Of these two are natives of India and a third extensively cultivated. It has been customary to speak of the Soy Bean of India as **Glycine Soja.** Maximowicz accepts **G. Soja,** *Sieb. et Zucc.,* as the wild form of the plant (**G. ussuriensis,** *Regel et Maack*)—a native of Japan and China, and reduces the cultivated state to a variety (=**Soja hispida,** *Mœnch,* **Dolichos hispida,** *Thbg.*). Forbes and Hemsley, in their enumeration of Chinese plants (*Journ. Linn. Soc., Vol. XXIII., 188*), accept these two forms as species under the names and synonyms given above. The cultivated plant differs chiefly from the wild, in its greater degree of hairiness, more erect stem, and larger legumes. Reference having been made to the authorities of the Calcutta Herbarium or the subject of **G. Soja,** *Sieb. et Zucc.,* being, as shown in the *Flora of British India,* a native of this country, **Dr. Prain** kindly went into the subject very carefully. He writes, "We have not, from any part of India, any specimens of **G. Soja** proper. The Khasia hill plant is more erect, more hispid, and has larger legumes than the Himálayan, and indeed resembles

The Soy Bean.

G. hispida, *Maxim.*, quite as much as it does the Indian cultivated " G. Soja," which, indeed, it connects with G. hispida. It is, in fact, the plant most like the wild G. Soja, *S. et Z.*, which no one ever professes to have found wild in India, while it is also the one most like G. hispida, *Maxim.* (which has never been found wild anywhere). It is the plant collected by Dr. Watt and myself in the Naga hills."

The writer noted on his Naga hill specimens that they were found in a semi-wild state, and that the plant was known to the Angami Nagas as *Tsu Dza*, a name not unlike Soja. Throughout India, the Soy Bean is cultivated, black and white-seeded forms being met with, which vary to some extent, but all preserve the specific characters of G. hispida. Plants raised at Saharanpur from Japanese seed have larger and broader leaves than the usual Indian forms. The fact that this cultivated plant possesses, even among the aborigınal tribes, names which are original, *i.e.*, in no way modern derivatives, points to an ancient cultivation, if, indeed, it may not be accepted as an indication of its indigenous nature. (*Editor.*)

[*et Zucc.* ; LEGUMINOSÆ.

263

Glycine hispida, *Maxim.* ; *Fl. Br. Ind., II., 184,* under G. Soja, *Sieb.*

THE SOY BEAN.

Syn.—DOLICHOS SOJA, *Linn.* ; SOJA HISPIDA, *Mœnch* ; S. ANGUSTIFOLIA, *Miq.*

Vern.—*Bhat, bhatwan, ram kurthi,* HIND. ; *Bhut,* PUNJ. ; *Gari-kulay,* BENG. ; *Hendedisom horec,* (black-seeded), *Pond disom, horec* (white-seeded variety), SANTAL ; *Tsu-dza,* NAGA. ; *Bhatnas, bhatwas,* NEPAL ; *Seta, kala botmas,* PARBAT. ; *Musa, gya,* NEWAR ; *Khajuwa,* EASTERN TERAI ; *Bhu t,* KUMAUN.

References.—*Roxb., Fl. Ind., Ed. C.B.C., 563 ; Stewart, Pb. Pl., 76 ; DC., Origin Cult. Pl., 330 ; Campbell, Ec. Prod., Chutia Nagpur, Nos. 8156, 8158 ; Atkinson, Him. Dist., 309, 696 ; Buchanan-Hamilton, Acct. of Nepal, 228 ; Church, Food Grains of India, 140 ; Spons' Encyclop., 1378, 1814 ; Smith, Dic., 386 ; Kew Reports, 1882, 42 ; Kew Off. Guide to the Mus. of Ec. Bot., 43 ; Trop. Agri., I., 567, IV., 695 ; Agri. Rep. Assam, 1882-83, No. 37 ; Special Reports, Director, Land Rev. and Agri., Bengal ; Rep. of Proc. of Rev. and Agri. Dept., 1882, 2 to 12 ; 1883, 1 to 7.*

Habitat.—Extensively cultivated throughout India and in Eastern Bengal, Khásia hills, Manipur, the Naga hills, and Burma, often found as a weed on fields or near cultivation.

Oil.—Large quantities of the SEED are annually used by the Chinese in the manufacture of an edible oil. " It is said that they obtain 17 per cent. of oil, by simple pressure. It bears a general resemblance to the ordinary edible oils of commerce, possessing an agreeable flavour and odour. It is useful for burning ; exposed to a low temperature it becomes pasty, and oxidizes rapidly on exposure to the air. As a drying oil it might replace linseed for some purposes. As an illuminator it is being rapidly replaced by American petroleum, but is still extensively used for food. It is an important article of Chinese commerce " (*Spons' Encyclopædia, 1378*).

Medicine.—A decoction of the ROOT is said to possess astringent properties.

Food and Fodder.—The Soy-bean forms an important article of food in China and Japan. Since 1873, it has been successfully grown in the warmer parts of Europe. It is also widely spread, in a cultivated state, over a great part of the Himálaya and the plains and lower hills of India. On the plains the crop is generally grown by itself, as a *kharif* crop ; the seeds are sown from June to September, and the harvesting takes place from November to January. Church gives the following information regarding the best methods of cultivation : " The seeds should be placed at a depth not

The Soy Bean.	(*J. Murray*.)	**GLYCINE hispida.**

exceeding 1 to 1½ inch; 18 plants may be left after weeding to the square yard. A peaty soil, or one rich in organic matter, suits the plant best; a calcareous soil is also favourable to its growth. Sulphate of potash is a good manure, nitrogen may be supplied either as nitrate of soda, or in the case of soils poor in organic matter, in the form of rape or mustard cake, but it is rarely needed, while large applications of nitrogenous manure exert a distinctly injurious effect upon the yield of beans. So far as we know this very important, vigorous, and productive pulse is not attacked by any insect or parasitic fungus." Two chief varieties of the cultivated Soy occur in India, one called " white," the other " black," but they are not distinguished by definite characters in chemical composition nor in properties.

Precise information cannot be given regarding the area under this crop in the various provinces of India. Attempts have been made by Government to extend its cultivation in Assam, but apparently without success. In 1882, **Professor Kinch** urged the advisability of renewed efforts in the Himálayan tracts, and, as a consequence, the Government of India directed the attention of local officials to the subject. Seed obtained from the Government Gardens, Saharanpur, were distributed to Madras, the Panjáb, Bengal, Bombay, Hyderabad, and Burma, for experimental cultivation. It appears to have been grown from seed obtained from China with a fair amount of success at the Saidapet Experimental Farm in 1882.

CHEMISTRY.—The chemical composition of the bean, according to **Professor Kinch**, places it above all other pulses as an albuminous food, while that of the straw also surpasses in nitrogenous value that of wheat, lentils, and even hay. The following composition is given by **Professor Church** : " In 100 parts of the bean, water 11, albumenoids 35·3, starch and sugar 26, fat 18·9, fibre 4·2, ash 4·6. The nutrient ratio is here about 1 : 2, while the nutrient value is 105."

The BEAN is eaten in India in the localities where cultivated. The Rev. A. Campbell states that in Chutia Nagpur it is generally used roasted and ground as *satú*, or simply roasted in the form of *átá*. In other parts of the country it is also eaten in the form of *dál*. In China and Japan three preparations are made from the soy-bean, namely, soy-sauce, soy-cheese, and a kind of paste, the two last of which are manufactured by crushing and pressing the seeds. The following description of the composition and preparation of the sauce is given in *Spons' Encyclopædia* :—" This useful condiment, said to form the basis of almost all the popular sauces made in Europe, is prepared by the Chinese and Japanese, by boiling the beans with an equal quantity of roughly-ground barley or wheat, and leaving it covered for 24 hours to ferment; salt is then added in quantity equal to the other ingredients; water is poured over, and the whole is stirred at least once daily for two months, when the liquid is poured and squeezed off, filtered, and preserved in wooden vessels, becoming brighter and clearer by long keeping. Its approximate value in the London market is 2s. 3d. to 3s. a gallon for Chinese and 2s. 4d. to 2s. 5d. for Japanese. It is not specified in the trade returns, but doubtless forms the chief item of the unenumerated species imported from China." As already mentioned, the OIL is extensively used in China and Japan as an article of food, and the cake left after the expression of the oil is also eaten by the poorer classes.

The soy-bean is an extremely valuable fodder-plant. If cut just when the pods are fully formed it makes most nutritious hay; and the residual cake above-mentioned, which contains, according to Church, 40 per cent. of flesh-forming materials and 7 per cent. of oil, is an extremely rich cattle-food.

Area.
267

Chemistry.
268

Bean.
269

Oil.
270

GLYCOSMIS, *Correa ; Gen. Pl., I., 303.*

271

Glycosmis pentaphylla, *Correa ; Fl. Br. Ind., I., 499 ;* RUTACEÆ.

Syn.—GLYCOSMIS CHYLOCARPA, *W. & A. ;* G. ARBOREA, *DC. ;* G. RETZII, *Roem. ;* LIMONIA PENTAPHYLLA, *Retz ;* L. ARBOREA, *Roxb. ;* MYXOSPERMUM CHYLOCARPUM, *Roem.*

Vern.—*Ban nimbu, potali, pilrupotala, girgitti, ban nimbu,* HIND. *; Ashshoura,* BENG. *; Kirmira,* BOMB. *; Kirmira menki,* GOA *; Gonji pandu, golugu, konda golugu,* TEL. *; Gúroda,* KAN. *; Tanshouk,* BURM.

References.—*Roxb., Fl. Ind., Ed. C.B.C., 364 ; Kurz, For. Fl. Burm., I., 185, 186 ; Beddome, Fl. Sylv. Anal. Gen., XLIII., t. 6, 66 ; Bedd. in Trans. Linn. Soc., XXV., 211 ; Gamble, Man. Timb., 59 ; Thwaites, En. Ceylon Pl., 45, 406 ; Dals. & Gibs., Bomb. Fl., 29 ; Elliot, Fl. Andhr., 61, 95 ; Atkinson, Him. Dist., 307 ; Lisboa, U. Pl. Bomb., 149, 274 ; Atkinson, Ec. Prod. of N.-W. P., Pt. V., 49 ; Indian Forester, X., 315, 325 ; XIV., 390 ; Gazetteers :—Mysore and Coorg, I., 169 ; N.-W. P., IV., lxix ; Bomb., XV., Pt. I., 429.*

Habitat.—A common evergreen shrub throughout the Tropical and Sub-tropical Himálaya, ascending to 7,000 feet in Sikkim. It extends from the Sutlej river in the North-West, eastwards and southwards to Upper Assam, Burma, Travancore, Malacca, and Ceylon.

MEDICINE.
Roots.
272

Medicine.—Mr. T. N. Mukharji states that the ROOTS pounded and mixed with sugar are given in cases of low fever by Native practitioners. Lisboa mentions that the WOOD bruised with water is administered internally as an antidote for snake-bite.

Wood.
273
FOOD.
Fruit.
274

Food.—The FRUIT, a white berry, about the size of a large pea, is commonly eaten.

Structure of the Wood.—White, hard, close-grained.

TIMBER.
275
DOMESTIC.
Twigs.
276
Leafy twigs.
277

Domestic Uses.—TWIGS used by the Bengalis to clean the teeth. The LEAFY TWIGS are in some of the rural parts of Bengal stuck into the walls and roofs of huts about the beginning of April to ward off lightning (see also **Euphorbia antiquorum**, p. 295).

GLYCYRRHIZA, *Linn. ; Gen. Pl., I., 508.*

278

Glycyrrhiza glabra, *Boiss. ; Fl. Or., II., 202 ; Linn. ;* LEGUMINOSÆ.

LIQUORICE ROOT.

Vern.—*Mulhatti, jethi-madh,* extract=*jathimadh-ká-ras, mulatthi-ká-ras,* HIND. *; Jashtimadhu, jai-shbomodhu,* BENG. *; Muraiti-ka-jur,* BEHAR *; Mulethi,* N.-W. P. *; Mitthi-lakri,* DEC. *; Bazar root=aslasús, jetimadh, muleti,* extract=*rabésús,* PB. *; Zaisi, makh, sús,* AFG. *; Jashti madhu,* BOMB *; Jéshtá madha,* MAR. *; Jethi madha,* GUZ. *; Anti-ma-duram, ati-maduram,* extract=*ati-maduram-pál,* TAM. *; Yashti-madhukam, ati-madhuramu,* extract=*yashti-maduram-pálu,* TEL. *; Yashti-madhuká, ati-madhurá,* KAN. *; Yashti-madhukam, ati madhuram, iratti-madhuram,* MALAY. *; Noe-khiyu, noe-khiyu-asui,* BURM *; Ati-maduram, velmí,* SING. *; Madhuka, yashti madhu, madhu-yashtikam,* SANS. *; Aslussús,* extract=*rubbussús,* ARAB. *; Bikhe-mahak,* extract=*asus, rob-a-sus, ausá-rahe-mahak,* PERS.

References.—*Stewart, Pb. Pl., 69 ; Aitchison, Botany of Afgh. Del. Comm., 56 ; Mason's Burma and Its People, 502 ; Pharm. Ind., 75 ; Ainslie, Mat. Ind., I., 199 ; O'Shaughnessy, Beng. Dispens., 293 ; Moodeen Sheriff, Supp. Pharm. Ind., 148 ; U. C. Dutt, Mat. Med. Hind., 143, 324 ; Dymock, Mat. Med. W. Ind., 2nd Ed., 244 ; Fleming, Med. Pl. and Drugs, as in As. Res., Vol. XI., I., 168 ; Flück. & Hanb., Pharmacog., 179 ; Bent. & Trim., Med. Pl., 74 ; S. Arjun, Bomb. Drugs, 41 ; Murray, Pl. and Drugs, Sind, 117 ; Med. Top. Ajmir, 146 ; Irvine, Mat. Med. Patna, 64 ; Baden Powell, Pb. Pr., 340 ; Birdwood, Bomb. Pr., 29 ; Buck, Dyes and Tans, N.-W. P., 44 ; Liotard, Dyes, 136 ; Smith, Dic., 247 ; Kew Off. Guide to the Mus. of Ec. Bot., 41 ; Report on the Settlement of the Hardoi District, Oude, 15 ; Indian Forester, XIII., 93.*

| Liquorice Root. | (*J. Murray.*) | **GLYCYRRHIZA glabra.** |

Habitat.—A perennial herb, of South Europe, Asia Minor, Armenia, Siberia, Persia, Turkistan, and Afghánistan. It is cultivated in Italy, France, Russia, Germany, Spain, and China, also to a small extent in England. Though neither wild nor cultivated in India it is an import of some consequence, and has been employed for medicine and in dyeing for many years. The root used in medicine is principally derived from two varieties, namely:—α typica, and γ glandulifera (*Boissier*).

Dye.—The wood, imported through the Panjáb from Afghánistan, is, in the North-West Provinces, employed in calico-printing, to perfume the fabric and give it a finish (*Sir E. C. Buck*).

DYE.
Wood.
279

Medicine.—Liquorice root has been used in Hindú medicine from a very remote period. U. C. Dutt states that "it is mentioned by Susruta, and is described as sweet, demulcent, cooling, and useful in inflammatory infections, cough, hoarseness, thirst, &c. It is much employed for flavouring medicinal decoctions, oils, and *ghritas*, and enters into the composition of numerous external cooling applications, along with red sandal wood, madder, **Andropogon muricatus**," &c. The drug also possesses a wide reputation in the works of Arabic and Persian physicians. Thus Dymock writes: "The author of the *Makhzan-el-Adwiya* gives a lengthy description of the plant, and directs the root to be decorticated before it is used. He says that the Egyptian is the best, next that of Irak, and then Syrian. The root is considered hot, dry, suppurative, demulcent, and lenitive, relieving thirst and cough, and removing unhealthy humours, also diuretic and emmenagogue, useful in asthma and irritable conditions of the bronchial passages. **Sheik-el-Ráis** recommends the decoction in cold colic; it is also dropped into the eyes to strengthen the sight. A poultice made of the leaves is said to be a cure for scald-head and stinking of the feet or arm-pits. **Muhammad bin Ahmad**, and **Yohanna bin Serapion** recommend the seeds as being the most active part of the plant, but remark that they are only produced in certain climates (*e.g.*, Basra)." In Europe also the medicinal value of Liquorice has long been known. It is unquestionably alluded to by **Theophrastus**, and by **Dioscorides**, who calls the plant γλυκιρρίζη; also by several Roman writers (**Ceerus**, **Scribonius**, **Largus**, and others), who describe it under the name of RADIX DULCIS. It appears to have originally enjoyed a reputation chiefly as a demulcent and sedative in diseases of the respiratory tract.

MEDICINE.
Root.
280

Leaves.
281
Seeds.
282

Characters and Chemical Composition.—The root occurs in long cylindrical branched pieces, an inch or less in diameter, tough and pliable; externally of a greyish-brown colour, yellow internally, with a somewhat disagreeable, earthy odour, and a sweet, mucilaginous, somewhat acrid taste (*Indian Pharm.*). Flückiger and Hanbury describe it as containing, in addition to sugar and albuminous matter, a peculiar sweet substance named *Glycyrrhizin*, which is precipitated from a strong decoction by the addition of an acid or a solution of cream of tartar, or by neutral or basic acetate of lead. When washed with dilute alcohol and dried, *Glycyrrhizin* is found to be an amorphous yellow powder with a strong bitter sweet taste, and an acid reaction. With hot water it forms a solution which gelatinizes on cooling, does not reduce alkaline tartrate of copper, is not fermentable, and does not rotate the plane of polarization. Gorup-Besanez (1876) found its composition to be represented by the formula $C_{16}H_{24}O_6$. By boiling with a dilute mineral acid a resinous, amorphous, bitter substance named *Glycyrretin*, the composition of which is undetermined, and an uncrystallizable sugar, are obtained. Other chemists have found *asparagin* and malic acid in the root, and the presence of starch, and a small amount of tannin in the outer layers is easily demonstrated.

Chemical
Composition.
283

Action and Uses.—Liquorice and its preparations are, in European

Action and
Uses.
284

| GMELINA arborea. | Liquorice; Gmelina. |

MEDICINE.

medicine, chiefly used for pharmaceutical purposes. It disguises the taste of many nauseous drugs, particularly senna, aloes, chloride of ammonium, senega, hyoscyamus, turpentine, and bitter sulphates, and is also, when powdered, a useful basis for pills. It has a pleasant taste, and, when slowly chewed or sucked, increases the flow of saliva and mucus. It is also a popular demulcent, and is largely employed to relieve sore-throats and coughs. It is used by Native practitioners as a demulcent in catarrh of the genito-urinary passages, and as a slight laxative.

TIMBER.
285

Structure of the Wood.—Bright yellow, tough and fibrous. Dymock writes : " In Persia glass bottle-makers use the wood for melting their materials, as they say it gives a greater heat than any other kind of fuel."

TRADE.
286

Trade.—The chief supply of the root in India is obtained from the Persian Gulf and Karáchi, and of the wood, for dyeing, from Afghánistan, *viâ* the Panjáb. Dymock states that the kind known as Karáchi liquorice is the best, and fetches from R50 to 80 per kandy of 5 cwt. Ordinary Persian liquorice is smaller, and not so sweet.

GMELINA, *Linn.; Gen. Pl., II., 1153.*

A genus of trees or shrubs belonging to the Natural Order VERBENACEÆ, and comprising eight species, of which five are natives of India.

287

Gmelina arborea, *Linn.; Fl. Br. Ind., IV., 581 ; Wight, Ic., t. 1470 ;*
[VERBENACEÆ,

Syn.—GMELINA RHEEDII, *Hook.;* PREMNA ARBOREA, *Roth. ;* P. TOMENTOSA, *Miq.*

Vern.—*Kumbhár, gumbhár, gamhar, gámbhár, khammara, kambhar, kúmár, gambari, sewan, shewan, gamári, khambhári,* HIND. ; *Gamari, gúmár, gúmbar,* BENG. ; *Gambari,* URIYA ; *Gumher, kasamar,* KOL. ; *Kasmár,* SANTAL ; *Gomari,* ASSAM ; *Gambari,* NEPAL ; *Numbon,* LEPCHA ; *Gumai,* CACHAR ; *Bolko bak,* GARO ; *Kurse,* GOND ; *Kúmhár, Gúmhár,* Bazár fruit=*kákódúmbári,* PB. ; *Sewan,* HAZARA ; *Kássamar,* KURKU ; *Gúmbhar, shíwun,* C. P. ; *Shewun,* BOMB. ; *Shewan, shivan,* MAR. ; *Chimman, sag,* BHIL ; *Gumudu téku, teggummadu, kasmaryamu, gumadi, cummi,* TAM. ; *Gúmar-tek, pedda gomru, tagumúda, gumudu, pedda gumudu téku, gumudu téku,* TEL. ; *Kasmiri, kuli, shewney, shivani,* KAN. ; *Kumbulu,* MALAY. ; *Ramani,* MAGH. ; *Yumanai, yémené, kyúnboc, kywon-pho,* BURM. ; *At-demmata,* SING. ; *Gumbhari, sripnari, Kásmari,* SANS.

References.—*Roxb., Fl. Ind., Ed. C.B.C., 486 ; Brandis, For. Fl., 364 ; Kurz, For. Fl. Burm., II., 264 ; Beddome, Fl. Sylv., t. 253 ; Gamble, Man. Timb., 295 ; Thwaites, En. Ceylon Pl., 244 ; Dals. & Gibs., Bomb. Fl., 201 ; Stewart, Pb. Pl., 166 ; Elliot, Fl. Andhr., 65, 88, 148, 174 ; Mason, Burma and Its People, 526, 793 ; Rev. A. Campbell, Ec. Prod., Chutia Nagpur, No. 9245 ; O'Shaughnessy, Beng. Dispens., 486 ; U. C. Dutt, Mat. Med. Hind., 218, 297, 304 ; Dymock, Mat. Med. W. Ind., 2nd Ed., 599 ; S. Arjun, Bomb. Drugs, 105 ; Baden Powell, Pb. Pr., 365, 581 ; Atkinson, Him. Dist., 315, 738 ; Drury, U. Pl., 228 ; Lisboa, U. Pl. Bomb., 107, 168 ; Birdwood, Bomb. Pr., 334 ; Balfour, Cyclop., I., 213 ; Treasury of Bot., I., 538 ; Aplin, Rep. on Shan States, 1887-88 : For. Adm. Rep., Chutia Nagpur, 1885, 6, 33 ; Buchanan, Statistics of Dinajpur, 151 ; Agri-Hort. Soc. of India, Journals (Old Series), VI., 26 ; VIII., Sel., 177 ; IX., 252, Sel., 44 ; XIII., 307 ; (New Series), VII., 276 ; Indian Forester, II., 19, 23 ; V., 190 ; VI., 101 ; VIII., 127, 128, 414, 438 ; IX., 238, 359, 607 ; X., 222, 325 ; XI., 354 ; XII., App. 19 ; XIII., 121 ; Gazetteers :—Mysore and Coorg, I., 48 ; Rajputana, 25 ; N.-W. P., IV., lxxvi ; Bombay, VI., 14 ; VII., 32, 36 ; XIII., 27 ; XV., 70 ; XVII., 26 ; XVIII., 52 ; Orissa, II., 179, App. VI. ; Settlement Reports :—Central Provs. :—Raipore District, 75 ; Chanda, App. VI. ; Manual of the Coimbatore District (Madras), 407.*

Habitat.—A large deciduous tree, sometimes attaining the height of 60 feet, met with in the Sub-Himálayan tract from the Chenab eastwards, also throughout India, Burma, and the Andaman Islands. **Mr. C. B. Clarke,**

in the *Flora of British India*, describes a variety—**glaucescens**,—which differs from the type species in having its leaves glaucous beneath, often nearly glabrous in the mature state. It is a native of the Sub-tropical Himálaya and the Khásia Mountains, at altitudes up to 2,000 feet.

Dye.—The Rev. A. Campbell states that the WOOD-ASHES and FRUIT are employed as dyes by the Santals. This fact is of interest, as the writer can find no reference to their being similarly utilised in other parts of India.

DYE.
Wood-Ashes.
288
Fruit.
289

Medicine.—The ROOT has long been an article of medicine with the Hindús. It is described as bitter, tonic, stomachic, laxative, and useful in fever, indigestion, anasarca, and various other complaints. U. C. Dutt writes: "It is an ingredient of *dasamula*" (a compound decoction of ten plants,—**Desmodium gangeticum, Tribulus terrestris**, and others) "and is thus much used in a variety of diseases. **Bangasena** says that *gambhári* root, taken with liquorice, honey, and sugar, increases the secretion of the milk. The FRUIT is sweetish, bitter, and cooling, and enters into the composition of several refrigerant decoctions for fever and bilious affections." The Kanára Gazetteer contains the information that the root, fruit, and BARK are all used medicinally in that district, and **Dymock** states that, in Bombay, the juice of the young LEAVES is used as a demulcent in gonorrhœa, cough, &c., either alone or combined with other drugs of similar properties. In other parts of India the root and fruit appear to be the parts generally employed medicinally, and, in Northern India, the former is believed to have anthelmintic properties.

MEDICINE.
Root.
290

Fruit.
291
Bark.
292
Leaves.
293

Food and Fodder.—This species flowers in the beginning of the hot season, and produces a FRUIT in April and May which is eaten by the Gonds and other hill tribes. The LEAVES are used as fodder, and are also much browsed by deer and other wild animals.

FOOD.
Fruit.
294
FODDER.
Leaves.
295

Structure of the Wood.—Yellowish, greyish or reddish-white, with a glossy lustre, close and even-grained, soft, strong, does not warp or crack in seasoning, weight from 28 to 35℔ per cubic foot, breaking weight of a bar, 6 feet × 2 inch × 2 inch, 580℔ (according to **Baker**). It is light, has a good surface, is very durable, is easily worked, and takes paint and varnish readily, and is, therefore, highly esteemed for planking, furniture, carriages, boat-decks, panelling and ornamental work of all kinds (*Gamble*). **Mason** states that it is largely employed by the Karens for canoes, and by the Burmans for clogs. Owing to its extreme durability, it has been recommended as an excellent timber for making tea-boxes, and has also attracted much attention as a very suitable wood for furniture, picture-frames, and similar work in which shrinking and warping have to be avoided. **Buchanan** states in his Statistics of Dinajpur that "it is much employed by the natives for making their instruments of music." The excellence of this timber for many purposes appears to have been first noticed and described by **Roxburgh**, who subjected it to various experiments which he describes as follows: "One of the experiments, and the most interesting, was made by placing part of an outside plank in the river, a little above low water mark, exactly where the worm is thought to exert its greatest powers. After remaining three years in this situation, though examined from time to time, the piece was cut, with the view of carrying a specimen of it to England, and to my great joy, I found it as sound and in every way as perfect throughout, as it was when first put into the river. Amongst other things, a valuable flood door was made of it, to keep the tides out of the Botanic Garden. It is now seven years and a half since the door (which is 4 feet square) was made, and though much exposed to the sun and water, yet it remains good; while similar doors, though much smaller, made of teak, were so much decayed a year ago, as

TIMBER.
296

The Asiatic Gmelina.

to render it necessary to replace them." Since the date of the publication of the above experiments, the wood has come permanently into notice and is in considerable demand in Calcutta for furniture-making.

**INDUSTRIAL
USE,
297**

Industrial Use.—The tree has been recommended as a good one on which to rear silkworms (*Agri.-Hort. Soc. of India Journ., VII. (New Series), 276*).

298

Gmelina asiatica, *Linn. ; Fl. Br. Ind., IV., 582 ; Wight, Ic., t. 174.*

Syn.—GMELINA COROMANDELIANA, *Burm.* ; G. LOBATA, *Gaertn.* ; *Fruct., I., 268, t. 56, excl. syn. Rumph.* ; G. PARVIFOLIA, *Roxb.* ; G. PARVIFLORA, *Roxb.* ; C. INERMIS, *Blanco* ; MICHELIA SPINOSA, *Amman.*

Vern.— *Badhára*, HIND.; *Bhedaira*, BEHAR; *Badhára*, PB.; *Láhán shivan*, MAR.; *Nilak-kumish*, TAM.; *Gamudu, gumudu, challa-gumudu, kavva-gumudu*, TEL.; *Láhán shivan, kal-shivani*, KAN.; *Nilak-kumash*, MALAY.; *Gatta-demmatta*, SING.; *Biddari*, SANS.

References.— *Roxb., Fl., Ind., Ed. C.B.C., 487* ; *Brandis, For. Fl., 365* ; *Kurz, For. Fl. Burm., II., 265* ; *Beddome, For. Man., 172* ; *Elliot, Flora Andhrica, 33, 65, 89* ; *Pharm. Ind., 164* ; *Ainslie, Mat. Ind., II., 240, 386* ; *O'Shaughnessy, Beng. Dispens., 486* ; *Dymock, Mat. Med. W. Ind., 2nd Ed., 599* ; *S. Arjun, Bomb. Drugs, 199* ; *Irvine, Mat. Med. Patna, 124* ; *Baden Powell, Pb. Pr., 364* ; *Drury, U. Pl., 229* ; *Balfour, Cyclop., I., 1214* ; *Treasury of Bot., I., 538* ; *Official Corresp. on proposed new Pharm. Ind., 240-1* ; *Gazetteers :—Mysore and Coorg, I., 64* ; *Bombay, XV., 70* ; *N.-W. P., Vol. I., 83* ; *IV., lxxvi.*

Habitat.—A large much-branching shrub of the forests of South India, Burma, and Ceylon; cultivated in Bengal.

**MEDICINE.
Root.
299**

Medicine.—The ROOT has been used as a demulcent by Hindú physicians from remote times. Rumphius mentions it under the name of "*jambusa sylvestris parviflora;*" Louriero speaks of its virtues in his *Flora of Cochin China*, commending it as of value in rheumatism and affections of the nerves; Dr. Horsfield in his Account of the Medicinal Plants of Java states that the plant was formerly in high esteem amongst the Portuguese, who called it *Rais Madre de Deos*. Ainslie also notices the plant, writing : "The root which, as it appears in the bazárs, is mucilaginous and demulcent, the Vytians reckon amongst those medicines which purify the blood, in cases of depraved habit of body; given in the form of electuary, to the quantity of a tea-spoonful twice daily." In another

**Leaves.
300**

passage he describes the virtues of the LEAVES as follows : "Its leaves would appear to have the quality of thickening water, and rendering it mucilaginous when agitated in it, so becoming a useful drink in gonorrhœa, and other maladies requiring demulcents. Certain other leaves have the same property, with this difference, that when our article is gently stirred in water, and the leaves at the same time a little bruised, the thickening of the water, by this means produced, does not pass away, as in the other instances, but remains; so it must be considered as a much more valuable medicine." Roxburgh and O'Shaughnessy comment on the same property of the leaves, and their observations are republished in the *Pharmacopœia of India*, which includes the drug in its non-officinal list. At the present time the root is principally employed as a demulcent for gonorrhœa and catarrh of the bladder, in doses of ʒii to ʒii in infusion, but it is also supposed to possess specific properties in the treatment of rheumatism and syphilis.

SPECIAL OPINIONS.—§ "Laxative and alterative. Useful in chronic rheumatism." (*Surgeon-Major J. McD. Houston, Travancore, and John Gomes, Esq., Medical Store-keeper, Trivandrum.*) "Useful in chronic rheumatism." (*Surgeon-Major J. J. L. Ratton, M.D., Medical College, Madras.*)

DOMESICS.

Domestic.—The Telegu names above given are said by Elliot to be

G   301

Gneiss Rocks.	(*J. Murray.*)	**GNEISS.**

derived from the fact that churning-sticks are made from the SHRUB :— *Challa* means butter-milk, and *Kavammu,* a churning-stick.

GNAPHALIUM, *Linn.; Gen. Pl., II., 305.*

Gnaphalium luteo-album, *Linn.; Fl. Br. Ind., III., 288;*
[COMPOSITÆ.

DOMESTIC.
Shrub.
301

302

> **Syn.**—GNAPHALIUM ORIXENSE, and G. ALBO-LUTEUM, *Roxb.;* SYNAN-THERA, *Wall., Cat.,* 7415.
> **Var. 1.**—MULTICEPS, heads golden yellow; G. MULTICEPS, *Wall.;* G. RAMIGERUM and CONFUSUM, *DC.;* G. AFFINE, *Don.;* G. MARTABANICUM, *Wall.*
> **Var. 2.**—PALLIDUM, heads pale brown; G. PALLIDUM, *Ham.*
> **Vern.**—*Bál ruksha,* PB.; *Byaing che piu,* BURM.
> **References.**—*Roxb., Fl. Ind., Ed. C.B.C.,* 600, 601; *Kurz, Prelim. For. Rep. on Pegu, App. C., xii.; Stewart, Pb. Pl.,* 127; *Gazetteer, N.-W. P., IV., lxxiii.*
> **Habitat.**—A very variable annual, common throughout India, from Kashmír to Burma and southwards to Martaban, ascending to 10,000 feet in Sikkim. *Var. 1,* **multiceps** is the rarer Indian form, seldom occurring on the plains, but fairly plentiful on the Sub-tropical and Tropical Himálaya, and the Khásia Mountains. *Var. 2,* **pallidum,** is very common all over the country.
> **Medicine.**—Stewart states that the LEAVES are sold as a medicine in the bazárs of the Panjáb, and quotes Madden to the effect that another unknown species is employed for tinder and moxas, in the region of the Sutlej.
> **Domestic.**—In Assam and the Naga Hills the leaves are rubbed in the hand to crumble away the cellular tissue, leaving behind the tomentum. This constitutes the tinder universally used on the eastern side of India.

MEDICINE.
Leaves.
303
Tinder.
304
DOMESTIC.
305

GNEISS, *Ball, Geology of India, III., 534.*

306

The following note has been kindly furnished by Mr. H. Medlicott, late Director of the Geological Survey.

Gneiss, *Eng.*

> GNEISS, GRANITE, *Fr.;* GNEISS, HOLZ GNEISS, GRANIT, *Ger.;* GRANITO *It.*

With the exception of a few comparatively small tracts of overlying strata, gneissic rocks extend, east of a line from Rotashgarh on the Son through Amarkantak to Goa, without a break from Cape Comorin to Colgong on the Ganges at the north-east corner of the peninsula, a distance of 1,400 miles, with a mean breadth of 350 miles. A continuation of this great exposure is found again in Assam and the Shillong plateau, where it also covers a considerable area, 250 miles in length, between the Dhansiri and Brahmaputra rivers. In Bundelkhand there is a large compact semi-circular area of gneiss. In the north-west quarter of Peninsular India, in the Arvali region, another area of gneiss occurs. In the Lower Himálayas, gneiss occurs over a considerable area in Sikkim in the neighbourhood of Darjiling, and more or less throughout the whole range to the Sutlej.

In the Himálayan Range proper, gneiss is the predominant rock for 300 miles to the west of Nepál, many of the highest peaks being formed of it. In Ladák a range of syenitic gneiss separates the Indus from its tributary the Shaiok and the Pang kong lake, and passes to the south-east on both sides of the Indus through Rupshu into Chinese Thibet. The Zánskár range in its central portion, and the Pir Panjál chain, consist

Building Stones.
307

| GNETUM scandens. | A Fibre used for making Fishing Nets. |

GNEISS.

to a great extent of this rock. Another gneissic ridge is the Dhauladhar range, extending north of the Kangra Valley in a north-west direction as far as Dalhousie.

In Burma the gneissic series consists to some extent of granitoid and hornblendic gneiss. Little attention has hitherto been paid to the metamorphic rocks of Burma; they occupy a large but unexplored area in Upper Burma; they form all the higher ranges in the neighbourhood of Ava and extend throughout a great portion of the country, extending thence to Salwin. Further north, they reach from Bhamo to the neighbourhood of Momein in Yunnan; the crystalline rocks then continue to the south, forming the Red Karen country and the hills between Sittang and Salwin, and extend into Tenasserim.

In the Nilghiri hills there are several places where excellent building stones could be obtained, but hitherto not much use has been made of them. In Mysore a variety is obtained which can be split into posts 20 feet long, which are used as supports for the telegraph wires. In the construction of walls, "bunds" of tanks, the beach groynes at Tranquebar, culverts, temples, bridges, &c., blocks of gneiss have been used. In Madras, beds of hornblendic gneiss are largely quarried at Palaveram, Cuddapary Choultry, and Pattandalum for the manufacture of articles of domestic use as well as for building purposes. In the Nellore-Kristna district it is used in the manufacture of cart-wheels.

Except for purely local purposes, the construction of bridges, &c., where the rock nearest at hand has, upon economical grounds, been made use of, this material has not commended itself for building purposes to English engineers. It is, however, peculiarly susceptible to fine carving, and, with the exception of some of the trap rocks, was the favourite stone for almost all the great temples in Southern India.

See publications of the Geological Survey of India and Journals of the Asiatic Societies of Bengal and Madras.

GNETUM, *Linn.; Gen. Pl., III., 419.*

308

Gnetum Gnemon, *Linn.; Fl. Br. Ind., V., 641;* GNETACEÆ.

Syn.—GNETUM BRUNONIANUM, *Griff.;* G. GRIFFITHII, *Parlat.*
References.—*Roxb., Fl. Ind., Ed. C.B.C., 632; Kurz, For. Fl. Burm., II., 497; in Flora, lv. (1872), 350; Gamble, Man. Timb., 293.*
Habitat.—An evergreen shrub-or small tree of the Khásia and Manipur Hills, extending southwards to Singapur, frequent in the dense forests of Southern Tenasserim.

FIBRE.
Bark.
309
Fibre.—"The BARK is made into strong cords at Sumatra" (*Roxburgh*).

FOOD.
Leaves.
310
Food.—"The LEAVES are eaten as spinach" (*Roxburgh*).

311

G. scandens, *Roxb.; Fl. Br. Ind., V., 642; Griff. in Trans. Linn. Soc., XXII., t. 55; f. 1-8, 22, 23, and t. 56, f. 39, 40, 42.*

Syn.—GNETUM EDULE, *Blume;* G. FUNICULARE, *Wight, Ic., t. 1955 (not of Blume);* G. PYRIFOLIUM, *Miq.;* THOA EDULIS, *Willd.*
Vern.—*Nanu-witi,* SYLHET; *Kúmbal, umble, úmbli,* BOMB.; *Umbrúth ballé,* KAN.; *Ula,* MALAY.; *Gyútnwé,* BURM.; *Pilita,* ANDAM.
References.—*Roxb., Fl. Ind., Ed. C.B.C., 632; Brandis, For. Fl., 502; Kurz, For. Fl. Burm., II., 495, in Flora, lv. (1872), 350; Gamble, Man. Timb., 393; Rheede, Hort. Malab., VII., t. 22; Grah. Cat. Bomb. Pl., l., 188; Dals. & Gibs., Bomb. Fl., 246; Lisboa, U. Pl. Bomb., 174, 273; Agri.-Hort. Soc. of India, Journal, IV., (Old Series), Sel., 264; Bombay Gazetteer, XV., 444.*
Habitat.—A lofty diœcious climbing shrub, met with in the Tropical Himálaya from Sikkim eastwards, to Assam, Singapore, and the Anda-

Gold.	(*J. Murray*).	GOLD.

man Islands; also in the hills of the Deccan from the Konkan to the Nilghiris.

Fibre.—The STEMS yield a fibre which is employed by the natives of
the Andaman Islands for the manufacture of hard fishing-nets called *Kud*.

FIBRE.
Stems.
312

Food.—The SHRUB, which flowers in March and April, yields an edible
fruit in September and October. It is rather larger than the largest
olive, and, when ripe, is smooth and orange-coloured. The outer succulent coat or PULP is commonly eaten by the Natives, and the SEEDS, when
roasted, are also employed as an article of food.

FOOD.
Shrub.
313
Pulp.
314
Seeds.
315

Structure of the Wood.—Dark-brown, soft, coarsely fibrous, porous,
rather heavy, but of no use except possibly for rough cordage (*Kurs*).

TIMBER.
316

Goa Bean; see **Psophocarpus tetragonolobus**, *DC.*; LEGUMINOSÆ.

Goats; see **Sheep & Goats**.

GOLD.

Gold, *Ball, Geology of India, III., 173-230, 608-610.*

317

The colour, lustre, power of resisting oxidation, extreme ductility and
malleability of this metal have caused it to be much valued from the
earliest ages. In the Bible mention is made of gold and silver ornaments,
cups, shields, &c., as abounding in the Court of Solomon, and of that king
having organised fleets of ships for obtaining these metals from Tarshish
and Ophir. It has been conjectured that the latter place may have been
some district or port of the Malabar coast. Whether this be so or not,
abundant evidence exists of the knowledge of gold in India from very
remote times. Pliny in A.D. 77 referred to the country of the Nareae,
as containing numerous mines of gold and silver; and that by the Nareae
were meant the Nairs of Malabar is now an established fact. Ancient
inscriptions shew that in the eleventh century gold existed at least in
Southern India in great abundance, and numerous and extensive, very
ancient mines have been described by various writers. In 1596 Linschoten
wrote of Ceylon: " It hath likewyse mynes of gold, silver, and other metals,"
but he makes no mention of having observed or heard of gold mines in
the Peninsula of India. In the *Ain-i-Akbari*, however, written at nearly
the same date, it is stated that "although gold is imported into Hindús-
tán it is to be found in abundance in the northern mountains of the
country, as also in Tibet. Gold may be obtained by the *Salóní* process" (washing) "from the sands of the Ganges, Indus, and several
other rivers, as most of the waters of this country are mixed with gold:
however, the labour and expense greatly exceed the profit. This last remark by Abul Fazl very correctly describes the condition of gold-washing
as an industry in most parts of India at the present day. Thus Ball wrote
in his Economic Geology, "The amount of gold brought down by the
rivers in a single year, gives him " (the gold-washer) "insignificant returns,"
. . . . "though in a country like India where a man can live for so small
a sum, it is possible to derive a subsistence, such as it is, from the washings of a few rivers year after year in succession." Recently, however,
gold-mining has been revived, especially in Southern India, with a fair
amount of success, and may develop into an industry of some importance. It may, accordingly, be of interest to give a short résumé of
the facts regarding the occurrence and supply of gold in India at the present day.

Vern.—*Sona*, HIND.; *Gser*, TIBET; *Sona, swarna*, MAR.; *Pwon, ponnú*,
TAM.; *Bungárum, bungárú*, TEL.; *Mas, amas, kanchana*, MALAY.;
Shwae, BURM.; *Run*, SING.; *Suvarna, swarna*, SANS.; *Tibr, sahab*,
ARAB.; *Tilla, thil, sir*, PERS.

References.—*Mallet, Geology of India (Mineralogy), IV., 1; Ainslie, Mat. Ind., I., 514-522; U. C. Dutt, Mat. Med. Hind., 57; Irvine, Med. Top. Ajmir, 169; Linschoten, Voyage to the East Indies, I., 27, 31, 109; II., 295; Abul Fasl, Ain-i-Akbari (Blochmann's Trans.), 17-30, 36-43; (Gladwin's Translation), II., 136; Buchanan, Journey through Mysore, &c., I., 441; Baden Powell, Pb. Pr., 12; Atkinson, Econ. Geol. of N.-W. P., 276; Mason, Burma and Its People, 560, 729; Oldham, Mission to Ava, 344; Forbes Watson Industrial Survey, II., 405; W. W. Hunter, Statistical Acct. of Bengal, II., 27. 75, App. 1; III., 39, 149; XIII., 228; XVII., 23, 167, 190, 202, 259; XIX., 203; Statistical Acct. of Assam, I., 106, 380; Balfour, Cyclop., I., 1220; Indian Agriculturist, Oct. 22nd, 1887; March 22nd, April 16th, July 13th, Nov. 9th and 11th, 1889; Bosworth-Smith, Rep. on the Kolar Gold Field, 1889; Proceedings of the Rev. & Agri. Dept. for March 1880, 19 and 20A; Brough Smith, Report on Wynaad, 1880; Bruce Foote, Auriferous Rock series in South India, Rec., G. S. I.; Gazetteers:—Mysore and Coorg, I., 17, 34, 432; Bhandara, Central Provs., 59; Bombay, V., 123; VII., 40; VIII., 261; Panjáb, Delhi, 133; Ambala, 11; Gurgaon, 14; Jhelam, 825; Rawal Pindi, 12; Bannu, 22; Peshawar, 24; Madras, Man. of Admin., II., App. VI., 33, 34; Admin. Rept., Central Provs., 124; Bombay, 1871-72, 373, 384; Settlement Reports:—Central Provs., Nagpur, Sup., 276; Seoni, 11; Upper Godavery Dist., 42; Chanda, 105; Panjáb, Hasara, 9; Peshawar, 12; Kohat, 32. Consult also the works quoted by Ball, Econ. Geology, pp. 608–611.*

OCCURRENCE
318

Occurrence.—The following account of the localities in which gold is chiefly to be found in India is abridged, for the most part, from the exhaustive article on the subject in *Ball's Economic Geology,* to which the reader is referred for more detailed information.

The ultimate derivation of most of the gold of Peninsular India is doubtless from the quartz reefs which occur traversing the metamorphic and sub-metamorphic series of rocks, but a certain quantity appears to exist in certain chlorotic schists and quartzites, and possibly also in some forms of gneiss. Existing evidence regarding the relative productiveness of the reefs in the different groups or series of metamorphosed rocks is conflicting, probably owing to the fact that a rule which holds good in one part of the country does not necessarily apply to other areas. The presence of gold has not yet been proved in any member of the Vindhyan formation, but in the next succeeding formation several of the groups included in the Gondwana system are believed to contain detrital gold. It is almost certain, also, that the gold obtained in the Godavari and in its tributary near Godalore, or Mungapet, is derived from rocks of Kamthi age, and the gold of the Ouli river in Talchir (Orissa) is derived from sandstones. The only other sources in Peninsular India are the recent and sub-recent alluvial deposits which rest on the metamorphic and sub-metamorphic rocks.

Passing to the extra-peninsular regions, gold is met with in rocks of several different periods. In Ladak it occurs in quartz reefs which traverse carboniferous rocks, in Kandahar it is found in cretaceous formations, as an original deposit connected with the intrusion of trap; while all along the foot of the Himálaya, the tertiary rocks which flank the bases of the hills are more or less auriferous. But the gold occurring in the last-mentioned area is all detrital, and is doubtless derived from the crystalline metamorphic rocks of the higher ranges which are, from other reasons, known to contain gold.

MADRAS.
319

I. MADRAS was in remote times famous for its gold mines, and has in recent years attracted much public attention and a large amount of capital in the endeavours that have been made to again open up a long dormant industry. Gold is known to exist in Travancore, Madura, Salem, Malabar, Wynaad, Mysore, and Bellary; but according to Ball its occurrence in Vizagapatam is as yet unproved.

| Gold. | (*J. Murray.*) | GOLD. |

In Travancore, it is found in outcrops of beds of quartzite including felspar, which run with the gneiss; but no real quartz reefs occur. Dr. W. King in a report to the Travancore Government (1881) stated that in only one case was the outcrop sufficiently large to promise a good tonnage of stone. In the *Madura District*, gold is found, according to Mr. J. H. Nelson, in two localities, namely, in Palakanuth and in the sands of the Veigei river. Ainslie mentions that an auriferous zinc blende was discovered in some part of the district by Mr. Mainwaring. At both these localities gold-washing is carried on by natives in a small way, barely affording a subsistence to those employed. In the *Salem district* gold used to be found
at the base of a hill called Kanjah Mallia, and was obtained from streams in that locality by washing. Heyne refers to some gold mines as existing at Sattergul, near Pangumpilly, in 1802, the exact locality of which does not appear to be now known.

Malabar district and the Wynaad.—As already stated, evidence exists
of gold having been obtained in this region as far back as the time of Pliny. In the report of a joint commission from Bengal and Bombay on the condition of Malabar in 1792-93, it is stated that at that time the Raja of Nilambar claimed a royalty on all gold found in his territory. Dr. Buchanan in his "Journey through Mysore, &c.," alludes to the existence of gold-mines at Malabar in 1801, and states that a Nair, who had the exclusive right to mine, paid a small annual tribute for the privilege. Ainslie includes Nilambar, Wynaad, and the sand of the Beypur rive at Calicut in his list of localities for gold. In 1830 a Mr. Baber stated before the Lords Committee on East Indian affairs, that in Coimbatore and the country west and south of the Nilghiri and Kunda hills, 2,000 square miles of soil were auriferous, and that at that time the Government derived a revenue from assessing the *puttis* or trays used to wash the gold. In 1831 the Collector of Malabar furnished a report to Government on the localities in which gold was then to be found. and in the same year Lieutenant Nicholson was appointed to prospect the gold-fields, and also to purchase on behalf of Government. His interesting report was on the whole favourable, but in many places referred to the evident jealousy with which his researches and enquiries were received by the natives. He stated, however, that in his opinion mines might be worked profitably by the British, and that the most promising localities appeared to be Cúpal and Carembat. After receipt of the report of a Committee in 1833, however, which condemned gold-working in the low country of Malabar as a European industry, the Governor General in Council decided that it would be inexpedient to work the mines. Nothing more appears to have been done for a quarter of a century, at the end of which time, in 1857-58, letters from the Collector of Malabar again attracted attention to the subject. In 1865 two Englishmen with experience of Australian gold-mining were attracted to the district, and soon afterwards machinery was erected to crush quartz at the Skull Reef—the first extensive attempt at British gold-working in India. Other applicants for the right to mine then came forward, and new mines were opened, but owing to many and (according to Mr. Brough Smith) preventible circumstances, all without success. In 1879-80 Mr. Brough Smith explored the Wynaad gold fields, and wrote an elaborate and exhaustive report of his investigations, in which it was stated that the tract was richly auriferous, the average yield of gold per ton, at ten reefs or workings, being from 6 dwt. 13 grains to 18 ounces 9 dwt. 1 grain. Omitting picked and exceptional samples which caused the latter very high figure, 88 samples from the ten sources yielded an average of 1 ounce 8 dwt. 22 grains per ton. Mr. Brough Smith deals fully with such important subjects as climate, water, and timber-supply,

GOLD.   Gold.

MADRAS.

&c., and, in his concluding remark, speaks with confidence as to the future of the industry, maintaining that failure can only result from want of care and forethought.

Professor Ball concludes his interesting account of the gold in this region, by giving an estimate of the cost of working a company on the authority of Mr. Ryan. As this is stated to be based on actual experience, it may prove both useful and interesting, and may be here quoted : "It being assumed that a concession of value cannot now be obtained at a less cost than £60,000, the following would represent the first year's expenditure : —

	£
Price paid for concession	60,000
Cost of machinery, 100 stamp-heads at £200 each .	20,000
One year's working expenses	12,000
Contingencies, law-expenses, &c.	8,000

Taking the value of gold at £3-15 per ounce, the return from 25,000 tons of stone, containing from 3 to 10 dwt. of gold per ton would be as follows : —

	Total ounces.	Value at £3-15.	Cost of production.*	Profit.	Percentage on capital of £100,000.
		£	£	£	£
3 dwt. per ton . .	3,750	14,062	11,875	2,187	2·19
4 ,, ,, ,, . .	5,000	18,750	11,875	6,875	6·87
5 ,, ,, ,, . .	6,250	23,437	11,875	11,562	11·56
6 ,, ,, ,, . .	7,500	28,135	11,875	16,250	16·25
7 ,, ,, ,, . .	8,750	38,812	11,875	20,937	20·93
8 ,, ,, ,, . .	10,000	37,50c	11,875	25,625	25·62
9 ,, ,, ,, . .	11,250	42,187	11,875	30,312	30·31
10 ,, ,, ,, . .	12,250	46,875	11,875	35,000	35·00

MYSORE.
324

II. MYSORE PROVINCE.—Captain Warren, in 1802, hearing of a rumour that gold had been found at the Yerra Baterine Hill, instituted enquiries which elicited the fact that there were gold-washings near the village of Wurigam (the modern Urigam or Ooregaum), and actual mining at Marcurpam. He proved the presence of gold in the surface soil and beds of the rivers over an extended area in the neighbourhood of the Manigatta, Wullur, and Yeldur hills, from Budikote to Ramasamudra. The people who washed were Dherus or Pariahs, and he appears to have thought that agriculture was for them a more profitable profession. He then described two mines, one at Kembly, 30 feet deep, having a gallery of 50 feet ; the other west of Surunpally, which was 45 feet deep and 56 feet in extent. From the sections given, Ball remarks "it is evident that these were not in solid rock, but that masses of quartz in an ochreous matrix had been taken out to be crushed." Later, Heyne alludes to Warren's researches, and various officers appear to have collected samples from the same region at subsequent dates. General Sir Mark Cubbon, when Commissioner of Mysore, is said to have prohibited more mines being sunk, in consequence of the frequency of accidents in those already existing. Subsequent to this date little attention appears to have been paid to the subject for nearly fifty years.

Of late years, however, the gold industry in this province has received a marked impetus, and its gradual growth can be traced through succes-

* This sum is arrived at as the average of several estimates of cost, 25,000 tons at 9s. 6d. = £11,875.

sive Administrative Reports. In 1868 it was stated that alluvial gold was occasionally found near Betmangla, but in too small quantities to repay labour; in 1870 washers were said to be able to earn 4 annas a day by working at the foot of the Hemagiri Hill, in the Huliyardurga taluk of the Nandidrug division; in 1872-73 it was recorded that five pounds weight of gold had been found in the Betmangla taluk, and in 1873-74 that six pounds weight had been obtained in Kolar. The same year an opinion was expressed that a proper system of working would disclose considerable quantities in certain districts; and permission was granted to a **Mr. Lavelle** to prospect for gold and other metals during a period of three years. He was informed that leases for a period of twenty years would be granted to him, of not more than ten blocks, each of two square miles or less in extent. As a result of this concession public attention began to be attracted to the Kolar gold-fields, and since the year 1880 several companies have started in the district, and have crushed and sent home gold. In 1889 **Mr. Bosworth Smith**, Government Mineralogist of Madras, issued a long and instructive report on the Kolar gold-field to which the reader is referred for a complete description of the Geology and Mineralogy of the district. His concluding remarks may be here quoted as they are of much interest and sum up, comparatively briefly, his opinions regarding the future of the industry. "There can be no doubt," he writes, "that the Kolar gold-field has a future before it. But that the expectations that were first started when gold-mining in India was revived in 1880 will ever be realised in this (or any other gold-field in any part of the globe) is very doubtful. Some of the mines are now paying expenses, and there can be no doubt that, managed economically and under scientific supervision, several others should easily pay their way at an early date. If regular dividends are to be paid, it will be found that prospecting work must be kept going side by side with the more pleasant task of stamping and crushing what pay-stone has already been found. It will not do, after finding a pay-shoot, to concentrate all the energies of the mine on getting out that shoot and rushing it through the stamps, to find, after taking all its quartz that has been left by the "old men" above 400 feet, that the rich shoot is getting out of control, and that it must practically remain untouched whilst a new shaft is sunk to cut the shoot lower down. It would be invidious to take each mine separately and write on its merits and demerits, but it can do no harm to mention the names of some of the best mines. That the oldest mines are the best is due to the fact that they have been more thoroughly prospected. and that when the field was started, the number of old workings on a block were taken (and very rightly too) as an indication of its value. The Oorghaum and Mysore mines contain a great number of large old workings, and without doubt these are the pick of the mines. Balaghat has a rich shoot opened out for over 200 feet, and Nundydrug has been returning an average of about 400 ounces per month for some time past. The mines that have crushed and sent home gold are the nine reefs, Balaghat, Nundydrug, Oorghaum, Mysore, Indian Consolidated (Kolar Section) Mining Companies, and the South-east Mysore Company is expected to crush very shortly." It may be remarked, however, that certain authorities in Madras hold a much higher opinion of the probable success of these mines than appears to have been entertained by **Mr. Bosworth Smith**, and that **Mr. Bruce Foote, F.G.S.**, in a recent paper contributed to the *Records of the Geological Survey of India*, has also taken a more favourable view of the subject. In one passage he writes, "the great success attained at a good number of the mines now being worked there has proved beyond all cavil, that gold does exist in richly paying quantity in many of the lodes running through the Dharwar schists" (the Kolar Gold-

GOLD.	Gold.

Field Band), " and I for one firmly believe that lodes of equal richness will be found in other tracts in which similar geological conditions prevail." In another passage he writes, "the results already obtained at Kolar are abundantly good enough to encourage sensible people to proceed with care and forethought to open other mines." In his opinion the gold-mining operations at present conducted, have only to a very small extent tapped the gold-bearing rocks of Mysore. Over the whole extent of the province from north to south, run well-marked bands of Dharwar schists, which all bear evidence of having been worked to a greater or lesser extent by Natives in remote times. The Kolar band does not belong to these well-marked great bands of Dharwar, but is an outlier of limited extent. Of the great bands traversing Mysore, the western is said by **Mr. Foote** to be the largest and least known, being covered by the dense forests and steep hills of the Western Ghâts.

HYDERABAD 325

III. HYDERABAD.—Gold-dust is found in the bed of the Godavari and its tributaries, and appears to have been fairly extensively worked up to the end of last century; at that time, however, operations ceased owing to an excessive rent charged by the Raja. According to **Dr. Walker** there was a gold mine about 1790 near the village of Goodloor or Godalore, in the vicinity of Mungapet, but **Ball** points out that, owing to the absence of crystalline rocks in the neighbourhood, it is improbable that there ever was a real mine there.

BENGAL. 326

IV. BENGAL.—Gold is obtained in Orissa, Midnapur, Bankura, and in the Province of Chutia Nagpur, the last-mentioned locality being apparently specially rich in the metal.

Orissa. 327

Orissa.—Ball states that "within the limits of the Province of Orissa gold-washing is or has been carried out in the Native States of Dhenkanal, Keonjhar, Pal Lahara, and Talchir. It is a poor pursuit, as in so many other parts of the country, but the fact is interesting as affording evidence of the existence of gold." At the present time gold-washing is carried on most actively in the Brahmini river, where it traverses Pal Lahara.

Midnapur. 328

Midnapur district contains a few professional gold-washers, who apparently carry on their industry in the beds of the Kasai river and its tributaries.

Bankura. 329

Bankura district.—Gold is reported to have been obtained in very small quantities in the sands of the Dalkissur at Bankura.

Chutia Nagpur. 330

Chutia Nagpur. —Ball writes, " From the characters of the rocks found in the sub-divisions of this province, it is not improbable that gold occurs in all of them, whether because it is less abundant in some, as is probable, or because it has never been properly searched for, the fact is certain that in others there is greater attraction for the indigenous gold-seeker. Judged by this standard, the richest tracts are situated in Manbhum, Singhbum, Gangpur, Jhashpur, and Udaipur. That these, or some of them, may yet be the scene of extensive operations, should the gold-mining in Southern India be successful, is very possible. The indications afforded by the alluvial deposits of sources of gold existing in the rocks over several large areas, are perhaps quite as striking in their way as those which led to the starting of the gold-mining industry in Southern India. Quartz or reef mining and crushing, however, can scarcely be said to have been tried in this area, but one solitary and not very expensive attempt having been made." It is stated that three companies have lately (1890) started for gold working in this province, and that a probability exists of two or three other companies being formed for the same purpose.

Manbhum. 331

In *Manbhum* the localities where gold-bearing sands exist are very numerous; indeed, in the southern half of the district, gold is to be found in nearly every stream. Ball discovered, by a systematic application of

G. 331

| Gold. | (*J Murray.*) | GOLD. |

the operations of two gold-washers, that the area in which gold was most abundant corresponded with a tract in which a particular series of rocks was found to occur. These rocks were sub-metamorphic, consisting chiefly of magnesian and mica schists, slates, and quartzites. They almost exclusively prevail " south of a line drawn from Simlapal on the east, through Bara Bazár, to a point a little north of Ichagarh on the west, and so on into the Chutia Nagpur highlands." In Manbhum, however, the metamorphic rocks also contain gold, but in much smaller quantity.

In *Singhbhum* the metal occurs in the same series of sub-metamorphic rocks, which runs continuously into this district from Manbhum. It is not found at all, however, in the metamorphic rocks. In this district, quartz reefs are more abundant than in Manbhum, and in all probability contain gold; indeed, Ball states that the only nugget seen by him from the district was in a quartz matrix, and that gold is said to have been obtained by quartz-crushing at Landu. The same writer enumerates the following as the most noteworthy gold-bearing localities in Singhbhum :—Kamerara, the Kapargadi Ghât in Dhalbhum, Landu in Seraikela, Asantoria in Kharsawan, Sonapet, Porahat, and Dhipa in Sarunda. Of these Sonapet, or the "mother of gold," is referred to by all writers on the district as the richest in the metal. Records, however, exist of gold-washing, to a greater or lesser extent, in the streams of all the localities.

In the *Lohardaga district*, the Kanchi river contains auriferous sands, probably derived from the same series of sub-metamorphic rocks, as that above described. As already mentioned gold occurs and is washed for in the Brahmini river in Bonai. In *Gangpur State*, gold-washing is carried on in the bed of the Ebe and in some of its tributaries particularly the Icha. Gold mines, in which large pieces of the pure metal were said to have been found, were also reported by Surgeon Breton to exist in the state (*Medico, Topography of Ceded Provinces, 1826*).

Many records exist of gold in *Jashpur State*, in some cases large nuggets having been found. In the early part of this century mines appear to have been worked by the Raja, but owing to an accident in one of the shafts operations were discontinued. In later years the ancient deposits have been considerably worked by gold-washers, who find them more profitable than the sands of the river beds. Ball writes of these, "On both sides of the river Ebe or Ib there are tracts at some distance from the banks, which are honey-combed with shafts sunk by successive generations of gold-seekers. These shafts are from 10 to 30 feet deep. The gold-bearing stratum is a layer of pebbles and fragments of quartz which underlies red soil and vegetable humus. The stuff selected is of a dirty drab or reddish colour with occasional balls of decomposed felspar, which latter are regarded as the surest indication of the presence of gold. The decomposed granitic rock on which this layer reposes is not generally washed, but Colonel Dalton found that it was likewise auriferous, but to a less degree." The outturn by the native method of simple washing was, according to Colonel Dalton, very uncertain, no mercury was used, only the visible gold being saved. Gold was sent by Colonel Ousely from Phrashabahal to the mint for assay, and a nugget from some other part of Jashpur was presented to the Geological Survey Museum by Colonel Dalton. The latter specimen weighed on receipt 221·87 grains, and after cleaning 199·6 grains, and contained 94·6 per cent. of the pure metal. Ball concludes his account of the Jashpur State with the following remarks : "The facts just given and those mentioned below, with reference to the states of Gangpur and Udaipur, establish, beyond a possibility of doubt, the existence of an ancient alluvial gold-bearing deposit at intervals throughout a tract of not far short of 2,000 square miles in area." "The principal rivers of this tract

GOLD.	**Gold.**

are the Mand and Ebe, with their numerous tributaries. As there is always water in the Ebe, it is possible that some system of hydraulic mining might be applicable. Be that as it may, there cannot but be gold-bearing reefs from which all this gold has been derived."

Udaipur.
333

In *Udaipur State*, also, the rivers contain auriferous sands. The first to call attention to the washings in this state was Colonel Ousely in 1847, at which time he reported that three families at Rabkob obtained a livelihood by the industry. In 1849 a Mr. Robinson took a lease of the village, with permission to work the mines, from Government, and found as the result of his trials that a man to whom he paid 1 anna could earn for him 3 to 4 annas worth of gold. The gold obtained was valued at the Calcutta Mint as worth R14¾ per tola. The unhealthiness of the district for Europeans appears, however, to have resulted in the cessation of the enterprise. In 1865 the number of native gold-washers was stated to have increased to six families, and the reporter (the late Colonel Dalton) wrote that the production of gold was only restricted by the number of washers.

CENTRAL
PROVINCES.
334

V. CENTRAL PROVINCES.—Gold-bearing sands occur in most parts of these provinces, wherever there are exposures of the older crystalline rocks. Judging by the census returns of 1872 Nagpur division is the richest, followed by Jabalpur and Chatisgarh, while in the Narbada division none of the inhabitants were returned as gold-washers.

Chatisgarh
335

Chatisgarh Division.—In the district of Sambalpur gold-washing is pursued as an industry at Sambalpur town on the Mahanadi, and at the village of Tahud on the Ebe. In the Bilaspur district gold is known to occur in the Jonk river at Sonakhan. In the Raipur district 12 gold-washers were returned in the 1872 census, though it is not known in what localities they pursue their avocation. It has been asserted, however, that gold is procurable in the Mahanadi at Rajoo (probably Rajim is meant by this name).

Nagpur.
336

Nagpur Division.—In the Bhandárá district gold-bearing sands occur in streams near Ambagarh and Thirora. In these waters gold-washing operations are carried on, and in some places mercury is employed in separating the finer particles. In the Chándá district the search for gold is said to be carried on in the eastern parts of the area, but there are no definite details as to the actual streams in which the metal is found. Gold is washed in several places in the Bálághát district, the auriferous streams being chiefly situated in the Lanji and Dhansua Parganas. Of these the Son and Deo are richest in the metal. The census returns of 1872 give 103 gold-washers in the Nagpur district, but it is probable that these men carry on their operations chiefly in the adjoining districts.

Jabalpur.
337

Jabalpur Division.—In the district of Wardhá, Ságar, and Dámoh returns are made of some 52 gold-washers, though there is no record of the occurrence of gold in these localities. The sands of the Parqudhur stream, in the Seoni district, however, produce gold. Balfour states that the washers of the sands of this river consider it unlucky to make more than 4 annas a day, as they believe that the goddess who makes the gold would leave the locality if they exceeded that amount.

Upper
Godávari.
438

In the *Upper Godávari District* gold is said to be found in two localities, namely, near Bhadrachellum and at Marigudem or Mariguram. The gold of the latter locality is of superior quality, being valued at R16 a tola, yet notwithstanding this fact, the work of washing is said in the *Central Provinces Gazetteer* to be "barely remunerative." It must consequently be inferred that the metal occurs in small quantity only. Gold-washing is also carried on in the Bastár State at Pratappur or Partabpur, and at Bharamgarh.

CENTRAL
INDIA.
339

VI. CENTRAL INDIA—*Ajmir-Merwara District.*—According to Dr.

G. 339

| Gold. | (*J. Murray.*) | GOLD. |

Irvine, gold-dust was at one time found in the sands of the Luni and Khari rivers, but the industry does not appear to be carried on at present.

VII. BOMBAY.—Auriferous rocks are reported to occur in the districts of Dharwar, Belgaum, Kaladgi in the Southern Mahratta Country, and in the province of Kathiawar.

Dharwar District.—Gold has been found at Chik Múlgúnd, Súrtúr, Dambal, Dhoni, and in the Hurti river near Guduk. **Mr. Foote**, in the *Records of the Geological Survey of India*, has given a *résumé* of the writings of other authors on the subject of gold in this district, together with his own observations. He considers that the rocks of the known gold-bearing area belong to three groups or series, each characterised by certain peculiarities. To these he has given the local names of Dhoni, Kappatgode and Súrtúr. The Dhoni series consists of a hematitic schist, accompanied by chloritic, hornblendic, and micaceous schists; and includes several beds of white and grey limestone, which might prove a valuable source of lime. The second group lies immediately above the first, and forms the Kappatgode hill. It also consists of hematitic schists, which, however, have associated with them argillaceous schists, and instead of having a green prevailing colour, as is the case with the first group, are reddish-buff or mottled white. The third group consists of hornblendic and chloritic schists intimately associated with a massive diorite. In all these series quartz reefs occur, but according to native opinion, only the streams arising from the Súrtúr series contain auriferous sands, and it is certain that the richest of all, the Súrtúr river, lies entirely within the area occupied by the chloritic schists and diorite. The quartz reefs in this section have, with few exceptions, been broken up by gold-seekers; and in the Kappatgode quartz reefs also, indications exist of workings at some past date. At the present time only a few families are engaged in gold-washing in Dharwar, and it appears probable that the unfavourable view taken by **Mr. Scholt** of the value of the alluvial deposits in the district was a just one. During the Bombay share mania, however, a Gold Company was started to work the locality, and apparently sank two shafts—one in the Dhoni, and one in the Kappatgode series.

Belgaum District. — Gold-dust is said to have been found within the limits of this district at or near the villages of Belowuddi, Byl Hongul, and Murgur. The quantity must, however, be small, since very few gold-washers pursue their calling in the district.

Kaladgi District. — **Mr. Foote** mentions a report of auriferous sands being found in the streams of this district, but adds that he has reason to doubt the accuracy of the statement.

Kathiawar.—Gold-dust in small quantities is said to be found in the Sourekha (a small river rising in the Girnar hills), also in the Aji, which passes Rájkot.

VIII. PANJÁB.—**Ball** writes, "It has been not unfrequently stated that all the rivers of the Panjáb, the Ravi alone excepted, contain auriferous sands. Probably there are some others which might be excluded from so general a statement; but the fact remains that the rivers and streams of the province, whether rising in the distant ranges of crystalline rocks forming the axis of the Himálayas, or merely having their sources in the outer and lower ranges of hills formed of detrital tertiary formations, do, as a general rule, contain gold. In the latter cases the gold must have a doubly derivative origin, and no veins, or other original deposits of it, can be expected to occur."

The practice of gold-washing in this province is probably of considerable antiquity; formerly it afforded a source of revenue; indeed during the Sikh predominance the tax amounted to one-fourth the gross produce. This

GOLD.	Gold.

PANJÁB.

revenue has, however, here, as in most other parts of India, dwindled down to very small proportions, or become totally extinct. In 1860–61 it was R444 and in 1861–62 R530. **Abul Fazl** mentions that in the time of Akbar gold was obtained by washing in rivers in the *subáh* of Lahore. **Ball** states that the districts it is at present found are, Bannu, Peshawar, Hazara, Rawal Pindi, Jhílam, Amballa, and certain Native States, and gives the following detailed information regarding each.

Bannu.
346

Bannu District.—Gold-dust is obtained from the Indus at and below Kalabagh to the annual value of about R200. It is doubtful whether the source of the metal is the low tertiary rocks or the older rocks higher up the valley.

Peshawar.
347

Peshawar District.—About 150 men wash for gold in the Indus above Attock and in the Kabul river, during part of the year, their regular avocation being that of boatmen. Each man is said to obtain on an average about 2 to 2½ tolas of gold, which sells for about R15 a tola. **Ball** calculates, from the time spent in collecting, that this amount only yields a daily wage of about 2 annas.

Hazara.
348

Hazara District.—Here as elsewhere the Indus yields a small quantity of gold-dust, which is similar in quality and value to that obtained in the Peshawar district.

Rawal Pindi.
349

Rawal Pindi District.—The sands of the Indus between Attock and Kalabagh are washed for the metal. **Dr. Jameson**, in 1843, stated that about 300 individuals used then to engage annually in the search for gold in this region, employing large wooden troughs and mercury, that one-fourth of the proceeds was claimed by the Sikh Government, and that the actual earnings of the men were estimated to be from 3 to 4 annas a day. Within the last few years, it is believed, endeavours have been made to establish washings on the Ravi and in other parts of the Rawal Pindi district on a large scale. The experiment was not, however, financially successful.

Jhílam.
350

Jhílam District contains most of the gold-washings of the Salt-Range. These are situated in the beds of rivers and streams arising from the lower Siwalik group, the detrital beds of which yield the metal. **Ball** states that "much of the gold is invisible or nearly so, and would be lost but for the employment of mercury." Under the Sikh Government about 160 cradles were worked, and afforded a revenue of over R500. **Baden Powell** quoting **Dr. Flemming** gives the annual production from these washings in 1848 as 1,013 tolas or about £1,600. The Bunhar river is specially mentioned by **Mr. Wynne** as gold-producing, and **Ball** states that from it westwards up to the Indus many of the streams which rise on the northern flank of the range contain gold.

Kangra.
351

Kangra District.—Gold is found in the Bias near Haripur, and also in Spiti, Kulu, and Lahul, but nowhere in large quantity.

Amballa.
352

Amballa District.—Specimens of gold from the Markunda river were exhibited at the Lahore Exhibition, and records exist of gold-washing having been carried in the neighbouring stream, the Gumti, from which the Raja of Nahan at one time derived a small revenue. **Balfour** mentions, but on what authority he does not state, that gold has been found in large quantities between Amballa and Kalka.

Gurgaon.
353
KASHMÍR.
354

Gurgaon District.—Gold is said to be found in the streams near Sonah.

IX. KASHMÍR.—**Abul Fazl** states in the *Aín-i-Akbari* that gold was found in the time of Akbar, in Padmatti, Puckely, and Gulkut (? Gilgit) of the Subah of Kashmír, and describes a peculiar process employed in obtaining it. This consisted in pegging down the skins of animals in the beds of gold-bearing streams. The hair on the skins acted, like the blanket used by miners in modern days, by arresting small particles of gold, which were

KASHMÍR.

shaken out after drying the skins. Though there is apparently little doubt but that gold was at one time obtained in Kashmír proper, few authentic records exist regarding it. At the present time, in the territories of the Maharajah of Kashmír, the industry appears to be almost confined to Ladak. Mention, however, is made by **Dr. Bellew** of an old deserted mine in auriferous sand at Kargil which had been given up in consequence of a portion of it having fallen in and killed some of the men employed. Gold-washing is said to be carried on in Ladak in the beds of the Indus and Shayok and at Kio on the Markha river.

Ladak.
355

X. TIBET.—Though this country is not within the limits of India a short account of the gold obtained from it may be here given, since there is every reason to believe that for many centuries it has been the source of a regular supply to this country. The survey parties of 1867–68 discovered the existence of large gold-fields at Thok Jalung (in the province of Nari Khorsam), Thok Nianmo, and Thok Sarlung, which were regularly worked by large encampments of Tibetan miners. One of the Pandits accompanying the expedition gave an interesting account of the habits and methods of work of these miners, one of the passages from which may be here quoted as throwing a light on the old story of **gold-digging ants.** "The cold is intense and the miners in winter are thickly clad with furs. They do not merely remain under ground when at work, but their small black tents, which are made of felt-like material, manufactured from the hair of the yak, are set in a series of pits, with steps leading down to them . . . . seven or eight feet below the surface of the ground" . . . . "Spite of the cold the diggers prefer working in winter, and the number of their tents, which in summer amounts to 300, rises to nearly 600 in winter. They prefer the winter, as the frozen soil then stands well, and is not likely to trouble them much by falling in." **Sir Henry Rawlinson** and **Professor Schiern,** commenting on these observations, arrive at the conclusion that the old tradition of gold-digging ants, mentioned in the writings of **Herodotus, Pliny,** &c., of the middle ages and of Arabian authors, owes its origin to these Tibetan miners. The latter learned writer remarks, "for us the story partakes no longer of the marvellous. The gold-digging ants were originally . . . . men of flesh and blood, and these men, Tibetan miners, whose mode of life and dress were in the remotest antiquity exactly what they are at the present day." The likelihood of this explanation being correct is strengthened by the fact that according to ancient writers the ants worked chiefly in winter. Further, **Pliny** states that the horns of the Indian ant were preserved in the temple of Hercules at Erythrae. **Professor Schiern** argues that these may have been horns taken from the fur dress of the miners. **Ball** thinks they may have been more probably the horns of *Ovis vignei,* which were probably in ancient times, as they are to this day, tipped with iron and employed as pick-axes by the miners.

TIBET.
356

The gold obtained by the Tibetan miners is tied up in little bags called *Sár-shu* weighing about 90 grains, which form the heavy currency of the country. It is chiefly given in exchange for grain or cloth and forms an important source of the metal in northern India. The mines are farmed or managed by a *Sár-pan* or gold commissioner, who holds a triennial contract direct from Lhassa. **Atkinson** states that the gold of the Thok Jalung mines has usually not more than 7·73 specific gravity, and that even the picked yellow grains have only a specific gravity of 11·96, showing that they are alloyed with some other metal.

XI. NORTH-WEST PROVINCES.—Gold-bearing sands occur in some of the rivers of Kumaon and Garhwál, also, as in Panjáb, in some of those which take their rise in the outer ranges of hills formed of tertiary rocks. Several of the rivers of the Moradabad district used formerly to be washed,

N. W. P.
357

GOLD.	Gold.

GOLD.

Gold.

N. W. P.

if they are not so still. Gold-washing was a source of revenue to the Gurkha Government, but when the country became British territory the smallness of the sum realised caused it to be remitted by the Commissioner. **Mr. Ravenshaw** states that in 1833 the gold-washers or *Nariyas* of Kot Kadir paid R50 a month, and those of Barapura R30 to the zamindar, while on the Dhela river a tax of R2-8 was levied by the Government on each washing trough.

Garhwál
358

Garhwál District.—The Alakananda, Benigunga, and Sona rivers contain auriferous sands, probably all doubly derivative, though an observer is said to have found a speck of gold in granite at Kedernath, near one of the sources of the first-mentioned stream. The Ganges, where it traverses the outer zone of tertiary rocks in Chandi, also contains gold.

Moradabad.
359

Moradabad District.—Gold-dust is to be found in the tributaries of the Ramgunga along the northern frontiers of the district, especially in the Koh and the Dhela.

NEPAL, &c.,
360

XII. NEPAL, DARJILING, & SIKKIM.—Though no definite information exists of gold being obtained in these localities, there is no reason for doubting that it exists under similar conditions to those prevailing in the North-West Himálaya. Gold imported from Tibet' is said, however, to be refined in Nepal to the value of 2 lakhs a year. It appears probable that the want of definite knowledge of gold in Sikkim and Nepal is at least partly due to the anxiety shown by Native Governments to conceal their wealth, a suspicion which is confirmed by the fact that gold does exist and is actually washed for in Champaran district at the foot of the hills.

Champaran.
361

Champaran District.—May be considered in this place, since from a geological point of view it is closely connected with the tract above described. A number of rivers and streams which rise from the outer ranges of tertiary rocks on the borders of this district and Nepal, are known to be auriferous, and their sands are annually washed at the commencement and termination of the rains, in the Pachnad, Hurha, Balui or Dhar, Achni and Kapan rivers. Notwithstanding the absence of actual knowledge of the occurrence of gold in Nepal, **Ball** holds that the metal in these outer Siwalik rocks must, as elsewhere in the Himálaya, be of detrital origin, derived from the higher ranges of crystalline rock. The gold-washers of Champaran are evidently of Mongolian origin. They earn, it is said, from 4 annas to 1 rupee a day, but this estimate, which gives a higher average than in almost any other part of India, may be too high.

ASSAM.
362

XIII. ASSAM.—**Ball** writes, "Assam has long been famous for the production of gold, and not a few authorities have stated that its rivers contain gold-bearing sands, some, however, limiting this general statement to those which rise on the hills to the North. Shorn of all exaggeration it would seem that there are few if any named rivers or streams in the districts of Darrang, Sibsagar, and Lakhimpur, which do not yield gold, while in eight other districts, namely, Goalpara, Kamrup, Nowgong, the Garo, Jaintia, and Naga Hills, Sylhet, and Cachar there is no gold as far as our sources of information go. That it is wholly absent in all is not likely, but it is not, and does not appear ever to have been sought for successfully in any of them." Most of the metal found in the first three localities is doubly derivative, coming from the disintegration of detrital rocks, but in the upper reaches of the Brahmaputra it is probably derived direct from the crystalline rocks. **Ball** gives a long and interesting account of the history of gold in Assam and the methods of washing employed in former times, to which the reader desiring such information may be referred. Suffice it to say in this place, that before British occupation, the *Sonwals* or gold-washers paid a yearly tribute of some R64,000, this sum, according to **Colonel Hannay,** representing at least 10,000 *Sonwals.*

G. 362

| Gold. | (*J. Murray.*) | GOLD. |

Sibsagar District.—The principal auriferous rivers of this district are the Dhaneswari, with its tributary the Pakerguri, the Desue, the Jangi, and the Buri Dihing. **Colonel Hannay** states that 15 men working 12 days, in each of the first three rivers, obtained 7¼ tolas of gold, while 24 men working for one month in the last obtained only 12 annas weight. The gold obtained in the Desue in this district and the Joglo in Lakhimpur had at one time the reputation of being the best in Assam, and the gold ornaments of the Assamese Royal Family are said to have been made entirely of the metal obtained from these sources.

ASSAM.
Sibsagar.
363

Lakhimpur District contains a greater number of named auriferous streams than the whole of the rest of Assam put together. In 1853 **Colonel Dalton** reported the total yield of the district to be about 20lb. per annum, worth, say, about £1,200. The chief auriferous streams of the district are the Brahmaputra with its tributaries, the Dikrang, Borpani, Subanshiri, Sisi, Dihong, Dibong, and Digara on the North, and on the South the Joglo, and Noa Dihing. The gold-washings of these streams were examined by **Colonels Dalton** and **Hannay** some years ago. The best results were obtained in the Soglo, from the alluvial deposit of which, 18 grains per ton of rubble washed, was obtained. The Noa Dihing was proved to be more productive than the Brahmaputra, and in this stream traces of platinum were found along with the gold.

Lakhimpur.
364

XIV. BURMA.—Gold is found in all the divisions of Burma, in some instances apparently directly derived from crystalline rocks, in others of doubly-derivative origin. In upper Burma, as in Assam, the latter is most frequently the case.

BURMA.
365

Pegu Division.—**Mr. Theobald** in the publications of the Geological Survey of India states that gold was at the time of his report occasionally washed for in the sand of the Irrawadi opposite Prome, but he himself only saw the operation being conducted at Shwe-Gyeng in coarse gravel.

Pegu.
366

Tenasserim Division.—In this area gold is reported by several observers to be found in the Shwe-Gyeng, Moot-ta-ma, and Tsit-toung rivers, in the streams falling from the granite ranges between Tay and Moungmagan, and in the waters of Henzai, Tavoy, and Tenasserim. Evidence exists of old gold-workings in many of these localities, and in 1867 an Australian Miner aided by Government attempted to obtain gold in the Moot-ta-ma and Baw-ga-ta, but without pecuniary success.

Tenasserim.
367

Upper Burma.—The use of gold in Burma, both for ornamenting buildings and as jewellery, is universal, but is perhaps more prominent in Upper Burma. Though a portion of the metal is obtained by washings in the country, by far the greater amount is imported from China. In 1855 the imports were estimated at an average of 1,100lb. and the indigenous gold which was brought to Mandalay at 300lb., making a total annual consumption of 1,460lb. The principal sources of native gold in Upper Burma are the Kapdup and Nam Kwan rivers in the Hukong Valley; the Kyendwen, and the Upper Irawadi. In the Kyendwen river platinum also occurs, and both metals are collected by a peculiar process. Horns of the wild cow, with the hair on, are fixed in the river, till charged with spangles, and are then sold.

Upper Burma.
368

Method of Collection.

It is unnecessary in an article such as the present to enter into the various methods employed in various parts of the world for obtaining gold by washing, quartz crushing, &c. It may be of interest, however, to give a short account

COLLECTION.
369

G. 369

| GOLD. | Gold. |

of the general method pursued by native gold-washers, with a few exceptions, in every part of India in which gold is to be found. The following short description of the practice followed in the Singhbhum District of Chutia Nagpur has been selected by Ball as typical, and may be here quoted :—" Each tribe occupies a distinct track, and poaching on one another's favourite streams is not indulged in to any great extent. The wooden dish used for washing, measures on an average about 28 by 18 inches for the men, smaller ones being used by the women and children amongst the Jhoras. The dish is hollowed somewhat eccentrically to a maximum depth of $2\frac{1}{2}$ inches. A scraper, formed of a flattened iron hook set in a handle, is used to collect the auriferous sand and gravel which accumulates in the angles formed by the rocks in the bed of the stream. The dish when filled is placed in shallow water, and the operator working with his hands soon separates and throws aside all the coarser gravel and stones, whilst the agitation of the water serves to carry away all the mud and lighter portions. The dish is then balanced on the palm of the left hand and oscillated to and fro with the right; this serves to throw off the greater portion of the remaining gravel, and the process is completed by a circular motion, which is communicated to the water in the hollow of the dish, by which even the smallest particles of foreign matter are separated, and the final result is a residue of black iron sand, in which the specks of gold are readily apparent ; but as mercury is not employed in this part of the country, all the very small and invisible gold is lost." As already stated, this process is supplemented in some parts of the country (*e.g.*, the Panjáb) by the employment of the amalgam method with mercury ; in others skins, horns, &c., &c., are placed in the stream to mechanically arrest fine particles of gold, and in Assam moss and slime scraped from the beds of the streams are similarly used. An idea also prevails in Assam that gold can be obtained by burning the leaves of a plant known as the *copat*. A somewhat peculiar system exists in the washings of the Ningthí river on the Burma-Manipúr border. "The sand and gravel is first placed on a sieve, the finer parts being allowed to fall through on to a hollowed plank 4 feet long and $2\frac{1}{2}$ feet wide at the upper end, and $1\frac{1}{2}$ feet at the lower which is open, the top and margins being protected by a rim $\frac{1}{2}$ inch high. The lower half is cut into grooves half an inch deep and the same in width. The fine sand caught in these grooves is washed in a wooden dish resembling a shield in shape which has a polished black internal surface and a receptacle in the centre. Placed floating in water it is revolved till all the sediment is removed and the mere sand and gold are alone left remaining."

Medicine.—Gold was in remote times employed as a medicine in Europe, and is to this day largely used by followers of Sanskrit medicine. Pliny informs us that in his time it was considered a sovereign remedy for 'green wounds,' that it was supposed to destroy warts, and that Roman mothers hung it round the necks of their children to ward off the evil effects of sorcery. By Sanskrit physicians it was supposed to be a valuable tonic and alterative, to increase strength and beauty, to improve the intellect and memory, to clear the voice, and to increase the sexual powers. These imaginary properties are still largely believed in, and gold is now, as it was centuries ago, much administered in Hindú medicine. Pure leaf gold is employed, purified by heating and cooling it alternately with *Kánjika*, oil, cow's urine, butter-milk, and a decoction of horse-gram. It is then reduced to powder by being rubbed with mercury and exposed to heat in a covered crucible with the addition of sulphur, and is in this form administered in doses of 1 to 2 grains. It enters into many complicated medicinal compounds, each of which is supposed to have some specific virtue. An exhaustive and interesting account of these will be found in *U. C. Dutt's Hindu Materia Medica*, from which the above abstract of the Indian methods of employment as a medicine has been mainly compiled.

G. 370

| Gold; Gordonia. | (*F. Murray.*) | **GORDONIA obtusa.** |

Domestic and Sacred.—Gold is largely employed by the richer classes in India for purposes of personal adornment, and also in the decoration of buildings. It is unnecessary in an article such as the present to enter into a consideration of the several art-industries, such as gold jewellery, filigree work, gold-wire, thread and lace, &c.; for interesting and exhaustive descriptions of which the reader may be specially referred to the volumes of the Journal of Indian Art.

DOMESTIC.
371

Trade.—The average imports of gold and bullion during the five years from 1883–84 to 1887–88 was R3,88,17,962; the average exports R26,83,717. The countries from which the metal was chiefly imported were the United Kingdom, China, Australasia, and Egypt. In 1887–88 R95½ lakhs were received from the first, 97 lakhs from the second, 54 lakhs from the third, and 20 lakhs from the last-mentioned country. The gold exported is almost entirely sent to the United Kingdom.

TRADE.
372

GOMPHIA, *Schreb.; Gen. Pl., I., 318.*

Gomphia angustifolia, *Vahl.; Fl. Br. Ind., I., 525;* OCHNACEÆ.

373

Syn.—GOMPHIA ZEYLANICA, *DC.;* G. MALABARICA, *DC.;* OCHNA ZEYLANICA, *Lam.;* WALKERA SERRATA, *Willd.;* MEESIA SERRATA, *Gærtn.*

Vern.—*Valermani,* MALAY.; *Bokaara-gass,* SING.

References.—*Gamble, Man. Timb., 65; Thwaites, En. Ceylon Pl., 71; W. and A., Prod., 152; Grah., Cat. Bomb. Pl., 36; Rheede, Hort. Mal., V., tt. 48, 52; O'Shaughnessy, Beng. Dispens., 269; Lisboa, U. Pl. Bomb., 37; Balfour, Cyclop., I., 1227.*

Habitat.—A small glabrous tree, native of South Western India from the South Konkan to Travancore, Singapore, and Ceylon. Distributed to China.

Medicine.—**O'Shaughnessy** states that the ROOT and LEAVES are bitter, and are employed in the form of a decoction in Malabar, as a tonic, stomachic, and anti-emetic.

MEDICINE.
Root.
374
Leaves.
375

Structure of the Wood.—Used for building purposes in Ceylon (*Thwaites*).

TIMBER.
376

Gomuti; see **Arenga saccharifera,** *Labill.;* VOL. I., 302.

GONIOTHALAMUS, *Blume; Gen. Pl., I., 26.*

Goniothalamus cardiopetalus, *H. f. & T.; Fl. Br. Ind., I., 75; Beddome, Ic.* [*Pl. Ind. Or., t. 62.;* ANONACEÆ.

377

Syn.—UVARIA OBOVATA, *Heyne;* POLYALTHIA CARDIOPETALA, *Dalz.;* ATRATEGIA THOMSONI, *Bedd.*

Habitat.—A small tree found on the mountains of Kanára.

Structure of the Wood.—Used for making posts (*Lisboa U., Pl. Bomb., 3*).

TIMBER.
378

Gooseberry; see **Ribes Grossularia,** *Linn.;* SAXIFRAGACEÆ.

Gooseberry, Cape; see **Physalis peruviana,** *Linn;* SOLANACEÆ.

GORDONIA, *Ellis; Gen. Pl., I., 186.*

Gordonia obtusa, *Wall.: Fl. Br. Ind., I., 291; Wight, Ill., I., 99;* [TERNSTRŒMIACEÆ.

379

Syn.—GORDONIA OBTUSIFOLIA, and G. PARVIFOLIA, *Wight;* SAURAUJA CRENULATA, *Wight in Wall., Cat., 1459 (not of DC.).*

Vern.—*Nagetta,* NILGHIRIS.

References.—*Beddome, Fl. Sylv., t. 83; Gamble, Man. Timb., 28; Drury, U. Pl. Ind., 229; Lisboa, U. Pl. Bomb., 14; Balfour, Cyclop., I., 1236; Ind. Forester, II., 22, 23; X., 35, 552.*

GORDONIA obtusa.	The Gordonia.

<table>
<tr><td>TIMBER.
380</td><td>Habitat.—An evergreen tree of the mountains of Western India from the Konkan to the Pulney hills, at altitudes of from 2,500 to 7,500 feet.
Structure of the Wood.—" White with a straw tint, even-grained and pleasant to work, not unlike beech; it is very generally used for planks, doors, rafters, and beams, but warps if not well seasoned " (Beddome).</td></tr>
</table>

O. S. G. P. I.—No. 215 R. & A. D.—11.9.90.—1,350.

Printed in the United States
By Bookmasters